Handbook of Experimental Pharmacology

Volume 218

Editor-in-Chief

F.B. Hofmann, München

Editorial Board

J.E. Barrett, Philadelphia
J. Buckingham, Uxbridge
V. Flockerzi, Homburg
P. Geppetti, Florence
M.C. Michel, Ingelheim
P. Moore, Singapore
C.P. Page, London
W. Rosenthal, Berlin

For further volumes:
http://www.springer.com/series/164

Marc Humbert • Oleg V. Evgenov •
Johannes-Peter Stasch
Editors

Pharmacotherapy of Pulmonary Hypertension

Editors
Marc Humbert
Faculté de Médecine
Univ Paris-Sud
Le Kremlin-Bicêtre
France

Oleg V. Evgenov
Harvard Medical School
Massachusetts General Hospital
Dep of Anasthesia and Critical Care
Boston, Massachusetts
USA

Johannes-Peter Stasch
Bayer Pharma AG
Cardiovascular Research
Wuppertal, Germany

ISSN 0171-2004 ISSN 1865-0325 (electronic)
ISBN 978-3-642-38663-3 ISBN 978-3-642-38664-0 (eBook)
DOI 10.1007/978-3-642-38664-0
Springer Heidelberg New York Dordrecht London

Library of Congress Control Number: 2013948127

© Springer-Verlag Berlin Heidelberg 2013

This work is subject to copyright. All rights are reserved by the Publisher, whether the whole or part of the material is concerned, specifically the rights of translation, reprinting, reuse of illustrations, recitation, broadcasting, reproduction on microfilms or in any other physical way, and transmission or information storage and retrieval, electronic adaptation, computer software, or by similar or dissimilar methodology now known or hereafter developed. Exempted from this legal reservation are brief excerpts in connection with reviews or scholarly analysis or material supplied specifically for the purpose of being entered and executed on a computer system, for exclusive use by the purchaser of the work. Duplication of this publication or parts thereof is permitted only under the provisions of the Copyright Law of the Publisher's location, in its current version, and permission for use must always be obtained from Springer. Permissions for use may be obtained through RightsLink at the Copyright Clearance Center. Violations are liable to prosecution under the respective Copyright Law.

The use of general descriptive names, registered names, trademarks, service marks, etc. in this publication does not imply, even in the absence of a specific statement, that such names are exempt from the relevant protective laws and regulations and therefore free for general use.

While the advice and information in this book are believed to be true and accurate at the date of publication, neither the authors nor the editors nor the publisher can accept any legal responsibility for any errors or omissions that may be made. The publisher makes no warranty, express or implied, with respect to the material contained herein.

Printed on acid-free paper

Springer is part of Springer Science+Business Media (www.springer.com)

Preface

Over the past decade, we have witnessed major advancements in the understanding of the molecular mechanisms and pathophysiology of pulmonary hypertension (PH) that occurred in parallel with the discovery and development of new therapies. Pharmacological agents that modulate the main pathophysiological pathways of pulmonary arterial hypertension, including endothelin-1, nitric oxide, and prostacyclin, have changed the course of this devastating disease by relieving symptoms and improving and prolonging patients' lives. Moreover, molecular biological tools have supplied a wealth of knowledge about the roles of established and emerging signaling pathways in PH. This book is an example of how pharmacotherapy can further expand our understanding of such a heterogeneous group of diseases as PH.

Written by leading experts in the field, this volume focuses on current, evidence-based pharmacological treatments of various forms of PH and also provides a comprehensive review of most recent developments in this area. The first part of the book covers definition, classification, pathophysiology, pathology, biomarkers, and animal models of the disease, thus laying the conceptual basis for what follows. The middle section provides an overview of the established therapies such as calcium channel blockers, prostanoids, endothelin receptor antagonists, phosphodiesterase-5 inhibitors, and inhaled nitric oxide. The last section explores novel pathways and emerging therapeutic approaches including soluble guanylate cyclase stimulators, Rho-kinase inhibitors, inhibitors of serotonin receptors and transporters, peptide growth factors, vasoactive peptides, modulators of redox equilibrium and cyclic nucleotides homeostasis, as well as immunosuppressive and anti-proliferative agents. Particular attention is given to clinical applications of these experimental therapies, which are on the horizon, thus spanning the continuum from basic science to clinical applications. The catalogue of signaling cascades in PH is rapidly growing and with it the wish to modulate these pathways for therapeutic reasons. We can expect to see further progress and achievements in developing novel therapeutics over the next decade, with new improved agents in established drug classes, new drug classes in established pathways, and novel drugs aimed at new molecular targets.

We hope that this book will help the reader appreciate the importance of elucidating the pathophysiology of PH and the research on new signaling pathways and their biochemical and genetic background. It also showcases the commitment of academia and industry to drug discovery and development focusing on this devastating condition. Unfortunately, despite recent advances in the understanding of the molecular mechanisms and pathology of PH, this disease remains incurable. There is still a significant need for identifying new therapeutic targets and developing new treatment options. Undoubtedly, exciting times lie ahead of us, for both the science and the patients. If this book helps to stimulate further research and improve patient care, the wishes of the authors, the editors, and the editorial manager, Susanne Dathe, will be fulfilled.

Paris, France	Marc Humbert
Boston, MA	Oleg V. Evgenov
Wuppertal, Germany	Johannes-Peter Stasch

Contents

Part I Pulmonary Hypertension: Conceptual Bases of the Disease

Definition and Classification of Pulmonary Hypertension 3
Marc Humbert, David Montani, Oleg V. Evgenov,
and Gérald Simonneau

Pulmonary Hypertension: Pathophysiology and Signaling Pathways ... 31
Bradley A. Maron and Joseph Loscalzo

Pulmonary Hypertension: Pathology 59
Peter Dorfmüller

Pulmonary Hypertension: Biomarkers 77
Christopher J. Rhodes, John Wharton, and Martin R. Wilkins

Rodent Models of Group 1 Pulmonary Hypertension 105
John J. Ryan, Glenn Marsboom, and Stephen L. Archer

Part II Pulmonary Hypertension: Established Therapies

General Supportive Care 153
Ioana R. Preston

Calcium-Channel Blockers in Pulmonary Arterial Hypertension 161
Marie-Camille Chaumais, Elise Artaud Macari, and Olivier Sitbon

Prostacyclins 177
Horst Olschewski

Endothelin Receptor Antagonists 199
Martine Clozel, Alessandro Maresta, and Marc Humbert

Phosphodiesterase-5 Inhibitors 229
Barbara A. Cockrill and Aaron B. Waxman

Inhaled Nitric Oxide for the Treatment of Pulmonary Arterial Hypertension 257
Steven H. Abman

Part III Pulmonary Hypertension: Novel Pathways and Emerging Therapies

Soluble Guanylate Cyclase Stimulators in Pulmonary Hypertension 279
Johannes-Peter Stasch and Oleg V. Evgenov

Therapeutics Targeting of Dysregulated Redox Equilibrium and Endothelial Dysfunction 315
Michael G. Risbano and Mark T. Gladwin

Rho-Kinase Inhibitors 351
Yoshihiro Fukumoto and Hiroaki Shimokawa

Serotonin Transporter and Serotonin Receptors 365
Serge Adnot, Amal Houssaini, Shariq Abid, Elisabeth Marcos, and Valérie Amsellem

Targeting of Platelet-Derived Growth Factor Signaling in Pulmonary Arterial Hypertension 381
Eva Berghausen, Henrik ten Freyhaus, and Stephan Rosenkranz

Emerging Molecular Targets for Anti-proliferative Strategies in Pulmonary Arterial Hypertension 409
Ly Tu and Christophe Guignabert

Anti-inflammatory and Immunosuppressive Agents in PAH 437
Jolyane Meloche, Sébastien Renard, Steeve Provencher, and Sébastien Bonnet

Vasoactive Peptides and the Pathogenesis of Pulmonary Hypertension: Role and Potential Therapeutic Application 477
Reshma S. Baliga, Raymond J. MacAllister, and Adrian J. Hobbs

Pulmonary Hypertension: Novel Pathways and Emerging Therapies Inhibitors of cGMP and cAMP Metabolism 513
Yassine Sassi and Jean-Sébastien Hulot

Pulmonary Hypertension: Old Targets Revisited (Statins, PPARs, Beta-Blockers) 531
Geoffrey Watson, Eduardo Oliver, Lan Zhao, and Martin R. Wilkins

Part IV Conclusions and Outlook

Pulmonary Hypertension: Current Management and Future Directions 551
Lewis J. Rubin

Index 557

Part I
Pulmonary Hypertension: Conceptual Bases of the Disease

Definition and Classification of Pulmonary Hypertension

Marc Humbert, David Montani, Oleg V. Evgenov, and Gérald Simonneau

Abstract Pulmonary hypertension is defined as an increase of mean pulmonary arterial pressure ≥25 mmHg at rest as assessed by right heart catheterization. According to different combinations of values of pulmonary wedge pressure, pulmonary vascular resistance and cardiac output, a hemodynamic classification of pulmonary hypertension has been proposed. Of major importance is the pulmonary wedge pressure which allows to distinguish pre-capillary (pulmonary wedge pressure ≤15 mmHg) and post-capillary (pulmonary wedge pressure >15 mmHg) pulmonary hypertension. Pre-capillary pulmonary hypertension includes the clinical groups 1 (pulmonary arterial hypertension), 3 (pulmonary hypertension due to lung diseases and/or hypoxia), 4 (chronic thrombo-embolic pulmonary hypertension) and 5 (pulmonary hypertension with unclear and/or multifactorial mechanisms). Post-capillary pulmonary hypertension corresponds to the clinical group 2 (pulmonary hypertension due to left heart diseases).

Keywords Pulmonary arterial hypertension • Pulmonary artery pressure • Pulmonary hypertension • Right heart catheterization

---

M. Humbert (✉) • D. Montani • G. Simonneau
Faculté de Médecine, Univ Paris-Sud, Le Kremlin-Bicêtre, France

Inserm U999, Centre Chirurgical Marie Lannelongue, Le Plessis-Robinson, France

Service de Pneumologie et Réanimation respiratoire, Hôpital Bicêtre, Assistance Publique – Hôpitaux de Paris, 78, rue Gabriel Péri, 94270 Le Kremlin-Bicêtre, France
e-mail: marc.humbert@abc.aphp.fr

O.V. Evgenov
Department of Anesthesia, Critical Care and Pain Medicine, Massachusetts General Hospital, Harvard Medical School, Boston, MA, USA

Contents

1 Introduction 4
2 Definition 5
3 Classification 6
4 Group 1: Pulmonary Arterial Hypertension 6
    4.1   Group 1.1/1.2: Idiopathic and Heritable PAH 7
    4.2   Group 1.3: Drug- and Toxin-Induced PAH 8
    4.3   Group 1.4.1: PAH Associated with Connective Tissue Diseases ... 10
    4.4   Group 1.4.2: Human Immunodeficiency Virus Infection 11
    4.5   Group 1.4.3: Portopulmonary Hypertension 11
    4.6   Group 1.4.4: Congenital Heart Diseases 12
    4.7   Group 1.4.5: Schistosomiasis 12
    4.8   Group 1.4.6: Chronic Hemolytic Anemia 14
    4.9   Lessons from the French PAH Registry in the Modern Management era 15
5 Group 1′: Pulmonary Veno-Occlusive Disease and Pulmonary Capillary Hemangiomatosis 16
6 Group 2: Pulmonary Hypertension due to Left Heart Disease 17
7 Group 3: Pulmonary Hypertension due to Lung Diseases and/or Hypoxia 18
8 Group 4: Chronic Thromboembolic Pulmonary Hypertension 18
9 Group 5: Pulmonary Hypertension with Unclear or Multifactorial Etiologies 19
    9.1   Group 5.1: Hematologic Disorders 19
    9.2   Group 5.2: Systemic Disorders 20
    9.3   Group 5.3: Metabolic Disorders 21
    9.4   Group 5.4: Miscellaneous Conditions 21
10 Conclusion 22
References 23

1 Introduction

Pulmonary hypertension (PH) has previously been called an orphan disease, that is, a condition which affects a few individuals and is overlooked by the medical profession and pharmaceutical industry (Humbert et al. 2007). Although undoubtedly rare, the concept that PH is overlooked cannot be considered to be the case today. Indeed, there have been a number of important discoveries in recent years that have significantly improved our understanding of the disease, helped guide patient management, and laid foundations for future research (Simonneau et al. 2009; Galiè et al. 2009). The fifth world symposium on PH has taken place in 2013 in Nice, France, 40 years after the World Health Organization sponsored the first international conference in Geneva, Switzerland, on a mysterious condition named primary pulmonary hypertension (PPH), spurred by the interest created by the sudden increase in the disease in patients who had used an anorectic pill (aminorex fumarate) (Humbert 2012). Significant achievements have been made in the field since then, as reflected by the results presented at the second world symposium on PPH (Evian, France, 1998), the third world symposium on pulmonary arterial hypertension (PAH) (Venice, Italy, 2003), and the fourth world symposium on PH (Dana Point, US, 2008) (Humbert 2012).

Table 1 Hemodynamic definitions of pulmonary hypertension. Adapted from Galiè et al. (2009)

Definition	Characteristics	Clinical groups(s)[a]
PH	$\bar{P}_{pa} \geq 25$ mmHg	All
Pre-capillary PH	$\bar{P}_{pa} \geq 25$ mmHg	1. Pulmonary arterial hypertension
	$P_{pcw} \leq 15$ mmHg	3. PH due to lung diseases
	CO normal or reduced[b]	4. Chronic thromboembolic PH
		5. PH with unclear and/or multifactorial mechanisms
Post-capillary PH	$\bar{P}_{pa} \geq 25$ mmHg	2. PH due to left heart disease
	$P_{pcw} > 15$ mmHg	
	CO normal or reduced[b]	
Passive	TPG ≤ 12 mmHg	
Reactive (out of proportion)	TPG >12 mmHg	

$\bar{P}_{pa}$ mean pulmonary arterial pressure; P_{pcw} pulmonary capillary wedge pressure; CO cardiac output; TPG transpulmonary pressure gradient ($\bar{P}_{pa} \bar{P}_{pcw}$)
[a]According to Table 2
[b]High CO can be present in cases of hyperkinetic conditions such as systemic-to-pulmonary shunts (only in the pulmonary circulation), anemia, hyperthyroidism, etc.
All values measured at rest

2 Definition

Pulmonary hypertension is defined as an increase of mean pulmonary arterial pressure (mPAP) ≥ 25 mmHg at rest as assessed by right heart catheterization (RHC) (Galiè et al. 2009). This value has been utilized for selecting patients in all randomized controlled trials on PAH and in PH registries. The normal mPAP at rest is 14 ± 3.3 mmHg and the upper limit of normal, according to these data, is 20.6 mmHg (Kovacs et al. 2009). It is currently undefined how to consider the values of mPAP between 21 and 24 mmHg (Galiè et al. 2009). In addition, the definition of PH during exercise as a mPAP > 30 mmHg as assessed by RHC is not supported by convincing published data, and healthy individuals can reach higher values (Galiè et al. 2009; Kovacs et al. 2009; Whyte et al. 2012). Therefore, no definition for PH during exercise is currently accepted (Galiè et al. 2009). According to different combinations of values of pulmonary wedge pressure (PWP), pulmonary vascular resistance (PVR), and cardiac output (CO), a hemodynamic classification of PH has been proposed (Table 1) (Galiè et al. 2009). Of major importance is the PWP, which allows to distinguish pre-capillary PH (PWP ≤ 15 mmHg) and post-capillary PH (PWP > 15 mmHg). Pre-capillary PH includes the following clinical groups: 1 (PAH), 3 (PH due to lung diseases and/or hypoxia), 4 (chronic thromboembolic PH, CTEPH), and 5 (PH with unclear and/or multifactorial mechanisms) (Simonneau et al. 2009; Galiè et al. 2009). Post-capillary PH includes the clinical group 2 (PH due to left heart diseases) (Simonneau et al. 2009; Galiè et al. 2009). In group 2, the transpulmonary gradient (TPG = mPAP − PWP) allows the distinction of so-called passive "proportionate" post-capillary PH (TPG ≤ 12 mmHg) and reactive "out of proportion" post-capillary PH (TPG > 12 mmHg) (Galiè et al. 2009). However, the concept of "proportionate" post-capillary PH is currently debated and the inclusion of other parameters such as the difference between mPAP and diastolic PAP may allow a more accurate analysis of the pulmonary hemodynamics. In this chapter, the features of each clinical PH group are discussed in specific sections with particular attention to PAH.

3 Classification

The PH classification has gone through a series of changes since the first classification proposed in 1973, which identified only two main categories, namely "primary" or "secondary" PH, depending on the presence or absence of identifiable causes or risk factors (Simonneau et al. 2009; Galiè et al. 2009; Hatano and Strasser 1975). During the second world symposium on PPH (Evian, France, 1998), a new classification was proposed and attempted to create categories of PH that shared similar pathogenesis, clinical features, and therapeutic options (Simonneau et al. 2009). This classification defined homogenous groups of patients to help in conducting clinical trials and obtaining approval for specific therapies. In 2003, the third PH conference (Venice, Italy) proposed only minor changes, with the exception of the introduction of the terms of idiopathic PAH (IPAH), familial PAH, or associated PAH defining three PAH subgroups sharing broadly similar physiopathology and response to therapy (Simonneau et al. 2009). The other prominent change was to move pulmonary veno-occlusive disease (PVOD) and pulmonary capillary hemangiomatosis (PCH) from separate categories into a single subcategory of PH associated with substantial venous or capillary involvement (Simonneau et al. 2009). Indeed PAH and PVOD share similarities with idiopathic PAH, including clinical presentation, hemodynamic characteristics, and risk factors, that justified placing them together in group 1. However, PVOD and PCH differ from PAH in terms of pathological changes and worst responses to vasodilator therapies (Simonneau et al. 2009).

In 2008, the fourth world PH conference held in Dana Point, California, proposed to revise previous classifications in order to accurately reflect information published over the past 5 years, as well as to clarify some areas that were ambiguous (Simonneau et al. 2009). The current Dana Point classification is presented in Table 2.

4 Group 1: Pulmonary Arterial Hypertension

The nomenclature of the PAH subgroups and associated conditions has evolved with time, and additional modifications were included in the 2008 revision (Simonneau et al. 2009; Galiè et al. 2009).

Table 2 Updated clinical classification of pulmonary hypertension. Adapted from Simonneau et al. (2009)

Updated clinical classification of pulmonary hypertension (Dana Point, 2008)
1. Pulmonary arterial hypertension (PAH)
1.1. Idiopathic PAH
1.2. Heritable
1.2.1. BMPR2
1.2.2. ALK1, endoglin (with or without hereditary hemorrhagic telangiectasia)
1.2.3. Unknown
1.3. Drug- and toxin-induced
1.4. Associated with
1.4.1. Connective tissue diseases
1.4.2. HIV infection
1.4.3. Portal hypertension
1.4.4. Congenital heart diseases
1.4.5. Schistosomiasis
1.4.6. Chronic hemolytic anemia
1.5. Persistent pulmonary hypertension of the newborn
1′. Pulmonary veno-occlusive disease (PVOD) and/or pulmonary capillary hemangiomatosis (PCH)
2. Pulmonary hypertension owing to left heart diseases
2.1. Systolic dysfunction
2.2. Diastolic dysfunction
2.3. Valvular disease
3. Pulmonary hypertension owing to lung diseases and/or hypoxia
3.1. Chronic obstructive pulmonary disease
3.2. Interstitial lung disease
3.3. Other pulmonary diseases with mixed restrictive and obstructive pattern
3.4. Sleep-disordered breathing
3.5. Alveolar hypoventilation disorders
3.6. Chronic exposure to high altitude
3.7. Developmental abnormalities
4. Chronic thromboembolic pulmonary hypertension (CTEPH)
5. Pulmonary hypertension with unclear multifactorial mechanisms
5.1. Hematologic disorders: myeloproliferative disorders, splenectomy
5.2. Systemic disorders: sarcoidosis, pulmonary Langerhans' cell histiocytosis: Lymphangioleiomyomatosis, neurofibromatosis, vasculitis
5.3. Metabolic disorders: glycogen storage disease, Gaucher's disease, thyroid disorders
5.4. Others: tumoral obstruction, fibrosing mediastinitis, chronic renal failure on dialysis

ALK1 activin receptor-like kinase type 1; *BMPR2* bone morphogenetic protein receptor type 2; *HIV* human immunodeficiency virus

4.1 Group 1.1/1.2: Idiopathic and Heritable PAH

IPAH corresponds to sporadic disease in which there is neither a familial history of PAH nor identified risk factors (Simonneau et al. 2009; Galiè et al. 2009). When PAH occurs in a familial context, germline mutations in the gene coding for the bone morphogenetic protein receptor 2 (*BMPR2*), a member of the transforming growth factor beta (TGF-β) signaling family, can be detected in more than 70 % of cases (Simonneau et al. 2009; Sztrymf et al. 2008). *BMPR2* mutations have also been detected in 11–40 % of apparently idiopathic cases with no familial history (Sztrymf et al. 2008; Girerd et al. 2010). Indeed, the distinction between idiopathic and familial PAH with *BMPR2* mutations is artificial, as all patients with a *BMPR2* mutation have heritable disease. Thus, it was decided to abandon the term "familial PAH" in favor of the term "heritable PAH" (Simonneau et al. 2009). Heritable forms of PAH include IPAH with germline mutations (mainly *BMPR2* but also *ACVRL1* or *endoglin*) and familial cases with or without identified mutations (Simonneau et al. 2009; Sztrymf et al. 2008; Girerd et al. 2010). Idiopathic and heritable PAH is more common in women than in men with a gender ratio around 2:1 (Sztrymf et al. 2008; Girerd et al. 2010).

Recent studies have shown that heritable PAH patients carrying a *BMPR2* mutation (irrespective of the familial history of other PAH cases) were younger at diagnosis, had more severe disease, and were less likely to demonstrate vasoreactivity than IPAH patients without a *BMPR2* mutation (Sztrymf et al. 2008; Girerd et al. 2010; Elliott et al. 2006). It is now commonly accepted that genetic testing should be performed as part of a comprehensive program that includes genetic counseling and discussion of the risks, benefits, and limitations of such testing (Girerd et al. 2010).

4.2 Group 1.3: Drug- and Toxin-Induced PAH

In the recent revision of the PAH classification aminorex, fenfluramine, dexfenfluramine, and toxic rapeseed oil represent the only identified "definite" risk factors for PAH (Simonneau et al. 2009; Galiè et al. 2009) (Table 3).

Aminorex fumarate structurally resembles adrenaline and ephedrine and is a potent appetite suppressant and central stimulant. Its use in the 1960s led to an outbreak of rapidly progressive PAH (then termed PPH) in Switzerland, Austria, and Germany, with a median exposure-to-onset time, when known, of 8 months (ranging from 3 weeks to over 1 year), as first described in a Swiss medical clinic (Follath et al. 1971; Gurtner 1979). The incidence of PAH in patients who had used aminorex was shown to be about 0.2 % overall, and proportional to the amount of drug taken. Furthermore, if discontinued early enough, a regression of PPH could be seen. Subsequently, aminorex was withdrawn from the market in 1968.

Both fenfluramine and dexfenfluramine have been marketed as appetite suppressants. An outbreak of drug-induced PAH has been identified with these two agents in the 1980s–1990s (Brenot et al. 1993; Abenhaim et al. 1996; Simonneau et al. 1998; Souza et al. 2008). PAH cases in patients exposed to fenfluramine derivatives share clinical, functional, hemodynamic, and genetic features with IPAH. This suggests that fenfluramine exposure represents a potential trigger for PAH without influencing its clinical course (Souza et al. 2008).

Table 3 Updated risk factors for pulmonary arterial hypertension. Adapted from Simonneau et al. (2009)

Definite	Possible
Aminorex	Cocaine
Fenfluramine	Phenylpropanolamine
Dexfenfluramine	St. John's Wort
Benfluorex	Chemotherapeutic agents
Toxic rapeseed oil	SSRI

Likely	Unlikely
Amphetamines	Oral contraceptives
L-tryptophan	Estrogen
Methamphetamines	Cigarette smoking

PAH pulmonary arterial hypertension; *SSRI* selective serotonin reuptake inhibitor

The association of fenfluramine and dexfenfluramine intake with the development of PAH was confirmed by the Surveillance of Pulmonary Hypertension in America (SOPHIA), which enrolled 1,335 subjects at tertiary PH centers in the USA between 1998 and 2001 (Walker et al. 2006). Of note, these agents have also been associated with an increased risk of valvular heart diseases, presumably because of their serotoninergic properties (Connolly et al. 1997). As a result, fenfluramine and dexfenfluramine were withdrawn from the market in the late 1990s (Souza et al. 2008; Humbert et al. 2006).

Benfluorex is a benzoate ester that shares similar structural and pharmacologic characteristics with dexfenfluramine and fenfluramine (Boutet et al. 2009; Savale et al. 2012; Frachon et al. 2010). The active and common metabolite of each of these molecules is norfenfluramine, which itself has a chemical structure similar to that of the amphetamines. Given its pharmacological properties, benfluorex would be expected to have toxic effects similar to the fenfluramine derivatives. An outbreak of valvular heart diseases and/or PAH induced by benfluorex use has been evidenced in France in the 2000s (Boutet et al. 2009; Savale et al. 2012; Frachon et al. 2010). Eighty-five cases of PH associated with benfluorex exposure were identified by the French PH network from June 1999 to March 2011 (Savale et al. 2012). Of these 85 cases, 70 patients had confirmed pre-capillary PH. Interestingly, 33 % of all patients also had prior exposure to fenfluramine or dexfenfluramine, and an additional risk factor for PH was identified in 30 % of the patients with pre-capillary PH (Savale et al. 2012). A quarter of patients in this series showed coexisting PH and mild-to-moderate cardiac valve involvement. These results, together with the accumulated data regarding the known toxic effects of fenfluramine and dexfenfluramine, strongly suggest that benfluorex exposure is a potent trigger for PAH. Benfluorex was withdrawn from the French market in 2009.

St John's Wort and over-the-counter antiobesity agents containing phenylpropanolamine also increase the risk of developing PAH (Walker et al. 2006). The SOPHIA study examined intake of a variety of nonselective monoamine reuptake inhibitors, selective serotonin reuptake inhibitors, antidepressants, and anxiolytics and found no increased risk for developing PAH (Walker et al. 2006). However, case–control studies of selective serotonin reuptake inhibitors used during pregnancy showed an increased risk of developing persistent PH of the newborn in the offspring (Chambers et al. 2006). Based on this study, selective serotonin reuptake inhibitors have been reclassified to the "possible" category.

Amphetamines use represents a "likely" risk factor, although they are frequently used in combination with fenfluramine. A recent comprehensive, retrospective study suggested a strong correlation between the use of methamphetamine (inhaled, smoked, oral, or intravenous) and the occurrence of PAH (Chin et al. 2006). Based primarily on the results of this study, methamphetamine use is now considered a "very likely" risk factor for the development of PAH.

Cases of pre-capillary PH fulfilling the criteria of drug-induced PAH have been reported in chronic myelogenous leukemia patients treated with the tyrosine kinase inhibitor dasatinib (Montani et al. 2012). At diagnosis, patients had moderate-to-severe pre-capillary PH with functional and hemodynamic impairment. Clinical, functional, and hemodynamic improvements were observed within a few months of dasatinib discontinuation in most patients (Montani et al. 2012). However, after a median follow-up of 9 months (min–max 3–36), the majority of patients failed to demonstrate complete clinical and hemodynamic recovery and no patients reached a normal value of mPAH (≤ 20 mmHg). The lowest estimate of incident PH occurring in patients exposed to dasatinib in France was 0.45 % (Montani et al. 2012). Thus, dasatinib may induce severe pre-capillary PH, fulfilling the criteria of PAH, suggesting a direct and specific effect of dasatinib on pulmonary vessels (Montani et al. 2012).

4.3 Group 1.4.1: PAH Associated with Connective Tissue Diseases

Pulmonary arterial hypertension associated with connective tissue diseases represents an important clinical subgroup (Humbert et al. 2006; Hachulla et al. 2005). The high prevalence of PAH in patients with systemic sclerosis has been well studied. Two prospective studies, using echocardiography as a screening method and RHC for diagnosis and confirmation, found a prevalence of PAH ranging from 8 % to 12 % (Hachulla et al. 2005; Mukerjee et al. 2003). Several long-term studies have showed that the outcomes of patients with PAH associated with systemic sclerosis are worse than those of IPAH patients. However, PAH does not represent the only cause of PH in systemic sclerosis. PH due to chronic lung disease such as pulmonary fibrosis and PH due to heart failure such as diastolic dysfunction are also frequent, emphasizing the importance of a complete evaluation to accurately classify and manage PH (Hachulla et al. 2009).

In systemic lupus erythematosus (Tanaka et al. 2002; Asherson et al. 1990) and mixed connective tissue disease (Burdt et al. 1999), the prevalence of PAH remains unknown, but likely occurs less frequently than in systemic sclerosis. Importantly, corticosteroids and immunosuppressants may improve outcomes in a subset of patients with PAH complicating the course of systemic lupus erythematosus or mixed connective tissue disease, while this is not the case in IPAH or PAH associated with systemic sclerosis (Jais et al. 2008; Sanchez et al. 2006). In the absence of chronic pulmonary parenchymal involvement, PAH has been reported infrequently in other connective tissue diseases such as Sjögren syndrome (Launay et al. 2007), polymyositis (Bunch et al. 1981), or rheumatoid arthritis (Dawson et al. 2000).

4.4 Group 1.4.2: Human Immunodeficiency Virus Infection

Pulmonary arterial hypertension is a rare complication of human immunodeficiency virus (HIV) infection (Speich et al. 1991; Opravil et al. 1997; Mehta et al. 2000; Nunes et al. 2003; Sitbon et al. 2008; Zuber et al. 2004; Opravil and Sereni 2008; Degano et al. 2010). This PAH subcategory is the strongest evidence of virus-induced PAH. The HIV-associated PAH has clinical, hemodynamic, and histologic characteristics similar to those of IPAH. Epidemiologic data in the early 1990s, a time when highly active antiretroviral therapy was not yet available, showed a PAH prevalence of 0.5 % and a dramatic early mortality (Speich et al. 1991; Opravil et al. 1997). Since the wide use of highly antiretroviral therapies, a stable prevalence of 0.46 % was confirmed, while the incidence of PAH has declined (due to the longer survival of the HIV-associated PAH patients) (Sitbon et al. 2008; Opravil and Sereni 2008). Uncontrolled studies show that patients with severe HIV-associated PAH benefit from specific PAH therapies (Degano et al. 2010). Interestingly, normalization of hemodynamics has been regularly reported in a substantial number of the HIV-associated PAH patients treated with specific PAH drugs and highly active antiretroviral therapy (Degano et al. 2010).

4.5 Group 1.4.3: Portopulmonary Hypertension

Portopulmonary hypertension (PoPH) is defined by the development of PAH associated with increased pressure in the portal circulation (Hadengue et al. 1991; Rodriguez-Roisin et al. 2004; Kawut et al. 2008; Le Pavec et al. 2008). Prospective hemodynamic studies have shown that 2–6 % of patients with portal hypertension have PH (Hadengue et al. 1991; Rodriguez-Roisin et al. 2004). As always, RHC is mandatory for the accurate diagnosis of PoPH, because several mechanisms may increase mPAP in the setting of advanced liver disease: hyperdynamic circulatory state with high CO, fluid overload, and diastolic dysfunction; in such circumstances, PVR is within the normal range. Pathologic changes in the small arteries appear identical to those seen in IPAH. A recent multicenter case–control study conducted in the USA identified that female gender and autoimmune hepatitis were independent risk factors for the development of PoPH and that hepatitis C infection was associated with a decreased risk (Kawut et al. 2008). Long-term prognosis of PoPH was related to the presence and severity of cirrhosis as well as to right heart function in a recently reported large French cohort study (Le Pavec et al. 2008).

4.6 Group 1.4.4: Congenital Heart Diseases

A significant proportion of patients with congenital heart disease, in particular those with systemic-to-pulmonary shunts, will develop PAH if left untreated (Simonneau et al. 2009; Galiè et al. 2009; Wood 1958; Hoffman and Rudolph 1965). Eisenmenger's syndrome is defined as congenital heart disease with an initial large systemic-to-pulmonary shunt that induces progressive pulmonary vascular disease and PAH, with resultant reversal of the shunt, hypoxemia, polycythemia, and cyanosis (Wood 1958; Hoffman and Rudolph 1965). It represents the most advanced form of PAH associated with congenital heart disease. The prevalence of PAH associated with congenital systemic-to-pulmonary shunts in Europe and North America has been estimated to be between 1.6 and 12.5 cases per million adults, with 25–50 % of this population affected by Eisenmenger's syndrome. These numbers are much higher in developing countries. The histological and pathophysiological changes seen in patients with PAH associated with congenital systemic-to-pulmonary shunts, in particular endothelial dysfunction, are similar to those observed in idiopathic or other associated forms of PAH. Following the fourth world symposium on PH, it was decided to update the pathologic and pathophysiological classification of congenital heart disease with systemic-to-pulmonary shunts (Table 4). In order to provide a more detailed description of each condition, four distinct phenotypes have been individualized (Table 5) (Simonneau et al. 2009; Galiè et al. 2009).

4.7 Group 1.4.5: Schistosomiasis

Embolic obstruction of pulmonary arteries by *Schistosoma* eggs was initially thought to be the main mechanism responsible for the development of PH in schistosomiasis (Simonneau et al. 2009; Chaves 1966; Lapa et al. 2009). However, it has been demonstrated that PH associated with schistosomiasis may have a similar clinical presentation and histologic findings than IPAH (Lapa et al. 2009). The mechanisms of PH in patients with schistosomiasis are probably multifactorial including PoPH and local vascular inflammation, whereas mechanical obstruction by *Schistosoma* eggs seems to play a minor role. Therefore, PH associated with schistosomiasis has been included in Group 1 (PAH) in the Dana Point updated classification (Simonneau et al. 2009; Galiè et al. 2009). More than 200 million people are infected by *Schistosoma* worldwide and 4–8 % of the infected individuals subsequently develop hepatosplenic disease. Therefore, PAH associated with schistosomiasis represents a frequent form of the disease,

Table 4 Anatomic-pathophysiological classification of congenital systemic-to-pulmonary shunts associated with PAH. Adapted from Simonneau et al. (2009)

---

1. Type
1.1. Single pre-tricuspid shunts
1.1.1. Atrial septal defect (ASD)
1.1.1.1. Ostium secundum
1.1.1.2. Sinus venosus
1.1.1.3. Ostium primum
1.1.2. Total or partial unobstructed anomalous pulmonary venous return
1.2. Simple post-tricuspid shunts
1.2.1. Ventricular septal defect (VSD)
1.2.2. Patent ductus arteriosus
1.3. Combined shunts (describe combination and define predominant defect)
1.4. Complex congenital heart disease
1.4.1. Complete atrioventricular septal defect
1.4.2. Truncus arteriosus
1.4.3. Single ventricle physiology with unobstructed pulmonary blood flow
1.4.4. Transposition of the great arteries with VSD (without pulmonary stenosis) and/or patent ductus arteriosus
1.4.5. Other
2. Dimension (specify for each defect if >1 congenital heart defect)
2.1. Hemodynamic (specify Qp/Qs)[a]
2.1.1. Restrictive (pressure gradient across the defect)
2.1.2. Nonrestrictive
2.2. Anatomic
2.2.1. Small to moderate (ASD ≤ 2.0 cm and VSD ≤ 1.0 cm)
2.2.2. Large (ASD >2.0 cm and VSD >1.0 cm)
3. Direction of shunt
3.1 Predominantly systemic-to-pulmonary
3.2 Predominantly pulmonary-to-systemic
3.3 Bidirectional
4. Associated cardiac and extracardiac abnormalities
5. Repair status
5.1. Unoperated
5.2. Palliated (specify type of operation(s), age at surgery)
5.3. Repaired (specify type of operation(s), age at surgery)

---

[a]Ratio of pulmonary (Op)-to-systemic (Qs) blood flow

especially in countries where the infection is endemic. Data from Brazil based on invasive hemodynamic measurements indicated the prevalence of PAH in patients with hepatosplenic disease of 4.6 % (Lapa et al. 2009) Of note, the prevalence of post-capillary PH was also important (3.0 %) reinforcing the need of RHC to appropriately characterize PH (Lapa et al. 2009).

Table 5 Clinical classification of congenital systemic-to-pulmonary shunts associated with PAH. Adapted from Simonneau et al. (2009)

A. Eisenmenger's syndrome	Includes all systemic-to-pulmonary shunts resulting from large defects and leading to a severe increase in PVR and a reversed (pulmonary-to-systemic) or bidirectional shunt; cyanosis, erythrocytosis, and multiple organ involvement are present
B. PAH associated with systemic-to-pulmonary shunts	Includes moderate to large defects; PVR is mildly to moderately increased, systemic-to-pulmonary shunt is still prevalent, and no cyanosis is present at rest
C. PAH with small defects	Small defects (usually ventricular septal defects <1 cm and atrial septal defects <2 cm of effective diameter assessed by echocardiography); clinical picture is very similar to idiopathic PAH
D. PAH after corrective cardiac surgery	Congenital heart disease has been corrected, but PAH is still present immediately after surgery or recurs several months or years after surgery in the absence of significant postoperative residual lesions

PAH pulmonary arterial hypertension; *PVR* pulmonary vascular resistance

4.8 Group 1.4.6: Chronic Hemolytic Anemia

There has been increasing evidence that PH is a complication of chronic hereditary and acquired hemolytic anemias, including sickle cell disease (Castro et al. 2003; Gladwin et al. 2004; Parent et al. 2011; Fonseca et al. 2012) and thalassemia (Aessopos et al. 1995). Pulmonary hypertension has been reported most frequently in patients with sickle cell disease; however, the prevalence of PAH is not clearly established. A large US study of patients with sickle cell disease, which screened PH by means of Doppler echocardiography by the presence of tricuspid regurgitation jet velocity (TRV) ≥ 2.5 m/s, found that 32 % of patients met this criteria (Gladwin et al. 2004). However, TRV ≥ 2.5 m/s is low threshold and leads to a substantial number of false-positive cases of PH if not confirmed by RHC (Parent et al. 2011; Fonseca et al. 2012). In this study, RHC was carried out in only 18 of 63 patients with TRV ≥ 2.5 m/s and PH defined by a mPAP ≥ 25 mmHg was confirmed in 17 patients (Gladwin et al. 2004). Nevertheless, PWP was elevated in a number of patients, emphasizing the complex mechanisms of PH in sickle cell patients. Indeed, a substantial proportion of sickle cell disease patients develop post-capillary PH (Parent et al. 2011). In addition, some patients present with a hyperkinetic state with moderate elevation in mPAP and normal PVR. Thus, the prevalence of pre-capillary PH in sickle cell disease is undoubtedly much lower than 32 %.

Two recent studies (Parent et al. 2011; Fonseca et al. 2012) have addressed the precise prevalence of pre- and post-capillary PH in sickle cell disease. In these two studies, patients with sickle cell disease underwent Doppler echocardiography with measurement of TRV. RHC had to be performed in all patients in whom PH was suspected on the basis of TRV ≥ 2.5 m/s. Pulmonary hypertension was defined as a mPAP ≥ 25 mmHg. The results of these two studies performed in France (Gladwin et al. 2004) and in Brazil (Parent et al. 2011) are remarkably similar showing a prevalence of PH of 6.2 % and 10 %, respectively, which was lower than previously reported. Interestingly, post-capillary PH was the most frequent cause with a prevalence of 3.3 % and 6.2 %, respectively, whereas the prevalence of pre-capillary PH was only 2.9 % and 3.8 %, respectively (Parent et al. 2011; Fonseca et al. 2012). In addition, these two studies clearly demonstrated that when a threshold TRV of ≥2.5 m/s was used to define PH, the positive predictive value of echocardiography for the detection of pulmonary hypertension was only 25 % and 32 %, respectively (Parent et al. 2011; Fonseca et al. 2012). Lastly, pre-capillary PH associated with sickle cell disease appears substantially different from other forms of PAH, in terms of both hemodynamic profile and response to specific PAH therapies. These observations call into question the rationale of keeping sickle cell disease in group 1 (PAH) of the clinical PH classification (Parent et al. 2011).

4.9 Lessons from the French PAH Registry in the Modern Management era

Pulmonary arterial hypertension is a rare and severe vascular disease for which the trend is to manage the patients in designated centers with multidisciplinary teams working in a shared-care approach (Humbert et al. 2006; Humbert et al. 2010a; Humbert et al. 2010b). The French registry was initiated in 17 university hospitals following at least 5 newly diagnosed patients per year. All consecutive adult (≥ 18 years) patients seen between October 2002 and October 2003 were to be included and followed up for at least 3 years: 674 patients (age of 50 ± 15 years (mean ± SD); range 18–85) were entered in the registry. Idiopathic, familial, anorexigen, connective tissue diseases, congenital heart diseases, portal hypertension, and HIV-associated PAH accounted for 39.2 %, 3.9 %, 9.5 %, 15.3 %, 11.3 %, 10.4 %, and 6.2 % of the population, respectively (Humbert et al. 2006). At the time of diagnosis, 75 % of patients were in the New York Heart Association (NYHA) functional class III or IV (Humbert et al. 2006). Six-minute walk distance was 329 ± 109 m. mPAP, cardiac index, and PVR index were 55 ± 15 mmHg, 2.5 ± 0.8 L/min/m^2, and 20.5 ± 10.2 mmHg/L/min/m^2, respectively (Humbert et al. 2006). Delay between symptom onset (mainly dyspnea on exercise) and PAH diagnosis was still ≥ 2 years (27 months), similar to that observed in the National Institutes of Health Registry in the 1980s, emphasizing the need for better PAH awareness and diagnostic strategy (Humbert et al. 2006). The low estimates of prevalence and incidence of PAH in France were 15.0 cases/million of adult population and 2.4 cases/million of adults per year, respectively (Humbert et al. 2006). Among the 354 consecutive adult patients with idiopathic, familial, or anorexigen-induced PAH who were prospectively enrolled, 56 had incident disease (PAH diagnosis was made at the time of recruitment in the registry) and

298 were prevalent cases (PAH diagnosis was made months or years before recruitment in the registry) (Humbert et al. 2010a; Humbert et al. 2010b). Patients were followed for 3 years and survival rates were analyzed. For incident idiopathic, familial, and anorexigen-induced PAH, estimated survival (95 % confidence intervals) at 1, 2, and 3 years was 85.7 % (76.5–94.9 %), 69.6 % (57.6–81.6 %), and 54.9 % (41.8–68.0 %), respectively (Humbert et al. 2010a). In a combined analysis population (incident patients and prevalent patients diagnosed within 3 years prior to study entry; $n = 190$), 1-, 2-, and 3-year survival estimates were 82.9 % (72.4 %–95.0 %), 67.1 % (57 · 1 %–78.8 %), and 58.2 % (49.0 %–69.3 %), respectively (Humbert et al. 2010a). Individual survival analysis identified the following factors as significantly and positively associated with survival: being female, NYHA functional class I/II, a greater six-minute walk distance, a lower right atrial pressure, and a higher cardiac output (Humbert et al. 2010a). Multivariable analysis showed that being female and having a greater six-minute walk distance and a higher cardiac output were jointly significantly associated with improved survival (Humbert et al. 2010a). This contemporary registry highlights current practice and shows that PAH is still detected late in the course of the disease with a majority of patients displaying severe functional and hemodynamic compromise. In the modern management era, idiopathic, familial, and anorexigen-induced PAH remains a progressive, fatal disease. Mortality is most closely associated with male gender, right ventricular hemodynamic function, and exercise limitation (Humbert et al. 2010a; Humbert et al. 2010b).

5 Group 1′: Pulmonary Veno-Occlusive Disease and Pulmonary Capillary Hemangiomatosis

Pulmonary veno-occlusive disease (PVOD) and pulmonary capillary hemangiomatosis (PCH) are uncommon conditions, but they are increasingly recognized as causes of PH (Montani et al. 2008). Similarities in pathological features and clinical presentation suggest that PVOD and PCH may in fact overlap, and it has been shown that PCH may be an angioproliferative process frequently associated with PVOD (Lantuejoul et al. 2006). Both conditions are difficult to categorize, as they share characteristics with PAH but also have a number of distinct differences. Given the current evidence, it has been decided at the Dana Point world symposium that PVOD/PCH should be a distinct category but not completely separated from PAH, and they are currently designated as a category 1′ in the updated classification (Simonneau et al. 2009; Galiè et al. 2009).

Pulmonary veno-occlusive disease and pulmonary capillary hemangiomatosis share a number of characteristics with PAH. First, some histologic changes in the small pulmonary arteries (intimal fibrosis and medial hypertrophy) observed in PAH are also found in PVOD/PCH. Second, clinical presentation and hemodynamic characteristics of PVOD/PCH and PAH are often indistinguishable (Montani et al. 2008; Montani et al. 2009). Third, PVOD/PCH and PAH share similar risk factors, including connective tissue diseases such as systemic sclerosis (Dorfmuller et al. 2007; Günther et al. 2012), HIV infection (Escamilla et al. 1995), and anorexigen intake (Montani et al. 2009). Last, familial occurrence has been reported in both PVOD/PCH, and mutations in the *BMPR2* gene have been documented in patients with PVOD (Montani et al. 2008; Montani et al. 2009). These findings suggest that PVOD, PCH, and PAH may represent different components of a single spectrum of disease.

Although PVOD/PCH may present similarly to PAH, there are also a number of important differences. These include the presence of crackles on clinical examination, radiologic abnormalities best seen on high-resolution computed tomography of the chest (nodular ground glass opacities, septal thickening, mediastinal lymph node enlargement) (Montani et al. 2008; Montani et al. 2009), hemosiderin-laden macrophages on bronchoalveolar lavage (Rabiller et al. 2006), a more pronounced hypoxemia, and a lower diffusing capacity for carbon monoxide (DLCO) in patients with PVOD/PCH (Montani et al. 2008; Montani et al. 2009). In addition, the response to PAH therapy and prognosis of PVOD/PCH are worse than in PAH, with an increased risk of pulmonary edema with continuous intravenous epoprostenol, calcium-channel blockers, endothelin receptor antagonists, and phosphodiesterase type 5 inhibitors (Montani et al. 2008; Montani et al. 2009; Humbert et al. 1998).

6 Group 2: Pulmonary Hypertension due to Left Heart Disease

Left-sided ventricular or valvular diseases may produce an increase in left atrial pressure, leading to a backward transmission of the pressure and a passive increase of pulmonary pressures. Left heart disease probably represents the most frequent cause of PH (Simonneau et al. 2009; Galiè et al. 2009; Oudiz 2007). In this situation, PVR is normal or near normal (< 3 mmHg/L/min) and there is no gradient between mPAP and PWP (TPG ≤ 12 mmHg) (Galiè et al. 2009). The increasing recognition of the left-sided heart dysfunction with a preserved ejection fraction has led to changes in the subcategories of Group 2, which now includes left heart systolic dysfunction, left heart diastolic dysfunction, and valvular diseases (Simonneau et al. 2009; Galiè et al. 2009). In some patients with left heart disease, the elevation of mPAP is out of proportion to that expected from the elevation of left arterial pressure (TPG > 12 mmHg) and PVR is higher than 3 mmHg/L/min (Galiè et al. 2009). Some patients with valvular disease or left heart dysfunction can develop severe PH of the same magnitude as that seen in PAH (Zener et al. 1972). The elevation of mPAP and PVR may be due to either the increase of pulmonary artery vasomotor tone and/or pulmonary vascular remodeling (Delgado et al. 2005; Moraes et al. 2000). No large, randomized controlled trial using medications approved for PAH have been performed in this patient population, and the efficacy and safety of PAH medications are currently unknown.

7 Group 3: Pulmonary Hypertension due to Lung Diseases and/or Hypoxia

In this group, the predominant cause of PH is alveolar hypoxia as a result of either chronic lung disease, impaired control of breathing, or living at high altitudes (Simonneau et al. 2009; Galiè et al. 2009; Weitzenblum et al. 1981; Thabut et al. 2005; Chaouat et al. 2005; Chaouat et al. 2008; Cottin et al. 2005). However, the precise prevalence of PH in all these conditions remains largely unknown. In the updated PH classification, a category of lung disease characterized by a mixed obstructive and restrictive pattern was added, including combined pulmonary fibrosis and emphysema (Simonneau et al. 2009; Cottin et al. 2005). In PH associated with parenchymal lung disease, the increase of mPAP is usually modest (mPAP < 35 mmHg) (Simonneau et al. 2009; Weitzenblum et al. 1981; Chaouat et al. 2008). In a retrospective study of 998 patients with chronic obstructive pulmonary disease who underwent RHC, only 1 % had severe PH defined as mPAP above 40 mmHg (Chaouat et al. 2005). These patients with severe PH had a mild-to-moderate airway obstruction, severe hypoxemia, hypocapnia, and a very low DLCO (Thabut et al. 2005; Chaouat et al. 2005). Large, randomized controlled trials of PAH therapies in patients with parenchymal lung disease and severe PH have not been performed, and there is no current recommendation supporting the use of PAH drugs in this group of patients (Humbert and Simonneau 2010).

8 Group 4: Chronic Thromboembolic Pulmonary Hypertension

Even if the exact incidence and prevalence of chronic thromboembolic pulmonary hypertension (CTEPH) are uncertain, it represents a frequent cause of PH and occurs in up to 4 % of patients after an acute pulmonary embolism (Tapson and Humbert 2006; Pengo et al. 2004). In the previous classification, CTEPH was divided into two subgroups: proximal CTEPH and distal CTEPH, depending on the technical possibility to manage PH by means of surgical pulmonary endarterectomy (Simonneau et al. 2009). However, there is no clear consensus about the exact definitions of proximal and distal CTEPH, and the decision regarding surgical treatment may vary depending on individual centers (Simonneau et al. 2009). Thus, in the new classification, it was decided to keep in Group 4 only a single category of CTEPH without distinction between proximal and distal forms (Simonneau et al. 2009; Galiè et al. 2009). Patients with suspected or confirmed CTEPH need to be referred to a center with expertise in the medical and surgical management of this condition, in order to consider the feasibility of performing surgery. The latter depends on the location of the obstruction, the correlation between hemodynamic findings, and the degree of mechanical obstruction assessed by pulmonary angiography, comorbidities, the willingness of the patient, and the experience of the medical and surgical teams

(Simonneau et al. 2009; Galiè et al. 2009; Kim 2006; Dartevelle et al. 2004; Jamieson et al. 2003). Patients who are not candidates for surgery may benefit from off-label use of PAH medical therapy (Suntharalingam et al. 2008; Rubin et al. 2006). However, further evaluation of these therapies in randomized control trials is needed (Rubin et al. 2006).

The clinical characteristics and management of patients enrolled in an international CTEPH registry have been recently reported (Pepke-Zaba et al. 2011). The international registry included 679 newly diagnosed ($\leq$ 6 months) consecutive patients with CTEPH (February 2007 to January 2009). Diagnosis was confirmed by RHC, ventilation–perfusion lung scintigraphy, computerized tomography, and/or pulmonary angiography. At diagnosis, a median of 14.1 months had passed since first symptoms; 427 patients (62.9 %) were considered operable, 247 (36.4 %) non-operable, and 5 (0.7 %) had no operability data; 386 patients (56.8 %, ranging 12.0–60.9 % across countries) underwent surgery (Pepke-Zaba et al. 2011). Operable patients did not differ from non-operable patients relative to symptoms, NYHA functional class, and hemodynamics. A history of acute pulmonary embolism was reported for 74.8 % of patients (77.5 % operable, 70 % non-operable) (Pepke-Zaba et al. 2011). Associated conditions included thrombophilic disorder in 31.9 % (37.1 % operable, 23.5 % non-operable) and splenectomy in 3.4 % of patients (1.9 % operable, 5.7 % non-operable) (Pepke-Zaba et al. 2011). At the time of CTEPH diagnosis, 37.7 % of patients initiated at least one PAH therapy (28.3 % operable, 53.8 % non-operable) (Pepke-Zaba et al. 2011). Pulmonary endarterectomy was performed with a 4.7 % documented mortality rate (Pepke-Zaba et al. 2011). Long-term follow-up of this CTEPH cohort is currently being studied with a goal to analyze outcomes according to clinical presentation and treatment strategy.

9 Group 5: Pulmonary Hypertension with Unclear or Multifactorial Etiologies

9.1 Group 5.1: Hematologic Disorders

Pulmonary hypertension has been reported in chronic myeloproliferative disorders including polycythemia vera, essential thrombocythemia, and chronic myelogenous leukemia (Guilpain et al. 2008; Adir and Humbert 2010). Several mechanisms may be implicated in PH associated with chronic myeloproliferative disorders including high cardiac output, spleen involvement (asplenia or spleen enlargement), direct obstruction of pulmonary arteries by circulating megakaryocytes, CTEPH, portopulmonary PH, drug-induced PAH (dasatinib), and congestive heart failure (Montani et al. 2012; Adir and Humbert 2010). Splenectomy, as a result of trauma or as a treatment for hematologic disorders, may increase the risk of developing PH (Adir and Humbert 2010). CTEPH and several cases of PAH with medial hypertrophy, intimal fibrosis, and plexiform lesions in the pulmonary vasculature have been reported in association with splenectomy (Hoeper et al. 1999; Jais et al. 2005).

9.2 Group 5.2: Systemic Disorders

The second subgroup includes systemic disorders, including sarcoidosis, pulmonary Langerhans' cell histiocytosis, lymphangioleiomyomatosis, neurofibromatosis, or vasculitis (Simonneau et al. 2009; Galiè et al. 2009).

Sarcoidosis is a common systemic granulomatous disease of an unknown etiology. PH is an increasingly recognized complication of sarcoidosis, with a reported prevalence of 1–28 % (Shorr et al. 2005; Nunes et al. 2006). PH is often attributed to the destruction of the capillary bed by the pulmonary fibrotic process and/or results from chronic hypoxia (Shorr et al. 2005; Nunes et al. 2006). However, the severity of PH does not occasionally correlate to the severity of parenchymal lung disease and blood gas abnormalities, suggesting that other mechanisms may contribute to the development of PH (Nunes et al. 2006). Among these mechanisms, one can consider extrinsic compression of large pulmonary vessels by lymph node enlargement or mediastinal fibrosis, granulomatous infiltration of the pulmonary vasculature, affecting especially the pulmonary veins (which sometimes mimics PVOD), cardiac sarcoidosis which may cause heart failure and post-capillary PH, and hepatic sarcoidosis which may cause portopulmonary PH (Shorr et al. 2005; Nunes et al. 2006).

Pulmonary Langerhans' cell histiocytosis is an uncommon lung disease that predominantly affects young adults and develops almost exclusively in those with a history of current or prior cigarette smoking. In pulmonary Langerhans' cell histiocytosis, pre-capillary PH is frequently observed in patients with advanced lung destruction, though no clear relationship exists between PH and the extent of parenchymal lung disease and/or hypoxia (Fartoukh et al. 2000). This observation suggests that alternative or additional mechanisms contribute to an intrinsic pulmonary vasculopathy that involves both the pre-capillary arterioles and post-capillary venous compartment, in addition to possible PVOD-like lesions (Fartoukh et al. 2000). Patients with pulmonary Langerhans' cell histiocytosis that develop PH have a particularly poor prognosis and an early referral for lung transplantation assessment is recommended (Fartoukh et al. 2000; Le Pavec et al. 2012). Recent encouraging data suggest that agents licensed for use in PAH confer improvements in pulmonary hemodynamics and are generally well tolerated. Further investigations into the use of PAH medical therapy in this population are warranted (Le Pavec et al. 2012).

Lymphangioleiomyomatosis is a rare multisystem disorder predominantly affecting women, characterized by cystic lung destruction, lymphatic abnormalities, and abdominal tumors. PH is relatively uncommon in patients with lymphangioleiomyomatosis (Cottin et al. 2012). A retrospective, multicenter study evaluated 20 patients with lymphangioleiomyomatosis and pre-capillary PH (Cottin et al. 2012). This study confirmed that PH of mild hemodynamic severity may occur in patients with lymphangioleiomyomatosis, even with mild pulmonary function impairment (Cottin et al. 2012). Off-label use of PAH therapies might improve hemodynamic parameters but requires further investigations (Cottin et al. 2012).

Neurofibromatosis type 1, also known as von Recklinghausen's disease, is an autosomal dominant disease that can be recognized by characteristic "café au lait" skin lesions and by cutaneous fibromas (Montani et al. 2011; Stewart et al. 2007). Neurofibromatosis type 1 is occasionally complicated by systemic vasculopathy. Several cases of PH have recently been reported in patients with von Recklinghausen's disease (Montani et al. 2011; Stewart et al. 2007). The mechanism of PH is unclear, and lung fibrosis and CTEPH may play a role in the development of PH. In rare cases, histological examination found both arteries and veins narrowed by medial and/or intimal hypertrophy and fibrosis (Montani et al. 2011; Stewart et al. 2007).

Last, some rare cases of PH have been observed in antineutrophil cytoplasmic antibody (ANCA)-associated vasculitis with clinical presentation similar to PAH; however, histological data are not available (Launay et al. 2006).

9.3 Group 5.3: Metabolic Disorders

PH has been reported in a few cases of type Ia glycogen storage disease, a rare autosomal recessive disorder caused by a deficiency of glucose-6-phosphatase (Hamaoka et al. 1990; Humbert et al. 2002; Pizzo 1980). The mechanisms of PH are uncertain, but portocaval shunts, atrial septal defects, severe restrictive pulmonary disease, or thromboembolic disease are thought to play a role. In one case, postmortem examination revealed the presence of plexiform lesions (Pizzo 1980).

Gaucher's disease is a rare disorder characterized by a deficiency of lysosomal B glucosidase, which results in an accumulation of glucocerebroside in reticuloendothelial cells. PH has been reported in Gaucher's disease with several potential mechanisms, including interstitial lung disease, chronic hypoxia, capillary plugging by Gaucher's cells, and splenectomy (Elstein et al. 1998; Theise and Ursell 1990).

The association between thyroid diseases and PH has been reported in a number of studies (Badesch et al. 1993; Chu et al. 2002; Li et al. 2007). High prevalence of autoimmune hypo- and hyperthyroidism suggests that these conditions may share a common (auto)immune susceptibility (Badesch et al. 1993; Chu et al. 2002; Li et al. 2007).

9.4 Group 5.4: Miscellaneous Conditions

The last subgroup includes a number of miscellaneous conditions, including tumoral extrinsic or intrinsic obstruction, fibrosing mediastinitis, and chronic renal failure with dialysis.

A progressive obstruction of proximal pulmonary arteries leading to PH may be observed when a tumor grows into the central pulmonary arteries with additional thrombosis. Such cases are mainly associated with pulmonary artery sarcomas (Anderson et al. 1995; Mayer et al. 2001). The differential diagnosis with CTEPH can be difficult and findings on angiography by computed tomography or magnetic resonance imaging, as well as 18 F-fluorodeoxyglucose positron emission tomography, may be useful to differentiate an obstruction by tumor or thrombotic material. Occlusion of the microvasculature by metastatic tumor emboli represents another rare cause of rapidly progressive PH (Roberts et al. 2003; Dot et al. 2007). The initial laboratory evaluation often shows severe hypoxemia. CT scanning does not reveal proximal thrombi but often shows thickening of septa. In contrast, the ventilation/perfusion lung scan is generally abnormal with multiple subsegmental perfusion defects. Pulmonary microvascular cytology sampling through a pulmonary artery catheter in the wedge position is an important diagnostic tool (Dot et al. 2007). The majority of reported cases occur in association with breast, lung, or gastric carcinoma (Roberts et al. 2003; Dot et al. 2007).

Fibrosing mediastinitis may be associated with severe PH due to compression of both pulmonary arteries and veins (Davis et al. 2001; Loyd et al. 1988; Goodwin et al. 1972). Ventilation/perfusion lung scan, computerized tomography of the chest, and pulmonary angiography are useful for accurate diagnosis; however, findings can mimic proximal thrombotic obstruction (Seferian et al. 2012). The predominant etiologies are histoplasmosis (Loyd et al. 1988; Goodwin et al. 1972), tuberculosis (Goodwin et al. 1972; Seferian et al. 2012), and sarcoidosis (Nunes et al. 2006).

Lastly, PH has been reported in patients with end-stage renal disease maintained on long-term hemodialysis (Yigla et al. 2003). Based on echocardiographic studies, elevated estimated PAP may be high in this patient population (Yigla et al. 2003). There are several potential explanations for the development of PH in these patients: mPAP may be increased by high CO (resulting from the arteriovenous access and anemia), as well as fluid overload. In addition, diastolic and systolic left heart dysfunctions are also frequent in this setting, leading to a significant proportion of post-capillary PH (Yigla et al. 2003; Nakhoul et al. 2005). Furthermore, hormonal and metabolic derangement associated with end-stage renal disease might promote dysfunction of pulmonary vascular tone.

10 Conclusion

Since the mid-twentieth century, a significant progress has been made in the PH field starting from the development of RHC techniques and the first description of PPH. The National Institutes of Health Registry and the world symposiums on PH have played an important role in this process. The most common causes of PH remain post-capillary PH due to left heart diseases (group 2) and pre-capillary PH due to chronic lung diseases and/or hypoxia (group 3). Besides these two common causes, CTEPH is of a major importance because it can be surgically managed. Lastly, PAH patients should be identified and classified early, as several targeted therapies have been approved in this group of diseases. The proceedings of the fifth world PH symposium will highlight current knowledge of PH that should help the international community to better manage PH patients (Humbert 2012).

References

Abenhaim L, Moride Y, Brenot F, Rich S, Benichou J, Kurz X, Higenbottam T, Oakley C, Wouters E, Aubier M, Simonneau G, Bégaud B (1996) Appetite-suppressant drugs and the risk of primary pulmonary hypertension. N Engl J Med 335:609–616

Adir Y, Humbert M (2010) Pulmonary hypertension in patients with chronic myeloproliferative disorders. Eur Respir J 35:1396–1406

Aessopos A, Stamatelos G, Skoumas V, Vassilopoulos G, Mantzourani M, Loukopoulos D (1995) Pulmonary hypertension and right heart failure in patients with beta-thalassemia intermedia. Chest 107:50–53

Anderson MB, Kriett JM, Kapelanski DP, Tarazi R, Jamieson SW (1995) Primary pulmonary artery sarcoma: a report of six cases. Ann Thorac Surg 59:1487–1490

Asherson RA, Higenbottam TW, Dinh Xuan AT, Khamashta MA, Hughes GR (1990) Pulmonary hypertension in a lupus clinic: experience with twenty-four patients. J Rheumatol 17:1292–1298

Badesch DB, Wynne KM, Bonvallet S, Voelkel NF, Ridgway C, Groves BM (1993) Hypothyroidism and primary pulmonary hypertension: an autoimmune pathogenetic link? Ann Intern Med 119:44–46

Boutet K, Frachon I, Jobic Y, Gut-Gobert C, Leroyer C, Carlhant-Kowalski D, Sitbon O, Simonneau G, Humbert M (2009) Fenfluramine-like cardiovascular side-effects of benfluorex. Eur Respir J 33:684–688

Brenot F, Hervé P, Petitpretz P, Parent F, Duroux P, Simonneau G (1993) Primary pulmonary hypertension and fenfluramine use. Br Heart J 70:537–541

Bunch TW, Tancredi RG, Lie JT (1981) Pulmonary hypertension in polymyositis. Chest 79:105–107

Burdt MA, Hoffman RW, Deutscher SL, Wang GS, Johnson JC, Sharp GC (1999) Long-term outcome in mixed connective tissue disease: longitudinal clinical and serologic findings. Arthritis Rheum 42:899–909

Castro O, Hoque M, Brown BD (2003) Pulmonary hypertension in sickle cell disease: cardiac catheterization results and survival. Blood 101:1257–1261

Chambers CD, Hernandez-Diaz S, Van Marter LJ, Werler MM, Louik C, Jones KL, Mitchell AA (2006) Selective serotonin-reuptake inhibitors and risk of persistent pulmonary hypertension of the newborn. N Engl J Med 354:579–587

Chaouat A, Bugnet AS, Kadaoui N, Schott R, Enache I, Ducolone A, Ehrhart M, Kessler R, Weitzenblum E (2005) Severe pulmonary hypertension and chronic obstructive pulmonary disease. Am J Respir Crit Care Med 172:189–194

Chaouat A, Naeije R, Weitzenblum E (2008) Pulmonary hypertension in COPD. Eur Respir J 32:1371–1385

Chaves E (1966) The pathology of the arterial pulmonary vasculature in Manson's schistosomiasis. Dis Chest 50:72–77

Chin KM, Channick RN, Rubin LJ (2006) Is methamphetamine use associated with idiopathic pulmonary arterial hypertension? Chest 130:1657–1663

Chu JW, Kao PN, Faul JL, Doyle RL (2002) High prevalence of autoimmune thyroid disease in pulmonary arterial hypertension. Chest 122:1668–1673

Connolly HM, Crary JL, McGoon MD, Hensrud DD, Edwards BS, Edwards WD, Schaff HV (1997) Valvular heart disease associated with fenfluramine-phentermine. N Engl J Med 337:581–588

Cottin V, Nunes H, Brillet PY, Delaval P, Devouassoux G, Tillie-Leblond I, Israel-Biet D, Court-Fortune I, Valeyre D, Cordier JF (2005) Combined pulmonary fibrosis and emphysema: a distinct underrecognised entity. Eur Respir J 26:586–593

Cottin V, Harari S, Humbert M, Mal H, Dorfmüller P, Jais X, Reynaud-Gaubert M, Prevot G, Lazor R, Taillé C, Lacronique J, Zeghmar S, Simonneau G, Cordier JF (2012) Pulmonary hypertension in lymphangioleiomyomatosis: characteristics in 20 patients. Eur Respir J 40(3):630–640

Dartevelle P, Fadel E, Mussot S, Chapelier A, Herve P, de Perrot M, Cerrina J, Ladurie FL, Lehouerou D, Humbert M, Sitbon O, Simonneau G (2004) Chronic thromboembolic pulmonary hypertension. Eur Respir J 23:637–648

Davis AM, Pierson RN, Loyd JE (2001) Mediastinal fibrosis. Semin Respir Infect 16:119–130

Dawson JK, Goodson NG, Graham DR, Lynch MP (2000) Raised pulmonary artery pressures measured with Doppler echocardiography in rheumatoid arthritis patients. Rheumatology (Oxford) 39:1320–1325

Degano B, Guillaume M, Savale L, Montani D, Jaïs X, Yaici A, Le Pavec J, Humbert M, Simonneau G, Sitbon O (2010) HIV-associated pulmonary arterial hypertension: survival and prognostic factors in the modern therapeutic era. AIDS 24:67–75

Delgado JF, Conde E, Sanchez V, Lopez-Rios F, Gomez-Sanchez MA, Escribano P, Sotelo T (2005) Gomez de la Camara A, Cortina J, de la Calzada CS. Pulmonary vascular remodeling in pulmonary hypertension due to chronic heart failure. Eur J Heart Fail 7:1011–1016

Dorfmuller P, Humbert M, Perros F, Sanchez O, Simonneau G, Muller KM, Capron F (2007) Fibrous remodeling of the pulmonary venous system in pulmonary arterial hypertension associated with connective tissue diseases. Hum Pathol 38:893–902

Dot JM, Sztrymf B, Yaici A, Dorfmuller P, Capron F, Parent F, Jais X, Sitbon O, Simonneau G, Humbert M (2007) Pulmonary arterial hypertension due to tumor emboli. Rev Mal Respir 24:359–366

Elliott CG, Glissmeyer EW, Havlena GT, Carlquist J, McKinney JT, Rich S, McGoon MD, Scholand MB, Kim M, Jensen RL, Schmidt JW, Ward K (2006) Relationship of BMPR2 mutations to vasoreactivity in pulmonary arterial hypertension. Circulation 113:2509–2515

Elstein D, Klutstein MW, Lahad A, Abrahamov A, Hadas-Halpern I, Zimran A (1998) Echocardiographic assessment of pulmonary hypertension in Gaucher's disease. Lancet 351:1544–1546

Escamilla R, Hermant C, Berjaud J, Mazerolles C, Daussy X (1995) Pulmonary veno-occlusive disease in a HIV-infected intravenous drug abuser. Eur Respir J 8:1982–1984

Fartoukh M, Humbert M, Capron F, Maitre S, Parent F, Le Gall C, Sitbon O, Herve P, Duroux P, Simonneau G (2000) Severe pulmonary hypertension in histiocytosis X. Am J Respir Crit Care Med 161:216–223

Follath F, Burkart F, Schweizer W (1971) Drug-induced pulmonary hypertension? Br Med J 5743:265–266

Fonseca GHH, Souza R, Salemi VC, Jardim CVP, Gualandro SFM (2012) Pulmonary hypertension diagnosed by right heart catheterization in sickle cell disease. Eur Respir J 39:112–118

Frachon I, Etienne Y, Jobic Y, Le Gal G, Humbert M, Leroyer C (2010) Benfluorex and unexplained valvular heart disease: a case-control study. PLoS One 5:e10128

Galiè N, Hoeper MM, Humbert M, Torbicki A, Vachiery JL, Barbera JA, Beghetti M, Corris P, Gaine S, Gibbs JS, Gomez-Sanchez MA, Jondeau G, Klepetko W, Opitz C, Peacock A, Rubin L, Zellweger M, Simonneau G (2009) Guidelines for the diagnosis and treatment of pulmonary hypertension. Eur Respir J 34:1219–1263

Girerd B, Montani D, Coulet F, Sztrymf B, Yaici A, Jaïs X, Tregouet D, Reis A, Drouin-Garraud V, Fraisse A, Sitbon O, O'Callaghan DS, Simonneau G, Soubrier F, Humbert M (2010) Clinical outcomes of pulmonary arterial hypertension in patients carrying an ACVRL1 (ALK1) mutation. Am J Respir Crit Care Med 181:851–861

Gladwin MT, Sachdev V, Jison ML, Shizukuda Y, Plehn JF, Minter K, Brown B, Coles WA, Nichols JS, Ernst I, Hunter LA, Blackwelder WC, Schechter AN, Rodgers GP, Castro O, Ognibene FP (2004) Pulmonary hypertension as a risk factor for death in patients with sickle cell disease. N Engl J Med 350:886–895

Goodwin RA, Nickell JA, Des Prez RM (1972) Mediastinal fibrosis complicating healed primary histoplasmosis and tuberculosis. Medicine (Baltimore) 51:227–246

Guilpain P, Montani D, Damaj G, Achouh L, Lefrère F, Marfaing-Koka A, Dartevelle P, Simonneau G, Humbert M, Hermine O (2008) Pulmonary hypertension associated with myeloproliferative disorders: a retrospective study of ten cases. Respiration 76:295–302

Günther S, Jaïs X, Maitre S, Bérezné A, Dorfmüller P, Seferian A, Savale L, Mercier O, Fadel E, Sitbon O, Mouthon L, Simonneau G, Humbert M, Montani D (2012) Computed tomography findings of pulmonary veno-occlusive disease in scleroderma patients presenting with precapillary pulmonary hypertension. Arthritis Rheum 64(9):2995–3005

Gurtner HP (1979) Pulmonary hypertension, "plexogenic pulmonary arteriopathy" and the appetite depressant drug aminorex: post or propter? Bull Eur Physiopathol Respir 15:897–923

Hachulla E, Gressin V, Guillevin L, Carpentier P, Diot E, Sibilia J, Kahan A, Cabane J, Frances C, Launay D, Mouthon L, Allanore Y, Tiev KP, Clerson P, de Groote P, Humbert M (2005) Early detection of pulmonary arterial hypertension in systemic sclerosis: a French nationwide prospective multicenter study. Arthritis Rheum 52:3792–3800

Hachulla E, de Groote P, Gressin V, Sibilia J, Diot E, Carpentier P, Mouthon L, Hatron PY, Jego P, Allanore Y, Tiev KP, Agard C, Cosnes A, Cirstea D, Constans J, Farge D, Viallard JF, Harle JR, Patat F, Imbert B, Kahan A, Cabane J, Clerson P, Guillevin L, Humbert M (2009) The 3-year incidence of pulmonary arterial hypertension associated with systemic sclerosis in a multicenter nationwide longitudinal study (ItinérAIR-Sclérodermie Study). Arthritis Rheum 60:1831–1839

Hadengue A, Benhayoun MK, Lebrec D, Benhamou JP (1991) Pulmonary hypertension complicating portal hypertension: prevalence and relation to splanchnic hemodynamics. Gastroenterology 100:520–528

Hamaoka K, Nakagawa M, Furukawa N, Sawada T (1990) Pulmonary hypertension in type I glycogen storage disease. Pediatr Cardiol 11:54–56

Hatano S, Strasser T (1975) Primary pulmonary hypertension. Report on a WHO meeting. October 15–17, 1973. WHO, Geneva

Hoeper MM, Niedermeyer J, Hoffmeyer F, Flemming P, Fabel H (1999) Pulmonary hypertension after splenectomy? Ann Intern Med 130:506–509

Hoffman JI, Rudolph AM (1965) The natural history of ventricular septal defects in infancy. Am J Cardiol 16:634–653

Humbert M (2012) The fifth world symposium on pulmonary hypertension will REVEAL the impact of registries. Eur Respir Rev 21:4–5

Humbert M, Simonneau G (2010) Vasodilators in patients with chronic obstructive pulmonary disease and pulmonary hypertension: not ready for prime time. Am J Respir Crit Care Med 181:202–203

Humbert M, Maitre S, Capron F, Rain B, Musset D, Simonneau G (1998) Pulmonary edema complicating continuous intravenous prostacyclin in pulmonary capillary hemangiomatosis. Am J Respir Crit Care Med 157:1681–1685

Humbert M, Labrune P, Sitbon O, Le Gall C, Callebert J, Herve P, Samuel D, Machado R, Trembath R, Drouet L, Launay JM, Simonneau G (2002) Pulmonary arterial hypertension and type-I glycogen-storage disease: the serotonin hypothesis. Eur Respir J 20:59–65

Humbert M, Sitbon O, Chaouat A, Bertocchi M, Habib G, Gressin V, Yaici A, Weitzenblum E, Cordier JF, Chabot F, Dromer C, Pison C, Reynaud-Gaubert M, Haloun A, Laurent M, Eric Hachulla E, Simonneau G (2006) Pulmonary arterial hypertension in France: results from a national registry. Am J Respir Crit Care Med 173:1023–1030

Humbert M, Khaltaev N, Bousquet J, Souza R (2007) Pulmonary hypertension: from an orphan disease to a public health problem. Chest 132:365–367

Humbert M, Sitbon O, Chaouat A, Bertocchi M, Habib G, Gressin V, Yaici A, Weitzenblum E, Cordier JF, Chabot F, Dromer C, Pison C, Reynaud-Gaubert M, Haloun A, Laurent M, Hachulla E, Cottin V, Degano B, Jaïs X, Montani D, Souza R, Simonneau G (2010a) Survival in patients with idiopathic, familial, and anorexigen-associated pulmonary arterial hypertension in the modern management era. Circulation 122:156–163

Humbert M, Sitbon O, Yaici A, Montani D, O'Callaghan DS, Jaïs X, Parent F, Savale L, Natali D, Günther S, Chaouat A, Chabot F, Cordier JF, Habib G, Gressin V, Jing ZC, Souza R, Simonneau G (2010b) Survival in incident and prevalent cohorts of patients with pulmonary arterial hypertension. Eur Respir J 36:549–555

Jais X, Ioos V, Jardim C, Sitbon O, Parent F, Hamid A, Fadel E, Dartevelle P, Simonneau G, Humbert M (2005) Splenectomy and chronic thromboembolic pulmonary hypertension. Thorax 60:1031–1034

Jais X, Launay D, Yaici A, Le Pavec J, Tcherakian C, Sitbon O, Simonneau G, Humbert M (2008) Immunosuppressive therapy in lupus- and mixed connective tissue disease-associated pulmonary arterial hypertension: a retrospective analysis of twenty-three cases. Arthritis Rheum 58:521–531

Jamieson SW, Kapelanski DP, Sakakibara N, Manecke GR, Thistlethwaite PA, Kerr KM, Channick RN, Fedullo PF, Auger WR (2003) Pulmonary endarterectomy: experience and lessons learned in 1,500 cases. Ann Thorac Surg 76:1457–1462, discussion 62–64

Kawut SM, Krowka MJ, Trotter JF, Roberts KE, Benza RL, Badesch DB, Taichman DB, Horn EM, Zacks S, Kaplowitz N, Brown RS Jr, Fallon MB (2008) Clinical risk factors for portopulmonary hypertension. Hepatology 48:196–203

Kim NH (2006) Assessment of operability in chronic thromboembolic pulmonary hypertension. Proc Am Thorac Soc 3:584–588

Kovacs G, Berghold A, Scheidl S, Olschewski H (2009) Pulmonary arterial pressure during rest and exercise in healthy subjects: a systematic review. Eur Respir J 34:888–894

Lantuejoul S, Sheppard MN, Corrin B, Burke MM, Nicholson AG (2006) Pulmonary veno-occlusive disease and pulmonary capillary hemangiomatosis: a clinicopathologic study of 35 cases. Am J Surg Pathol 30:850–857

Lapa M, Dias B, Jardim C, Fernandes CJ, Dourado PM, Figueiredo M, Farias A, Tsutsui J, Terra-Filho M, Humbert M, Souza R (2009) Cardiopulmonary manifestations of hepatosplenic schistosomiasis. Circulation 119:1518–1523

Launay D, Souza R, Guillevin L, Hachulla E, Pouchot J, Simonneau G, Humbert M (2006) Pulmonary arterial hypertension in ANCA-associated vasculitis. Sarcoidosis Vasc Diffuse Lung Dis 23:223–228

Launay D, Hachulla E, Hatron PY, Jais X, Simonneau G, Humbert M (2007) Pulmonary arterial hypertension: a rare complication of primary Sjogren syndrome: report of 9 new cases and review of the literature. Medicine (Baltimore) 86:299–315

Le Pavec J, Souza R, Herve P, Lebrec D, Savale L, Tcherakian C, Jais X, Yaici A, Humbert M, Simonneau G, Sitbon O (2008) Portopulmonary hypertension: survival and prognostic factors. Am J Respir Crit Care Med 178:637–643

Le Pavec J, Lorillon G, Jaïs X, Tcherakian C, Feuillet S, Dorfmüller P, Simonneau G, Humbert M, Tazi A (2012) Pulmonary Langerhans cell histiocytosis associated pulmonary hypertension: clinical characteristics and impact of pulmonary arterial hypertension therapies. Chest 142(5):1150–1157

Li JH, Safford RE, Aduen JF, Heckman MG, Crook JE, Burger CD (2007) Pulmonary hypertension and thyroid disease. Chest 132:793–797

Loyd JE, Tillman BF, Atkinson JB, Des Prez RM (1988) Mediastinal fibrosis complicating histoplasmosis. Medicine (Baltimore) 67:295–310

Mayer E, Kriegsmann J, Gaumann A, Kauczor HU, Dahm M, Hake U, Schmid FX, Oelert H (2001) Surgical treatment of pulmonary artery sarcoma. J Thorac Cardiovasc Surg 121:77–82

Mehta NJ, Khan IA, Mehta RN, Sepkowitz DA (2000) HIV-Related pulmonary hypertension: analytic review of 131 cases. Chest 118:1133–1141

Montani D, Achouh L, Dorfmuller P, Le Pavec J, Sztrymf B, Tcherakian C, Rabiller A, Haque R, Sitbon O, Jais X, Dartevelle P, Maitre S, Capron F, Musset D, Simonneau G, Humbert M (2008) Pulmonary veno-occlusive disease: clinical, functional, radiologic, and hemodynamic characteristics and outcome of 24 cases confirmed by histology. Medicine (Baltimore) 87:220–233

Montani D, Price LC, Dorfmuller P, Achouh L, Jais X, Yaici A, Sitbon O, Musset D, Simonneau G, Humbert M (2009) Pulmonary veno-occlusive disease. Eur Respir J 33:189–200

Montani D, Coulet F, Girerd B, Eyries M, Bergot E, Mal H, Biondi G, Dromer C, Hugues T, Marquette C, O'Connell C, O'Callaghan DS, Savale L, Jaïs X, Dorfmüller P, Begueret H,

Bertoletti L, Sitbon O, Bellanné-Chantelot C, Zalcman G, Simonneau G, Humbert M, Soubrier F (2011) Pulmonary hypertension in patients with neurofibromatosis type I. Medicine (Baltimore) 90:201–211

Montani D, Bergot E, Günther S, Savale L, Bergeron A, Bourdin A, Bouvaist H, Canuet M, Pison C, Macro M, Poubeau P, Girerd B, Natali D, Guignabert C, Perros F, O'Callaghan DS, Jaïs X, Tubert-Bitter P, Zalcman G, Sitbon O, Simonneau G, Humbert M (2012) Pulmonary arterial hypertension in patients treated by dasatinib. Circulation 125:2128–2137

Moraes DL, Colucci WS, Givertz MM (2000) Secondary pulmonary hypertension in chronic heart failure: the role of the endothelium in pathophysiology and management. Circulation 102:1718–1723

Mukerjee D, St George D, Coleiro B, Knight C, Denton CP, Davar J, Black CM, Coghlan JG (2003) Prevalence and outcome in systemic sclerosis associated pulmonary arterial hypertension: application of a registry approach. Ann Rheum Dis 62:1088–1093

Nakhoul F, Yigla M, Gilman R, Reisner SA, Abassi Z (2005) The pathogenesis of pulmonary hypertension in haemodialysis patients via arterio-venous access. Nephrol Dial Transplant 20:1686–1692

Nunes H, Humbert M, Sitbon O, Morse JH, Deng Z, Knowles JA, Le Gall C, Parent F, Garcia G, Herve P, Barst RJ, Simonneau G (2003) Prognostic factors for survival in human immunodeficiency virus-associated pulmonary arterial hypertension. Am J Respir Crit Care Med 167:1433–1439

Nunes H, Humbert M, Capron F, Brauner M, Sitbon O, Battesti JP, Simonneau G, Valeyre D (2006) Pulmonary hypertension associated with sarcoidosis: mechanisms, haemodynamics and prognosis. Thorax 61:68–74

Opravil M, Sereni D (2008) Natural history of HIV-associated pulmonary arterial hypertension: trends in the HAART era. AIDS 22:S35–S40

Opravil M, Pechere M, Speich R, Joller-Jemelka HI, Jenni R, Russi EW, Hirschel B, Luthy R (1997) HIV-associated primary pulmonary hypertension: a case control study. Am J Respir Crit Care Med 155:990–995

Oudiz RJ (2007) Pulmonary hypertension associated with left-sided heart disease. Clin Chest Med 28:233–241

Parent F, Bachir D, Inamo J, Lionnet F, Driss F, Loko G, Habibi A, Bennani S, Savale L, Adnot S, Maitre B, Yaici A, Hajji L, O'Callaghan DS, Clerson P, Girot R, Galacteros F, Simonneau G (2011) A hemodynamic study of pulmonary hypertension in sickle cell disease. N Engl J Med 365:44–53

Pengo V, Lensing AW, Prins MH, Marchiori A, Davidson BL, Tiozzo F, Albanese P, Biasiolo A, Pegoraro C, Iliceto S, Prandoni P (2004) Incidence of chronic thromboembolic pulmonary hypertension after pulmonary embolism. N Engl J Med 350:2257–2264

Pepke-Zaba J, Delcroix M, Lang I, Mayer E, Jansa P, Ambroz D, Treacy C, D'Armini AM, Morsolini M, Snijder R, Bresser P, Torbicki A, Kristensen B, Lewczuk J, Simkova I, Barberà JA, de Perrot M, Hoeper MM, Gaine S, Speich R, Gomez-Sanchez MA, Kovacs G, Hamid AM, Jaïs X, Simonneau G (2011) Chronic thromboembolic pulmonary hypertension (CTEPH): results from an international prospective registry. Circulation 124:1973–1981

Pizzo CJ (1980) Type I, glycogen storage disease with focal nodular hyperplasia of the liver and vasoconstrictive pulmonary hypertension. Pediatrics 65:341–343

Rabiller A, Jais X, Hamid A, Resten A, Parent F, Haque R, Capron F, Sitbon O, Simonneau G, Humbert M (2006) Occult alveolar haemorrhage in pulmonary veno-occlusive disease. Eur Respir J 27:108–113

Roberts KE, Hamele-Bena D, Saqi A, Stein CA, Cole RP (2003) Pulmonary tumor embolism: a review of the literature. Am J Med 115:228–232

Rodriguez-Roisin R, Krowka MJ, Herve P, Fallon MB (2004) Pulmonary-hepatic vascular disorders. Eur Respir J 24:861–880

Rubin LJ, Hoeper MM, Klepetko W, Galie N, Lang IM, Simonneau G (2006) Current and future management of chronic thromboembolic pulmonary hypertension: from diagnosis to treatment responses. Proc Am Thorac Soc 3:601–607

Sanchez O, Sitbon O, Jaïs X, Simonneau G, Humbert M (2006) Immunosuppressive therapy in connective tissue diseases-associated pulmonary arterial hypertension. Chest 130:182–189

Savale L, Chaumais MC, Cottin V, Bergot E, Frachon I, Prevot G, Pison C, Dromer C, Poubeau P, Lamblin N, Habib G, Reynaud-Gaubert M, Bourdin A, Sanchez O, Tubert-Bitter P, Jaïs X, Montani D, Sitbon O, Simonneau G, Humbert M (2012) Pulmonary hypertension associated with benfluorex exposure. Eur Respir J 40(5):1164–1172

Seferian A, Jaïs X, Creuze N, Savale L, Humbert M, Sitbon O, Simonneau G, Montani D (2012) Mediastinal fibrosis mimicking proximal chronic thromboembolic disease. Circulation 125: 2045–2047

Shorr AF, Helman DL, Davies DB, Nathan SD (2005) Pulmonary hypertension in advanced sarcoidosis: epidemiology and clinical characteristics. Eur Respir J 25:783–788

Simonneau G, Fartoukh M, Sitbon O, Humbert M, Jagot JL, Herve P (1998) Primary pulmonary hypertension associated with the use of fenfluramine derivatives. Chest 114:195S–199S

Simonneau G, Robbins IM, Beghetti M, Channick RN, Delcroix M, Denton CP, Elliott CG, Gaine SP, Gladwin MT, Jing ZC, Krowka MJ, Langleben D, Nakanishi N, Souza R (2009) Updated clinical classification of pulmonary hypertension. J Am Coll Cardiol 54:S43–S54

Sitbon O, Lascoux-Combe C, Delfraissy JF, Yeni PG, Raffi F, De Zuttere D, Gressin V, Clerson P, Sereni D, Simonneau G (2008) Prevalence of HIV-related pulmonary arterial hypertension in the current antiretroviral therapy era. Am J Respir Crit Care Med 177:108–113

Souza R, Humbert M, Sztrymf B, Jais X, Yaici A, Le Pavec J, Parent F, Herve P, Soubrier F, Sitbon O, Simonneau G (2008) Pulmonary arterial hypertension associated with fenfluramine exposure: report of 109 cases. Eur Respir J 31:343–348

Speich R, Jenni R, Opravil M, Pfab M, Russi EW (1991) Primary pulmonary hypertension in HIV infection. Chest 100:1268–1271

Stewart DR, Cogan JD, Kramer MR, Miller WT Jr, Christiansen LE, Pauciulo MW, Messiaen LM, Tu GS, Thompson WH, Pyeritz RE, Ryu JH, Nichols WC, Kodama M, Meyrick BO, Ross DJ (2007) Is pulmonary arterial hypertension in neurofibromatosis type 1 secondary to a plexogenic arteriopathy? Chest 132:798–808

Suntharalingam J, Treacy CM, Doughty NJ, Goldsmith K, Soon E, Toshner MR, Sheares KK, Hughes R, Morrell NW, Pepke-Zaba J (2008) Long-term use of sildenafil in inoperable chronic thromboembolic pulmonary hypertension. Chest 134:229–236

Sztrymf B, Coulet F, Girerd B, Yaici A, Jais X, Sitbon O, Montani D, Souza R, Simonneau G, Soubrier F, Humbert M (2008) Clinical outcomes of pulmonary arterial hypertension in carriers of BMPR2 mutation. Am J Respir Crit Care Med 177:1377–1383

Tanaka E, Harigai M, Tanaka M, Kawaguchi Y, Hara M, Kamatani N (2002) Pulmonary hypertension in systemic lupus erythematosus: evaluation of clinical characteristics and response to immunosuppressive treatment. J Rheumatol 29:282–287

Tapson VF, Humbert M (2006) Incidence and prevalence of chronic thromboembolic pulmonary hypertension: from acute to chronic pulmonary embolism. Proc Am Thorac Soc 3:564–567

Thabut G, Dauriat G, Stern JB, Logeart D, Levy A, Marrash-Chahla R, Mal H (2005) Pulmonary hemodynamics in advanced COPD candidates for lung volume reduction surgery or lung transplantation. Chest 127:1531–1536

Theise ND, Ursell PC (1990) Pulmonary hypertension and Gaucher's disease: logical association or mere coincidence? Am J Pediatr Hematol Oncol 12:74–76

Walker AM, Langleben D, Korelitz JJ, Rich S, Rubin LJ, Strom BL, Gonin R, Keast S, Badesch D, Barst RJ, Bourge RC, Channick R, Frost A, Gaine S, McGoon M, McLaughlin V, Murali S, Oudiz RJ, Robbins IM, Tapson V, Abenhaim L, Constantine G (2006) Temporal trends and drug exposures in pulmonary hypertension: an American experience. Am Heart J 152:521–526

Weitzenblum E, Hirth C, Ducolone A, Mirhom R, Rasaholinjanahary J, Ehrhart M (1981) Prognostic value of pulmonary artery pressure in chronic obstructive pulmonary disease. Thorax 36:752–758

Whyte K, Hoette S, Hervé P, Montani D, Jaïs X, Parent F, Savale L, Natali D, O'Callaghan DS, Garcia G, Sitbon O, Simonneau G, Humbert M, Chemla D (2012) The association between resting and mild-to-moderate exercise pulmonary artery pressure. Eur Respir J 39:313–318

Wood P (1958) The Eisenmenger syndrome or pulmonary hypertension with reversed central shunt. I. Br Med J 2:701–709

Yigla M, Nakhoul F, Sabag A, Tov N, Gorevich B, Abassi Z, Reisner SA (2003) Pulmonary hypertension in patients with end-stage renal disease. Chest 123:1577–1582

Zener JC, Hancock EW, Shumway NE, Harrison DC (1972) Regression of extreme pulmonary hypertension after mitral valve surgery. Am J Cardiol 30:820–826

Zuber JP, Calmy A, Evison JM, Hasse B, Schiffer V, Wagels T, Nuesch R, Magenta L, Ledergerber B, Jenni R, Speich R, Opravil M (2004) Pulmonary arterial hypertension related to HIV infection: improved hemodynamics and survival associated with antiretroviral therapy. Clin Infect Dis 38:1178–1185

Pulmonary Hypertension: Pathophysiology and Signaling Pathways

Bradley A. Maron and Joseph Loscalzo

Abstract Pulmonary hypertension (PH) is characterized by pathological changes to cell signaling pathways within the alveolar-pulmonary arteriole–right ventricular axis that results in increases in pulmonary vascular resistance and, ultimately, the development of right ventricular (RV) dysfunction. Cornerstone histopathological features of the PH vasculopathy include intimal thickening, concentric hypertrophy, and perivascular fibrosis of distal pulmonary arterioles. The presence of plexogenic lesions is pathognomonic of pulmonary arterial hypertension (PAH); when present, this severe form of remodeling is associated with subtotal obliteration of the blood vessel lumen. The extent of RV remodeling in PH correlates with clinical symptom severity and portends a poor outcome. Currently available PH-specific pharmacotherapies that aim to improve symptom burden by targeting pulmonary vasodilatory/vasoconstrictor cell signaling pathways do not fully reverse pulmonary vascular remodeling and, thus, are largely unsuccessful at maintaining normal cardiopulmonary hemodynamics long term. Thus, determining the molecular mechanisms that are responsible for pulmonary vascular remodeling in PH is of great potential therapeutic value, particularly pathways that promote apoptosis-resistant cellular proliferation, disrupt normal cellular bioenergetics to alter cell function, and/or modulate severely abnormal responses to pulmonary vascular injury. This chapter reviews current insights into PH pathophysiology and disease mechanisms, and discusses novel cell signaling pathways that implicate microRNAs and mitochondrial dysfunction in the development of the PH phenotype.

B.A. Maron
Cardiovascular Division, Department of Medicine, Brigham and Women's Hospital, Harvard Medical School, 77 Avenue Louis Pasteur, NRB Room 0630-AO, Boston, MA 02115, USA

J. Loscalzo (✉)
Cardiovascular Division, Department of Medicine, Brigham and Women's Hospital, Harvard Medical School, 77 Avenue Louis Pasteur, NRB Room 0630-AO, Boston, MA 02115, USA
e-mail: jloscalzo@partners.org

Keywords Network medicine • Systems pharmacology • Complex diseases • Pharmacogenetics

Contents

1	Introduction	33
2	PH Pathophysiology	34
3	Cell Signaling Mechanisms in the Pathobiology of PH	37
	3.1 Endothelial Nitric Oxide Synthase in PH	37
	3.2 Endothelin-1 System	40
	3.3 Soluble Guanylyl Cyclase and Phosphodiesterase Inhibition in PH	42
	3.4 Phosphodiesterase Inhibition in PH	43
	3.5 Prostacyclin Signaling in PH	44
	3.6 Mitochondrial Dysfunction	46
	3.7 Peroxisome Proliferator-Activated Receptor-γ	48
	3.8 MicroRNA-Mediated Regulation of Cellular Responses to Hypoxia	50
4	Conclusions	52
References		52

Abbreviations

BH_4	Tetrahydrobiopterin
BMP-RII	Bone morphogenetic protein receptor II
cAMP	Cyclic adenosine monophosphate
cGMP	Cyclic guanosine monophosphate
COX	Cyclooxygenase
EGFR	Epidermal growth factor receptor
eNOS	Endothelial nitric oxide synthase
ET-1	Endothelin-1
ET_A	Endothelin-type A receptor
ET_B	Endothelin-type B receptor
FeNO	Iron-nitrosyl
GTP	Guanosine triphosphate
HAPE	High altitude pulmonary edema syndrome
HHT	Hereditary hemorrhagic telangiectasia
HIF	Hypoxia-inducible factor
HIV	Human immunodeficiency virus
IL	Interleukin
ISCU1/2	Iron–sulfur cluster assembly proteins
KL	Kruppel-like factor
LO	Lipooxygenases
LV	Left ventricle
MAPK	Mitogen-activated protein kinase
miR	MicroRNA
NO•	Nitric oxide
NOX	NADPH oxidase

•O₂⁻	Superoxide
O₂NOO⁻	Peroxynitrate
ONOO⁻	Peroxynitrite
PAEC	Pulmonary artery endothelial cells
PAH	Pulmonary arterial hypertension
PDE	Phosphodiesterase inhibitor
PDGF	Platelet-derived growth factor
PDK	Pyruvate dehydrogenase kinase
PG	Prostaglandin
PH	Pulmonary hypertension
PKG	Protein kinase G
PPAR γ	Peroxisome proliferator-activated receptor
PSMC	Pulmonary artery smooth muscle cells
PTPC	Permeability transition pore complex
ROS	Reactive oxygen species
RV	Right ventricle
sGC	Soluble guanylyl cyclase
SOD	Superoxide dismutase
TAPSE	Tricuspid annular plane systolic excursion
TGF	Transforming growth factor
TXA₂	Thromboxane
VEGF	Vascular endothelial growth factor

1 Introduction

Maladaptive changes to the phenotype of pulmonary arterioles resulting in pulmonary vascular dysfunction, right ventricular (RV) pressure loading, and, ultimately, right heart failure are a central pathophysiological mechanism leading to the development of clinically evident pulmonary hypertension (PH). The "two-hit" hypothesis of PH proposes that in the presence of a predisposing genetic and/or molecular substrate, exposure to certain environmental or biological mediators of vascular injury initiates a cascade of adverse cell signaling events culminating in gross structural malformation and functional deterioration to pulmonary arterioles. Although no single inciting event is known to trigger universally the development of PH, pulmonary endothelial dysfunction and decreased levels of bioavailable nitric oxide (NO•) are observed in early stages of many PH disease forms. Importantly, the pulmonary vascular bed hosts the greatest density of vascular tissue within the human circulatory system (Barst and Rubin 2011); thus, even subtle perturbations to signaling pathways that regulate structure and function of cells within the alveolar-pulmonary circulation interface may translate into meaningful changes to cardiopulmonary performance.

The cornerstone histopathological feature of PH is adverse remodeling of distal pulmonary arterioles that is characterized by intimal thickening (Farber and

Loscalzo 2004), dysregulated proliferation of apoptosis-resistant pulmonary artery endothelial cells (PAECs) and pulmonary vascular smooth muscle cells (PSMCs) (Abe et al. 2010), increased perivascular fibrosis, and, in certain forms of PH, the genesis of plexogenic lesions (Archer et al. 2010). Subtotal luminal obliteration of small- and medium-sized pulmonary arterioles, abnormal pulmonary vascular reactivity, and increased pulmonary blood vessel tone contribute to elevations in pulmonary vascular resistance and uncoupling of RV-pulmonary circulatory function (Rondelet et al. 2010). Enhanced understanding of cross talk between signaling pathways in PAECs, PSMCs, lung fibroblasts, and RV myocytes that occurs in response to injury has led to the development of PH-specific pharmacotherapies. These treatments aim to improve pulmonary vascular tone by restoring nitric oxide (NO$^{\bullet}$)- or prostacyclin-mediated signaling pathways, or through inhibition of endothelin-1 (ET-1)-dependent and -independent activation of vascular calcium channels that promotes vascular mitogenesis and vasoconstriction (Schneider et al. 2007; McLaughlin et al. 2009). Despite this progress, however, clinical outcome in PH remains poor, particularly among patients afflicted with pulmonary arterial hypertension (PAH), in which mortality rates approach 10 % within 1 year of diagnosis (Benza et al. 2010). This observation has stimulated novel dimensions of investigation that emphasize abnormalities in mitochondrial function, cellular metabolism, and microRNA (miR)-dependent responses to hypoxia as potentially under-recognized mechanisms involved in the pathogenesis of PH.

2 PH Pathophysiology

In PH, pulmonary circulatory performance is impaired as a consequence of adverse changes to the compliance of medium- and small-sized pulmonary arterioles that occur in response to chronic pulmonary vascular injury. In the majority of patients, these changes occur owing to hypoxic pulmonary vasoconstriction; vascular congestion in the setting of left atrial hypertension (i.e., impaired left ventricular [LV] function, mitral valve disease); or impedance to pulmonary blood flow as a consequence of primary lung, cardiac, pulmonary, or vascular thromboembolic disease (Maron and Loscalzo 2013). In PAH, the interplay between specific molecular and genetic factors induces the effacement of pulmonary arterioles and disrupts homeostatic mechanisms that control normal blood vessel tone and platelet function. This results in the classic PAH phenotypic triad of microvascular thrombosis, increased pulmonary vascular reactivity, and plexiform lesions (Fig. 1).

The contemporary definition of PH stipulates that the following hemodynamic criteria be met: a sustained elevation in mean pulmonary artery pressure (>25 mmHg) and pulmonary vascular resistance (>3 Wood units) in the setting of a normal pulmonary capillary wedge pressure. These measures emphasize *pulmonary vascular dysfunction* as the central determinate mitigating the diagnosis of PH. This distinction departs from previous iterations of this definition by identifying the pulmonary circulatory system as a specific entity within the larger

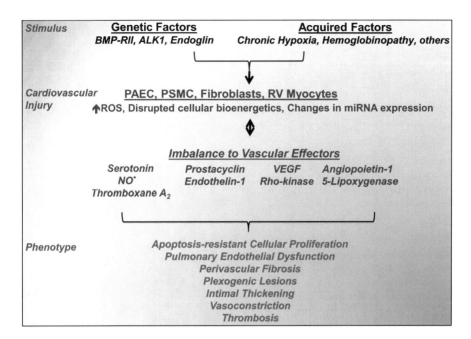

Fig. 1 Pathobiology of pulmonary arterial hypertension. The "two-hit" hypothesis of PAH contends that in the presence of certain genetic and/or molecular risk factors, exposure to environmental and/or biological stimulators of pulmonary/cardiovascular injury increases oxidant stress levels, promotes mitochondrial dysfunction, and upregulates specific microRNAs, which, in turn, dysregulate cell signaling pathways responsible for maintaining normal pulmonary vascular structure and function. Ultimately, these changes are associated with the classical PAH vasculopathy that is characterized by pulmonary endothelial dysfunction, plexogenic lesions, vasoconstriction, and microvascular thrombosis. *BMP-RII* bone morphogenetic protein receptor II, *PAEC* pulmonary artery endothelial cell, *PSMC* pulmonary smooth muscle cell, *RV* right ventricle, *ROS* reactive oxygen species, *NO•* nitric oxide, *VEGF* vascular endothelial growth factor, *miRNA* microRNA

cardiopulmonary apparatus. This approach furthermore reflects the fact that traditional PH treatment strategies, which emphasize PH-associated comorbidities (i.e., hypoxic lung disease, impaired left ventricular diastolic function) to alleviate symptoms, are often unsuccessful at providing patients with sufficient and sustained improvements to cardiopulmonary hemodynamics. Analyzing PH pathophysiology and, thus, the pursuit of novel therapies in the modern era must be predicated upon an understanding of biological/molecular factors that drive disease progression.

Increases in pulmonary vascular resistance are tolerated poorly by the RV, which, compared to the LV, is a thin-walled and non-compacted structure. Chronic changes to RV volume- and/or pressure-loading conditions result in adverse remodeling of the RV that is characterized by increased end-diastolic volume, geometric conformational changes from a normal tetrahedron to a crescentic trapezoid, and RV free wall hypertrophy (Voelkel et al. 2006) (Fig. 2). Eventual RV systolic dysfunction may be accelerated or compounded in severity by

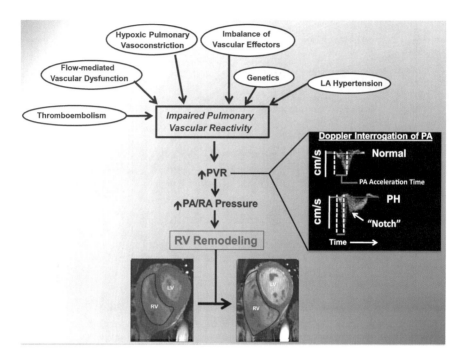

Fig. 2 Pathophysiology of pulmonary hypertension. Pulmonary hypertension (PH) is characterized by impaired pulmonary vascular reactivity due to a heterogeneous range of pathological cellular/molecular processes that mediate pulmonary vascular injury. Pulmonary vascular dysfunction in PH is detected clinically by elevations in pulmonary vascular reactivity, which may be assessed invasively by cardiac catheterization or noninvasively by transthoracic echocardiography (*inset*). Under normal conditions, Doppler interrogation of the pulmonary artery outflow tract results in a broad-based triangular signal envelope. In the presence of decreased vascular compliance, as is the case in moderate to severe forms of PH, the time to peak blood flow acceleration is decreased (PA acceleration time) and a reflected (i.e., retrograde) Doppler signal is detected as a mid- or late-systolic 'notch.' Increased right ventricular (RV) pressure loading results in geometric conformational changes to the RV cavity (outlined in black) as well as RV hypertrophy and systolic dysfunction. *LA* left atrial, *PA* pulmonary artery, *RA* right atrial, *cm/s* centimeters/second. Cardiac magnetic resonance images are reproduced with permission from Fernandez-Friera L, Alvarez-Garcia A, Guzman G et al. (2011) Apical right ventricular dysfunction in patients with pulmonary hypertension demonstrated with magnetic resonance. Heart 97:1250–1256

progressive tricuspid valve regurgitation that increases RV end-diastolic volume and enhances cavitary dilation. The pathobiological mechanisms involved in the development of frank, irreversible RV failure (i.e., cor pulmonale) are unresolved, but likely involve RV (subendocardial) ischemia (Gautier et al. 2007), strain/stress-induced intramural replacement fibrosis (Umar et al. 2012), and torsional effects on RV myocytes that are mediated by global changes to RV shape (Puwanant et al. 2010).

Pulmonary artery pressure is dependent partly upon RV systolic function; thus, in the setting of diminished RV contractility, pulmonary artery pressure may be normal despite severe pulmonary vascular disease. Along these lines, decremental changes in RV systolic function in patients with PH are associated with worsening symptomatology (e.g., dyspnea, fatigue, abdominal/peripheral edema), decreased functional capacity, and increased mortality. This is the case in patients with even *mild* heart failure (New York Heart Association Class II) and left atrial hypertension-associated PH due to LV systolic dysfunction, in which an RV ejection fraction ≤ 39 % is an independent predictor of early mortality (de Groote et al. 1998). Similarly, in patients with PAH, decreases in tricuspid annular plane systolic excursion (TAPSE), an echocardiographic measurement of RV systolic function, correlate inversely with 1-year mortality rates (Forfia et al. 2006). In turn, clinical benefits afforded to PAH patients by endothelin receptor antagonists and prostacyclin replacement therapy (see Part II, Olschewski 2013; Clozel et al. 2013) occur by virtue of their favorable effect on RV loading conditions, which promotes reverse RV remodeling and restores RV-pulmonary vascular coupling (Oikawa et al. 2005; Chin et al. 2008).

3 Cell Signaling Mechanisms in the Pathobiology of PH

3.1 Endothelial Nitric Oxide Synthase in PH

Nitric oxide (NO•) is a 30 Da lipophilic gaseous molecule, which may diffuse through PAEC/PSMC membranes to participate in intercellular signaling. Nitric oxide is synthesized in mammalian tissues via activation of three nitric oxide synthase (NOS) isoforms, each of which are homodimeric enzymes containing a calmodulin-binding domain that separates an N-terminal heme-binding domain and a C-terminal reductase domain (Porter et al. 1990). Nitric oxide synthases catalyze the formation of NO• from L-arginine in a reaction that consists of two distinct monooxygenation steps. In the first monooxygenation step, two moles of electrons are donated by one mole of NADPH to a heme-bound oxygen via flavin adenine dinucleotide (FAD) and flavin mononucleotide (FMN). This allows for the two-electron oxidation of a guanidine nitrogen of L-arginine to form one mole each of omega-*N*-hydroxy-L-arginine and water (Delker et al. 2010). In the second monooxygenation step, one-half mole of NADPH transfers one electron to a second heme-bound oxygen, and omega-*N*-hydroxy-L-arginine undergoes a three-electron oxidation to form one mole each of NO• and L-citrulline (Griffith and Stuehr 1995).

Activation of endothelial NOS (eNOS), such as in response to vascular endothelial shear stress, is modulated by various intracellular posttranslational modifications, including *S*-nitrosylation (e.g., Cys94, Cys99), phosphorylation (e.g., Ser1177, Ser65, Thr495), and palmitoylation, among others (Dudzinski et al. 2006). The classical extracellular signaling pathways involved in eNOS activation include G-protein-coupled receptor signal transduction, which increases intracellular Ca^{2+} levels and, subsequently, levels of Ca^{2+}-calmodulin; Akt signaling via sphingosine 1-phosphate; vascular endothelial growth factor (VEGF) via phosphatase calcineurin; and hormonal stimuli (e.g., estrogen and insulin) (Murata et al. 2002; Egom et al. 2011; Cerqueira et al. 2012). Decreased pulmonary vascular eNOS activity is observed in numerous animal models of PH in vivo and in humans with this disease (Steudel et al. 1998; Gangopahyay et al. 2011). Specifically, loss of NO˙ bioavailability is linked to impaired endothelium-dependent and -independent vasodilation, increased PSMC mitogenesis, and platelet aggregation. Proposed mechanisms to account for diminished levels of functional eNOS in PH are provided below.

3.1.1 Hypoxia and eNOS in PH

The mechanism(s) by which hypoxia influences eNOS gene expression is (are) controversial, as PAEC exposure to $PaO_2 < 70$ mmHg has been associated with both increased *and* decreased eNOS protein expression levels. Fish and colleagues demonstrated that hypoxia induces a decrease in acetylation and lysine 4 methylation of eNOS proximal promoter histones to decrease eNOS gene transcription (Fish et al. 2010). In contrast, others have suggested that hypoxia-inducible factor-1α (HIF-1α), a master transcription factor that modulates a wide range of cellular processes in response to hypoxia, binds to a HIF response element near the promoter region of eNOS to increase eNOS gene expression (Coulet et al. 2003). However, in this scenario, hypoxia-mediated upregulation of eNOS expression does not necessarily imply increased eNOS *activity*. To the contrary, tonic stimulation of eNOS is associated with a paradoxical *decrease* in eNOS activity, likely owing to the consumption and subsequent depletion of key cofactors (i.e., 5,6,7,8-tetrahydrobiopterin [BH_4]) necessary for normal eNOS function. Under these conditions, eNOS is 'uncoupled,' resulting in the preferential generation of superoxide ($˙O_2^-$) over NO˙ (see Sect. 3.3). Data from PH experiments in vivo support this claim: eNOS deficiency and/or impaired eNOS function is a key factor in disease pathogenesis. For example, eNOS knockout mice ($eNOS^{-/-}$) exposed to *mild* hypoxia demonstrate significantly increased RV systolic pressure and diminished markers of eNOS bioactivity as compared to wild-type controls (Fagan et al. 1999). Diminished eNOS activity is also implicated in inflammatory (monocrotaline), genetic (bone morphogenetic protein receptor II [BMP-RII] deficient), and angioproliferative (VEGF inhibition with SU-5416) experimental models of PAH in vivo (Tang et al. 2004).

Hypoxia may also decrease eNOS activity by inducing posttranslational modification(s) of eNOS and/or caveolin-1, which decreases Ca^{2+} sensing by eNOS and results in dissociation of eNOS from its regulatory proteins, heat shock protein 90 and calmodulin (Murata et al. 2002). Alternatively, hypoxia may decrease levels of bioavailable NO˙ through eNOS-independent mechanisms. In red blood cells, for example, hypoxia promotes increased levels of heme iron-nitrosyl (FeNO) that limits hemoglobin S-nitrosylation, which, in turn, is a key PaO_2-sensitive mechanism implicated in the regulation of pulmonary vascular tone (McMahon et al. 2005).

3.1.2 Oxidant Stress and eNOS

Perturbations to the redox status of PAECs, PSMCs, RV myocytes, and lung fibroblasts due to activation of reactive oxygen species-generating (ROS) enzymes, such as NADPH oxidase (NOX), xanthine oxidase, and uncoupled eNOS, or via disrupted electron transport chain function in mitochondria promote pulmonary vasculopathy characterized by impaired NO•-dependent vasodilation, intimal thickening, and perivascular fibrosis (Mittal et al. 2012). In humans, increases in pulmonary vascular ROS generation may occur as a pathological response to chronic hypoxia, or increased pulmonary vascular blood flow (e.g., secondary to intracardiac shunt); or due to impaired antioxidant enzyme function, as is the case in sickle cell anemia-associated PH in which glutathione peroxidase deficiency is observed (Gizi et al. 2011). ROS may impair eNOS activity through the oxidation of enzyme cofactors (i.e., BH_4), or inactivate NO• such as in the case of •O_2^- which reacts with NO• to generate peroxynitrite ($ONOO^-$). Additionally, the interaction of •O_2^- with the stable NO• by-product nitrite (NO_2^-) forms peroxynitrate (O_2NOO^-) and, thus, decreases levels of NO_2^-, which is a key substrate for NOS-independent synthesis of NO• ($2HNO_2 \rightarrow N_2O_3 + H_2O$; $N_2O_3 \rightarrow NO• + NO_2•$) (Lundberg et al. 2011); (Spiegelhalder et al. 1976).

3.1.3 Genetic Mediators of eNOS in PH

BMP-RII is a serine–threonine kinase and member of the transforming growth factor-β (TGF-β) superfamily of receptors (Rosenzweig et al. 1995). Approximately 70 % of familial PAH cases involve mutations in BMP-RII, and receptor dysfunction is increasingly recognized as a contributor to non-PAH forms of PH (Machado et al. 2006). Although BMP-RII is believed to contribute to remodeling of pulmonary blood vessels through a wide range of signaling pathways, including SMAD-dependent PSMC migration (Long et al. 2009), it was recently demonstrated that two BMP-RII ligands, BMP2 and BMP4, are involved in BMP-RII-dependent phosphorylation of eNOS at Ser-1177 to upregulate eNOS activity (Gangopahyay et al. 2011). Similarly, abnormalities in the function of endoglin, a key BMP receptor accessory protein in human PAECs, are linked to the development of PAH when present in the setting of the clinical syndromes hereditary hemorrhagic telangiectasia (HHT) type 1 and type 2. Mice heterozygous for vascular endothelial endoglin expression ($Eng^{+/-}$) develop PH spontaneously in vivo due, in part, to increased pulmonary vascular ROS generation, eNOS uncoupling, and decreased NOS-inhibitable NO• production (Toporsian et al. 2010).

Several eNOS polymorphisms are implicated in the development of PH and other vascular diseases. For example, a single nucleotide polymorphism leading to a substitution of aspartic acid for glutamic acid at position 298 (Glu298Asp) of eNOS and an increased NOS4a allelic frequency of 27-bp variable number of repeats increase susceptibility to developing the high altitude pulmonary edema (HAPE) syndrome, including elevations in pulmonary artery pressure and pulmonary vascular resistance. These changes may occur owing to function-limiting changes in the conformation of eNOS, although the precise mechanism by which to account for the phenomenon is unknown (Miyamoto et al. 1998; Droma et al. 2002; McDonald et al. 2004).

3.2 Endothelin-1 System

Endothelin-1 (ET-1) is a 21-amino acid vasoactive peptide that contains two disulfide bridges between Cys1-Cys15 and Cys3-Cys11 (Yeager et al. 2012), which are necessary for endothelin converting enzyme-mediated proteolytic cleavage of ET-1 from its precursor, 'Big ET-1.' Endothelin-1 is constitutively expressed in a wide range of mammalian cell types, including hepatic sinusoidal cells, renal epithelial cells, and PAECs (Huggins et al. 1993). Endothelin-1 gene expression levels are upregulated significantly in RV myocytes, PAECs, PSMCs, and lung fibroblasts in the presence of stimuli associated with pulmonary vascular injury in PH, including cytokines that mediate vascular inflammation (i.e., TGF-β, IL-6) (Olave et al. 2012), increased levels of pulmonary vascular ROS (An et al. 2007), hypoxia (Yamashita et al. 2001), and decreased levels of bioavailable NO$^\cdot$ (Kourembanas et al. 1993). In fact, plasma ET-1 levels may be increased fourfold in patients with PAH or PH due to left atrial hypertension, and anti-ET-1 immunohistochemical analysis demonstrates significantly increased immunoreactivity in PAECs and PSMCs of plexiform lesions compared to blood vessels harvested from normal controls (Giaid et al. 1993). Endothelin-1 is also released from sickled red blood cells and interacts with the blood vessel wall to promote vasoconstriction in a process that contributes to the systemic and pulmonary vasculopathy of sickle cell anemia (Gladwin and Vichinsky 2008).

Endothelin-1 regulates pulmonary vascular tone through its interaction with the vasoconstrictor endothelin-type A (ET$_A$) and -type B (ET$_B$) receptors in PSMCs and vasodilatory ET$_B$ receptors in PAECs, which do not constitutively express ET$_A$. Endothelin-type A and ET$_B$ receptors are members of the superfamily of G-protein-coupled receptors and are overall highly homologous (55 %), with the exception of the cysteine-rich 35-amino acid sequence distal to the seventh transmembrane domain, in which homology between receptors is only 75 % (Doi et al. 1999). Since cysteine(s) in this region are believed to regulate G-protein coupling to both ET$_A$ and ET$_B$ receptors, and, thus, are integral to receptor signal transduction, it has been postulated that differences in this region between receptor subtypes may account, in part, for their differential functions (Okamoto et al. 1997).

In PSMCs, stimulation of $ET_{A/B}$ receptors by ET-1 induces G_i and G_q coupling to modulate phospholipase C-mediated hydrolysis of phosphatidylinositol 4,5-bisphosphate to inositol 1,4,5-triphosphate (IP_3). In turn, opening of IP_3-sensitive calcium (Ca^{2+}) channels as well as ET-1-mediated opening of the store-operated and nonselective Ca^{2+} channels induces an increase in intracellular Ca^{2+} flux ($[Ca^{2+}]_i$), Ca^{2+} waves, and Ca^{2+} oscillations that promotes vasoconstriction (Liu et al. 2012). Importantly, ET-1-induced vasoconstriction persists following ET-1 dissociation from the ET_A receptor, indicating that the $[Ca^{2+}]_i$ flux response mediated by ET-1 is robust, Ca^{2+}-dependent hyperpolarization is delayed during ET-1 signaling, or both (Zhang et al. 2003; Liu et al. 2012). The functional consequence of ET_A receptor signaling on vascular tone is noteworthy: relative to norepinephrine, the concentration of ET-1 required to induce 50 % blood vessel contraction (i.e., EC_{50}) in pig coronary arteries, rat aorta, and rat pulmonary artery is 0.52, 1.4, and 0.68, respectively (Huggins et al. 1993). ET-1 binding to ET_A receptors ($K_i = 0.6$ nmol l^{-1}) also promotes vascular smooth muscle cell mitogenesis by activating various signaling intermediaries that regulate protein synthesis, including protein kinase C; mitogen-activated protein kinase (MAPK); p70S6K, which targets the ribosomal protein S6K to increase cellular protein synthesis; and epidermal growth factor receptor (EGFR) via tyrosine phosphorylation (Iwasaki et al. 1999; Kapakos et al. 2010). Interestingly, upregulation of the proto-oncogene transcription factor *c-fos* by ET-1 (or hypoxia) is linked to cellular proliferation and fibrosis of PSMCs, lung fibroblasts, and myocytes in experimental animal models of PH (Rothman et al. 1994; Nishimura et al. 2003; Recchia et al. 2009), providing molecular evidence to account for the proliferative phenotypic overlap between plexogenic lesions of PAH and various solid tumors.

In contrast to PSMCs, ET-1 binding to the ET_B receptor (K_i of 0.12 nmol l^{-1}) in PAECs results in the activation of eNOS and synthesis of $NO^•$, which is required to maintain normal pulmonary vascular tone and prevent vascular remodeling. Endothelin-type B receptor-dependent activation of eNOS is believed to occur via G-protein coupling to the ET_B receptor that stimulates $[Ca^{2+}]_i$ flux and subsequent elevations in Ca^{2+} binding to calmodulin, which is a key allosteric modulator of eNOS activity. Recent work from our laboratory has demonstrated that pathophysiological levels of the mineralocorticoid hormone aldosterone akin to those observed in humans with PH increase NOX4-dependent ROS generation in PAECs in vitro, which is associated with ET_B receptor dysfunction, impaired ET_B receptor-dependent activation of eNOS, and oxidation of $NO^•$ to $ONOO^-$ (Maron et al. 2012).

Endothelin-type B receptor signal transduction also results in the synthesis of vasodilatory prostaglandins (PG). In isolated guinea pig lungs exposed to ET-1, a ~50-fold increase in ET_B receptor-dependent PGI_2 synthesis is observed (D'Orleans-Juste et al. 1991), although the mechanism by which ET_B receptor activation stimulates PGI_2 synthesis is not well characterized. Internalization of the ET_B receptor/ET-1 complex and subsequent proteasomal degradation is the chief mechanism by which ET-1 elimination occurs. This conclusion is supported in vivo by experiments involving the transgenic spotting lethal rat (sl/sl), which lacks constitutively expressed vascular ET_B receptors. Compared to wild-type rats, these rats demonstrate significantly higher circulating levels of ET-1 and more severe PH following monocrotaline injection to induce pulmonary vascular injury (Nishida et al. 2004).

3.3 Soluble Guanylyl Cyclase and Phosphodiesterase Inhibition in PH

Nitric oxide is the primary biological activator of the heterodimeric (α_1/β_1 or α_1/β_2) enzyme soluble guanylyl cyclase (sGC) that catalyzes the conversion of cytosolic GTP to cGMP, which is a critical secondary signaling molecule necessary for activation of cGMP-dependent protein kinase (i.e., protein kinase G [PKG]) to promote PSMC relaxation and inhibit platelet aggregation and thrombosis. Nitric oxide binding to the heme (Fe^{2+}) prosthetic group of sGC results in the formation of a hexa-coordinated histidine–heme–NO• intermediate. Subsequent cleavage of the heme–histidine bond leads to a dramatic upregulation of enzyme activity: nanomolar concentrations of NO• may induce an appropriate 100-fold increase in sGC activation (Evgenov et al. 2006). In PH, elevated levels of ROS accumulation may impair sGC activity through the oxidation of heme from the ferrous (Fe^{2+}) to ferric (Fe^{3+}) state that converts sGC to an NO•-insensitive state, presumably owing to decreased affinity of NO• for oxidized heme. Alternatively, others have demonstrated that sGC activity is influenced by the redox status of functional sGC cysteinyl thiol(s) in a manner that is independent of the heme redox state (Fernhoff et al. 2009; Yoo et al. 2012). Work from our laboratory has demonstrated that pathophysiological concentrations of H_2O_2 induce the formation of higher cysteinyl thiol oxidative states of Cys-122 on the β_1 subunit of sGC in vascular smooth muscle cells, including sulfenic acid, sulfinic acid, and the disulfide form. Posttranslational oxidative modification of Cys122, in turn, functions as a redox 'switch' that regulates enzyme function, resulting in decreased NO•-sensing by sGC and impaired enzyme activity (Maron et al. 2009). The importance of abnormal sGC function in the pathogenesis of PH is well established. Transgenic mice deficient in the sGC α_1-subunit develop exaggerated elevations in RV systolic pressure and muscularization of intraacinar pulmonary arterioles following exposure to chronic hypoxia compared to wild-type mice (Vermeersch et al. 2007). Moreover, hypoxia alone is associated with decreased mRNA and protein levels of sGC as well as sGC-dependent cGMP formation (Hassoun et al. 2004).

Collectively, these observations implicate the pharmacotherapeutic potential of heme-independent sGC activators in PH. Work from Ko and colleagues and drug discovery experiments performed at Bayer HealthCare in the early 1990s identified YC-1 [3-5′-hydroxymethyl-2′furyl)-1-benzyl indazole] and 5-substituted-2-furaldehyde-hydrazone derivative compounds (i.e., BAY compounds), respectively, as synthetic heme-(in)dependent activators of sGC (Ko et al. 1994; Stasch et al. 2006). BAY 58-2667, perhaps the best studied among these compounds, activates sGC with a K_m of 74 μM and a V_{max} of 0.134 μmol min^{-1} mg^{-1} (Schmidt et al. 2003), and although the precise mechanism by which this (and other) BAY compounds activates heme-oxidized sGC is not fully resolved, one leading hypothesis contends that BAY 58-2667 competes with the oxidized heme moiety for binding to the sGC-activating motif to induce enzyme activation (Pellicena et al. 2004) (see Part III, Sect. 1). The effect of these compounds on sGC-NO$^•$ vasodilatory signaling and NO$^•$-dependent vascular remodeling has been assessed in PH in vivo. In one study, the administration of YC-1 to hypoxic mice decreased PSMC proliferation and pulmonary artery pressure (Huh et al. 2011). The effect of BAY 63-2521 (Riociguat™) on cardiopulmonary hemodynamics was also assessed in a small cohort of patients with PAH, chronic thromboembolic PH, or PH from interstitial lung disease. Drug therapy (3.0–7.5 mg day^{-1}) over 12 weeks decreased pulmonary vascular resistance by 215 dyne s cm^{-5}, which was associated with an increase in the median 6-min walk distance by 55.0 m from baseline (Ghofrani et al. 2010).

3.4 Phosphodiesterase Inhibition in PH

In 1962, Butcher and Sutherland implicated phosphodiesterase enzymatic activity in endogenous degradation of adenosine 3', 5' phosphate (cAMP) (Butcher and Sutherland 1962). Eleven PDE isoforms have since been detected in mammalian tissue (Table 1). The fields of PDE biochemistry and PH intersected following the identification of cGMP-specific PDE type-5, at elevated concentrations in PSMCs, platelets, and myocytes. Phosphodiesterase type-5 regulates cGMP bioactivity via (1) the hydrolysis of cGMP to 5'-GMP, and (2) allosteric binding of cGMP to PDE-5 GAF[1] domains, which induces a conformational change to the structure of PDE-5 and positively feeds back to promote cGMP metabolism (Fig. 3). PDE-5-cGMP binding ranges from K_d of 2.4 μM (pH = 5.2) to 0.15 μM (pH = 9.5) (Turko et al. 1999). The pH-dependent manner by which this occurs is in concert with reports demonstrating a regulatory role for the cyclic nucleotide ionizing residues (i.e., pH-sensitive) Asp-289 and Asp-478 in modulating PDE-5 function (McAllister-Lucas et al. 1995). The intracellular concentration of cGMP may also be influenced by flux through membrane-bound cGMP-gated channels and the multidrug transporter, although the contribution of these to total levels of bioactive cGMP is negligible in pulmonary vascular tissue (Serre et al. 1995).

---

[1] GAF is an acronym of the various tissues in which these domains were originally described: cGMP-dependent phosphodiesterases (PDEs), *nabaena* adenylyl cyclases, and *E. coli* FhlA (Francis et al. 2010).

Table 1 cGMP-specific kinetic properties of phosphodiesterases and their tissue distribution

Isoenzyme	K_m cGMP (μM)	V_{max} cGMP (μM)	Tissue distribution
PDE1A	3–4	50–300	VSMC, cardiomyocyte, brain
PDE1B	1–6	30	VSMC, cardiomyocyte, brain
PDE1C	1–2	Not determined	VSMC, cardiomyocyte, brain
PDE2A	10	123	VSMC, cardiomyocyte, brain, corpus cavernosum
PDE3A	0.02–0.2	0.3	VSMC, cardiomyocyte, brain, corpus cavernosum, platelets
PDE3B	0.3	2	VSMC, cardiomyocytes
PDE5A	1–6	1–3	Corpus cavernosum, PSMC, skeletal muscle, platelets, cardiomyocytes
PDE6A/B	15	2,300	Retina
PDE 6C	17	1,400	Retina
PDE9A	0.2–0.7	Not determined	Various
PDE 10A	13	Not determined	Various
PDE 11A	0.4–0.2	Not determined	Skeletal muscle, heart, VSMC

PDE phosphodiesterase, *VSMC* vascular smooth muscle cell, *PSMC* pulmonary smooth muscle cell. Adapted from Francis SH et al. (2010) cGMP-dependent protein kinase and cGMP phosphodiesterases in nitric oxide and cGMP action. Pharm Rev 62:525–563 and Reffelman T, Kloner RA (2003) Therapeutic role of phosphodiesterase 5 inhibition for cardiovascular disease. Circulation 108:239–244

In PAH, expression of PDE-5 is increased in PSMCs and in RV myocytes (Wharton et al. 2005; Nagendran et al. 2007), which is associated with decreased levels of bioactive NO$^{\bullet}$, pulmonary vascular dysfunction, and impaired RV lusitropy (Waxman 2011). In cultured PSMCs, PDE-5 inhibition attenuates key indices of adverse remodeling, including DNA synthesis/cell growth, cellular proliferation, and suppression of apoptosis (Wharton et al. 2005). Phosphodiesterase type-5 is also linked to decreased thrombotic burden in chronic thromboembolic PH, presumably by increasing bioactive cGMP levels in platelets to inhibit platelet aggregation (Suntharalingam et al. 2007).

3.5 Prostacyclin Signaling in PH

Arachidonic acid, or 5,8,11,14-eicosatetraenoic acid, is released from membrane phospholipids in response to mechanical or chemical stimuli, resulting in the synthesis of two major classes of eicosanoids: prostanoids, via cyclooxygenase (COX) pathway, and leukotrienes, via the lipooxygenase (LO) pathway. Cyclooxygenase (COX) exists in at least two isoforms. The constitutively expressed form, COX-1, is present in various cell types including the vascular endothelium, gastric mucosa, and platelets. In contrast, COX-2, which is the inducible form of COX, is present in cells involved in inflammation, particularly macrophages. COX-2 is also present in normal vascular endothelial cells and appears to be upregulated in

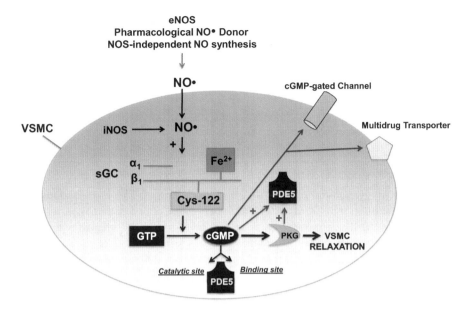

Fig. 3 Influence of phosphodiesterase type-5 on the nitric oxide signaling axis. Nitric oxide (NO•) derived from endothelial nitric oxide synthase (eNOS), pharmacological NO• donors, or via induction of inducible NOS in vascular smooth muscle cells (VSMCs) activates soluble guanylyl cyclase (sGC) to generate cGMP. Normal NO•–sGC signaling is regulated by the redox status of the prosthetic heme ligand and functional cysteinyl thiol(s), such as Cys122, which are present near the catalytically active region of the β_1-subunit of sGC. The interaction of cGMP with its most relevant biological target, protein kinase G (PKG), results in VSMC relaxation. This vasodilatory effect of this pathway is offset through the interaction of cGMP with phosphodiesterases (PDEs), which decreased bioactive cGMP levels through hydrolysis of cGMP to form 5′GMP, or by allosteric binding of cGMP to PDE. In the case of PDE type-5 (PDE5), increased levels of cGMP and/or PKG upregulates PDE5 activity. The contribution of gated cGMP channels and the multidrug transported to the regulation of bioactive cGMP levels is negligible in pulmonary vascular tissue. Adapted from Francis SH et al. (2010) cGMP-dependent protein kinase and cGMP phosphodiesterases in nitric oxide and cGMP action. Pharm Rev 62:525–563, with permission

response to stimuli associated with vascular injury, such as shear stress (Topper et al. 1996). Key products of the COX pathway that are pertinent to the pathophysiology of PH include the potent vasoconstrictor and stimulus for platelet aggregation, thromboxane A_2 (TXA_2), and prostaglandin I_2 (prostacyclin), which exerts opposing effects to TXA_2, including vasodilation and inhibition of platelet activation. Prostacyclin is synthesized from PGH_2 by prostacyclin synthase in a reaction that occurs primarily in vascular endothelial cells. Early investigators speculated that the pulmonary vasoconstriction phenotype in PAH was a consequence of imbalanced TXA_2/PGI_2 synthesis. This hypothesis was supported by the observation that, compared to normal pulmonary blood vessels, pre-capillary pulmonary arterioles harvested from patients with PAH, hepato-pulmonary PH, and HIV-associated PH demonstrate significantly decreased PGI_2 synthase mRNA and protein expression levels (Tuder et al. 1999). In contrast, elevated levels of

TXA$_2$ and increased pulmonary vascular *sensitivity* to TXA$_2$ have been reported in PH, such as in the lamb model of hypoxia-induced neonatal persistent pulmonary hypertension (Hinton et al. 2007). These observations are consistent with other reports in adults demonstrating that, compared to healthy controls, patients with PAH or PH due to hypoxic lung disease demonstrate elevated levels of urinary excretion of 11-dehydro-thromboxane B$_2$, a stable metabolite of TXA$_2$, and decreased levels of the stable prostacyclin metabolite, 2,3-dinor-6-keto-prostaglandin F1α (Christman et al. 1992).

Shear stress, ET-1, hypoxia, and BMP-RII dysfunction are each associated with overactivation of the TXA$_2$ synthesis pathway relative to PGI$_2$ in vascular tissue (Zaugg et al. 1996; Song et al. 2005; Racz et al. 2010). Although the precise mechanism by which to account for this imbalance is unresolved, abnormal COX-2 function in the setting of vascular injury may play a role. It was recently demonstrated that compared to control mice, COX-2 knockdown mice administered monocrotaline to induce vascular inflammation exhibit a robust increase in NOX4 gene expression, dihydroethidium fluorescence (indicative of ROS accumulation), and ET$_A$ receptor expression in pulmonary arterioles, whereas prostacyclin levels were decreased significantly (Seta et al. 2011). These findings are consistent with reports in cultured COX-2-deficient PSMCs, in which hypoxia results in a hypertrophic remodeling response and a vasoconstrictor phenotype (Fredenburgh et al. 2008).

5-lipooxygenase (5-LO) catalyzes the conversion of arachidonic acid ultimately into various leukotrienes that mediate cellular processes involved in vascular remodeling and cellular responses to injury. Leukotriene B$_4$, for example, exerts both chemotactic and chemokinetic activity on polymorphonuclear leukocytes and eosinophils (Ford-Hutchinson et al. 1980), and leukotrienes C$_4$, D$_4$, and E$_4$ (each of which contain a cysteine) are implicated in pulmonary vasoconstriction and increased pulmonary vascular permeability. Although rats that overexpress 5-LO do not develop PAH spontaneously, pulmonary vascular dysfunction and abnormal cardiopulmonary hemodynamics are accelerated in the presence of pulmonary vascular inflammation (Jones et al. 2004). This supports other observations indicating that an inflammatory milieu is conducive to 5-LO-dependent synthesis of the vasoactive cysteinyl leukotrienes (Listi et al. 2008). Molecular inhibition of the 5-LO-activating protein (FLAP) has, in turn, been shown to prevent pulmonary hypertension in rats exposed to chronic hypoxia (Voelkel et al. 1996).

3.6 Mitochondrial Dysfunction

Mitochondria regulate bioenergetics, cellular respiration, and the intracellular redox status and, thus, have the potential to regulate PAEC/PSMC signaling pathways linked to cell survival, proliferation, and ROS production. Hydrogen (hydride) derived from dietary carbohydrate and fats is oxidized by molecular oxygen (O$_2$) via the tricarboxylic acid (TCA) cycle and β-oxidation pathways, respectively, to generate adenosine triphosphate (ATP). These biochemical events occur via the electron transport chain, in which two electrons donated by NADH + H$^+$ flow sequentially from complex I to ubiquinone (coenzyme Q) to complex III (ubiquinol: cytochrome c oxidoreductase) and then to cytochrome c. Electrons are then transferred to complex IV to reduce ½O$_2$ and generate H$_2$O (Wallace 2005). Protons are pumped across the inner mitochondrial membrane to establish the significantly negative electrochemical gradient ($\Delta\psi$m: ~ −200 mV) across that membrane, which provides the electromotive force necessary for ATP synthesis.

Changes to mitochondrial membrane permeability, and, hence, the normal $\Delta\psi$m, are antecedent to reversible structural and functional changes in mitochondria and, if unchecked, commit the cell to apoptosis (Kroemer and Reed 2000; Michelakis et al. 2008). Numerous mechanisms to account for the relationship between mitochondrial membrane permeability and changes to cell survival have been proposed and include increased permeability of the voltage-sensitive permeability transition pore complex (PTPC), alkalinization of the local pH, and perturbations to the intramitochondrial redox status that results in oxidation of a key thiol involved in regulating PTPC opening and/or oxidation of pyridine nucleotides (i.e., NADH/NAD$^+$) to favor PTPC opening (Woodfield et al. 1998; Zamzami and Kroemer 2001). Collectively, these changes afford egress of apoptosis-associated proteins (e.g., Bax, Bcl-2, others) from the intramitochondrial to extramitochondrial space, thereby activating programmed cell death signaling pathways (Mossalam et al. 2012).

Pathological disruptions to mitochondria-dependent regulation of cell survival are a central mechanism in the pathobiology of various angioproliferative diseases, including solid tumor cancers. Apoptosis-resistant proliferation of PAECs/PSMCs is likewise a prominent pathophenotypic feature of PH, particularly with respect to plexigenic lesions in PAH. This observation has raised attention to the possibility that mitochondrial dysfunction is an under-recognized pathobiological factor by which to account for the phenotypic overlap between these two broad categories of disease. Evidence in support of this concept is derived partly from observations made in the fawn-hooded rat, a unique animal strain that develops PAH spontaneously. Pulmonary vascular smooth muscle cells harvested from these animals demonstrate mitochondria that are decreased in size and fragmented prior to the development of pulmonary vascular remodeling (Bonnet et al. 2006). The functional effects of these changes are linked to a shift in mitochondrial metabolism from oxidative phosphorylation toward glycolysis, impairment to electron flux, and subsequent activation of hypoxia-inducible factor (HIF)-1α (Archer et al. 2008). In turn, HIF-1α has been shown in endothelial cells cultured from patients with idiopathic PAH to target carbonic anhydrase IX, which decreases levels of the antioxidant enzyme manganese superoxide dismutase (SOD2) to increase vascular ROS generation and decrease levels of NO$^{\bullet}$ (Fijalkowska et al. 2010). Interestingly, in these experiments, increased HIF-1α expression correlated inversely with low numbers of mitochondria, indicating that negative control of mitochondrial biogenesis by HIF-1α may be one mechanism by which to account for abnormal cellular respiration patterns observed in in vivo models of PAH. Conventional factors associated with PAH may also influence mitochondrial dysfunction directly. For example, compared to healthy controls, stimulation of PSMCs with ET-1, platelet-derived growth factor (PDGF), or IL-6 harvested from PAH patients results in Kruppel-like factor 5 (KL-5)-mediated activation of cyclin B1 that hyperpolarizes the mitochondrial inner membrane to inhibit apoptosis (Courboulin et al. 2011b).

Under normoxic conditions, electron transport chain complexes I or II generate $^{\bullet}O_2^-$ that is dismutated to form H_2O_2, which is a key signaling molecule required for activation of Kv channels necessary to maintain the negative electrochemical gradient of the mitochondria (Bonnet et al. 2006). At $PaO_2 < 70$ mmHg, there is decreased intramitochondrial H_2O_2 generation, opening of O_2-sensitive Kv1.5 channels, and subsequent activation of L-type Ca^{2+} channels that promotes pulmonary vasoconstriction (Archer et al. 2004). Human PSMCs in PAH, however, are deficient in Kv1.5 channels, and data from experimental animal models of PAH suggest that the effect of this deficiency is mitochondrial hyperpolarization, and, consequently, tonic activation of L-type Ca^{2+} channels associated with vasoconstriction and proliferation of PSMCs (Reeve et al. 2001).

Less well established is the role of abnormal mitochondrial bioenergetics in the development of pulmonary vascular dysfunction and/or RV hypertrophy in PH. There is increasing evidence suggesting that in cardiomyocytes, an abnormal shift in cellular fuel utilization vis-à-vis the glucose–fatty acid cycle (i.e., *Randle's cycle*) (Randle et al. 1963) accounts for changes to myocardial structure (i.e., hypertrophy) and function (i.e., impaired contractility) (Fig. 4). In the monocrotaline and pulmonary artery banding rat models of PAH, for example, decreased RV O_2 consumption is observed and modulates a shift from oxidative phosphorylation to glycolysis by a mechanism involving increased Glut-1 expression and upregulation of pyruvate dehydrogenase kinase (PDK) expression with consequent increased phosphorylation of pyruvate dehydrogenase leading to its inhibition (Piao et al. 2010). The functional effects of this process include impaired RV systolic function and prolongation of the QT interval, which can be reversed by PDK inhibition or through inhibition of fatty acid oxidation to induce an indirect reciprocal shift in the mitochondrial fuel source back to glucose (oxidation) (Fang et al. 2012).

3.7 Peroxisome Proliferator-Activated Receptor-γ

Peroxisome proliferator-activated receptor (PPAR-γ) is a transcription factor most commonly associated with its regulatory effect on genes involved in fatty acid storage and glucose metabolism in adipocytes (Kilroy et al. 2012); PPAR-γ, and its transcription target apoE, are also key downstream targets of BMP-RII signal transduction. In turn, loss of function to BMP-RII via somatic mutation or dissociation of BMP-RII-interacting proteins is associated with PSMC proliferation in vitro and the development of PAH in vivo (Merklinger et al. 2005; Chan et al. 2007; Song et al. 2008). In PSMCs, the antiproliferative effect of BMP-RII

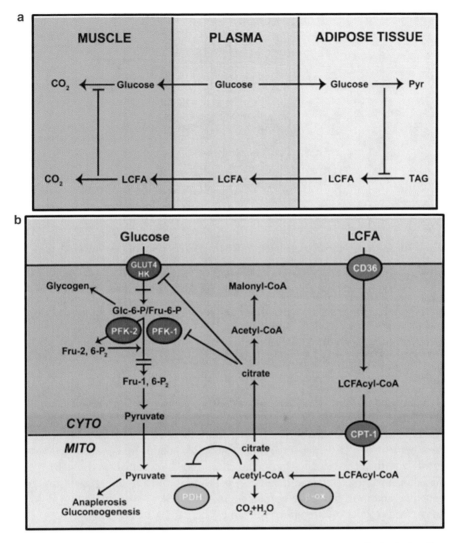

Fig. 4 Mitochondrial bioenergetics. (a) The "glucose–fatty acid cycle" or *Randle Cycle* describes a mechanism to maintain homeostatic control of circulating levels of glucose and fatty acids. (b) Inhibition of glucose utilization by fatty acid oxidation is regulated most strongly by pyruvate dehydrogenase (PDH) and, to a lesser extent, by 6-phosphofructo-1-kinase (PFK) and glucose uptake. Phosphofructokinase inhibition owing to citrate accumulation or via pharmacological/molecular strategies shifts glucose toward glycogen synthesis and pyruvate to gluconeogenesis or the synthesis of TCA intermediates (i.e., anaplerosis). Overactivation of PDK in right ventricular myocytes in an in vivo model of pulmonary hypertension has been associated with a shift from glucose oxidation to glycolysis and subsequent myocardial dysfunction (Piao et al. 2010). *LCFA* long-chain fatty acids, *TAG* triacylglycerol, *Pyr* pyruvate, *Cyto* cytosol, *MITO* mitochondria, *GLUT4* glucose transporter 4, *HK* hexokinase, *Glc-6-P* glucose-6-phosphate, *Fru-6-P* fructose 6-phosphate, *CPT I* carnitine palmitoyltransferase I, *β-ox* β-oxidation. Reproduced with permission from Hue et al. (2009) Am J Physiol Endocrinol Metab 297:E578–E591

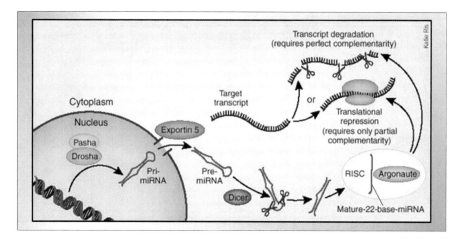

Fig. 5 MicroRNA processing. MircoRNA (miRNA) transcription in the nucleus is facilitated by the RNase III enzymes, drosha and pasha, to generate hairpin-looped molecules known as primary miRNAs (pri-mRNA), which are processed further to form miRNA precursors (pre-miRNA). Once exported from the nucleus to the cytoplasm, the RNA endonuclease, dicer, facilitates the synthesis of the mature miRNA duplex by removing the hairpin loop. The catalytic component to the RNA-induced silencing complex (RISC) is referred to as argonaute and facilitates RISC-mediated miRNA incorporation. miRNA then interacts with 3′ untranslated regions of specific mRNA targets to negatively regulate gene expression. Reproduced with permission from Mack GS (2007) MicroRNA gets down to business. Nat Biotechnol 25:631–638

is modulated by phospho-ERK and PPAR-γ binding to DNA, which, in turn, stimulates apoE synthesis and secretion (Hansmann et al. 2007). Transgenic ApoE knockout mice (ApoE$^{-/-}$) fed a high-fat diet demonstrate spontaneous development of PH, which is reversible through pharmacological stimulation of PPAR-γ with pioglitazone (Hansmann et al. 2007). In human PAECs, BMP-RII signaling appears to induce a PPAR-γ/β–catenin complex that targets the gene encoding apelin to modulate normal cellular responses to injury and, in PSMCs, suppresses cellular proliferation (Falcetti et al. 2010; Alastalo et al. 2011).

3.8 MicroRNA-Mediated Regulation of Cellular Responses to Hypoxia

MicroRNA (miRNA) are non-canonical and highly conserved noncoding ribonucleic acid molecules (~20 nucleotides) that participate in a heterogeneous range of cellular processes and are believed to regulate over 30 % of all mRNA transcripts (Berezikov et al. 2005). MicroRNA transcription generates hairpin-looped molecules known as primary miRNAs (pri-mRNA), which are processed in the cell nucleus to form miRNA precursors (pre-miRNA). Once exported from the nucleus to the cytoplasm, the RNA endonuclease, dicer, facilitates the synthesis of the mature double-stranded miRNA by removing the hairpin loop (Fig. 5). miRNAs interact with 3′ untranslated regions of specific mRNA targets to regulate negatively gene expression (Chan and Loscalzo 2010). More than 90 miRNAs have been identified to be upregulated in response to hypoxia, although only a select few (miR-210, miR-424, miR-17, miR-328) have been studied in detail with respect to PH disease pathophysiology (Fasanaro et al. 2008).

Hypoxia-inducible factor-1α-dependent upregulation of miR-210 targets iron–sulfur cluster assembly proteins (ISCU1/2) to repress mitochondrial respiration. Under hypoxic conditions, miR-210 levels are increased in PAECs in vitro, which results in miR-210-dependent downregulation of ISCU1/2 that inhibits mitochondrial electron transport (i.e., Complex I) and the tricarboxylic acid cycle. In this way, miR-210 is a critical molecular intermediate that accounts, in part, for the effect of hypoxia on HIF-1α-dependent disruptions to electron transport chain function (Chan et al. 2009). Importantly, HIF-1α itself is likely to be under miRNA-dependent regulation. In human vascular endothelial cells, hypoxia-induced upregulation of miR-424 and subsequent targeting of the scaffolding protein, cullin 2, by miR-424 appears to be an important regulatory mechanism stabilizing HIF-1α (Ghosh et al. 2010). Moreover, the observation that miR-424 promotes angiogenesis in peripheral blood vessels following ischemia (i.e., locally hypoxic environment) in mice in vivo raises speculation that this particular miRNA may be relevant in the angioproliferative pattern observed in pulmonary arterioles under hypoxic conditions in PH.

Along these lines, miR-17 is also implicated in hypoxia-mediated vascular endothelial cell proliferation through the negative regulation of the cell cycle inhibitor p21. In one study, overexpression of miR-17 increased PDGF-stimulated cellular proliferation in cultured PSMCs. Administration of a miR-17 antagomir to mice exposed to chronic hypoxia, however, was shown to protect against increases in pulmonary artery pressure and pulmonary arterial muscularization (Courboulin et al. 2011a, b; Pullamsetti et al. 2011).

Recently, downregulation of miR-328 by hypoxia was linked to hypoxic pulmonary vasoconstriction and negative pulmonary vascular remodeling in rats with moderate PH (Guo et al. 2012). In these experiments, hypoxia-induced suppression of miR-328-dependent inhibition of L-type calcium channel-α1C expression through a mechanism involving the interaction of miR-328 with the 3′ untranslated region of the L-type calcium channel-α1C was associated with increased RV systolic pressure. Furthermore, miR-328 signaling suppressed insulin-like growth factor 1, and was proposed by the authors of that study as a potential mechanism by which to account for the relationship between hypoxia, miR-328, and decreased PSMC apoptosis.

Parikh and colleagues performed a network bioinformatics analysis, which predicted miR-21 to participate in PH pathobiology by regulating BMP-, BMP-RII-, inflammation-, and hypoxia-associated signaling pathways (Parikh et al. 2012). This analysis was consistent with previous observations in vitro implicating miR-21 in negative vascular remodeling (Ji et al. 2007). Moreover, hypoxia-mediated miR-21 upregulation in PAECs appears to contribute to the PH vascular phenotype by decreasing BMP-RII, RhoB, and Rho kinase, which, under normal conditions, are involved in pulmonary vasodilatory signaling. miR-21 was likewise linked to disease expression in various PH animal models in vivo, and was observed to be highly expressed in pulmonary vascular tissue in humans with PH.

4 Conclusions

Pulmonary hypertension describes a complex disorder characterized by dysregulation of cell signaling pathways that maintain normal structure and function to distal pulmonary blood vessels. In severe forms of PH, this may result in an obliterative vasculopathy, severely elevated pulmonary artery pressure and pulmonary vascular resistance, and adverse RV remodeling. The development of successful PH pharmacotherapies in the future that aim to modify disease progression will likely hinge on the identification of novel molecular mechanisms that modulate pulmonary vascular remodeling. This pursuit is expected to require enhanced understanding of the processes by which miRNAs, mitochondria, and other molecular factors regulate cellular bioenergetics, survival, and proliferation to contribute to PH disease expression.

References

Abe K, Toba M et al (2010) Formation of plexiform lesions in experimental severe pulmonary arterial hypertension. Circulation 121(25):2747–2754

Alastalo TP, Li M et al (2011) Disruption of PPARgamma/beta-catenin-mediated regulation of apelin impairs BMP-induced mouse and human pulmonary arterial EC survival. J Clin Invest 121(9):3735–3746

An SJ, Boyd R et al (2007) NADPH oxidase mediates angiotensin II-induced endothelin-1 expression in vascular adventitial fibroblasts. Cardiovasc Res 75(4):702–709

Archer SL, Wu XC et al (2004) O2 sensing in the human ductus arteriosus: redox-sensitive K+ channels are regulated by mitochondria-derived hydrogen peroxide. Biol Chem 385 (3–4):205–216

Archer SL, Gomberg-Maitland M et al (2008) Mitochondrial metabolism, redox signaling, and fusion: a mitochondria-ROS-HIF-1alpha-Kv1.5 O2-sensing pathway at the intersection of pulmonary hypertension and cancer. Am J Physiol Heart Circ Physiol 294(2):H570–H578

Archer SL, Weir EK et al (2010) Basic science of pulmonary arterial hypertension for clinicians: new concepts and experimental therapies. Circulation 121(18):2045–2066

Barst RJ, Rubin LJ (2011) Pulmonary hypertension. In: Fuster V, Walsh RA, Harrington RA (eds) Hurst's the heart, 13th edn. McGraw-Hill, New York

Benza RL, Miller DP et al (2010) Predicting survival in pulmonary arterial hypertension: insights from the Registry to Evaluate Early and Long-Term Pulmonary Arterial Hypertension Disease Management (REVEAL). Circulation 122(2):164–172

Berezikov E, Guryev V et al (2005) Phylogenetic shadowing and computational identification of human microRNA genes. Cell 120(1):21–24

Bonnet S, Michelakis ED et al (2006) An abnormal mitochondrial-hypoxia inducible factor-1alpha-Kv channel pathway disrupts oxygen sensing and triggers pulmonary arterial hypertension in fawn hooded rats: similarities to human pulmonary arterial hypertension. Circulation 113(22):2630–2641

Butcher RW, Sutherland EW (1962) Adenosine 3′,5′-phosphate in biological materials. I. Purification and properties of cyclic 3′,5′-nucleotide phosphodiesterase and use of this enzyme to characterize adenosine 3′,5′-phosphate in human urine. J Biol Chem 237:1244–1250

Cerqueira FM, Brandizzi LI et al (2012) Serum from calorie-restricted rats activates vascular cell eNOS through enhanced insulin signaling mediated by adiponectin. PLoS One 7(2):e31155

Chan SY, Loscalzo J (2010) MicroRNA-210: a unique and pleiotropic hypoxamir. Cell Cycle 9 (6):1072–1083

Chan MC, Nguyen PH et al (2007) A novel regulatory mechanism of the bone morphogenetic protein (BMP) signaling pathway involving the carboxyl-terminal tail domain of BMP type II receptor. Mol Cell Biol 27(16):5776–5789

Chan SY, Zhang YY et al (2009) MicroRNA-210 controls mitochondrial metabolism during hypoxia by repressing the iron-sulfur cluster assembly proteins ISCU1/2. Cell Metab 10 (4):273–284

Chin KM, Kingman M et al (2008) Changes in right ventricular structure and function assessed using cardiac magnetic resonance imaging in bosentan-treated patients with pulmonary arterial hypertension. Am J Cardiol 101(11):1669–1672

Christman BW, McPherson CD et al (1992) An imbalance between the excretion of thromboxane and prostacyclin metabolites in pulmonary hypertension. N Engl J Med 327(2):70–75

Clozel M, Maresta A, Humbert M (2013) Endothelin receptor antagonists. In: Humbert M, Evgenov OV, Stasch JP (eds) Pharmacotherapy of pulmonary hypertension. Springer, Heidelberg

Coulet F, Nadaud S et al (2003) Identification of hypoxia-response element in the human endothelial nitric-oxide synthase gene promoter. J Biol Chem 278(47):46230–46240

Courboulin A, Paulin R et al (2011a) Role for miR-204 in human pulmonary arterial hypertension. J Exp Med 208(3):535–548

Courboulin A, Tremblay VL et al (2011b) Kruppel-like Factor 5 contributes to pulmonary artery smooth muscle proliferation and resistance to apoptosis in human pulmonary arterial hypertension. Respir Res 12:128

D'Orleans-Juste P, Telemaque S et al (1991) Different pharmacological profiles of big-endothelin-3 and big-endothelin-1 in vivo and in vitro. Br J Pharmacol 104(2):440–444

de Groote P, Millaire A et al (1998) Right ventricular ejection fraction is an independent predictor of survival in patients with moderate heart failure. J Am Coll Cardiol 32(4):948–954

Delker SL, Xue F et al (2010) Role of zinc in isoform-selective inhibitor binding to neuronal nitric oxide synthase. Biochemistry 49(51):10803–10810

Doi T, Sugimoto H et al (1999) Interactions of endothelin receptor subtypes A and B with Gi, Go, and Gq in reconstituted phospholipid vesicles. Biochemistry 38(10):3090–3099

Droma Y, Hanaoka M et al (2002) Positive association of the endothelial nitric oxide synthase gene polymorphisms with high-altitude pulmonary edema. Circulation 106(7):826–830

Dudzinski DM, Igarashi J et al (2006) The regulation and pharmacology of endothelial nitric oxide synthase. Annu Rev Pharmacol Toxicol 46:235–276

Egom EE, Mohamed TM et al (2011) Activation of Pak1/Akt/eNOS signaling following sphingosine-1-phosphate release as part of a mechanism protecting cardiomyocytes against ischemic cell injury. Am J Physiol Heart Circ Physiol 301(4):H1487–H1495

Evgenov OV, Pacher P et al (2006) NO-independent stimulators and activators of soluble guanylate cyclase: discovery and therapeutic potential. Nat Rev Drug Discov 5(9):755–768

Fagan KA, Fouty BW et al (1999) The pulmonary circulation of homozygous or heterozygous eNOS-null mice is hyperresponsive to mild hypoxia. J Clin Invest 103(2):291–299

Falcetti E, Hall SM et al (2010) Smooth muscle proliferation and role of the prostacyclin (IP) receptor in idiopathic pulmonary arterial hypertension. Am J Respir Crit Care Med 182 (9):1161–1170

Fang YH, Piao L et al (2012) Therapeutic inhibition of fatty acid oxidation in right ventricular hypertrophy: exploiting Randle's cycle. J Mol Med (Berl) 90(1):31–43

Farber HW, Loscalzo J (2004) Pulmonary arterial hypertension. N Engl J Med 351(16):1655–1665

Fasanaro P, D'Alessandra Y et al (2008) MicroRNA-210 modulates endothelial cell response to hypoxia and inhibits the receptor tyrosine kinase ligand Ephrin-A3. J Biol Chem 283 (23):15878–15883

Fernhoff NB, Derbyshire ER et al (2009) A nitric oxide/cysteine interaction mediates the activation of soluble guanylate cyclase. Proc Natl Acad Sci USA 106(51):21602–21607

Fijalkowska I, Xu W et al (2010) Hypoxia inducible-factor1alpha regulates the metabolic shift of pulmonary hypertensive endothelial cells. Am J Pathol 176(3):1130–1138

Fish JE, Yan MS et al (2010) Hypoxic repression of endothelial nitric-oxide synthase transcription is coupled with eviction of promoter histones. J Biol Chem 285(2):810–826

Ford-Hutchinson AW, Bray MA et al (1980) Leukotriene B, a potent chemokinetic and aggregating substance released from polymorphonuclear leukocytes. Nature 286 (5770):264–265

Forfia PR, Fisher MR et al (2006) Tricuspid annular displacement predicts survival in pulmonary hypertension. Am J Respir Crit Care Med 174(9):1034–1041

Francis SH et al (2010) cGMP-dependent protein kinases and cGMP phosphodiesterases in nitric oxide and cGMP action. Pharmacol Rev 62(3):525–563

Fredenburgh LE, Liang OD et al (2008) Absence of cyclooxygenase-2 exacerbates hypoxia-induced pulmonary hypertension and enhances contractility of vascular smooth muscle cells. Circulation 117(16):2114–2122

Gangopahyay A, Oran M et al (2011) Bone morphogenetic protein receptor II is a novel mediator of endothelial nitric-oxide synthase activation. J Biol Chem 286(38):33134–33140

Gautier M, Antier D et al (2007) Continuous inhalation of carbon monoxide induces right ventricle ischemia and dysfunction in rats with hypoxic pulmonary hypertension. Am J Physiol Heart Circ Physiol 293(2):H1046–H1052

Ghofrani HA, Hoeper MM et al (2010) Riociguat for chronic thromboembolic pulmonary hypertension and pulmonary arterial hypertension: a phase II study. Eur Respir J 36(4):792–799

Ghosh G, Subramanian IV et al (2010) Hypoxia-induced microRNA-424 expression in human endothelial cells regulates HIF-alpha isoforms and promotes angiogenesis. J Clin Invest 120 (11):4141–4154

Giaid A, Yanagisawa M et al (1993) Expression of endothelin-1 in the lungs of patients with pulmonary hypertension. N Engl J Med 328(24):1732–1739

Gizi A, Papassotiriou I et al (2011) Assessment of oxidative stress in patients with sickle cell disease: the glutathione system and the oxidant-antioxidant status. Blood Cells Mol Dis 46 (3):220–225

Gladwin MT, Vichinsky E (2008) Pulmonary complications of sickle cell disease. N Engl J Med 359(21):2254–2265

Griffith OW, Stuehr DJ (1995) Nitric oxide synthases: properties and catalytic mechanism. Annu Rev Physiol 57:707–736

Guo L, Qiu Z et al (2012) The microRNA-328 regulates hypoxic pulmonary hypertension by targeting at insulin growth factor 1 receptor and L-type calcium channel-alpha1C. Hypertension 59(5):1006–1013

Hansmann G, Wagner RA et al (2007) Pulmonary arterial hypertension is linked to insulin resistance and reversed by peroxisome proliferator-activated receptor-gamma activation. Circulation 115(10):1275–1284

Hassoun PM, Filippov G et al (2004) Hypoxia decreases expression of soluble guanylate cyclase in cultured rat pulmonary artery smooth muscle cells. Am J Respir Cell Mol Biol 30(6):908–913

Hinton M, Gutsol A et al (2007) Thromboxane hypersensitivity in hypoxic pulmonary artery myocytes: altered TP receptor localization and kinetics. Am J Physiol Lung Cell Mol Physiol 292(3):L654–L663

Huggins JP, Pelton JT et al (1993) The structure and specificity of endothelin receptors: their importance in physiology and medicine. Pharmacol Ther 59(1):55–123

Huh JW, Kim SY et al (2011) YC-1 attenuates hypoxia-induced pulmonary arterial hypertension in mice. Pulm Pharmacol Ther 24(6):638–646

Iwasaki H, Eguchi S et al (1999) Endothelin-mediated vascular growth requires p42/p44 mitogen-activated protein kinase and p70 S6 kinase cascades via transactivation of epidermal growth factor receptor. Endocrinology 140(10):4659–4668

Ji R, Cheng Y et al (2007) MicroRNA expression signature and antisense-mediated depletion reveal an essential role of MicroRNA in vascular neointimal lesion formation. Circ Res 100(11):1579–1588

Jones JE, Walker JL et al (2004) Effect of 5-lipoxygenase on the development of pulmonary hypertension in rats. Am J Physiol Heart Circ Physiol 286(5):H1775–H1784

Kapakos G, Bouallegue A et al (2010) Modulatory role of nitric oxide/cGMP system in endothelin-1-induced signaling responses in vascular smooth muscle cells. Curr Cardiol Rev 6(4):247–254

Kilroy G, Kirk-Ballard H et al (2012) The ubiquitin ligase Siah2 regulates PPARgamma activity in adipocytes. Endocrinology 153(3):1206–1218

Ko FN, Wu CC et al (1994) YC-1, a novel activator of platelet guanylate cyclase. Blood 84(12):4226–4233

Kourembanas S, McQuillan LP et al (1993) Nitric oxide regulates the expression of vasoconstrictors and growth factors by vascular endothelium under both normoxia and hypoxia. J Clin Invest 92(1):99–104

Kroemer G, Reed JC (2000) Mitochondrial control of cell death. Nat Med 6(5):513–519

Listi F, Caruso M et al (2008) Pro-inflammatory gene variants in myocardial infarction and longevity: implications for pharmacogenomics. Curr Pharm Des 14(26):2678–2685

Liu XR, Zhang MF et al (2012) Enhanced store-operated Ca(2)+ entry and TRPC channel expression in pulmonary arteries of monocrotaline-induced pulmonary hypertensive rats. Am J Physiol Cell Physiol 302(1):C77–C87

Long L, Crosby A et al (2009) Altered bone morphogenetic protein and transforming growth factor-beta signaling in rat models of pulmonary hypertension: potential for activin receptor-like kinase-5 inhibition in prevention and progression of disease. Circulation 119(4):566–576

Lundberg JO, Weitzberg E et al (2011) The nitrate-nitrite-nitric oxide pathway in mammals. In: Bryan NS, Loscalzo J (eds) Nitrite and nitrate in human health and disease. Humana, New York

Machado RD, Aldred MA et al (2006) Mutations of the TGF-beta type II receptor BMPR2 in pulmonary arterial hypertension. Hum Mutat 27(2):121–132

Maron BA, Loscalzo J (2013) Pulmonary hypertension in non-pulmonary arterial hypertension patients. In: Creager MA, Beckman JA, Loscalzo J (eds) Vascular medicine: a companion to Braunwald's heart disease, 2nd edn. Elsevevier, Philadelphia, PA, pp 419–432

Maron BA, Zhang YY et al (2009) Aldosterone increases oxidant stress to impair guanylyl cyclase activity by cysteinyl thiol oxidation in vascular smooth muscle cells. J Biol Chem 284(12):7665–7672

Maron BA, Zhang YY et al (2012) Aldosterone inactivates the endothelin-B receptor via a cysteinyl thiol redox switch to decrease pulmonary endothelial nitric oxide levels and modulate pulmonary arterial hypertension. Circulation 126(8):963–974

McAllister-Lucas LM, Haik TL et al (1995) An essential aspartic acid at each of two allosteric cGMP-binding sites of a cGMP-specific phosphodiesterase. J Biol Chem 270(51):30671–30679

McDonald DM, Alp NJ et al (2004) Functional comparison of the endothelial nitric oxide synthase Glu298Asp polymorphic variants in human endothelial cells. Pharmacogenetics 14(12):831–839

McLaughlin VV, Archer SL et al (2009) ACCF/AHA 2009 expert consensus document on pulmonary hypertension a report of the American College of Cardiology Foundation Task Force on Expert Consensus Documents and the American Heart Association developed in collaboration with the American College of Chest Physicians; American Thoracic Society, Inc.; and the Pulmonary Hypertension Association. J Am Coll Cardiol 53(17):1573–1619

McMahon TJ, Ahearn GS et al (2005) A nitric oxide processing defect of red blood cells created by hypoxia: deficiency of S-nitrosohemoglobin in pulmonary hypertension. Proc Natl Acad Sci USA 102(41):14801–14806

Merklinger SL, Jones PL et al (2005) Epidermal growth factor receptor blockade mediates smooth muscle cell apoptosis and improves survival in rats with pulmonary hypertension. Circulation 112(3):423–431

Michelakis ED, Wilkins MR et al (2008) Emerging concepts and translational priorities in pulmonary arterial hypertension. Circulation 118(14):1486–1495

Mittal M, Gu XQ et al (2012) Hypoxia induces K(v) channel current inhibition by increased NADPH oxidase-derived reactive oxygen species. Free Radic Biol Med 52(6):1033–1042

Miyamoto Y, Saito Y et al (1998) Endothelial nitric oxide synthase gene is positively associated with essential hypertension. Hypertension 32(1):3–8

Mossalam M, Matissek KJ et al (2012) Direct induction of apoptosis using an optimal mitochondrially targeted P53. Mol Pharm 9(5):1449–1458

Murata T, Sato K et al (2002) Decreased endothelial nitric-oxide synthase (eNOS) activity resulting from abnormal interaction between eNOS and its regulatory proteins in hypoxia-induced pulmonary hypertension. J Biol Chem 277(46):44085–44092

Nagendran J, Archer SL et al (2007) Phosphodiesterase type 5 is highly expressed in the hypertrophied human right ventricle, and acute inhibition of phosphodiesterase type 5 improves contractility. Circulation 116(3):238–248

Nishida M, Okada Y et al (2004) Role of endothelin ETB receptor in the pathogenesis of monocrotaline-induced pulmonary hypertension in rats. Eur J Pharmacol 496(1–3):159–165

Nishimura T, Vaszar LT et al (2003) Simvastatin rescues rats from fatal pulmonary hypertension by inducing apoptosis of neointimal smooth muscle cells. Circulation 108(13):1640–1645

Oikawa M, Kagaya Y et al (2005) Increased [18F]fluorodeoxyglucose accumulation in right ventricular free wall in patients with pulmonary hypertension and the effect of epoprostenol. J Am Coll Cardiol 45(11):1849–1855

Okamoto Y, Ninomiya H et al (1997) Palmitoylation of human endothelinB. Its critical role in G protein coupling and a differential requirement for the cytoplasmic tail by G protein subtypes. J Biol Chem 272(34):21589–21596

Olave N, Nicola T et al (2012) Transforming growth factor-beta regulates endothelin-1 signaling in the newborn mouse lung during hypoxia exposure. Am J Physiol Lung Cell Mol Physiol 302(9):L857–L865

Olschewski H (2013) Prostacyclins. In: Humbert M, Evgenov OV, Stasch JP (eds) Pharmacotherapy of pulmonary hypertension. Springer, Heidelberg

Parikh VN, Jin RC et al (2012) MicroRNA-21 integrates pathogenic signaling to control pulmonary hypertension: results of a network bioinformatics approach. Circulation 125(12):1520–1532

Pellicena P, Karow DS et al (2004) Crystal structure of an oxygen-binding heme domain related to soluble guanylate cyclases. Proc Natl Acad Sci USA 101(35):12854–12859

Piao L, Fang YH et al (2010) The inhibition of pyruvate dehydrogenase kinase improves impaired cardiac function and electrical remodeling in two models of right ventricular hypertrophy: resuscitating the hibernating right ventricle. J Mol Med (Berl) 88(1):47–60

Porter TD, Beck TW et al (1990) NADPH-cytochrome P-450 oxidoreductase gene organization correlates with structural domains of the protein. Biochemistry 29(42):9814–9818

Pullamsetti SS, Doebele C et al (2011) Inhibition of microRNA-17 improves lung and heart function in experimental pulmonary hypertension. Am J Respir Crit Care Med 185(4):409–419

Puwanant S, Park M et al (2010) Ventricular geometry, strain, and rotational mechanics in pulmonary hypertension. Circulation 121(2):259–266

Racz A, Veresh Z et al (2010) Cyclooxygenase-2 derived thromboxane A(2) and reactive oxygen species mediate flow-induced constrictions of venules in hyperhomocysteinemia. Atherosclerosis 208(1):43–49

Randle PJ, Garland PB et al (1963) The glucose fatty-acid cycle. Its role in insulin sensitivity and the metabolic disturbances of diabetes mellitus. Lancet 1(7285):785–789

Recchia AG, Filice E et al (2009) Endothelin-1 induces connective tissue growth factor expression in cardiomyocytes. J Mol Cell Cardiol 46(3):352–359

Reeve HL, Michelakis E et al (2001) Alterations in a redox oxygen sensing mechanism in chronic hypoxia. J Appl Physiol 90(6):2249–2256

Rondelet B, Dewachter L et al (2010) Sildenafil added to sitaxsentan in overcirculation-induced pulmonary arterial hypertension. Am J Physiol Heart Circ Physiol 299(4):H1118–H1123

Rosenzweig BL, Imamura T et al (1995) Cloning and characterization of a human type II receptor for bone morphogenetic proteins. Proc Natl Acad Sci USA 92(17):7632–7636

Rothman A, Wolner B et al (1994) Immediate-early gene expression in response to hypertrophic and proliferative stimuli in pulmonary arterial smooth muscle cells. J Biol Chem 269(9):6399–6404

Schmidt P, Schramm M et al (2003) Mechanisms of nitric oxide independent activation of soluble guanylyl cyclase. Eur J Pharmacol 468(3):167–174

Schneider MP, Boesen EI et al (2007) Contrasting actions of endothelin ET(A) and ET(B) receptors in cardiovascular disease. Annu Rev Pharmacol Toxicol 47:731–759

Serre V, Ildefonse M et al (1995) Effects of cysteine modification on the activity of the cGMP-gated channel from retinal rods. J Membr Biol 146(2):145–162

Seta F, Rahmani M et al (2011) Pulmonary oxidative stress is increased in cyclooxygenase-2 knockdown mice with mild pulmonary hypertension induced by monocrotaline. PLoS One 6(8):e23439

Song Y, Jones JE et al (2005) Increased susceptibility to pulmonary hypertension in heterozygous BMPR2-mutant mice. Circulation 112(4):553–562

Song Y, Coleman L et al (2008) Inflammation, endothelial injury, and persistent pulmonary hypertension in heterozygous BMPR2-mutant mice. Am J Physiol Heart Circ Physiol 295(2):H677–H690

Spiegelhalder B, Eisenbrand G et al (1976) Influence of dietary nitrate on nitrite content of human saliva: possible relevance to in vivo formation of N-nitroso compounds. Food Cosmet Toxicol 14(6):545–548

Stasch JP, Schmidt PM et al (2006) Targeting the heme-oxidized nitric oxide receptor for selective vasodilatation of diseased blood vessels. J Clin Invest 116(9):2552–2561

Steudel W, Scherrer-Crosbie M et al (1998) Sustained pulmonary hypertension and right ventricular hypertrophy after chronic hypoxia in mice with congenital deficiency of nitric oxide synthase 3. J Clin Invest 101(11):2468–2477

Suntharalingam J, Hughes RJ et al (2007) Acute haemodynamic responses to inhaled nitric oxide and intravenous sildenafil in distal chronic thromboembolic pulmonary hypertension (CTEPH). Vascul Pharmacol 46(6):449–455

Tang JR, Markham NE et al (2004) Inhaled nitric oxide attenuates pulmonary hypertension and improves lung growth in infant rats after neonatal treatment with a VEGF receptor inhibitor. Am J Physiol Lung Cell Mol Physiol 287(2):L344–L351

Toporsian M, Jerkic M et al (2010) Spontaneous adult-onset pulmonary arterial hypertension attributable to increased endothelial oxidative stress in a murine model of hereditary hemorrhagic telangiectasia. Arterioscler Thromb Vasc Biol 30(3):509–517

Topper JN, Cai J et al (1996) Identification of vascular endothelial genes differentially responsive to fluid mechanical stimuli: cyclooxygenase-2, manganese superoxide dismutase, and endothelial cell nitric oxide synthase are selectively up-regulated by steady laminar shear stress. Proc Natl Acad Sci USA 93(19):10417–10422

Tuder RM, Cool CD et al (1999) Prostacyclin synthase expression is decreased in lungs from patients with severe pulmonary hypertension. Am J Respir Crit Care Med 159(6):1925–1932

Turko IV, Francis SH et al (1999) Studies of the molecular mechanism of discrimination between cGMP and cAMP in the allosteric sites of the cGMP-binding cGMP-specific phosphodiesterase (PDE5). J Biol Chem 274(41):29038–29041

Umar S, Lee JH et al (2012) Spontaneous ventricular fibrillation in right ventricular failure secondary to chronic pulmonary hypertension. Circ Arrhythm Electrophysiol 5(1):181–190

Vermeersch P, Buys E et al (2007) Soluble guanylate cyclase-alpha1 deficiency selectively inhibits the pulmonary vasodilator response to nitric oxide and increases the pulmonary vascular remodeling response to chronic hypoxia. Circulation 116(8):936–943

Voelkel NF, Tuder RM et al (1996) Inhibition of 5-lipoxygenase-activating protein (FLAP) reduces pulmonary vascular reactivity and pulmonary hypertension in hypoxic rats. J Clin Invest 97(11):2491–2498

Voelkel NF, Quaife RA et al (2006) Right ventricular function and failure: report of a National Heart, Lung, and Blood Institute working group on cellular and molecular mechanisms of right heart failure. Circulation 114(17):1883–1891

Wallace DC (2005) A mitochondrial paradigm of metabolic and degenerative diseases, aging, and cancer: a dawn for evolutionary medicine. Annu Rev Genet 39:359–407

Waxman AB (2011) Pulmonary hypertension in heart failure with preserved ejection fraction: a target for therapy? Circulation 124(2):133–135

Wharton J, Strange JW et al (2005) Antiproliferative effects of phosphodiesterase type 5 inhibition in human pulmonary artery cells. Am J Respir Crit Care Med 172(1):105–113

Woodfield K, Ruck A et al (1998) Direct demonstration of a specific interaction between cyclophilin-D and the adenine nucleotide translocase confirms their role in the mitochondrial permeability transition. Biochem J 336(Pt 2):287–290

Yamashita K, Discher DJ et al (2001) Molecular regulation of the endothelin-1 gene by hypoxia. Contributions of hypoxia-inducible factor-1, activator protein-1, GATA-2, AND p300/CBP. J Biol Chem 276(16):12645–12653

Yeager ME, Belchenko DD et al (2012) Endothelin-1, the unfolded protein response, and persistent inflammation: role of pulmonary artery smooth muscle cells. Am J Respir Cell Mol Biol 46(1):14–22

Yoo BK, Lamarre I et al (2012) Quaternary structure controls ligand dynamics in soluble guanylate cyclase. J Biol Chem 287(9):6851–6859

Zamzami N, Kroemer G (2001) The mitochondrion in apoptosis: how Pandora's box opens. Nat Rev Mol Cell Biol 2(1):67–71

Zaugg CE, Hornstein PS et al (1996) Endothelin-1-induced release of thromboxane A2 increases the vasoconstrictor effect of endothelin-1 in postischemic reperfused rat hearts. Circulation 94(4):742–747

Zhang WM, Yip KP et al (2003) ET-1 activates Ca2+ sparks in PASMC: local Ca2+ signaling between inositol trisphosphate and ryanodine receptors. Am J Physiol Lung Cell Mol Physiol 285(3):L680–L690

Pulmonary Hypertension: Pathology

Peter Dorfmüller

Abstract Pulmonary hypertension (PH) is a life-threatening and often fatal disease, characterized by elevated pulmonary vascular resistance and secondary right ventricular failure. Since etiologies of PH are multiple and its pathogenesis is complex, histology from lungs of patients with PH may help us to determine different etiological factors of the disease. The degree of involvement of various cell types and structures within the lung tissue represents an important indicator of the pathophysiologal process. So even if the role for pathologists in routine management of PH is limited, lessons can be learned from morphology. The present chapter outlines the current understanding of this disease from the pathologist's point of view.

Keywords Pulmonary hypertension • Histology • Intimal fibrosis • Plexiform lesion • Medial hypertrophy • Pulmonary veno-occlusive disease

Contents

1 Introduction 60
2 Arterial Lesions 60
   2.1 Medial Hypertrophy 61
   2.2 Intimal Fibrosis 63
   2.3 Complex Lesions 64
   2.4 Perivascular Inflammation and Tertiary Lymphoid Tissue 66
3 Venous and Venular Lesions 68
   3.1 Pulmonary Veno-occlusive Disease/Pulmonary Capillary Hemangiomatosis 68

P. Dorfmüller (✉)
Service d'Anatomie et de Cytologie Pathologiques, Hôpital Marie Lannelongue, 133, Avenue de la Résistance, Le Plessis Robinson, France

INSERM U999 "Hypertension Artérielle Pulmonaire: Physiopathologie et Innovation Thérapeutique", 133, Avenue de la Résistance, Le Plessis Robinson, France
e-mail: peter.dorfmuller@u-psud.fr

4	Forms of Pulmonary Hypertension with Significant Arterial and Venous Remodeling...	70
4.1	PAH Associated with Connective Tissue Disease...	70
5	Conclusion...	71
References...		72

1 Introduction

The morphologic study of diseased lungs from patients with PH represents an important landmark on the chart of pathophysiologic concepts: a real status and probably the histopathologic correlate to increased pulmonary vascular resistance. The knowledge of pathological anatomy in PH must be of importance to the researcher, as well as to the clinical physician, since observation and interpretation always precede deduction and effective action. Therefore, the histologic inventory of PH vasculopathy has been addressed in the consensus papers of the past World Symposia (Simonneau et al. 2004, 2009). Despite these publications being enriched with up-to-date scientific knowledge from the scientific community, the foundations of descriptive histopathology reach back to the 1960s and 1970s, when such historical figures as Heath, Edwards, or Wagenvoort performed their meticulous studies on large case numbers (Heath and Edwards 1958; Wagenvoort and Wagenvoort 1970). Nonetheless, recent histology-oriented papers have helped to differentiate morphological phenotypes and peculiarities in PH subgroups, stimulating the ongoing discussion within the PH community. The following account of histological observations is paralleled with historical and actual concepts of PH pathophysiology.

2 Arterial Lesions

The pulmonary arterial tree comprises at least three levels with different anatomical/histological architectural pattern. Arteries from the main arterial trunk (27 mm in diameter) to the segmentary level (0.5 mm in diameter) belong to the elastic type, characterized by an important proportion of elastic fibers within the medial layer of the vessel wall. Since their architecture is similar to large systemic arteries, chronic vessel wall stress through increased pressure will manifest in atheromatous lesions as found, e.g., in the aorta of patients with systemic hypertension. In patients displaying PH all stages from subendothelial foamy macrophages to large atheroma plaques can be encountered, even if such lesions will never reach the significance and severity of aortic plaques. Also, occlusive or near-occlusive lesions as they can be seen in carotid arteries are usually not found in PH. The increase in pulmonary vascular resistances is most probably due to the distal arterial level, comprising arteries of the muscular type:

the proportion of elastic fibers within the medial layer decreases and the share of smooth muscle cells will increase in vessels beneath 0.5 mm in diameter. From here on, peculiar lesions of the vessel wall can be found in patients with PH, lesions with different microscopic appearance, owing to the difference of the prevailing cell type. Those different cell types are physiologically present and some of them even constitute the three arterial wall compartments. In consequence, they also determine the pathological anatomy of the artery. The principal cells involved in pulmonary arterial lesions of patients suffering from PH are smooth muscle cells, fibroblasts/myofibroblasts, endothelial cells, and mononuclear cells. Figure 1 shows all different arterial lesions and the cell type prevailing (Dorfmüller and Humbert 2012).

2.1 Medial Hypertrophy

This abnormality of the vessel wall can be observed in all subgroups of pulmonary arterial hypertension (PAH, Group 1, Dana Point) and may be encountered in other forms of PH, e.g., Group 2 or 3 (PH owing to left heart disease and PH owing to lung diseases and/or hypoxia). The lesion corresponds to a proliferation of smooth muscle cells within the tunica media (Fig. 2a). The histological criterion of medial hypertrophy (which corresponds more precisely to hypertrophy and hyperplasia, i.e., increase in volume and number of smooth muscle cells, respectively) is fulfilled when the diameter of a single medial layer, delineated by its internal and external elastic lamina, exceeds 10 % of the artery's cross-sectional diameter. In the PH lung, hyperplasia of smooth muscle cells is virtually always observed at the level of pre-capillary arterioles; it should be avoided to use the term "medial hypertrophy" in this context, since human arterioles of the lung are physiologically quasi non-muscularized and do not show the classic arterial layers.

Isolated hypertrophy of the medial layer may be considered as an early and even reversible event, as has been shown for pulmonary hypertension due to hypoxia at high altitude (Heath and Williams 1989). Also, as recently stated by Sakao et al. (2010), new therapeutic concepts that appear to work in experimental PH, e.g., the tyrosine kinase receptor inhibitor imatinib, solely reduce the pulmonary arterial/arteriolar muscularization, an observation which underlines the reversibility of muscular remodeling. In a recent study and review of the literature, Penaloza and Arias-Stella (2007) noted that healthy natives of high altitude regions over 3,500 m above sea level have pulmonary hypertension, right ventricular hypertrophy and increased amount of smooth muscle cells in the distal pulmonary arterial branches. Clinical and histological findings from 30 healthy high altitude natives and 30 sea level natives revealed that the main factor responsible for pulmonary hypertension in healthy highlanders is the increased amount of smooth muscle cells in the distal pulmonary arteries and arterioles. Vasoconstriction is a secondary factor in this setting, because the administration of oxygen decreases the pulmonary arterial pressure only by 15–20 %. The adaptive increase in cardiac and smooth muscle

Diversity of Lesions in PAH - Variation of the Cell Type Prevailing

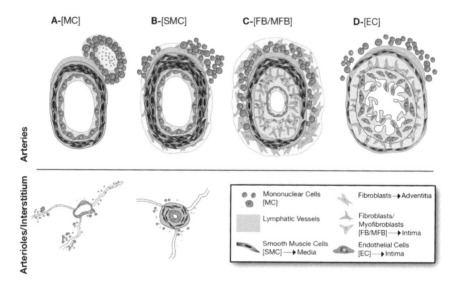

Fig. 1 Different lesions or different snap shots of one process in time? Typical histological lesions encountered in lungs from patients with PAH. (**a**) Tertiary lymphoid tissue (lymphoid follicles) can be observed in the vicinity of inconspicuous and remodeled arteries of patients with PAH, frequently associated with dilated lymphatic vessels (*top*); interstitial inflammatory infiltrates, mostly mononuclear cells, are frequently seen (*bottom*). (**b**) Hypertrophy/hyperplasia of the media is encountered in different forms of PH and corresponds to a proliferation of smooth muscle cells (SMC) (*top*). SMC are also the prevailing cell type involved in the muscularization of normally non-muscularized arterioles (*bottom*); perivascular lymphocytic infiltrates are regularly observed; most animal models of PH display this reversible lesion type with prevailing SMC. Fibrosis of the adventitia may be present. (**c**) Concentric intimal fibrosis (non-laminar or laminar) is constantly found in PAH; fibroblasts and myofibroblasts prevail and occlude the lumen through deposition of collagen. The lesion is often combined with hypertrophy of the media and fibrous broadening of the adventitia; perivascular inflammation is frequently observed. (**d**) Plexiform lesions are considered as a hallmark of PAH histology and correspond to an exuberant proliferation of endothelial cells within a dilated, thin-walled arterial lumen. Inflammation is classically described in the range of plexiform lesions. Figure reprinted with permission of the American Thoracic Society. Copyright © 2012 American Thoracic Society. Dorfmüller P, Humbert M (2012) Progress in pulmonary arterial hypertension pathology: relighting a torch inside the tunnel. Am J Respir Crit Care Med 186(3):210–212. Official journal of the American Thoracic Society

cell mass reverses after a prolonged residence at sea level. In PAH, medial hypertrophy is commonly associated to the remodeling of other vascular compartments, which are discussed below.

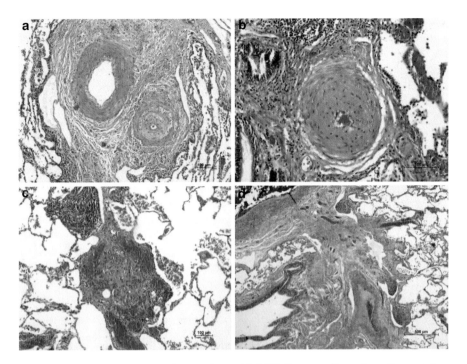

Fig. 2 Pulmonary arterial lesions from patients with PAH. Hematoxylin–eosin–saffron staining. (**a**) Two arterial walls displaying different histological patterns; the left artery shows important medial hypertrophy, while the right artery predominantly presents with intimal fibrosis (concentric, partially laminar). (**b**) Small pulmonary artery with important concentric, non-laminar thickening of the intima. Note the discrete perivascular lymphocytic infiltrate. (**c**) Complex lesion with central plexiform pattern and peripheral dilation lesions. Perivascular inflammatory infiltrate and a lymphoid follicle are present in the range of the diseased vessel. (**d**) Remodeled artery with adjacent bronchus. Note the tertiary lymphoid tissue in the vicinity of the vessel

2.2 Intimal Fibrosis

Fibrotic lesions of the intimal layer are frequent in PH lungs. The intima may be thickened by proliferation and recruitment of fibroblasts, myofibroblasts, and other connective tissue cells, and consequently by the interstitial deposition of collagen (Fig. 2a, b). In a purely descriptive approach, this thickening may be uniform and concentric, or focally predominating and eccentric. Both forms can lead to a complete occlusion of the artery. The gain in intimal cellularity is generally understood as a reaction of the inner arterial layer to a luminal stimulus, e.g., chronic pressure and shear stress or as a scarring process after primary endothelial damage (Voelkel and Tuder 1995). In many cases, adventitial fibrosis is associated with intimal changes but remains difficult to evaluate, due to the lack of a clear anatomical delimitation.

Eccentric intimal thickening is frequently observed in cases with thrombotic events and probably represents residues of wall-adherent, organized thrombi. Thrombotic lesions, or so-called in situ thrombosis, are a characteristic pattern of chronic thromboembolic pulmonary hypertension (CTEPH) but also regularly encountered in different PAH subgroups and in PVOD; organization and recanalization of totally occluding thrombotic material may lead to bizarre, fibrotic multichannel lesions (so-called colander-like lesions) which can easily be confounded with proliferative complex lesions (see below).

A conspicuous phenotype of fibrosis with laminar arrangement of collagen layers within the intima of the diseased vessel may be observed in PAH (Fig. 2a). This lesion is also known as "onion-skin" or "onion-bulb" lesion. The scar-like, collagen-rich arterial lesions may be found in lungs of patients suffering from different forms of PAH, including PAH associated with connective tissue disease (Cool et al. 1997). Nevertheless, the observation that intimal thickening proximal to plexiform lesions in supernumerary arteries usually displays a concentric laminar phenotype seems to closely associate these two lesions. Immunohistochemical analysis reveals fibroblasts, myofibroblasts, and smooth muscle cells.

2.3 Complex Lesions

Complex pulmonary arterial lesions comprise different patterns, which are commonly observed in close topographic association. However, the pathophysiologic significance of this typical PAH lesion has yet to be elucidated and its peculiar structure hitherto remains a mystery. It is still unclear if this arterial anomaly represents a cause for increased vascular resistances or if it is simply a secondary phenomenon resulting from chronic pulmonary hypertension. In this context, it should be mentioned that Stacher et al. (2012) have recently published an exhaustive histological analysis of 62 cases of PAH, taking into account associated PAH-specific treatment. Interestingly, the authors found a significant positive correlation between the quantity of plexiform lesions and case history of prostacyclin treatment, a finding which puts too mechanistic conclusions of primary luminal obstruction through endothelial proliferation into question.

The plexiform pattern is the main criterion and at the core of complex lesions (Fig. 2c). It has been first described in congenital heart disease-associated PAH as well as in idiopathic PAH (formerly called primary pulmonary hypertension) (Heath and Edwards 1958; Wagenvoort and Wagenvoort 1970). Eventually, plexiform lesions have been shown to occur in other PAH subgroups such as anorexigen intake associated PAH or portopulmonary hypertension (Widgren 1986; Krowka and Edwards 2000). The peculiar lesion concerns various vascular compartments. A focal intimal thickening of small pulmonary arteries, preferably beyond branching points, is followed by an aneurysm-like widening of the vessel: the lumen is obstructed by an exuberant proliferation of endothelial cells, leading to the formation of capillary-like, sinusoidal channels on a fibroblast-rich background (Tuder et al. 2009). This intra-arterial capillary convolute, resembling sometimes to the glomerulum of a kidney, appears to feed into dilated, vein-like congestive vessels, which are perceivable at low magnification and may be helpful as sentinel lesions when tracing plexiform lesions. The latter anomaly is called dilation lesion and is virtually always associated with plexiform lesions (Fig. 2c). Arteritis with transmural inflammation and fibrinoid necrosis, as firstly described by Heath and Edwards (1958), maybe observed, but has become a rather infrequent phenomenon in PAH, possibly due to recent advances in medical therapy and/or associated prolonged survival of patients. However, perivascular inflammatory infiltrates in the vicinity of complex lesions and other diseased pulmonary arteries in PAH patients are commonly found (Stacher et al. 2012). The infiltrates mainly consist of T and B lymphocytes, to a lesser extent macrophages, as well as scattered mast cells. This inflammatory phenotype, in most cases, is evaluated as "mild" to "moderate".

From all phenotypes of arterial remodeling encountered in pulmonary hypertension, complex lesions appear to be found, if not specifically, than at least typically in PAH (Group 1 of the Dana Point Classification) (Tuder et al. 2009). There have been controversial publications in the past, questioning this specificity: in a publication from 1993 (Moser and Bloor 1993), plexiform lesions were detected in 20 out of 31 patients with confirmed chronic major vessel thromboembolic pulmonary hypertension, suggesting a possible relation between organizing thromboembolic and complex lesions, which mirrors an old theory first inspired by Harrison in 1958 (Harrison 1958). Recently, this finding has been confirmed in a review by Piazza and Goldhaber (2011). Another historic hypothesis explaining the pathogenesis of the glomerulum-like exuberant endothelial cell proliferation within the diseased pulmonary artery is the formation of arteriovenous shunts (Kucsko 1953). Wagenvoort explained the generation of plexiform lesions in cardiac left-to-right shunting by an increased blood-flow eliciting reflective vasoconstriction with subsequent development of endothelial alteration (Wagenvoort and Wagenvoort 1970). He speculated that vascular necrosis with arteritis might result from intense vasospasm, as seen in the systemic circulation. The pathogenetic plot, here, would be that plexiform lesions develop in the focal areas of fibrinoid necrosis by active cellular reorganization and recanalization of the thrombus composed of fibrin and platelets, usually present in this setting and observed within plexiform lesions. In fact, the only animal model leading to obvious plexiform lesions was achieved by producing necrosis of pulmonary arterioles in dogs after 2 weeks of severe pulmonary hypertension due to the creation of a shunt between pulmonary and systemic circulations (Saldaña et al. 1968). This explanation would take into consideration that elements of inflammation are consistently present in the range of plexiform and other vascular lesions in PAH (Tuder et al. 1994; Dorfmuller et al. 2003). On the other hand, Lee et al. (1998) have proposed a neoplastic approach to the hardly understandable proliferation of intraluminal neovessels at the core of plexiform lesions: they found that a large proportion of the endothelial cells in this area show monoclonality, raising the question of possible tumor-like growth. These different observations of arterial wall alteration within plexiform lesions are probably connected in a temporary line. However, they could fit into a broader concept of pathogenesis, first mentioned by Wagenvoort, but interestingly developed by Yaginuma et al. (1990). They submitted histological data from 11 patients with pulmonary hypertension due to congenital heart disease to a computer-based three-dimensional image reconstruction and found that plexiform lesions mostly occurred in supernumerary arteries branching apart from larger pulmonary arteries, proximal-to-arterial lesions with intimal fibrosis and medial hypertrophy. They also gathered evidence for generation of indirect anastomoses between the post-plexiform arterial segment to bronchial arteries running along the close bronchiole, passing via arterioles and the capillary bed and thereby creating the thin-walled congestive dilation lesions. In this view, the generation of proximal complex lesions might be the mere attempt of the pulmonary vasculature to bypass the primary downstream obstruction and to ensure capillary oxygenation through overt contact with arterial blood from proximal pulmonary and bronchial arteries. On the other hand, plexiform lesions are not restricted to supernumerary branching and can be observed after distal dichotomous branching of pulmonary arteries. Cool and coworkers have come to different conclusions after a computerized three-dimensional study on five patients with severe pulmonary hypertension of different cause. In their view, plexiform lesions are functionally important because blood flow is severely obstructed along the entire length of a vessel affected by a single lesion (Cool et al. 1999). This, in fact, puts at least a working shunt concept into question. They hypothesize that the plexiform lesion could be an early vascular alteration in severe pulmonary hypertension, independent of a component of medial smooth muscle cell hypertrophy. At a later time point in disease evolution, the plexiform lesion could transform into an intraluminal concentric obstruction composed of endothelial cells and recruited myofibroblasts, following the path of the plexiform lesion and thus representing a fibrous scar of the latter. The close association that can be observed between concentric laminar intimal fibrosis and plexiform lesions in lungs of patients suffering from PAH, indeed, makes this thought a tempting assumption. Nevertheless, a shunt hypothesis would not be contradictory if seen in the light of a failed attempt to shortcut other correlates of obstruction such as medial hypertrophy, intimal non-laminar fibrosis, and thrombotic lesions.

2.4 Perivascular Inflammation and Tertiary Lymphoid Tissue

It has not yet been elucidated whether the inflammatory pattern seen in plexiform lesions and other intimal lesions is of pathogenetic importance, or if it represents a pure epiphenomenon within disease evolution. The reported evidence of proinflammatory mediators, so-called chemokines, released by altered endothelial cells of PAH lungs strongly indicates a self-supporting and self-amplifying process (Balabanian et al. 2002; Dorfmuller et al. 2002; Sanchez et al. 2007). Although elements of inflammation seem to be present in diseased arteries of patients displaying PAH in various associated conditions, as well as in idiopathic PAH, the specific role of immune cells within installation and/or maintenance of obstructive lesions remains unclear. It seems unlikely that intimal proliferation and medial hypertrophy of pulmonary arteries could be the sole result of "scarring" vasculitis-like lesions, because lymphocytic and macrophagic infiltrates observed in the setting of PAH remain perivascular and are less abundant than in pulmonary forms of vasculitis, as seen e.g., in Wegener's granulomatosis (Dorfmüller et al. 2003). Nonetheless, Wagenvoort and coworkers had discussed such a possibility 40 years ago, and it cannot be excluded that a phase of intense inflammatory activity precedes a clinically symptomatic arterial remodeling with proliferation and recruitment of smooth muscle cells, endothelial cells, and fibroblasts/myofibroblasts. In fact, recent investigations provide evidence that chemokines with specific chemotactic activity for T lymphocytes and macrophages can increase smooth muscle proliferation (Perros et al. 2007).

A recent study from Perros et al. (2012) provides evidence for lymphoid neogenesis in lungs from patients with idiopathic PAH (IPAH). The authors analyzed lung histology of 21 patients with IPAH, of 5 patients with Eisenmenger syndrome associated PAH (ESPAH), and of 21 control patients (healthy lung areas from lobectomies of patients with adenocarcinoma or squamous carcinoma). They found pulmonary tTLs ranging from small lymphoid aggregates to large accumulations of lymphocytes resembling highly organized lymphoid follicles (Fig. 2d). tLTs were distributed throughout the pulmonary vasculature, from small distal remodeled arteries to larger pulmonary arteries; a 20-fold increase in pulmonary tLTs was measured by morphometrical means in patients with IPAH as compared with control subjects. Patients with ESPAH were not different from controls regarding lung tLT density. Lymphoid neogenesis in the target organ is considered to be a hallmark of autoimmune diseases. As such, tertiary lymphoid tissues tLTs or lymphoid follicles are commonly found in salivary glands of patients suffering from Sjögren syndrome, in thyroid gland of patients with Hashimoto thyroiditis, in joints and lungs of patients with rheumatoid arthritis, and around airways of patients with chronic obstructive pulmonary disease (Hogg et al. 2004; Aloisi and Pujol-Borrell 2006; Carragher et al. 2008). Several studies suggest that lymphoid neogenesis correlates with local autoantibody production (Carragher et al. 2008). Interestingly, Perros and coworkers also showed that lymphoid follicles in IPAH patients display canonical cellularity and structure of bona fide tLTs and lymphorganogenic chemokines such as CXCL13 and CCL19/CCL21 are overexpressed. Also, the study revealed that an IL-17-mediated response may promote the formation of tLTs in IPAH. Indeed, Rangel-Moreno et al. (2011) provide evidences that this signaling is essential for lymphoid neogenesis in lung. The presence of pulmonary tLTs in IPAH lungs could provide a structural basis for a local autoimmune response occurring in this disease. In the initially cited study, Perros and colleagues found accumulation of long-lived antibody-secreting CD^{138+} plasma cells together with immunoglobulin deposits around pulmonary vascular lesions, which appears to be consistent with locally produced autoantibodies, a prominent feature of IPAH (Tamby et al. 2005; Terrier et al. 2008; Dib et al. 2012).

3 Venous and Venular Lesions

During the Dana Point meeting in 2008, a consensus was reached, formally separating Group 1 (pulmonary arterial hypertension) of the classification of pulmonary hypertension from the rare entities of PVOD and PCH, now categorized as the new Group 1' (one prime): pulmonary venous hypertension. This differentiation seemed necessary because of particularities regarding therapeutic strategies and the cautious, if not restrictive use of potent vasodilating drugs, such as intravenous epoprostenol, in the case of pulmonary venous involvement. From the pathologist's standpoint of view, this separation is comprehensible, but not ideal; in fact some cases of PAH (Group 1) show PVOD-like pattern (see beneath), and the vast majority of PVOD and PCH cases (Group 1') present at least mild arterial alteration. The frequent occurrence of mixed vascular involvement seems to demand a less absolutistic evaluation into pulmonary vascular lesions either predominating the arterial or the venous system.

3.1 Pulmonary Veno-occlusive Disease/Pulmonary Capillary Hemangiomatosis

PVOD is a rare pulmonary vascular disease causing pulmonary hypertension and has been considered together with pulmonary capillary hemangiomatosis (PCH) a subgroup of pulmonary arterial hypertension (PAH) until recently, according to the Venice classification of 2003. It shows an estimated prevalence of 0.1–0.2 per million persons and per year (Mandel et al. 2000). Historical reports and large case studies have extrapolated a proportion of PVOD in PAH ranging from 5 to 25 % (Rich et al. 1987; Mandel et al. 2000). In PVOD as in PCH, vascular lesions predominate on the post-capillary level of pulmonary vasculature. However, lesions frequently involve both veins *and* arteries in lungs of patients with PVOD. Interestingly, a recent report indicates that certain subgroups of PAH regarded as pre-capillary forms simultaneously display a PVOD-like pattern (see beneath). In PVOD, the observed post-capillary lesions involves septal veins and pre-septal venules and frequently consist of loose, fibrous remodeling of the intima, which may totally occlude the lumen (Fig. 3a, b). The involvement of pre-septal venules should be considered as necessary for the histological diagnosis of PVOD; fibrous occlusion of large septal veins may be seen in many forms of pulmonary venous hypertension, including a frequently reported obstruction of large pulmonary veins following catheter ablation for cardiac atrial fibrillation (Di Biase et al. 2006). While septal veins usually display a paucicellular, cushion-like fibrous obstruction, intimal thickening of pre-septal venules can present with a dense pattern and increased cellularity. Anti-α-actin staining may reveal involvement of smooth muscle cells and/or myofibroblasts within such venular lesions. Also, thrombotic occlusion of small post-capillary microvessels has been observed, corresponding to

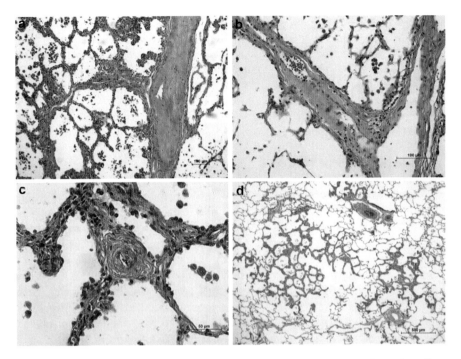

Fig. 3 Histological pattern in lungs from patients with PVOD. Hematoxylin–eosin–saffron staining. (**a**) Septal vein with fibrous, collagen-rich occlusion. The adjacent alveolar septa show thickening with discrete interstitial fibrosis and multiplication of capillaries. Note numerous intraalveolar macrophages. (**b**) Fibrous remodeling of the intima in a smaller, pre-septal venule. (**c**) Microvessels (either arterioles or venules) frequently display muscularization. (**d**) Typical patchy pattern of capillary hemangiomatosis-like foci. Alveolar septa are thickened by capillary proliferation. Note a remodeled pulmonary artery (*top right*) and a pulmonary vein with fibrous intimal thickening (*bottom right*)

"colander-like" lesions, which can be seen otherwise in small pulmonary arteries. The tunica media may be muscularized in both septal veins and pre-septal venules. The latter can be difficult to distinguish from muscularized arterioles of less than 80 μm in diameter, usually present in lungs from patients with PVOD (Fig. 3c). Pleural and pulmonary lymphatic vessels are usually dilated (Pietra 1997). The presence of calcium-encrusting elastic fibers in the vessel wall or the perivascular space, and inflammatory activation through a foreign body giant cell response is considered as an argument in favor of PVOD as compared to secondary venous hypertension (Pietra et al. 2004). Importantly, occult pulmonary hemorrhage regularly occurs in patients displaying PVOD. This particularity, which is certainly due to the post-capillary block, is of diagnostic interest, as bronchoalveolar lavage can reveal occult hemorrhage. The degree of hemorrhage can be evaluated semiquantitatively and qualitatively using the Golde score, which assesses intra-alveolar siderin-laden macrophages by Perls Prussian blue staining (Golde et al. 1975; Capron 1999). In addition to an increased number of siderophages, large amounts of hemosiderin can be found in type II pneumocytes, and within the interstitial space. Moreover, post-capillary obstruction may frequently lead to capillary angiectasia and even capillary angioproliferation; in PVOD cases, doubling and tripling of the alveolar septal capillary layers may be focally present (Fig. 3a, d). Lately, this histological peculiarity has raised questions concerning a possible overlap between PVOD and cases of PCH, a disease classically characterized by an aggressive patch-like angioproliferation of capillaries; indeed, Lantuéjoul et al. (2006) recently reported 35 cases of PVOD and PCH with more or less similar pattern and evoke the possibility of a same disease entity. Historically, PCH was described by Wagenvoort and other authors as an aggressive capillary proliferation with patchy to nodular distribution; several rows of capillaries along alveolar walls progressed to nodules and sheets of back-to-back capillaries in advanced lesions (Wagenvoort et al. 1978). A malignant disorder which had been evoked with the term of "angiomatous growth" however seems unlikely, because cytological atypia and mitoses are usually absent. An infiltration of bronchiolar structures had also been described. A possible explanation to this angiomatoid expansion might be a hemodynamically relevant post-capillary block. Occult hemorrhage or hemosiderosis, therefore, is frequently found. As in PVOD, these characteristics lead to compensatory muscularization of arterioles and medial hypertrophy in pulmonary arteries (Fig. 3c). As already mentioned, the similarities in clinical and histological presentation indeed suggest that PVOD and PCH are a same disease entity with either a vein-predominating or a capillary-predominating phenotype (Lantuéjoul et al. 2006).

4 Forms of Pulmonary Hypertension with Significant Arterial and Venous Remodeling

4.1 PAH Associated with Connective Tissue Disease

Until recently, lesions of the pulmonary arterial system, more or less similar to those occurring in idiopathic PAH, have been held responsible for pulmonary hypertension in patients with connective tissue-associated PAH. In a published analysis of eight patients suffering from CTD-associated PAH, we observed that six out of eight patients (75 %) exhibited occlusive lesions of pulmonary veins and venules, as it can be typically seen in PVOD. In contrast, only 5 out of 29 non-CTD control patients with the primary diagnosis PAH displayed venous involvement (Dorfmüller et al. 2007). Though all investigated CTD-PAH patients exhibited pulmonary arterial changes, venous and venular fibrous remodeling, when present, was more pronounced than arterial changes, concerning the quantity of diseased vessels. All cases of CTD-associated PAH revealed pulmonary arterial lesions, involving small muscular vessels on the pre- and intra-acinar level. The vessel wall remodeling corresponded to arterial changes found in PAH, ranging from intimal constrictive and non-constrictive lesions to associated medial hypertrophy and adventitial thickening. In another report on this subject, Overbeek et al. (2009) analyzed lung tissue of eight PAH patients with limited cutaneous SSc and compared it to samples of 11 idiopathic PAH patients. They found that all SSc-PAH patients displayed arterial and venous remodeling, while venous lesions were present in only 3 of the 11 IPAH patients. It is notable that four SSc-PAH cases displayed a PVOD-like pattern with patchy capillary congestion and signs of occult alveolar hemorrhage.

Systemic sclerosis represents one of the leading pathological conditions associated with pulmonary hypertension and CTD-associated pulmonary hypertension belongs to Group 1 of the Dana Point Classification. Prevalence of PAH in certain forms of CTD has been estimated to up to 50 % of cases (Ungerer et al. 1983). In a recent cross-sectional national screening, at least 8 % of scleroderma patients displayed moderate-to-severe PAH (Hachulla et al. 2005). In patients with CTD, PAH is the leading cause of mortality and necessitates intensive medical treatment, which frequently proves difficult and with mixed results (Sanchez et al. 1999). Treatment with vasodilators like continuous intravenous epoprostenol has shown improvement of exercise capacity and cardiopulmonary hemodynamics, but the response is less effective than in IPAH and survival remains poor among patients with associated CTD (Badesch et al. 2000; Ramirez and Varga 2004). Equally, endothelin receptor antagonists seem to show less impressive effects in systemic sclerosis patients than in other forms of PAH (Humbert and Simonneau 2005). In addition, adverse effects of vasoactive treatment in PAH associated with CTD can occur and may lead to severe pulmonary edema (Humbert et al. 1999). Considering this background, the described findings of a PVOD-like setting in CTD-associated PAH highlight the important hemodynamic effect of post-capillary occlusion on the pulmonary vasculature. Resistance of CTD-associated PAH to common vasodilator therapies and complications, such as pulmonary edema, could be the mere consequence of a higher prevalence of veno-occlusive remodeling in these patients, as compared to other forms of PAH.

5 Conclusion

Characteristic vascular lesions in pulmonary hypertension may concern the lung vasculature and microvasculature from the pre- to the post-capillary level. Differentiation of histological phenotypes has helped to categorize pulmonary hypertension into different groups and subgroups, now summed up under the actual Dana Point Classification (which will be shortly revisited in the upcoming Fifth World Symposium on PH in 2013). It is important to stress that the differentiation of lesions and their topography within the vascular bed of the lung can be of significant importance considering outcome and specific therapy. However, a clear-cut separation of different forms of pulmonary hypertension through recognition of a histological phenotype will always be difficult or even impossible; different etiological factors (e.g., hypoxia versus anorexigen intake) may trigger a final common step in the cascade of pathologic events (e.g., oxidative stress and consecutive growth factor expression) and finally lead to the same morphologic pattern (e.g., medial hypertrophy). Nonetheless, the correct interpretation of the pathological vascular anatomy remains an indispensable roadmap to the successful coordination of clinical, biological, and pharmaceutical research dedicated to pulmonary hypertension.

References

Aloisi F, Pujol-Borrell R (2006) Lymphoid neogenesis in chronic inflammatory diseases. Nat Rev Immunol 6(3):205–217

Badesch DB, Tapson VF, McGoon MD, Brundage BH, Rubin LJ, Wigley FM, Rich S, Barst RJ, Barrett PS, Kral KM, Jöbsis MM, Loyd JE, Murali S, Frost A, Girgis R, Bourge RC, Ralph DD, Elliott CG, Hill NS, Langleben D, Schilz RJ, McLaughlin VV, Robbins IM, Groves BM, Shapiro S, Medsger TA (2000) Continuous intravenous epoprostenol for pulmonary hypertension due to the scleroderma spectrum of disease. A randomized, controlled trial. Ann Intern Med 132(6):425–434

Balabanian K, Foussat A, Dorfmüller P, Durand-Gasselin I, Capel F, Bouchet-Delbos L, Portier A, Marfaing-Koka A, Krzysiek R, Rimaniol A, Simonneau G, Emilie D, Humbert M (2002) CX (3)C chemokine fractalkine in pulmonary arterial hypertension. Am J Respir Crit Care Med 165(10):1419–1425

Capron F (1999) Bronchoalveolar lavage and alveolar hemorrhage. Ann Pathol 19(5):395–400

Carragher DM, Rangel-Moreno J, Randall TD (2008) Ectopic lymphoid tissues and local immunity. Semin Immunol 20(1):26–42

Cool C, Kennedy D, Voelkel N, Tuder R (1997) Pathogenesis and evolution of plexiform lesions in pulmonary hypertension associated with scleroderma and human immunodeficiency virus infection. Hum Pathol 28(4):434–442

Cool CD, Stewart JS, Werahera P, Miller GJ, Williams RL, Voelkel NF, Tuder RM (1999) Three-dimensional reconstruction of pulmonary arteries in plexiform pulmonary hypertension using cell-specific markers. Evidence for a dynamic and heterogeneous process of pulmonary endothelial cell growth. Am J Pathol 155(2):411–419

Di Biase L, Fahmy TS, Wazni OM, Bai R, Patel D, Lakkireddy D, Cummings JE, Schweikert RA, Burkhardt JD, Elayi CS, Kanj M, Popova L, Prasad S, Martin DO, Prieto L, Saliba W, Tchou P, Arruda M, Natale A (2006) Pulmonary vein total occlusion following catheter ablation for atrial fibrillation: clinical implications after long-term follow-up. J Am Coll Cardiol 48 (12):2493–2499

Dib H, Tamby MC, Bussone G, Regent A, Berezné A, Lafine C, Broussard C, Simonneau G, Guillevin L, Witko-Sarsat V, Humbert M, Mouthon L (2012) Targets of anti-endothelial cell antibodies in pulmonary hypertension and scleroderma. Eur Respir J 39(6):1405–1414

Dorfmüller P, Humbert M (2012) Progress in pulmonary arterial hypertension pathology: relighting a torch inside the tunnel. Am J Respir Crit Care Med 186(3):210–212

Dorfmuller P, Zarka V, Durand-Gasselin I, Monti G, Balabanian K, Garcia G, Capron F, Coulomb-Lhermine A, Marfaing-Koka A, Simonneau G, Emilie D, Humbert M (2002) Chemokine RANTES in severe pulmonary arterial hypertension. Am J Respir Crit Care Med 165 (4):534–539

Dorfmuller P, Perros F, Balabanian K, Humbert M (2003) Inflammation in pulmonary arterial hypertension. Eur Respir J 22(2):358–363

Dorfmüller P, Humbert M, Capron F, Müller KM (2003) Pathology and aspects of pathogenesis in pulmonary arterial hypertension. Sarcoidosis Vasc Diffuse Lung Dis 20(1):9–19

Dorfmüller P, Humbert M, Perros F, Sanchez O, Simonneau G, Müller KM, Capron F (2007) Fibrous remodeling of the pulmonary venous system in pulmonary arterial hypertension associated with connective tissue diseases. Hum Pathol 38(6):893–902

Golde DW, Drew WL, Klein HZ, Finley TN, Cline MJ (1975) Occult pulmonary haemorrhage in leukaemia. Br Med J 2(5964):166–168

Hachulla E, Gressin V, Guillevin L, Carpentier P, Diot E, Sibilia J, Kahan A, Cabane J, Francès C, Launay D, Mouthon L, Allanore Y, Tiev KP, Clerson P, de Groote P, Humbert M (2005) Early detection of pulmonary arterial hypertension in systemic sclerosis: a French nationwide prospective multicenter study. Arthritis Rheum 52(12):3792–3800

Harrison CV (1958) IV. The pathology of the pulmonary vessels in pulmonary hypertension. Br J Radiol 31(364):217–226

Heath D, Edwards JE (1958) The pathology of hypertensive pulmonary vascular disease; a description of six grades of structural changes in the pulmonary arteries with special reference to congenital cardiac septal defects. Circulation 18(4 Part 1):533–547

Heath D, Williams D (1989) High-altitude medicine and pathology. Butterworths, London

Hogg JC, Chu F, Utokaparch S, Woods R, Elliott WM, Buzatu L, Cherniack RM, Rogers RM, Sciurba FC, Coxson HO, Paré PD (2004) The nature of small-airway obstruction in chronic obstructive pulmonary disease. N Engl J Med 350(26):2645–2653

Humbert M, Simonneau G (2005) Drug Insight: endothelin-receptor antagonists for pulmonary arterial hypertension in systemic rheumatic diseases. Nat Clin Pract Rheumatol 1(2):93–101

Humbert M, Sanchez O, Fartoukh M, Jagot JL, Le Gall C, Sitbon O, Parent F, Simonneau G (1999) Short-term and long-term epoprostenol (prostacyclin) therapy in pulmonary hypertension secondary to connective tissue diseases: results of a pilot study. Eur Respir J 13(6):1351–1356

Krowka MJ, Edwards WD (2000) A spectrum of pulmonary vascular pathology in portopulmonary hypertension. Liver Transpl 6(2):241–242

Kucsko L (1953) Arteriovenous communications in the human lung and their functional significance. Frankf Z Pathol 64(1):54–83

Lantuéjoul S, Sheppard MN, Corrin B, Burke MM, Nicholson AG (2006) Pulmonary veno-occlusive disease and pulmonary capillary hemangiomatosis: a clinicopathologic study of 35 cases. Am J Surg Pathol 30(7):850–857

Lee SD, Shroyer KR, Markham NE, Cool CD, Voelkel NF, Tuder RM (1998) Monoclonal endothelial cell proliferation is present in primary but not secondary pulmonary hypertension. J Clin Invest 101(5):927–934

Mandel J, Mark EJ, Hales CA (2000) Pulmonary veno-occlusive disease. Am J Respir Crit Care Med 162(5):1964–1973

Moser KM, Bloor CM (1993) Pulmonary vascular lesions occurring in patients with chronic major vessel thromboembolic pulmonary hypertension. Chest 103(3):685–692

Overbeek MJ, Vonk MC, Boonstra A, Voskuyl AE, Vonk-Noordegraaf A, Smit EF, Dijkmans BA, Postmus PE, Mooi WJ, Heijdra Y, Grünberg K (2009) Pulmonary arterial hypertension in limited cutaneous systemic sclerosis: a distinctive vasculopathy. Eur Respir J 34(2):371–379

Penaloza D, Arias-Stella J (2007) The heart and pulmonary circulation at high altitudes: healthy highlanders and chronic mountain sickness. Circulation 115(9):1132–1146

Perros F, Dorfmüller P, Souza R, Durand-Gasselin I, Godot V, Capel F, Adnot S, Eddahibi S, Mazmanian M, Fadel E, Hervé P, Simonneau G, Emilie D, Humbert M (2007) Fractalkine-induced smooth muscle cell proliferation in pulmonary hypertension. Eur Respir J 29(5):937–943

Perros F, Dorfmüller P, Montani D, Hammad H, Waelput W, Girerd B, Raymond N, Mercier O, Mussot S, Cohen-Kaminsky S, Humbert M, Lambrecht BN (2012) Pulmonary lymphoid neogenesis in idiopathic pulmonary arterial hypertension. Am J Respir Crit Care Med 185(3):311–321

Piazza G, Goldhaber SZ (2011) Chronic thromboembolic pulmonary hypertension. N Engl J Med 364(4):351–360

Pietra G (1997) Pathology of primary pulmonary hypertension. Marcel Dekker, New York

Pietra GG, Capron F, Stewart S, Leone O, Humbert M, Robbins IM, Reid LM, Tuder RM (2004) Pathologic assessment of vasculopathies in pulmonary hypertension. J Am Coll Cardiol 43 (12 Suppl S):25S–32S

Ramirez A, Varga J (2004) Pulmonary arterial hypertension in systemic sclerosis: clinical manifestations, pathophysiology, evaluation, and management. Treat Respir Med 3 (6):339–352

Rangel-Moreno J, Carragher DM, de la Luz Garcia-Hernandez M, Hwang JY, Kusser K, Hartson L, Kolls JK, Khader SA, Randall TD (2011) The development of inducible bronchus-associated lymphoid tissue depends on IL-17. Nat Immunol 12(7):639–646

Rich S, Dantzker DR, Ayres SM, Bergofsky EH, Brundage BH, Detre KM, Fishman AP, Goldring RM, Groves BM, Koerner SK (1987) Primary pulmonary hypertension. A national prospective study. Ann Intern Med 107(2):216–223

Sakao S, Tatsumi K, Voelkel NF (2010) Reversible or irreversible remodeling in pulmonary arterial hypertension. Am J Respir Cell Mol Biol 43(6):629–634

Saldaña ME, Harley RA, Liebow AA, Carrington CB (1968) Experimental extreme pulmonary hypertension and vascular disease in relation to polycythemia. Am J Pathol 52(5):935–981

Sanchez O, Humbert M, Sitbon O, Simonneau G (1999) Treatment of pulmonary hypertension secondary to connective tissue diseases. Thorax 54(3):273–277

Sanchez O, Marcos E, Perros F, Fadel E, Tu L, Humbert M, Dartevelle P, Simonneau G, Adnot S, Eddahibi S (2007) Role of endothelium-derived CC chemokine ligand 2 in idiopathic pulmonary arterial hypertension. Am J Respir Crit Care Med 176(10):1041–1047

Simonneau G, Galie N, Rubin LJ, Langleben D, Seeger W, Domenighetti G, Gibbs S, Lebrec D, Speich R, Beghetti M, Rich S, Fishman A (2004) Clinical classification of pulmonary hypertension. J Am Coll Cardiol 43(12):5S–12S

Simonneau G, Robbins IM, Beghetti M, Channick RN, Delcroix M, Denton CP, Elliott CG, Gaine SP, Gladwin MT, Jing ZC, Krowka MJ, Langleben D, Nakanishi N, Souza R (2009) Updated clinical classification of pulmonary hypertension. J Am Coll Cardiol 54(1):S43–S54

Stacher E, Graham BB, Hunt JM, Gandjeva A, Groshong SD, McLaughlin VV, Jessup M, Grizzle WE, Aldred MA, Cool CD, Tuder RM (2012) Modern age pathology of pulmonary arterial hypertension. Am J Respir Crit Care Med 186(3):261–272

Tamby MC, Chanseaud Y, Humbert M, Fermanian J, Guilpain P, Garcia-de-la-Peña-Lefebvre P, Brunet S, Servettaz A, Weill B, Simonneau G, Guillevin L, Boissier MC, Mouthon L (2005) Anti-endothelial cell antibodies in idiopathic and systemic sclerosis associated pulmonary arterial hypertension. Thorax 60(9):765–772

Terrier B, Tamby MC, Camoin L, Guilpain P, Broussard C, Bussone G, Yaïci A, Hotellier F, Simonneau G, Guillevin L, Humbert M, Mouthon L (2008) Identification of target antigens of antifibroblast antibodies in pulmonary arterial hypertension. Am J Respir Crit Care Med 177 (10):1128–1134

Tuder R, Groves B, Badesch D, Voelkel N (1994) Exuberant endothelial cell growth and elements of inflammation are present in plexiform lesions of pulmonary hypertension. Am J Pathol 144 (2):275–285

Tuder RM, Abman SH, Braun T, Capron F, Stevens T, Thistlethwaite PA, Haworth SG (2009) Development and pathology of pulmonary hypertension. J Am Coll Cardiol 54(1 Suppl):S3–S9

Ungerer RG, Tashkin DP, Furst D, Clements PJ, Gong H, Bein M, Smith JW, Roberts N, Cabeen W (1983) Prevalence and clinical correlates of pulmonary arterial hypertension in progressive systemic sclerosis. Am J Med 75(1):65–74

Voelkel NF, Tuder RM (1995) Cellular and molecular mechanisms in the pathogenesis of severe pulmonary hypertension. Eur Respir J 8(12):2129–2138

Wagenvoort C, Wagenvoort N (1970) Primary pulmonary hypertension: a pathologic study of the lung vessels in 156 clinically diagnosed cases. Circulation 42(6):1163–1184

Wagenvoort CA, Beetstra A, Spijker J (1978) Capillary haemangiomatosis of the lungs. Histopathology 2(6):401–406
Widgren S (1986) Prolonged survey of cases of pulmonary hypertension in relation to consumption of aminorex. Histological, quantitative and morphometric study of 9 cases. Schweiz Med Wochenschr 116(27–28):918–924
Yaginuma G, Mohri H, Takahashi T (1990) Distribution of arterial lesions and collateral pathways in the pulmonary hypertension of congenital heart disease: a computer aided reconstruction study. Thorax 45(8):586–590

Pulmonary Hypertension: Biomarkers

Christopher J. Rhodes, John Wharton, and Martin R. Wilkins

Abstract Physicians look to biomarkers to inform the management of pulmonary hypertension (PH) at all stages, from assessing susceptibility through screening, diagnosis, and risk stratification to drug selection and monitoring. PH is a heterogeneous disorder and currently there are no accepted blood biomarkers specific to any manifestation of the condition. Brain natriuretic peptide and its N-terminal peptide have been most widely studied. Other candidate prognostic biomarkers in patients with pulmonary arterial hypertension (PAH) include growth and differentiation factor-15, red cell distribution width, uric acid, creatinine, inflammatory markers such as interleukin-6, angiopoietins, and microRNAs. Combining the measurement of biomarkers reflecting different components of the pathology with other modalities may enable better molecular characterisation of PH subtypes and permit improved targeting of therapeutic strategies and disease monitoring.

Keywords Pulmonary hypertension • Biomarkers • Pulmonary arterial hypertension • Brain natriuretic peptide • Cytokines • Inflammation • Iron • microRNA

Contents

1	Biomarkers	78
	1.1 Biomarkers of Susceptibility to PH	78
	1.2 Biomarkers for Screening for PH	79
	1.3 Biomarkers for Making the Diagnosis of PH	79
	1.4 Biomarkers for Risk Stratification in PH	79
	1.5 Biomarkers for Therapeutic Selection and Monitoring	80
	1.6 Statistical Evaluation of Novel Biomarkers	81
2	Individual Biomarkers for PH	85
	2.1 Genetic Biomarkers	85
	2.2 Clinical and Haemodynamic Indices	86

C.J. Rhodes • J. Wharton • M.R. Wilkins (✉)
Imperial College London, Hammersmith Hospital, London W12 0NN, UK
e-mail: m.wilkins@imperial.ac.uk

2.3   Natriuretic Peptides 87
2.4   Growth and Differentiation Factor-15 87
2.5   Red Cell Distribution Width and Other Iron Parameters 88
2.6   Inflammatory Markers 89
2.7   Creatinine 89
2.8   Uric Acid 90
2.9   Angiopoietins 90
2.10 MicroRNAs 91
2.11 Other Potential Biomarkers in PH 91
2.12 Comments on Biomarkers in PH 92
3 Future Directions and Conclusions 94
References 95

1 Biomarkers

The US National Institutes of Health Biomarkers Definitions Working Group have defined a biomarker as "any characteristic that is objectively measured and evaluated as an indicator of normal biological processes, pathogenic processes, or pharmacologic responses to a therapeutic intervention" (Atkinson et al. 2001). That is to say, a biomarker is any objective measure that informs on the state of health of an individual.

To be useful it has to satisfy a number of criteria. The strength and consistency of the relationship between the biomarker and the disease parameter it is reporting are an important consideration; associations must be convincing (statistically significant) and reproducible across different cohorts before a relationship can be considered true. A biomarker's *specificity* is the proportion of true negatives over the true negatives and false positives—and so a measure of the ability of a biomarker to accurately rule out disease. *Sensitivity*, conversely, is the proportion of true positives over the true positives and false negatives—and so the ability of a biomarker to predict disease. The *temporality* of a biomarker is also important; some biomarkers respond rapidly to events, while others may accumulate and integrate information. The *biological gradient* refers to the dynamic range of the measurement and together with the quality of the assay used must be sufficient to differentiate accurately between groups.

Biomarkers may help in the management of PH at several stages, from susceptibility, screening, diagnosis, and risk stratification to therapeutic selection and monitoring.

1.1 Biomarkers of Susceptibility to PH

Genomic information holds the greatest potential for identifying susceptible individuals, i.e. individuals at risk but with no manifestations of the disease. Family members of patients with PAH can be tested for mutations in known risk genes, namely *BMPR2*, *ALK1*, or *ENG* (see Sect. 2.1). Following appropriate counselling, those harbouring mutations can be seen at regular intervals to detect and manage the disease at its earliest occurrence.

Other scenarios in which biomarkers for susceptibility to PH may be identified include identifying those individuals most likely to develop PH following pulmonary embolism/thromboendarterectomy, or in other at-risk populations including but not limited to those where PAH is associated with connective tissue disorders, HIV, or schistosomiasis.

1.2 Biomarkers for Screening for PH

A considerable proportion of the pulmonary vascular bed needs to be 'lost' before pulmonary vascular disease becomes symptomatic (Lau et al. 2011), and even in patients with mildly symptomatic PAH, there is already a marked increase in pulmonary vascular resistance (Galië et al. 2008). Screening individuals for subclinical disease requires a non-invasive test and the most practical tool for this purpose is the echocardiogram. No blood biomarker currently identified is sensitive enough to detect subclinical disease.

1.3 Biomarkers for Making the Diagnosis of PH

The insidious onset and non-specific nature of presenting symptoms in PH means that the diagnosis is delayed while other pathologies are excluded. The definitive diagnosis of PH is dependent on demonstrating a raised resting mean pulmonary artery pressure at cardiac catheterization. While a normal level of the cardiac stress-marker brain natriuretic peptide (BNP) or its N-terminal peptide (NT-proBNP) may help rule out PH, there is no blood biomarker that is diagnostic of the condition. The cost and invasive nature of cardiac catheterization drive the search for alternative non-invasive tests. Further developments in other imaging technologies such as cardiomagnetic resonance (CMR) imaging or 3D echocardiography may provide a solution but remain research tools at present. Cardiac output is a prognostic marker in PH patients and inert-gas rebreathing may represent a non-invasive means of measuring this important parameter (Clinical Trials.gov Identifier NCT01606839), detecting for example treatment responses in patients with PAH (Lee et al. 2011).

1.4 Biomarkers for Risk Stratification in PH

Risk stratification can inform not just prognosis but may help optimization of treatment and the achievement of specific goals aimed at improving longer-term survival. Multicentre patient registries—National Institutes of Health (NIH), French PAH registry, and US-based Registry to Evaluate Early and Long-term Pulmonary Arterial Hypertension Disease Management (REVEAL)—have been very useful in identifying independent factors that are associated with prognosis of PAH, including age, sex, disease aetiology, World Health Organization (WHO) functional class, 6 min walk distance (6MWD), and haemodynamic parameters (D'Alonzo et al. 1991; Rich et al. 2000; Humbert et al. 2010; Benza et al. 2010). But these are blunt instruments and we remain poorly equipped to make judgements at the individual patient level (Thenappan et al. 2012). That is, to predict exactly which individuals will survive for years with their condition versus those who will succumb within months or weeks of their original diagnosis and may have benefitted from earlier intervention with combination therapy or more aggressive treatment (Hoeper et al. 2005; Nickel et al. 2011a).

Measurements of circulating BNP and NT-proBNP levels remain the most commonly used blood biomarkers for stratifying PH patients and are recommended in the European guidelines for the diagnosis and treatment of PH (Galie et al. 2009). It is recognised, however, that some patients with advanced disease have circulating levels within the normal range and the levels for an individual patient have to be interpreted in the context of all the clinical information available. Repetitive measurements may add information and improve their utility in disease management (Nickel et al. 2011a).

Novel candidates for prognostic biomarkers in PH, which will be described in more detail below, include growth and differentiation factor-15 (Nickel et al. 2008; Rhodes et al. 2011a), red cell distribution width (Hampole et al. 2009; Rhodes et al. 2011a), inflammatory markers such as interleukin-6 (Humbert et al. 1995; Soon et al. 2010), creatinine (Leuchte et al. 2007; Shah et al. 2008), uric acid (Nagaya et al. 1999), angiopoietins (Kumpers et al. 2010), and microRNAs (Caruso et al. 2010; Courboulin et al. 2011).

1.5 Biomarkers for Therapeutic Selection and Monitoring

There is interest in the PH community in goal-orientated treatment strategies, whereby treatments are titrated according to response. The biomarkers used to assess response are 6 min walk distance, cardiopulmonary exercise testing, and BNP levels (Hoeper et al. 2005).

With the increase in number of drugs available to treat PH, the possibilities for individualising treatment have increased. Biomarkers that can identify which patients will respond best to particular therapies enable expensive treatments to be used more cost effectively and reduce unnecessary drug exposures. This point is well illustrated by the use of acute vasodilator testing at cardiac catheterization to identify patients suitable for treatment with calcium antagonists. A small number (<10 %) of patients with PAH who show an acute reduction in pulmonary artery pressures, when exposed to nitric oxide or adenosine, respond well to calcium channel blockers with excellent haemodynamic recovery and long-term survival (Sitbon et al. 2005; Tonelli et al. 2010). The recent finding that over 60 % of patients with PAH are iron deficient (in the absence of anaemia) may identify a subgroup of patients who will benefit from iron, but this awaits formal testing (Rhodes et al. 2011b).

While there is most interest in biomarkers of efficacy it is important not to forget toxicity and the role of safety biomarkers. For PH patients on warfarin, the International Normalised Ratio (INR) is used to monitor the clotting tendency of blood and thereby ensure the correct dosage. Patients on endothelin antagonists also have their liver enzymes monitored for early signs of liver toxicity, which is generally reversible upon discontinuation of the drug. A new protein kinase inhibitor-induced toxicity has been identified, pre-capillary PH having been associated with dasatinib therapy (Montani et al. 2012). Cardiac biomarkers such as Troponin I and BNP may represent an early means of detecting cardiac toxicity in patients exposed to protein kinase inhibitors (ClinicalTrials.gov Identifier NCT00532064).

1.6 Statistical Evaluation of Novel Biomarkers

Key to the discovery and validation of biomarkers is access to well-phenotyped patient populations from cohort studies and clinical trials. The data and/or samples are mined based on plausible candidates or using unbiased screens. These datasets can be very large, particularly if fed from platform technologies, such as proteomics or metabolomics.

Before an analysis can be started, general requirements for statistical testing must be met. Whether the biomarker of interest is normally distributed (i.e. fits to a "normal" bell-shaped curve) will determine whether parametric (for normal or Guassian data) or non-parametric tests can be deployed. Normality can be assessed informally by box and whisker plots/histograms and formally by the Kolmogorov–Smirnov test. If required, distribution of a marker can be transformed by logging, square-rooting, or inverting all numbers to achieve normality.

Tests for two groups (e.g. control versus patient) of continuous data (i.e. not categories) include the Student's t-test and Mann–Whitney U-test. Multiple groups require analysis of variance (ANOVA or Kruskal–Wallis) followed by post hoc testing. Correlations can be assessed by Pearson's test or Spearman's rank test. Chi-squared analysis is required when assessing relationships between categorical variables, for example, levels of a marker above or below a cut-off level against WHO functional class, or diagnosis (see Fig. 1).

The first step in evaluating a novel biomarker is to understand its association with the disease of interest. This usually involves comparing levels measured in a patient population with appropriate control subjects. In PH, appropriate controls may simply be healthy subjects, but could include non-PH patients with a chronic disease (e.g. asthma, COPD without PH) or different diagnostic classes of PH to that being investigated (e.g. chronic thromboembolic PH versus idiopathic PAH).

Once an association with a disease state has been demonstrated, the meaning and importance of this association can be further investigated by assessing correlations

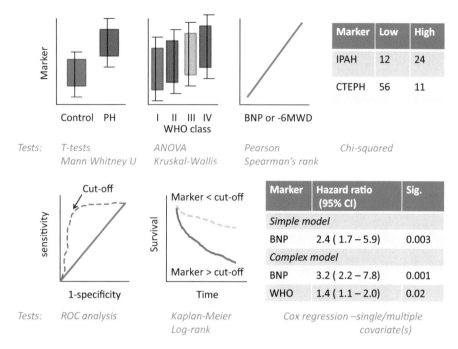

Fig. 1 Common data types and statistical analyses employed in biomarker assessment

between markers of disease severity (e.g. 6MWD, WHO functional class, pulmonary haemodynamic measurements) and circulating levels of markers such as BNP. Demonstrating the relevance of a marker to disease severity can also be hypothesis-generating; for example a marker that strongly correlates to BNP may be more likely to reflect myocardial stress than one that did not. Unless these investigations are hypothesis-driven, considerations for multiple testing may be necessary (see below).

Ultimately, if a biomarker truly reflects the severity and progression of disease, then its levels should reflect the prognosis of an individual patient. For this it is necessary to censor the survival/mortality of patients in a cohort at one given time point, and to calculate the time between biomarker measurement and the censor date. Receiver operating characteristic (ROC) analysis can then be used to assess the relationship between potential cut-off levels of a biomarker and survival over a set period, for example 1-, 2-, or 3-year survival. It plots the sensitivity versus 1-specificity for each possible cut-off level of a marker, and the area under the curve represents the power of the marker to predict survival over this time period. ROC analysis also allows for direct comparisons of performance between candidates, as those with larger area-under-the-curve (AUC) values (also known as c-statistic) will best discriminate survivors and non-survivors. The cut-off furthest deviating from the line x=y will represent an optimum cut-off for this purpose (Fig. 1) and may be determined by plotting biomarker concentrations against their respective sensitivity and specificity data (Schumann et al. 2010). Cut-offs can also be derived from distributions in the general/healthy populous or simply by taking median/quartile values.

The relationships of biomarker cut-offs to survival are frequently illustrated by Kaplan–Meier survival estimates and tested statistically by the log-rank test. The predictive value of stepwise changes in biomarker levels (i.e. 4–5, or 5–6, or 7.4–8.4) to the hazard rate (relative risk of mortality in a set period of time, i.e. year 1–2, year 2–3, year 3.4–4.4) can be assessed by Cox regression analysis. If a biomarker performs well by ROC, Kaplan–Meier, and Cox analysis then it can be said to have prognostic power in the population studied.

To determine whether a prognostic biomarker might add to current clinical practice, it must be compared to established gold standards, and multiple variable Cox regression analysis is the test of choice for this purpose. If a biomarker is still significantly related to survival in a Cox model including the best current prognostic measures (e.g. haemodynamic values, exercise capacity, BNP levels), then it can be said to be an "independent" prognostic marker which adds to the information currently available.

Important assumptions must be met for simple Cox analysis including that the hazard rate stays constant over time (so a high BNP would imply the same increased risk from year 1 to 2 as from year 3 to 4) and the same amplitude of change has the same hazard whatever the starting level of biomarker is (so an increase in BNP from 300 to 400 is as bad as an increase from 700 to 800). Normalisation of biomarkers prior to Cox analysis is standard and most likely aids the meeting of the latter criteria.

If a biomarker appears useful following the above tests, then the reproducibility of any findings must then be validated in further independent cohorts (see below), and eventually, its use in a clinical setting tested in specifically designed prospective clinical trials.

1.6.1 Cautionary Points

Over-fitting. An important consideration in any regression model, including Cox analysis, is not to over-fit the model. A general rule of thumb in regression is to only consider 1 covariate for each 10 data points and in Cox analysis this applies to the events (e.g. death, transplantation, hospitalisation) being considered. So a study with 100 patients with only 10 deaths would not be sufficient to assess the predictive value of a novel biomarker in comparison to BNP on death by Cox regression analysis.

Multiple Testing. This is a particular issue with the advent of high-throughput methodologies, such as genomics, proteomics, and metabolomics. If corrections are severe (such as the simplistic Bonferroni method) they may mask true differences but if ignored then false positives will arise. Estimations of the false discovery rate of a methodology may help gauge confidence in hits, but there remains no substitute for validation of hypotheses in independent cohorts. Intrinsic to a discovery study is that identified candidates will be those that best perform, leading to over-estimation of their value compared to established markers previously identified. Hence discovery and validation cohorts are required to prove the value of a novel biomarker, and as the validation cohort will be used to assess an a priori defined hypothesis, then corrections for multiple testing will no longer be required.

Missing Data. Another problem commonly faced in biomarker discovery and development is that of missing data. This may reflect the retrospective nature of many studies which are considered secondary in importance to more orthodox clinical studies of novel therapeutics (Jenkins et al. 2011).

Incident and Prevalent Disease. It is notable that the recent REVEAL and French registry studies included prevalent as well as treatment naïve patients with incident disease (Benza et al. 2010; Humbert et al. 2010). Survival modelling with both groups of patients has its limitations and statistical techniques were used to take account of potential survivor bias arising from the inclusion of patients with prevalent disease. These methodological issues may still have resulted in overoptimistic survival rates due to an underestimation of the proportion of patients with severe disease who died soon after diagnosis (McLaughlin and Suissa 2010). Nonetheless, it is the response to treatment, captured through clinical observations and biochemical measurements, rather than the presence of treatment itself that is important in determining patient prognosis. For biomarker studies, careful analysis of survival from the date of blood sample collection is required in order to stratify risk at clinical follow-up appointments (Rhodes et al. 2011a).

1.6.2 Additional Criteria

Several groups have proposed criteria for evaluating new biomarkers and emphasised the need for prospective validation (Hlatky et al. 2009; Pletcher and Pignone 2011; Wang 2011). The lack of prospective clinical studies directly designed to assess the efficacy of adding in biomarker measurements to treatment decisions demonstrates that we are still some distance from incorporating any of the novel biomarkers proposed into standard guidelines. An appropriate trial design could be the randomization of front-line therapeutic strategies following grouping of subjects by biomarker measurement, or the randomization of patients to either normal treatment algorithms or algorithms that incorporate biomarker measurements into treatment decisions (Fig. 2).

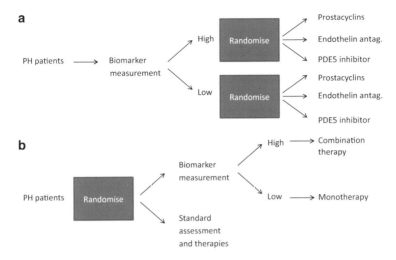

Fig. 2 Potential trial designs to test the value of biomarkers in treatment decisions in PH. (**a**) Comparison of treatment efficacies in "high" and "low" biomarker groups. (**b**) Comparison of example treatment algorithms with and without biomarker measurements. *Antag.* antagonists, *PDE5* phosphodiesterase type 5. Modified from Sargent et al. (2005)

2 Individual Biomarkers for PH

2.1 Genetic Biomarkers

Mutations in the bone morphogenetic receptor type 2 gene (*BMPR2*) are found in 50–80% of hereditary PAH (Deng et al. 2000; Lane et al. 2000; Machado et al. 2001; Morisaki et al. 2004), 10–20 % of idiopathic PAH patients (Newman et al. 2008; Girerd et al. 2010), and rarely in secondary PAH associated with anorexic drugs or congenital heart disease (Humbert et al. 2002; Roberts et al. 2004). Most predict loss of function (Machado et al. 2006). Patients with *BMPR2* mutations present approximately 10 years earlier than PAH patients without a mutation and have more severe haemodynamic disturbance leading to death at a younger age (Sztrymf et al. 2008). These patients are also less likely to respond to vasodilator therapy (Rosenzweig et al. 2008).

Two other genes in the tumour growth factor (TGF)-β receptor superfamily, activin receptor-like kinase 1 (*ALK1*) and endoglin (*ENG*), have also been causally linked to the development of PAH in hereditary haemorrhagic telangiectasia (Harrison et al. 2003). *ALK1* mutations are also found in hereditary PAH patients, leading to earlier presentation and rapid clinical decline even in comparison to *BMPR2* mutation carriers (Girerd et al. 2010). In addition, gene mutations in downstream intermediaries of the TGF-β/BMP-signalling pathway, namely Smads 1, 4, 8, and 9, have been identified as an infrequent cause of PAH (Nasim et al. 2011; Shintani et al. 2009).

Penetrance of PAH in *BMPR2* mutation carriers is generally around 20% and can vary depending on TGF-β polymorphisms (Phillips et al. 2008). Reduced penetrance suggests the presence of modifying factors that influence disease risk and the need for a so-called second hit to make the disease clinically expressed. Several genetic variants have been investigated, including polymorphisms of serotonin transporter, prostacyclin synthase, endothelial nitric oxide synthase, angiotensin-converting enzyme type 1, angiotensin II receptor type 1, carbamoyl phosphate synthetase 1, smooth muscle potassium channels (KCNQ, KCNA5), and cytochrome P450 1B1 (Fessel et al. 2011). With the cost of genome sequencing falling, the opportunities to investigate the genomic influence on PAH will increase. The sizable datasets that accompany these studies will however require careful analysis.

2.2 Clinical and Haemodynamic Indices

Before a novel marker can be considered useful it must be compared against the current gold standard. In PH, definitive diagnosis is dependent on invasive haemodynamic measurements by right heart catheterization—mean pulmonary artery pressure, mean right atrial pressure, pulmonary vascular resistance, and cardiac index all reflect disease severity and have been shown to hold prognostic information (D'Alonzo et al. 1991; McLaughlin et al. 2002; Sitbon et al. 2002). The current therapies for PAH have been approved on the basis of improvements in 6MWD as a surrogate for survival (Galie et al. 2005, 2008; Rubin et al. 2002). Indeed this measure has been shown to predict survival in PH (Miyamoto et al. 2000; Sitbon et al. 2002), although its variability between centres limits its use. WHO functional class is a simple clinical assessment of the symptoms of right heart failure that has been consistently demonstrated to predict survival in PH (McLaughlin et al. 2002; Sitbon et al. 2002; Humbert et al. 2010). All these measures should be considered, together with parameters such as age, diagnosis, sex, disease duration, and therapies, when assessing the potential value of a putative new biomarker in PH.

An empirically derived equation, based on baseline haemodynamic measurements, was developed to estimate survival of patients with primary PAH in the NIH registry (D'Alonzo et al. 1991). Clinical trials have used this equation to compare observed and predicted survival, but the management and treatment of PAH have changed dramatically over the last decade and contemporary registries show that the NIH equation significantly underestimates current survival (Benza et al. 2010; Humbert et al. 2010; Thenappan et al. 2010). New risk prediction equations have been produced in order to better model survival in patients with category 1.1–1.3 PAH (Humbert et al. 2010; Thenappan et al. 2010) or all groups of PAH (Benza et al. 2010). These equations have since been applied to prospectively collected data from PAH cohorts and a simplified risk score calculator developed that may be more applicable to general clinical practice (Benza et al. 2012; Thenappan et al. 2012).

2.3 Natriuretic Peptides

The natriuretic peptides—atrial and brain, ANP and BNP—are synthesised by myocardium and secreted into the circulation, regulating natriuresis, diuresis, and vasomotor tone (Yoshimura et al. 1991; Nakao et al. 1992). Pressure and volume overload increases secretion and circulating levels of ANP and BNP and acts to reduce the workload of the heart, albeit ineffectively in PH.

BNP is produced as a 108 amino acid prohormone, which is cleaved and secreted as a biologically active sequence (BNP-32) and inactive N-terminal proBNP (NT-proBNP) peptide. The former has a short half-life of around 20 min and NT-proBNP a longer half-life of several hours, but both fragments are cleared by the kidney and influenced by renal function (Holmes et al. 1993; Yandle et al. 1993). Other non-cardiac factors (e.g. age, obesity, diabetes mellitus) are also known to influence NT-proBNP/BNP plasma levels.

Circulating levels of ANP, BNP-32, and NT-proBNP correlate with haemodynamic measures in PAH (Nagaya et al. 1998; Leuchte et al. 2005; Souza et al. 2005) and reduced BNP levels following long-term vasodilator therapy also correlate strongly ($p < 0.001$) with falls in PVR, suggesting that BNP measurements may reflect treatment efficacy (Nagaya et al. 1998).

Natriuretic peptide levels are elevated by left as well as right ventricular dysfunction and therefore not specific to PH. Nonetheless, the right ventricle is a major determinant of prognosis in PH (van de Veerdonk et al. 2011) and circulating BNP/NT-proBNP levels predict survival and can be used to risk stratify patients (Nagaya et al. 2000; Fijalkowska et al. 2006; Nickel et al. 2008; Hampole et al. 2009; Rhodes et al. 2011a). In a study of 55 PAH patients with a mean follow-up time of 36 months, a cut-off NT-proBNP level of 1,400 pg/ml predicted mortality with 88 % sensitivity and 53 % specificity (log rank, $p < 0.01$) (Fijalkowska et al. 2006). The influences of renal insufficiency were also examined in a later study of 118 patients, which showed in a multiple covariate analysis that NT-proBNP levels and creatinine clearance outperformed BNP, and were independent predictors of mortality (Leuchte et al. 2007).

2.4 Growth and Differentiation Factor-15

Growth and differentiation factor (GDF)-15 is a member of the TGF-β superfamily (Bootcov et al. 1997) that protects mice from developing ventricular hypertrophy following pressure overload (Xu et al. 2006). It is ubiquitously expressed at low levels in most tissues and is induced in response to tissue injury and inflammation.

Mechanical stretch, as well as other stimuli, triggers its release from cardiomyocytes and it is thought that GDF-15 has a role in protecting myocytes against apoptosis and induction of hypertrophy (Heger et al. 2010).

Circulating GDF-15 levels have prognostic power in several cardiovascular diseases (Kempf and Wollert 2009; Khan et al. 2009) and hence this biomarker is not specific to PH. GDF-15 expression has also been localised to plexigenic lesions in PAH patients (Nickel et al. 2011b) but, as it expressed elsewhere, the source and major stimulus for GDF-15 production in PAH are unclear. This may in part explain why GDF-15 levels added prognostic information to BNP in one study (Nickel et al. 2008), but appeared redundant in another where other markers of inflammation and general cardiovascular health were also measured (Rhodes et al. 2011a).

2.5 Red Cell Distribution Width and Other Iron Parameters

Anaemia is a recognised predictor of survival in both chronic heart failure and PAH (Rhodes et al. 2011b). Iron deficiency, measured using a number of indicators—red cell distribution width (RDW), serum ferritin, and soluble transferrin receptor—predicts a poor survival in PAH even in the absence of anaemia (Rhodes et al. 2011c). Inappropriately raised hepcidin levels suggest impaired absorption of iron from the gut. Interestingly, hepcidin is regulated by BMP6 and the possibility that dysregulated hepcidin production and PAH may be linked through a perturbation of BMP signalling merits further investigation. Nevertheless, hepcidin is an acute phase protein and hepatic expression is induced by circulating pro-inflammatory cytokines such as interleukin-6 (IL-6, an inflammatory cytokine linked to the development of PH, see below).

Inflammation can induce iron deficiency, but the relationship of iron deficiency to survival in PAH seems independent of measures of inflammation. In 139 IPAH patients (Rhodes et al. 2011a), RDW was assessed as a prognostic biomarker alongside several other previously identified plasma biomarkers: NT-proBNP, GDF-15, creatinine (a marker of renal dysfunction, see below), and IL-6, as well as baseline haemodynamics and coincident clinical measurements. All 5 plasma markers predicted survival, though RDW was the strongest performer in ROC analysis, with an area under the curve of 0.82 (GDF-15 and 6 min walk distance 0.76/0.75, NT-proBNP, and IL-6 0.67/0.66). Again, a "normal" RDW level of 15.7 % was the best cut-off, levels above identifying a group of patients with around 40 % 3-year survival compared to over 70 % in those with lower levels. Haemodynamic measurements (mPAP, mRAP, CI, and PVR) also predicted survival, as did WHO functional class, but a Cox regression model comprising just 6 min walk distance, RDW, and NT-proBNP levels could not be improved by the addition of any further measure identified as prognostic in single variable regressions.

2.6 Inflammatory Markers

Inflammation has long been appreciated as an element of PH pathology (Hassoun et al. 2009) and patients with PAH associated with connective tissue disorders generally exhibit a poorer prognosis and increased mortality risk when compared with other forms of the disease (Condliffe et al. 2009; Benza et al. 2010). The inflammatory marker C-reactive protein (CRP) has been shown to predict outcome, the normalisation of CRP levels being associated with improved haemodynamic and functional capacity as well as long-term survival (Quarck et al. 2009; Sztrymf et al. 2010). The levels of interleukins (ILs) such as IL-1β and IL-6 are also raised in PAH patients (Humbert et al. 1995) and a study characterising circulating endothelial progenitor cells in PAH demonstrated elevated tumour necrosis factor (TNF)-α, as well as the levels of IL-6 (CRP) in both IPAH and Eisenmenger congenital heart disease PAH patients (Diller et al. 2008). Increased IL-6 has also been demonstrated in inoperable congenital heart disease patients (Lopes et al. 2011). Interestingly, IL-6 can be induced by loss of BMPR2 expression, interacts with BMP signalling (Hagen et al. 2007), and has been implicated in the development of PH in experimental models (Savale et al. 2009; Steiner et al. 2009). Furthermore, high IL-6 levels may contribute to the reduced BMPR2 expression seen in IPAH through activation of the microRNA (miR)-17/92 cluster (Brock et al. 2009).

A recent screen of serum TNF-α, interferon-gamma, and several interleukins (IL-1β, -2, -4, -5, -6, -8, -10, -12p70, and -13) in 60 idiopathic and heritable PAH patients showed that all the cytokines measured (except IL-5 and IL-13) were elevated when compared with healthy volunteers (Soon et al. 2010). In addition, Kaplan–Meier analysis indicated that levels of IL-6, -8, -10, and -12p70 predicted survival. Another study of 139 IPAH patients also found that raised IL-6 was a good predictor of survival, but may not offer prognostic information above that provided by NT-proBNP, RDW, and 6MWD (Rhodes et al. 2011a). However, a direct measure of the inflammatory component in PH may yet prove useful in guiding therapeutic decisions.

2.7 Creatinine

Failure of the right ventricle is associated not only with reduced cardiac output but also back-pressure on the liver and kidney. In a study highlighting the dependence of NT-proBNP levels on renal function, estimates of creatinine clearance were shown to predict survival in over 100 PH patients of mixed aetiology (Leuchte et al. 2007). In one of the largest studies of biomarkers in PH to date, serum creatinine levels not only correlated with pulmonary haemodynamics, functional class, and exercise capacity but also predicted survival, particularly in a subgroup of patients with lower mRAP (<10 mmHg) measurements (Shah et al. 2008). In a more recent study of idiopathic PAH patients, creatinine levels correlated with baseline mRAP and CI, predicting survival in single variable analyses but not independently of NT-proBNP, RDW, and 6MWD (Rhodes et al. 2011a).

2.8 Uric Acid

Uric acid is the final degradation product of purines and an endogenous free radical scavenger that is elevated following tissue ischaemia and hypoxia. Raised levels have been described in chronic obstructive pulmonary disease (Braghiroli et al. 1993), chronic heart failure (Leyva et al. 1998), and cyanotic congenital heart disease (Hayabuchi et al. 1993; Leyva et al. 1998). Circulating uric acid levels are also widely reported to be elevated in PAH, with values around 50 % higher than seen in controls (Hoeper et al. 1999; Nagaya et al. 1999; Bendayan et al. 2003; Voelkel et al. 2000). Correlations of uric acid serum levels with cardiac output, PVR, WHO functional class, and mortality suggested its potential usefulness as a biomarker in PH (Nagaya et al. 1999). This was also underlined by reductions of uric acid levels in response to chronic vasodilator therapy. Potential confounding factors however include renal dysfunction and diuretic use and hence uric acid levels must be interpreted in the context of these factors.

2.9 Angiopoietins

Angiopoietin (Ang)-1 and its antagonist Ang-2 are ligands for the Tie2 tyrosine kinase receptor on endothelial cells and, in balance with vascular endothelial growth factor (VEGF), they act to control vascular development and maturation (Augustin et al. 2009). Overexpression of Ang-1 has been proposed as a murine model of PH (Chu et al. 2004), whereas other studies have suggested that the pathway is protective (Kugathasan et al. 2009). Circulating levels of Ang-1, Ang-2, VEGF, and soluble Tie2 (which inhibits circulating Ang-1) were all shown to be elevated in a study of IPAH patients (Kumpers et al. 2010). Of these four markers only Ang-2 levels correlated with disease severity (as assessed by mRAP, PVR, CI, and WHO functional class) and also correlated with elevated NT-proBNP levels. Furthermore, Ang-2 levels predicted disease outcomes; the best cut-off for predicting death or transplantation at 3 years was 2.9 ng/ml (sensitivity 85 % and specificity 58 %), a level rarely reached in controls measured. A small second cohort also demonstrated changes in Ang-2 following the initiation of treatment that correlated with alterations in parameters such as mRAP, PVR, and 6MWD (Kumpers et al. 2010). The usefulness of this interesting new biomarker, which might potentially reflect the dysfunctional endothelium in PAH, must be validated in independent studies.

2.10 MicroRNAs

Recently, several non-protein coding microRNAs (miRs) have been identified that modulate gene expression at the post-transcriptional level, regulating a wide range of physiological and pathological processes in the cardiovascular system, and circulate in the bloodstream (Creemers et al. 2012). Several miRs have been implicated in the pathogenesis of PH and could have utility as biomarkers as well as therapeutic targets in PAH (McDonald et al. 2012). These include the miR-17/92 cluster—mainly miR-20a and miR-17—(Brock et al. 2009, 2012; Pullamsetti et al. 2012), miR-21 (Caruso et al. 2010), and miR-204 (Courboulin et al. 2011). Most of these studies focused on changes in miRs in hypoxia- and monocrotaline-induced experimental models of PH, but miR-21 was shown to be reduced in human lung tissue and serum from patients with IPAH (Caruso et al. 2010). Downregulation of miR-204 has also been described in lung tissue, cultured pulmonary artery smooth muscle cells, and circulating buffy coat cells from patients with PAH (Courboulin et al. 2011). Interestingly, a recent in silico investigation, using a network-based bioinformatics approach, identified these miRs as ones predicted to be disease modifying in PH (Parikh et al. 2012). In addition to their potential role as biomarkers, circulating miRs are now thought to have a role in mediating intercellular communication. For example, miR-150 released from monocytic cells affects endothelial cell function and miR-143/145 from endothelial cells modulates smooth muscle cell phenotypes (Hergenreider et al. 2012). Circulating miRs might provide new insights into the pathogenesis of PH, particularly as they are accessible, relatively stable, and in some cases tissue-specific. Research in circulating miRs is however still at an early stage and the methodology is challenging, requiring careful consideration of study design, standards, statistical analysis, and validation in independent cohorts (Zampetaki et al. 2012).

2.11 Other Potential Biomarkers in PH

While the focus of this chapter has been primarily on blood biomarkers which represent possibly the simplest objective measure of disease at relatively low cost, other methodologies relevant to PH are being improved and may in future become more commonly adopted in disease management. *Cardiopulmonary exercise testing (CPET)* is an improvement upon more standard exercise measures recommended in Europe for risk stratification, which can discern between PH of different aetiologies (Zhai et al. 2011) and has prognostic value (Ferrazza et al. 2009; Deboeck et al. 2012), but requires expertise and specialist equipment that are not available at all centres. *Radionuclide imaging*, such as positron emission tomography (PET), and the availability of novel radiotracers for studies of relevant signal transduction pathways (Jakobsen et al. 2006) offer the potential to monitor disease progression and response to therapy in PH. There has been particular interest in the use of ^{18}flurodeoxyglucose PET in PAH (Xu et al. 2007), providing a surrogate marker of changes in glycolytic activity, vascular structure, and immune/inflammatory activation in the lung (Marsboom et al. 2012) as well as a marker of right ventricular dysfunction in PAH (Bokhari et al. 2011). Imaging techniques such as *3D echocardiography* (Amaki et al. 2009; Grapsa et al. 2012a, b) and *cardiomagnetic resonance (CMR) imaging* (Vonk Noordegraaf and Galie 2011; Bradlow et al. 2012) also offer greater accuracy and reproducibility in the assessment of the right ventricle and can provide volumetric measurements that reflect its function. CMR in particular has been utilised to assess the efficacy of therapeutics in recent clinical trials (Wilkins et al. 2005, 2010) and provides prognostic information in PAH (van Wolferen et al. 2007; van de Veerdonk et al. 2011), but the expense and the expertise required to operate the machinery and interpret results may have inhibited its more widespread use.

Many other blood biomarkers have also been associated with PH and are summarised in Table 1.

2.12 Comments on Biomarkers in PH

While many different biomarkers have been proposed for use in the management of PH, most have been inadequately validated in larger, independent cohorts and few have met the criteria to support routine measurement and use in clinical practice. The evidence for the utility of BNP and NT-proBNP as indices of right ventricular stress in PH is strong and their measurement is recommended by European guidelines (Galie et al. 2009). Despite this, the lack of validated specific cut-offs and an understanding of the meaning of proportional changes in BNP levels over time limits their use in practice. GDF-15 and RDW have been particularly strong performers in recent biomarker studies (Nickel et al. 2008; Rhodes et al. 2011a), possibly because they report on multiple facets of the disease pathology. Other markers such as IL-6, creatinine, uric acid, and angiopoietins may reflect more specific components, inflammation, renal dysfunction, oxidative stress, and endothelial cell dysfunction, respectively. The relatively recent discovery of microRNAs opens up a new avenue for biomarker discovery in PH. But before any of these biomarkers will be used routinely, circulating levels have to be shown to provide additional prognostic information that aids clinical decision-making and the necessary laboratory tests are readily available and reliable. The use of multiple markers, reflecting different aspects of PH pathology and incorporating imaging, may best represent the overall status of a patient and help guide future therapeutic decisions as more specific therapies become available. A critical question however is whether such a panel of biomarkers can identify not just a cohort but an individual patient with PH who is likely to derive benefit from a particular treatment?

Table 1 Other potential biomarkers in pulmonary hypertension

Biomarker	Biology	Evidence
Insulin resistance	Control of blood glucose levels	Insulin resistance twice as common in females with PAH (Zamanian et al. 2009)
Soluble TWEAK	Member of tumour necrosis factor (TNF) receptor superfamily	Predicts hemodynamic impairment and functional capacity in patients with PAH (Filusch et al. 2011)
Osteoprotegerin (OPG)	TNF/apoptotic pathway	Serum OPG elevated in IPAH patients and predicts survival (Lawrie et al. 2008; Condliffe et al. 2012)
Apelin	Vasodilator	Plasma levels decreased in IPAH, chronic lung disease, and heart failure (Goetze et al. 2006)
Endothelial progenitor cells (EPCs)	Endothelial repair/dysfunction	Heterogeneous population; altered circulating cell number and/or function of cultured cells in PAH (Diller et al. 2008, 2010; Asosingh et al. 2008; Toshner et al. 2009)
Circulating microvesicles	Endothelial function	Increased endothelial microvesicles predict haemodynamic severity and outcome in PH (Amabile et al. 2008, 2009)
Von Willebrand factor (vWf)	Endothelial function	Elevated levels prognostic in a small study of PAH associated with congenital heart disease (Lopes et al. 2011)
Cyclic GMP	Nitric oxide (NO) pathway	cGMP levels increased in PH (Bogdan et al. 1998), decreased by iloprost (Wiedemann et al. 2001), and correlate with PVR (Ghofrani et al. 2002)
Asymmetric dimethyl arginine (ADMA)	Inhibitor of NO synthase	Upregulated in PAH associated with congenital heart disease (Gorenflo et al. 2001). Levels correlate with haemodynamics and mortality in IPAH (Kielstein et al. 2005) and CTEPH (Skoro-Sajer et al. 2007)
Osteopontin	Inflammatory cytokine	Raised levels in IPAH correlate with 6MWD, mPAP, and functional class and predict survival (Lorenzen et al. 2011)
Isoprostanes	Oxidative stress	Increased in PAH and urinary levels associated with survival (Cracowski et al. 2001, 2012)
Nitric oxide (NO)	Vasodilator	Exhaled NO and urine NO metabolites reduced in IPAH and reversed by bosentan treatment (Girgis et al. 2005)
Serotonin	Vasoconstrictor	Raised levels in PAH correlate with PVR, but unresponsive to vasodilator treatment (Kereveur et al. 2000)
Endothelin-1	Vasoconstrictor	Raised plasma levels correlate with RAP, PVR, and disease severity in PAH (Stewart et al. 1991; Cacoub et al. 1993; Nootens et al. 1995; Rubens et al. 2001)
Cardiac troponins T & I	Myocardial injury	Troponin T & I levels predict disease severity and survival in PAH (Torbicki et al. 2003; Heresi et al. 2012)

(continued)

Table 1 (continued)

Biomarker	Biology	Evidence
Fatty acid binding protein	Myocardial injury	Negative prognostic marker in CTEPH (Lankeit et al. 2008) and pulmonary embolism (Puls et al. 2007)
Complement C4a des Arg	Complement cascade	Elevated levels of C4a des Arg (C4a activation fragment) in IPAH (Abdul-Salam et al. 2006)
Matrix metalloproteinases (MMPs)	Tissue remodelling	Elevated plasma levels of MMP-2, MMP-9, and tissue inhibitor of MMP (TIMP)-4 in PH (Elias et al. 2008; Hiremath et al. 2010; Schumann et al. 2010)

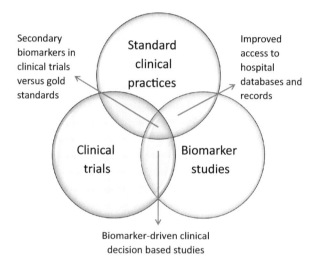

Fig. 3 Areas where improvements are required to take biomarker science forward in PH

3 Future Directions and Conclusions

The advent of "unbiased" genomic, metabolomic, and proteomic screening methodologies has not yet provided a major improvement in our understanding of disease biology or identified many new biomarkers. As the technologies are refined and improved upon, however, they will inevitably generate novel targets for therapy or biomarker use. MicroRNA screens in particular may generate novel hits, with panels of miRNAs being used to assess disease progression and response to therapy. The replication of findings using independent cohorts from other centres is vital in order to validate a potential biomarker and comparisons against current gold standards may be accelerated using hospital databases and registries in research protocols.

Genetic factors in PH can be useful in estimating susceptibility but currently apply only to a relatively small proportion of the overall PH population. More biomarker-focused clinical studies and better integration of pre-clinical biomarker studies into standard practices and clinical therapy trials are needed to take biomarkers in PH to the next level (Fig. 3).

References

Abdul-Salam VB, Paul GA, Ali JO, Gibbs SR, Rahman D, Taylor GW, Wilkins MR, Edwards RJ (2006) Identification of plasma protein biomarkers associated with idiopathic pulmonary arterial hypertension. Proteomics 6:2286–2294

Amabile N, Heiss C, Real WM, Minasi P, McGlothlin D, Rame EJ, Grossman W, De Marco T, Yeghiazarians Y (2008) Circulating endothelial microparticle levels predict hemodynamic severity of pulmonary hypertension. Am J Respir Crit Care Med 177:1268–1275

Amabile N, Heiss C, Chang V, Angeli FS, Damon L, Rame EJ, McGlothlin D, Grossman W, De Marco T, Yeghiazarians Y (2009) Increased CD62e+ endothelial microparticle levels predict poor outcome in pulmonary hypertension patients. J Heart Lung Transplant 28:1081–1086

Amaki M, Nakatani S, Kanzaki H, Kyotani S, Nakanishi N, Shigemasa C, Hisatome I, Kitakaze M (2009) Usefulness of three-dimensional echocardiography in assessing right ventricular function in patients with primary pulmonary hypertension. Hypertens Res 32:419–422

Asosingh K, Aldred MA, Vasanji A, Drazba J, Sharp J, Farver C, Comhair SA, Xu W, Licina L, Huang L, Anand-Apte B, Yoder MC, Tuder RM, Erzurum SC (2008) Circulating angiogenic precursors in idiopathic pulmonary arterial hypertension. Am J Pathol 172:615–627

Atkinson AJ, Colburn WA, DeGruttola VG, DeMets DL, Downing GJ, Hoth DF, Oates JA, Peck CC, Schooley RT, Spilker BA, Woodcock J, Zeger SL (2001) Biomarkers and surrogate endpoints: preferred definitions and conceptual framework. Clin Pharmacol Ther 69:89–95

Augustin HG, Koh GY, Thurston G, Alitalo K (2009) Control of vascular morphogenesis and homeostasis through the angiopoietin-Tie system. Nat Rev Mol Cell Biol 10:165–177

Bendayan D, Shitrit D, Ygla M, Huerta M, Fink G, Kramer MR (2003) Hyperuricemia as a prognostic factor in pulmonary arterial hypertension. Respir Med 97:130–133

Benza RL, Miller DP, Gomberg-Maitland M, Frantz RP, Foreman AJ, Coffey CS, Frost A, Barst RJ, Badesch DB, Elliott CG, Liou TG, McGoon MD (2010) Predicting survival in pulmonary arterial hypertension. Circulation 122:164–172

Benza RL, Gomberg-Maitland M, Miller DP, Frost A, Frantz RP, Foreman AJ, Badesch DB, McGoon MD (2012) The REVEAL Registry risk score calculator in patients newly diagnosed with pulmonary arterial hypertension. Chest 141:354–362

Bogdan M, Humbert M, Francoual J, Claise C, Duroux P, Simonneau G, Lindenbaum A (1998) Urinary cGMP concentrations in severe primary pulmonary hypertension. Thorax 53:1059–1062

Bokhari S, Raina A, Berman Rosenweig E, Schulze PC, Bokhari J, Einstein AJ, Barst RJ, Johnson LL (2011) PET imaging may provide a novel biomarker and understanding of right ventricular dysfunction in patients with idiopathic pulmonary arterial hypertension/clinical perspective. Circ Cardiovasc Imaging 4:641–647

Bootcov MR, Bauskin AR, Valenzuela SM, Moore AG, Bansal M, He XY, Zhang HP, Donnellan M, Mahler S, Pryor K, Walsh BJ, Nicholson RC, Fairlie WD, Por SB, Robbins JM, Breit SN (1997) MIC-1, a novel macrophage inhibitory cytokine, is a divergent member of the TGF-beta superfamily. Proc Natl Acad Sci USA 94:11514–11519

Bradlow WM, Gibbs JSR, Mohiaddin RH (2012) Cardiovascular magnetic resonance in pulmonary hypertension. J Cardiovasc Magn Reson 14:6

Braghiroli A, Sacco C, Erbetta M, Ruga V, Donner CF (1993) Overnight urinary uric acid: creatinine ratio for detection of sleep hypoxemia. Validation study in chronic obstructive pulmonary disease and obstructive sleep apnea before and after treatment with nasal continuous positive airway pressure. Am Rev Respir Dis 148:173–178

Brock M, Trenkmann M, Gay RE, Michel BA, Gay S, Fischler M, Ulrich S, Speich R, Huber LC (2009) Interleukin-6 modulates the expression of the bone morphogenic protein receptor type II through a novel STAT3-microRNA cluster 17/92 pathway. Circ Res 104:1184–1191

Brock M, Samillan VJ, Trenkmann M, Schwarzwald C, Ulrich S, Gay RE, Gassmann M, Ostergaard L, Gay S, Speich R, Huber LC (2012) AntagomiR directed against miR-20a restores functional BMPR2 signalling and prevents vascular remodelling in hypoxia-induced pulmonary hypertension. Eur Heart J. doi:10.1093/eurheartj/ehs060

Cacoub P, Dorent R, Maistre G, Nataf P, Carayon A, Piette C, Godeau P, Cabrol C, Gandjbakhch I (1993) Endothelin-1 in primary pulmonary hypertension and the Eisenmenger syndrome. Am J Cardiol 71:448–450

Caruso P, MacLean MR, Khanin R, McClure J, Soon E, Southgate M, MacDonald RA, Greig JA, Robertson KE, Masson R, Denby L, Dempsie Y, Long L, Morrell NW, Baker AH (2010) Dynamic changes in lung microRNA profiles during the development of pulmonary hypertension due to chronic hypoxia and monocrotaline. Arterioscler Thromb Vasc Biol 30:716–723

Chu D, Sullivan CC, Du L, Cho AJ, Kido M, Wolf PL, Weitzman MD, Jamieson SW, Thistlethwaite PA (2004) A new animal model for pulmonary hypertension based on the overexpression of a single gene, angiopoietin-1. Ann Thorac Surg 77:449–456

Condliffe R, Kiely DG, Peacock AJ, Corris PA, Gibbs JS, Vrapi F, Das C, Elliot CA, Johnson M, DeSoyza J, Torpy C, Goldsmith K, Hodgkins D, Hughes RJ, Pepke-Zaba J, Coghlan JG (2009) Connective tissue disease-associated pulmonary arterial hypertension in the modern treatment era. Am J Respir Crit Care Med 179:151–157

Condliffe R, Pickworth J, Hopkinson K, Walker S, Hameed A, Suntharaligam J, Soon E, Treacy C, Pepke-Zaba J, Francis S, Crossman D, Newman C, Elliot C, Morton A, Morrell N, Lawrie A (2012) Serum osteoprotegerin is increased and predicts survival in idiopathic pulmonary arterial hypertension. Pulm Circ 2:21–27

Courboulin A, Paulin R, Giguere NJ, Saksouk N, Perreault T, Meloche J, Paquet ER, Biardel S, Provencher S, Cote J, Simard MJ, Bonnet S (2011) Role for miR-204 in human pulmonary arterial hypertension. J Exp Med 208:535–548

Cracowski JL, Cracowski C, Bessard G, Pepin JL, Bessard J, Schwebel C, Stanke-Labesque F, Pison C (2001) Increased lipid peroxidation in patients with pulmonary hypertension. Am J Respir Crit Care Med 164:1038–1042

Cracowski JL, Degano B, Chabot F, Labarere J, Schwedhelm E, Monneret D, Iuliano L, Schwebel C, Chaouat A, Reynaud-Gaubert M, Faure P, Maas R, Renversez JC, Cracowski C, Sitbon O, Yaici A, Simonneau G, Humbert M (2012) Independent association of urinary F2-isoprostanes with survival in pulmonary arterial hypertension. Chest DOI. doi:10.1378/chest.11-1267, prepublished online 8 Mar 2012

Creemers EE, Tijsen AJ, Pinto YM (2012) Circulating microRNAs: novel biomarkers and extracellular communicators in cardiovascular disease? Circ Res 110:483–495

D'Alonzo GE, Barst RJ, Ayres SM, Bergofsky EH, Brundage BH, Detre KM, Fishman AP, Goldring RM, Groves BM, Kernis JT (1991) Survival in patients with primary pulmonary hypertension. Results from a national prospective registry. Ann Intern Med 115:343–349

Deboeck G, Coditti C, Uez S, Luc Vachiéry J, Amotte M, Harples L, Elot C, Naeije R (2012) Exercise to predict outcome in idiopathic vs associated pulmonary arterial hypertension. Eur Respir J. doi:10.1183/09031936.00217911, published on 22 Mar 2012

Deng Z, Morse JH, Slager SL, Cuervo N, Moore KJ, Venetos G, Kalachikov S, Cayanis E, Fischer SG, Barst RJ, Hodge SE, Knowles JA (2000) Familial primary pulmonary hypertension (gene PPH1) is caused by mutations in the bone morphogenetic protein receptor-II gene. Am J Hum Genet 67:737–744

Diller GP, van Eijl S, Okonko DO, Howard LS, Ali O, Thum T, Wort SJ, Bedard E, Gibbs JS, Bauersachs J, Hobbs AJ, Wilkins MR, Gatzoulis MA, Wharton J (2008) Circulating endothelial progenitor cells in patients with Eisenmenger syndrome and idiopathic pulmonary arterial hypertension. Circulation 117:3020–3030

Diller GP, Thum T, Wilkins MR, Wharton J (2010) Endothelial progenitor cells in pulmonary arterial hypertension. Trends Cardiovasc Med 20:22–29

Elias GJ, Ioannis M, Theodora P, Dimitrios PP, Despoina P, Kostantinos V, Charalampos K, Vassilios V, Petros SP (2008) Circulating tissue inhibitor of matrix metalloproteinase-4 (TIMP-4) in systemic sclerosis patients with elevated pulmonary arterial pressure. Mediators Inflamm 2008:164134

Ferrazza AM, Martolini D, Valli G, Palange P (2009) Cardiopulmonary exercise testing in the functional and prognostic evaluation of patients with pulmonary diseases. Respiration 77:3–17

Fessel JP, Loyd JE, Austin ED (2011) The genetics of pulmonary arterial hypertension in the post-BMPR2 era. Pulm Circ 1:305 319

Fijalkowska A, Kurzyna M, Torbicki A, Szewczyk G, Florczyk M, Pruszczyk P, Szturmowicz M (2006) Serum N-terminal brain natriuretic peptide as a prognostic parameter in patients with pulmonary hypertension. Chest 129:1313–1321

Filusch A, Zelniker T, Baumgartner C, Eschricht S, Frey N, Katus HA, Chorianopoulos E (2011) Soluble TWEAK predicts hemodynamic impairment and functional capacity in patients with pulmonary arterial hypertension. Clin Res Cardiol 100:879–885

Galie N, Ghofrani HA, Torbicki A, Barst RJ, Rubin LJ, Badesch D, Fleming T, Parpia T, Burgess G, Branzi A, Grimminger F, Kurzyna M, Simonneau G (2005) Sildenafil citrate therapy for pulmonary arterial hypertension. N Engl J Med 353:2148–2157

Galië N, Rubin LJ, Hoeper MM, Jansa P, Al-Hiti H, Meyer GMB, Chiossi E, Kusic-Pajic A, Simonneau G (2008) Treatment of patients with mildly symptomatic pulmonary arterial hypertension with bosentan (EARLY study): a double-blind, randomised controlled trial. Lancet 371:2093–2100

Galie N, Olschewski H, Oudiz RJ, Torres F, Frost A, Ghofrani HA, Badesch DB, McGoon MD, McLaughlin VV, Roecker EB, Gerber MJ, Dufton C, Wiens BL, Rubin LJ, for the Ambrisentan in Pulmonary Arterial Hypertension RD-BP-CMESAG (2008) Ambrisentan for the treatment of pulmonary arterial hypertension. Circulation 117:3010–3019

Galie N, Hoeper MM, Humbert M, Torbicki A, Vachiery JL, Barbera JA, Beghetti M, Corris P, Gaine S, Gibbs JS, Gomez-Sanchez MA, Jondeau G, Klepetko W, Opitz C, Peacock A, Rubin L, Zellweger M, Simonneau G (2009) Guidelines for the diagnosis and treatment of pulmonary hypertension. Eur Respir J 34:1219–1263

Ghofrani HA, Wiedemann R, Rose F, Weissmann N, Schermuly RT, Quanz K, Grimminger F, Seeger W, Olschewski H (2002) Lung cGMP release subsequent to NO inhalation in pulmonary hypertension: responders versus nonresponders. Eur Respir J 19:664–671

Girerd B, Montani D, Coulet F, Sztrymf B, Yaici A, Jais X, Tregouet D, Reis A, Drouin-Garraud V, Fraisse A, Sitbon O, O'Callaghan DS, Simonneau G, Soubrier F, Humbert M (2010) Clinical outcomes of pulmonary arterial hypertension in patients carrying an ACVRL1 (ALK1) mutation. Am J Respir Crit Care Med 181:851–861

Girgis RE, Champion HC, Diette GB, Johns RA, Permutt S, Sylvester JT (2005) Decreased exhaled nitric oxide in pulmonary arterial hypertension: response to bosentan therapy. Am J Respir Crit Care Med 172:352–357

Goetze JP, Rehfeld JF, Carlsen J, Videbaek R, Andersen CB, Boesgaard S, Friis-Hansen L (2006) Apelin: a new plasma marker of cardiopulmonary disease. Regul Pept 133:134–138

Gorenflo M, Zheng C, Werle E, Fiehn W, Ulmer HE (2001) Plasma levels of asymmetrical dimethyl-L-arginine in patients with congenital heart disease and pulmonary hypertension. J Cardiovasc Pharmacol 37:489–492

Grapsa J, Gibbs JS, Cabrita IZ, Watson GF, Pavlopoulos H, Dawson D, Gin-Sing W, Howard LS, Nihoyannopoulos P (2012a) The association of clinical outcome with right atrial and ventricular remodelling in patients with pulmonary arterial hypertension: study with real-time three-dimensional echocardiography. Eur Heart J Cardiovasc Imaging 13(8):666–672

Grapsa J, Gibbs JS, Dawson D, Watson G, Patni R, Athanasiou T, Punjabi PP, Howard LS, Nihoyannopoulos P (2012b) Morphologic and functional remodeling of the right ventricle in pulmonary hypertension by real time three dimensional echocardiography. Am J Cardiol 109:906–913

Hagen M, Fagan K, Steudel W, Carr M, Lane K, Rodman DM, West J (2007) Interaction of interleukin-6 and the BMP pathway in pulmonary smooth muscle. Am J Physiol Lung Cell Mol Physiol 292:L1473–L1479

Hampole CV, Mehrotra AK, Thenappan T, Gomberg-Maitland M, Shah SJ (2009) Usefulness of red cell distribution width as a prognostic marker in pulmonary hypertension. Am J Cardiol 104:868–872

Harrison RE, Flanagan JA, Sankelo M, Abdalla SA, Rowell J, Machado RD, Elliott CG, Robbins IM, Olschewski H, McLaughlin V, Gruenig E, Kermeen F, Halme M, Raisanen-Sokolowski A, Laitinen T, Morrell NW, Trembath RC (2003) Molecular and functional analysis identifies ALK-1 as the predominant cause of pulmonary hypertension related to hereditary haemorrhagic telangiectasia. J Med Genet 40:865–871

Hassoun PM, Mouthon L, Barbera JA, Eddahibi S, Flores SC, Grimminger F, Jones PL, Maitland ML, Michelakis ED, Morrell NW, Newman JH, Rabinovitch M, Schermuly R, Stenmark KR, Voelkel NF, Yuan JX, Humbert M (2009) Inflammation, growth factors, and pulmonary vascular remodeling. J Am Coll Cardiol 54:S10–S19

Hayabuchi Y, Matsuoka S, Akita H, Kuroda Y (1993) Hyperuricaemia in cyanotic congenital heart disease. Eur J Pediatr 152:873–876

Heger J, Schiegnitz E, von Waldthausen D, Anwar MM, Piper HM, Euler G (2010) Growth differentiation factor 15 acts anti-apoptotic and pro-hypertrophic in adult cardiomyocytes. J Cell Physiol 224:120–126

Heresi GA, Tang WH, Aytekin M, Hammel J, Hazen SL, Dweik RA (2012) Sensitive cardiac troponin I predicts poor outcomes in pulmonary arterial hypertension. Eur Respir J 39:939–944

Hergenreider E, Heydt S, Treguer K, Boettger T, Horrevoets AJ, Zeiher AM, Scheffer MP, Frangakis AS, Yin X, Mayr M, Braun T, Urbich C, Boon RA, Dimmeler S (2012) Atheroprotective communication between endothelial cells and smooth muscle cells through miRNAs. Nat Cell Biol 14:249–256

Hiremath J, Thanikachalam S, Parikh K, Shanmugasundaram S, Bangera S, Shapiro L, Pott GB, Vnencak-Jones CL, Arneson C, Wade M, White RJ (2010) Exercise improvement and plasma biomarker changes with intravenous treprostinil therapy for pulmonary arterial hypertension: a placebo-controlled trial. J Heart Lung Transplant 29:137–149

Hlatky MA, Greenland P, Arnett DK, Ballantyne CM, Criqui MH, Elkind MSV, Go AS, Harrell FE, Hong Y, Howard BV, Howard VJ, Hsue PY, Kramer CM, McConnell JP, Normand SL, O'Donnell CJ, Smith SC, Wilson PWF, on behalf of the American Heart Association Expert Panel on Subclinical Atherosclerotic Diseases and Emerging Risk Factors and the Stroke Council (2009) Criteria for evaluation of novel markers of cardiovascular risk. Circulation 119:2408–2416

Hoeper MM, Hohlfeld JM, Fabel H (1999) Hyperuricaemia in patients with right or left heart failure. Eur Respir J 13:682–685

Hoeper MM, Markevych I, Spiekerkoetter E, Welte T, Niedermeyer J (2005) Goal-oriented treatment and combination therapy for pulmonary arterial hypertension. Eur Respir J 26:858–863

Holmes SJ, Espiner EA, Richards AM, Yandle TG, Frampton C (1993) Renal, endocrine, and hemodynamic effects of human brain natriuretic peptide in normal man. J Clin Endocrinol Metab 76:91–96

Humbert M, Monti G, Brenot F, Sitbon O, Portier A, Grangeot-Keros L, Duroux P, Galanaud P, Simonneau G, Emilie D (1995) Increased interleukin-1 and interleukin-6 serum concentrations in severe primary pulmonary hypertension. Am J Respir Crit Care Med 151:1628–1631

Humbert M, Deng Z, Simonneau G, Barst RJ, Sitbon O, Wolf M, Cuervo N, Moore KJ, Hodge SE, Knowles JA, Morse JH (2002) BMPR2 germline mutations in pulmonary hypertension associated with fenfluramine derivatives. Eur Respir J 20:518–523

Humbert M, Sitbon O, Chaouat A, Bertocchi M, Habib G, Gressin V, Yaici A, Weitzenblum E, Cordier JF, Chabot F, Dromer C, Pison C, Reynaud-Gaubert M, Haloun A, Laurent M, Hachulla E, Cottin V, Degano B, Jais X, Montani D, Souza R, Simonneau G (2010) Survival in patients with idiopathic, familial, and anorexigen-associated pulmonary arterial hypertension in the modern management era. Circulation 122:156–163

Jakobsen S, Kodahl GM, Olsen AK, Cumming P (2006) Synthesis, radiolabeling and in vivo evaluation of [11C]RAL-01, a potential phosphodiesterase 5 radioligand. Nucl Med Biol 33:593–597

Jenkins M, Flynn A, Smart T, Harbron C, Sabin T, Ratnayake J, Delmar P, Herath A, Jarvis P, Matcham J (2011) A statistician's perspective on biomarkers in drug development. Pharm Stat 10:494–507

Kempf T, Wollert KC (2009) Growth differentiation factor-15: a new biomarker in cardiovascular disease. Herz 34:594–599

Kereveur A, Callebert J, Humbert M, Herve P, Simonneau G, Launay JM, Drouet L (2000) High plasma serotonin levels in primary pulmonary hypertension. Effect of long-term epoprostenol (prostacyclin) therapy. Arterioscler Thromb Vasc Biol 20:2233–2239

Khan SQ, Ng K, Dhillon O, Kelly D, Quinn P, Squire IB, Davies JE, Ng LL (2009) Growth differentiation factor-15 as a prognostic marker in patients with acute myocardial infarction. Eur Heart J 30:1057–1065

Kielstein JT, Bode-Boger SM, Hesse G, Martens-Lobenhoffer J, Takacs A, Fliser D, Hoeper MM (2005) Asymmetrical dimethylarginine in idiopathic pulmonary arterial hypertension. Arterioscler Thromb Vasc Biol 25:1414–1418

Kugathasan L, Ray JB, Deng Y, Rezaei E, Dumont DJ, Stewart DJ (2009) The angiopietin-1-Tie2 pathway prevents rather than promotes pulmonary arterial hypertension in transgenic mice. J Exp Med 206:2221–2234

Kumpers P, Nickel N, Lukasz A, Golpon H, Westerkamp V, Olsson KM, Jonigk D, Maegel L, Bockmeyer CL, David S, Hoeper MM (2010) Circulating angiopoietins in idiopathic pulmonary arterial hypertension. Eur Heart J 31:2291–2300

Lane KB, Machado RD, Pauciulo MW, Thomson JR, Phillips JA III, Loyd JE, Nichols WC, Trembath RC (2000) Heterozygous germline mutations in BMPR2, encoding a TGF-beta receptor, cause familial primary pulmonary hypertension. Nat Genet 26:81–84

Lankeit M, Dellas C, Panzenbock A, Skoro-Sajer N, Bonderman D, Olschewski M, Schafer K, Puls M, Konstantinides S, Lang IM (2008) Heart-type fatty acid-binding protein for risk assessment of chronic thromboembolic pulmonary hypertension. Eur Respir J 31:1024–1029

Lau EM, Manes A, Celermajer DS, Galie N (2011) Early detection of pulmonary vascular disease in pulmonary arterial hypertension: time to move forward. Eur Heart J 32:2489–2498

Lawrie A, Waterman E, Southwood M, Evans D, Suntharalingam J, Francis S, Crossman D, Croucher P, Morrell N, Newman C (2008) Evidence of a role for osteoprotegerin in the pathogenesis of pulmonary arterial hypertension. Am J Pathol 172:256–264

Lee WTN, Brown A, Peacock AJ, Johnson MK (2011) Use of non-invasive haemodynamic measurements to detect treatment response in precapillary pulmonary hypertension. Thorax 66:810–814

Leuchte HH, Holzapfel M, Baumgartner RA, Neurohr C, Vogeser M, Behr J (2005) Characterization of brain natriuretic peptide in long-term follow-up of pulmonary arterial hypertension. Chest 128:2368–2374

Leuchte HH, El NM, Tuerpe JC, Hartmann B, Baumgartner RA, Vogeser M, Muehling O, Behr J (2007) N-terminal pro-brain natriuretic peptide and renal insufficiency as predictors of mortality in pulmonary hypertension. Chest 131:402–409

Leyva F, Anker SD, Godsland IF, Teixeira M, Hellewell PG, Kox WJ, Poole-Wilson PA, Coats AJ (1998) Uric acid in chronic heart failure: a marker of chronic inflammation. Eur Heart J 19:1814–1822

Lopes AA, Barreto AC, Maeda NY, Cicero C, Soares RP, Bydlowski SP, Rich S (2011) Plasma von Willebrand factor as a predictor of survival in pulmonary arterial hypertension associated with congenital heart disease. Braz J Med Biol Res 44:1269–1275

Lorenzen JM, Nickel N, Kramer R, Golpon H, Westerkamp V, Olsson KM, Haller H, Hoeper MM (2011) Osteopontin in patients with idiopathic pulmonary hypertension. Chest 139:1010–1017

Machado RD, Pauciulo MW, Thomson JR, Lane KB, Morgan NV, Wheeler L, Phillips JA III, Newman J, Williams D, Galiè N, Manes A, McNeil K, Yacoub M, Mikhail G, Rogers P, Corris P, Humbert M, Donnai D, Martensson G, Tranebjaerg L, Loyd JE, Trembath RC, Nichols WC (2001) BMPR2 haploinsufficiency as the inherited molecular mechanism for primary pulmonary hypertension. Am J Hum Genet 68:92–102

Machado RD, Aldred MA, James V, Harrison RE, Patel B, Schwalbe EC, Gruenig E, Janssen B, Koehler R, Seeger W, Eickelberg O, Olschewski H, Elliott CG, Glissmeyer E, Carlquist J, Kim M, Torbicki A, Fijalkowska A, Szewczyk G, Parma J, Abramowicz MJ, Galie N, Morisaki H, Kyotani S, Nakanishi N, Morisaki T, Humbert M, Simonneau G, Sitbon O, Soubrier F, Coulet F, Morrell NW, Trembath RC (2006) Mutations of the TGF-beta type II receptor BMPR2 in pulmonary arterial hypertension. Hum Mutat 27:121–132

Marsboom G, Wietholt C, Haney CR, Toth PT, Ryan JJ, Morrow E, Thenappan T, Bache-Wiig P, Piao L, Paul J, Chen CT, Archer SL (2012) Lung 18F-fluorodeoxyglucose positron emission tomography for diagnosis and monitoring of pulmonary arterial hypertension. Am J Respir Crit Care Med 185:670–679

McDonald RA, Hata A, MacLean MR, Morrell NW, Baker AH (2012) MicroRNA and vascular remodelling in acute vascular injury and pulmonary vascular remodelling. Cardiovasc Res 93:594–604

McLaughlin VV, Suissa S (2010) Prognosis of pulmonary arterial hypertension: the power of clinical registries of rare diseases. Circulation 122:106–108

McLaughlin VV, Shillington A, Rich S (2002) Survival in primary pulmonary hypertension. Circulation 106:1477–1482

Miyamoto S, Nagaya N, Satoh T, Kyotani S, Sakamaki F, Fujita M, Nakanishi N, Miyatake K (2000) Clinical correlates and prognostic significance of six-minute walk test in patients with primary pulmonary hypertension. Comparison with cardiopulmonary exercise testing. Am J Respir Crit Care Med 161:487–492

Montani D, Bergot E, Gunther S, Savale L, Bergeron A, Bourdin A, Bouvaist H, Canuet M, Pison C, Macro M, Poubeau P, Girerd B, Natali D, Guignabert C, Perros F, O'Callaghan DS, Jais X, Tubert-Bitter P, Zalcman G, Sitbon O, Simonneau G, Humbert M (2012) Pulmonary arterial hypertension in patients treated by dasatinib. Circulation 125:2128–2137

Morisaki H, Nakanishi N, Kyotani S, Takashima A, Tomoike H, Morisaki T (2004) BMPR2 mutations found in Japanese patients with familial and sporadic primary pulmonary hypertension. Hum Mutat 23:632

Nagaya N, Nishikimi T, Okano Y, Uematsu M, Satoh T, Kyotani S, Kuribayashi S, Hamada S, Kakishita M, Nakanishi N, Takamiya M, Kunieda T, Matsuo H, Kangawa K (1998) Plasma brain natriuretic peptide levels increase in proportion to the extent of right ventricular dysfunction in pulmonary hypertension. J Am Coll Cardiol 31:202–208

Nagaya N, Uematsu M, Satoh T, Kyotani S, Sakamaki F, Nakanishi N, Yamagishi M, Kunieda T, Miyatake K (1999) Serum uric acid levels correlate with the severity and the mortality of primary pulmonary hypertension. Am J Respir Crit Care Med 160:487–492

Nagaya N, Nishikimi T, Uematsu M, Satoh T, Kyotani S, Sakamaki F, Kakishita M, Fukushima K, Okano Y, Nakanishi N, Miyatake K, Kangawa K (2000) Plasma brain natriuretic peptide as a prognostic indicator in patients with primary pulmonary hypertension. Circulation 102:865–870

Nakao K, Ogawa Y, Suga S, Imura H (1992) Molecular biology and biochemistry of the natriuretic peptide system. I: Natriuretic peptides. J Hypertens 10:907–912

Nasim MT, Ogo T, Ahmed M, Randall R, Chowdhury HM, Snape KM, Bradshaw TY, Southgate L, Lee GJ, Jackson I, Lord GM, Gibbs JS, Wilkins MR, Ohta-Ogo K, Nakamura K, Girerd B, Coulet F, Soubrier F, Humbert M, Morrell NW, Trembath RC, Machado RD (2011) Molecular genetic characterization of SMAD signaling molecules in pulmonary arterial hypertension. Hum Mutat 32:1385–1389

Newman JH, Phillips JA III, Loyd JE (2008) Narrative review: the enigma of pulmonary arterial hypertension: new insights from genetic studies. Ann Intern Med 148:278–283

Nickel N, Kempf T, Tapken H, Tongers J, Laenger F, Lehmann U, Golpon H, Olsson K, Wilkins MR, Gibbs JS, Hoeper MM, Wollert KC (2008) Growth differentiation factor-15 in idiopathic pulmonary arterial hypertension. Am J Respir Crit Care Med 178:534–541

Nickel N, Golpon H, Greer M, Knudsen L, Olsson K, Westerkamp V, Welte T, Hoeper MM (2011a) The prognostic impact of follow-up assessments in patients with idiopathic pulmonary arterial hypertension. Eur Respir J 39:589–596

Nickel N, Jonigk D, Kempf T, Bockmeyer CL, Maegel L, Rische J, Laenger F, Lehmann U, Sauer C, Greer M, Welte T, Hoeper MM, Golpon HA (2011b) GDF-15 is abundantly expressed in plexiform lesions in patients with pulmonary arterial hypertension and affects proliferation and apoptosis of pulmonary endothelial cells. Respir Res 12:62

Nootens M, Kaufmann E, Rector T, Toher C, Judd D, Francis GS, Rich S (1995) Neurohormonal activation in patients with right ventricular failure from pulmonary hypertension: relation to hemodynamic variables and endothelin levels. J Am Coll Cardiol 26:1581–1585

Parikh VN, Jin RC, Rabello S, Gulbahce N, White K, Hale A, Cottrill KA, Shaik RS, Waxman AB, Zhang YY, Maron BA, Hartner JC, Fujiwara Y, Orkin SH, Haley KJ, Barabasi AL, Loscalzo J, Chan SY (2012) MicroRNA-21 integrates pathogenic signaling to control pulmonary hypertension: results of a network bioinformatics approach. Circulation 125:1520–1532

Phillips JA III, Poling JS, Phillips CA, Stanton KC, Austin ED, Cogan JD, Wheeler L, Yu C, Newman JH, Dietz HC, Loyd JE (2008) Synergistic heterozygosity for TGFbeta1 SNPs and BMPR2 mutations modulates the age at diagnosis and penetrance of familial pulmonary arterial hypertension. Genet Med 10:359–365

Pletcher MJ, Pignone M (2011) Evaluating the clinical utility of a biomarker: a review of methods for estimating health impact. Circulation 123:1116–1124

Pullamsetti SS, Doebele C, Fischer A, Savai R, Kojonazarov B, Dahal BK, Ghofrani HA, Weissmann N, Grimminger F, Bonauer A, Seeger W, Zeiher AM, Dimmeler S, Schermuly RT (2012) Inhibition of microRNA-17 improves lung and heart function in experimental pulmonary hypertension. Am J Respir Crit Care Med 185:409–419

Puls M, Dellas C, Lankeit M, Olschewski M, Binder L, Geibel A, Reiner C, Schafer K, Hasenfuss G, Konstantinides S (2007) Heart-type fatty acid-binding protein permits early risk stratification of pulmonary embolism. Eur Heart J 28:224–229

Quarck R, Nawrot T, Meyns B, Delcroix M (2009) C-reactive protein: a new predictor of adverse outcome in pulmonary arterial hypertension. J Am Coll Cardiol 53:1211–1218

Rhodes CJ, Wharton J, Howard LS, Gibbs JS, Wilkins MR (2011a) Red cell distribution width outperforms other potential circulating biomarkers in predicting survival in idiopathic pulmonary arterial hypertension. Heart 97:1054–1060

Rhodes CJ, Wharton J, Howard L, Gibbs JS, Vonk-Noordegraaf A, Wilkins MR (2011b) Iron deficiency in pulmonary arterial hypertension: a potential therapeutic target. Eur Respir J 38:1453–1460

Rhodes CJ, Howard LS, Busbridge M, Ashby D, Kondili E, Gibbs JS, Wharton J, Wilkins MR (2011c) Iron deficiency and raised hepcidin in idiopathic pulmonary arterial hypertension: clinical prevalence, outcomes, and mechanistic insights. J Am Coll Cardiol 58:300–309

Rich S, Rubin L, Walker AM, Schneeweiss S, Abenhaim L (2000) Anorexigens and pulmonary hypertension in the United States. Chest 117:870–874

Roberts KE, McElroy JJ, Wong WP, Yen E, Widlitz A, Barst RJ, Knowles JA, Morse JH (2004) BMPR2 mutations in pulmonary arterial hypertension with congenital heart disease. Eur Respir J 24:371–374

Rosenzweig EB, Morse JH, Knowles JA, Chada KK, Khan AM, Roberts KE, McElroy JJ, Juskiw NK, Mallory NC, Rich S, Diamond B, Barst RJ (2008) Clinical implications of determining BMPR2 mutation status in a large cohort of children and adults with pulmonary arterial hypertension. J Heart Lung Transplant 27:668–674

Rubens C, Ewert R, Halank M, Wensel R, Orzechowski HD, Schultheiss HP, Hoeffken G (2001) Big endothelin-1 and endothelin-1 plasma levels are correlated with the severity of primary pulmonary hypertension. Chest 120:1562–1569

Rubin LJ, Badesch DB, Barst RJ, Galiè N, Black CM, Keogh A, Pulido T, Frost A, Roux S, Leconte I, Landzberg M, Simonneau G (2002) Bosentan therapy for pulmonary arterial hypertension. N Engl J Med 346:896–903

Sargent DJ, Conley BA, Allegra C, Collette L (2005) Clinical trial designs for predictive marker validation in cancer treatment trials. J Clin Oncol 23:2020–2027

Savale L, Tu L, Rideau D, Izziki M, Maitre B, Adnot S, Eddahibi S (2009) Impact of interleukin-6 on hypoxia-induced pulmonary hypertension and lung inflammation in mice. Respir Res 10:6

Schumann C, Lepper PM, Frank H, Schneiderbauer R, Wibmer T, Kropf C, Stoiber KM, Rudiger S, Kruska L, Krahn T, Kramer F (2010) Circulating biomarkers of tissue remodelling in pulmonary hypertension. Biomarkers 15:523–532

Shah SJ, Thenappan T, Rich S, Tian L, Archer SL, Gomberg-Maitland M (2008) Association of serum creatinine with abnormal hemodynamics and mortality in pulmonary arterial hypertension. Circulation 117:2475–2483

Shintani M, Yagi H, Nakayama T, Saji T, Matsuoka R (2009) A new nonsense mutation of SMAD8 associated with pulmonary arterial hypertension. J Med Genet 46:331–337

Sitbon O, Humbert M, Nunes H, Parent F, Garcia G, Herve P, Rainisio M, Simonneau G (2002) Long-term intravenous epoprostenol infusion in primary pulmonary hypertension: prognostic factors and survival. J Am Coll Cardiol 40:780–788

Sitbon O, Humbert M, Jais X, Ioos V, Hamid AM, Provencher S, Garcia G, Parent F, Herve P, Simonneau G (2005) Long-term response to calcium channel blockers in idiopathic pulmonary arterial hypertension. Circulation 111:3105–3111

Skoro-Sajer N, Mittermayer F, Panzenboeck A, Bonderman D, Sadushi R, Hitsch R, Jakowitsch J, Klepetko W, Kneussl MP, Wolzt M, Lang IM (2007) Asymmetric dimethylarginine is increased in chronic thromboembolic pulmonary hypertension. Am J Respir Crit Care Med 176:1154–1160

Soon E, Holmes AM, Treacy CM, Doughty NJ, Southgate L, Machado RD, Trembath RC, Jennings S, Barker L, Nicklin P, Walker C, Budd DC, Pepke-Zaba J, Morrell NW (2010) Elevated levels of inflammatory cytokines predict survival in idiopathic and familial pulmonary arterial hypertension. Circulation 122:920–927

Souza R, Bogossian HB, Humbert M, Jardim C, Rabelo R, Amato MB, Carvalho CR (2005) N-terminal-pro-brain natriuretic peptide as a haemodynamic marker in idiopathic pulmonary arterial hypertension. Eur Respir J 25:509–513

Steiner MK, Syrkina OL, Kolliputi N, Mark EJ, Hales CA, Waxman AB (2009) Interleukin-6 overexpression induces pulmonary hypertension. Circ Res 104:236–244

Stewart DJ, Levy RD, Cernacek P, Langleben D (1991) Increased plasma endothelin-1 in pulmonary hypertension: marker or mediator of disease? Ann Intern Med 114:464–469

Sztrymf B, Coulet F, Girerd B, Yaici A, Jais X, Sitbon O, Montani D, Souza R, Simonneau G, Soubrier F, Humbert M (2008) Clinical outcomes of pulmonary arterial hypertension in carriers of BMPR2 mutation. Am J Respir Crit Care Med 177:1377–1383

Sztrymf B, Souza R, Bertoletti L, Jais X, Sitbon O, Price LC, Simonneau G, Humbert M (2010) Prognostic factors of acute heart failure in patients with pulmonary arterial hypertension. Eur Respir J 35:1286–1293

Thenappan T, Shah SJ, Rich S, Tian L, Archer SL, Gomberg-Maitland M (2010) Survival in pulmonary arterial hypertension: a reappraisal of the NIH risk stratification equation. Eur Respir J 35:1079–1087

Thenappan T, Glassner C, Gomberg-Maitland M (2012) Validation of the pulmonary hypertension connection equation for survival prediction in pulmonary arterial hypertension. Chest 141:642–650

Tonelli AR, Alnuaimat H, Mubarak K (2010) Pulmonary vasodilator testing and use of calcium channel blockers in pulmonary arterial hypertension. Respir Med 104:481–496

Torbicki A, Kurzyna M, Kuca P, Fijalkowska A, Sikora J, Florczyk M, Pruszczyk P, Burakowski J, Wawrzynska L (2003) Detectable serum cardiac troponin T as a marker of poor prognosis among patients with chronic precapillary pulmonary hypertension. Circulation 108:844–848

Toshner M, Voswinckel R, Southwood M, Al-Lamki R, Howard LS, Marchesan D, Yang J, Suntharalingam J, Soon E, Exley A, Stewart S, Hecker M, Zhu Z, Gehling U, Seeger W, Pepke-Zaba J, Morrell NW (2009) Evidence of dysfunction of endothelial progenitors in pulmonary arterial hypertension. Am J Respir Crit Care Med 180:780–787

van de Veerdonk MC, Kind T, Marcus JT, Mauritz GJ, Heymans MW, Bogaard HJ, Boonstra A, Marques KMJ, Westerhof N, Vonk-Noordegraaf A (2011) Progressive right ventricular dysfunction in patients with pulmonary arterial hypertension responding to therapy. J Am Coll Cardiol 58:2511–2519

van Wolferen SA, Marcus JT, Boonstra A, Marques KMJ, Bronzwaer JGF, Spreeuwenberg MD, Postmus PE, Vonk-Noordegraaf A (2007) Prognostic value of right ventricular mass, volume, and function in idiopathic pulmonary arterial hypertension. Eur Heart J 28:1250–1257

Voelkel MA, Wynne KM, Badesch DB, Groves BM, Voelkel NF (2000) Hyperuricemia in severe pulmonary hypertension. Chest 117:19–24

Vonk Noordegraaf A, Galie N (2011) The role of the right ventricle in pulmonary arterial hypertension. Eur Respir Rev 20:243–253

Wang TJ (2011) Assessing the role of circulating, genetic, and imaging biomarkers in cardiovascular risk prediction. Circulation 123:551–565

Wiedemann R, Ghofrani HA, Weissmann N, Schermuly R, Quanz K, Grimminger F, Seeger W, Olschewski H (2001) Atrial natriuretic peptide in severe primary and nonprimary pulmonary hypertension: response to iloprost inhalation. J Am Coll Cardiol 38:1130–1136

Wilkins MR, Paul GA, Strange JW, Tunariu N, Gin-Sing W, Banya WA, Westwood MA, Stefanidis A, Ng LL, Pennell DJ, Mohiaddin RH, Nihoyannopoulos P, Gibbs JS (2005) Sildenafil versus Endothelin Receptor Antagonist For Pulmonary Hypertension (SERAPH) study. Am J Respir Crit Care Med 171:1292–1297

Wilkins MR, Ali O, Bradlow W, Wharton J, Taegtmeyer A, Rhodes CJ, Ghofrani HA, Howard L, Nihoyannopoulos P, Mohiaddin RH, Gibbs JS, for the Simvastatin Pulmonary Hypertension Trial (SiPHT) Study Group (2010) Simvastatin as a treatment for pulmonary hypertension trial. Am J Respir Crit Care Med 181:1106–1113

Xu J, Kimball TR, Lorenz JN, Brown DA, Bauskin AR, Klevitsky R, Hewett TE, Breit SN, Molkentin JD (2006) GDF15/MIC-1 functions as a protective and antihypertrophic factor released from the myocardium in association with SMAD protein activation. Circ Res 98:342–350

Xu W, Koeck T, Lara AR, Neumann D, DiFilippo FP, Koo M, Janocha AJ, Masri FA, Arroliga AC, Jennings C, Dweik RA, Tuder RM, Stuehr DJ, Erzurum SC (2007) Alterations of cellular bioenergetics in pulmonary artery endothelial cells. Proc Natl Acad Sci USA 104:1342–1347

Yandle TG, Richards AM, Gilbert A, Fisher S, Holmes S, Espiner EA (1993) Assay of brain natriuretic peptide (BNP) in human plasma: evidence for high molecular weight BNP as a major plasma component in heart failure. J Clin Endocrinol Metab 76:832–838

Yoshimura M, Yasue H, Morita E, Sakaino N, Jougasaki M, Kurose M, Mukoyama M, Saito Y, Nakao K, Imura H (1991) Hemodynamic, renal, and hormonal responses to brain natriuretic peptide infusion in patients with congestive heart failure. Circulation 84:1581–1588

Zamanian RT, Hansmann G, Snook S, Lilienfeld D, Rappaport KM, Reaven GM, Rabinovitch M, Doyle RL (2009) Insulin resistance in pulmonary arterial hypertension. Eur Respir J 33:318–324

Zampetaki A, Willeit P, Drozdov I, Kiechl S, Mayr M (2012) Profiling of circulating microRNAs: from single biomarkers to re-wired networks. Cardiovasc Res 93:555–562

Zhai Z, Murphy K, Tighe H, Wang C, Wilkins MR, Gibbs JS, Howard LS (2011) Differences in ventilatory inefficiency between pulmonary arterial hypertension and chronic thromboembolic pulmonary hypertension. Chest 140:1284–1291

Rodent Models of Group 1 Pulmonary Hypertension

John J. Ryan, Glenn Marsboom, and Stephen L. Archer

Abstract World Health Organization category 1 pulmonary hypertension (PH) is a heterogeneous syndrome in which PH originates in the small pulmonary arteries and is therefore also referred to as pulmonary arterial hypertension (PAH). Common pathophysiologic features include endothelial dysfunction, excessive proliferation and impaired apoptosis of vascular cells, and mitochondrial fragmentation. The proliferation/apoptosis imbalance relates in part to activation of the transcription factors hypoxia-inducible factor-1α (HIF-1α) and nuclear factor of activated T-cells (NFAT) and apoptosis repressors, such as survivin. Perivascular inflammation, disruption of adventitial connective tissue, and a glycolytic metabolic shift in vascular cells and right ventricular myocytes also occur in PAH. There are important genetic and epigenetic predispositions to PAH. This review assesses the fidelity of existing animal models to human PAH. No single model can perfectly recapitulate the many diverse forms of PH in Category 1; however, acceptable models exist. PAH induced by monocrotaline and chronic hypoxia plus SU-5416 (CH+SU) in rats display endothelial dysfunction, proliferation/apoptosis imbalance, and develop the glycolytic metabolic profile of human PAH. Histologically, CH+SU best conforms to PAH in that it develops complex vascular lesions, including plexiform lesions. However, the monocrotaline model can be induced to manifest complex vascular lesions and does manifest the tendency of PAH patients to die of right ventricular (RV) failure. Murine models offer greater molecular certainty than

J.J. Ryan
Cardiovascular Center, University Hospital, University of Utah, 50 N Medical Dr, Salt Lake City, UT 84132, USA
e-mail: john.ryan@hsc.utah.edu

G. Marsboom
University of Chicago, 5841 South Maryland Avenue, MC 6080, Chicago, IL 60637, USA

S.L. Archer (✉)
Department of Medicine, Queen's University, Etherington Hall, Rm. 3041, 94 Stuart Street, Kingston, ON, Canada, K7L 3N6
e-mail: archers@queensu.ca

rat models but rarely develop significant PH, have less right ventricular hypertrophy (RVH) and pulmonary artery (PA) remodeling, and are harder to image and catheterize. The use of high fidelity catheterization and advanced imaging (microPET-CT, high frequency echocardiography, high field strength MRI) and functional testing (treadmill) permit accurate phenotyping of experimental models of PAH. Preclinical trial design is an important aspect of testing experimental PAH therapies. The use of multiple complementary models with adequate sample size and trial duration and appropriate endpoints are required for preclinical assessment of experimental PAH therapies.

Keywords Hypoxia • Monocrotaline • Sugen 5416 • Right ventricular hypertrophy • Plexiform lesions

Contents

1	Introduction	107
2	Molecular Signaling	110
3	Outcome Assessments in Animal Models	114
4	Exercise Capacity	115
5	Catheterization and Echocardiography	115
	5.1 Monocrotaline-Induced PAH	119
	5.2 Hypoxia Models	122
	5.3 Chronic Hypoxia Rats	123
	5.4 Fawn-Hooded Rats	125
	5.5 Chronic Hypoxia and Sugen-5416	128
6	Transgenic Models	129
	6.1 BMPR-2 Mutation Models	129
7	Global Knockout Mice	129
8	Conditional Knockout Mice	130
9	BMPR-2-Dominant Negative Transgenic Mice	130
10	Short Hairpin (shRNA) BMPR-2 Knockdown Mice	131
	10.1 5-HTT Overexpressing Mice	131
	10.2 Models That Require Further Study	132
	10.3 Broiler Fowl Disease	132
	10.4 Endothelin B Receptor Deficiency in Rats	132
	10.5 Angiopoietin-1 Overexpression in Rats	133
	10.6 High Fat Diet to Apo E Knockout Mice	133
11	Inflammatory Models	134
	11.1 γ-Herpesvirus Reactivation Model	134
	11.2 Schistosomiasis	136
	11.3 *Stachybotrys chartarum*	136
	11.4 HIV-Associated Pulmonary Hypertension	137
	11.5 Radiation Exposure	137
12	Surgical Models	139
	12.1 Pulmonary Artery Banding Model	139
	12.2 Aorto-Caval Shunt	140
13	Summary	140
	References	141

Non-standard Abbreviations and Acronyms

5-HTT	Serotonin transporter
Ang-1	Angiopoietin-1
BMPR-2	Bone morphogenetic protein receptor-2
CH + SU	Chronic hypoxia plus Sugen-5416
CO	Cardiac output
DRP-1	Dynamin-related protein 1
ET-1	Endothelin-1
FHR	Fawn-Hooded rat
HHV	Human herpes virus
HIF-1α	Hypoxia-inducible factor-1α
HIV	Human immunodeficiency virus
HPV	Hypoxic pulmonary vasoconstriction
LVEDP	Left ventricular end diastolic pressure
mPAP	mean pulmonary artery pressure
MRI	Magnetic resonance imaging
NFAT	Nuclear factor of activated T cells
PA	Pulmonary artery
PAAT	Pulmonary artery acceleration time
PAH	Pulmonary arterial hypertension
PASMC	Pulmonary artery smooth muscle cell
PDH	Pyruvate dehydrogenase
PDK	Pyruvate dehydrogenase kinase
PH	Pulmonary hypertension
PVR	Pulmonary vascular resistance
RVH	Right ventricular hypertrophy
SOD2	Superoxide dismutase 2
SU-5416	Sugen-5416
TAPSE	Tricuspid annulus plane systolic excursion
VEGF	Vascular endothelial growth factor
VTI	Velocity time integral

1 Introduction

In Group 1 PH (or pulmonary arterial hypertension, PAH) not only is mPAP greater than 25 mmHg, but the designation requires the exclusion of thromboembolic disease, lung disease, and left heart disease (the left ventricular end diastolic pressure should be <15 mmHg). Group 1 PH occurs in a rare idiopathic form and an even less common familial form. Most Group 1 PH is associated with an identifiable risk factor (such as connective tissue disease, congenital heart disease, or anorexigen consumption). The female/male ratio in PAH is >4/1 (Archer and Rich 2000; Pugh and Hemnes 2010). There is considerable heterogeneity within

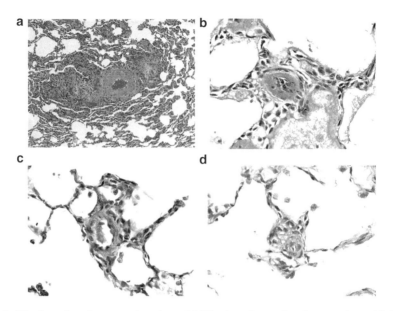

Fig. 1 Histology from human and rat lung. (**a**) Histology from scleroderma patient with PAH showing medial hypertrophy. (**b**) Vascular obliteration in monocrotaline rat. (**c**) Chronic hypoxia rat with mild medial hypertrophy. (**d**) Plexiform lesions in Sugen-Hypoxia rat

Group 1 with associated conditions as diverse as congenital heart disease, hemolytic anemia, schistosomiasis, and HIV. PAH has a significant mortality with 5-year survival of only 60 % (Thenappan et al. 2010), despite advances in therapies. The estimated incidence of PAH is 2.4–7.6 cases/million annually (Humbert et al. 2006).

Common histological features of Group 1 include medial hypertrophy and adventitial hyperplasia of small pulmonary arteries. Often there are complex vascular lesions (including intimal hyperplasia and plexiform arteriopathy) (Fig. 1a) (reviewed in Tuder et al. 2007). Plexiform lesions are reminiscent of glomeruli, containing multiple, endothelial-lined, slit-like channels interspersed with myofibroblasts and pulmonary artery smooth muscle cells (PASMCs). Clusters of small, dilated pulmonary arteries, often found distal to these plexiform lesions, are called angiomatoid lesions. Other common features of Group 1 PH are platelet deposition, thrombus in situ, and pulmonary hemorrhage (Tuder et al. 2007; Pogoriler et al. 2012). The histological features of human PAH that have been hardest to reproduce in animal models include the intimal proliferation, plexiform lesions, and pulmonary veno-occlusive lesions. These histologic factors (combined with an element of exaggerated vasoconstriction) increase right ventricular afterload (measured as increased pulmonary vascular resistance, PVR > 3 Wood units) and reduce vascular compliance (evident as shortened pulmonary artery acceleration time, PAAT).

Even in human PAH there is histological diversity. Plexiform lesions are uncommon in scleroderma-associated PAH (Pogoriler et al. 2012). Within the lungs of a single patient, there can be considerable regional heterogeneity in the localization of complex vascular lesions. As in human PAH, the pathology in most animal models of Group 1 PH, is confined to the lung vasculature, although here again there is diversity among forms of Group 1 PH. Scleroderma patients more often manifest pulmonary veno-occlusive disease and may have concomitant interstitial lung fibrosis, systemic vascular disease (notably Raynaud's phenomenon,) and left ventricular disease (Sweiss et al. 2010). This is perhaps not surprising, since scleroderma is a systemic disease and circulating autoantibodies that target endothelium and smooth muscle cells (SMCs) are not specific for the lung vasculature. There are currently no good models of Category 1' PH (patients with pulmonary capillary hemangiomatosis and pulmonary veno-occlusive disease). Thus, no single animal model of PH will accurately capture all the facets of Group 1 PH, perhaps because Group 1 is itself a heterogeneous collection of PH syndromes.

The other groups classified by the WHO system are not the focus of this review, but they too are best studied using models that closely replicate their central disease process. In Group 2 PH, the PH is secondary to an increase in left atrial pressure caused by left heart disease, including systolic or diastolic left ventricular dysfunction, or valvular diseases, such as mitral stenosis (Califf et al. 1997). There are no convenient and readily reproducible small animal models for Group 2 PH, although transaortic banding can induce this syndrome (Chen et al. 2012). Group 3 PH results from chronic hypoxia exposure or chronic lung disease resulting in mild elevations in PAP (Elwing and Panos 2008). The chronic hypoxic rat model is reflective of Category 3 PH (Ryan et al. 2011). Group 4 PH is secondary to chronic thromboembolic disease and small animal models are lacking. Group 5 is a collection of PH syndromes associated with miscellaneous diseases, such as extrinsic pulmonary artery (PA) obstruction and sarcoidosis (and is thus difficult to model).

Many PAH patients have multiple, identifiable pathogenic stimuli, such as a component of hypoxia and/or increased LVEDP, that contribute to their PH syndrome. Models that offer molecular purity (e.g., knockout mice) may not recapitulate the complex clinical picture of PH. Interestingly, the models induced by multiple insults, such as monocrotaline plus pneumonectomy or chronic hypoxia plus Sugen-5416 (CH + SU), best mimic human PAH.

Many experimental therapies studied in animal models have demonstrated a benefit, which fails to translate into clinical benefit in patients (Stenmark et al. 2009). In a review of 91 studies that used the monocrotaline model, therapies were demonstrated to be beneficial in 88 of the studies (Stenmark et al. 2009). The excess optimism from these animal studies originates partially from publication bias (which favors positive findings) and partly from flaws in the models (i.e., monocrotaline imperfectly mimics human PAH). However, poor study design often contributes to the confusion. Many of these "positive" studies reflect the short duration of the trial, the reliance on surrogate end points instead of mortality, and the rather poor phenotypic characterization of hemodynamics and right ventricular function. For example, by extending study duration in a rodent trial of statins for prevention of PAH, we found the same lack of efficacy (McMurtry et al. 2007a) that was ultimately noted in patients (Kawut et al. 2011). Several positive rodent studies of statins in PAH were of shorter duration, perhaps contributing to an overly optimistic conclusion.

The ideal model would reflect the distinct clinical characteristics of Group 1 PH, namely a female predominance, isolated pulmonary vascular disease, right ventricular hypertrophy, a high mortality from right heart failure, and the classical histopathological features of human PAH. Such a model has been challenging to create. In addition, for those studying pediatric Group 1 PH, a model would need to be created in neonatal animals, since the pulmonary vascular biology of neonates and adults differs substantially.

In this chapter, we review the models of Group 1 PH, emphasizing common rodent models, summarized in Table 1. We briefly discuss large animal models, including dogs, sheep, cows, and primates. These large animals oftentimes more successfully replicate human disease than do rodent models; however, their utility is limited by their size, the expense of the models, the slower disease progression, as well as less scientific emotional/ethical caveats, relating to large animal studies.

2 Molecular Signaling

The animal models of PAH should replicate the molecular and cellular signaling disorders see in PAH patients. Plasma serotonin is increased in PAH (Herve et al. 1995), as is the ratio of endothelial-derived vasoconstrictors/dilators (Christman et al. 1992). Increased expression and activity of serotonin transporter (5-HTT) in PASMC contributes to the proliferative excess in PAH (Marcos et al. 2003), as does increased expression of platelet-derived growth factor receptor (Perros et al. 2008). The decreased PASMC apoptosis and increased proliferation seen in PAH raises comparisons to cancer (Merklinger et al. 2005). Apoptosis resistance is driven by several factors, including de novo expression of the anti-apoptotic protein survivin (McMurtry et al. 2005). Apoptosis resistance is seen in multiple rodent models, including Fawn-Hooded rats and 5-HTT overexpressing mice (Archer et al. 2010).

Mutations in bone morphogenetic protein receptor-2 (BMPR-2) (Morrell et al. 2001), the genetic basis for familial PAH (Thomson et al. 2000), also promote proliferation and impair apoptosis. The resulting loss of BMPR-2 receptor function contributes to PASMC proliferation (Zhang et al. 2003). BMPR-2 also affects endothelial cells (Yang et al. 2011). Although most rodent models of PAH have no genetic BMPR-2 abnormality, impaired BMP signaling and acquired downregulation of BMPR-2 expression is commonly observed (McMurtry et al. 2007b; Morrell 2006). In addition, the BMP antagonist gremlin is increased in hypoxic PH and gremlin is also upregulated in the vessels of PAH patients (Cahill et al. 2012). Increased gremlin secretion by endothelial cells blocks BMP signaling thereby contributing to PASMC proliferation and PAH.

Normoxic activation of HIF-1α promotes PASMC proliferation and creates a glycolytic phenotype, which can be detected by increased pulmonary uptake of fluorodeoxyglucose (FDG) on Positron Emission Tomography (PET) scans (Marsboom et al. 2012a; Fijalkowska et al. 2010). HIF-1α activation occurs despite

Table 1 Animal models of Group 1 PAH

Animal model	Method	Medial hypertrophy	Plexiform lesions	mPAP (mmHg)	RVH	HIF-1α activity	Kv1.5	Spontaneity	Mortality
Human PAH	Spontaneous or hereditary	+++	+++	~30–60	Varies	+	↓	+++	30 % mortality at 3 years
Monocrotaline rat	60–100 mg/kg sc Can be combined with pneumonectomy	+++	+	40–60	Moderate	+	No change	–	100 % ~2 months
Chronic hypoxia rat	10 % O_2 for 3–4 weeks Also studied in mice	+	–	30–40	Mild	+	↓	–	Minor
Chronic hypoxia plus SU-5416 rat	20 mg/kg of SU-5416 coupled with 3 weeks of hypoxia at 10 %	++	+	40–60	Moderate	+	↓	–	10 % at 6 weeks then stable until ~20 weeks
Fawn-Hooded rat	Hereditary	++	–	30–50	Moderate	+	↓	+++	Survives up to 1 year
BMPR-2 mutation mouse	Overexpression of mutant BMPR-2 alleles	++	–	35–45	Mild	+	↓	+++	Stable up to 3 months
Endothelin-B receptor deficiency	Monocrotaline administered to ET_B-deficient rats	+	–	40–45	Mild	–	No change	–	Stable up to 3 months
Angiopoietin-1 overexpression	AAV-Ang-1 is injected into RV outflow track	+	–	30–40	Mild	–	No change	–	Stable up to 3 months
5-HTT overexpressing mouse	Transgenic mice exposed to hypoxia for 2–4 weeks	+	–	45–55	Mild	–	No change	+	Stable up to 3 months
γ-Herpesvirus reactivation mice	Virus given to 1 year old S100A4/Mts mice	–	+	20–30	None	–	?	–	Stable up to 1 year
Schistosomiasis mice	Female C57/BL6 adult mice infected sc with *S. mansoni*	+	+	20–30	None	Undefined	?	–	Stable up to 20 weeks

(continued)

Table 1 (continued)

Animal model	Method	Medial hypertrophy	Plexiform lesions	mPAP (mmHg)	RVH	HIF-1α activity	Kv1.5	Spontaneity	Mortality
Stachybotrys chartarum mice	Intratracheal injection of fungal spores into mice over 4–12 weeks	+	–	20–30	Moderate	Undefined	?	–	Stable up to 12 weeks
High fat diet to ApoE knockout mice	Starting at 4 weeks of age, mice are fed high fat chow for 11 weeks	+	+	20–30	Mild	Undefined	?	–	Stable up to 3 months
Monocrotaline in primates	Administered interscapularly 1/month for 4 months	+	–	40–60	Moderate	+	?	–	Stable up to 6 months
HIV-associated PH in primates	SHIV-*nef* infection	+	+	Limited data	Moderate	Undefined	?	–	Stable up to 6 months
Radiation exposure to Wister rats	Exposed to protons with shoot-through technique	+	–	25–35	Mild	Undefined	?	–	Stable up to 2 months
Pulmonary artery banding rodent	Main PA ligated with a silk suture	–	–	Normal	Severe	–	No change	–	Stable up to 12 weeks
Aorto-caval shunt rodent	Needle advanced through abdominal aorta into IVC. Can be combined with monocrotaline	+	–	25–35	Severe	–	No change	–	Stable up to 12 weeks
Brisket disease in cattle	Cattle reared at 7,000 ft	+	–	60–80	Severe	+	?	–	Stable up to 6 months

? represents the unknown effect of the animal model on Kv1.5

5-HTT serotonin transporter, *Ang-1* angiopoietin-1, *BMPR-2* bone morphogenetic protein receptor-2, *CH + SU* chronic hypoxia plus Sugen-5416, *ET-1* endothelin-1, *FHR* Fawn-Hooded rat, *HHV* human herpes virus, *HIF-1α* hypoxia-inducible factor-1α, *HIV* human Immunodeficiency virus, *mPAP* mean pulmonary artery pressure, *PA* pulmonary artery, *PAH* pulmonary arterial hypertension, *PH* pulmonary hypertension, *RVH* right ventricular hypertrophy, *SU-5416* Sugen-5416 normal PO_2 and (at least in some cases) is the result of epigenetic silencing of superoxide dismutase 2 (SOD2) (Archer et al. 2010). Decreased SOD2 expression impairs mitochondrial redox signaling and permits normoxic activation of HIF-1α (Archer et al. 2010). HIF-1α activation increases the transcription of the pyruvate dehydrogenase kinases (PDKs), through interaction of HIF-1α with the PDK gene's hypoxia-responsive element (HRE) (Kim et al. 2006). PDKs phosphorylate and inhibit the pyruvate dehydrogenase (PDH) complex, thereby shifting PASMC metabolism to glycolysis (McMurtry et al. 2004; Michelakis et al. 2002; Guignabert et al. 2009). This abnormal mitochondrial-SOD2-HIF-1α pathway is seen in Fawn-Hooded rats, as well as in the monocrotaline and chronic hypoxic models. The glycolytic phenotype is reminiscent of the Warburg phenomenon seen in cancer cells (Bonnet et al. 2007a). Recently it has been noted that the mitochondria in PAH PASMC are not only functionally impaired, but also they are structurally abnormal—specifically they are fragmented (Archer et al. 2008). Mitochondrial fragmentation reflects a disorder in two tightly regulated processes: mitochondrial fission (which is increased) and mitochondrial fusion (which is depressed). The excessive fission is caused by activation of dynamin-related protein 1 (DRP-1) (Marsboom et al. 2012b). HIF-1α activation increases the activity of DRP-1 by phosphorylation of DRP-1 at serine 616. This phosphorylation is coupled to mitosis because the same regulatory kinase (cyclin B1/CDK1) phosphorylates DRP-1 and permits mitosis. For the cell to undergo mitosis, the mitochondria must divide and DRP-1 inhibition prevents both mitotic fission and cell division. For example, the DRP-1 inhibitor, mdivi-1, arrests PAH PASMCs at the G2/M interphase. Thus the mitochondria looms large in the pathogenesis of PAH, which is interesting, since mitochondria are also the redox sensor for hypoxia and are central to hypoxic pulmonary vasoconstriction (Weir et al. 2005).

Kv1.5 and other voltage-gated potassium channels are decreased in human PAH and rodent models of PAH. Loss of Kv1.5 contributes to constriction and proliferation of PASMCs (Michelakis et al. 2001). Loss of Kv channels depolarizes PASMC membrane potential, thereby increasing influx of Ca^{2+} via the activation of voltage-gated calcium channels (Fig. 2) (Archer and Rich 2000; Reeve et al. 2001; Yuan et al. 1998). This raises cytosolic calcium promoting vasoconstriction and proliferation. If the increase in cytosolic calcium (which results from loss of Kv channels and activation of TrpC6 channels (Zhang et al. 2007)) is sustained, this activates nuclear factor of activated T cells (NFAT), a Ca^{2+}-calcineurin-dependent transcription factor that increases PASMC proliferation (Bonnet et al. 2007b). Loss of Kv channels also increases cytosolic K^+, which inhibits caspases and promotes apoptosis resistance. Downregulation of Kv1.5 seen in human PAH (Yuan et al. 1998; Weir et al. 1998; Geraci et al. 2001) is recapitulated in many rodent models including those induced by chronic hypoxia (Pozeg et al. 2003) ($\pm$ Sugen-5416) and the PH in Fawn-Hooded rats, BMPR-2 dominant negative transgenic mice (Young et al. 2006), and serotonin transporter (5-HTT) transgenic mice (Guignabert et al. 2006).

The role of inflammation in PAH is evident from the strong association with connective tissue disease and the histological presence of perivascular mononuclear

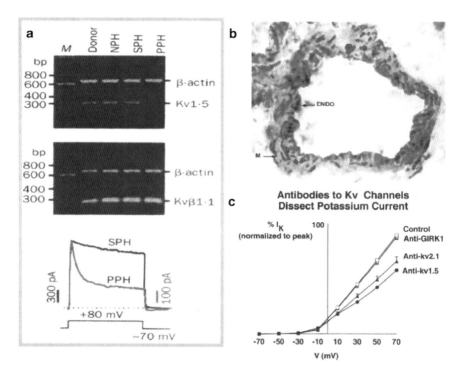

Fig. 2 Potassium channel abnormalities in PPH. (a) Reverse-transcriptase polymerase chain reaction shows that patients with PPH have less mRNA for Kv1.5 than donors, chronic lung disease patients with normal PVR (NPH), or secondary PHT patient (SPH). Expression of other mRNA is unaffected (such as B-actin and Kvb1.1 subunit). (b) Immunohistochemistry from rat pulmonary arteries with PH shows Kv1.5 channel in PASMCs (*light gray*). Kv1.5 is also present in endothelium (ENDO, *arrow*) but is not seen because slide is double-stained with an antibody to endothelial NO synthase (*dark gray*). M indicates media. (c) Whole-cell potassium current (IK) is inhibited by internal application of an antibody directed against Kv1.5 and Kv2.1 on whole-cell patch clamping of normal rat PASMCs but not by an antibody against a K1 channel that is not expressed in PASMCs (GIRK-1). This demonstrates that Kv1.5 and Kv2.1 contribute to electrical activity of rat PASMCs. Reprinted with permission from Archer et al. (1998)

cell infiltrates adjacent to pulmonary arteries in PAH. Mononuclear cells have also been seen around the characteristic plexiform lesions (Cool et al. 1997). Reproducing this inflammatory infiltration is important in animal models of PAH. Inflammation is present in the monocrotaline model but is notably absent in the CH + SU model (Zhao 2010).

3 Outcome Assessments in Animal Models

A brief summary of the techniques required to comprehensively assess PAH in rodent models is provided. Exercise capacity, accurate hemodynamic measurements, and advanced imaging tools are necessary to properly categorize, quantify, and monitor disease progression and regression of PH in animal models. Rodent imaging has been revolutionized by miniaturization of micromanometer-tipped catheters, high field-strength magnetic resonance imaging (MRI), high-frequency echocardiography, and advanced application of nuclear imaging (Urboniene et al. 2010; Thibault et al. 2010; Marsboom et al. 2012a).

4 Exercise Capacity

Exercise capacity can be assessed using treadmill testing. However, there are multiple determinants of maximal exercise capacity, including not only RV function and the status of the pulmonary circulation but also pulmonary function, left ventricular performance, skeletal muscle function, and activity of the peripheral and central nervous systems. Typical rodent treadmills are motorized and one varies the speed and slope following a standardized protocol (summarized in Fig. 3). The animal is motivated to run by a shock bar that delivers a mild noxious stimulus should they cease running (Fig. 3a). Maximal exercise capacity is defined as the distance covered before the animal fatigues and comes to rest on the shock grid (at which time they are immediately removed from the device). Concern has been raised about the reproducibility and reliability of this metric (Knab et al. 2009). Despite its limitations, assessment of exercise capacity is an important outcome. It is helpful as a surrogate for cardiac output and, when interpreted in light of findings on catheterization and echocardiography, is useful in assessing the response to therapy. Indeed, exercise capacity is the parameter used by the FDA to approve medications for patients with PAH (Miyamoto et al. 2000).

5 Catheterization and Echocardiography

Anesthesia. Both catheterization and echocardiograpy are performed using inhaled isoflurane mixed with medical air or 100 % O_2. To induce anesthesia in rats, the rodent is placed in an anesthesia induction chamber with gas flow at 2–3 l/min. Anesthesia is induced by administration of isoflurane at 3–4 %. After induction, anesthesia is maintained using isoflurane 1.5–2 % (for echocardiography) or 3 % (for more invasive procedures, such as cardiac catheterization). Anesthesia should be sufficient to prevent responses to noxious stimuli (like paw pinch) but sufficiently light that the heart rate is >300 bpm in rats or >500 bmp in mice.

Hemodynamics. Complete hemodynamic assessment needs to be performed to fully understand the PH models, including measurement of PVR. Without a left and right heart catheterization one cannot exclude Group 2 PH. In rodents and other animals used to study PH, the normal PAP is the same as in normal humans. However, in rats and mice the heart rate is 5–10 times faster than in humans,

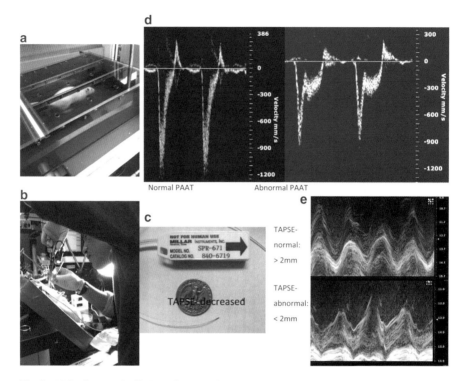

Fig. 3 (**a**) Rodent treadmill. Exercise capacity is tested by measuring maximal distance run on a motorized treadmill (Simplex II Instrument; Columbus Instruments, Columbus, OH). The initial treadmill speed is 5 m/min and increased 5 m/min every 5 min for 30 min or until the rat fatigued. (**b**) Right heart catheterization on rodent. (**c**) Microtipped catheter compared to 25 cents piece. (**d**) Pulmonary artery acceleration time as measured by Doppler echocardiogram. (**e**) Tricuspid annular plane systolic excursion measured on M mode echocardiogram

respectively. Hemodynamics in a group of normal and pulmonary hypertensive rats studied in our laboratory are shown in Table 2. A complete summary of the technique for rat catheterization is provided in ref (Urboniene et al. 2010).

Pulmonary hemodynamics are optimally measured in lightly anesthetized, close-chested animals. Although cannulating the PA is faster and requires less expertise in the open-chest preparation, opening the thorax reduces cardiac output, heart rate, and PA pressures (particularly end diastolic pressures). It is acceptable to use the open chest technique but this technical detail and its limitations should be noted. When performing pressure–volume assessments, we have found that, hemodynamics are more precisely obtained in prone or upright positions (rather than in the commonly used supine position) (Fig. 3b). Measurement of mPAP, systolic PAP, and LVEDP are best performed using micromanometer-tipped catheters that have sufficient frequency response to measure pressures at heart rates of greater than 300 bpm (Fig. 3c). Fluid-filled catheters produce damped (and thus inaccurate) pressure tracings at these heart rates. In addition to the right heart catheterization, it is important to measure LVEDP by retrograde catheterization of the LV (performed

Table 2 Rat hemodynamics

Rat hemodynamics	Normal Sprague-Dawley rat ($n = 8$)	Pulmonary hypertensive rat: Monocrotaline rat ($n = 6$)
HR (beats/min)	372 ± 16	311 ± 9
PAAT (ms)	33.4 ± 1.1	22.1 ± 1.4
sPAP (mmHg)	20.7 ± 1.5	51.2 ± 3.8
mPAP (mmHg)	14.6 ± 1.6	32.8 ± 2.0
PVR (dyn/s/cm^5)	0.15 ± 0.02	0.52 ± 0.06
SV (ml)	0.31 ± 0.02	0.22 ± 0.01
CO (ml/min)	111 ± 8	66 ± 6
CI (ml/min/g)	0.36 ± 0.01	0.20 ± 0.01
Treadmill distance (m)	405 ± 75	125 ± 54

BW body weight, *CI* cardiac index, *CO* cardiac output, *HR* heart rate, *PAAT* pulmonary artery acceleration time, *PAcsa* pulmonary artery cross sectional area, *SV* stroke volume, *VTI* velocity time integral

via a carotid arteriotomy). For right heart measurements, the catheter is placed into the right internal jugular vein via a venotomy. A fluid-filled catheter is advanced through a curved sheath into the RV and subsequently the PA. Once PA pressures are obtained, the more fragile Millar catheter replaces the fluid filled catheter. This is challenging to perform in mice and open-chested preparations using Millar catheters may be required.

Measurement of cardiac output using thermodilution techniques is reliable and easily performed in rodents. The technique involves bolus infusion of cold saline in the internal jugular vein with temperature recording by a thermistor in the aorta (Physitemp Instruments, Clifton, NJ). Changes in temperature at the thermistor are inversely related to cardiac output. One must be mindful of the small volumes of fluid that rodents can safely tolerate to avoid perturbing hemodynamics. (we use boluses of 0.1 ml of iced saline). Normal cardiac output in adult male rats is >110 ml/min.

Echocardiography. Because catheterization is performed by the cut-down technique, (rather than percutaneously) rodents are typically euthanized at the end of the procedure. Serial studies are more easily performed using echocardiography. The challenge in measuring PAP by echocardiogram in rodent models is the absence of tricuspid regurgitation. This precludes the use of Bernoulli's equation, the technique commonly used to assess PA pressure in patients (Rich et al. 2011), to provide a noninvasive estimate of PA systolic pressure. However, in rodents the PA Doppler signal is easily obtained and this permits assessment of PA compliance and PA pressure using the PA acceleration time (PAAT) (Fig. 3d).

For echocardiography the anesthetized rat is placed in a supine position. After shaving the chest and applying gel to improve acoustic coupling, two-dimensional images are obtained (we use the Vevo 2100 (VisualSonics, Toronto, ON) equipped with a 25 MHz transducer). The two-dimensional echocardiogram allows measurement of PA diameter and with color guidance allows optimal placement of the

Table 3 Measurement of hemodynamics

Rat hemodynamic measurements	Formula
Stroke volume	$PA_{CSA} \times VTI$
Cardiac output	$SV \times HR$
Cardiac index	CO/BW
Mean PA pressure	$58.7 - (1.21 \times PAAT)$

BW body weight, *CI* cardiac index, *CO* cardiac output, *HR* heart rate, *PAAT* pulmonary artery acceleration time, PA_{CSA} pulmonary artery cross sectional area, *SV* stroke volume, *VTI* velocity time integral

pulsed-wave Doppler signal in the main PA. The PA pulsed Doppler envelope is analyzed to determine the PAAT (defined as the time to peak flow velocity) and flow velocity time integral (VTI). Two-dimensionally guided M-mode echocardiography is used to measure RV wall thickness

PA pressure is best measured using PAAT, which is inversely related to mean PA pressure. By performing simultaneous Doppler and right heart catheterization studies, one can create a regression equation that accurately relates PAAT to mPAP (Urboniene et al. 2010). The normal PAAT in an adult rat is 30–35 ms and a PAAT < 20 ms indicates significant PH. PAAT correlates inversely with mPAP on catheterization ($r^2 = 0.76$) and our laboratory's regression equation is mPAP = $58.7 - (1.21 \times PAAT)$ (Urboniene et al. 2010). Ideally, each research laboratory should create its own regression equation.

In the rat monocrotaline PAH model, good correlation has been found between echocardiography and MRI assessment of PA pressure by PAAT versus the gold standard of right heart catheterization, offering the potential for noninvasive techniques to ultimately replace cardiac catheterization in rat models (Urboniene et al. 2010; Jones et al. 2002). PAAT also correlates well with mPAP in mice. A decrease in the ratio of acceleration time to ejection time (AT/ET) correlates closely with increases in murine RVSP (Thibault et al. 2010).

In PH, there is also mid-systolic notching of the PA Doppler envelope due to a retrograde cancellation wave that reflects back from the noncompliant distal vasculature (Fig. 3d). Echocardiographic estimates of CO correlate well with results of thermodilution studies. RV stroke volume is calculated through multiplying PA VTI by PA cross-sectional area (πr^2, where r = the radius of the main PA). CO is calculated by multiplying stroke volume by heart rate (Table 3).

The RV undergoes systolic shortening due to a deep layer of vertically arrayed RV fibers and this shortening is diminished when RV function is impaired. Shortening of the RV can be measured by orienting the M-mode cursor longitudinally across the tricuspid annulus in the direction of the RV apex. The excursion of the annulus (base to apex) is measured in mm and is called the tricuspid annular plane systolic excursion (TAPSE) (Fig. 3e). TAPSE is a useful noninvasive assessment of RV function (Forfia et al. 2006). TAPSE is measured in rodents and can assess RV function. The normal value in our laboratory is >2 mm. In rats with severe PAH this value decreases to <1 mm.

We will now review the most established and reproduced models of Group 1 PH and subsequently briefly review the less frequently used animal models, including transgenic mice and surgical models of congenital heart disease.

5.1 Monocrotaline-Induced PAH

Although the monocrotaline rat model has recently been subject to considerable criticism (Gomez-Arroyo et al. 2012), it does capture many of the cardinal features of human PH, such as endothelial cell damage, medial hypertrophy of small PAs, vascular inflammation, RVH, and RV failure. Monocrotaline is derived from the seeds of *Crotalaria spectabilis* and has been shown to induce pulmonary vascular injury primarily after hepatic generation of monocrotaline pyrrole. In a 1954 study of Jamaicans who used herbal medicines derived from *Crotalaria spectabilis*, Bras et al. (1954)implicated monocrotaline in the development of pulmonary veno-occlusive disease. Pulmonary vasculitis was observed in rats fed with monocrotaline in 1961 by Lalich (Lalich and Merkow 1961). Lalich then demonstrated that these rats develop right ventricular hypertrophy. A reliable and reproducible model for monocrotaline-induced PH was developed by Heath in 1967 (Kay et al. 1967).

Method. A single administration of monocrotaline (60–100 mg/kg, subcutaneously) to rats predictably results in PH within approximately 3 weeks and mortality in most rats after 6–8 weeks (Urboniene et al. 2010). The quick advance to death, secondary to RV failure likely decreases the ability of this model to develop compensatory mechanisms. There is a dose–response to monocrotaline and lower dose of monocrotaline (20 mg/kg) result in milder PH and a nonlethal, compensated model with RV hypertrophy rather than RV failure (Buermans et al. 2005).

Features of model. Monocrotaline is converted to the active metabolite dehydromonocrotaline in the liver by cytochrome P-450 enzyme CYP3A4. It exerts multiple toxicological effects on the pulmonary endothelium including induction of DNA damage (Wagner et al. 1993), abnormalities in nitric oxide signaling, and activation of proliferation factors (Huang et al. 2010). Mitochondrial swelling and increased production of reactive oxygen species in the lung have been observed in monocrotaline rats (Rosenberg and Rabinovitch 1988; Archer et al. 1990). Monocrotaline also causes cell-cycle arrest to pulmonary artery endothelial cells, as well as to glial cells, pneumocytes, and hepatocytes. These effects on pneumocytes likely contribute to the extravascular effects of monocrotaline, namely pulmonary edema and alveolar cell septal hyperplasia (Woods et al. 1999). Some of the toxic effects of monocrotaline are also likely mediated through hematological effects. Dehydromonocrotaline accumulates in the erythrocytes and circulates systemically in red blood cells (Pan et al. 1991). Platelet-mediated thrombosis and fibrin activation are observed (Schulze and Roth 1998). The role played by platelets in monocrotaline PAH is highlighted by the protective effect of thrombocytopenia (Schultze and Roth 1998). Inflammation plays an important role in this model, with mononuclear cells accumulating in the adventitia early in the evolution of PAH (Wilson et al. 1989). Macrophages accumulate and contribute to the proliferation of smooth muscle cells (Wilson et al. 1989). The liver toxicity observed in monocrotaline model contributes to development of portal hypertension and this may predispose to PH (Epstein et al. 1992).

The initial visible insult in monocrotaline PAH is endothelial injury. This endothelial damage is seen on electron microscopy (Fig. 4) (Rosenberg and Rabinovitch 1988). The endothelial injury can also result in edema in the adventitia (Wilson and Segall 1990). In an alternate model, rats that received monocrotaline in their drinking water had decreased angiotensin converting enzyme (ACE) activity early in the development of PH, again reflective of endothelial damage (Molteni et al. 1984). Within 8 days of administration of monocrotaline subcutaneously, extension of SMCs peripherally into the small PAs is observed with changes occurring at a later stage in the larger PAs (approximately 2 weeks after monocrotaline administration). The predominant histology on light microscopy is medial hypertrophy of small (20–100 µm) PAs (Fig. 1b). There is also obliteration of these arteries, which reduces vascular density. The first hemodynamic changes are not observed until ~2 weeks after administration of monocrotaline. Mean PA pressures in monocrotaline rats increase to greater than 35 mmHg by 4–6 weeks. The development of RVH follows the development of PH, being evident at 3–4 weeks (Marsboom et al. 2012a; Pan et al. 1991).

When monocrotaline is combined with pneumonectomy, mean PA pressures are increased even further (>45 mmHg) (White et al. 2007). To create this model, 12-week-old Sprague-Dawley rats are subjected to pneumonectomy and 1 week later, monocrotaline is administered (60 mg/kg subcutaneously). The histological findings are also more severe than with monocrotaline alone, with advanced medial hypertrophy and neointimal lesions observed, including plexiform-like lesions. This exacerbation is likely secondary to increased pulmonary blood flow or hypoxia in the residual lung (Okada et al. 1997). An unexplained finding is that the plexiform lesions seen in this model develop within 1 week of administration of monocrotaline—before the onset of PH. These lesions resemble human plexiform lesions and also have a similar molecular profile expressing vascular endothelial growth factor receptor-2 and contain PASMC rich in tissue factor (White et al. 2007).

The right ventricle. The RV in monocrotaline PAH is hypertrophied and has a glycolytic shift in metabolism characterized by PDK activation and inhibition of PDH. This RV has a phenotype similar to hibernating myocardium (Piao et al. 2010a).

Limitations of model. The monocrotaline model has been subject to criticism, such as the development of concomitant myocarditis in some animals (Gomez-Arroyo et al. 2012), the lack of plexiform lesions, and damage to airways [pulmonary fibrosis and alveolar septal cell hyperplasia (Dumitrascu et al. 2008)].

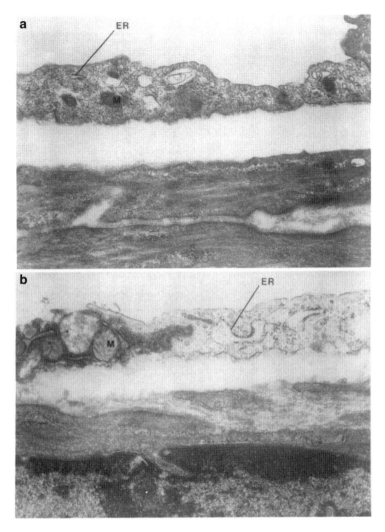

Fig. 4 Photomicrographs of small preacinar artery 4 days after injection with monocrotaline (**b**) or normal saline (**a**). Endothelium from monocrotaline-injected animals appears swollen and less dense. Note swollen mitochondria (M) and dilated endoplasmic reticulum (ER). Reproduced with permission from Rosenberg and Rabinovitch (1988)

Monocrotaline has also been found to cause obstructive pulmonary vein thrombosis (Lalich et al. 1977). However, this post-capillary obstruction is similar to the pathology observed in pulmonary veno-occlusive disease (PVOD) and thus the monocrotaline model reflects aspects of Group 1' PH.

Monocrotaline has been studied in other animals. Importantly, monocrotaline does not lead to the development of pulmonary hypertension in mice for an unknown reason (Dumitrascu et al. 2008). PAH arteriopathy has been shown in young beagles who have been exposed to dehydromonocrotaline (Gust and Schuster 2001). Eight weeks after venous injection of dehydromonocrotaline (3 mg/kg) in 12-week-old beagles, pulmonary artery pressures increase to 30–35 mmHg. Medial hypertrophy of small PAs develops and an increase in neointimal proliferation is seen. However this model is infrequently used, primarily because this canine model offers minimal advantage over the corresponding rodent model. Likewise, monocrotaline (30 mg/kg) causes PAH in baboons and macaca monkeys leading to RV disease (Raczniak et al. 1978). Monthly administration of monocrotaline for 4 months leads to clinical manifestations and histological changes very similar to those found in humans with PAH (Chesney and Allen 1973). Although, limited hemodynamics are available in this model (Raczniak et al. 1978), these monkeys live 220 days after injection of monocrotaline on average (range 100–355 days). RV dilatation and RVH is pronounced, with wide separation of the RV myocytes and an increase in RV collagen. The authors report that these myocytes had an increase in the number of mitochondria (Raczniak et al. 1978). However, the use of monocrotaline in monkeys offers very little advantage (if any) over the rodent monocrotaline model and introduces economic and ethical challenges. It does suggest that ingestion of monocrotaline could cause PAH in humans and is a reminder about the need for caution in the ingestion of biologicals, which may occur as health supplements or herbal teas.

Despite limitations, the monocrotaline rat remains a convenient, reliable, and reproducible form of PH (Stenmark et al. 2009) that mimics many of the cellular and molecular components of PAH. The fact that RV failure seems to be important to the death of these animals is a strength of the model. There appears to be some poorly characterized individual variability in susceptibility to monocrotaline since we find that not all exposed animals develop PAH (approximately 10 % of exposed animals fail to develop PH in our protocols).

5.2 Hypoxia Models

Chronic hypoxia is a model of Group 3 PH that is often used to make inferences regarding Group 1 PH. The PH and the vascular remodeling are mild and animals usually do not die of RV failure. We discuss the chronic hypoxic model because it is relevant to the CH + SU model (injection of Sugen 5416 + 3 weeks of chronic hypoxia 10 %). Moreover, HIF-1α activation (seen in chronic hypoxia) is also seen in Group 1 PH. Upregulation of HIF-1α occurs in the setting of hypoxia and results in mitochondrial fragmentation and subsequent smooth muscle cell proliferation (which is more easily demonstrated in vivo than in cell culture). Mice that are haploinsufficient for HIF-1α are relatively protected from this form of PH (Shimoda et al. 2001), implicating a central role for HIF-1α in the pathogenesis of this form of PH.

Of note, there is remarkable interspecies variation in the response to hypoxia. Animals that are genetically acclimated to high altitude such as llamas or yaks have no demonstrable PH or vascular remodeling upon exposure to chronic hypoxia (Stenmark et al. 2009; Rabinovitch et al. 1979). Rats are intermediate responders to chronic hypoxia, however, not all breeds of rats respond equally. Fawn-Hooded rats (FHR) respond most significantly, with mPAP 45–50 mmHg (Bonnet et al. 2006). After Fawn-Hooded rats, Sprague-Dawley rats are the most susceptible to hypoxia-induced pulmonary hypertension with mPAP 30–35 mmHg (Kentera et al. 1988). Fischer rats are relatively resistant to hypoxia-induced PH and develop minimal PH (mPAP 25–30 mmHg) or vascular remodeling and thus should preferentially not be used to create this model of PH (Zhao et al. 2001). It has also been observed that younger rats are also more sensitive to hypoxia (Rabinovitch et al. 1981; Stenmark et al. 2006). Mice are very poor responders to hypoxia (Stenmark et al. 2009). The difference between mice and rats in response to hypoxia may be due to differences in the up- and downregulation of various genes. For example, chronic hypoxia in the mice results in downregulation of genes associated with vascular smooth muscle cell proliferation (Hoshikawa et al. 2003), whereas in rats hypoxia causes an increase in genes associated with proliferation of endothelial cells and a decrease in genes involved in apoptosis (Bull et al. 2007).

5.3 Chronic Hypoxia Rats

Method. In chronic hypoxic rats, most experimental protocols gradually decrease FiO_2 over 3–5 days to acclimate rodents and then maintain the animals at 10 % O_2 (approximately P_aO_2 40 mmHg, which simulates an altitude of 18,000 ft) for up to 4 weeks. This model is tolerated well with minimal mortality, provided time for acclimatization is allowed. Hypobaric and normobaric hypoxia yield similar PH; although normobaric hypoxia is often used, because it is not complicated by the noise of the vacuum pump, which is required for the hypobaric model. However, the hypobaric model is cheaper as it does not require premixed gases. Importantly, to rule out any effects of HPV on measured pressures, animals should be kept in normal oxygen tensions for at least 1 h prior to catheterization.

Features of model. There is usually little evidence of vascular obstruction or rarefaction. The hallmark is medial hypertrophy of small PAs, which develops after ~2 weeks of hypoxic exposure (Zhao 2010) (Fig. 1c). Hypoxic pulmonary vasoconstriction (HPV) plays a role early in the disease process with vascular remodeling becoming more important over time. HPV is selectively decreased in chronic hypoxia-induced pulmonary hypertension (whereas constriction to other stimuli is enhanced) (Reeve et al. 2001; McMurtry et al. 1976). Within 2 days of exposure to hypoxia, HPV is no longer marked as impaired redox signaling develops in the mitochondria and vascular remodeling plays a more critical role in elevating PA pressures (Fig. 5) (Reeve et al. 2001).

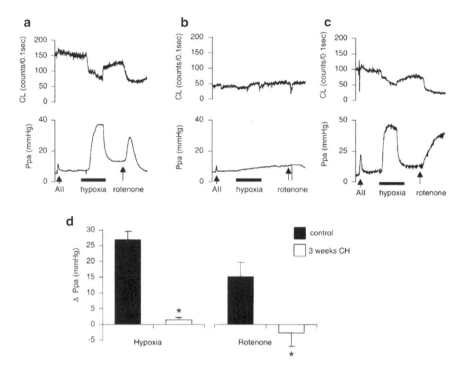

Fig. 5 Representative traces from isolated, perfused lungs showing simultaneous recording of chemiluminescence (CL) and pulmonary arterial pressure (Ppa). (a) Response in control lung to angiotension II (AII), hypoxia, and rotenone (10 mM). (b) Response in 3-week chronic hypoxia (CH) lung to AII, hypoxia, and rotenone. (c) Response in 3-week CH returned to normoxia for 2 days to AII, hypoxia, and rotenone. *Bar* indicates period of acute hypoxic challenge. Rotenone added at *arrow*. (d) Means ± SE changes (Δ) in Ppa in control and CH rat lungs to acute hypoxia and rotenone (10 µM). *$P < 0.05$, different from control. Reproduced with permission from Reeve et al. (2001)

The vascular remodeling in chronic hypoxia increases PVR. It has been observed that luminal area is preserved in the rat model of hypoxia in vessels between 30 and 200 µm (Hyvelin et al. 2005). Potentially the medial thickening can grow in an outward direction in some vessels, thereby not affecting the luminal area (Stenmark and McMurtry 2005). Histologically, the increased medial hypertrophy is characterized by an increase in α-SM-actin. Fibroblasts and myofibroblasts increase deposition of elastin, collagen, and fibronectin thereby contributing to adventitial thickening (Davie et al. 2006). Distal extension of smooth muscle into normally nonmuscularized pulmonary blood vessels is a characteristic feature of chronic hypoxic PH (Stenmark et al. 2006). Inflammation plays an important role in the chronic hypoxia model and both an early and a late stage of inflammation is observed (Burke et al. 2009). Mesenchymal precursor cells and mononuclear cells that infiltrate the vessel wall may contribute to the vascular remodeling in chronic hypoxia (Frid et al. 2006).

The RV. The RV in this model undergoes mild hypertrophy. A local increase in ACE activity has been observed in the RV of chronic hypoxic rats, similar to the increase reported in monocrotaline rats (Molteni et al. 1984; Morrell et al. 1997).

Limitations of model. Limitations from this model include its failure to reflect the severe obstructive remodeling and complex vascular lesions in Group 1 PH. In addition, the hemodynamic abnormalities are modest and RV failure is not a significant feature of the model in rodents. Also this model is accompanied by polycythemia (which can increase PH), which is not a feature of PAH. Finally, many experimental treatments (including many of which are ineffective in more severe models) are "effective" in regressing chronic hypoxic PH, suggesting the model is insufficiently robust for preclinical assessment of PAH therapies (Girgis et al. 2003).

Brisket disease is a form of chronic hypoxic PH induced by exposure of lowland cattle to high altitude (>7,000 ft). This form of hypoxic PH is severe and does lead to right heart failure, evident by fluid accumulation in the dependent "brisket" region of the cow (Grover and Reeves 1962). This disease is extensively reviewed in ref (Rhodes 2005). We mention Brisket disease in this review to highlight the ability to study chronic hypoxia in cows, which are robust responders to hypoxia (unlike rodents). Medial hypertrophy of the small pulmonary arteries is observed in these animals and is proportional to the extent of the increase in mPAP (Alexander and Jensen 1963). Some cattle such as *Bos taurus* (Astrup et al. 1968) have a heritable disposition to develop severe PH secondary to high altitude, although the disease typically occurs in about 20 % of steers brought to high altitude (Grover 1965). Systolic PAP in the affected animals is typically 80–85 mmHg, whereas it is 30 mmHg in the resistant animals. Recently genetic studies performed on herds at high altitude have identified single nucleotide polymorphisms and candidate genes which cause and protect against high altitude pulmonary hypertension (Newman et al. 2012).

5.4 Fawn-Hooded Rats

Three major FHR strains are known: (a) PAH prone FHR/EurMcwiCrl, (b) a systemically hypertensive FHR strain (prone to glomerulonephritis,) and (c) a strain prone to depression/substance addiction (Cowley et al. 2004; Brown et al. 1998; Kuijpers and de Jong 1987; van Rodijnen et al. 2002; Kuijpers and Gruys 1984). PAH-prone FHR have normal renal function and no systemic hypertension. FHR spontaneously develop PAH, largely through an increase in HIF-1α, and mitochondrial-metabolic abnormalities reflecting a complex, heritable etiology. The FHR was originally an outbred strain using "German brown" albino rats and Long Evan's rats at the University of Michigan (Tschopp and Weiss 1974; Tobach et al. 1984). In FHR the PAH is heritable, occurring in 68 % of male and female FHR offspring (Kentera et al. 1988).

Method. FHR are allowed to age and monitored periodically using echocardiography for the development of PH. The time to onset of PH varies with altitude. FHR are sensitive to hypoxia and when raised at high altitude demonstrate marked vasoconstrictor response (Nagaoka et al. 2006). The development of PH in FHR is accelerated by mild hypoxia, even at levels that would not cause PH in normal rodents (Le Cras et al. 1999, 2000). In Denver at an elevation of 5,200 ft, FHR animals develop PAH within 1 month of birth (Sato et al. 1992). For FHR raised in Denver, the risk of PH can be decreased by the administration of mild, neonatal hyperbaria, which simulates sea-level PO_2 (Le Cras et al. 2000). In Edmonton, Alberta, at half the altitude, PAH develops much later in life, approximately between 20 and 40 weeks (Bonnet et al. 2006). In Chicago Illinois (sea level), PH does not develop until age 40–60 weeks (Piao et al. 2012).

Features of disease. FHR have increased expression of lung endothelin (ET-1) and decreased endothelial nitric oxide synthase, as is also seen in human PAH (Le Cras et al. 2000; Nagaoka et al. 2001). Platelet abnormalities observed in human PAH are also seen in FHR with PH, along with increased vasoconstrictor response to serotonin (Ashmore et al. 1991). FHR have a normal serotonin transporter gene, despite abnormal serotonin storage (Gonzalez et al. 1998). FHR and Tester-Moriyama rats share pigmentation and platelet disorders, but differ at 62 % of microsatellite markers, indicating their genetic differences (Hamada et al. 1997). Although Tester-Moriyama rats share a similar serotonin platelet disorder, they have no PAH, suggesting that the serotonin pathway does not cause FHR PAH (Nagaoka et al. 2001). The characteristic FHR coat color is also observed to be localized to chromosome 1, as is the defect in the platelet storage pool (Brown et al. 1996). FHR PASMC, like PASMC from human PAH, has an exaggerated predilection to proliferate (Morrell et al. 2009), particularly in response to growth factors, such as Epidermal Growth Factor. This may contribute to their pulmonary vascular remodeling and PAH (Janakidevi et al. 1995). In addition, FHR are hypoxia sensitive, being prone to develop PAH and alveolar simplification (Le Cras et al. 1999, 2000) when exposed to mild hypoxia [at levels that do not affect other rodents (Sato et al. 1992)]. Alveolar simplification (a phenotype common in perinatal lung injury) manifests as fewer and larger alveoli with reduced numbers of septae, loss of small PAs, and decreased capillary density. In neonatal FHR, mild hypoxia that does not adversely affect other rat strains, causes severe PAH that is associated with increased lung endothelin (ET-1) expression (Nagaoka et al. 2001). However, the ET_A receptor-antagonist, BQ-123, only inhibits excessive FHR PASMC proliferation partially and under specific culture conditions [0.3 but not 0.1 or 10 %serum (Zamora et al. 1996)], indicating that excess ET-1 is also not the fundamental cause of FHR's PAH. Also, as in human PAH, the PASMCs in FHR are hyperproliferative and the resulting histologic abnormality is medial hypertrophy of small PAs (Bonnet et al. 2006). FHR have not been reported to spontaneously develop complex vascular lesions.

We believe that much of the hypoxia sensitivity and hyperproliferative phenotype of FHR relates to abnormalities in mitochondrial structure and defects in oxygen sensing that they share with human PAH (Bonnet et al. 2006). Activation

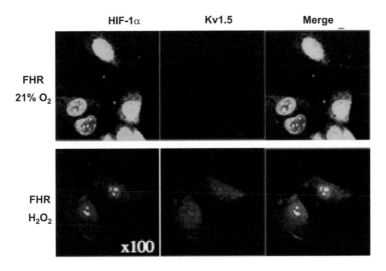

Fig. 6 Disruption in mitochondrial fusion in PAH. 48-h incubation in H_2O_2 reverses nuclear translocation of HIF-1 and restores Kv1.5 (*red*) in FHR PASMCs, consistent with the hypothesis that a loss of mitochondrial ROS production causes the FHR's abnormalities. *iPAH* idiopathic PAH, *BN1* consomic rat. Reproduced with permission from Archer et al. (2008)

of HIF-1α is observed in the PASMC of FHR, even when cell cultures are maintained at 120 mmHg (Archer et al. 2010; Marsboom et al. 2012b; Bonnet et al. 2006) (Fig. 6). This normoxic activation of HIF-1α is associated with fragmentation of mitochondria as well as a decrease in mitochondrial expression of SOD2 (Archer et al. 2010). SOD2 downregulation in FHR is the result of epigenetic silencing of the SOD2 gene. Increased activation of DNA methyltransferases 1 and 3B lead to methylation of two key CpG islands in the promotor and intron 2 (Archer et al. 2010). 5-azacytidine, a DNA methyltransferase inhibitor, restores SOD2 expression by demethylating the CpG islands. With return of SOD2 function, proliferation of FHR PASMC decreases. The mitochondrial fragmentation in FHR (and human PAH) relates in a large part to activation of dynamin-related protein (DRP-1) (Marsboom et al. 2012b). Inhibition of mitochondrial fission in PAH PASMC prevents mitotic division of mitochondria and arrests the cells in G2/M phase of the cell cycle. The resulting growth inhibition plus induction of apoptosis suggest this recently identified abnormality may contribute to PAH in the FHR model (and humans).

It is important to remember that when using the FHR model, one needs to use an appropriate control. The FHR-BN1 consomic rat can be used as this control—these rats are made through introgression of chromosome 1 from the hypoxia resistant Brown Norway rats into an isogenic FHR background (Cowley et al. 2004). Although these animals are genetically identical to FHR except for chromosome 1, there is neither activation of HIF-1α or PDK nor any mitochondrial abnormality and they do not develop PAH. Thus the PAH-initiating gene in FHR is likely present on chromosome 1 (Bonnet et al. 2006).

The RV. The right ventricle in FHR undergoes a glycolytic shift secondary to elevated PA pressures as well as RVH (Piao et al. 2009).

Limitations. The main disadvantage of this model is that the animals develop PH slowly, although this can be expedited by exposing the animals to mild hypoxia. Also, the FHR manifest some degree of alveolar simplification, raising the issue that FHR are not a pure model of PH (Le Cras et al. 1999). Finally, the availability of consomic rats is limited.

5.5 Chronic Hypoxia and Sugen-5416

Sugen-5416 (SU-5416) is a small molecule inhibitor of vascular endothelial growth factor receptor 2 (VEGFR-2) that was developed as a potential anticancer drug, semaxinib. Taraseviciene-Stewart et al. (2001) developed this rat model of PH by combining SU-5416 with chronic hypoxia. The original model was created by subcutaneous injection of 200 mg/kg of SU-5416, three times a week for 3 weeks at Denver's altitude, although subsequently it was found that a lower dose of SU-5416 was sufficient and that more prolonged observation in normoxia (after the hypoxic exposure) lead to the development of complex vascular lesions (Abe et al. 2010).

Method. This model has now evolved and is typically made by one subcutaneous injection of 20 mg/kg of SU-5416 coupled with 3 weeks of hypoxia at 10 % with a return to normoxia for 2–9 weeks. The pulmonary hypertension continues to progress after removal from the hypoxic stimulus. Thus, 5–6 weeks after subcutaneous injection these animals have severe and progressive PAH, with mean PAP ~50 mmHg. By 5–6 weeks, medial hypertrophy develops in all pulmonary arteries and neointimal proliferation is seen in small pulmonary arteries (those <50 μm). A concentric laminar neointima is seen 8 weeks after administration of SU-5416. Plexiform lesions, similar to those in human PAH, develop at approximately 10–12 weeks (Abe et al. 2010).

Features of disease. The model is well tolerated with a much lower mortality than that seen in monocrotaline rats despite the fact that the PA pressures are typically higher in SU-5416 rats (Sakao and Tatsumi 2011). This model has minimal response to vasodilators such as iloprost and thus bears a similarity to human PAH (Oka et al. 2007). The low mortality, the simplicity of creating the model plus its replication of the key features of human PH, particularly plexiform lesions, recommends the Chronic Hypoxia-SU-5416 (CH + SU) model (Fig. 1d).

SU-5416 causes death of PA endothelial cells and this apoptosis is followed by obliteration of precapillary arterial lumen by the emergence of apoptosis-resistant proliferating endothelial cells. This subset of cells with a predisposition to survival is evident by an increase in IGF-1, Bcl-2, and Bcl-X expression (Taraseviciene-Stewart et al. 2001).

The CH + SU model has recently been adapted for mice (Ciuclan et al. 2011). The murine protocol uses a weekly dose of 20 mg/kg SU-5416 subcutaneously for 3 weeks, administered while the mice are in hypoxia (FiO_2 10 %). Although used by a limited number of investigators thus far, this model has identified abnormalities in TGF-β, BMP, and SMAD signaling as well as increases in HIF-1α and IL-6 typical of PAH. CH + SU in mice does result in a more robust increase in RVSP (>45 mmHg) than is seen in other murine models.

Limitations. By its nature, this model depends on VEGF inhibition to achieve its features. However, the role of VEGF inhibition in human PAH is unclear. Additionally, the SU-5416 model has a notable absence of perivascular inflammation, which is observed in other models [monocrotaline (Wilson et al. 1989)] and in human PAH (Price et al. 2012). This may be an important deficiency since monocytes and macrophages and other inflammatory cells may mediate pulmonary vascular remodeling in hypoxic models of PH (Zhao 2010).

6 Transgenic Models

6.1 BMPR-2 Mutation Models

Familial PAH is associated with multiple mutations in the BMPR-2 gene (Lane et al. 2000), but the abnormal genotype is not sufficient to cause PAH in most cases. Carriers of BMPR-2 mutations have only a 20 % risk of developing PAH, indicating that other factors are required to cause PH (Newman et al. 2001). While a complete deletion of BMPR-2 is lethal (see, for example, Beppu et al. 2000), mice with a BMPR-2 gene haploinsufficiency do not develop PH (Liu et al. 2007). To circumvent embryonic lethality of a complete deletion, several genetically modified mice have been developed. We have previously reviewed animal models of BMPR2 extensively in Ryan et al. (2011) and this discussion is based on that review.

7 Global Knockout Mice

Beppu et al. (2000) generated mice in which the BMPR-2 exons 4 and 5 were deleted globally. Homozygous mutant mice died during gastrulation. Beppu et al. (2004) found that the heterozygous mice had mild PH and pulmonary vascular remodeling. However, other investigators did not find PH at baseline in these mice, although increased pulmonary vascular remodeling and increased tone in response to serotonin or overexpression of 5-lipoxygenase are observed (Song et al. 2005; Long et al. 2006).

8 Conditional Knockout Mice

Beppu et al. (2005) also generated mice containing loxP sequences flanking exons 4 and 5, allowing conditional deletion of the transmembrane domain and rendering the resulting protein incapable of initiating downstream signaling. It is feasible to eliminate BMPR-2 signaling in a cell type-specific manner by crossing these mice with mice expressing Cre recombinase in specific cell types. These conditional knockout mice have been used together with a Cre transgene that is selectively expressed in endothelial cells to generate another mouse model of PH (Hong et al. 2008). PH develops in a subset of mice when the BMPR-2 is deleted in mature animals. In those mice with RSVP >30 mmHg, there is medial hypertrophy of small PAs and RVH. When Beppu et al. (2009) used the TIE2-Cre transgene to knockout BMPR-2 in endothelial cells during development, the mice died in utero with extensive cardiac defects.

9 BMPR-2-Dominant Negative Transgenic Mice

These are mice in which human mutant BMPR-2 alleles are overexpressed in a tissue-specific and inducible manner. West et al. (2004) induced expression of a kinase-dead, dominant-negative BMPR-2 mutation in mice. This mutation leads to a loss of the entire intracellular domain. By using a SMC-specific promotor, transgene activity is limited to the SMCs after treatment with the activator doxycycline. Doxycycline was fed to the mothers after giving birth and was continued in the diet until mice reached adulthood. In the transgenic mice, the RVSP was increased to 55–60 mmHg. Modest increased medial thickness of the PAs is observed in this model.

West et al. (2008) then developed a second series of mice with a transgene containing a BMPR-2 with a different mutation observed in familial PAH (a carboxyl terminal stop codon at position 899). Interestingly, when expression of this transgene was induced in SMC, the mice developed PH and extensive pulmonary vascular pruning. Compared to mice with a mutation in the kinase domain, this study suggests that it is worse to have a dysfunctional BMPR-2 (only missing the regulatory carboxy terminal domain) (West et al. 2008) than have an almost complete deletion of BMPR-2 (West et al. 2004). Overexpression of this dominant-negative BMPR-2 gene is associated with downregulation of the voltage-gated potassium channel, Kv1.5, and causes PA vasoconstriction. Nifedipine, a selective L-type calcium channel blocker, reduces PH in these mice. This suggests that activation of L-type calcium channels, caused by reduced expression of Kv1.5 channels and membrane depolarization, mediates PH in these mice (Young et al. 2006). A concern with these transgenic models is that mutated

BMPR-2 isoforms are expressed at supraphysiological levels, thereby potentially disrupting the entire BMP signaling system.

10 Short Hairpin (shRNA) BMPR-2 Knockdown Mice

Liu et al. (2007) generated mice in which BMPR-2 gene expression was globally reduced, using a shRNA-based strategy. Silencing BMPR-2 expression by RNA interference does not cause PAH (although the knockdown of BMPR-2 is approximately 90 %) but instead causes gastrointestinal hyperplasia, incomplete mural cell coverage of vessel walls, and mucosal hemorrhage. It appears that BMP receptor signaling regulates angiogenesis by maintaining the expression of endothelial guidance molecules and if this is disrupted, vessels are malformed. However, there remain moderate amounts of BMPR2 in the lungs of these animals, thereby suggesting that downregulation of BMPR2 alone is not sufficient to induce PAH, in line with the aforementioned conditional knockout mice (Beppu et al. 2005).

10.1 5-HTT Overexpressing Mice

Serotonin (5-HT) increases proliferation of PASMC and migration of fibroblasts and causes pulmonary vasoconstriction. Expression of the serotonin transporter (5-HTT) is increased in Group 1 PH (Eddahibi et al. 2001). Transgenic overexpression of 5-HTT has been achieved in mice both globally (MacLean et al. 2004) and in SMC-specific (Guignabert et al. 2006) models to assess the role this transporter plays in PAH.

Method. When 5-HTT transgenic mice are exposed to hypobaric hypoxia for 14–28 days, the RVSP increases to mPAP 45–50 mmHg. When bred in normoxia, these 5-HTT transgenic mice have been found to have elevated RVSP (mPAP 30–35 mmHg) (MacLean et al. 2004). The SMC-specific, 5-HTT transgenic mouse, is created using a SM22 promoter, and RVSP increases to 30–35 mmHg in this model after 15 days of 10 % O_2 (Guignabert et al. 2006).

Features of model. Hypoxic pulmonary vasoconstriction is decreased in this model. In the hypoxia treated transgenic mice, the distal PAs become more muscularized and there is an increase in proliferation of PASMCs. Curiously, monocrotaline, which as mentioned earlier does not normally cause PH in mice, causes an increase in RVSP to 25–35 mmHg when administered to these transgenic mice (5 mg/kg subcutaneously).

Limitations. The main limitation of this model is that both the PH and histological changes are not severe.

10.2 Models That Require Further Study

Whereas the animal models discussed heretofore have been studied by many investigators and are quite reproducible, there are a number of intriguing, but less widely studied, models that merit mention. These models require further study to determine their relevance as PH models.

10.3 Broiler Fowl Disease

Although not a transgenic model, PAH spontaneously develops in 3 % of young domestic fowl bred for meat production (broiler chickens, broilers) when raised in conditions that promote maximal growth. In this setting the rapid growth is accompanied by increases in cardiac output that enters lungs which have not increased in size as quickly as the increase in cardiac output (Wideman 2000). Relevance of this model to 5-HTT overexpressing mice is suggested by the observed increase serotonin in these birds (Chapman et al. 2008). Although too sporadic to make a robust model of PH without some modification, broiler fowl remains the only animal "model" that spontaneously develops plexiform lesions (Wideman and Hamal 2011).

10.4 Endothelin B Receptor Deficiency in Rats

Endothelin-1 (ET-1) interacts with endothelin B (ET_B) receptors in the endothelium, which causes pulmonary vasodilatation through increased prostacyclin and nitric oxide production. ET-1 is increased in human PAH (Giaid et al. 1993) and in the chronic hypoxic PH model, ET_B receptors and endothelin-1 levels are increased in the lungs (Li et al. 1994). The ET_B receptor deficiency model is based on the molecular elimination for a compensatory vasodilator pathway (ET_B-mediated vasodilatation).

Method. Exposing ET_B-deficient rats [ET_B (sl/sl)] to monocrotaline (60 mg/kg subcutaneously at 4–6 weeks of age) results in neointimal lesions, similar to those observed in human PAH (Ivy et al. 2005).

Features of model. In ET_B-deficient rats exposed to monocrotaline, mean PAP increases to 40–45 mmHg, with a reduction in cardiac output to 30 ml/min. The total pulmonary resistance in monocrotaline-treated, ET_B-receptor deficient rats increases considerably from a baseline of 400 mmHg/l/min to 1,400 mmHg/ml/min. Thus, deficiency of the ET_B receptor accelerates progression of monocrotaline-induced PAH in rats. This suggests an antiproliferative effect of the ET_B receptor in pulmonary vascular homeostasis and is another example of how multiple insults (removal of a compensatory mechanism) exacerbate experimental PH. Significant perivascular inflammation is observed in this model.

Limitations of model. The biggest limitation of this model is that these rats are not readily available for purchase. Also, plexiform lesions are not seen in this model (Ivy et al. 2002).

10.5 Angiopoietin-1 Overexpression in Rats

Angiopoietin-1 (Ang-1) plays a controversial role in PAH with two groups reporting opposite findings. Thistlewaithe et al. reported that expression of angiopoietin-1 and phosphorylation of Tie2 (the endothelial-specific receptor of Ang-1) are upregulated in the lungs of patients with a variety of forms of PH (Chu et al. 2004). It is noted that Ang-1 expression correlated directly with the severity of PH (Du et al. 2003), supporting their contention that overexpression of Ang leads to PH in rodents. In contrast, Stewart et al. found that decreased activity of the Ang-Tie2 pathway worsens PAH by decreasing endothelial cell-survival signaling (Zhao et al. 2003). They note that transfer of the Ang-1 gene improves PAH (Kugathasan et al. 2005). This controversy needs to be resolved before Ang-overexpression can be used widely as a PAH model.

Method. To create a rat model of PH through overexpression of Ang-1, 2×10^{10} genomic particles of adeno-associated virus-angiopoietin-1 (AAV-Ang-1) are injected into the RV outflow track of anesthetized Fischer rats at 12 weeks of age (Chu et al. 2004). Rats are subsequently sacrificed 1–2 months later. The average survival of these rats, if not sacrificed, is approximately 5–6 months after injection.

Features of model. In the animals that express high levels of Ang-1 protein, PH is observed and histological changes such as medial hypertrophy and muscular hyperplasia of the small PAs. RVSP typically reaches 50 mmHg in the Ang-1 overexpressing rats. There are no complex lesions and the endothelium does not seem to be adversely affected.

Limitations. The inflammatory changes, which are often observed in the monocrotaline model and in human PAH, are not seen in this model. The complex histologic lesions seen in the CH + SU5416 model and in human PAH are not seen. Also, the role of Ang-1 in human PAH remains undefined and the model has not been replicated, thus limiting its applicability at this time.

10.6 High Fat Diet to Apo E Knockout Mice

Metabolic syndrome in humans has been identified as a risk factor for PH (Robbins et al. 2009) and thus ApoE knockout mice have been studied to determine if the high fat diet can also cause a PH-syndrome (Hansmann et al. 2007). In this model, starting at 4 weeks of age, ApoE knockout mice are fed high fat chow for 11 weeks (Pendse et al. 2009). RVSP increases to 30 mmHg and the pulmonary arteries are observed to develop mild medial hypertrophy, however, no plexiform lesions are observed. Another limitation of this model is that the male rats develop the bulk of the PH, which contrasts with the female preponderance observed in humans.

11 Inflammatory Models

Several lines of evidence indicate that inflammation plays an important role in the development of PH (Hassoun et al. 2009). For example, autoimmune diseases such as scleroderma are a common cause of PAH and HIV infection is also associated with PAH. In addition, schistosomiasis contains an important inflammatory component. While inflammation is induced as a component of the monocrotaline model, a number of models exist that specifically look at how inflammation influences the development of PAH.

11.1 γ-Herpesvirus Reactivation Model

The importance of viral infections in the development of PAH has been suggested by the identification of human herpes virus 8 (HHV-8) in the lungs of PAH patients (Cool et al. 2003), but this finding is somewhat controversial because other authors have not found signs of HHV-8 infection in PAH (Valmary et al. 2011). Nevertheless, the murine γ-Herpesvirus 68, which resembles HHV-8, leads to increased perivascular inflammation and vascular remodeling. However, this vascular remodeling only occurs in transgenic mice overexpressing the small calcium binding protein S100A4 and even in these mice no significant increase in RV pressures was observed (Greenway et al. 2004; Spiekerkoetter et al. 2008).

Method. γ-Herpesvirus 68 (γHV68) is administered to S100A4/Mts mice at 1 year of age. The model does not result in PH if the animals are inoculated at less than 3 months of age.

Features of model. Using this late inoculation, the animals develop neointimal hyperplasia, which is maintained (Fig. 7A). However, the increase in RVSP, which is mild 6 weeks after administration of virus (RVPS ~ 30 mmHg) is no longer evident 3 months later. In order to cause persistent PH, a mutant M1-γHV68 strain can be administered, which has a fivefold higher efficiency in reactivation and causes neointimal proliferation in all S100A4/Mts1 mice.

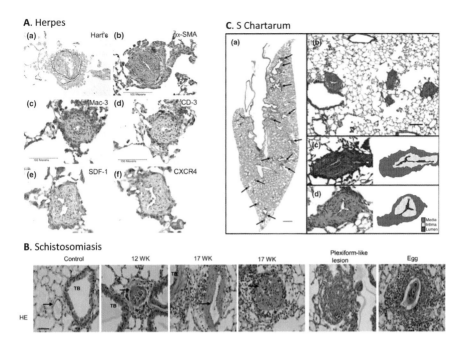

Fig. 7 (**A**) Herpes-γ: the PVD lesion is characterized by fragmented elastin, a neointima composed largely of α-smooth muscle actin (α-SMA)-positive cells, and a perivascular inflammatory infiltrate consisting of macrophages and T cells. Representative PVD lesions from 1.5-year-old S100A4/Mts1 mice were characterized by Hart's stain showing fragmented elastin fibers (**a**) and positive immunohistochemistry for α-SMA (**b**). In the perivascular region surrounding the PVD lesion and occasionally within the vessel wall, cells show positive immunoreactivity for the macrophage marker Mac-3 (**c**), the T lymphocyte marker CD-3 (**d**), stromal cell-derived factor-1 (SDF-1; **e**), and its cognate receptor, CXCR4 (**f**). Original magnification ×400. *Bars*: 100 μm. Reproduced with permission from Spiekerkoetter et al. (2008). (**B**) Schistosomiasis—representative photomicrographs of serial lung sections from control mice and mice chronically infected with *S. mansoni* at 12, 17, and 20 weeks. Reproduced with permission from Crosby et al. (2010). (**C**) Histopathological findings in the lungs of a mouse from the 12-week exposure of *Stachybotrys chartarum* at 1×10^4 spores/mouse (18 times) and were sacrificed 1 day after the last injection. (**a**) Pulmonary arteries with thickened walls were diffusely distributed in the lung (haematoxylin and eosin stain). *Arrows* indicate representative pulmonary arteries with stenotic changes. (**b**) Symmetrical thickening of the intima and media of the vascular wall is demonstrated at the periphery of the pulmonary arteries (elastica Masson's double stain). (**c**) Intima and media thickening of the vascular wall caused pulmonary artery stenosis (elastica Masson's double stain). (**d**) Myointimal cell proliferation in the intima caused symmetric thickening of the arterial wall and extensive stenosis (haematoxylin and eosin stain). *Scale bars*, 500 μm (**a**), 100 μm (**b**), 10 μm (**c, d**). Reproduced with permission from Ochiai et al. (2008)

Limitations of model. The hemodynamic changes observed in this model remain modest and the complexity required to generate PH suggest that much is left to understand before this could be used widely as a murine PH model.

11.2 Schistosomiasis

Parasitic worms from the *Schistosoma* genus have a complicated life cycle which needs two different hosts to complete (one being a mammal or bird and another being a specific kind of snail). Eggs released in the urine or feces of infected humans or animals will infect snails, which in turn leads to the formation of cercariae, which are able to swim and penetrate the skin of humans. Therefore, poor sanitation is responsible for maintaining this cycle, explaining why schistosomiasis almost exclusively occurs in developing countries (Graham et al. 2010). From a global perspective, Schistosomiasis is a common cause of PH, especially in underdeveloped countries. While the mechanism of the disease process is still not fully defined, it has been suggested that egg embolism, increased pulmonary blood flow, and especially inflammation can contribute to the onset of PAH (Graham et al. 2010). Currently, there is only limited knowledge about whether therapies used in other patients with PAH will have therapeutic benefit in PAH associated with schistosomiasis.

Method. The disease process and the subsequent PH can be mimicked in an animal model (Crosby et al. 2010) by transcutaneously infecting female C57/BL6 adult mice with a suspension of approximately 30 cercariae of the Puerto Rican strain of *S. mansoni*.

Features of model. After approximately 5 weeks, the worms mature and start making eggs. Since worms preferentially inhabit the portal veins, initially eggs will accumulate in the liver. However, eggs ultimately appear in the lungs and lead to extensive inflammation with severe thickening of the media of the small PA and plexiform lesions from 12 weeks onwards (Fig. 7B).

Limitations of model. Despite the success in generating the classical histological features of human PH, these animals do not develop PH or RVH. Further study is required to understand the lack of PH in this model.

11.3 Stachybotrys chartarum

Although it is unclear if *Stachybotrys chartarum* causes PH in humans, repeated intratracheal injected of 10,000 spores of this fungus into mice 6–18 times over 4–12 weeks has been shown after 4 weeks to cause mild increases in PA pressures as well as medial hypertrophy and intimal thickness (Fig. 7C) (Ochiai et al. 2008). RVH is observed at 3 months. The RVSP measured in this mouse model is mild (20 mmHg compared to normal values at approximately 10 mmHg). However, the combination of chronic inflammation with histological findings of pulmonary vascular remodeling makes this potentially an attractive model, worthy of further study.

11.4 HIV-Associated Pulmonary Hypertension

HIV associated with pulmonary hypertension has been observed since the late 1980s (Kim and Factor 1987), but it remains unclear how infection leads to an increased risk for PAH. In humans with HIV the prevalence is low (0.46 %) (Sitbon et al. 2008).

Method. A nonhuman primate models of HIV-1 infection can be created with simian immunodeficiency virus (SIV) or using a HIV-1 nef gene, which is crucial for maintaining a high viral load by downregulating CD4 (Marecki et al. 2006). Alternatively, HIV-1 transgenic rats can be studied which develop increased RVSP and PA remodeling (Lund et al. 2011).

Features of model. Both SIV (Chalifoux et al. 1992) and SIV modified to overexpress nef (Marecki et al. 2006) can recapitulate many of the lung pathologies observed in patients with PH secondary to HIV, although with limited or no hemodynamics changes (Marecki et al. 2006). In this model, complex, plexiform lesions, coupled with luminal obliteration, medial hypertrophy, and thrombosis have been observed in SIV-*nef* animals, but not in SIV infected animals. There is a high prevalence of mononuclear cells in this model supportive of the inflammatory role played by HIV (Humbert 2008).

Limitations of model. Monkeys are very costly to house. Moreover, the exact hemodynamics in this model remain undefined. A fundamental limitation is likely the low prevalence of PAH even in humans with HIV (Sitbon et al. 2008).

11.5 Radiation Exposure

Recently a novel animal model of PAH was developed in the form of a lung injury from radiation (Ghobadi et al. 2012).

Method. Male adult Wister rats were exposed to 150 MeV protons from a cyclotron, using a shoot-through technique so that rats' lungs were irradiated at different doses (Fig. 8). Eight weeks after lung irradiation, the animals were sacrificed and hemodynamics were measured.

Features. Lung irradiation caused PH (systolic PAP ranged from 30 to 60 mmHg). The severity of the hemodynamic changes of PH was proportional to the irradiated lung volume. Histologically, lung irradiation caused medial hypertrophy, adventitial thickening, and obliteration of small pulmonary vessels. These features were also seen out of the field of radiation. This is an interesting model of Group 1 PH, although replication by other groups is required.

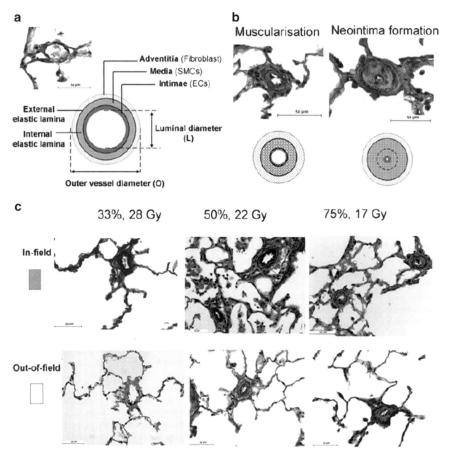

Fig. 8 Early radiation response of lung vasculature leads to global pulmonary vascular remodeling. (**a**) Anatomy of nonirradiated control pulmonary intra-acinar vessel (<50 mm). Pulmonary vascular remodeling was assessed by immunohistochemical staining (Verhoeff's elastica stain) of pulmonary intra-acinar vessels 8 weeks postirradiation of 33, 50, and 75 % of rats' lungs to 28, 22, and 17 Gy, respectively. (**b**) Lung irradiation induces vascular remodeling such as muscularization, adventitia thickening, and neointima formation. (**c**) Vascular remodeling was present infield and out-of-field showing a global pulmonary vascular response in all irradiated volumes. *L* luminal diameter, *O* outer vessel diameter, *SMC* smooth muscle cell. Magnification ×400. Ghobadi G, Bartelds B, van der Veen SJ et al (2011) Thorax

Limitations. Irradiation has an effect on both vasculature as well as airway tissue thus this is not a pure Group 1 PH model. The exact mechanism through which it achieves it changes remain unclear as this model has not been studied extensively, but primarily the changes appear mediated through endothelial cell damage.

12 Surgical Models

Surgical models provide insight in particular to congenital cardiac disease, which are an important cause of Group 1 PH. They provide valuable hemodynamic information as well as insight into adaptive and maladaptive response of the right ventricle (Piao et al. 2010b). Increased pulmonary blood flow leads to increased shear stress on the vascular wall and smooth muscle cell hypertrophy and proliferation (Buus et al. 2001). Since right ventricular failure is an important cause of death in PAH patients, it is noteworthy that different models have been developed to induce RVH in the absence of pulmonary disease. Both pressure overload (by pulmonary artery banding) and volume overload (by shunting blood from the aorta to the vena cava) have been used to study RV hypertrophy.

12.1 Pulmonary Artery Banding Model

Prognosis and functional status in congenital heart disease is determined to a large extent by right ventricular function. Pulmonary artery banding (PAB) replicates pulmonic stenosis (either isolated or after tetralogy of Fallot repair) and does not cause pulmonary hypertension. As such it is a useful control model to separate pulmonary disease from cardiac disease.

Methods of model. To create this model, under anesthesia, via a limited median sternotomy, the main pulmonary artery is dissected from the ascending aorta. A small needle [16G in rats (Fang et al. 2012), 23G in mice (Bartelds et al. 2011)] is then placed parallel to the main PA and ligated with a 4-0 silk suture (7-0 in mouse model). Withdrawal of the needle then creates a fixed PA stenosis.

Features of model. The PAB model demonstrates a decrease in cardiac output and exercise performance. Right ventricular hypertrophy develops 3–4 weeks after banding. This model has been useful to study right ventricular metabolism in the pressure-overloaded right ventricle (Fang et al. 2012). Using this model, it has been observed that fatty acid oxidation (FAO) is increased in the right ventricle and inhibiting FAO in PAB rats increased right ventricle ATP production and improved exercise capacity, as well as RV performance.

Limitations of model. As with all the surgical models, technical expertise is required to develop a consistent disease in animals.

12.2 Aorto-Caval Shunt

Aorto-caval shunt is a successful model of RV volume overload.

Methods of model. To create this model in mice, through a midline laparotomy, a 25G needle is advanced through the abdominal aorta toward the inferior vena cava. The distal vein is compressed throughout the procedure. Upon removal of the needle, tissue glue is used to close the aortic puncture site.

Features of model. When studied in mice, aorto-caval shunting develop a similar extent of RVH to PAB (Bartelds et al. 2011). Aorto-caval shunting can also be combined with monocrotaline to further accentuate the PA pressures in addition to PA flow. To create this model, 7 days after monocrotaline is administered (60 mg/kg), aorto-caval shunt surgery is performed (van Albada et al. 2005). Compared to aorto-caval alone or monocrotaline alone, MCT plus aorto-caval shunt develop functionally more severe PH with increased mortality over 5–6 weeks, although the pressures observed in MCT and MCT plus aorto-caval shunt are similar (systolic PAP 55–60 mmHg) as is the extent of neo-intimal proliferation and muscularization. The combined insult does result in a greater amount of RVH and RV dilatation. This further highlights the vital impact of right ventricular function on morbidity and mortality.

Limitations of model. This technique, although reproducible and rapid, has been associated with a 10 % mortality at 24 h postsurgery (Garcia and Diebold 1990).

13 Summary

When determining which animal model of Group 1 PH to study, the following characteristics are important to consider:

1. Accurate representation of human PAH
2. Hemodynamic changes
3. Pathologic changes, including inflammation
4. Ease, cost and reproducibility of the model
5. Cellular or molecular question being addressed
6. Local expertise and resources

Group 1 PH is not a uniform disease and a single model cannot successfully capture the all the features of disparate diseases within this group. Comprehensive assessment using advanced imaging, catheterization, and molecular characterization are essential. Adequate trial design and selection of appropriate endpoints are as important as the model itself in obtaining results that are relevant to human PAH.

References

Abe K, Toba M, Alzoubi A et al (2010) Formation of plexiform lesions in experimental severe pulmonary arterial hypertension. Circulation 121:2747–2754

Alexander AF, Jensen R (1963) Pulmonary vascular pathology of high altitude-induced pulmonary hypertension in cattle. Am J Vet Res 24:1112–1122

Archer S, Rich S (2000) Primary pulmonary hypertension: a vascular biology and translational research "work in progress". Circulation 102:2781–2791

Archer SL, Gebhard R, Levine A, Prigge W, Weir EK (1990) Exacerbation of monocrotaline pulmonary hypertension by a high fat diet: association with increased production lung levels of activated oxygen species. J Vasc Med Biol 2:125–135

Archer SL, Souil E, Dinh-Xuan AT et al (1998) Molecular identification of the role of voltage-gated K+ channels, Kv1.5 and Kv2.1, in hypoxic pulmonary vasoconstriction and control of resting membrane potential in rat pulmonary artery myocytes. J Clin Invest 101:2319–2330

Archer SL, Gomberg-Maitland M, Maitland ML, Rich S, Garcia JG, Weir EK (2008) Mitochondrial metabolism, redox signaling, and fusion: a mitochondria-ROS-HIF-1alpha-Kv1.5 O2-sensing pathway at the intersection of pulmonary hypertension and cancer. Am J Physiol Heart Circ Physiol 294:H570–H578

Archer SL, Marsboom G, Kim GH et al (2010) Epigenetic attenuation of mitochondrial superoxide dismutase 2 in pulmonary arterial hypertension: a basis for excessive cell proliferation and a new therapeutic target. Circulation 121:2661–2671

Ashmore RC, Rodman DM, Sato K et al (1991) Paradoxical constriction to platelets by arteries from rats with pulmonary hypertension. Am J Physiol 260:H1929–H1934

Astrup T, Glas P, Kok P (1968) Lung fibrinolytic activity and bovine high mountain disease. Proc Soc Exp Biol Med 127:373–377

Bartelds B, Borgdorff MA, Smit-van Oosten A (2011) Differential responses of the right ventricle to abnormal loading conditions in mice: pressure vs. volume load. Eur J Heart Fail 13:1275–1282

Beppu H, Kawabata M, Hamamoto T et al (2000) BMP type II receptor is required for gastrulation and early development of mouse embryos. Dev Biol 221:249–258

Beppu H, Ichinose F, Kawai N et al (2004) BMPR-II heterozygous mice have mild pulmonary hypertension and an impaired pulmonary vascular remodeling response to prolonged hypoxia. Am J Physiol Lung Cell Mol Physiol 287:L1241–L1247

Beppu H, Lei H, Bloch KD, Li E (2005) Generation of a floxed allele of the mouse BMP type II receptor gene. Genesis 41:133–137

Beppu H, Malhotra R, Beppu Y, Lepore JJ, Parmacek MS, Bloch KD (2009) BMP type II receptor regulates positioning of outflow tract and remodeling of atrioventricular cushion during cardiogenesis. Dev Biol 331:167–175

Bonnet S, Michelakis ED, Porter CJ et al (2006) An abnormal mitochondrial-hypoxia inducible factor-1alpha-Kv channel pathway disrupts oxygen sensing and triggers pulmonary arterial hypertension in fawn hooded rats: similarities to human pulmonary arterial hypertension. Circulation 113:2630–2641

Bonnet S, Archer SL, Allalunis-Turner J et al (2007a) A mitochondria-K+ channel axis is suppressed in cancer and its normalization promotes apoptosis and inhibits cancer growth. Cancer Cell 11:37–51

Bonnet S, Rochefort G, Sutendra G et al (2007b) The nuclear factor of activated T cells in pulmonary arterial hypertension can be therapeutically targeted. Proc Natl Acad Sci USA 104:11418–11423

Bras G, Jelliffe DB, Stuart KL (1954) Veno-occlusive disease of liver with nonportal type of cirrhosis, occurring in Jamaica. AMA Arch Pathol 57:285–300

Brown DM, Provoost AP, Daly MJ, Lander ES, Jacob HJ (1996) Renal disease susceptibility and hypertension are under independent genetic control in the fawn-hooded rat. Nat Genet 12:44–51

Brown DM, Van Dokkum RP, Korte MR et al (1998) Genetic control of susceptibility for renal damage in hypertensive fawn-hooded rats. Ren Fail 20:407–411

Buermans HP, Redout EM, Schiel AE et al (2005) Microarray analysis reveals pivotal divergent mRNA expression profiles early in the development of either compensated ventricular hypertrophy or heart failure. Physiol Genomics 21:314–323

Bull TM, Coldren CD, Geraci MW, Voelkel NF (2007) Gene expression profiling in pulmonary hypertension. Proc Am Thorac Soc 4:117–120

Burke DL, Frid MG, Kunrath CL et al (2009) Sustained hypoxia promotes the development of a pulmonary artery-specific chronic inflammatory microenvironment. Am J Physiol Lung Cell Mol Physiol 297:L238–L250

Buus CL, Pourageaud F, Fazzi GE, Janssen G, Mulvany MJ, De Mey JG (2001) Smooth muscle cell changes during flow-related remodeling of rat mesenteric resistance arteries. Circ Res 89:180–186

Cahill E, Costello CM, Rowan SC et al (2012) Gremlin plays a key role in the pathogenesis of pulmonary hypertension. Circulation 125:920–930

Califf RM, Adams KF, McKenna WJ et al (1997) A randomized controlled trial of epoprostenol therapy for severe congestive heart failure: the Flolan International Randomized Survival Trial (FIRST). Am Heart J 134:44–54

Chalifoux LV, Simon MA, Pauley DR, MacKey JJ, Wyand MS, Ringler DJ (1992) Arteriopathy in macaques infected with simian immunodeficiency virus. Lab Invest 67:338–349

Chapman ME, Taylor RL, Wideman RF Jr (2008) Analysis of plasma serotonin levels and hemodynamic responses following chronic serotonin infusion in broilers challenged with bacterial lipopolysaccharide and microparticles. Poult Sci 87:116–124

Chen Y, Guo H, Xu D et al (2012) Left ventricular failure produces profound lung remodeling and pulmonary hypertension in mice: heart failure causes severe lung disease. Hypertension 59 (6):1170–1178

Chesney CF, Allen JR (1973) Animal model: pulmonary hypertension, cor pulmonale and endocardial fibroelastosis in monocrotaline-intoxicated nonhuman primates. Am J Pathol 70:489–492

Christman BW, McPherson CD, Newman JH et al (1992) An imbalance between the excretion of thromboxane and prostacyclin metabolites in pulmonary hypertension. N Engl J Med 327:70–75

Chu D, Sullivan CC, Du L et al (2004) A new animal model for pulmonary hypertension based on the overexpression of a single gene, angiopoietin-1. Ann Thorac Surg 77:449–456, discussion 56–57

Ciuclan L, Bonneau O, Hussey M et al (2011) A novel murine model of severe pulmonary arterial hypertension. Am J Respir Crit Care Med 184:1171–1182

Cool CD, Kennedy D, Voelkel NF, Tuder RM (1997) Pathogenesis and evolution of plexiform lesions in pulmonary hypertension associated with scleroderma and human immunodeficiency virus infection. Hum Pathol 28:434–442

Cool CD, Rai PR, Yeager ME et al (2003) Expression of human herpesvirus 8 in primary pulmonary hypertension. N Engl J Med 349:1113–1122

Cowley AW Jr, Liang M, Roman RJ, Greene AS, Jacob HJ (2004) Consomic rat model systems for physiological genomics. Acta Physiol Scand 181:585–592

Crosby A, Jones FM, Southwood M et al (2010) Pulmonary vascular remodeling correlates with lung eggs and cytokines in murine schistosomiasis. Am J Respir Crit Care Med 181:279–288

Davie NJ, Gerasimovskaya EV, Hofmeister SE et al (2006) Pulmonary artery adventitial fibroblasts cooperate with vasa vasorum endothelial cells to regulate vasa vasorum neovascularization: a process mediated by hypoxia and endothelin-1. Am J Pathol 168:1793–1807

Du L, Sullivan CC, Chu D et al (2003) Signaling molecules in nonfamilial pulmonary hypertension. N Engl J Med 348:500–509

Dumitrascu R, Koebrich S, Dony E et al (2008) Characterization of a murine model of monocrotaline pyrrole-induced acute lung injury. BMC Pulm Med 8:25

Eddahibi S, Humbert M, Fadel E et al (2001) Serotonin transporter overexpression is responsible for pulmonary artery smooth muscle hyperplasia in primary pulmonary hypertension. J Clin Invest 108:1141–1150

Elwing J, Panos RJ (2008) Pulmonary hypertension associated with COPD. Int J Chron Obstruct Pulmon Dis 3:55–70

Epstein RB, Min KW, Anderson SL, Syzek L (1992) A canine model for hepatic venoocclusive disease. Transplantation 54:12–16

Fang YH, Piao L, Hong Z et al (2012) Therapeutic inhibition of fatty acid oxidation in right ventricular hypertrophy: exploiting Randle's cycle. J Mol Med (Berl) 90:31–43

Fijalkowska I, Xu W, Comhair SA et al (2010) Hypoxia inducible-factor1alpha regulates the metabolic shift of pulmonary hypertensive endothelial cells. Am J Pathol 176:1130–1138

Forfia PR, Fisher MR, Mathai SC et al (2006) Tricuspid annular displacement predicts survival in pulmonary hypertension. Am J Respir Crit Care Med 174:1034–1041

Frid MG, Brunetti JA, Burke DL et al (2006) Hypoxia-induced pulmonary vascular remodeling requires recruitment of circulating mesenchymal precursors of a monocyte/macrophage lineage. Am J Pathol 168:659–669

Garcia R, Diebold S (1990) Simple, rapid, and effective method of producing aortocaval shunts in the rat. Cardiovasc Res 24:430–432

Geraci MW, Moore M, Gesell T et al (2001) Gene expression patterns in the lungs of patients with primary pulmonary hypertension: a gene microarray analysis. Circ Res 88:555–562

Ghobadi G, Bartelds B, van der Veen SJ et al (2012) Lung irradiation induces pulmonary vascular remodelling resembling pulmonary arterial hypertension. Thorax 67:334–341

Giaid A, Yanagisawa M, Langleben D et al (1993) Expression of endothelin-1 in the lungs of patients with pulmonary hypertension. N Engl J Med 328:1732–1739

Girgis RE, Li D, Zhan X et al (2003) Attenuation of chronic hypoxic pulmonary hypertension by simvastatin. Am J Physiol Heart Circ Physiol 285:H938–H945

Gomez-Arroyo JG, Farkas L, Alhussaini AA et al (2012) The monocrotaline model of pulmonary hypertension in perspective. Am J Physiol Lung Cell Mol Physiol 302:L363–L369

Gonzalez AM, Smith AP, Emery CJ, Higenbottam TW (1998) The pulmonary hypertensive fawn-hooded rat has a normal serotonin transporter coding sequence. Am J Respir Cell Mol Biol 19:245–249

Graham BB, Bandeira AP, Morrell NW, Butrous G, Tuder RM (2010) Schistosomiasis-associated pulmonary hypertension: pulmonary vascular disease: the global perspective. Chest 137:20S–29S

Greenway S, van Suylen RJ, Du Marchie Sarvaas G et al (2004) S100A4/Mts1 produces murine pulmonary artery changes resembling plexogenic arteriopathy and is increased in human plexogenic arteriopathy. Am J Pathol 164:253–262

Grover RF (1965) Pulmonary circulation in animals and man at high altitude. Ann NY Acad Sci 127:632–639

Grover RF, Reeves JT (1962) Experimental induction of pulmonary hypertension in normal steers at high altitude. Med Thorac 19:543–550

Guignabert C, Izikki M, Tu LI et al (2006) Transgenic mice overexpressing the 5-hydroxytryptamine transporter gene in smooth muscle develop pulmonary hypertension. Circ Res 98:1323–1330

Guignabert C, Tu L, Izikki M et al (2009) Dichloroacetate treatment partially regresses established pulmonary hypertension in mice with SM22alpha-targeted overexpression of the serotonin transporter. FASEB J 23:4135–4147

Gust R, Schuster DP (2001) Vascular remodeling in experimentally induced subacute canine pulmonary hypertension. Exp Lung Res 27:1–12

Hamada S, Nishikawa T, Yokoi N, Serikawa T (1997) TM rats: a model for platelet storage pool deficiency. Exp Anim 46:235–239

Hansmann G, Wagner RA, Schellong S et al (2007) Pulmonary arterial hypertension is linked to insulin resistance and reversed by peroxisome proliferator-activated receptor-gamma activation. Circulation 115:1275–1284

Hassoun PM, Mouthon L, Barbera JA et al (2009) Inflammation, growth factors, and pulmonary vascular remodeling. J Am Coll Cardiol 54:S10–S19

Herve P, Launay JM, Scrobohaci ML et al (1995) Increased plasma serotonin in primary pulmonary hypertension. Am J Med 99:249–254

Hong KH, Lee YJ, Lee E et al (2008) Genetic ablation of the BMPR2 gene in pulmonary endothelium is sufficient to predispose to pulmonary arterial hypertension. Circulation 118:722–730

Hoshikawa Y, Nana-Sinkam P, Moore MD et al (2003) Hypoxia induces different genes in the lungs of rats compared with mice. Physiol Genomics 12:209–219

Huang J, Wolk JH, Gewitz MH, Mathew R (2010) Progressive endothelial cell damage in an inflammatory model of pulmonary hypertension. Exp Lung Res 36:57–66

Humbert M (2008) Mediators involved in HIV-related pulmonary arterial hypertension. AIDS 22 (Suppl 3):S41–S47

Humbert M, Sitbon O, Chaouat A et al (2006) Pulmonary arterial hypertension in France: results from a national registry. Am J Respir Crit Care Med 173:1023–1030

Hyvelin JM, Howell K, Nichol A, Costello CM, Preston RJ, McLoughlin P (2005) Inhibition of Rho-kinase attenuates hypoxia-induced angiogenesis in the pulmonary circulation. Circ Res 97:185–191

Ivy DD, Yanagisawa M, Gariepy CE, Gebb SA, Colvin KL, McMurtry IF (2002) Exaggerated hypoxic pulmonary hypertension in endothelin B receptor-deficient rats. Am J Physiol Lung Cell Mol Physiol 282:L703–L712

Ivy DD, McMurtry IF, Colvin K et al (2005) Development of occlusive neointimal lesions in distal pulmonary arteries of endothelin B receptor-deficient rats: a new model of severe pulmonary arterial hypertension. Circulation 111:2988–2996

Janakidevi K, Tiruppathi C, Del Vecchio PJ, Pinheiro JM, Malik AB (1995) Growth characteristics of pulmonary artery smooth muscle cells from fawn-hooded rats. Am J Physiol 268:L465–L470

Jones JE, Mendes L, Rudd MA, Russo G, Loscalzo J, Zhang YY (2002) Serial noninvasive assessment of progressive pulmonary hypertension in a rat model. Am J Physiol Heart Circ Physiol 283:H364–H371

Kawut SM, Bagiella E, Lederer DJ et al (2011) Randomized clinical trial of aspirin and simvastatin for pulmonary arterial hypertension: ASA-STAT. Circulation 123:2985–2993

Kay JM, Harris P, Heath D (1967) Pulmonary hypertension produced in rats by ingestion of Crotalaria spectabilis seeds. Thorax 22:176–179

Kentera D, Susic D, Veljkovic V, Tucakovic G, Koko V (1988) Pulmonary artery pressure in rats with hereditary platelet function defect. Respiration 54:110–114

Kim KK, Factor SM (1987) Membranoproliferative glomerulonephritis and plexogenic pulmonary arteriopathy in a homosexual man with acquired immunodeficiency syndrome. Hum Pathol 18:1293–1296

Kim JW, Tchernyshyov I, Semenza GL, Dang CV (2006) HIF-1-mediated expression of pyruvate dehydrogenase kinase: a metabolic switch required for cellular adaptation to hypoxia. Cell Metab 3:177–185

Knab AM, Bowen RS, Moore-Harrison T, Hamilton AT, Turner MJ, Lightfoot JT (2009) Repeatability of exercise behaviors in mice. Physiol Behav 98:433–440

Kugathasan L, Dutly AE, Zhao YD et al (2005) Role of angiopoietin-1 in experimental and human pulmonary arterial hypertension. Chest 128:633S–642S

Kuijpers MH, de Jong W (1987) Relationship between blood pressure level, renal histopathological lesions and plasma renin activity in fawn-hooded rats. Br J Exp Pathol 68:179–187

Kuijpers MH, Gruys E (1984) Spontaneous hypertension and hypertensive renal disease in the fawn-hooded rat. Br J Exp Pathol 65:181–190

Lalich JJ, Merkow L (1961) Pulmonary arteritis produced in rat by feeding Crotalaria spectabilis. Lab Invest 10:744–750

Lalich JL, Johnson WD, Raczniak TJ, Shumaker RC (1977) Fibrin thrombosis in monocrotaline pyrrole-induced cor pulmonale in rats. Arch Pathol Lab Med 101:69–73

Lane KB, Machado RD, Pauciulo MW et al (2000) Heterozygous germline mutations in BMPR2, encoding a TGF-beta receptor, cause familial primary pulmonary hypertension. Nat Genet 26:81–84

Le Cras TD, Kim DH, Gebb S et al (1999) Abnormal lung growth and the development of pulmonary hypertension in the Fawn-Hooded rat. Am J Physiol 277:L709–L718

Le Cras TD, Kim DH, Markham NE, Abman AS (2000) Early abnormalities of pulmonary vascular development in the Fawn-Hooded rat raised at Denver's altitude. Am J Physiol Lung Cell Mol Physiol 279:L283–L291

Li H, Elton TS, Chen YF, Oparil S (1994) Increased endothelin receptor gene expression in hypoxic rat lung. Am J Physiol 266:L553–L560

Liu D, Wang J, Kinzel B et al (2007) Dosage-dependent requirement of BMP type II receptor for maintenance of vascular integrity. Blood 110:1502–1510

Long L, MacLean MR, Jeffery TK et al (2006) Serotonin increases susceptibility to pulmonary hypertension in BMPR2-deficient mice. Circ Res 98:818–827

Lund AK, Lucero J, Herbert L, Liu Y, Naik JS (2011) Human immunodeficiency virus transgenic rats exhibit pulmonary hypertension. Am J Physiol Lung Cell Mol Physiol 301:L315–L326

MacLean MR, Deuchar GA, Hicks MN et al (2004) Overexpression of the 5-hydroxytryptamine transporter gene: effect on pulmonary hemodynamics and hypoxia-induced pulmonary hypertension. Circulation 109:2150–2155

Marcos E, Adnot S, Pham MH et al (2003) Serotonin transporter inhibitors protect against hypoxic pulmonary hypertension. Am J Respir Crit Care Med 168:487–493

Marecki JC, Cool CD, Parr JE et al (2006) HIV-1 Nef is associated with complex pulmonary vascular lesions in SHIV-nef-infected macaques. Am J Respir Crit Care Med 174:437–445

Marsboom G, Wietholt C, Haney CR et al (2012a) Lung ^{18}F-fluorodeoxyglucose positron emission tomography for diagnosis and monitoring of pulmonary arterial hypertension. Am J Respir Crit Care Med 185(6):670–679

Marsboom G, Toth PT, Ryan JJ et al (2012b) Dynamin-related protein 1-mediated mitochondrial mitotic fission permits hyperproliferation of vascular smooth muscle cells and offers a novel therapeutic target in pulmonary hypertension. Circ Res 110(11):1484–1497

McMurtry IF, Davidson AB, Reeves JT, Grover RF (1976) Inhibition of hypoxic pulmonary vasoconstriction by calcium antagonists in isolated rat lungs. Circ Res 38:99–104

McMurtry MS, Bonnet S, Wu X et al (2004) Dichloroacetate prevents and reverses pulmonary hypertension by inducing pulmonary artery smooth muscle cell apoptosis. Circ Res 95:830–840

McMurtry MS, Archer SL, Altieri DC et al (2005) Gene therapy targeting survivin selectively induces pulmonary vascular apoptosis and reverses pulmonary arterial hypertension. J Clin Invest 115:1479–1491

McMurtry MS, Bonnet S, Michelakis ED, Haromy A, Archer SL (2007a) Statin therapy, alone or with rapamycin, does not reverse monocrotaline pulmonary arterial hypertension: the rapamcyin-atorvastatin-simvastatin study. Am J Physiol Lung Cell Mol Physiol 293: L933–L940

McMurtry MS, Moudgil R, Hashimoto K, Bonnet S, Michelakis ED, Archer SL (2007b) Overexpression of human bone morphogenetic protein receptor 2 does not ameliorate monocrotaline pulmonary arterial hypertension. Am J Physiol Lung Cell Mol Physiol 292: L872–L878

Merklinger SL, Jones PL, Martinez EC, Rabinovitch M (2005) Epidermal growth factor receptor blockade mediates smooth muscle cell apoptosis and improves survival in rats with pulmonary hypertension. Circulation 112:423–431

Michelakis ED, Dyck JR, McMurtry MS et al (2001) Gene transfer and metabolic modulators as new therapies for pulmonary hypertension. Increasing expression and activity of potassium channels in rat and human models. Adv Exp Med Biol 502:401–418

Michelakis ED, McMurtry MS, Wu XC et al (2002) Dichloroacetate, a metabolic modulator, prevents and reverses chronic hypoxic pulmonary hypertension in rats: role of increased expression and activity of voltage-gated potassium channels. Circulation 105:244–250

Miyamoto S, Nagaya N, Satoh T et al (2000) Clinical correlates and prognostic significance of six-minute walk test in patients with primary pulmonary hypertension. Comparison with cardiopulmonary exercise testing. Am J Respir Crit Care Med 161:487–492

Molteni A, Ward WF, Ts'ao CH, Port CD, Solliday NH (1984) Monocrotaline-induced pulmonary endothelial dysfunction in rats. Proc Soc Exp Biol Med 176:88–94

Morrell NW (2006) Pulmonary hypertension due to BMPR2 mutation: a new paradigm for tissue remodeling? Proc Am Thorac Soc 3:680–686

Morrell NW, Danilov SM, Satyan KB, Morris KG, Stenmark KR (1997) Right ventricular angiotensin converting enzyme activity and expression is increased during hypoxic pulmonary hypertension. Cardiovasc Res 34:393–403

Morrell NW, Yang X, Upton PD et al (2001) Altered growth responses of pulmonary artery smooth muscle cells from patients with primary pulmonary hypertension to transforming growth factor-beta(1) and bone morphogenetic proteins. Circulation 104:790–795

Morrell NW, Adnot S, Archer SL et al (2009) Cellular and molecular basis of pulmonary arterial hypertension. J Am Coll Cardiol 54:S20–S31

Nagaoka T, Muramatsu M, Sato K, McMurtry I, Oka M, Fukuchi Y (2001) Mild hypoxia causes severe pulmonary hypertension in fawn-hooded but not in Tester Moriyama rats. Respir Physiol 127:53–60

Nagaoka T, Gebb SA, Karoor V et al (2006) Involvement of RhoA/Rho kinase signaling in pulmonary hypertension of the fawn-hooded rat. J Appl Physiol 100:996–1002

Newman JH, Wheeler L, Lane KB et al (2001) Mutation in the gene for bone morphogenetic protein receptor II as a cause of primary pulmonary hypertension in a large kindred. N Engl J Med 345:319–324

Newman JH, Holt TN, Hedges LK et al (2012) High-altitude pulmonary hypertension in cattle (brisket disease): candidate genes and gene expression profiling of peripheral blood mononuclear cells. Pulm Circ 1:462–469

Ochiai E, Kamei K, Watanabe A et al (2008) Inhalation of Stachybotrys chartarum causes pulmonary arterial hypertension in mice. Int J Exp Pathol 89:201–208

Oka M, Homma N, Taraseviciene-Stewart L et al (2007) Rho kinase-mediated vasoconstriction is important in severe occlusive pulmonary arterial hypertension in rats. Circ Res 100:923–929

Okada K, Tanaka Y, Bernstein M, Zhang W, Patterson GA, Botney MD (1997) Pulmonary hemodynamics modify the rat pulmonary artery response to injury. A neointimal model of pulmonary hypertension. Am J Pathol 151:1019–1025

Pan LC, Lame MW, Morin D, Wilson DW, Segall HJ (1991) Red blood cells augment transport of reactive metabolites of monocrotaline from liver to lung in isolated and tandem liver and lung preparations. Toxicol Appl Pharmacol 110:336–346

Pendse AA, Arbones-Mainar JM, Johnson LA, Altenburg MK, Maeda N (2009) Apolipoprotein E knock-out and knock-in mice: atherosclerosis, metabolic syndrome, and beyond. J Lipid Res 50(Suppl):S178–S182

Perros F, Montani D, Dorfmuller P et al (2008) Platelet-derived growth factor expression and function in idiopathic pulmonary arterial hypertension. Am J Respir Crit Care Med 178:81–88

Piao L, Toth PT, Urboniene D, Archer SL (2009) Impaired oxidative metabolism and enhanced glycolysis in right ventricular hypertrophy: the warburg effect. PVRI Rev 1:163–166

Piao L, Fang YH, Cadete VJ et al (2010a) The inhibition of pyruvate dehydrogenase kinase improves impaired cardiac function and electrical remodeling in two models of right ventricular hypertrophy: resuscitating the hibernating right ventricle. J Mol Med (Berl) 88:47–60

Piao L, Marsboom G, Archer SL (2010b) Mitochondrial metabolic adaptation in right ventricular hypertrophy and failure. J Mol Med (Berl) 88:1011–1020

Piao L, Sidhu V, Fang Y, Thenappan T, Lopaschuk GD, Archer S (2012) Chronic inhibition of pyruvate dehydrogenase kinase with dichloroacetate improves cardiac metabolism and function in right ventricular hypertrophy in fawn-hooded rats. J Am Coll Cardiol 59:E1604

Pogoriler J, Rich S, Archer S, Husain A (2012) Persistence of complex vascular lesions despite prolonged prostacyclin therapy of pulmonary arterial hypertension. Histopathology 61 (4):597–609

Pozeg ZI, Michelakis ED, McMurtry MS et al (2003) In vivo gene transfer of the O2-sensitive potassium channel Kv1.5 reduces pulmonary hypertension and restores hypoxic pulmonary vasoconstriction in chronically hypoxic rats. Circulation 107:2037–2044

Price LC, Wort SJ, Perros F et al (2012) Inflammation in pulmonary arterial hypertension. Chest 141:210–221

Pugh ME, Hemnes AR (2010) Development of pulmonary arterial hypertension in women: interplay of sex hormones and pulmonary vascular disease. Womens Health (Lond Engl) 6:285–296

Rabinovitch M, Fisher K, Gamble W, Reid L, Treves S (1979) Thallium-201: quantitation of right ventricular hypertrophy in chronically hypoxic rats. Radiology 130:223–225

Rabinovitch M, Gamble WJ, Miettinen OS, Reid L (1981) Age and sex influence on pulmonary hypertension of chronic hypoxia and on recovery. Am J Physiol 240:H62–H72

Raczniak TJ, Chesney CF, Allen JR (1978) Ultrastructure of the right ventricle after monocrotaline-induced cor pulmonale in the nonhuman primate (Macaca arctoides). Exp Mol Pathol 28:107–118

Reeve HL, Michelakis E, Nelson DP, Weir EK, Archer SL (2001) Alterations in a redox oxygen sensing mechanism in chronic hypoxia. J Appl Physiol 90:2249–2256

Rhodes J (2005) Comparative physiology of hypoxic pulmonary hypertension: historical clues from brisket disease. J Appl Physiol 98:1092–1100

Rich JD, Shah SJ, Swamy RS, Kamp A, Rich S (2011) Inaccuracy of Doppler echocardiographic estimates of pulmonary artery pressures in patients with pulmonary hypertension: implications for clinical practice. Chest 139:988–993

Robbins IM, Newman JH, Johnson RF et al (2009) Association of the metabolic syndrome with pulmonary venous hypertension. Chest 136:31–36

Rosenberg HC, Rabinovitch M (1988) Endothelial injury and vascular reactivity in monocrotaline pulmonary hypertension. Am J Physiol 255:H1484–H1491

Ryan J, Bloch K, Archer SL (2011) Rodent models of pulmonary hypertension: harmonisation with the world health organisation's categorisation of human PH. Int J Clin Pract Suppl (172):15–34

Sakao S, Tatsumi K (2011) The effects of antiangiogenic compound SU5416 in a rat model of pulmonary arterial hypertension. Respiration 81:253–261

Sato K, Webb S, Tucker A et al (1992) Factors influencing the idiopathic development of pulmonary hypertension in the fawn hooded rat. Am Rev Respir Dis 145:793–797

Schultze AE, Roth RA (1998) Chronic pulmonary hypertension – the monocrotaline model and involvement of the hemostatic system. J Toxicol Environ Health B Crit Rev 1:271–346

Schulze A, Roth R (1998) Chronic pulmonary hypertension – the monocrotaline model and involvement of the hemostatic system. J Toxicol Environ Health B Crit Rev 1:271–346

Shimoda LA, Manalo DJ, Sham JS, Semenza GL, Sylvester JT (2001) Partial HIF-1alpha deficiency impairs pulmonary arterial myocyte electrophysiological responses to hypoxia. Am J Physiol Lung Cell Mol Physiol 281:L202–L208

Sitbon O, Lascoux-Combe C, Delfraissy JF et al (2008) Prevalence of HIV-related pulmonary arterial hypertension in the current antiretroviral therapy era. Am J Respir Crit Care Med 177:108–113

Song Y, Jones JE, Beppu H, Keaney JF Jr, Loscalzo J, Zhang YY (2005) Increased susceptibility to pulmonary hypertension in heterozygous BMPR2-mutant mice. Circulation 112:553–562

Spiekerkoetter E, Alvira CM, Kim YM et al (2008) Reactivation of gammaHV68 induces neointimal lesions in pulmonary arteries of S100A4/Mts1-overexpressing mice in association with degradation of elastin. Am J Physiol Lung Cell Mol Physiol 294:L276–L289

Stenmark KR, McMurtry IF (2005) Vascular remodeling versus vasoconstriction in chronic hypoxic pulmonary hypertension: a time for reappraisal? Circ Res 97:95–98

Stenmark KR, Fagan KA, Frid MG (2006) Hypoxia-induced pulmonary vascular remodeling: cellular and molecular mechanisms. Circ Res 99:675–691

Stenmark KR, Meyrick B, Galie N, Mooi WJ, McMurtry IF (2009) Animal models of pulmonary arterial hypertension: the hope for etiological discovery and pharmacological cure. Am J Physiol Lung Cell Mol Physiol 297:L1013–L1032

Sweiss NJ, Hushaw L, Thenappan T et al (2010) Diagnosis and management of pulmonary hypertension in systemic sclerosis. Curr Rheumatol Rep 12:8–18

Taraseviciene-Stewart L, Kasahara Y, Alger L et al (2001) Inhibition of the VEGF receptor 2 combined with chronic hypoxia causes cell death-dependent pulmonary endothelial cell proliferation and severe pulmonary hypertension. FASEB J 15:427–438

Thenappan T, Shah SJ, Rich S, Tian L, Archer SL, Gomberg-Maitland M (2010) Survival in pulmonary arterial hypertension: a reappraisal of the NIH risk stratification equation. Eur Respir J 35:1079–1087

Thibault HB, Kurtz B, Raher MJ et al (2010) Noninvasive assessment of murine pulmonary arterial pressure: validation and application to models of pulmonary hypertension. Circ Cardiovasc Imaging 3:157–163

Thomson JR, Machado RD, Pauciulo MW et al (2000) Sporadic primary pulmonary hypertension is associated with germline mutations of the gene encoding BMPR-II, a receptor member of the TGF-beta family. J Med Genet 37:741–745

Tobach E, DeSantis JL, Zucker MB (1984) Platelet storage pool disease in hybrid rats. F1 fawn-hooded rats derived from crosses with their putative ancestors (Rattus norvegicus). J Hered 75:15–18

Tschopp B, Weiss HJ (1974) Decreased ATP, ADP and serotonin in young platelets of fawn-hooded rats with storage pool disease. Thromb Diath Haemorrh 2:670–677

Tuder RM, Marecki JC, Richter A, Fijalkowska I, Flores S (2007) Pathology of pulmonary hypertension. Clin Chest Med 28:23–42, vii

Urboniene D, Haber I, Fang YH, Thenappan T, Archer SL (2010) Validation of high-resolution echocardiography and magnetic resonance imaging vs. high-fidelity catheterization in experimental pulmonary hypertension. Am J Physiol Lung Cell Mol Physiol 299:L401–L412

Valmary S, Dorfmuller P, Montani D, Humbert M, Brousset P, Degano B (2011) Human gamma-herpesviruses Epstein-Barr virus and human herpesvirus-8 are not detected in the lungs of patients with severe pulmonary arterial hypertension. Chest 139:1310–1316

van Albada ME, Schoemaker RG, Kemna MS, Cromme-Dijkhuis AH, van Veghel R, Berger RM (2005) The role of increased pulmonary blood flow in pulmonary arterial hypertension. Eur Respir J 26:487–493

van Rodijnen WF, van Lambalgen TA, Tangelder GJ, van Dokkum RP, Provoost AP, ter Wee PM (2002) Reduced reactivity of renal microvessels to pressure and angiotensin II in fawn-hooded rats. Hypertension 39:111–115

Wagner JG, Petry TW, Roth RA (1993) Characterization of monocrotaline pyrrole-induced DNA cross-linking in pulmonary artery endothelium. Am J Physiol 264:L517–L522

Weir EK, Reeve HL, Johnson G, Michelakis ED, Nelson DP, Archer SL (1998) A role for potassium channels in smooth muscle cells and platelets in the etiology of primary pulmonary hypertension. Chest 114:200S–204S

Weir EK, Lopez-Barneo J, Buckler KJ, Archer SL (2005) Acute oxygen-sensing mechanisms. N Engl J Med 353:2042–2055

West J, Fagan K, Steudel W et al (2004) Pulmonary hypertension in transgenic mice expressing a dominant-negative BMPRII gene in smooth muscle. Circ Res 94:1109–1114

West J, Harral J, Lane K et al (2008) Mice expressing BMPR2R899X transgene in smooth muscle develop pulmonary vascular lesions. Am J Physiol Lung Cell Mol Physiol 295:L744–L755

White RJ, Meoli DF, Swarthout RF et al (2007) Plexiform-like lesions and increased tissue factor expression in a rat model of severe pulmonary arterial hypertension. Am J Physiol Lung Cell Mol Physiol 293:L583–L590

Wideman R (2000) Cardio-pulmonary hemodynamics and ascites in broiler chickens. Poult Avian Biol Rev 11:21–43

Wideman RF Jr, Hamal KR (2011) Idiopathic pulmonary arterial hypertension: an avian model for plexogenic arteriopathy and serotonergic vasoconstriction. J Pharmacol Toxicol Methods 63:283–295

Wilson DW, Segall HJ (1990) Changes in type II cell populations in monocrotaline pneumotoxicity. Am J Pathol 136:1293–1299

Wilson DW, Segall HJ, Pan LC, Dunston SK (1989) Progressive inflammatory and structural changes in the pulmonary vasculature of monocrotaline-treated rats. Microvasc Res 38:57–80

Woods LW, Wilson DW, Segall HJ (1999) Manipulation of injury and repair of the alveolar epithelium using two pneumotoxicants: 3-methylindole and monocrotaline. Exp Lung Res 25:165–181

Yang X, Long L, Reynolds PN, Morrell NW (2011) Expression of mutant BMPR-II in pulmonary endothelial cells promotes apoptosis and a release of factors that stimulate proliferation of pulmonary arterial smooth muscle cells. Pulm Circ 1:103–110

Young KA, Ivester C, West J, Carr M, Rodman DM (2006) BMP signaling controls PASMC KV channel expression in vitro and in vivo. Am J Physiol Lung Cell Mol Physiol 290:L841–L848

Yuan JX, Aldinger AM, Juhaszova M et al (1998) Dysfunctional voltage-gated K+ channels in pulmonary artery smooth muscle cells of patients with primary pulmonary hypertension. Circulation 98:1400–1406

Zamora MR, Stelzner TJ, Webb S, Panos RJ, Ruff LJ, Dempsey EC (1996) Overexpression of endothelin-1 and enhanced growth of pulmonary artery smooth muscle cells from fawn-hooded rats. Am J Physiol 270:L101–L109

Zhang S, Fantozzi I, Tigno DD et al (2003) Bone morphogenetic proteins induce apoptosis in human pulmonary vascular smooth muscle cells. Am J Physiol Lung Cell Mol Physiol 285:L740–L754

Zhang S, Patel HH, Murray F et al (2007) Pulmonary artery smooth muscle cells from normal subjects and IPAH patients show divergent cAMP-mediated effects on TRPC expression and capacitative Ca2+ entry. Am J Physiol Lung Cell Mol Physiol 292:L1202–L1210

Zhao L (2010) Chronic hypoxia-induced pulmonary hypertension in rat: the best animal model for studying pulmonary vasoconstriction and vascular medial hypertrophy. Drug Discov Today Dis Models 7:83–88

Zhao L, Sebkhi A, Nunez DJ et al (2001) Right ventricular hypertrophy secondary to pulmonary hypertension is linked to rat chromosome 17: evaluation of cardiac ryanodine Ryr2 receptor as a candidate. Circulation 103:442–447

Zhao YD, Campbell AI, Robb M, Ng D, Stewart DJ (2003) Protective role of angiopoietin-1 in experimental pulmonary hypertension. Circ Res 92:984–991

Part II
Pulmonary Hypertension: Established Therapies

General Supportive Care

Ioana R. Preston

Abstract In addition to PAH-specific therapies proven to be effective in achieving a better outcome and/or symptomatic improvement, the care of patients with pulmonary arterial hypertension (PAH) requires a complex strategy that includes general pharmacological and non-pharmacological measures. This chapter will detail the general supportive measures beneficial for PAH patients as adjuvant treatment to targeted therapies.

Keywords Diuretics • Anticoagulation • Pulmonary rehabilitation • Travel • Pulmonary arterial hypertension

Contents

1 Introduction	154
2 General Pharmacological Measures	154
2.1 Diuretics	154
2.2 Oxygen	154
2.3 Anticoagulants	156
2.4 Digoxin	156
3 Non-pharmacological Measures	157
3.1 Physical Exercise and Pulmonary Rehabilitation	157
3.2 Pregnancy and Birth Control	157
3.3 Screening for Depression and Psychosocial Support	158
3.4 Infection Control	158
3.5 Elective Surgery	159
References	159

I.R. Preston (✉)
Pulmonary, Critical Care and Sleep Division, Tufts Medical Center, Tufts University School of Medicine, 800 Washington Street, Box #257, Boston, MA 02111, USA
e-mail: ipreston@tuftsmedicalcenter.org

M. Humbert et al. (eds.), *Pharmacotherapy of Pulmonary Hypertension*,
Handbook of Experimental Pharmacology 218, DOI 10.1007/978-3-642-38664-0_6,
© Springer-Verlag Berlin Heidelberg 2013

1 Introduction

Pulmonary arterial hypertension (PAH) is a complex disease that requires an integrated approach between multiple disciplines. In addition to targeted therapies, general measures have been recommended to prevent complications that may lead to disease decompensation and to treat nonspecifically certain aspects of PAH. A summary of recommendations for general care is presented in Table 1. Among those, only supervised exercise therapy has been formally studied in randomized trials in PAH. Therefore, the recommendations in this chapter are mostly based on small, nonrandomized reports and clinical experience. The principles of general care described below aid in maximizing the clinical care of PAH patients. General supportive care measures can be divided into pharmacological and non-pharmacological measures.

2 General Pharmacological Measures

2.1 Diuretics

Diuretics are often used in decompensated right heart failure in clinical practice. Although there are no clinical trials testing the effectiveness of diuretics in PAH, clinical experience suggests that a tight volume control achieved with diuretics improves symptoms. There are no guidelines regarding the choice or dose of diuretics. Unlike the experience in left heart failure, where aldosterone antagonists have shown to improve survival in patients with systolic left ventricular failure (Hamaguchi et al. 2010), there are no data of their efficacy in right heart failure. Frequent monitoring of renal function and electrolytes, as well as close weight control, is advisable. In randomized trials of left heart failure, adjustment of cardiovascular medications based on regular monitoring of brain natriuretic peptide (BNP) or its precursor, N-terminal pro-BNP (NT-pro-BNP) led to better control of heart failure symptoms and fewer hospitalizations (Januzzi et al. 2011). Prospective studies on PAH patients showed that a decrease in NT-pro-BNP levels with therapy was associated with improved outcome (Mauritz et al. 2011). Although there are no studies proving the benefit of actively adjusting medical therapy (i.e., diuretics) based on BNP levels in PAH, BNP or NT-pro-BNP levels can be monitored as a measure of compensated right heart failure.

2.2 Oxygen

Most PAH patients have a minor degree of hypoxemia and many present with nocturnal desaturations, even in the absence of obstructive sleep apnea (Minai et al. 2007). There are few exceptions where a more severe degree of hypoxemia

Table 1 Summary of recommendations for general supportive care of PAH patients

Pharmacological measures	Considerations
Diuretics	The goal is to achieve a tight control of volume status. The choice of diuretic is variable
Oxygen	In hypoxemic patients, oxygen supplementation can be considered. Traveling requires assessment before the flight for possible oxygen supplementation
Anticoagulants	Only warfarin has been studied to date. Long-term anticoagulation should be considered in idiopathic PAH, heritable PAH, and PAH associated with anorexigen use
Digoxin	Goal to treat tachyarrhythmias

Non-pharmacological measures	Considerations
Physical exercise and pulmonary rehabilitation	Exercise is encouraged in patients stable on targeted therapies. Patients benefit from supervised exercise programs
Pregnancy and birth control	Pregnancy is contraindicated and birth control measures are encouraged
Screening for depression and psychosocial support	Screening for depression is important and patients may benefit from a multidisciplinary approach
Infection control	Preventive vaccinations are recommended and catheter care using sterile techniques are important in avoiding infections
Elective surgery	The increased risk of complications requires careful assessment and planning prior to surgery

is encountered: patients with congenital systemic to pulmonary shunts and right to left physiology (Eisenmenger's syndrome), those with patent foramen ovale and right to left shunts, and those with scleroderma-associated PAH who have a more severe gas exchange abnormality and a low diffusion capacity for carbon monoxide. Oxygen was shown to be an acute vasodilator (Atz et al. 1999), but long-term treatment with oxygen has not been studied in PAH patients. One exception constitutes patients with Eisenmenger syndrome, in whom long-term chronic oxygen therapy did not bring any benefit (Sandoval et al. 2001). Currently, the recommendations for oxygen therapy in PAH are extrapolated from the chronic obstructive lung disease data. Long-term oxygen can be considered in PAH when partial pressure of oxygen is <60 mmHg, or oxygen saturation <88 % (Weitzenblum et al. 1985). In addition, oxygen supplementation can be used for exertional and nocturnal hypoxemia. Most stable PAH patients can safely travel by plane, although patients who require oxygen replacement, either continuously, or with exertion will most likely require oxygen during a flight. In a retrospective study in which 159 out of 430 PAH patients traveled at least once (Thamm et al. 2011), 57 patients were on long-term ambulatory oxygen therapy, and 29 used supplemental oxygen while traveling. There were 20 adverse events, most of them mild to moderate, and only 7 required medical intervention. A more recent study investigated prospectively 34 PAH patients and found that hypoxemia is common during flight and is associated with cabin pressures equivalent to

>6,000 ft above sea level, ambulation, and flight duration (Roubinian et al. 2012). Therefore, while most stable PAH patients can safely travel by plane, we recommend evaluation for supplemental in-flight oxygen for patients taking long flights, if not already on supplemental therapy.

2.3 Anticoagulants

The role of chronic thrombosis in PAH is controversial (Wagenvoort 1980; Fuster et al. 1984). It is not clear whether the thrombotic arteriopathy described in pathological samples from PAH patients are an epiphenomenon or an integral part of the pulmonary vascular remodeling. The experience with anticoagulation is derived from one retrospective cohort and two prospective studies that showed that the use of warfarin was associated with a better outcome in PAH (Fuster et al. 1984; Rich et al. 1992; Kawut et al. 2005). These studies looked only at patients with idiopathic PAH, heritable PAH, and PAH due to anorexigen use. To date, there is no information on the effects of anticoagulation in other forms of PAH. Therefore, current guidelines for chronic anticoagulation include these three categories of PAH. The consideration for anticoagulation must be weighed against the potential risk of bleeding. In particular, patients with PAH associated with scleroderma, those with congenital heart disease, and portopulmonary hypertension patients may be at increased risk due to the presence of gastrointestinal, or pulmonary arteriovenous malformations, or an abnormal coagulation profile, respectively. When anticoagulation is considered, the goal is to maintain an international normalized ratio between 1.5 and 2.5 (as it is practiced in the USA) or between 2.0 and 3.0 (the European experience). Anticoagulation can be stopped for invasive procedures without the need to bridge with heparins, unless patients carry the diagnosis of chronic thromboembolic pulmonary hypertension, or have other reasons for anticoagulation.

2.4 Digoxin

Digoxin was shown to improve cardiac output acutely in PAH patients (Rich et al. 1998), but long-term studies are lacking. The current recommendations include its use in tachyarrhythmias. As the therapeutic window of digoxin is narrow, caution needs to be exerted in patients with prerenal azotemia, in order to avoid toxicity.

3 Non-pharmacological Measures

3.1 Physical Exercise and Pulmonary Rehabilitation

Once started on targeted therapies, PAH patients are encouraged to gradually increase their activity to a level that brings on mild to moderate dyspnea. Activities that elicit presyncope, syncope, or chest pain should be discouraged. In addition, patients who are on chronic anticoagulation should avoid contact sports. Patients with both early and advanced disease benefit from supervised rehabilitation programs. To date, four studies showed that supervised rehabilitation programs both in inpatient and outpatient settings improve exercise capacity and quality of life, as well as skeletal muscle function, when added to pharmacological therapy (Mereles et al. 2006; de Man et al. 2009; Fox et al. 2011; Grunig et al. 2012). The most recent study evaluated the effect of exercise training in an inpatient setting for 3 weeks followed by outpatient training for 15 weeks in a cohort of patients with different forms of pulmonary hypertension (Grunig et al. 2012). Compared with baseline, 3- and 15-week assessments showed improvement in 6-min walk distance (6MWD), scores of quality of life, functional class, peak oxygen consumption, oxygen pulse, heart rate and systolic pulmonary artery pressure at rest, and maximal workload. The improvement in 6MWD was similar across the different forms and functional classes of pulmonary hypertension. Adverse events such as respiratory infections, syncope, or presyncope occurred in 13 % of patients, suggesting that exercise training in pulmonary hypertension is an effective but not a completely harmless add-on therapy and patients and should be closely monitored. Once patients complete the supervised program, they are encouraged to continue their maintenance exercises regularly.

3.2 Pregnancy and Birth Control

PAH is considered a contraindication to pregnancy because it has been previously associated with 30–50 % mortality in the peripartum period, although recent reports show more favorable outcomes, especially in long-term responders to calcium channel blockers (Jais et al. 2012). Barrier contraceptive methods are safe, but not very reliable. Hormonal replacement therapy is a suitable option, if high doses of estrogen are avoided due to their association with increased risk of thrombosis. Alternatively, progesterone-only preparations such as medroxyprogesterone acetate and etonogestrel are effective approaches to contraception. Intrauterine devices (IUDs) are a safe and reliable method of long-term contraception, although special care needs to be taken during the implantation. We recommend the procedure be done in the hospital setting with placement of two peripheral intravenous lines and atropine readily available in case of vasovagal reactions. PAH patients who become pregnant should be informed of the high risk of pregnancy and termination of pregnancy discussed. Patients who choose to continue pregnancy should be closely monitored by a team involving high-risk obstetricians and PAH specialists, preferably in a PAH center. Treatment with PAH-specific therapies should be initiated or modified, taking into account the risk profile of PAH therapies, and a planned elective delivery should be scheduled (Bendayan et al. 2005; Bonnin et al. 2005).

3.3 Screening for Depression and Psychosocial Support

Similar to other chronic conditions, many patients with PAH are diagnosed and/or being treated for depression. Some patients are reluctant to discuss their psychosocial limitations with their PAH specialists and many PAH specialists do not have formal training to recognize and treat depression. The type of therapy chosen is a complex matter. Serotonin receptor antagonists (SSRIs) are commonly used for depression. Nevertheless, the role of serotonin in PAH remains controversial (see Chap. 4.4) and the decision of what pharmacological therapy, should one be indicated, needs to be weighted and discussed among the various healthcare professionals providing for the care of the PAH patient. A report that surveyed two cohorts of patients showed that the use of SSRIs was associated with decreased mortality after adjustment for age, gender, etiology of PAH, and obesity (Shah et al. 2009), while in the large Regisitry to Evaluate Early And Long-term PAH disease management (REVEAL) incident SSRI use was associated with increased mortaily and a greater risk of clinical worsening (Sadoughi et al. 2013). Conversely, the use of SSRIs in pregnant women was associated with an increased risk for primary pulmonary hypertension of the newborn in the offspring (Chambers et al. 2006; Kieler et al. 2012). Therefore, an interdisciplinary approach involving the PAH treating physician, the primary care provider, a psychologist, social worker, and/or a psychiatrist should be considered in determining the best approach to screening and treating depression in PAH. In addition to pharmacological intervention, patient support groups may provide psychosocial and emotional support.

3.4 Infection Control

Similar to other patients with chronic cardiopulmonary conditions, vaccinations against influenza and pneumococcal pneumonia are recommended. Special catheter care, as well as maintenance of sterile techniques while mixing the drug are advised in patients on long-term intravenous infusion with prostacyclin analogues. The involvement of specialized nurses to train the patient in the technique of sterile handling of the parenteral drug and catheter site care is extremely important to ensure a good practice to avoid blood stream infections.

3.5 Elective Surgery

Patients with pulmonary hypertension, in particular those with PAH, are at high risk for complications when undergoing anesthesia and major surgery (Kaw et al. 2011). The perioperative management can be challenging as it may be complicated by a systemic inflammatory response, hypoxemia, and the development of decompensated right heart failure. The mortality rate in the perioperative period was reported between 7 and 18 % (Minai et al. 2006; Kaw et al. 2011; Meyer et al. 2013). Risk factors for complications included emergency surgery, major surgery, and a long operative time (Kaw et al. 2011; Meyer et al. 2013). Patients on oral PAH-specific therapies may require temporary conversion to parenteral therapies (inhaled, intravenous, or subcutaneous). It is not clear what type of anesthesia is better tolerated, as both epidural and general anesthesia cause peripheral vasodilation with a possible decrease in cardiac output. Careful assessment and planning prior to surgery by the treating physicians and anesthesiologist team are advised.

In conclusion, in the intricate strategy of PAH management, general supportive measures play an important role in aiding to the pharmacological targeted therapy in ensuring stability, avoiding the development of complications, and providing symptomatic improvement. These measures need to be integrated in a complex approach, best achieved with the collaboration between the PAH treating physician and multiple other specialties.

References

Atz AM, Adatia I et al (1999) Combined effects of nitric oxide and oxygen during acute pulmonary vasodilator testing. J Am Coll Cardiol 33(3):813–819
Bendayan D, Hod M et al (2005) Pregnancy outcome in patients with pulmonary arterial hypertension receiving prostacyclin therapy. Obstet Gynecol 106(5 Pt 2):1206–1210
Bonnin M, Mercier FJ et al (2005) Severe pulmonary hypertension during pregnancy: mode of delivery and anesthetic management of 15 consecutive cases. Anesthesiology 102(6): 1133–1137, discussion 1135A–1136A
Chambers CD, Hernandez-Diaz S et al (2006) Selective serotonin-reuptake inhibitors and risk of persistent pulmonary hypertension of the newborn. N Engl J Med 354(6):579–587
de Man FS, Handoko ML et al (2009) Effects of exercise training in patients with idiopathic pulmonary arterial hypertension. Eur Respir J 34(3):669–675
Fox BD, Kassirer M et al (2011) Ambulatory rehabilitation improves exercise capacity in patients with pulmonary hypertension. J Card Fail 17(3):196–200
Fuster V, Steele PM et al (1984) Primary pulmonary hypertension: natural history and the importance of thrombosis. Circulation 70(4):580–587
Grunig E, Lichtblau M et al (2012) Safety and efficacy of exercise training in various forms of pulmonary hypertension. Eur Respir J 40(1):84–92
Hamaguchi S, Kinugawa S et al (2010) Spironolactone use at discharge was associated with improved survival in hospitalized patients with systolic heart failure. Am Heart J 160(6): 1156–1162
Jais X, Olsson KM et al (2012) Pregnancy outcomes in pulmonary arterial hypertension in the modern management era. Eur Respir J 40(4):881–885

Januzzi JL Jr, Rehman SU et al (2011) Use of amino-terminal pro-B-type natriuretic peptide to guide outpatient therapy of patients with chronic left ventricular systolic dysfunction. J Am Coll Cardiol 58(18):1881–1889

Kaw R, Pasupuleti V et al (2011) Pulmonary hypertension: an important predictor of outcomes in patients undergoing non-cardiac surgery. Respir Med 105(4):619–624

Kawut SM, Horn EM et al (2005) New predictors of outcome in idiopathic pulmonary arterial hypertension. Am J Cardiol 95(2):199–203

Kieler H, Artama M et al (2012) Selective serotonin reuptake inhibitors during pregnancy and risk of persistent pulmonary hypertension in the newborn: population based cohort study from the five Nordic countries. BMJ 344:d8012

Mauritz GJ, Rizopoulos D et al (2011) Usefulness of serial N-terminal pro-B-type natriuretic peptide measurements for determining prognosis in patients with pulmonary arterial hypertension. Am J Cardiol 108(11):1645–1650

Mereles D, Ehlken N et al (2006) Exercise and respiratory training improve exercise capacity and quality of life in patients with severe chronic pulmonary hypertension. Circulation 114(14):1482–1489

Meyer S, McLaughlin VV et al (2013) Outcomes of noncardiac, nonobstetric surgery in patients with PAH: an international prospective survey. Eur Respir J 41(6):1302–1307

Minai OA, Venkateshiah SB et al (2006) Surgical intervention in patients with moderate to severe pulmonary arterial hypertension. Conn Med 70(4):239–243

Minai OA, Pandya CM et al (2007) Predictors of nocturnal oxygen desaturation in pulmonary arterial hypertension. Chest 131(1):109–117

Rich S, Kaufmann E et al (1992) The effect of high doses of calcium-channel blockers on survival in primary pulmonary hypertension. N Engl J Med 327(2):76–81

Rich S, Seidlitz M et al (1998) The short-term effects of digoxin in patients with right ventricular dysfunction from pulmonary hypertension. Chest 114(3):787–792

Roubinian N, Elliott CG et al (2012) Effects of commercial air travel on patients with pulmonary hypertension. Chest 142(4):885–892

Sadoughi A, Roberts KE et al (2013) Use of selective serotonin uptake inhibitors and outcomes in pulmonary arterial hypertension. Chest. doi: 10.1378/chest.12-2081

Sandoval J, Aguirre JS et al (2001) Nocturnal oxygen therapy in patients with the Eisenmenger syndrome. Am J Respir Crit Care Med 164(9):1682–1687

Shah SJ, Gomberg-Maitland M et al (2009) Selective serotonin reuptake inhibitors and the incidence and outcome of pulmonary hypertension. Chest 136(3):694–700

Thamm M, Voswinckel R et al (2011) Air travel can be safe and well tolerated in patients with clinically stable pulmonary hypertension. Pulm Circ 1(2):239–243

Wagenvoort CA (1980) Lung biopsy specimens in the evaluation of pulmonary vascular disease. Chest 77(5):614–625

Weitzenblum E, Sautegeau A et al (1985) Long-term oxygen therapy can reverse the progression of pulmonary hypertension in patients with chronic obstructive pulmonary disease. Am Rev Respir Dis 131(4):493–498

Calcium-Channel Blockers in Pulmonary Arterial Hypertension

Marie-Camille Chaumais, Elise Artaud Macari, and Olivier Sitbon

Abstract Voltage-activated calcium channels are a family of membrane proteins that provide the major influx pathway for calcium in many different types of cells. Calcium-channel blockers inhibit the calcium influx into vascular cells leading to relaxation of smooth muscle cells and vasodilatation. Vasoconstriction of small pulmonary arteries is recognized as a component of the pathogenesis of pulmonary arterial hypertension and treatment with calcium-channel blockers appears to be rational in this setting. No randomized controlled trial has been performed to demonstrate the beneficial effects of calcium-channel blockers in the treatment of patients with pulmonary arterial hypertension. However, uncontrolled studies have suggested that long-term administration of high-dose calcium antagonists dramatically improves survival in a small subset of patients who respond acutely to those drugs, compared with unresponsive patients. The initial response to an acute vasodilator test with inhaled nitric oxide or intravenous prostacyclin or adenosine accurately identifies patients with pulmonary arterial hypertension who are likely to respond to long-term treatment with calcium-channel blockers.

M.-C. Chaumais
Faculté de Pharmacie, Université Paris-Sud, Chatenay-Malabry, France

Service de Pharmacie, Hôpital Antoine Béclère, Clamart, France

Centre de Référence de l'Hypertension Pulmonaire Sévère, Service de Pneumologie et Soins Intensifs, CHU de Bicêtre, 78 rue du Général Leclerc, 94275 Le Kremlin-Bicêtre, France

E. Artaud Macari • O. Sitbon (✉)
Centre de Référence de l'Hypertension Pulmonaire Sévère, Service de Pneumologie et Soins Intensifs, CHU de Bicêtre, 78 rue du Général Leclerc, 94275 Le Kremlin-Bicêtre, France

Faculté de Médecine, Université Paris-Sud, Le Kremlin-Bicêtre, France

INSERM U999, Centre Chirurgical Marie-Lannelongue, Le Plessis-Robinson, France
e-mail: olivier.sitbon@bct.aphp.fr

Keywords Calcium-channel blockers • Vasodilators • Acute vasodilator testing • Nitric oxide • Pulmonary arterial hypertension

Contents

1	Pharmacology of calcium channel blockers	163
	1.1 Mechanism of Action	163
	1.2 Classification of Calcium-Channel Blockers	164
	1.3 Usual Indications	166
	1.4 Side Effects	166
	1.5 Drug Interactions	167
2	Calcium-Channel Blockers in Pulmonary Arterial Hypertension	167
	2.1 Acute Vasodilator Testing	167
	2.2 Drugs Used for Acute Vasodilator Testing	168
	2.3 Place of Calcium-Channel Blockers in PAH	169
	2.4 Clinical Use of Calcium-Channel Blockers in PAH	170
3	Conclusion	172
References		172

Abbreviations

Ca^{2+}	Calcium
CCBs	Calcium-channel blockers
CO	Cardiac output
DHP	Dihydropyridine
NO	Nitric oxide
NYHA	New York Heart Association
PAH	Pulmonary arterial hypertension
PAP	Pulmonary artery pressure
PVR	Pulmonary vascular resistance
SMC	Smooth muscle cells

Calcium channel antagonists or calcium channels blockers (CCBs) inhibit the calcium influx into vascular cells leading to relaxation of smooth muscle cells (SMC) and vasodilatation. For more than 25 years, development of CCBs led to the management of hypertension, angina pectoris or cardiac dysrhythmia and these drugs represent nowadays a major pharmacological approach.

Beside proliferation of endothelial and SMC, vasoconstriction of small muscular pulmonary arteries is recognized as a component of the pathogenesis of pulmonary arterial hypertension (PAH) and treatment with CCBs appears to be rational in this setting. Unfortunately, CCBs can be used only in a minority of patients with PAH. This chapter focuses on pharmacology of CCBs and their use in the management of patients with PAH

1 Pharmacology of calcium channel blockers

1.1 Mechanism of Action

Voltage-activated calcium channels are a family of membrane proteins that provide the major influx pathway for calcium in many different types of cells. Long-lasting voltage-gated calcium channels (L-type Ca^{2+} channels) are a major subclass of voltage-activated calcium channels present in cardiac, skeletal, and SMC, in various types of cells in the brain, and in many nonexcitable cells. L-type Ca^{2+} channels open in response to a change in membrane potential and are regulated by a variety of receptor-activated signaling pathways. In cardiac and SMC, activation of L-type Ca^{2+} channels initiates contraction directly by increasing cytosolic calcium concentration and indirectly by activating calcium-dependent calcium release by ryanodine-sensitive calcium release channels in the sarcoplasmic reticulum (Tsien 1983; Bers 2002).

Four subunits make up the SMC and cardiac L-type Ca^{2+} channels. The α-1 subunit associated with the disulfide-linked α-2δ subunit and the intracellular β subunit (Fig. 1). The α-1 subunit of 190–250 kDa is the largest subunit, and it incorporates the conduction pore, the voltage sensor and gating apparatus, and most of the known sites of channel regulation by second messengers, drugs, and toxins (Catterall et al. 2005). Several α-1 subunit genes that give rise to L-type currents have been identified, including α-1S which is found in skeletal muscle, α-1C which is expressed in heart, smooth muscle, and brain, as well as α-1D which is found in neuroendocrine cells (De Waard et al. 1996). Diversity in the α-1 subunits also arises owing to alternative splicing of the DNAs; for example, the α1C-subunit gene undergoes alternative splicing to give rise to isoforms found in heart (α1c-a), smooth muscle (α1c-b), and brain (α1c-c) (Hosey et al. 1996). In 2000, a rational nomenclature was adopted (Ertel et al. 2000) based on the well-defined potassium channel nomenclature (Chandy and Gutman 1993). Calcium channels were named using the chemical symbol of the principal permeating ion (Ca^{2+}) with the principal physiological regulator (voltage) indicated as a subscript (CaV). The numerical identifier corresponds to the CaV channel α-1 subunit gene subfamily (1–3 at present) and the order of discovery of the alpha-1 subunit within that subfamily (1 through n). According to this nomenclature, the CaV1 subfamily (CaV1.1–CaV1.4) includes channels containing _1S, _1C, _1D, and _1F, which mediate L-type Ca^{2+} channels and where CaV1.2 is the isoform in the heart and SMC. α-1 subunit is organized in four homologous domains (I–IV), with six transmembrane segments (S1–S6) in each (Fig. 1). The S4 segment serves as the voltage sensor (Catterall et al. 2005). The pore loop between transmembrane segments S5 and S6 in each domain determines ion conductance and selectivity, and changes of only three amino acids in the pore loops in domains I, III, and IV will convert a sodium channel to calcium selectivity. Although auxiliary subunits modulate the properties of the channel complex, the pharmacological and electrophysiological diversity of calcium channels arises primarily from the existence of multiple alpha-1

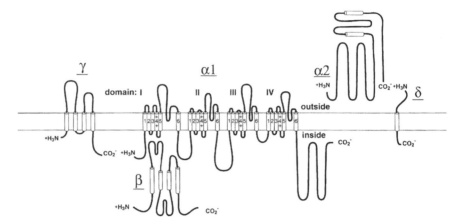

Fig. 1 L-voltage Ca^{2+} channels' structure (from Hockerman et al. 1997). *Cylinders* represent alpha helical transmembrane segments. *Bold lines* represent the polypeptide chains of each subunit

subunits (Hofmann et al. 1994). Phosphorylation of channel-forming subunits by different kinases is one of the most important ways to change the activity of L-type Ca^{2+} channels. L-type Ca^{2+} channels in cardiac cells, neurons, skeletal muscle, and chromaffin cells are regulated by a protein kinase A-dependent phosphorylation (McDonald et al. 1994). In addition, other protein kinases (for example, protein kinase C and calcium-dependent kinases) and phosphoprotein phosphatases have been shown to modify channel activity (Catterall 2011; Brette et al. 2006).

The α-1 subunit is the binding site for CCBs. By inhibition of L-type Ca^{2+} channel, CCBs decrease calcium influx into SMC of arterial wall and myocardium cells leading to vasodilatation and a decrease of systemic blood pressure.

1.2 Classification of Calcium-Channel Blockers

The three major classes of CCBs are chemically distinct and exhibit different functional effects depending on their biophysical, conformation-dependent interactions with the L-type Ca^{2+} channel. Dihydropyridines (DHP) have exclusively vascular effects, whereas non-dihydropyridines (regrouping phenylalkylamines and benzothiazepines) have both vascular and cardiac effects. Drugs in these three subclasses bind to separate sites on the L-type calcium channel: Bound phenylalkylamine (verapamil-like) makes both hydrophobic contacts with residues in transmembrane segments IIIS6 and IVS6 and ionic interactions with the glutamate residues in the pore region of the α-1 subunit. The benzothiazepine (diltiazem-like) calcium antagonists bind to the cytoplasmic bridge between motif III (IIIS) and motif IV (IVS) and more specifically the S6 segments of these domains. The benzothiazepine site is therefore located in close proximity to the phenylalkylamine receptor site (Kimball et al. 1993). The DHP (nifedipine-like) calcium antagonists bind to transmembrane segment 6 of

Table 1 Characteristics of available calcium-channel blockers on cardiac and vascular smooth muscle cells

CCB group		Agent	Vascular effects	Cardiac effects	Indications
Dihydropyridines		Nifedipine	++++	−	Hypertension/angina
		Nicardipine	++++	−	Hypertension
		Felodipine	++++	−	Hypertension
		Nitrendipine	++++	−	Hypertension
		Isradipine	++++	−	Hypertension
		Nisoldipine	++++	−	Hypertension
		Amlodipine	++++	−	Hypertension/angina
		Lercanidipine	++++	−	Hypertension
Non-dihydropyridines	Benzothiazepines	Diltiazem	+++	++	Hypertension/angina/dysrhythmia
	Phenylalkylamines	Verapamil	++++	++++	Hypertension/angina/dysrhythmia

CCBs have different effects on vascular and cardiac tissues. The non-dihydropyridine agents have inotropic and chronotropic negative effects leading to their use in angina pectoris and dysrhythmia, whereas dihydropyridine agents have only peripheral vasodilator effect and are prescribed mainly for hypertension

both motif III (IIIS6) and motif IV (IVS6) and segment 5 of motif III (IIIS5) (Hockerman et al. 1997). Whereas the three major classes of CCBs all interact with receptor sites formed by the same general region of the α-1 subunit, binding with specific separate sites on this α-1 subunit allows a pharmacologic rationale for their combination, especially for diltiazem and dihydropyridines (Elliott and Ram 2011).

All currently available CCBs cause vasodilation, with lowering of blood pressure. The DHPs are more potent vasodilators and generally have less cardiodepressant activity than representatives of other classes of CCBs due to their high selectivity on the vasculature. On the opposite, verapamil has a high cardiac selectivity for the L-type Ca^{2+} channel and diltiazem has an intermediary profile (Lee and Tsien 1983; Herzig et al. 1992). Associated with the decrease in blood pressure, non-DHP induces a depression of cardiac contractility and conduction. Verapamil and diltiazem slow conduction through the atrioventricular node and increase the atrioventricular nodal refractory period, which, in turn, results in the slowing of the ventricular response rate in atrial fibrillation or flutter or in the conversion of atrioventricular nodal reentry tachyarrhythmias to sinus rhythm by disruption of the timing of the reentry circuit (Abernethy and Schwartz 1999). All the available CCBs are described in Table 1.

1.3 Usual Indications

All CCBs have been approved by the Food and Drug Administration (FDA) for the treatment of hypertension, alone or in combination with other antihypertensive drugs. Specific CCBs could also be indicated in angina pectoris prevention or cardiac dysrhythmia (Table 1). CCBs may also be beneficial in patients with left ventricular diastolic dysfunction, Raynaud's phenomenon, migraine, preterm labor, esophageal spasm, and bipolar disorders (Abernethy and Schwartz 1999). Verapamil and diltiazem are the only non-dihydropyridine drugs (phenylalkylamine and benzothiazepine, respectively) currently used in clinical practice whereas several DHPs are approved. The newer DHP calcium antagonists are structurally related to nifedipine, but may provide greater vascular selectivity and wider clinical utility. CCBs are usually used long term by oral route. Most are available in long-acting formulations that permit once-daily administration. Only diltiazem, nicardipine, and verapamil are available in intravenous formulations.

In addition to their role in the acute changes in ion flux associated with changes in membrane potential, L-type calcium channels seem to have a role in cellular growth and proliferation. Long-term treatment with CCBs for hypertension has been reported to decrease arteriolar and left ventricular hypertrophy at different levels regarding models and drugs (Jesmin et al. 2002; Kyselovic et al. 2001).

1.4 Side Effects

The most common side effects with CCB treatment are related to their vasodilator properties, with headache, flushing, dizziness, and nausea. Leg edema may occur in patients on long-term CCBs, the mechanism of this side effect being however unknown. Other side effects such as constipation (principally with verapamil), rash, and drowsiness may also occur. For short-acting DHPs, a reflex increase in heart rate has been reported (Meredith and Elliott 2004). Bradycardia is usually observed with diltiazem. Gingival hyperplasia may occur in patients who are given a DHP, but it is rare in those given verapamil or diltiazem.

CCBs are generally not recommended for patients with or high risk for heart failure due to reduction of left ventricular function. Verapamil and diltiazem are contraindicated in patients with hypotension, sick sinus syndrome, and second- or third-degree atrioventricular block and in patients with atrial flutter or atrial fibrillation. In addition, verapamil is contraindicated in patients with severe left ventricular dysfunction, whereas diltiazem is contraindicated in patients with acute myocardial infarction and pulmonary congestion (Elliott and Ram 2011).

1.5 Drug Interactions

All CCBs have drug interactions due to their metabolism by the CYP 3A4. Indeed, strong enzyme inducer as rifampicin or phenytoin or enzyme inhibitor as ketoconazole or cimetidine could significantly alter CCBs concentrations. Verapamil and diltiazem inhibit also the clearance of other substrates of CYP 3A4 (e.g., cyclosporine, carbamazepine, cyclosporine, tacrolimus, simvastatin, midazolam, and HIV-protease inhibitors), whereas the dihydropyridine drugs do not. Moreover, verapamil inhibits P-glycoprotein-mediated drug transport, which may alter the intestinal absorption of several drugs and their distribution into peripheral tissues and the central nervous system (Doppenschmitt et al. 1999). Associated with inhibition of P-glycoprotein-mediated transport of digoxin into peripheral tissues, verapamil also decreases its clearance (Kuhlmann 1987). After an initiation with verapamil in a patient who is receiving digoxin, the cardiac glycoside should be measured in serum after it has reached a new steady state to determine whether the dose should be reduced. Finally, due to their antiarrhythmic properties and additive negative inotropic effects, verapamil and diltiazem have to be used with caution when combined with β-blockers.

2 Calcium-Channel Blockers in Pulmonary Arterial Hypertension

Vasoconstriction of small pulmonary arteries is recognized as a component of the pathogenesis of PAH (Rich et al. 1992). Pure vasodilators alleviate vasoconstriction with little effect on fibrotic and proliferative changes that usually predominate over vasoconstriction in PAH (Rubin 1997). The goal of vasodilator therapy is to reduce pulmonary arterial pressure (PAP), increase cardiac output (CO), and thus decrease pulmonary vascular resistance (PVR) without symptomatic systemic hypotension. Before availability of PAH-targeted therapies (prostacyclin analogues, endothelin-receptor antagonists, PDE5 inhibitors), the question of the overall efficacy of administering vasodilators was of concern, as well as the way of safely identifying the patients who may benefit from long-term vasodilator therapy. Many vasodilators have been studied in PAH with opposite effects (Rubin and Peter 1980; Packer 1985; Hughes and Rubin 1986; Packer et al. 1982; McGoon and Vlietstra 1984). Among these, only CCBs have been proved to be effective long term in rare carefully selected PAH patients (Rich et al. 1992; Packer 1985; Sitbon et al. 1998, 2005).

2.1 Acute Vasodilator Testing

Pulmonary vasodilator testing establishes the relative contribution of reversible vasoconstriction in patients with PAH. It is generally accepted that the initial response to an acute pulmonary vasodilator testing accurately identified patients with PAH who were likely to respond to chronic vasodilator therapy (Reeves et al. 1986; Sitbon et al. 1998, 2005; Morales-Blanhir et al. 2004; Galie et al. 2009). Multiple criteria for a positive acute vasodilator test in patients with PAH have been suggested: a decrease in both mean PAP of at least 20 % and PVR of more than 20 or 30 % (Barst et al. 1999; Rich et al. 1992); a decrease in PVR of more than 30 % with a decrease in mPAP of more than 10 % (Groves et al. 1993); a decrease in PVR of 50 % with a mean PAP lower than 30 mmHg (Raffy et al. 1996). Conclusions from the third World Symposium on PAH (Barst et al. 2004) and guidelines from the European Society of Cardiology/European Respiratory Society (Galie et al. 2004) and from the American College of Chest Physicians (Badesch et al. 2004) proposed a decrease in mean PAP of at least 10 mmHg to reach an absolute value of 40 mmHg or less without decrease in CO, which is currently recommended to consider a pulmonary vasodilator test as positive (Fig. 2). The current criteria derive from a retrospective analysis of 557 idiopathic PAH patients who received long-term treatment with CCBs after a positive acute pulmonary vasodilator test by older criteria (fall in both mean PAP and PVR > 20 %). Of 12.6 % of the patients who exhibited acute pulmonary vasoreactivity, only half (6.8 %) improved long term with the use of CCBs (Sitbon et al. 2005). The mean PAP reached during acute vasodilator testing was less than 40 mmHg in the majority of patients who were considered long-term CCB responders (Sitbon et al. 2005).

2.2 Drugs Used for Acute Vasodilator Testing

Drugs used for acute vasodilator testing have been chosen based on selectively for the pulmonary circulation and short duration of action to minimize systemic adverse effects (Naeije and Vachiery 2001). The agents currently recommended to evaluate the pulmonary vasoresponsiveness include inhaled nitric oxide, intravenous adenosine, or epoprostenol (Galie et al. 2004; Badesch et al. 2009; Zuo et al. 2012; Nootens et al. 1995; Sitbon et al. 1995). In addition, inhaled iloprost has acquired substantial evidence for this use, although this agent is not currently recommended by the most updated guidelines (Jing et al. 2009; Hoeper et al. 2000). There is a lack of consensus on the preferred agent for determining acute pulmonary vasoreactivity (Ghofrani et al. 2008). NO has the advantages to be most selective agent for pulmonary vascular bed and to have an easy route of administration, good safety, and low cost. The ACCP/AHA 2009 Expert Consensus Document on Pulmonary Hypertension recommends inhaled NO as the preferred vasodilator agent and considers IV epoprostenol and adenosine acceptable alternatives (McLaughlin et al. 2009). NO doses of 10–80 ppm for up to 15 min are usually given, even though two studies have shown that the vasodilator response to NO was not dependent upon the NO concentration (Sitbon et al. 1995; Ricciardi et al. 1998). Doses of 10–20 ppm for 5 min are usually used in clinical practice.

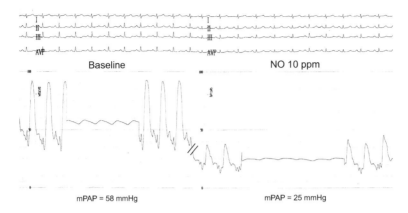

Fig. 2 Acute vasodilator testing with inhaled NO in a 45-year-old male with idiopathic PAH. Mean PAP dropped from 58 to 25 mmHg during acute vasodilator testing. Patient subsequently received long-term treatment with diltiazem (360 mg daily) leading to major and sustained clinical and hemodynamic improvement

2.3 Place of Calcium-Channel Blockers in PAH

A small uncontrolled study has suggested that long-term administration of high-dose CCBs dramatically improves survival in a small subset of patients who respond acutely to those drugs, compared with unresponsive patients (Rich et al. 1992). A large retrospective study from the French reference center for PAH has shown that less than 10 % of patients with idiopathic and anorexigen-associated PAH may respond to long-term CCB therapy (Sitbon et al. 2005). Patients with heritable PAH are less likely to exhibit a significant response to acute vasodilator testing (Elliott et al. 2006; Rosenzweig et al. 2008; Sztrymf et al. 2008; Montani et al. 2010). In forms of PAH other than idiopathic PAH and PAH associated with anorexigens (i.e., PAH associated with connective tissue diseases, HIV infection, portal hypertension, or congenital cardiac shunts), the rate of acute responders is even rare, and long-term response to CCBs is exceptional in these patients (Montani et al. 2010).

The updated evidence-based treatment algorithm for patients with PAH recommends to perform an acute vasodilator testing in all patients with PAH at the time of first evaluation by right heart catheterization, acknowledging that patients with iPAH, and anorexigen-associated PAH are more likely to respond to CCB treatment (Barst et al. 2009; Galie et al. 2009). In the ESC/ERS guidelines, the degree of recommendation to perform an acute vasodilator testing is strong (I-C) for patients with idiopathic and anorexigen-associated PAH and weak (IIb-C) for patients with other forms of PAH (Galie et al. 2009).

2.4 Clinical Use of Calcium-Channel Blockers in PAH

No randomized controlled trial has been performed to confirm the beneficial effects of CCBs in the treatment of PAH, predominantly because of the rarity of the condition (Badesch et al. 2004), and CCB therapy in PAH has not received regulatory approval anywhere because of this. However, open-label prospective studies have shown significant improvement outcome in selected PAH patients treated with CCBs. In 1987, Rich et al. have shown a decrease in PAP, PVR, and right ventricular hypertrophy in more than half of patients with idiopathic PAH (formerly primary pulmonary hypertension) who responded favorably to acute vasodilator testing and long-term treatment with CCBs (Rich and Brundage 1987). In this study, high doses of CCBs were required to produce marked hemodynamic response (Rich and Brundage 1987). These results were confirmed by the same group in 1992 in a larger group of patients (Rich et al. 1992). They found that 27 % of patients with idiopathic PAH responded favorably to an acute vasodilator testing with high doses of CCBs and the majority experienced clinical and hemodynamic improvement on long-term treatment with such drugs. In addition, there was a major difference in survival favoring responders versus nonresponders (Rich et al. 1992). This benefit in survival in CCB responders was confirmed in a large retrospective study performed in France with a survival rate of 97 % at 5 years in responders versus 48 % in patients who failed on long-term CCBs (Sitbon et al. 2005). In that study, a sustained benefit with CCBs was defined as the improvement or being maintained in NYHA functional class I or II with near-normalization in pulmonary haemodynamics after a year of treatment (Sitbon et al. 2005).

The current approach of treatment with CCBs in PAH is to initiate the drug if the patient displays a positive vasodilator response, an adequate systemic blood pressure (above 90 mmHg) and a stable NYHA functional class I–III prior to initiation of therapy (Barst et al. 2009; Galie et al. 2009). The agents commonly recommended are diltiazem, nifedipine, or amlodipine (Barst et al. 2009; Galie et al. 2009). The choice between nifedipine/amlodipine and diltiazem is guided by heart rate at rest: diltiazem is preferred in patients with resting HR > 80 bpm (Rich et al. 1992). Verapamil is not recommended due to its prominent negative inotropic effect (Packer et al. 1984). Moreover, CCBs are contraindicated in patients with right heart failure or hemodynamic instability (McLaughlin et al. 2009). Finally, high doses of CCBs must not be used in patients with negative pulmonary vasodilator test as these agents can lead to systemic side effects, worsening of right ventricular function, pulmonary edema, and ultimately death in those patients (Partanen et al. 1993; Farber et al. 1983).

Acute testing with CCBs as proposed by Rich (Rich et al. 1992) has been progressively abandoned for a progressive up-titration of the dose over a few weeks (Gaine 2000). Nifedipine is usually initiated at a dose of 30 mg/day and up-titrated to 60 mg/day and sometimes over up to 240 mg/day. Diltiazem is started at 180 mg/day and up-titrated to 360 mg/day, up to 720 mg/day (Rich et al. 1992).

Table 2 Use of calcium-channel blockers in PAH

CCB agent	Started dosage (mg/day)	Usual dosage (mg/day)	Most frequent side effects
Nifedipine	30	60–240	Systemic hypotension, edema, headache, nausea
Amlodipine	5	20–40	Not reported
Diltiazem	180	360–720	Systemic hypotension, bradycardia, edema, headache, nausea

Nifedipine, diltiazem, and to a less extent amlodipine are the three CCBs used in PAH patients who are responders to acute vasodilator testing. CCBs are given at high dosages with a progressive up-titration. Side effects are related to their vasodilator properties (Sitbon et al. 1998)

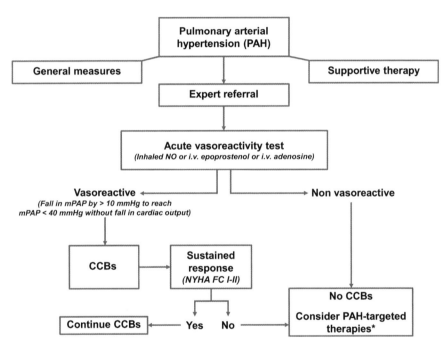

Fig. 3 Place of calcium-channel blockers in the PAH treatment algorithm (adapted from Galie et al. 2009). Abbreviations: *CCBs* calcium-channel blockers, *mPAP* mean pulmonary artery pressure, *NO* nitric oxide, *NYHA FC* New York Heart Association functional class. *Asterisk*: PAH-targeted therapies: prostacyclin analogues, endothelin-receptor antagonists, phosphodiesterase type-5 inhibitors

The target dose of amlodipine is 20 mg/day (Woodmansey et al. 1996; Barst et al. 2009) (Table 2). The side effects most frequently reported with CCBs in PAH are systemic hypotension and edema of the lower limbs (Sitbon et al. 1998). Association with low dose of diuretics and elastic stockings will help to reverse these side effects.

Patients with iPAH who meet the criteria for a positive vasodilator response and receive treatment with CCBs should be followed closely for both safety and efficacy with an initial reassessment after 3–4 months on therapy including clinical and hemodynamic evaluation. We recommend repeating acute vasodilator testing during the first reassessment on CCBs in order to determine the possibility for further increase in CCB daily dose in case of demonstration of an additional pulmonary vasodilator response. Finally, if NYHA functional class I or II has not been achieved, then additional PAH-targeted therapies are suggested (Fig. 3).

3 Conclusion

CCBs are the first drugs that have been shown to improves long-term survival in patients with PAH. Unfortunately, treatment with those agents should be restricted to rare patients with idiopathic or anorexigen-related PAH who demonstrate an acute pulmonary vasodilator response during first assessment with right heart catheterization. It is widely accepted that the initial response to an acute vasodilator test with inhaled nitric oxide or intravenous prostacyclin or adenosine accurately identify patients with PAH who are likely to respond to chronic CCBs therapy. Calcium antagonists must not ultimately be used in patients with PAH in the absence of positive vasodilator response.

References

Abernethy DR, Schwartz JB (1999) Calcium-antagonist drugs. N Engl J Med 341:1447–1457
Badesch DB, Abman SH, Ahearn GS, Barst RJ, McCrory DC, Simonneau G, McLaughlin VV (2004) Medical therapy for pulmonary arterial hypertension: ACCP evidence-based clinical practice guidelines. Chest 126:35S–62S
Badesch DB, Champion HC, Sanchez MA, Hoeper MM, Loyd JE, Manes A, McGoon M, Naeije R, Olschewski H, Oudiz RJ, Torbicki A (2009) Diagnosis and assessment of pulmonary arterial hypertension. J Am Coll Cardiol 54:S55–S66
Barst RJ, Maislin G, Fishman AP (1999) Vasodilator therapy for primary pulmonary hypertension in children. Circulation 99:1197–1208
Barst RJ, McGoon M, Torbicki A, Sitbon O, Krowka MJ, Olschewski H, Gaine S (2004) Diagnosis and differential assessment of pulmonary arterial hypertension. J Am Coll Cardiol 43:40S–47S
Barst RJ, Gibbs JS, Ghofrani HA, Hoeper MM, McLaughlin VV, Rubin LJ, Sitbon O, Tapson VF, Galie N (2009) Updated evidence-based treatment algorithm in pulmonary arterial hypertension. J Am Coll Cardiol 54:S78–S84
Bers DM (2002) Sarcoplasmic reticulum Ca release in intact ventricular myocytes. Front Biosci 7: d1697–d1711
Brette F, Leroy J, Le Guennec JY, Salle L (2006) Ca^{2+} currents in cardiac myocytes: old story, new insights. Prog Biophys Mol Biol 91:1–82
Catterall WA (2011) Voltage-gated calcium channels. Cold Spring Harb Perspect Biol 3:a003947
Catterall WA, Perez-Reyes E, Snutch TP, Striessnig J (2005) International Union of Pharmacology. XLVIII. Nomenclature and structure-function relationships of voltage-gated calcium channels. Pharmacol Rev 57:411–425
Chandy KG, Gutman GA (1993) Nomenclature for mammalian potassium channel genes. Trends Pharmacol Sci 14:434

De Waard M, Gurnett CA, Campbell KP (1996) Structural and functional diversity of voltage-activated calcium channels. Ion Channels 4:41–87

Doppenschmitt S, Langguth P, Regardh CG, Andersson TB, Hilgendorf C, Spahn-Langguth H (1999) Characterization of binding properties to human P-glycoprotein: development of a [3H] verapamil radioligand-binding assay. J Pharmacol Exp Ther 288:348–357

Elliott WJ, Ram CV (2011) Calcium channel blockers. J Clin Hypertens (Greenwich) 13:687–689

Elliott CG, Glissmeyer EW, Havlena GT, Carlquist J, McKinney JT, Rich S, McGoon MD, Scholand MB, Kim M, Jensen RL, Schmidt JW, Ward K (2006) Relationship of BMPR2 mutations to vasoreactivity in pulmonary arterial hypertension. Circulation 113:2509–2515

Ertel EA, Campbell KP, Harpold MM, Hofmann F, Mori Y, Perez-Reyes E, Schwartz A, Snutch TP, Tanabe T, Birnbaumer L, Tsien RW, Catterall WA (2000) Nomenclature of voltage-gated calcium channels. Neuron 25:533–535

Farber HW, Karlinsky JB, Faling LJ (1983) Fatal outcome following nifedipine for pulmonary hypertension. Chest 83:708–709

Gaine S (2000) Pulmonary hypertension. JAMA 284:3160–3168

Galie N, Torbicki A, Barst R, Dartevelle P, Haworth S, Higenbottam T, Olschewski H, Peacock A, Pietra G, Rubin LJ, Simonneau G, Priori SG, Garcia MA, Blanc JJ, Budaj A, Cowie M, Dean V, Deckers J, Burgos EF, Lekakis J, Lindahl B, Mazzotta G, McGregor K, Morais J, Oto A, Smiseth OA, Barbera JA, Gibbs S, Hoeper M, Humbert M, Naeije R, Pepke-Zaba J (2004) Guidelines on diagnosis and treatment of pulmonary arterial hypertension. The Task Force on Diagnosis and Treatment of Pulmonary Arterial Hypertension of the European Society of Cardiology. Eur Heart J 25:2243–2278

Galie N, Hoeper MM, Humbert M, Torbicki A, Vachiery JL, Barbera JA, Beghetti M, Corris P, Gaine S, Gibbs JS, Gomez-Sanchez MA, Jondeau G, Klepetko W, Opitz C, Peacock A, Rubin L, Zellweger M, Simonneau G (2009) Guidelines for the diagnosis and treatment of pulmonary hypertension: the Task Force for the Diagnosis and Treatment of Pulmonary Hypertension of the European Society of Cardiology (ESC) and the European Respiratory Society (ERS), endorsed by the International Society of Heart and Lung Transplantation (ISHLT). Eur Heart J 30:2493–2537

Ghofrani HA, Wilkins MW, Rich S (2008) Uncertainties in the diagnosis and treatment of pulmonary arterial hypertension. Circulation 118:1195–1201

Groves BM, Badesch DB, Turkevich D et al (1993) Correlation of acute prostacyclin response in primary (unexplained) pulmonary hypertension with efficacy of treatment with calcium channel blockers and survival. In: Weir K (ed) Ion flux in pulmonary vascular control. Plenum, New York

Herzig S, Lullmann H, Sieg H (1992) Frequency- and potential-dependency of the negative inotropic action of various dihydropyridine and non-dihydropyridine calcium antagonists. Pharmacol Toxicol 71:229–235

Hockerman GH, Peterson BZ, Johnson BD, Catterall WA (1997) Molecular determinants of drug binding and action on L-type calcium channels. Annu Rev Pharmacol Toxicol 37:361–396

Hoeper MM, Olschewski H, Ghofrani HA, Wilkens H, Winkler J, Borst MM, Niedermeyer J, Fabel H, Seeger W (2000) A comparison of the acute hemodynamic effects of inhaled nitric oxide and aerosolized iloprost in primary pulmonary hypertension. German PPH study group. J Am Coll Cardiol 35:176–182

Hofmann F, Biel M, Flockerzi V (1994) Molecular basis for Ca^{2+} channel diversity. Annu Rev Neurosci 17:399–418

Hosey MM, Chien AJ, Puri TS (1996) Structure and regulation of L-type calcium channels a current assessment of the properties and roles of channel subunits. Trends Cardiovasc Med 6:265–273

Hughes JD, Rubin LJ (1986) Primary pulmonary hypertension. An analysis of 28 cases and a review of the literature. Medicine (Baltimore) 65:56–72

Jesmin S, Sakuma I, Hattori Y, Fujii S, Kitabatake A (2002) Long-acting calcium channel blocker benidipine suppresses expression of angiogenic growth factors and prevents cardiac remodelling in a Type II diabetic rat model. Diabetologia 45:402–415

Jing ZC, Jiang X, Han ZY, Xu XQ, Wang Y, Wu Y, Lv H, Ma CR, Yang YJ, Pu JL (2009) Iloprost for pulmonary vasodilator testing in idiopathic pulmonary arterial hypertension. Eur Respir J 33:1354–1360

Kimball SD, Hunt JT, Barrish JC, Das J, Floyd DM, Lago MW, Lee VG, Spergel SH, Moreland S, Hedberg SA et al (1993) 1-Benzazepin-2-one calcium channel blockers – VI. Receptor-binding model and possible relationship to desmethoxyverapamil. Bioorg Med Chem 1:285–307

Kuhlmann J (1987) Effects of quinidine, verapamil and nifedipine on the pharmacokinetics and pharmacodynamics of digitoxin during steady state conditions. Arzneimittelforschung 37: 545–548

Kyselovic J, Krenek P, Wibo M, Godfraind T (2001) Effects of amlodipine and lacidipine on cardiac remodelling and renin production in salt-loaded stroke-prone hypertensive rats. Br J Pharmacol 134:1516–1522

Lee KS, Tsien RW (1983) Mechanism of calcium channel blockade by verapamil, D600, diltiazem and nitrendipine in single dialysed heart cells. Nature 302:790–794

McDonald TF, Pelzer S, Trautwein W, Pelzer DJ (1994) Regulation and modulation of calcium channels in cardiac, skeletal, and smooth muscle cells. Physiol Rev 74:365–507

McGoon MD, Vlietstra RE (1984) Vasodilator therapy for primary pulmonary hypertension. Mayo Clin Proc 59:672–677

McLaughlin VV, Archer SL, Badesch DB, Barst RJ, Farber HW, Lindner JR, Mathier MA, McGoon MD, Park MH, Rosenson RS, Rubin LJ, Tapson VF, Varga J (2009) ACCF/AHA 2009 expert consensus document on pulmonary hypertension a report of the American College of Cardiology Foundation Task Force on Expert Consensus Documents and the American Heart Association developed in collaboration with the American College of Chest Physicians; American Thoracic Society, Inc.; and the Pulmonary Hypertension Association. J Am Coll Cardiol 53:1573–1619

Meredith PA, Elliott HL (2004) Dihydropyridine calcium channel blockers: basic pharmacological similarities but fundamental therapeutic differences. J Hypertens 22:1641–1648

Montani D, Savale L, Natali D, Jais X, Herve P, Garcia G, Humbert M, Simonneau G, Sitbon O (2010) Long-term response to calcium-channel blockers in non-idiopathic pulmonary arterial hypertension. Eur Heart J 31:1898–1907

Morales-Blanhir J, Santos S, de Jover L, Sala E, Pare C, Roca J, Rodriguez-Roisin R, Barbera JA (2004) Clinical value of vasodilator test with inhaled nitric oxide for predicting long-term response to oral vasodilators in pulmonary hypertension. Respir Med 98:225–234

Naeije R, Vachiery JL (2001) Medical therapy of pulmonary hypertension. Conventional therapies. Clin Chest Med 22:517–527

Nootens M, Schrader B, Kaufmann E, Vestal R, Long W, Rich S (1995) Comparative acute effects of adenosine and prostacyclin in primary pulmonary hypertension. Chest 107:54–57

Packer M (1985) Therapeutic application of calcium-channel antagonists for pulmonary hypertension. Am J Cardiol 55:196B–201B

Packer M, Greenberg B, Massie B, Dash H (1982) Deleterious effects of hydralazine in patients with pulmonary hypertension. N Engl J Med 306:1326–1331

Packer M, Medina N, Yushak M, Wiener I (1984) Detrimental effects of verapamil in patients with primary pulmonary hypertension. Br Heart J 52:106–111

Partanen J, Nieminen MS, Luomanmaki K (1993) Death in a patient with primary pulmonary hypertension after 20 mg of nifedipine. N Engl J Med 329:812, author reply 812–813

Raffy O, Azarian R, Brenot F, Parent F, Sitbon O, Petitpretz P, Hervé P, Duroux P, Dinh-Xuan AT, Simonneau G (1996) Clinical significance of the pulmonary vasodilator response during short-term infusion of prostacyclin in primary pulmonary hypertension. Circulation 93:484–488

Reeves JT, Groves BM, Turkevich D (1986) The case for treatment of selected patients with primary pulmonary hypertension. Am Rev Respir Dis 134:342–346

Ricciardi MJ, Knight BP, Martinez FJ, Rubenfire M (1998) Inhaled nitric oxide in primary pulmonary hypertension: a safe and effective agent for predicting response to nifedipine. J Am Coll Cardiol 32:1068–1073

Rich S, Brundage BH (1987) High-dose calcium channel-blocking therapy for primary pulmonary hypertension: evidence for long-term reduction in pulmonary arterial pressure and regression of right ventricular hypertrophy. Circulation 76:135–141

Rich S, Kaufmann E, Levy PS (1992) The effect of high doses of calcium-channel blockers on survival in primary pulmonary hypertension. N Engl J Med 327:76–81

Rosenzweig EB, Morse JH, Knowles JA, Chada KK, Khan AM, Roberts KE, McElroy JJ, Juskiw NK, Mallory NC, Rich S, Diamond B, Barst RJ (2008) Clinical implications of determining BMPR2 mutation status in a large cohort of children and adults with pulmonary arterial hypertension. J Heart Lung Transplant 27:668–674

Rubin LJ (1997) Primary pulmonary hypertension. N Engl J Med 336:111–117

Rubin LJ, Peter RH (1980) Primary pulmonary hypertension: new approaches to therapy. Am Heart J 100:757–759

Sitbon O, Brenot F, Denjean A, Bergeron A, Parent F, Azarian R, Herve P, Raffestin B, Simonneau G (1995) Inhaled nitric oxide as a screening vasodilator agent in primary pulmonary hypertension. A dose-response study and comparison with prostacyclin. Am J Respir Crit Care Med 151: 384–389

Sitbon O, Humbert M, Jagot JL, Taravella O, Fartoukh M, Parent F, Herve P, Simonneau G (1998) Inhaled nitric oxide as a screening agent for safely identifying responders to oral calcium-channel blockers in primary pulmonary hypertension. Eur Respir J 12:265–270

Sitbon O, Humbert M, Jais X, Ioos V, Hamid AM, Provencher S, Garcia G, Parent F, Herve P, Simonneau G (2005) Long-term response to calcium channel blockers in idiopathic pulmonary arterial hypertension. Circulation 111:3105–3111

Sztrymf B, Coulet F, Girerd B, Yaici A, Jais X, Sitbon O, Montani D, Souza R, Simonneau G, Soubrier F, Humbert M (2008) Clinical outcomes of pulmonary arterial hypertension in carriers of BMPR2 mutation. Am J Respir Crit Care Med 177:1377–1383

Tsien RW (1983) Calcium channels in excitable cell membranes. Annu Rev Physiol 45:341–358

Woodmansey PA, O'Toole L, Channer KS, Morice AH (1996) Acute pulmonary vasodilatory properties of amlodipine in humans with pulmonary hypertension. Heart 75:171–173

Zuo XR, Zhang R, Jiang X, Li XL, Zong F, Xie WP, Wang H, Jing ZC (2012) Usefulness of intravenous adenosine in idiopathic pulmonary arterial hypertension as a screening agent for identifying long-term responders to calcium channel blockers. Am J Cardiol 109(12): 1801–1806

Prostacyclins

Horst Olschewski

Abstract Prostacyclins have a favourable pharmacological profile for treatment of pulmonary hypertension as they possess vasodilative, antiproliferative, anti-aggregatory, and anti-inflammatory properties that may compensate the main pathologic changes in the small pulmonary arteries. In severe pulmonary hypertension these vessels show a deficit in the endogenous prostacyclin secretion. The therapeutic potential of prostacyclin for pulmonary hypertension has been known since 30 years, and since nearly 20 years prostacyclin has been approved for idiopathic PAH. There are intravenous, subcutaneous, and inhaled approaches of different substances who share many but not all pharmacologic properties. However, none of these approaches are easy and free of adverse effects. Long-term experience and careful decision-making are instrumental to achieve favourable clinical long-term results.

Keywords Prostacyclin • Epoprostenol • Iloprost • Treprostinil • Beraprost • Selexipag

Contents

1	Introduction	178
2	History of Prostacyclin and Analogues	178
3	Pharmacologic Properties	179
	3.1 Differences Between Prostacyclin Analogues	181
	3.2 Hemodynamic Effects	182
4	Clinical Application	183
	4.1 Epoprostenol: Intravenous	183
	4.2 Iloprost: Intravenous	185
	4.3 Treprostinil: Intravenous	186

H. Olschewski (✉)
Division of Pulmonology, Department of Internal Medicine, Medical University of Graz, Graz, Austria
e-mail: Horst.Olschewski@klinikum-graz.at

4.4	Treprostinil: Subcutaneous	186
4.5	Beraprost: Oral	187
4.6	Selexipag	188
4.7	Iloprost: Inhaled	188
4.8	Treprostinil: Inhaled	190
5	Management of Prostacyclin Therapy	190
5.1	Decision for a Prostacyclin or Its Analogues	190
5.2	Potential Treatment Hazards	191
5.3	Practical Issues of Prostacyclin Treatment	191
5.4	Switch from iv Prostacyclins to Other Forms of Application	192
5.5	Switch from Inhaled Prostacyclins to Continuous Infusion	192
6	Conclusion	193
References		193

1 Introduction

Prostacyclin and its stable analogues possess a chemical prostaglandin structure and preferentially bind to the prostaglandin I (IP) receptor. Prostacyclins are potent vasodilators and possess antithrombotic, antiproliferative, and anti-inflammatory properties. They also reduce matrix secretion in smooth muscle cells (SMC), endothelial cells, and fibroblasts (Fig. 1). The endothelial cells are the major source of endogenous prostacyclin. Due to the short half-life, the prostacyclin action will be directed both on the local vascular wall and on passing blood cells that get in contact with the endothelium. Pulmonary arterial hypertension (PAH) is associated with vasoconstriction, thrombosis, proliferation, inflammation, and a lack of endogenous prostacyclin secretion. This provides a strong rationale for prostacyclin use as therapy for PAH.

2 History of Prostacyclin and Analogues

Endogenous prostacyclin was discovered in 1976 (Moncada et al. 1976). In the same year, prostacyclin was chemically synthesised (epoprostenol) and investigated in animals and humans (Gryglewski 1980). Chemically stable analogues were synthesised in the 1980s. Clinical studies in PAH patients were performed with epoprostenol, iloprost, beraprost, and treprostinil while cicaprost, a highly specific IP-receptor agonist, was not clinically developed.

The first applications of prostacyclin in pulmonary hypertension were published in 1980 (O'Grady et al. 1980) and the first long-term therapy for severe idiopathic pulmonary arterial hypertension (PAH) was applied in 1984 (Higenbottam et al. 1984). Despite many new developments in PAH therapy, prostacyclins remain a mainstay in the treatment of PAH.

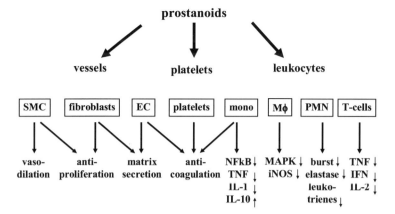

Fig. 1 Effects of prostacyclins on the vessel wall and adherent blood cells. *SMC* smooth muscle cells; *EC* endothelial cells; *mono* mononuclear cells; *MΦ* macrophages; *PMN* polymorphonuclear neutrophils; *T-cells* T-lymphocytes; *TNF* transforming nuclear factor α; *IL* interleukin; *MAPK* mitogen-activated protein kinase, burst, generation of reactive oxygen species, elastase, elastase secretion; *IFN* interferon γ. From Olschewski et al. (2001b)

3 Pharmacologic Properties

The main target of prostacyclin and its analogues is the human IP receptor which is abundantly expressed in blood vessels, leucocytes, and thrombocytes and is rapidly activated by binding of the ligand. The IP receptor in small animals is much less sensitive to prostacyclin as the human receptor. Therefore predictions about dose-response relationships based on such animal models may not be reliable. The IP receptor is coupled with Gs proteins and activates adenylate cyclase, leading to increased cAMP levels in the target cells which explain most of the biologic effects. Principally, it also couples with Gq proteins and might activate vasoconstrictive pathways under certain instances (Chow et al. 2003; Wise 2003). In addition, prostacyclin and analogues are not highly specific to the IP receptor but may also activate EP receptors (Abramovitz et al. 2000) which are located on the cell surface as well as in the nucleus (Bhattacharya et al. 1998, 1999). They may further activate the peroxisome proliferator-activated receptor delta (PPARδ) (Gupta et al. 2000). Both PPARα and PPARδ may also be activated via IP receptor-dependent PKA activation, but the intracellular prostacyclin produced by the endogenous PGI synthase seems to specifically activate the apoptosis pathway by activation of PPARδ (Hatae et al. 2001) (Fig. 2). There is evidence that PPARδ is also involved in the acute signalling in prostacyclin-induced vasodilatation (Li et al. 2012). Specifically, prostacyclins activate the most important potassium channels, the TWIK-related acid-sensitive potassium channel (TASK1) and the calcium-activated potassium channel (KCA) and thereby cause membrane hyperpolarization and inhibition of L-type calcium channels in SMCs. Figure 3 represents the current knowledge about the mechanisms underlying the powerful activation of potassium current

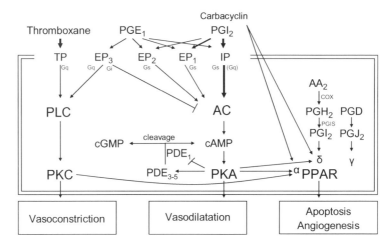

Fig. 2 Intracellular targets of prostacyclin. Extracellular PGI_2 mainly but not specifically acts on the IP receptor and the EP1 receptor while PGE1 is more specific to EP receptors. *Gs, Gq, and Gi* G-proteins, *PLC* phospholipase C, *PKC* proteinkinase C, *AC* adenylyl-cyclase, *PKA* protein kinase A, AA_2 arachidonic acid, *COX* cyclooxygenases 1 and 2, PGH_2 protaglandin H_2, *PGIS* prostacyclin synthase, *PPAR* peroxisome proliferator-activated receptors, *PGD* prostaglandin D, PGJ_2 prostaglandin J_2, *PDE* phosphodiesterases. From Gomberg-Maitland and Olschewski (2008)

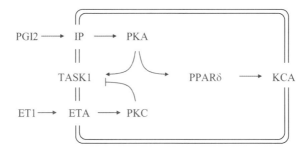

Fig. 3 Prostacyclin-induced activation of potassium channels prostacyclin (PGI2) activates proteinkinase A (PKA) via IP receptors of the cell membrane. PKA activates both the TWIK-related acid-sensitive K channel (TASK1) and the calcium-activated potassium channel (KCA). The activation of KCA is critically dependent on activation of PPARδ, while TASK1 activation is independent of PPARδ. Endothelin1 (ET1) counteracts the activation of TASK1 via proteinkinase C (PKC) activation. Data from Li et al. (2012) and Tang et al. (2009)

from TASK1 and KCA channels by prostacyclin and the opposite effects of endothelin-1 (Tang et al. 2009). In human pulmonary arterial smooth muscle cells, TASK1 is the only active potassium channel at physiologic membrane potential and is therefore mainly responsible for the negative membrane potential which provides a low vascular tone (Olschewski et al. 2006). It has recently been shown that TASK1 activity depends on both oxygen and src tyrosine kinase activity (Nagaraj et al. 2012).

3.1 Differences Between Prostacyclin Analogues

Epoprostenol is provided as a dry powder with a highly basic glycine buffer as a solvent drug. After mixing drug powder with the solvent, the solution has to be used within 12–24 h due to spontaneous degradation of the compound. A new formulation that was FDA approved in 2009 (Veletri®) allows for 48 h use at 25 °C and 5 days of storage at 2–8 °C. All other prostanoids are chemically stable in solution and their plasma half-life is much longer. It is about 30 min with iloprost and beraprost and up to 4.5 h with treprostinil. Vein irritation is common to all prostanoids. All stable prostanoids have been provided as oral preparations; however, only beraprost has received approval for PAH in Japan and some Eastern countries (see below). Iloprost has been approved as inhalative therapy and treprostinil as subcutaneous infusion or intravenous infusion or inhalation (see below). The doses that are needed for strong vasodilative effects differ considerably among prostanoids. A continuous intravenous dose of 50 ng/kg/min of epoprostenol is, on average, translated into 100 ng/kg/min of treprostinil but just 3 ng/kg/min of iloprost. The reasons for these huge differences are not known.

Apart from differences in pharmacokinetics and pharmacodynamics, there may be other specific differences between the prostacyclin analogues. The antiproliferative effects of treprostinil appeared more impressive than those of other compounds (Clapp et al. 2002), but due to differences in chemical stability such results must be interpreted with caution. Non-IP-receptor effects of prostacyclin analogues have attracted some interest in tumour biology (Keith and Geraci 2006) and might be interesting for pulmonary hypertension as well. Iloprost and cicaprost were found to have different effects in the murine corneal model of angiogenesis. While iloprost caused significant angiogenesis, comparable to VEGF, cicaprost had no such effects (Pola et al. 2004). The explanation may be that iloprost and carbacyclin but not cicaprost activate peroxisome proliferator-activated receptors (PPAR) (Reginato et al. 1998), which results in VEGF secretion (Forman et al. 1995; Barger 2002). VEGF increase may antagonise endothelial dysfunction and represents a potential beneficial factor (le Cras et al. 2002; Taraseviciene-Stewart et al. 2001).

Some effects have only been described in one of the compounds but may be common to all prostacyclin analogues. Treprostinil augments the positive inotropic effects of catecholamines in isolated ventricular myocytes (Fontana et al. 2007), although it has no positive inotropic effects of its own. This effect may have clinical relevance because during right heart failure there is an increased catecholamine drive. These effects may add on indirect effects that originate from systemic vasodilation and subsequent baroreflex activation (Fig. 4) and improved ventriculoarterial coupling (Kerbaul et al. 2007).

Iloprost suppresses neutrophil adhesion, respiratory burst, and elastase secretion (Rose et al. 2003). This may be important because inflammation appears to play a role among the pathologic mechanisms of pulmonary arterial hypertension (Stacher et al. 2012). In SMC, a number of beneficial changes in gene expression

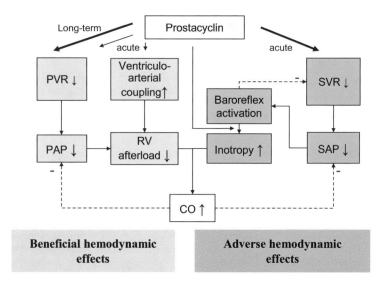

Fig. 4 Hemodynamic effects of systemically applied prostacyclins or its analogues. Within minutes, prostacyclins cause a strong vasodilatation which in most patients is more prominent on the systemic (SVR) compared to the pulmonary vascular resistance (PVR). The decrease in systemic pressure (SAP) activates the arterial baroreflex which increases SVR and has positive inotropic effects. The reduction in PVR would decrease pulmonary arterial pressure (PAP); however, in the acute phase, the increase in cardiac output counteracts this resulting in an unchanged PAP. Within weeks, PVR decreases more prominently and SVR does not further decrease which may result in a significant PAP decrease. For ventriculoarterial coupling and inotropic effects of prostacyclin, see text. Adapted from Olschewski and Olschewski (2011)

(hyaluronidonic acid, COX 2, and VEGF upregulation; monocyte chemotactic protein and plasminogen activator inhibitor-1 downregulation) were shown after incubation with iloprost (Meyer-Kirchrath et al. 2004; Sussmann et al. 2004). If we consider that endothelial dysfunction or damage may play a major role in the pathologic development of pulmonary artery remodelling, this translates into a number of beneficial effects of iloprost (Fig. 5).

3.2 Hemodynamic Effects

The hemodynamic effects and the associated side effects of all prostacyclin analogues are similar while the mode of application and the applied doses differ. In the case of continuous intravenous infusion there is a relatively fast dose increase with systemic prostacyclin analogues over the first weeks, followed by a gradual dose increase over many months. Interestingly, there is no loss of pharmacological effect if the infusion rate is adequately adapted. The doses for epoprostenol typically start at 4 ng/kg/min and increase to about 20–60 ng/kg/min after a year.

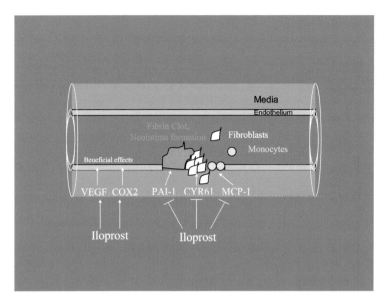

Fig. 5 Hypothesis for neointimal formation and possible role of prostacyclins. After endothelial damage, monocytes, fibroblasts, and the coagulation system will form a neointima, which leads to intima fibrosis. Iloprost has been shown to inhibit the mRNA expression of important mediators of these processes. *VEGF* vascular endothelial growth factor, *COX2* cyclooxygenase 2, *PAI-1* plasminogen activator inhibitor, *Cyr 61* cysteine-rich angiogenic protein, *MCP-1* monocyte chemotactic protein. From Gomberg-Maitland and Olschewski (2008)

In the majority of patients, the beneficial hemodynamic effects, characterized by a significant decrease in pulmonary vascular resistance, will occur after some days of therapy; however, they may be obscured by systemic side effects. It takes several weeks until the patient will fully anticipate the beneficial effect of the drug because the pulmonary effects become more and more prominent while the systemic side effects fade away.

If the prostacyclin analogues are applied by inhalation every 2–3 h, there is no need for dose increase in the majority of patients (Olschewski et al. 1996, 2000, 2003, 2010), suggesting that this mode of application may not cause receptor desensitisation.

4 Clinical Application

4.1 Epoprostenol: Intravenous

Continuous intravenous prostacyclin (epoprostenol) was the first therapy for idiopathic and heritable PAH. Case reports and small case series (Higenbottam et al. 1984; Weir et al. 1989; Rubin et al. 1982) demonstrated improvement in symptoms and hemodynamics and were followed by a prospective, randomised open-label controlled trial (Barst et al. 1996) for 12 weeks. The 6 min walk distance (6MWD) improved by 32 m in the epoprostenol group compared with a decrease of 15 m in the control group. The mean pulmonary vascular resistance decreased by 21 % in the epoprostenol group vs. an increase of 9 % in the conventional group. This study also suggested a survival benefit as those eight patients who died during the 12-week trial were all in the conventional therapy group.

Long-term effects of epoprostenol have never been compared with placebo because this is considered unethical due to the substantial treatment benefit after 3 months of therapy. Observational cohort analyses comparing the survival of epoprostenol-treated IPAH and HPAH patients with historical controls suggested long-term benefits (McLaughlin et al. 2002; Sitbon et al. 2002).

In scleroderma-associated PAH, intravenous epoprostenol improved exercise capacity (Badesch et al. 2000). Similar improvements in functional capacity and functional class have been seen in subjects with congenital left-to-right cardiac shunts (Rosenzweig et al. 1999) and infection with the human immunodeficiency virus (HIV) (Nunes et al. 2003; Aguilar and Farber 2000) while results are ambiguous in portopulmonary hypertension (Krowka and Swanson 2006) and show unfavourable effects in pulmonary hypertension secondary to left heart failure (Califf et al. 1997).

Increasing use of epoprostenol was associated with a considerable decrease in the number of lung transplantations for pulmonary hypertension. A US survey indicated that treatment with epoprostenol allowed two-thirds of subjects to deactivate from the transplant list (Robbins et al. 1998). It is unclear how long the disease will remain stable and therefore frequent clinical follow-up with examination, exercise, and hemodynamics is recommended (Badesch et al. 2007; Galie et al. 2009).

4.1.1 Combination with Other PAH Medication

In the BREATHE 2 study, subjects with PAH were started on epoprostenol with uptitration for 16 weeks and randomised in a 2:1 ratio to bosentan or placebo (Humbert et al. 2004). Addition of bosentan caused a trend, but no significant benefit in clinical or hemodynamic measurements.

Simonneau et al. reported the results of a 16-week multinational, double-blind, placebo-controlled trial assessing the safety and efficacy of sildenafil in addition to epoprostenol (PACES) (Simonneau et al. 2008). Patients had improvements in exercise capacity (adjusted increase 26 m, $p < 0.001$), hemodynamics, and time to clinical worsening in favour of sildenafil. The long-term open-label extension at 1 year maintained this benefit.

4.1.2 Limitations of Intravenous Epoprostenol

Limitations are mainly based on the pharmacology of epoprostenol with its chemical instability and extremely short half-life after infusion. Long-term administration of epoprostenol requires a permanent central venous catheter and a portable infusion pump. Conventional medication needs to be prepared daily and to be kept cold, requiring ice packs to be worn with medication in 24 h cassettes. The new formulation provides more stability after dissolving the drug, but the plasma half-life remains unchanged.

Patients need education of sterile techniques, operation of the pump, taking care of the catheter, and a strong support system to help ensure a safe environment prior to initiation. A "caregiver," a family member, or a friend who lives in close proximity to the patient is strongly recommended to obviate problems if the subject is too ill or unable to prepare medication on a given day. Serious complications include infection and thrombosis of the catheter and temporary interruption of the infusion because of inadvertent disconnection or pump malfunction. The incidence of catheter-related sepsis ranges from 0.04 to 0.6 cases per patient-year in experienced centres (Sitbon et al. 2002; Humbert et al. 1999; Kitterman et al. 2012) but may be much higher in less experienced centres.

Side effects are usually well tolerated, provided the dose is uptitrated at an adequate rate. The most common side effects include flushing, headache, nausea, loose stool, jaw pain, and foot pain and ascites (Barst et al. 1996; McLaughlin et al. 2002; Sitbon et al. 2002). While jaw pain is mostly transient and well tolerated, foot pain and ascites may deteriorate physical activity and quality of life.

4.2 Iloprost: Intravenous

Intravenous iloprost has been approved for pulmonary hypertension in New Zealand, but it has also been used in other countries like Germany, Switzerland, England, Australia, Thailand, Israel, Argentina, and Brazil. Intravenous iloprost has similar acute hemodynamic effects as epoprostenol (Higenbottam et al. 1998). It comes as concentrated solution in glass ampoules and does not require cooling. Due to its longer half-life it is less risky in the case of accidental therapy disruption. In Europe, the typical starting dose is 0.5 ng/kg/min and the long-term dose of infused iloprost is mostly 3 ng/kg/min (Ewert et al. 2007). In a few cases doses up to 10 ng/kg/min were applied. It has not been formally tested if infused iloprost has the same clinical efficacy as infused epoprostenol or treprostinil.

4.3 Treprostinil: Intravenous

The FDA approved the use of intravenous treprostinil for patients who do not tolerate subcutaneous treprostinil therapy. The advantage over epoprostenol is the chemical stability and the longer half-life that may also decrease the risk of rebound pulmonary hypertension in case of inadvertent cessation of the infusion. When patients transitioned from IV epoprostenol to IV treprostinil they finished on more than twice the dose of treprostinil compared with epoprostenol (Gomberg-Maitland et al. 2005a; Sitbon et al. 2007). It is unclear why a higher treprostinil dose is required compared to epoprostenol as, theoretically, its longer half-life would suggest the opposite. There were significantly more septic events with treprostinil as compared to epoprostenol in the USA; particularly the number of gram-negative sepsis episodes was increased (Kitterman et al. 2012). This might be associated with the different pump system, the slower infusion rate, or pharmacologic effects of the drug. There was no indication that the drug itself was contaminated.

4.4 Treprostinil: Subcutaneous

Treprostinil was originally approved as subcutaneous (sc) infusion. Subcutaneous therapy avoids a permanent central venous catheter. Medication is infused via a small self-inserted sc needle with continuous injection by syringe by a battery-driven pump. The infusion site should be changed every 3–4 days according to the package insert. In clinical practice patients often leave sites for 2–4 weeks (Lang et al. 2006). The drug comes as a ready-to-use solution and there is no need for cooling. The most common side effect occurring in up to 85 % of the patients is local infusion site pain (Simonneau et al. 2002). Patients are treated with variable response with a combination of local anaesthetic solutions, nonsteroidal anti-inflammatory agents, gabapentin, pregabalin, or low-dose narcotics. The pain does not appear to be dose related and cannot be predicted before therapy starts.

Class IV patients and patients receiving higher doses of treprostinil had a more significant improvement. There was also an improvement in hemodynamic parameters, including right atrial pressure, mean pulmonary artery pressure, pulmonary vascular resistance, and cardiac output. The drug was FDA approved for the treatment of NYHA class II–IV PAH patients in 2002 and EMA approved for PAH NYHA III in 2005.

Recent retrospective and observational studies of sc treprostinil suggest long-term clinical improvement and survival benefits. Lang and colleagues evaluated clinical outcomes in 99 PAH patients and 23 patients with inoperable chronic thromboembolic pulmonary hypertension for a mean follow-up of 26 ± 17 months in the open-label phase of a randomised clinical trial (Lang et al. 2006). The mean dose was 40 ± 3 ng/kg/min (range 16–84 ng/kg/min). Patients maintained their improved exercise tolerance (6MWD) and FC at 3 years. Event-free survival was

83 % at 1 year and 69 % at 3 years and overall survival was 89 % at 1 year and 71 % at 3 years, similar rates to those observed on iv epoprostenol.

Evaluation of all patients who had been enrolled in randomised controlled trials with open-label extension and available long-term observations revealed 860 patients who were treated for up to 4 years (Barst et al. 2006). Survival in patients on monotherapy censoring patients with the addition of targeted PAH therapy (130 patients) or premature discontinuation due to adverse events (199 patients) was 88, 79, 73, and 70 % at 1, 2, 3, and 4 years, respectively; however, only few patients remained on therapy for more than 2 years. The most frequently reported side effect was site pain.

Addition of sildenafil to sc treprostinil improved exercise capacity (6MWD) and functional class (Gomberg-Maitland et al. 2005b) and the same was found for the addition of sildenafil to inhaled treprostinil (Voswinckel et al. 2008).

4.4.1 Treprostinil Oral

The oral form of treprostinil is an extended release formulation. This therapy had a significant benefit in treatment-naïve PAH patients (FREEDOM-M) but no significant effect in patients on other targeted PAH medications (FREEDOM-C). Recently, the FDA decided not to approve oral treprostinil based on the moderate effects on 6 min walk distance and no evidence for a delay in clinical worsening achieved in the FREEDOM trials.

4.5 Beraprost: Oral

Beraprost is absorbed rapidly after oral administration and has an elimination half-life of 20–40 min. In Europe, 130 FC II and III subjects were enrolled in a 12-week randomised double-blind placebo-controlled trial of beraprost (Galie et al. 2002). Subjects had pulmonary arterial hypertension caused by IPAH, connective tissue diseases, congenital left-to-right shunts, portal hypertension, and HIV. At a median dose of 80 μg, four times daily, the 6MWD improved by 25 m, compared to placebo ($p = 0.04$). Subjects with IPAH had a mean increase of 46 m whereas PAH from other causes had no significant improvement. In the USA, PAH patients in functional class II and III were included in a 12-month study. The 6 min walk distance improved at 3 and 6 months compared with placebo, but this benefit was not sustained at 9 and 12 months (Barst et al. 2003). Beraprost has not been approved in the USA and Europe but is approved in Japan and several other Far East countries.

4.6 Selexipag

Selexipag is a non-prostanoid substance that specifically activates the IP receptor (Morrison et al. 2012). It has been investigated in a phase II study and had significant hemodynamic effects and side effects as expected for a prostacyclin analogue (Simonneau et al. 2012). Currently, a phase III study is performed.

4.7 Iloprost: Inhaled

The lung anatomy allows direct access to the precapillary vessels of the lung via drug deposition in the alveolar space or the surface of the bronchiolus terminalis (Fig. 6). Inhalation of prostacyclin as compared to infusion was shown to provide intrapulmonary selectivity and pulmonary selectivity with less systemic side effects (Olschewski et al. 1996, 2003; Walmrath et al. 1993, 1996). The inhalative approach was used with epoprostenol from the early 1990s, with iloprost from 1994 and with treprostinil from 2003.

When iloprost is delivered by different appropriate nebulizers, the pharmacodynamic and pharmacokinetics applied by the same dose are very similar (Olschewski et al. 2003). To ensure alveolar deposition, the delivery system produces small aerosolized particles with a median diameter of 3.0–5.0 µm. Iloprost must be inhaled six to nine times a day to achieve good clinical results. The inhaled doses for significant clinical effects were much smaller with inhaled iloprost (about 0.31 ng/kg/min) (Olschewski et al. 2002), compared to the intravenous infusion of iloprost for PAH.

Inhaled iloprost was approved for PAH, NYHA III + IV in the USA; for PAH, NYHA III and IV as well as inoperable CTEPH in Australia; and for IPAH, NYHA III in Europe. In the European study patients with IPAH, HPAH, PAH associated with connective tissue diseases, or inoperable CTEPH in FC III or IV were enrolled in a 12-week multi-centre placebo-controlled trial (Olschewski et al. 2002). The primary endpoint was a combined clinical endpoint where four criteria had to be met to be counted as a responder: a 10 % increase in 6MWD, an improvement in NYHA functional class, no deterioration, and no death. Seventeen percent of patients on iloprost reached this endpoint compared with 4 % in the placebo group ($p = 0.007$). The mean increase in 6MWD was 37 m ($p = 0.004$) and 59 m amongst subjects with primary pulmonary hypertension. Subjects' well-being improved as evidenced by quality of life scores and the Mahler dyspnea index. Hemodynamic trough values (before inhalation) at 12 weeks improved slightly but significantly in the treatment group compared with placebo ($p < 0.001$). There was a major improvement when post-inhalation values were considered. Side effects consisted of symptoms related to systemic vasodilation. More syncopal episodes (8 vs. 5, NS) occurred in the iloprost group, although they were not associated with clinical deterioration.

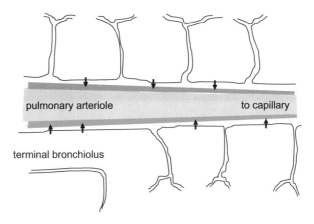

Fig. 6 The inhaled route of application for prostacyclin or its analogues. *Black arrows* mark areas where locally deposited drug penetrates the airway wall and diffuses into the pulmonary artery wall. From Gomberg-Maitland and Olschewski (2008)

Iloprost also improved hemodynamics and physical capacity in a small series of lung fibrosis patients (Olschewski et al. 1999), decompensated right heart failure (Olschewski et al. 2000), and HIV patients (Ghofrani et al. 2004). In a long-term observational study from Germany, the survival after 1 and 2 years was 92 % and 87 % and event-free survival was 84 % and 74 % in the IPAH patients (Olschewski et al. 2010). These results are comparable with the recent long-term data with endothelin receptor antagonists.

4.7.1 Combination Studies with Inhaled Iloprost

Hoeper et al. studied 20 patients with IPAH that were on either inhaled iloprost or oral beraprost. Three months of therapy with bosentan added to the prostacyclin resulted in improvement in exercise capacity (6MWD) and maximal oxygen consumption (Hoeper et al. 2003).

The reverse order, addition of inhaled prostacylin to oral endothelin antagonist, was used in the STEP trial (Iloprost inhalation solution safety and pilot efficacy trial in combination with bosentan for evaluation in pulmonary arterial hypertension), a double-blind placebo-controlled trial of 67 FC III and IV patients on a stable dose of bosentan for 3 months (McLaughlin et al. 2006). Subjects were randomised to either iloprost 5 µg six to nine times daily or placebo with 6MWD distance post-inhalation as the primary endpoint. The mean improvement of 26 m vs. placebo did not meet statistical significance ($p = 0.051$), but improvements in secondary endpoints including functional class and time to clinical worsening favoured inhaled iloprost.

Hoeper et al. in their 12-week study randomised stable IPAH patients FC III, currently on bosentan (125 mg twice daily) to either inhaled iloprost 5 µg six times daily or no additional therapy. According to slow recruitment and the results of the interim analysis (36 completed and 40 enrolled patients), this trial was early terminated (Hoeper et al. 2006).

Sildenafil may act synergistically with prostanoids due to its interaction with cAMP breakdown via inhibition of phosphodiesterase 3 via cGMP (Olschewski et al. 2001a). Ghofrani et al. (2002) demonstrated impressive additive effects of inhaled iloprost on sildenafil-induced pulmonary vasodilatation and beneficial clinical effects of adjunct therapy with sildenafil in patients on inhaled iloprost. Over 9–12 months of follow-up, exercise capacity and hemodynamics improved (Ghofrani et al. 2003).

4.8 Treprostinil: Inhaled

Inhaled treprostinil has effects similar to inhaled iloprost; however, there are significant differences in the pharmacodynamics (Voswinckel et al. 2006). This allows for a shorter inhalation time and less inhalations per day. It may be even possible to inhale the full dose with one single breath (Voswinckel et al. 2009). The maximally tolerated dose with a near maximal acute decrease in pulmonary vascular resistance occurred with the 30 µg dose.

In a small cohort, the addition of inhaled treprostinil to oral PAH therapy found significant improvements in 6MWD with nearly all patients improving physical capacity and functional class (Channick et al. 2006). The same was found in a large multinational study (TRIUMPH-1). Study drug or placebo was inhaled with a special ultrasonic nebulizer four times daily. Patients were on a stable treatment with bosentan or sildenafil before enrollment and this therapy was not changed during the study period. Patients on inhaled treprostinil improved significantly as compared to placebo ($p < 0.001$) both post-inhalation and pre-inhalation. Side effects were mostly mild and did not involve toxic effects (McLaughlin et al. 2010). Based on these results inhaled treprostinil was approved by the FDA, while the EMA declined approval based on formal issues.

Combination of sildenafil with inhaled treprostinil had synergistic hemodynamic effects (Voswinckel et al. 2008).

5 Management of Prostacyclin Therapy

5.1 Decision for a Prostacyclin or Its Analogues

The decision for a prostacyclin therapy is—in most of the cases—equivalent with a lifelong commitment for this kind of drug. Because all approved drugs employ demanding application modes and are prone to side effects, this decision must be based on solid ground.

5.2 Potential Treatment Hazards

Left ventricular dysfunction, either systolic or diastolic, represents a contraindication for prostacyclins due to the results of the FIRST study (Califf et al. 1997). In this study, 471 patients with left ventricular failure and mild pulmonary hypertension were randomised to continuous intravenous epoprostenol or placebo. After 6 months there was a significantly higher mortality in the epoprostenol group compared to control.

There was no study in patients with mild left ventricular failure and severe pulmonary hypertension and there is no consensus how to treat "out of proportion" pulmonary hypertension in left heart failure. However, there is consensus, although not formally studied, not to employ prostacyclin or its analogues in patients with a PAWP > 15 mmHg. As a consequence, all candidates for this therapy should have undergone right heart catheterization and a number of further diagnostic tests before a prostacyclin therapy is started.

There are several diseases and conditions that may amplify adverse effects of prostacyclins even if they are not clinically manifest before application. This is the case in pulmonary veno-occlusive disease (PVOD) (Gunther et al. 2012; Rabiller et al. 2006), where prostacyclins can have beneficial effects but are prone to lung edema and hypoxemia (Montani et al. 2009), and in chronic thromboembolic pulmonary hypertension (CTEPH), where any vasodilators will primarily increase the blood flow of the hyper-perfused areas, resulting in worsening of hypoxemia. Patients with obstructive lung disease or lung fibrosis have both heterogeneous ventilation and heterogeneous perfusion and are prone to ventilation/perfusion mismatch and shunt blood flow with any kind of pulmonary vasodilator.

The inhalative application as compared to the systemic application of prostacyclins may be advantageous in patients who are prone to V/Q mismatch (Olschewski et al. 1999). This also applies for liver diseases (Krowka and Swanson 2006; Reichenberger et al. 2006). However, prostacyclin or its analogues have not been approved in any of these conditions.

5.3 Practical Issues of Prostacyclin Treatment

As prostacyclin patients represent a small population with a life-threatening disease and a risky medication, a strong partnership between the patient and the health professionals is very important. In most specialised centres there is a long-lasting personal relationship between individual patients and nurses and/or physicians.

Most patients experience some degree of side effects. They must learn to accept this for the sake of their physical capacity and survival. They often need to talk with an experienced professional about these issues. Both water retention and overnutrition can cause significant clinical deterioration. Patients must learn to control their water, salt, and calorie intake and to adapt the diuretic dose to their needs. Infection management is very important, particularly in patients on iv or sc medication. This may be very demanding for both the patient and the caregiver (Hall et al. 2012).

Patients profit from a written plan for the correct response to a given situation including antibiotic treatment and eventually removal of the intravenous line or change of the sc line.

It seems that an oral-first treatment strategy is not associated with a disadvantage in patients with pulmonary hypertension (Cornwell et al. 2011). Actually, most of the patients who start a prostacyclin treatment have been treated with non-prostanoid medication like endothelin receptor antagonists or phosphodiesterase-5 inhibitors. If they are in WHO class III or IV, they receive a prostacyclin or analogue on top of their oral therapy. A complete switch is mostly due to severe side effects from the oral drugs.

Patients who are primarily started on prostacyclin therapy are mostly in NYHA III or IV with signs of right heart decompensation and severely reduced cardiac index and central venous oxygen saturation. Sometimes these patients need to stay in an intensive care unit. It may be advantageous to start these patients on upfront combination therapy, but the evidence for such an approach is based on small uncontrolled case series (Kemp et al. 2012).

Before starting the therapy with prostacyclin or its analogues, water retention is treated with iv diuretics. If systemic pressure is too low, patients receive low-dose catecholamines in parallel. There is some experience with dobutamine, dopamine, adrenaline, and noradrenalin in such a setting. Noradrenalin should only be used in parallel with a continuous prostacyclin because it may further constrict the pulmonary vessels and promote right heart decompensation and death. The prostacyclin dose is then uptitrated while the resting medication is downtitrated as soon as possible.

If a patient is recompensated on prostacyclin treatment, it may be possible to switch to a less demanding therapy like an endothelin receptor antagonist, a phosphodiesterase inhibitor, or even a calcium channel blocker if the patient is a responder.

5.4 Switch from iv Prostacyclins to Other Forms of Application

If a patient is on intravenous medication, it may be considered to switch to a less demanding application form. Transition from iv epoprostenol to inhaled iloprost or from iv treprostinil to inhaled treprostinil is possible. The advantage is to have less systemic side effects. There are small series of patients where this transition worked in most of the cases (Ivy et al. 2008; Perez et al. 2012).

5.5 Switch from Inhaled Prostacyclins to Continuous Infusion

There are three major reasons to consider this switch: (1) the inhalations are too time-consuming and demanding, (2) there are specific side effects (mostly coughing), and (3) the effect is not sufficient. The first reason normally becomes evident within the first days or weeks of therapy and the second normally goes away after the first days of therapy, but the third may become evident after months or years. Although the dose of inhaled prostacyclin analogues is quite stable over long periods of time, the needed dose may gradually increase.

6 Conclusion

In pulmonary hypertension, a lack of prostacyclin is an important feature of the pathologic changes. Therapy with prostacyclin or its analogues restores physiological pathways involving the pulmonary arterial vessel wall, the platelets, and inflammatory cells and exerts beneficial clinical effects antagonising vasoconstriction, inflammation, and aggregation of circulating cells in the pulmonary vessels. Prostacyclin therapy has been shown to be efficacious, particularly in advanced disease. Combination of other targeted therapies with prostacyclin or its analogues appears to be effective and safe. Despite demanding application methods, this class of drugs remains a mainstay in the treatment of PAH.

References

Abramovitz M, Adam M, Boie Y, Carriere M, Denis D, Godbout C, Lamontagne S, Rochette C, Sawyer N, Tremblay NM, Belley M, Gallant M, Dufresne C, Gareau Y, Ruel R, Juteau H, Labelle M, Ouimet N, Metters KM (2000) The utilization of recombinant prostanoid receptors to determine the affinities and selectivities of prostaglandins and related analogs. Biochim Biophys Acta 1483:285–293

Aguilar RV, Farber HW (2000) Epoprostenol (prostacyclin) therapy in HIV-associated pulmonary hypertension. Am J Respir Crit Care Med 162:1846–1850

Badesch DB, Tapson VF, McGoon MD, Brundage BH, Rubin LJ, Wigley FM, Rich S, Barst RJ, Barrett PS, Kral KM, Jobsis MM, Loyd JE, Murali S, Frost A, Girgis R, Bourge RC, Ralph DD, Elliott CG, Hill NS, Langleben D, Schilz RJ, McLaughlin VV, Robbins IM, Groves BM, Shapiro S, Medsger TA Jr (2000) Continuous intravenous epoprostenol for pulmonary hypertension due to the scleroderma spectrum of disease. A randomized, controlled trial. Ann Intern Med 132:425–434

Badesch DB, Abman SH, Simonneau G, Rubin LJ, McLaughlin VV (2007) Medical therapy for pulmonary arterial hypertension: updated ACCP evidence-based clinical practice guidelines. Chest 131:1917–1928

Barger PM (2002) Has angiogenesis been invited to the PPARty? J Mol Cell Cardiol 34:713–716

Barst RJ, Rubin LJ, Long WA, McGoon MD, Rich S, Badesch DB, Groves BM, Tapson VF, Bourge RC, Brundage BH (1996) A comparison of continuous intravenous epoprostenol (prostacyclin) with conventional therapy for primary pulmonary hypertension. The Primary Pulmonary Hypertension Study Group. N Engl J Med 334:296–302

Barst RJ, McGoon M, McLaughlin V, Tapson V, Rich S, Rubin L, Wasserman K, Oudiz R, Shapiro S, Robbins IM, Channick R, Badesch D, Rayburn BK, Flinchbaugh R, Sigman J, Arneson C, Jeffs R (2003) Beraprost therapy for pulmonary arterial hypertension. J Am Coll Cardiol 41:2119–2125

Barst RJ, Galie N, Naeije R, Simonneau G, Jeffs R, Arneson C, Rubin LJ (2006) Long-term outcome in pulmonary arterial hypertension patients treated with subcutaneous treprostinil. Eur Respir J 28:1195–1203

Bhattacharya M, Peri KG, Almazan G, Ribeiro-da-Silva A, Shichi H, Durocher Y, Abramovitz M, Hou X, Varma DR, Chemtob S (1998) Nuclear localization of prostaglandin E2 receptors. Proc Natl Acad Sci USA 95:15792–15797

Bhattacharya M, Peri K, Ribeiro-da-Silva A, Almazan G, Shichi H, Hou X, Varma DR, Chemtob S (1999) Localization of functional prostaglandin E2 receptors EP3 and EP4 in the nuclear envelope. J Biol Chem 274:15719–15724

Califf RM, Adams KF, McKenna WJ, Gheorghiade M, Uretsky BF, McNulty SE, Darius H, Schulman K, Zannad F, Handberg-Thurmond E, Harrell FE Jr, Wheeler W, Soler-Soler J, Swedberg K (1997) A randomized controlled trial of epoprostenol therapy for severe congestive heart failure: the Flolan International Randomized Survival Trial (FIRST). Am Heart J 134:44–54

Channick RN, Olschewski H, Seeger W, Staub T, Voswinckel R, Rubin LJ (2006) Safety and efficacy of inhaled treprostinil as add-on therapy to bosentan in pulmonary arterial hypertension. J Am Coll Cardiol 48:1433–1437

Chow KB, Jones RL, Wise H (2003) Protein kinase A-dependent coupling of mouse prostacyclin receptors to Gi is cell-type dependent. Eur J Pharmacol 474:7–13

Clapp LH, Finney P, Turcato S, Tran S, Rubin LJ, Tinker A (2002) Differential effects of stable prostacyclin analogs on smooth muscle proliferation and cyclic AMP generation in human pulmonary artery. Am J Respir Cell Mol Biol 26:194–201

Cornwell WK, McLaughlin VV, Krishnan SM, Rubenfire M (2011) Does the outcome justify an oral-first treatment strategy for management of pulmonary arterial hypertension? Chest 140:697–705

Ewert R, Opitz CF, Wensel R, Winkler J, Halank M, Felix SB (2007) Continuous intravenous iloprost to revert treatment failure of first-line inhaled iloprost therapy in patients with idiopathic pulmonary arterial hypertension. Clin Res Cardiol 96:211–217

Fontana M, Olschewski H, Olschewski A, Schluter KD (2007) Treprostinil potentiates the positive inotropic effect of catecholamines in adult rat ventricular cardiomyocytes. Br J Pharmacol 151:779–786

Forman BM, Umesono K, Chen J, Evans RM (1995) Unique response pathways are established by allosteric interactions among nuclear hormone receptors. Cell 81:541–550

Galie N, Humbert M, Vachiery JL, Vizza CD, Kneussl M, Manes A, Sitbon O, Torbicki A, Delcroix M, Naeije R, Hoeper M, Chaouat A, Morand S, Besse B, Simonneau G (2002) Effects of beraprost sodium, an oral prostacyclin analogue, in patients with pulmonary arterial hypertension: a randomized, double-blind, placebo-controlled trial. J Am Coll Cardiol 39:1496–1502

Galie N, Hoeper MM, Humbert M, Torbicki A, Vachiery JL, Barbera JA, Beghetti M, Corris P, Gaine S, Gibbs JS, Gomez-Sanchez MA, Jondeau G, Klepetko W, Opitz C, Peacock A, Rubin L, Zellweger M, Simonneau G, Vahanian A, Auricchio A, Bax J, Ceconi C, Dean V, Filippatos G, Funck-Brentano C, Hobbs R, Kearney P, McDonagh T, McGregor K, Popescu BA, Reiner Z, Sechtem U, Sirnes PA, Tendera M, Vardas P, Widimsky P, Sechtem U, Al AN, Andreotti F, Aschermann M, Asteggiano R, Benza R, Berger R, Bonnet D, Delcroix M, Howard L, Kitsiou AN, Lang I, Maggioni A, Nielsen-Kudsk JE, Park M, Perrone-Filardi P, Price S, Domenech MT, Vonk-Noordegraaf A, Zamorano JL (2009) Guidelines for the diagnosis and treatment of pulmonary hypertension: The Task Force for the Diagnosis and Treatment of Pulmonary Hypertension of the European Society of Cardiology (ESC) and the European Respiratory Society (ERS), endorsed by the International Society of Heart and Lung Transplantation (ISHLT). Eur Heart J 30:2493–2537

Ghofrani HA, Wiedemann R, Rose F, Olschewski H, Schermuly RT, Weissmann N, Seeger W, Grimminger F (2002) Combination therapy with oral sildenafil and inhaled iloprost for severe pulmonary hypertension. Ann Intern Med 136:515–522

Ghofrani HA, Rose F, Schermuly RT, Olschewski H, Wiedemann R, Kreckel A, Weissmann N, Ghofrani S, Enke B, Seeger W, Grimminger F (2003) Oral sildenafil as long-term adjunct therapy to inhaled iloprost in severe pulmonary arterial hypertension. J Am Coll Cardiol 42:158–164

Ghofrani HA, Friese G, Discher T, Olschewski H, Schermuly RT, Weissmann N, Seeger W, Grimminger F, Lohmeyer J (2004) Inhaled iloprost is a potent acute pulmonary vasodilator in HIV-related severe pulmonary hypertension. Eur Respir J 23:321–326

Gomberg-Maitland M, Olschewski H (2008) Prostacyclin therapies for the treatment of pulmonary arterial hypertension. Eur Respir J 31:891–901

Gomberg-Maitland M, Tapson VF, Benza RL, McLaughlin VV, Krichman A, Widlitz AC, Barst RJ (2005a) Transition from intravenous epoprostenol to intravenous treprostinil in pulmonary hypertension. Am J Respir Crit Care Med 172:1586–1589

Gomberg-Maitland M, McLaughlin V, Gulati M, Rich S (2005b) Efficacy and safety of sildenafil added to treprostinil in pulmonary hypertension. Am J Cardiol 96:1334–1336

Gryglewski RJ (1980) Prostaglandins, platelets, and atherosclerosis. CRC Crit Rev Biochem 7:291–338

Gunther S, Jais X, Maitre S, Berezne A, Dorfmuller P, Seferian A, Savale L, Mercier O, Fadel E, Sitbon O, Mouthon L, Simonneau G, Humbert M, Montani D (2012) Computed tomography findings of pulmonary venoocclusive disease in scleroderma patients presenting with precapillary pulmonary hypertension. Arthritis Rheum 64:2995–3005

Gupta RA, Tan J, Krause WF, Geraci MW, Willson TM, Dey SK, DuBois RN (2000) Prostacyclin-mediated activation of peroxisome proliferator-activated receptor delta in colorectal cancer. Proc Natl Acad Sci USA 97:13275–13280

Hall H, Cote J, McBean A, Purden M (2012) The experiences of patients with pulmonary arterial hypertension receiving continuous intravenous infusion of epoprostenol (Flolan) and their support persons. Heart Lung 41:35–43

Hatae T, Wada M, Yokoyama C, Shimonishi M, Tanabe T (2001) Prostacyclin-dependent apoptosis mediated by PPAR delta. J Biol Chem 276:46260–46267

Higenbottam T, Wheeldon D, Wells F, Wallwork J (1984) Long-term treatment of primary pulmonary hypertension with continuous intravenous epoprostenol (prostacyclin). Lancet 1:1046–1047

Higenbottam TW, Butt AY, Dinh-Xuan AT, Takao M, Cremona G, Akamine S (1998) Treatment of pulmonary hypertension with the continuous infusion of a prostacyclin analogue, iloprost. Heart 79:175–179

Hoeper MM, Taha N, Bekjarova A, Gatzke R, Spiekerkoetter E (2003) Bosentan treatment in patients with primary pulmonary hypertension receiving nonparenteral prostanoids. Eur Respir J 22:330–334

Hoeper MM, Leuchte H, Halank M, Wilkens H, Meyer FJ, Seyfarth HJ, Wensel R, Ripken F, Bremer H, Kluge S, Hoeffken G, Behr J (2006) Combining inhaled iloprost with bosentan in patients with idiopathic pulmonary arterial hypertension. Eur Respir J 28:691–694

Humbert M, Sanchez O, Fartoukh M, Jagot JL, Le Gall C, Sitbon O, Parent F, Simonneau G (1999) Short-term and long-term epoprostenol (prostacyclin) therapy in pulmonary hypertension secondary to connective tissue diseases: results of a pilot study. Eur Respir J 13:1351–1356

Humbert M, Barst RJ, Robbins IM, Channick RN, Galie N, Boonstra A, Rubin LJ, Horn EM, Manes A, Simonneau G (2004) Combination of bosentan with epoprostenol in pulmonary arterial hypertension: BREATHE-2. Eur Respir J 24:353–359

Ivy DD, Doran AK, Smith KJ, Mallory GB Jr, Beghetti M, Barst RJ, Brady D, Law Y, Parker D, Claussen L, Abman SH (2008) Short- and long-term effects of inhaled iloprost therapy in children with pulmonary arterial hypertension. J Am Coll Cardiol 51:161–169

Keith RL, Geraci MW (2006) Prostacyclin in lung cancer. J Thorac Oncol 1:503–505

Kemp K, Savale L, O'Callaghan DS, Jais X, Montani D, Humbert M, Simonneau G, Sitbon O (2012) Usefulness of first-line combination therapy with epoprostenol and bosentan in pulmonary arterial hypertension: an observational study. J Heart Lung Transplant 31:150–158

Kerbaul F, Brimioulle S, Rondelet B, Dewachter C, Hubloue I, Naeije R (2007) How prostacyclin improves cardiac output in right heart failure in conjunction with pulmonary hypertension. Am J Respir Crit Care Med 175:846–850

Kitterman N, Poms A, Miller DP, Lombardi S, Farber HW, Barst RJ (2012) Bloodstream infections in patients with pulmonary arterial hypertension treated with intravenous prostanoids: insights from the REVEAL REGISTRY(R). Mayo Clin Proc 87:825–834

Krowka MJ, Swanson KL (2006) How should we treat portopulmonary hypertension? Eur Respir J 28:466–467

Lang I, Gomez-Sanchez M, Kneussl M, Naeije R, Escribano P, Skoro-Sajer N, Vachiery JL (2006) Efficacy of long-term subcutaneous treprostinil sodium therapy in pulmonary hypertension. Chest 129:1636–1643

le Cras TD, Markham NE, Tuder RM, Voelkel NF, Abman SH (2002) Treatment of newborn rats with a VEGF receptor inhibitor causes pulmonary hypertension and abnormal lung structure. Am J Physiol Lung Cell Mol Physiol 283:L555–L562

Li Y, Connolly M, Nagaraj C, Tang B, Balint Z, Popper H, Smolle-Juettner FM, Lindenmann J, Kwapiszewska G, Aaronson PI, Wohlkoenig C, Leithner K, Olschewski H, Olschewski A (2012) Peroxisome proliferator-activated receptor-beta/delta, the acute signaling factor in prostacyclin-induced pulmonary vasodilation. Am J Respir Cell Mol Biol 46:372–379

McLaughlin VV, Shillington A, Rich S (2002) Survival in primary pulmonary hypertension: the impact of epoprostenol therapy. Circulation 106:1477–1482

McLaughlin VV, Oudiz RJ, Frost A, Tapson VF, Murali S, Channick RN, Badesch DB, Barst RJ, Hsu HH, Rubin LJ (2006) Randomized study of adding inhaled iloprost to existing bosentan in pulmonary arterial hypertension. Am J Respir Crit Care Med 174:1257–1263

McLaughlin VV, Benza RL, Rubin LJ, Channick RN, Voswinckel R, Tapson VF, Robbins IM, Olschewski H, Rubenfire M, Seeger W (2010) Addition of inhaled treprostinil to oral therapy for pulmonary arterial hypertension: a randomized controlled clinical trial. J Am Coll Cardiol 55:1915–1922

Meyer-Kirchrath J, Debey S, Glandorff C, Kirchrath L, Schror K (2004) Gene expression profile of the G(s)-coupled prostacyclin receptor in human vascular smooth muscle cells. Biochem Pharmacol 67:757–765

Moncada S, Gryglewski R, Bunting S, Vane JR (1976) An enzyme isolated from arteries transforms prostaglandin endoperoxides to an unstable substance that inhibits platelet aggregation. Nature 263:663–665

Montani D, Price LC, Dorfmuller P, Achouh L, Jais X, Yaici A, Sitbon O, Musset D, Simonneau G, Humbert M (2009) Pulmonary veno-occlusive disease. Eur Respir J 33:189–200

Morrison K, Studer R, Ernst R, Haag F, Kauser K, Clozel M (2012) Differential effects of selexipag and prostacyclin analogs in rat pulmonary artery. J Pharmacol Exp Ther 343:547–555

Nagaraj C, Tang B, Balint Z, Wygrecka M, Hrzenjak A, Kwapiszewska G, Stacher E, Lindenmann J, Weir EK, Olschewski H, Olschewski A (2012) Src tyrosine kinase is crucial for potassium channel function in human pulmonary arteries. Eur Respir J 41(1):85–95

Nunes H, Humbert M, Sitbon O, Morse JH, Deng Z, Knowles JA, Le Gall C, Parent F, Garcia G, Herve P, Barst RJ, Simonneau G (2003) Prognostic factors for survival in HIV-associated pulmonary arterial hypertension. Am J Respir Crit Care Med 167:1433–1439

O'Grady J, Warrington S, Moti MJ, Bunting S, Flower R, Fowle AS, Higgs EA, Moncada S (1980) Effects of intravenous infusion of prostacyclin (PGI2) in man. Prostaglandins 19:319–332

Olschewski H, Olschewski A (2011) Pulmonale hypertonie, 2nd edn. Uni-Med Science, Bremen

Olschewski H, Walmrath D, Schermuly R, Ghofrani A, Grimminger F, Seeger W (1996) Aerosolized prostacyclin and iloprost in severe pulmonary hypertension. Ann Intern Med 124:820–824

Olschewski H, Ghofrani HA, Walmrath D, Schermuly R, Temmesfeld-Wollbruck B, Grimminger F, Seeger W (1999) Inhaled prostacyclin and iloprost in severe pulmonary hypertension secondary to lung fibrosis. Am J Respir Crit Care Med 160:600–607

Olschewski H, Ghofrani HA, Schmehl T, Winkler J, Wilkens H, Hoper MM, Behr J, Kleber FX, Seeger W (2000) Inhaled iloprost to treat severe pulmonary hypertension. An uncontrolled trial. German PPH Study Group. Ann Intern Med 132:435–443

Olschewski H, Rose F, Grunig E, Ghofrani HA, Walmrath D, Schulz R, Schermuly R, Grimminger F, Seeger W (2001a) Cellular pathophysiology and therapy of pulmonary hypertension. J Lab Clin Med 138:367–377

Olschewski H, Olschewski A, Rose F, Schermuly R, Schutte H, Weissmann N, Seeger W, Grimminger F (2001b) Physiologic basis for the treatment of pulmonary hypertension. J Lab Clin Med 138:287–297

Olschewski H, Simonneau G, Galie N, Higenbottam T, Naeije R, Rubin LJ, Nikkho S, Speich R, Hoeper MM, Behr J, Winkler J, Sitbon O, Popov W, Ghofrani HA, Manes A, Kiely DG, Ewert R, Meyer A, Corris PA, Delcroix M, Gomez-Sanchez M, Siedentop H, Seeger W (2002) Inhaled iloprost for severe pulmonary hypertension. N Engl J Med 347:322–329

Olschewski H, Rohde B, Behr J, Ewert R, Gessler T, Ghofrani HA, Schmehl T (2003) Pharmacodynamics and pharmacokinetics of inhaled iloprost, aerosolized by three different devices, in severe pulmonary hypertension. Chest 124:1294–1304

Olschewski A, Li Y, Tang B, Hanze J, Eul B, Bohle RM, Wilhelm J, Morty RE, Brau ME, Weir EK, Kwapiszewska G, Klepetko W, Seeger W, Olschewski H (2006) Impact of TASK-1 in human pulmonary artery smooth muscle cells. Circ Res 98:1072–1080

Olschewski H, Hoeper MM, Behr J, Ewert R, Meyer A, Borst MM, Winkler J, Pfeifer M, Wilkens H, Ghofrani HA, Nikkho S, Seeger W (2010) Long-term therapy with inhaled iloprost in patients with pulmonary hypertension. Respir Med 104:731–740

Perez VA, Rosenzweig E, Rubin LJ, Poch D, Bajwa A, Park M, Jain M, Bourge RC, Kudelko K, Spiekerkoetter E, Liu J, Hsi A, Zamanian RT (2012) Safety and efficacy of transition from systemic prostanoids to inhaled treprostinil in pulmonary arterial hypertension. Am J Cardiol 110(10):1546–1550

Pola R, Gaetani E, Flex A, Aprahamian TR, Bosch-Marce M, Losordo DW, Smith RC, Pola P (2004) Comparative analysis of the in vivo angiogenic properties of stable prostacyclin analogs: a possible role for peroxisome proliferator-activated receptors. J Mol Cell Cardiol 36:363–370

Rabiller A, Jais X, Hamid A, Resten A, Parent F, Haque R, Capron F, Sitbon O, Simonneau G, Humbert M (2006) Occult alveolar haemorrhage in pulmonary veno-occlusive disease. Eur Respir J 27:108–113

Reginato MJ, Krakow SL, Bailey ST, Lazar MA (1998) Prostaglandins promote and block adipogenesis through opposing effects on peroxisome proliferator-activated receptor gamma. J Biol Chem 273:1855–1858

Reichenberger F, Voswinckel R, Steveling E, Enke B, Kreckel A, Olschewski H, Grimminger F, Seeger W, Ghofrani HA (2006) Sildenafil treatment for portopulmonary hypertension. Eur Respir J 28:563–567

Robbins IM, Christman BW, Newman JH, Matlock R, Loyd JE (1998) A survey of diagnostic practices and the use of epoprostenol in patients with primary pulmonary hypertension. Chest 114:1269–1275

Rose F, Hattar K, Gakisch S, Grimminger F, Olschewski H, Seeger W, Tschuschner A, Schermuly RT, Weissmann N, Hanze J, Sibelius U, Ghofrani HA (2003) Increased neutrophil mediator release in patients with pulmonary hypertension – suppression by inhaled iloprost. Thromb Haemost 90:1141–1149

Rosenzweig EB, Kerstein D, Barst RJ (1999) Long-term prostacyclin for pulmonary hypertension with associated congenital heart defects. Circulation 99:1858–1865

Rubin LJ, Groves BM, Reeves JT, Frosolono M, Handel F, Cato AE (1982) Prostacyclin-induced acute pulmonary vasodilation in primary pulmonary hypertension. Circulation 66:334–338

Simonneau G, Barst RJ, Galie N, Naeije R, Rich S, Bourge RC, Keogh A, Oudiz R, Frost A, Blackburn SD, Crow JW, Rubin LJ (2002) Continuous subcutaneous infusion of treprostinil, a prostacyclin analogue, in patients with pulmonary arterial hypertension. A double-blind, randomized, placebo-controlled trial. Am J Respir Crit Care Med 165:800–804

Simonneau G, Rubin LJ, Galie N, Barst RJ, Fleming TR, Frost AE, Engel PJ, Kramer MR, Burgess G, Collings L, Cossons N, Sitbon O, Badesch DB (2008) Addition of sildenafil to long-term intravenous epoprostenol therapy in patients with pulmonary arterial hypertension: a randomized trial. Ann Intern Med 149:521–530

Simonneau G, Torbicki A, Hoeper MM, Delcroix M, Karlocai K, Galie N, Degano B, Bonderman D, Kurzyna M, Efficace M, Giorgino R, Lang IM (2012) Selexipag: an oral, selective prostacyclin receptor agonist for the treatment of pulmonary arterial hypertension. Eur Respir J 40:874–880

Sitbon O, Humbert M, Nunes H, Parent F, Garcia G, Herve P, Rainisio M, Simonneau G (2002) Long-term intravenous epoprostenol infusion in primary pulmonary hypertension: prognostic factors and survival. J Am Coll Cardiol 40:780–788

Sitbon O, Manes A, Jais X, Pallazini M, Humbert M, Presotto L, Paillette L, Zaccardelli D, Davis G, Jeffs R, Simonneau G, Galie N (2007) Rapid switch from intravenous epoprostenol to intravenous treprostinil in patients with pulmonary arterial hypertension. J Cardiovasc Pharmacol 49:1–5

Stacher E, Graham BB, Hunt JM, Gandjeva A, Groshong SD, McLaughlin VV, Jessup M, Grizzle WE, Aldred MA, Cool CD, Tuder RM (2012) Modern age pathology of pulmonary arterial hypertension. Am J Respir Crit Care Med 186:261–272

Sussmann M, Sarbia M, Meyer-Kirchrath J, Nusing RM, Schror K, Fischer JW (2004) Induction of hyaluronic acid synthase 2 (HAS2) in human vascular smooth muscle cells by vasodilatory prostaglandins. Circ Res 94:592–600

Tang B, Li Y, Nagaraj C, Morty RE, Gabor S, Stacher E, Voswinckel R, Weissmann N, Leithner K, Olschewski H, Olschewski A (2009) Endothelin-1 inhibits background two-pore domain channel TASK-1 in primary human pulmonary artery smooth muscle cells. Am J Respir Cell Mol Biol 41:476–483

Taraseviciene-Stewart L, Kasahara Y, Alger L, Hirth P, Mc MG, Waltenberger J, Voelkel NF, Tuder RM (2001) Inhibition of the VEGF receptor 2 combined with chronic hypoxia causes cell death-dependent pulmonary endothelial cell proliferation and severe pulmonary hypertension. FASEB J 15:427–438

Voswinckel R, Enke B, Reichenberger F, Kohstall M, Kreckel A, Krick S, Gall H, Gessler T, Schmehl T, Ghofrani HA, Schermuly RT, Grimminger F, Rubin LJ, Seeger W, Olschewski H (2006) Favorable effects of inhaled treprostinil in severe pulmonary hypertension: results from randomized controlled pilot studies. J Am Coll Cardiol 48:1672–1681

Voswinckel R, Reichenberger F, Enke B, Kreckel A, Krick S, Gall H, Schermuly RT, Grimminger F, Rubin LJ, Olschewski H, Seeger W, Ghofrani HA (2008) Acute effects of the combination of sildenafil and inhaled treprostinil on haemodynamics and gas exchange in pulmonary hypertension. Pulm Pharmacol Ther 25:824–832

Voswinckel R, Reichenberger F, Gall H, Schmehl T, Gessler T, Schermuly RT, Grimminger F, Rubin LJ, Seeger W, Ghofrani HA, Olschewski H (2009) Metered dose inhaler delivery of treprostinil for the treatment of pulmonary hypertension. Pulm Pharmacol Ther 22:50–56

Walmrath D, Schneider T, Pilch J, Grimminger F, Seeger W (1993) Aerosolised prostacyclin in adult respiratory distress syndrome. Lancet 342:961–962

Walmrath D, Schneider T, Schermuly R, Olschewski H, Grimminger F, Seeger W (1996) Direct comparison of inhaled nitric oxide and aerosolized prostacyclin in acute respiratory distress syndrome. Am J Respir Crit Care Med 153:991–996

Weir EK, Rubin LJ, Ayres SM, Bergofsky EH, Brundage BH, Detre KM, Elliott CG, Fishman AP, Goldring RM, Groves BM (1989) The acute administration of vasodilators in primary pulmonary hypertension. Experience from the National Institutes of Health Registry on Primary Pulmonary Hypertension. Am Rev Respir Dis 140:1623–1630

Wise H (2003) Multiple signalling options for prostacyclin. Acta Pharmacol Sin 24:625–630

Endothelin Receptor Antagonists

Martine Clozel, Alessandro Maresta, and Marc Humbert

Abstract Three pathways have been identified in the pathogenesis of pulmonary arterial hypertension (PAH): the endothelin (ET), nitric oxide (NO) and prostacyclin pathways. These pathways represent the targets of approved PAH therapies and their discovery has facilitated significant progress in the understanding and treatment of PAH. The ET system is well established as a key player in the pathophysiology of PAH, with deleterious effects mediated by both the ET_A and ET_B receptors. Endothelin receptor antagonists (ERAs) are an important part of PAH therapy, with two ERAs currently approved for the treatment of PAH and a novel ERA that has recently been investigated in a Phase III clinical trial. This chapter describes the role of ET in the pathogenesis of PAH, reviews experimental data and examines the clinical status of ERAs in PAH treatment.

Keywords Endothelin • Endothelin receptor antagonist • Pulmonary arterial hypertension

Contents

1	The ET System	200
	1.1 Endothelins	200
	1.2 ET Receptors	201
	1.3 The ET System and the Pulmonary Circulation	202

M. Clozel • A. Maresta
Actelion Pharmaceuticals Ltd, Gewerbestrasse 16, 4123 Allschwil, Switzerland
e-mail: martine.clozel@actelion.com

M. Humbert (✉)
Université Paris-Sud, Le Kremlin-Bicêtre, France

Inserm U999, Centre Chirurgical Marie Lannelongue, Le Plessis-Robinson, France

Assistance Publique Hôpitaux de Paris, Service de Pneumologie, Hôpital Bicêtre, Le Kremlin-Bicêtre, France
e-mail: marc.humbert@abc.aphp.fr; mjc.humbert@gmail.com

2 The Role of the ET System in the Pathogenesis of PH 202
   2.1 Imbalance of Endothelial Mediators 202
   2.2 Effects of ET-1 Contributing to PAH 203
   2.3 Upregulation of the ET System in PH 204
3 ET Receptor Antagonism: Preclinical Studies 204
   3.1 ET Receptor Antagonism and Vasoconstriction/Vasodilation 205
   3.2 ET Receptor Antagonism and Vascular/Cardiac Hypertrophy 206
   3.3 ET Receptor Antagonism and Fibrosis 208
4 ET Receptor Antagonism for the Treatment of PAH: Several Molecules with Different
   Characteristics 209
   4.1 Pharmacological Profile 209
   4.2 Safety Profile 211
   4.3 Drug Interaction Profile 213
5 ET Receptor Antagonism in PAH: Human Studies 214
   5.1 Bosentan 214
   5.2 Sitaxentan 217
   5.3 Ambrisentan 217
   5.4 Macitentan 218
References 219

1 The ET System

1.1 Endothelins

ETs represent a group of three distinctive 21-amino acid peptides: ET-1, ET-2, and ET-3 (Inoue et al. 1989), with ET-1 being the main isoform produced in the cardiovascular system. Each ET isoform is synthesized as a large pre-proendothelin (pre-pro-ET) that is cleaved to pro-ET (big ET-1, -2, and -3) and then further processed by ET-converting enzymes (ECE-1 and ECE-2) to yield mature ET (Kedzierski and Yanagisawa 2001).

ET-1 is one of the most potent vasoconstrictor ever identified (Yanagisawa et al. 1988). It is probably also one of the most potent co-proliferative agents. ET-1 is released, predominantly from vascular endothelial cells, but also by every other cell type. Its secretion can be constitutive or regulated, and it contributes to the maintenance of vascular tone (Russell and Davenport 1999). In endothelial cells, ET-1 is not stored, but secreted predominantly toward the interstitium and smooth muscle (80 %) and to a lesser extent into the circulation (20 %) (Wagner et al. 1992), supporting a paracrine/autocrine role of ET-1, rather than hormonal. The paracrine activity of ET-1 is also corroborated by a strong affinity for its receptors combined with a high tissue receptor density (e.g., 10–100 nM in the vascular wall), resulting in newly secreted ET-1 being immediately trapped by receptor binding (Frelin and Guedin 1994). These unique features explain why plasma ET-1 concentrations are a poor marker of activity of the ET system.

The production of ET-1 is upregulated in response to a range of stimuli including mechanical stress (Yoshizumi et al. 1989), hypoxia (Rakugi et al. 1990), thrombin (Golden et al. 1998), angiotensin II (Imai et al. 1992), vasopressin (Imai et al. 1992), human immunodeficiency virus glycoprotein 120 (HIV gp120) (Ehrenreich et al. 1990), and growth factors (Olave et al. 2012). Conversely it is downregulated by NO (Bourque et al. 2011), prostacyclin (Prins et al. 1994), natriuretic peptides (Emori et al. 1993), and estrogens (Akishita et al. 1998). ET-1 can also be produced, although to a lower extent, by a large number of different cell types including smooth muscle cells (Markewitz et al. 2001), cardiac myocytes (Ito et al. 1993), lung fibroblasts (Davie et al. 2006), macrophages (Ehrenreich et al. 1990), and leukocytes (Sessa et al. 1991).

1.2 ET Receptors

ET-1 exerts its action through two pharmacologically distinct G protein-coupled receptors, namely, ET_A and ET_B receptors (Davenport 2002). The ET_A receptor is mostly expressed on vascular smooth muscle cells, fibroblasts, and cardiomyocytes and mediates the vasoconstrictor and proliferative effects of ETs (Hosoda et al. 1991; MacLean et al. 1994; Davie et al. 2002). The ET_B receptor is expressed on vascular endothelial cells where it induces endothelial-dependent vasodilatation through NO production (Hosoda et al. 1991; Hirata et al. 1993) and clearance of circulating ET-1 (Dupuis et al. 1996a). The ET_B receptor is also present on vascular smooth muscle cells, where its activation induces vasoconstriction and cell proliferation (Davie et al. 2002; Clozel et al. 1992), and on fibroblasts, where it mediates cell proliferation, matrix contraction, and collagen production (Shi-Wen et al. 2001, 2004). In the kidney, the ET_B receptor modulates sodium and water handling (Kohan 1997). The ET_B receptors on smooth muscle cells and fibroblasts are inducible (Adner et al. 1998). In a number of pathological situations, there is a shift in the expression of ET_B receptors, which experience a downregulation on endothelial cells and a marked upregulation on smooth muscle cells (Iglarz and Clozel 2010). In rats with systemic hypertension, diabetes mellitus, or hypercholesterolemia, ET_B receptor immunostaining is reduced in endothelial cells and increased in vascular smooth muscle (Kakoki et al. 1999). Upregulation of smooth muscle ET_B receptors has been observed in animal models of heart failure (Gray et al. 2000) and subarachnoid hemorrhage (Roux et al. 1995) and in human diseases including chronic thromboembolic pulmonary hypertension (PH) (Bauer et al. 2002), atherosclerosis (Dagassan et al. 1996), ischemic heart disease (Dimitrijevic et al. 2009), and scleroderma (Shi-Wen et al. 2001, 2004). In such pathological situations, both ET_A and ET_B receptors mediate the detrimental actions of ET-1. Because of a "cross talk" between ET_A and ET_B receptors (Clozel and Flores 2006; Clozel and Gray 1995), selective antagonism of only one receptor subtype may result in compensation by the other receptor subtype and may be insufficient, whereas dual antagonism may be necessary for complete inhibition.

The molecular basis of this cross talk between receptors could be the formation of ET_A–ET_B heterodimers in cells expressing both receptor subtypes, which may further modulate the functionality of the receptors (Sauvageau et al. 2007; Gregan et al. 2004).

1.3 The ET System and the Pulmonary Circulation

The pulmonary circulation is the major site of production of ET-1. In rats, the lung displays the highest immunoreactive ET concentrations (Matsumoto et al. 1989), with mRNA expression levels five times higher than in any other organ studied, including heart, kidney, and brain (Firth and Ratcliffe 1992). The human lung possesses high-affinity binding sites for ET which are variably distributed (Zhao et al. 1995). In human pulmonary arteries, the ET_A receptor is predominant in large vessels, with an increasing proportion of ET_B receptor in more distal pulmonary arteries (Davie et al. 2002).

The lung has also the capacity to clear plasma ET-1 from the circulation and removes about 50 % of circulating ET-1 through the ET_B receptors (Dupuis et al. 1996b). In the normal lung, the amount of ET-1 extracted is quantitatively similar to the amount produced, resulting in a physiological balance of ET-1 across the pulmonary circulation.

2 The Role of the ET System in the Pathogenesis of PH

2.1 Imbalance of Endothelial Mediators

The pathogenesis of PH is a multifactorial process that involves endothelial cell dysfunction and altered production of endothelial cell-derived vasoactive mediators (Tuder et al. 2001). The result is an imbalance between vasodilators/antimitotics, such as prostacyclin and NO, and vasoconstrictors/mitogens, such as thromboxane A2, serotonin, angiopoietin-2, and ET-1, in favor of vasoconstriction and proliferation (Archer and Rich 2000; Kumpers et al. 2010). ET-1 seems to play a major role in endothelial dysfunction and in the pathogenesis of PH, because of its abundance in pulmonary endothelial cells and its increased production in response to many insults. ET-1 is a major initiator of endothelial dysfunction and its production is inversely correlated with that of NO. ET-1 decreases NO bioavailability either by decreasing its production or by increasing its degradation (Iglarz and Clozel 2007). Impairment of these signaling pathways, which regulate normal vascular tone, can lead to the development of PH.

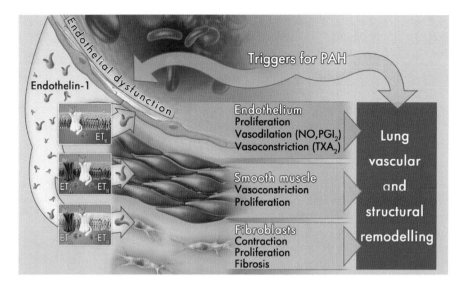

Fig. 1 The pathological roles of ET-1 in PAH

2.2 Effects of ET-1 Contributing to PAH

The effects of ET-1 are in line with the pathological changes that are typical of PAH: PAH is characterized by pulmonary arterial wall remodeling leading to formation of fibrotic neointimal and plexiform lesions, endothelial injury associated with vasoconstriction and in situ thrombosis, and infiltration of inflammatory cells (Stenmark et al. 2009). ET-1 may contribute to each mechanism involved in the pathogenesis of PAH (Fig. 1). In addition to its pressor effects, ET-1 can cause both an abnormal growth pattern in smooth muscle cells (Davie et al. 2002), endothelial cells (Morbidelli et al. 1995), and fibroblasts (Peacock et al. 1992) and an inhibition of cell apoptosis (Kulasekaran et al. 2009; Jankov et al. 2006; Shichiri et al. 1997). ET-1 significantly increases the chemotaxis of endothelial cells and fibroblasts (Morbidelli et al. 1995; Peacock et al. 1992). It also induces fibrosis by stimulating the activation and contraction of fibroblasts (Shi-Wen et al. 2001, 2004), and the synthesis of extracellular matrix compounds (Shi-Wen et al. 2001). Another effect of ET-1 is the induction of inflammation through various mechanisms, including the increase in vascular permeability (Filep et al. 1995), the stimulation of pro-inflammatory cytokine production (Browatzki et al. 2000), the activation of platelet and neutrophils, and the promotion of cellular adhesion (Helset et al. 1996). Finally, there is increasing evidence suggesting that ET-1 increases reactive oxygen species production in the vasculature (Loomis et al. 2005). This was confirmed in a transgenic mouse model where endothelium-restricted ET-1 overexpression led to endothelial dysfunction, vascular hypertrophy, and increased activity and expression of NADPH oxidase (Amiri et al. 2004).

Taken together, these enhanced biological activities shifted in favor of constriction, proliferation, fibrosis, and inflammation suggest that disturbances in the control of ET-1 production can contribute to the pathogenesis of PAH.

2.3 Upregulation of the ET System in PH

The most compelling evidence for activation of the ET system in PH comes from the observation that ET-1 is overexpressed in a number of conditions associated with PH, including idiopathic PAH (Stewart et al. 1991; Giaid et al. 1993; Rubens et al. 2001), PAH associated with connective tissue diseases (Zamora et al. 1990; Julkunen et al. 1991; Vancheeswaran et al. 1994), HIV infection (Rolinski et al. 1994, 1999), portal hypertension (Benjaminov et al. 2003), congenital heart diseases (Yoshibayashi et al. 1991), sickle cell disease (Benza 2008), but also persistent PH of the newborn (Rosenberg et al. 1993), chronic thromboembolic PH (Bauer et al. 2002), and PH complicating chronic obstructive pulmonary disease (Carratu et al. 2008), and interstitial lung disease (Trakada et al. 2003). One can speculate that the release of ET-1 is reflecting a common endothelial injury in response to a variety of stimulants, such as HIV gp120, anti-endothelial cell antibodies, shear stress, and *BMPR2* mutation, and that the common response will be a sequence of pathological alterations leading to PAH. Furthermore, the increase in plasma ET-1 concentrations tends to correlate with disease severity. In patients with PAH, the increase in plasma ET-1 concentrations is correlated with right atrial pressure, mixed venous oxygen saturation, and pulmonary vascular resistance (Cacoub et al. 1997; Nootens et al. 1995). The elevated circulating ET-1 concentrations associated with PH are not related to a decrease in ET_B receptor-mediated clearance, as indicated by normal or near-normal clearance in patients with PAH, suggesting that increased synthesis could be a primary abnormality (Langleben et al. 2006).

Changes in the ET signaling system in PH are not only limited to an increased production of ET-1 but also affect the ET receptor gene expression. ET receptors are upregulated in PAH. PAH is associated with a significant upregulation of ET binding sites in the distal pulmonary artery circulation, and the relative proportions of both receptor subtypes (60 % ET_A: 40 % ET_B) generally remain the same in normal and remodeled vessels (Davie et al. 2002). In patients with chronic thromboembolic PH, the ET_B receptor is overexpressed in pulmonary arterial smooth muscle cells where it mediates contraction and proliferation, whereas the mRNA levels of ET_A receptor in both endothelium and media are unchanged (Bauer et al. 2002).

3 ET Receptor Antagonism: Preclinical Studies

ET receptor blockade is the most efficient way to antagonize the ET system. Inhibition of ECE is also possible but less effective (Davenport and Maguire 2006). ERAs are classified as ET_A-selective, ET_B-selective, or dual ET_A/ET_B

antagonists depending on their relative affinities for the ET_A and ET_B receptors, as determined either in in vitro studies in cell culture, and/or in vivo by measurement of ET-1 plasma concentrations since they are elevated by ET_B antagonism (Attina et al. 2005). No pharmacological definition of selectivity has been clearly established and ET_A-selective antagonists can bind to both receptors at high concentrations (Goddard and Webb 2002), whereas dual antagonists usually demonstrate a higher affinity for the ET_A receptor. The majority of preclinical studies describing ET receptor antagonism in PH have been performed using in vitro cell culture assays (animal and human), ex vivo tissue preparations (animal and human), and animal models.

3.1 ET Receptor Antagonism and Vasoconstriction/ Vasodilation

Given the role of the ET_B receptor in vasodilation (Hirata et al. 1993), the beneficial effects of combining ET_A and ET_B antagonism have been challenged (Hirata et al. 1993; Dupuis et al. 1996a; Sato et al. 1999). Whether ET_B antagonism results in arterial vasoconstriction or vasodilation appears to depend upon the model and vessels investigated. The ET_B receptor antagonist BQ-788 significantly inhibited ET-1-induced contraction in human internal mammary arteries, but not in radial arteries, whereas the ET_A receptor antagonist BQ-610 produced a potent inhibitory effect in both arteries (Liu et al. 1996). Dilatory effects have been observed with both selective ET_A receptor blockade and dual antagonism. Pretreatment with the ET_A-selective antagonists sitaxentan and BQ-123 or with the dual antagonist bosentan completely inhibited the pulmonary vasoconstrictor response to acute hypoxia in rats (Tilton et al. 2000; Oparil et al. 1995; Chen et al. 1995). Other dual ET_A/ ET_B receptor antagonists (BSF 420627, SB 217242 and SB 209670) also demonstrated vasodilatory properties in hypoxia models of PH in rats and dogs (Jasmin et al. 2001; Underwood et al. 1998; Willette et al. 1997). In isolated perfused rat lungs, blockade of both ET_A and ET_B receptors with the ET_A-selective BQ-123 and ET_B-selective BQ-788 antagonists was required for maximal inhibition of ET-1-induced contraction (Sato et al. 1995). Dual blockade was also necessary in isolated human bronchi (Fukuroda et al. 1996) and in isolated small pulmonary arteries from rats with induced myocardial infarction (Sauvageau et al. 2006) to inhibit the contractive effect of ET-1. In arteries from rats with monocrotaline-induced PH, the dual antagonist bosentan and the ET_A receptor antagonist A-147627 potently reduced the contraction induced by ET-1, while bosentan also reduced vascular sensitivity to ET-1 (Sauvageau et al. 2009). These studies suggest that combined ET_A/ ET_B receptor blockade may be necessary for complete inhibition of ET-1-induced vasoconstriction. This synergistic effect may be explained by a cross talk between the ET_A and ET_B receptors and the formation of functional heterodimers (Sauvageau et al. 2007).

3.2 ET Receptor Antagonism and Vascular/Cardiac Hypertrophy

Beyond vasodilation, the impact of ET-1 antagonists on vascular hypertrophy and right ventricle afterload is essential for pharmacological evaluation, especially in conditions where vascular remodeling is already present. Experimental animal models have been used to reproduce the pathophysiology of PH and its cardiac consequences such as right ventricular hypertrophy and failure. These models mimic some of the clinical conditions that may lead to PH, such as hypoxia, endothelial injury/inflammation, left-to-right shunt, or vascular obstruction.

In several models of acute PH, such as acute hypoxia in rats (Chen et al. 1995; Eddahibi et al. 1995), pigs (Holm et al. 1996), and piglets (Pearl et al. 1999), cardiopulmonary bypass in pigs (Carteaux et al. 1999), and endotoxic shock in pigs (Weitzberg et al. 1996), bosentan blocked the progression of PH and increased cardiac output. The ET_A antagonist, sitaxentan, demonstrated similar effects in hypoxic rats (Tilton et al. 2000).

In chronic models of hypoxia-induced PH in rats, sitaxentan (Tilton et al. 2000) and bosentan (Chen et al. 1995) prevented the development of PH. Furthermore, both drugs, when initiated after the hypoxic injury, produced a significant reversal of established hypoxia-induced PH with reduction of right ventricular hypertrophy and pulmonary vascular remodeling, despite continuous hypoxic exposure. Likewise, selective ET_A antagonists [BQ-123 (DiCarlo et al. 1995) and A-127722 (Chen et al. 1997)] have been examined in this model and found to consistently attenuate or even prevent right ventricular hypertrophy and pulmonary artery medial thickening.

Mixed results have been reported in the model of monocrotaline-induced PH in rats. Sakai et al. (2000) reported that dual antagonism with SB 209670 had the potential to worsen PH compared with selective ET_A blockade with BMS 193884, because blockade of the ET_B receptor inhibited the hypotensive effect of ET-1 on pulmonary circulation. However, the dual antagonist bosentan was found to reduce right ventricular hypertrophy and medial hypertrophy in this model (Hill et al. 1997; Clozel et al. 2006). Using different time courses of administration, Hill et al. (1997) evidenced that bosentan acts mainly during the remodeling phase rather than during the acute lung injury following monocrotaline injection. ET_A-selective antagonists, including sitaxentan (Tilton et al. 2000), BQ-123 (Miyauchi et al. 1993), LU 135252 (Prie et al. 1997), ABT-627 (Nishida et al. 2004), and YM598 (Yuyama et al. 2004), have similarly demonstrated a protective effect in this model although a reversal of medial hypertrophy was not systematically observed (Prie et al. 1997). In contrast, ET_B-selective antagonism with A-192621 aggravated PH, whereas combination with the ET_A-selective antagonist ABT-627 improved PH to the same extent as ET_A antagonism alone (Nishida et al. 2004). This exaggerated response to monocrotaline during ET_B receptor blockade was attributed to ET_A receptor activation, suggesting an exclusive contribution of ET_A receptor-mediated action to the pathogenesis of monocrotaline-induced PH.

This interpretation was challenged by a comparative study in rats with monocrotaline-induced PH, where the dual ET receptor antagonist BSF 420627 doubled survival compared with the untreated animals and increased it by 10 % compared with the animals receiving the ET_A-selective antagonist LU 135252 (Jasmin et al. 2001). Reduction of right ventricular hypertrophy was only seen in the animals receiving the dual ET receptor antagonist, suggesting that blockade of both receptors was necessary to prevent the deleterious effects of ET-1 in this model.

A direct comparison between the dual ET receptor antagonist bosentan and the ET_A-selective antagonist ambrisentan in the monocrotaline-induced PH model (Schroll et al. 2008) indicated that ambrisentan significantly increased prostacyclin synthase I expression compared with bosentan. However, no benefit on pulmonary hemodynamics was observed: both antagonists had similar effects on right ventricular systolic pressure, pulmonary vascular remodeling, and right ventricular hypertrophy.

More recently, macitentan, a novel ET receptor dual antagonist, was also compared with bosentan in the monocrotaline-induced PH rat model (Iglarz et al. 2008). Macitentan prevented right ventricle hypertrophy and the development of PH at a ten times lower dose than bosentan. Prior studies have shown that dual antagonists could significantly reduce mortality in experimental PH (Jasmin et al. 2001; Clozel et al. 2006). Macitentan had a major effect of decreasing mortality in this animal model. In the Dahl-salt rat model of systemic hypertension, macitentan administered on top of the maximally effective dose of bosentan further reduced mean arterial pressure (MAP). In contrast, bosentan administered on top of the maximally effective dose of macitentan failed to elicit any further reduction in MAP. The add-on effect of macitentan on top of bosentan suggests that macitentan is able to achieve a more complete blockade of ET receptors (Iglarz et al. 2012). The potential for increased potency and efficacy of macitentan has been attributed to physicochemical and binding properties promoting tissue penetration (Iglarz et al. 2011). Overall, both selective and dual ERAs have proven to be efficacious in monocrotaline-induced PH in rats and the Dahl-salt rat model of systemic hypertension.

In models of overcirculation-induced PH, the objective is to experimentally mimic PH caused by congenital heart diseases with left-to-right shunts. This situation can be reproduced by anastomosis of the left subclavian artery to the pulmonary arterial trunk in piglets and results in pronounced chronic PH, with medial hypertrophy, and increased pulmonary vascular resistance (Morin 1989; Rondelet et al. 2003). In this model, bosentan and sitaxentan have been shown to prevent and reverse pulmonary vascular remodeling and right heart hypertrophy (Rondelet et al. 2003, 2007).

In models of chronic thromboembolic PH, experimental vascular obstruction or occlusion results in a sustained increase in pulmonary vascular resistance and remodeling of the pulmonary vascular bed. In a canine model where PH was induced by repeated embolization with ceramic beads, treatment with bosentan not only reduced medial thickening of pulmonary arteries, but also reduced adventitial thickening and prevented intimal fibrosis and peripheral neomuscularization (Kim et al. 2000). The lack of effect on hemodynamics was attributed to the fact that hypertension was due to mechanical obstruction in this model.

The above studies document that dual and ET_A-selective ERAs prevented, attenuated, or even reversed vascular remodeling and/or hypertrophy associated with experimental PH.

3.3 ET Receptor Antagonism and Fibrosis

The evaluation of ERAs in PH also includes the benefit derived from their antiproliferative and anti-fibrotic effects. Smooth muscle cell proliferation, adventitial thickening, altered angiogenesis, and fibrosis are generally observed in PH, especially in PH associated with connective tissue disease or PH due to chronic lung disease, which are characterized by a wide range of vascular inflammatory and fibrotic complications (Crystal et al. 1984).

At the cellular level, proliferation of distal human pulmonary artery smooth muscle cells was found to be modulated by both the ET_A and ET_B receptors and to be inhibited by the dual ET_A/ ET_B receptor antagonist PD145065 (Davie et al. 2002). In contrast, human lung fibroblast growth was stopped by the ET_A-selective receptor antagonist BQ-123 but not by the specific ET_B receptor antagonist BQ-788 (Gallelli et al. 2005). In human skin fibroblast cultures exposed to ET-1, interstitial collagenase downregulation and collagen matrix contraction were inhibited by bosentan and by the ET_A-selective receptor antagonist PD156707, but dual antagonism was necessary to prevent the production of collagen type I and III (Shi-Wen et al. 2001). Both antagonists inhibited ET-1-induced collagen matrix contraction in fibroblasts from patients with scleroderma showing an enhanced contractile phenotype (Shi-Wen et al. 2004). In human pulmonary endothelial cells exposed to cigarette smoke extract, ET_B receptor blockade, either with bosentan or with BQ-788, attenuated cell dysfunction (Milara et al. 2010). In these sensitized endothelial cells, the ET_B receptor is overexpressed and may mediate vasoconstriction via reduced NO synthesis and increased thromboxane A2 productions, and potentiate endothelial cell dysfunction induced by ET-1.

The above in vitro results, suggesting a role for ET antagonism in countering the pro-fibrotic and pro-inflammatory effects of ET-1, are supported by several experimental models. Using the bleomycin-induced pulmonary fibrosis model in rats, Park et al. (1997) reported that bosentan reduced lung fibrosis, as measured by tissue morphometry. However, this finding was not confirmed by Mutsaers et al. (1998), who observed that neither dual ET receptor antagonism with bosentan nor single ET_A blockade with BQ-485 prevented the accumulation of collagen, despite the inhibition of ET-1 expression in lung tissue. A more recent study, however, confirmed the beneficial effects of bosentan in this model both on the development of PH and of pulmonary fibrosis. Bosentan-treated rats had improved exercise capacity, less right ventricular hypertrophy, and lower hydroxyproline content in lung tissue (Schroll et al. 2010). The novel dual antogonist macitentan improved right ventricular hypertrophy and hydroxyproline content at a dose three times lower than bosentan (Iglarz et al. 2011).

In a model of cigarette smoke-induced chronic obstructive disease associated with PH in rats, BQ-123 and bosentan both suppressed the development of emphysema, and the benefit of specifically blocking the ET_B receptor suggested by in vitro studies (Milara et al. 2010) was not confirmed. Both ERAs similarly inhibited pulmonary apoptosis, reduced matrix metalloproteinases (MMP-2 and MMP-9) activities, decreased inflammatory cytokine levels, and increased serum antioxidant activity in the cigarette smoke extract-treated rats.

4 ET Receptor Antagonism for the Treatment of PAH: Several Molecules with Different Characteristics

Two ERAs are currently approved for the treatment of PAH: bosentan (Tracleer®, Actelion Pharmaceuticals Ltd, Allschwil, Switzerland) and ambrisentan (Letairis®, Gilead Sciences, Foster City, California, USA and Volibris®, GlaxoSmithKline, London, UK). Sitaxentan (Thelin®, Pfizer Ld, Kent, UK) was withdrawn from the market in December 2010 for safety concerns (Galie et al. 2011). Macitentan (Actelion Pharmaceuticals Ltd, Allschwil, Switzerland) is a novel ERA that has completed the Phase III clinical trial in PAH [SERAPHIN: Study with an Endothelin Receptor Antagonist in Pulmonary arterial Hypertension to Improve cliNical outcome (ClinicalTrials.gov)].

4.1 Pharmacological Profile

These ERAs differ in several aspects, including chemical structure, receptor affinity, and pharmacokinetics (Table 1). Ambrisentan is ET_A selective, whereas bosentan and macitentan are dual ERAs, with a slightly higher affinity for the ET_A receptor. The specificity of macitentan combines several unique features: it has demonstrated an effect on survival in animal models of PAH; it is a slow-offset competitive antagonist with high receptor occupancy; it is optimized for tissue penetration, including an optimal shape, a high distribution coefficient for lipid versus aqueous solutions, and a minimal ionization state. In vitro experiments investigating the receptor affinity and receptor binding kinetics of macitentan and other ERAs, demonstrated that macitentan has a receptor occupancy half-life 15-times greater than bosentan and ambrisentan (Gatfield et al. 2012) (Fig. 2). All of these features are expected to result in improved and enhanced targeting of the autocrine–paracrine ET-1 system (Fig. 3), and especially in pathological situations associated with upregulation of ET receptors, such as PAH (Iglarz et al. 2008;

Table 1 Profiles of the endothelin receptor antagonists

	Bosentan	Ambrisentan	Macitentan
Status	Approved in 2001 in USA, 2002 in Europe	Approved in 2007 in USA, 2008 in Europe	Phase III completed
Chemical structure	Pyrimidinyl sulfonamide	Diphenylpropanoic acid	Propylsulfamide
Receptor selectivity	Dual	ETA-selective	Dual
Tissue selectivity	No	No	Yes
Log D	20:1	1:2.5	800:1
pKa	5.1	3.5	6.2
Peak plasma concentration (hours)	3–5	2	8
Bioavailability (%)	50	90	30–90[a]
Elimination half life	5.4	9	16
Protein binding (%)	>98	99	99
Metabolism			
Route	Hepatic + biliary excretion	Hepatic (glucuronidation) + biliary excretion	Hepatic + biliary excretion
Enzyme involved	CYP 3A4, 2C9	CYP 3A4, 2C19, UGTs (1A9S, 2B7S, 1A3S)	CYP 3A4
Excretion in urine (%)	<3	Low	49.7 %
Effect on CYP and hepatobiliary transporters			
CYP	↑CYP 3A4, ↑2C9, (↑2C19)	↑CYP 3A4	↑CYP 3A4
NTCP/OATP inhibition	Yes/no	No/no	No/no
BSEP inhibition	Yes	No	No
Adverse effects			
Liver enzyme	Yes (11 %)	No (0 %)	Yes (3 %)
Peripheral edema	Yes (13.2 %)	Yes (17 %)	Yes (18 %)
Anemia/hemoglobin decrease	Yes (9.9 %)	Yes (7 %)	Yes (13 %)
Teratogenicity	Yes	Yes	Yes
Significant drug–drug interactions			
Cyclosporine A	Yes	Yes	No
Warfarin	Yes	No	No
Sildenafil	Yes	No	No
Ketoconazole	Yes	No	No
Fluconazole	Yes		
Rifampicin	Yes	No	
Glibenclamide	Yes		No
References	Actelion Pharmaceuticals Ltd. (2011)	Gilead Sciences, Inc. (2011), Raja (2010a)	Iglarz et al. (2008), Raja (2010a), Sidharta et al. (2011), Sidharta et al. (2012a), Sidharta et al. (2012b),

(continued)

Table 1 (continued)

	Bosentan	Ambrisentan	Macitentan
			Bruderer et al. (2012a), Bruderer et al. (2012b), Sidharta et al. (2010), Raja (2010b)

↑ drug induces
CYP: cytochrome P450, Log D: distribution coefficient, *NTCP*: sodium-dependent taurocholate cotransporter, *OATP*: organic anion transporting polypeptide, p*K*a: ionization dissociation constant, *UGTs*: uridine glucuronosyltransferases
[a]In dogs and rats (dose-dependent)

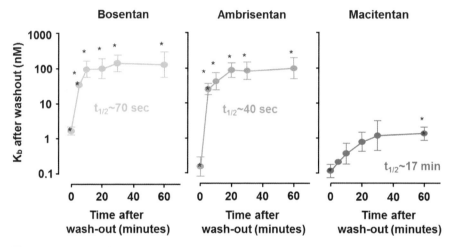

Fig. 2 Receptor binding half-lives of bosentan, ambrisentan and macitentan

Iglarz et al. 2011). Macitentan has a long elimination half-life allowing for a once-a-day dosing regimen, as ambrisentan, whereas bosentan is dosed twice daily. All three ERAs are metabolized by cytochrome P450 (CYP) enzymes in the liver, but the major hepatic metabolism pathway for ambrisentan is glucuronidation by hepatic uridine glucuronosyltransferases (UGTs). Subsequent excretion is primarily in the bile for all three agents.

4.2 Safety Profile

The main difference between ERAs appears to be in safety profiles, with higher rates of peripheral edema observed with the ET_A-selective ambrisentan and a greater frequency of serum liver function abnormalities occurring with bosentan.

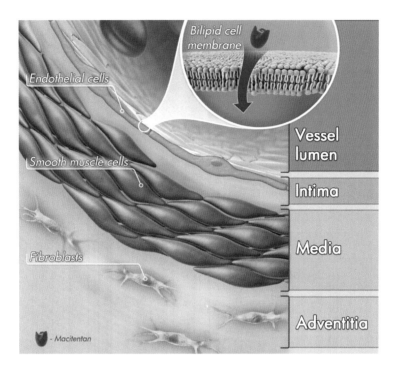

Fig. 3 Tissue-targeting potential of macitentan

Peripheral edema is a clinical consequence of worsening PH, but it is also a known class effect of ERAs (Battistini et al. 2006; Kohan et al. 2011). Even though ET_B receptors have been suggested to play a role in sodium excretion, ET_A-selective antagonists pose a greater risk of fluid overload than dual ET_A/ET_B antagonists (Barton and Kohan 2011; Trow and Taichman 2009). The mechanism of this fluid retention has been studied in rodents and the results show that ET_A-selective receptor antagonists cause water retention by stimulating, via the activation of un-antagonized ET_B receptors, the release of vasopressin and the expression of V2 receptors.

Liver abnormalities have also been associated with ERAs but may vary widely among individual agents (Battistini et al. 2006). In pivotal clinical trials, the incidence of serum aminotransferase elevations was lower with ambrisentan [0 % (Gilead Sciences, Inc. 2011)] than that with bosentan [11 % (Actelion Pharmaceuticals Ltd. 2011)]. In the long-term extension trial with ambrisentan, the incidence of serum aminotransferase elevations observed with ambrisentan was of 3.9 % (Oudiz et al. 2009). In the SERAPHIN trial, elevations of liver aminotransferases were observed in 4 % of patients on 3 mg macitentan and in 3 % of patients on 10 mg macitentan (Pulido et al. 2013).

The mechanism of hepatotoxicity of ERAs has been evaluated in sandwich-cultured human hepatocytes (Hartman et al. 2010) and in vivo (Fattinger et al. 2001; Mano et al. 2007). It has been shown that the mechanism of liver enzyme elevations with bosentan was due—at least in part—to an inhibition of the biliary efflux of taurocholate mediated by the bile salt export pump (BSEP). The observation that in humans an elevation in bile salts preceded an elevation of aminotransferases and a potentiation by glibenclamide, also an inhibitor of BSEP, suggests that this hypothesis also stands in man. Macitentan has been designed to avoid an interaction with bile salt elimination.

Reductions in hemoglobin concentration and hematocrit have been reported with the use of ERAs (Battistini et al. 2006). The cause of this reduction is not a direct anemia but is secondary to hemodilution. As a class, ERAs have shown teratogenic effects in animals, and pregnancy should be avoided.

4.3 Drug Interaction Profile

Trough concentrations of bosentan are increased 30-fold by cyclosporine and bosentan should therefore not be co-administered with cyclosporine A. The mechanism of the interaction has been suggested to be an inhibition of OATP-mediated uptake of bosentan into hepatocytes (Treiber et al. 2007). Bosentan should also not be used concomitantly with glibenclamide, due to an increased risk of elevated liver aminotransferases. Bosentan is metabolized by CYP2C9 and CYP3A4 and inhibition of these isoenzymes may increase the plasma concentration of bosentan. Therefore administration with fluconazole, which inhibits CYP2C9 and also CYP3A4, and administration with both a potent CYP3A4 inhibitor (such as ketoconazole, itraconazole, or ritonavir) and a CYP2C9 inhibitor (such as voriconazole) are not recommended. Bosentan is also a mild inducer of CYP2C9 and CYP3A4, and possibly CYP2C19. Consequently, plasma concentrations of substances metabolized by these isoenzymes such as sildenafil, simvastatin, hormonal contraceptives (ethinyl estradiol and norethindrone), and to a lower extent warfarin, are decreased by bosentan coadministration. Conversely, bosentan concentration rises by up to 50 % with concomitant sildenafil therapy due to inhibition of OATP-mediated transport (Treiber et al. 2007). Inducers of CYP2C9 and CYP3A4 are expected to decrease the concentration of bosentan. As such, rifampicin reduces bosentan steady-state concentrations, although an initial concentration increase has been attributed to inhibition OATP-mediated uptake of bosentan into hepatocytes (Treiber et al. 2007).

Ambrisentan is sensitive to drug–drug interaction with cyclosporine, as its concentrations are increased two-fold by cyclosporine A. If coadministration is necessary, a reduction of the dose of ambrisentan should be considered.

Macitentan has a very low propensity for drug–drug interactions, as shown by clinical pharmacology studies, which showed no significant interactions between macitentan and sildenafil, warfarin, cyclosporine, and only a two-fold increase with ketoconazole (Raja 2010b). Macitentan and its metabolites do not interfere with the BSEP-mediated elimination of bile salts (Raja 2010b, Bolli et al. 2012), which may mean that there is no drug–drug interaction with glibenclamide. Therefore, concomitant use of macitentan and glibenclamide was permitted in the completed SERAPHIN trial.

Thus, both ET_A-selective and dual ERAs have demonstrated improvements in hemodynamics, right ventricular hypertrophy, pulmonary arterial remodeling, pulmonary vessel endothelial function, and survival in experimental PH. The registration of bosentan for the treatment of PAH has constituted a first step forward in the treatment of patients with PAH and has been followed by the registration of ambrisentan. Understanding how to improve efficacy and the mechanisms of edema and liver enzyme changes have led to the development of macitentan, a dual ERA with maximal tissue-targeting properties, enhanced affinity for the ET receptors and long-lasting receptor occupancy, and minimal risk of fluid retention and interaction with bile salts. Only clinical trials can appropriately describe the efficacy and safety of each of these molecules. Together with other classes of drugs, phosphodiesterase type 5 inhibitors, and prostacyclin receptor agonists, the treatment paradigm will continue to evolve. The goal of 'standard of care treatment' will shift from improving quality of life and symptoms, to slowing the progression of this very severe disease.

5 ET Receptor Antagonism in PAH: Human Studies

5.1 Bosentan

Bosentan was the first oral therapy approved for PAH. Studies have demonstrated improvements in exercise capacity, functional class, and hemodynamic parameters and an increase in time to clinical worsening in PAH patients. The first randomized, double-blind, placebo-controlled, multicenter trial (study 351) with bosentan was designed to assess its effects on exercise capacity and cardiopulmonary hemodynamics, as well as safety and tolerability (Channick et al. 2001). Eligible patients had severe idiopathic PAH or PAH associated with scleroderma (in functional classes III or IV, according to the New York Heart Association (NYHA) classification), despite prior treatment including vasodilators, anticoagulants, diuretics, cardiac glycosides, or supplemental oxygen as clinically indicated. Thirty-two patients were randomized to receive bosentan or placebo (2:1 ratio). Patients randomized to bosentan received 62.5 mg bid for the first 4 weeks, followed by the target dose (125 mg bid) unless drug-related adverse events were observed. After 12 weeks, the 6-min walk distance (6MWD) increased by 70 m (from 360 ± 19 m [m $\pm$ SE] at baseline to 430 ± 14 m at week 12; $P < 0.05$) in the bosentan arm, whereas no improvement was seen with placebo (355 ± 25 m at baseline and 349 ± 44 m at week 12). The median change from baseline was +51 m with bosentan versus −6 m with placebo. The difference between treatment arms in the mean change in 6MWD was 76 ± 31 m (mean $\pm$ SEM) in favor of bosentan (95 % CI, 12–139; $P = 0.021$). Bosentan improved cardiac index, with the difference between treatment groups in the mean change at week 12 being 1.0 ± 0.2 L/min/m^2 (mean $\pm$ SEM) in favor of bosentan (95 % CI, 0.6–1.4 L/min/m^2, $P < 0.001$). Bosentan decreased pulmonary vascular resistance, whereas it increased with placebo ($P < 0.001$). Asymptomatic increases in hepatic aminotransferases were observed in two bosentan-treated patients, but these normalized without discontinuation or change of dose.

A larger double-blind, placebo-controlled study (BREATHE-1) evaluated bosentan in 213 patients with PAH (idiopathic or associated with connective tissue diseases) who were randomized to placebo, bosentan 125 or 250 mg bid for a minimum of 16 weeks (62.5 mg bid for 4 weeks and then target dose) (Rubin et al. 2002). The primary endpoint was the change in exercise capacity assessed by 6MWD, and secondary endpoints included changes in Borg dyspnea index, NYHA functional class, and time to clinical worsening. Patients had symptomatic, severe PAH (NYHA functional class III or IV) despite treatment with anticoagulants, vasodilators, diuretics, cardiac glycosides, or supplemental oxygen as clinically indicated. Glibenclamide was contraindicated since it was increasing the risk of liver enzyme elevations. Patients were randomized to receive either bosentan 62.5 mg bid for 4 weeks followed by the target dose (125 or 250 mg bid) or matching doses of placebo (144 patients received bosentan and 69 received placebo). After 16 weeks, bosentan improved the 6MWD by 36 m whereas a deterioration (-8 m) was seen with placebo, and the difference between treatment groups in the mean change in 6MWD was 44 m in favor of bosentan ($P = 0.0002$). A dose response for efficacy could not be ascertained. The risk of clinical worsening was reduced by bosentan compared with placebo ($P = 0.0015$). The most frequent adverse event in both the bosentan and placebo treatment groups was headache. Abnormal hepatic function [as indicated by elevated levels of alanine aminotransferase (ALT) and/or aspartate aminotransferase (AST)], syncope, and flushing occurred more frequently in the bosentan groups. Increases in hepatic enzymes over three times the upper limit of normal occurred in 12 % of the bosentan 125 mg group and in 14 % of the bosentan 250 mg group; two patients (3 %) in the 125 mg bid group and five (7 %) in the 250 mg bid group had elevations $> 8 \times$ ULN. Hepatic function abnormalities were transient except for three patients (all in the 250 mg dosage bosentan group); these patients were withdrawn prematurely from the study. All hepatic transaminases increases have been reported to be reversible with continued treatment, decrease in dose, or discontinuation. Three patients died during the course of the study: two on placebo and one on bosentan.

With respect to longer term data on bosentan therapy, McLaughlin et al. (2005) reported that first-line therapy with bosentan monotherapy, with the addition of or transition to other therapy as needed, resulted in Kaplan–Meier survival estimates of 96 % at 12 months and 89 % at 24 months. At 12 and 24 months, 85 % and 70 % of patients, respectively, remained alive and on bosentan monotherapy. Sitbon et al. (2005) compared survival in 139 NYHA functional class III idiopathic PAH treated with bosentan with historical data from 346 similar idiopathic PAH patients treated with continuous intravenous epoprostenol. Baseline characteristics suggested that the epoprostenol cohort had more severe disease. Kaplan–Meier survival estimates after 1 and 2 years were 97 % and 91 %, respectively, in the bosentan-treated cohort and 91 % and 84 %, respectively, in the epoprostenol cohort.

In a retrospective study of 86 children with idiopathic PAH and PAH associated with congenital heart disease or connective tissue diseases (Rosenzweig et al. 2005), bosentan was used with or without concomitant intravenous epoprostenol or subcutaneous treprostinil. At data cutoff, 68 patients (79 %) were still treated with bosentan, 13 (15 %) were discontinued, and 5 (6 %) had died. Median exposure to bosentan was 14 months. In 90 % of the patients ($n = 78$), NYHA functional class improved (46 %) or was unchanged (44 %) with bosentan treatment. Kaplan–Meier survival estimates at 1 and 2 years were 98 % and 91 %, respectively. Galiè et al. (2006) evaluated bosentan therapy in a multicenter, double-blind, randomized, placebo-controlled study in patients with functional class III Eisenmenger syndrome (BREATHE-5). Fifty-four patients were randomized 2:1 to bosentan vs. placebo for 16 weeks. Resting systemic arterial oxygen saturation did not worsen with bosentan therapy, and, compared with placebo, bosentan decreased pulmonary vascular resistance ($P < 0.05$) and mean pulmonary arterial pressure ($P < 0.05$), and increased exercise capacity ($P < 0.01$). Four patients discontinued due to adverse events: 2 (5 %) in the bosentan arm and 2 (12 %) in the placebo arm. An open-label multicenter study with bosentan therapy in 16 patients with HIV-associated PAH similarly demonstrated favorable clinical, functional, and hemodynamic effects (Sitbon et al. 2004).

While bosentan is used relatively widely in patients with PAH, close follow-up over time of both efficacy and safety is encouraged. After approval of bosentan for treatment of PAH in the European Union, European authorities required the introduction of a post-marketing surveillance system to obtain further data on its safety profile (Humbert et al. 2007). For that purpose, a prospective, Internet-based post-marketing surveillance system was developed which solicited reports on elevated aminotransferases, medical reasons for bosentan discontinuation, and other serious adverse events requiring hospitalization. Captured data included demographics, PAH etiology, baseline functional status, and concomitant PAH-specific medications. Safety signals captured included death, hospitalization, serious adverse events, unexpected adverse events, and elevated aminotransferases. Within 30 months, 4,994 patients were included, representing 79 % of patients receiving bosentan in Europe. 4,623 patients were naïve to treatment; of these 352 had elevated aminotransferases, corresponding to a crude incidence of 8 % and an annualized rate of 10 %. Bosentan was discontinued due to elevated aminotransferases in 150 bosentan-naïve patients (3 %) with increased hepatic transaminases reported to normalize in all patients after stopping bosentan. Safety results were consistent across subgroups and etiologies. Therefore, liver function tests should be checked monthly. In addition to hepatotoxicity, other side effects include decrease in hemoglobin (hematocrit may be checked every 3 months) and peripheral edema.

Women treated with bosentan should use effective contraception because pregnancy and labor increase the demand on the heart-pulmonary system and is contraindicated in patients with PH; in addition bosentan is also potentially teratogenic. Of note, bosentan may reduce the effects of hormonal contraception and it is recommended that women who are taking bosentan use a second method of contraception. In males, there is concern that the ERAs, as a class may be capable of causing testicular tubular atrophy and male infertility. Younger males who may consider conceiving in the future should be informed regarding this possibility prior to taking these drugs.

5.2 Sitaxentan

Sitaxentan, a selective ETA receptor antagonist, when given at a dose of 100 mg/day orally, was approved in the European Union and other countries, but not in the USA, for the treatment of PAH in adult patients. Indeed, multicenter, randomized placebo-controlled clinical trials demonstrated that sitaxentan has beneficial effects on exercise capacity (i.e., 6MWD), NYHA functional class, and hemodynamic parameters in PAH patients (Barst et al. 2004, 2006). However, sitaxentan had to be removed from the world market because of the occurrence of acute severe idiosyncratic hepatotoxicity (Barst et al. 2006).

5.3 Ambrisentan

Ambrisentan is another oral, once-daily, ETA-selective ERA approved for the treatment of PAH. In Phase III clinical trials in patients with PAH, ambrisentan (2.5–10 mg orally once daily) improved exercise capacity, Borg dyspnea index, time to clinical worsening, NYHA functional class, and quality of life compared with placebo (Galiè et al. 2008). Ambrisentan in Pulmonary Arterial Hypertension, Randomized, Double-Blind, Placebo-Controlled, Multicenter, Efficacy Study 1 and 2 (ARIES-1 and ARIES-2) were concurrent, double-blind, placebo-controlled studies that randomized 202 and 192 patients with PAH, respectively, to placebo or ambrisentan (ARIES-1, 5 or 10 mg; ARIES-2, 2.5 or 5 mg) orally once daily for 12 weeks (Galiè et al. 2008). The primary endpoint for each study was change in 6MWD from baseline to week 12. Clinical worsening, NYHA functional class, Short Form-36 Health Survey score, Borg dyspnea score, and B-type natriuretic peptide plasma concentrations were also assessed. In addition, a long-term extension study was performed. The 6MWD increased in all ambrisentan groups; mean placebo-corrected treatment effects were 31 m ($P = 0.008$) and 51 m ($P < 0.001$) in ARIES-1 for 5 and 10 mg ambrisentan, respectively, and 32 m ($P = 0.022$) and 59 m ($P < 0.001$) in ARIES-2 for 2.5 and 5 mg ambrisentan, respectively. Improvements in time to clinical worsening (ARIES-2), NYHA functional class

(ARIES-1), Short Form-36 score (ARIES-2), Borg dyspnea score (both studies), and B-type natriuretic peptide (both studies) were observed. Elevation of liver aminotransferases have been reported with ambrisentan use. In most cases alternative causes of liver injury could be identified. Cases consistent with autoimmune hepatitis, including possible exacerbation of underlying autoimmune hepatitis, hepatic injury, and hepatic enzyme elevations, which could be potentially related to therapy, have been mentioned in the Summary of Product Information labeling text of Volibris/Letairis). In 280 patients completing 48 weeks of treatment with ambrisentan monotherapy, the improvement from baseline in 6MWD at 48 weeks was 39 m.

In light of marked aggravation in patients with pulmonary hypertension associated with idiopathic pulmonary fibrosis and pulmonary fibrosis without pulmonary hypertension, ambrisentan is now contraindicated in EU for these patients.

5.4 Macitentan

The dual ERA macitentan has been tested in a large pivotal, long-term, event-driven randomized, placebo-controlled study (SERAPHIN, Study with an Endothelin Receptor Antagonist in Pulmonary arterial Hypertension to Improve cliNical outcome) (Pulido et al. 2013). This study was designed to evaluate the efficacy and safety of macitentan through the primary endpoint of time to first morbidity and all-cause mortality event in 742 patients with symptomatic PAH with a mean duration of treatment of 85, 100 and 104 weeks for placebo, macitentan 3 mg and macitentan 10 mg respectively. Macitentan, at both the 3 and 10 mg dose, has met its primary endpoint, decreasing the risk of a morbidity and mortality event over the treatment period versus placebo (Pulido et al. 2013). This risk was reduced by 45 % in the 10 mg dose group ($P < 0.001$). At 3 mg, the observed risk reduction was 30 % ($P = 0.01$). The statistically significant effect of macitentan 10 mg on this endpoint was observed irrespective of background PAH therapy. Treatment with macitentan in the SERAPHIN study was safe and well tolerated. Secondary efficacy endpoints, including change from baseline to month 6 in 6MWD and change from baseline to month 6 in NYHA functional class (FC) and time—over the whole treatment period—to either death due to PAH or hospitalization due to PAH, also showed a dose-dependent effect ($P < 0.05$ for either dose). Macitentan had a significant effect on the baseline-adjusted mean change in 6MWD ($P = 0.04$ for macitentan 3 mg and $P = 0.007$ for macitentan 10 mg). The effect of macitentan on change from baseline to month 6 in 6MWD was also examined by baseline FC. For patients in FC I/II, the baseline-adjusted mean change in 6-MWD was 6.3 m (97.5 % CL, −19.0 to 31.6 m) for macitentan 3 mg and 12.3 m (97.5 % CL, −8.1 to 32.7 m) for macitentan 10 mg. For patients in FC III/IV the baseline-adjusted mean change in 6-MWD was 32.8 m (97.5 % CL, 2.4 to 63.1 m) for macitentan 3 mg and 37.0 m (97.5 % CL, 5.4 to 68.6 m) for macitentan 10 mg.

The safety set comprised 741 patients (randomized 1:1:1), who received at least one dose of study treatment. Macitentan in this patient population was safe and well tolerated. The number of adverse events reported and patients discontinuing treatment due to adverse events was similar across all groups. Elevations of liver aminotransferases greater than three times the upper limit of normal were observed in 4 % of patients receiving placebo, in 4 % of patients on 3 mg of macitentan, and in 3 % of patients on 10 mg of macitentan. In addition, no difference was observed between macitentan and placebo on fluid retention (edema). A decrease in hemoglobin—reported as an adverse event—was observed more frequently on macitentan than placebo, with no difference in treatment discontinuation between groups.

In conclusion, several ERAs have been successfully evaluated in randomized controlled clinical trials in patients with symptomatic PAH. Compounds have been shown to be different from each other and the differentiation is not only related to whether the compound belongs to the class of selective ET_A receptor antagonists or dual ET_A/ET_B receptor antagonists. Progress in diagnostic and treatment in PAH has been markedly influenced by the ERAs; however, many questions remain. Whether various PAH subgroups such as PAH associated with connective tissue diseases or associated with congenital heart diseases may respond more or less favorably to one or another molecule is currently unknown. Also, whether an ERA should be first-line monotherapy, i.e., as opposed to a phosphodiesterase type 5 inhibitor or a prostacyclin analogue, is also unclear. Whether there is an additive or synergistic effect of an ERA with a prostanoid (e.g., epoprostenol, iloprost, or treprostinil) and/or a phosphodiesterase type 5 inhibitor (e.g., sildenafil or tadalafil) is yet to be determined. As observed in many chronic diseases, combination therapy (an area of active investigation) may further improve the overall efficacy for treating patients with PAH.

Acknowledgements The authors thank Sylvie I. Ertel (Sundgau Medical Writer, France) for editorial assistance

References

Actelion Pharmaceuticals Ltd. (2011) Summary of product characteristics Tracleer®. Actelion Pharmaceuticals Ltd., Allschwil, Switzerland

Actelion press release, 30th April 2012

Adner M, Uddman E, Cardell LO, Edvinsson L (1998) Regional variation in appearance of vascular contractile endothelin-B receptors following organ culture. Cardiovasc Res 37(1):254–262

Akishita M, Kozaki K, Eto M, Yoshizumi M, Ishikawa M, Toba K et al (1998) Estrogen attenuates endothelin-1 production by bovine endothelial cells via estrogen receptor. Biochem Biophys Res Commun 251(1):17–21

Amiri F, Virdis A, Neves MF, Iglarz M, Seidah NG, Touyz RM et al (2004) Endothelium-restricted overexpression of human endothelin-1 causes vascular remodeling and endothelial dysfunction. Circulation 110(15):2233–2240

Archer S, Rich S (2000) Primary pulmonary hypertension: a vascular biology and translational research "work in progress". Circulation 102(22):2781–2791

Attina T, Camidge R, Newby DE, Webb DJ (2005) Endothelin antagonism in pulmonary hypertension, heart failure, and beyond. Heart 91(6):825–831

Barst RJ, Langleben D, Frost A et al (2004) Sitaxsentan therapy for pulmonary arterial hypertension. Am J Respir Crit Care Med 169:441–447

Barst RJ, Langleben D, Badesch D et al (2006) Treatment of pulmonary arterial hypertension with the selective endothelin-A receptor antagonist sitaxsentan. J Am Coll Cardiol 47:2049–2056

Barton M, Kohan DE (2011) Endothelin antagonists in clinical trials: lessons learned. Contrib Nephrol 172:255–260

Battistini B, Berthiaume N, Kelland NF, Webb DJ, Kohan DE (2006) Profile of past and current clinical trials involving endothelin receptor antagonists: the novel "-sentan" class of drug. Exp Biol Med (Maywood) 231(6):653–695

Bauer M, Wilkens H, Langer F, Schneider SO, Lausberg H, Schafers HJ (2002) Selective upregulation of endothelin B receptor gene expression in severe pulmonary hypertension. Circulation 105(9):1034–1036

Benjaminov FS, Prentice M, Sniderman KW, Siu S, Liu P, Wong F (2003) Portopulmonary hypertension in decompensated cirrhosis with refractory ascites. Gut 52(9):1355–1362

Benza RL (2008) Pulmonary hypertension associated with sickle cell disease: pathophysiology and rationale for treatment. Lung 186(4):247–254

Bolli MH, Boss C, Binkert C et al (2012) The discovery of N-[5-(4-bromophenyl)-6-[2-[(5-bromo-2-pyrimidinyl)oxy]ethoxy]-4-pyrimidinyl]-N'-propylsulfamide (Macitentan), an orally active, potent dual endothelin receptor antagonist. J Med Chem 55(17):7849–7861

Bourque SL, Davidge ST, Adams MA (2011) The interaction between endothelin-1 and nitric oxide in the vasculature: new perspectives. Am J Physiol Regul Integr Comp Physiol 300(6):R1288–R1295

Browatzki M, Schmidt J, Kubler W, Kranzhofer R (2000) Endothelin-1 induces interleukin-6 release via activation of the transcription factor NF-kappaB in human vascular smooth muscle cells. Basic Res Cardiol 95(2):98–105

Bruderer S, Aanismaa P, Homery MC, Hausler S, Landskroner K, Sidharta PN, Treiber A, Dingemanse J (2012a) Effect of cyclosporine and rifampin on the pharmacokinetics of macitentan, a tissue-targeting dual endothelin receptor antagonist. AAPS J 14:68–78

Bruderer S, Hopfgartner G, Seiberling M, Wank J, Sidharta PN, Treiber A, Dingemanse J (2012b) Absorption, distribution, metabolism, and excretion of macitentan, a dual endothelin receptor antagonist, in humans. Xenobiotica 42:901

Cacoub P, Dorent R, Nataf P, Carayon A, Riquet M, Noe E et al (1997) Endothelin-1 in the lungs of patients with pulmonary hypertension. Cardiovasc Res 33(1):196–200

Carratu P, Scoditti C, Maniscalco M, Seccia TM, Di Gioia G, Gadaleta F et al (2008) Exhaled and arterial levels of endothelin-1 are increased and correlate with pulmonary systolic pressure in COPD with pulmonary hypertension. BMC Pulm Med 8:20

Carteaux JP, Roux S, Siaghy M, Schjoth B, Dolofon P, Bechamps Y et al (1999) Acute pulmonary hypertension after cardiopulmonary bypass in pig: the role of endogenous endothelin. Eur J Cardiothorac Surg 15(3):346–352

Channick RN, Simonneau G, Sitbon O et al (2001) Effects of the dual endothelin-receptor antagonist bosentan in patients with pulmonary hypertension: a randomised placebo-controlled study. Lancet 358:1119–1123

Chen SJ, Chen YF, Meng QC, Durand J, Dicarlo VS, Oparil S (1995) Endothelin-receptor antagonist bosentan prevents and reverses hypoxic pulmonary hypertension in rats. J Appl Physiol 79(6):2122–2131

Chen SJ, Chen YF, Opgenorth TJ, Wessale JL, Meng QC, Durand J et al (1997) The orally active nonpeptide endothelin A-receptor antagonist A-127722 prevents and reverses hypoxia-induced pulmonary hypertension and pulmonary vascular remodeling in Sprague-Dawley rats. J Cardiovasc Pharmacol 29(6):713–725

ClinicalTrials.gov. Study of ACT-064992 on morbidity and mortality in patients with symptomatic pulmonary arterial hypertension. NCT00660179 [updated July 26, 2012]

Clozel M, Flores S (2006) Endothelin receptors as drug targets in chronic cardiovascular diseases: the rationale for dual antagonism. Drug Dev Res 67(11):825–834

Clozel M, Gray GA (1995) Are there different ETB receptors mediating constriction and relaxation? J Cardiovasc Pharmacol 26(Suppl 3):S262–S264

Clozel M, Gray GA, Breu V, Loffler BM, Osterwalder R (1992) The endothelin ETB receptor mediates both vasodilation and vasoconstriction in vivo. Biochem Biophys Res Commun 186(2):867–873

Clozel M, Hess P, Rey M, Iglarz M, Binkert C, Qiu C (2006) Bosentan, sildenafil, and their combination in the monocrotaline model of pulmonary hypertension in rats. Exp Biol Med (Maywood) 231(6):967–973

Crystal RG, Bitterman PB, Rennard SI, Hance AJ, Keogh BA (1984) Interstitial lung diseases of unknown cause. Disorders characterized by chronic inflammation of the lower respiratory tract (first of two parts). N Engl J Med 310(3):154–166

Dagassan PH, Breu V, Clozel M, Kunzli A, Vogt P, Turina M et al (1996) Up-regulation of endothelin-B receptors in atherosclerotic human coronary arteries. J Cardiovasc Pharmacol 27(1):147–153

Davenport AP (2002) International Union of Pharmacology. XXIX. Update on endothelin receptor nomenclature. Pharmacol Rev 54(2):219–226

Davenport AP, Maguire JJ (2006) Endothelin. Handb Exp Pharmacol (176 Pt 1):295–329

Davie N, Haleen SJ, Upton PD, Polak JM, Yacoub MH, Morrell NW et al (2002) ET(A) and ET(B) receptors modulate the proliferation of human pulmonary artery smooth muscle cells. Am J Respir Crit Care Med 165(3):398–405

Davie NJ, Gerasimovskaya EV, Hofmeister SE, Richman AP, Jones PL, Reeves JT et al (2006) Pulmonary artery adventitial fibroblasts cooperate with vasa vasorum endothelial cells to regulate vasa vasorum neovascularization: a process mediated by hypoxia and endothelin-1. Am J Pathol 168(6):1793–1807

DiCarlo VS, Chen SJ, Meng QC, Durand J, Yano M, Chen YF et al (1995) ETA-receptor antagonist prevents and reverses chronic hypoxia-induced pulmonary hypertension in rat. Am J Physiol 269(5 Pt 1):L690–L697

Dimitrijevic I, Edvinsson ML, Chen Q, Malmsjo M, Kimblad PO, Edvinsson L (2009) Increased expression of vascular endothelin type B and angiotensin type 1 receptors in patients with ischemic heart disease. BMC Cardiovasc Disord 9:40

Dupuis J, Goresky CA, Fournier A (1996a) Pulmonary clearance of circulating endothelin-1 in dogs in vivo: exclusive role of ETB receptors. J Appl Physiol 81(4):1510–1515

Dupuis J, Stewart DJ, Cernacek P, Gosselin G (1996b) Human pulmonary circulation is an important site for both clearance and production of endothelin-1. Circulation 94(7):1578–1584

Eddahibi S, Raffestin B, Clozel M, Levame M, Adnot S (1995) Protection from pulmonary hypertension with an orally active endothelin receptor antagonist in hypoxic rats. Am J Physiol 268(2 Pt 2):H828–H835

Ehrenreich H, Anderson RW, Fox CH, Rieckmann P, Hoffman GS, Travis WD et al (1990) Endothelins, peptides with potent vasoactive properties, are produced by human macrophages. J Exp Med 172(6):1741–1748

Emori T, Hirata Y, Imai T, Eguchi S, Kanno K, Marumo F (1993) Cellular mechanism of natriuretic peptides-induced inhibition of endothelin-1 biosynthesis in rat endothelial cells. Endocrinology 133(6):2474–2480

Fattinger K, Funk C, Pantze M, Weber C, Reichen J, Stieger B et al (2001) The endothelin antagonist bosentan inhibits the canalicular bile salt export pump: a potential mechanism for hepatic adverse reactions. Clin Pharmacol Ther 69(4):223–231

Filep JG, Fournier A, Foldes-Filep E (1995) Acute pro-inflammatory actions of endothelin-1 in the guinea-pig lung: involvement of ETA and ETB receptors. Br J Pharmacol 115(2):227–236

Firth JD, Ratcliffe PJ (1992) Organ distribution of the three rat endothelin messenger RNAs and the effects of ischemia on renal gene expression. J Clin Invest 90(3):1023–1031

Frelin C, Guedin D (1994) Why are circulating concentrations of endothelin-1 so low? Cardiovasc Res 28(11):1613–1622

Fukuroda T, Ozaki S, Ihara M, Ishikawa K, Yano M, Miyauchi T et al (1996) Necessity of dual blockade of endothelin ETA and ETB receptor subtypes for antagonism of endothelin-1-induced contraction in human bronchi. Br J Pharmacol 117(6):995–999

Galiè N, Beghetti M, Gatzoulis MA et al (2006) Bosentan therapy in patients with Eisenmenger syndrome: a multicenter, double-blind, randomized, placebo-controlled study. Circulation 114:48–54

Galiè N, Olschewski H, Oudiz RJ et al (2008) Ambrisentan for the treatment of pulmonary arterial hypertension. Circulation 117:3010–3019

Galie N, Hoeper MM, Gibbs JS, Simonneau G (2011) Liver toxicity of sitaxentan in pulmonary arterial hypertension. Eur Respir J 37(2):475–476

Gallelli L, Pelaia G, D'Agostino B, Cuda G, Vatrella A, Fratto D et al (2005) Endothelin-1 induces proliferation of human lung fibroblasts and IL-11 secretion through an ET(A) receptor-dependent activation of MAP kinases. J Cell Biochem 96(4):858–868

Gatfield J, Mueller Grandjean C, Sasse T, Clozel M, Nayler O (2012) Slow receptor dissociation kinetics differentiate macitentan from other endothelin receptor antagonists in pulmonary arterial smooth muscle cells. PLoS One 7:e47662

Giaid A, Yanagisawa M, Langleben D, Michell RP, Levy R, Shennib H et al (1993) Expression of endothelin-1 in the lungs of patients with pulmonary hypertension. N Engl J Med 328:1732–1739

Gilead Sciences, Inc. (2011) Prescribing Information Letairis®. Gilead Sciences, Inc., Foster City, CA, USA

Goddard J, Webb DJ (2002) Endothelin antagonists and hypertension: a question of dose? Hypertension 40(3):e1–e2, author reply e1–e2

Golden CG, Nick HS, Visner GA (1998) Thrombin regulation of endothelin-1 gene in isolated human pulmonary endothelial cells. Chest 114(1 Suppl):63S–64S

Gray GA, Mickley EJ, Webb DJ, McEwan PE (2000) Localization and function of ET-1 and ET receptors in small arteries post-myocardial infarction: upregulation of smooth muscle ET (B) receptors that modulate contraction. Br J Pharmacol 130(8):1735–1744

Gregan B, Schaefer M, Rosenthal W, Oksche A (2004) Fluorescence resonance energy transfer analysis reveals the existence of endothelin-A and endothelin-B receptor homodimers. J Cardiovasc Pharmacol 44(Suppl 1):S30–S33

Hartman JC, Brouwer K, Mandagere A, Melvin L, Gorczynski R (2010) Evaluation of the endothelin receptor antagonists ambrisentan, darusentan, bosentan, and sitaxsentan as substrates and inhibitors of hepatobiliary transporters in sandwich-cultured human hepatocytes. Can J Physiol Pharmacol 88(6):682–691

Helset E, Lindal S, Olsen R, Myklebust R, Jorgensen L (1996) Endothelin-1 causes sequential trapping of platelets and neutrophils in pulmonary microcirculation in rats. Am J Physiol 271(4 Pt 1):L538–L546

Hill NS, Warburton RR, Pietras L, Klinger JR (1997) Nonspecific endothelin-receptor antagonist blunts monocrotaline-induced pulmonary hypertension in rats. J Appl Physiol 83(4):1209–1215

Hirata Y, Emori T, Eguchi S, Kanno K, Imai T, Ohta K et al (1993) Endothelin receptor subtype B mediates synthesis of nitric oxide by cultured bovine endothelial cells. J Clin Invest 91(4):1367–1373

Holm P, Liska J, Clozel M, Franco-Cereceda A (1996) The endothelin antagonist bosentan: hemodynamic effects during normoxia and hypoxic pulmonary hypertension in pigs. J Thorac Cardiovasc Surg 112(4):890–897

Hosoda K, Nakao K, Hiroshi A, Suga S, Ogawa Y, Mukoyama M et al (1991) Cloning and expression of human endothelin-1 receptor cDNA. FEBS Lett 287(1–2):23–26

Humbert M, Seagal ES, Kiely DG, Carlsen J, Schwerin B, Hoeper MM (2007) Results of European post-marketing surveillance of bosentan in pulmonary hypertension. Eur Resp J 30:338–344

Iglarz M, Clozel M (2007) Mechanisms of ET-1-induced endothelial dysfunction. J Cardiovasc Pharmacol 50(6):621–628
Iglarz M, Clozel M (2010) At the heart of tissue: endothelin system and end-organ damage. Clin Sci (Lond) 119(11):453–463
Iglarz M, Binkert C, Morrison K, Fischli W, Gatfield J, Treiber A et al (2008) Pharmacology of macitentan, an orally active tissue-targeting dual endothelin receptor antagonist. J Pharmacol Exp Ther 327(3):736–745
Iglarz M, Landskroner K, Rey M, Wanner D, Hess P, Clozel M (2011) Optimization of tissue targeting properties of macitentan, a new dual endothelin receptor antagonist, improves its efficacy in a rat model of pulmonary fibrosis associated with pulmonary arterial hypertension. Am J Respir Crit Care Med 183:A6445
Iglarz M, Rey M, Hess P, Kauser K, Clozel M (2012) Superior in vivo efficacy of macitentan: comparison to other endothelin receptor antagonists. Eur Respir J 40(Suppl 56):717s
Imai T, Hirata Y, Emori T, Yanagisawa M, Masaki T, Marumo F (1992) Induction of endothelin-1 gene by angiotensin and vasopressin in endothelial cells. Hypertension 19(6 Pt 2):753–757
Inoue A, Yanagisawa M, Kimura S, Kasuya Y, Miyauchi T, Goto K et al (1989) The human endothelin family: three structurally and pharmacologically distinct isopeptides predicted by three separate genes. Proc Natl Acad Sci USA 86(8):2863–2867
Ito H, Hirata Y, Adachi S, Tanaka M, Tsujino M, Koike A et al (1993) Endothelin-1 is an autocrine/paracrine factor in the mechanism of angiotensin II-induced hypertrophy in cultured rat cardiomyocytes. J Clin Invest 92(1):398–403
Jankov RP, Kantores C, Belcastro R, Yi M, Tanswell AK (2006) Endothelin-1 inhibits apoptosis of pulmonary arterial smooth muscle in the neonatal rat. Pediatr Res 60(3):245–251
Jasmin JF, Lucas M, Cernacek P, Dupuis J (2001) Effectiveness of a nonselective ET(A/B) and a selective ET(A) antagonist in rats with monocrotaline-induced pulmonary hypertension. Circulation 103(2):314–318
Julkunen H, Saijonmaa O, Gronhagen-Riska C, Teppo AM, Fyhrquist F (1991) Raised plasma concentrations of endothelin-1 in systemic lupus erythematosus. Ann Rheum Dis 50(7):526–527
Kakoki M, Hirata Y, Hayakawa H, Tojo A, Nagata D, Suzuki E et al (1999) Effects of hypertension, diabetes mellitus, and hypercholesterolemia on endothelin type B receptor-mediated nitric oxide release from rat kidney. Circulation 99(9):1242–1248
Kedzierski RM, Yanagisawa M (2001) Endothelin system: the double-edged sword in health and disease. Annu Rev Pharmacol Toxicol 41:851–876
Kim H, Yung GL, Marsh JJ, Konopka RG, Pedersen CA, Chiles PG et al (2000) Endothelin mediates pulmonary vascular remodelling in a canine model of chronic embolic pulmonary hypertension. Eur Respir J 15(4):640–648
Kohan DE (1997) Endothelins in the normal and diseased kidney. Am J Kidney Dis 29(1):2–26
Kohan DE, Pritchett Y, Molitch M, Wen S, Garimella T, Audhya P et al (2011) Addition of atrasentan to renin-angiotensin system blockade reduces albuminuria in diabetic nephropathy. J Am Soc Nephrol 22(4):763–772
Kulasekaran P, Scavone CA, Rogers DS, Arenberg DA, Thannickal VJ, Horowitz JC (2009) Endothelin-1 and transforming growth factor-beta1 independently induce fibroblast resistance to apoptosis via AKT activation. Am J Respir Cell Mol Biol 41(4):484–493
Kumpers P, Nickel N, Lukasz A, Golpon H, Westerkamp V, Olsson KM et al (2010) Circulating angiopoietins in idiopathic pulmonary arterial hypertension. Eur Heart J 31(18):2291–2300
Langleben D, Dupuis J, Langleben I, Hirsch AM, Baron M, Senecal JL et al (2006) Etiology-specific endothelin-1 clearance in human precapillary pulmonary hypertension. Chest 129(3):689–695
Liu JJ, Chen JR, Buxton BF (1996) Unique response of human arteries to endothelin B receptor agonist and antagonist. Clin Sci (Lond) 90(2):91–96

Loomis ED, Sullivan JC, Osmond DA, Pollock DM, Pollock JS (2005) Endothelin mediates superoxide production and vasoconstriction through activation of NADPH oxidase and uncoupled nitric-oxide synthase in the rat aorta. J Pharmacol Exp Ther 315(3):1058–1064

MacLean MR, McCulloch KM, Baird M (1994) Endothelin ETA- and ETB-receptor-mediated vasoconstriction in rat pulmonary arteries and arterioles. J Cardiovasc Pharmacol 23(5):838–845

Mano Y, Usui T, Kamimura H (2007) Effects of bosentan, an endothelin receptor antagonist, on bile salt export pump and multidrug resistance-associated protein 2. Biopharm Drug Dispos 28(1):13–18

Markewitz BA, Farrukh IS, Chen Y, Li Y, Michael JR (2001) Regulation of endothelin-1 synthesis in human pulmonary arterial smooth muscle cells. Effects of transforming growth factor-beta and hypoxia. Cardiovasc Res 49(1):200–206

Matsumoto H, Suzuki N, Onda H, Fujino M (1989) Abundance of endothelin-3 in rat intestine, pituitary gland and brain. Biochem Biophys Res Commun 164(1):74–80

McLaughlin VV, Sitbon O, Badesch DB et al (2005) Survival with first-line bosentan in patients with primary pulmonary hypertension. Eur Respir J 25:244–249

Milara J, Ortiz JL, Juan G, Guijarro R, Almudever P, Martorell M et al (2010) Cigarette smoke exposure up-regulates endothelin receptor B in human pulmonary artery endothelial cells: molecular and functional consequences. Br J Pharmacol 161(7):1599–1615

Miyauchi T, Yorikane R, Sakai S, Sakurai T, Okada M, Nishikibe M et al (1993) Contribution of endogenous endothelin-1 to the progression of cardiopulmonary alterations in rats with monocrotaline-induced pulmonary hypertension. Circ Res 73(5):887–897

Morbidelli L, Orlando C, Maggi CA, Ledda F, Ziche M (1995) Proliferation and migration of endothelial cells is promoted by endothelins via activation of ETB receptors. Am J Physiol 269 (2 Pt 2):H686–H695

Morin FC 3rd (1989) Ligating the ductus arteriosus before birth causes persistent pulmonary hypertension in the newborn lamb. Pediatr Res 25(3):245–250

Mutsaers SE, Marshall RP, Goldsack NR, Laurent GJ, McAnulty RJ (1998) Effect of endothelin receptor antagonists (BQ-485, Ro 47-0203) on collagen deposition during the development of bleomycin-induced pulmonary fibrosis in rats. Pulm Pharmacol Ther 11(2–3):221–225

Nishida M, Eshiro K, Okada Y, Takaoka M, Matsumura Y (2004) Roles of endothelin ETA and ETB receptors in the pathogenesis of monocrotaline-induced pulmonary hypertension. J Cardiovasc Pharmacol 44(2):187–191

Nootens M, Kaufmann E, Rector T, Toher C, Judd D, Francis GS et al (1995) Neurohormonal activation in patients with right ventricular failure from pulmonary hypertension: relation to hemodynamic variables and endothelin levels. J Am Coll Cardiol 26(7):1581–1585

Olave N, Nicola T, Zhang W, Bulger A, James ML, Oparil S et al (2012) Transforming growth factor-beta regulates endothelin B signaling in the newborn mouse lung during hypoxia exposure. Am J Physiol Lung Cell Mol Physiol 302(9):L857–L865

Oparil S, Chen SJ, Meng QC, Elton TS, Yano M, Chen YF (1995) Endothelin-A receptor antagonist prevents acute hypoxia-induced pulmonary hypertension in the rat. Am J Physiol 268(1 Pt 1):L95–L100

Oudiz RJ, Galie N, Olschewski H, Torres F, Frost A, Ghofrani HA et al (2009) Long-term ambrisentan therapy for the treatment of pulmonary arterial hypertension. J Am Coll Cardiol 54(21):1971–1981

Park SH, Saleh D, Giaid A, Michel RP (1997) Increased endothelin-1 in bleomycin-induced pulmonary fibrosis and the effect of an endothelin receptor antagonist. Am J Respir Crit Care Med 156(2 Pt 1):600–608

Peacock AJ, Dawes KE, Shock A, Gray AJ, Reeves JT, Laurent GJ (1992) Endothelin-1 and endothelin-3 induce chemotaxis and replication of pulmonary artery fibroblasts. Am J Respir Cell Mol Biol 7(5):492–499

Pearl JM, Wellmann SA, McNamara JL, Lombardi JP, Wagner CJ, Raake JL et al (1999) Bosentan prevents hypoxia-reoxygenation-induced pulmonary hypertension and improves pulmonary function. Ann Thorac Surg 68(5):1714–1721, discussion 21–22

Prie S, Leung TK, Cernacek P, Ryan JW, Dupuis J (1997) The orally active ET(A) receptor antagonist (+)-(S)-2-(4,6-dimethoxy-pyrimidin-2-yloxy)-3-methoxy-3,3-diphe nyl-propionic acid (LU 135252) prevents the development of pulmonary hypertension and endothelial metabolic dysfunction in monocrotaline-treated rats. J Pharmacol Exp Ther 282(3):1312–1318

Prins BA, Hu RM, Nazario B, Pedram A, Frank HJ, Weber MA et al (1994) Prostaglandin E2 and prostacyclin inhibit the production and secretion of endothelin from cultured endothelial cells. J Biol Chem 269(16):11938–11944

Pulido T, Adzerikho I, Channick RN et al (2013) Macitentan and morbidity and mortality in pulmonary arterial hypertension. N Engl J Med 369:809–818

Raja SG (2010a) Endothelin receptor antagonists for pulmonary arterial hypertension: an overview. Cardiovasc Ther 28(5):e65–e71

Raja SG (2010b) Macitentan, a tissue-targeting endothelin receptor antagonist for the potential oral treatment of pulmonary arterial hypertension and idiopathic pulmonary fibrosis. Curr Opin Investig Drugs 11(9):1066–1073

Rakugi H, Tabuchi Y, Nakamaru M, Nagano M, Higashimori K, Mikami H et al (1990) Evidence for endothelin-1 release from resistance vessels of rats in response to hypoxia. Biochem Biophys Res Commun 169(3):973–977

Rolinski B, Geier SA, Sadri I, Klauss V, Bogner JR, Ehrenreich H et al (1994) Endothelin-1 immunoreactivity in plasma is elevated in HIV-1 infected patients with retinal microangiopathic syndrome. Clin Investig 72(4):288–293

Rolinski B, Heigermoser A, Lederer E, Bogner JR, Loch O, Goebel FD (1999) Endothelin-1 is elevated in the cerebrospinal fluid of HIV-infected patients with encephalopathy. Infection 27(4–5):244–247

Rondelet B, Kerbaul F, Motte S, van Beneden R, Remmelink M, Brimioulle S et al (2003) Bosentan for the prevention of overcirculation-induced experimental pulmonary arterial hypertension. Circulation 107(9):1329–1335

Rondelet B, Kerbaul F, Vivian GF, Hubloue I, Huez S, Fesler P et al (2007) Sitaxsentan for the prevention of experimental shunt-induced pulmonary hypertension. Pediatr Res 61(3):284–288

Rosenberg AA, Kennaugh J, Koppenhafer SL, Loomis M, Chatfield BA, Abman SH (1993) Elevated immunoreactive endothelin-1 levels in newborn infants with persistent pulmonary hypertension. J Pediatr 123(1):109–114

Rosenzweig EB, Ivy DD, Widlitz A et al (2005) Effects of long-term bosentan in children with pulmonary arterial hypertension. J Am Coll Cardiol 46:697–704

Roux S, Löffler BM, Gray GA, Sprecher U, Clozel M, Clozel JP (1995) The role of endothelin in experimental cerebral vasospasm. Neurosurgery 37(1):78–85

Rubens C, Ewert R, Halank M, Wensel R, Orzechowski HD, Schultheiss HP et al (2001) Big endothelin-1 and endothelin-1 plasma levels are correlated with the severity of primary pulmonary hypertension. Chest 120(5):1562–1569

Rubin LJ, Badesch DB, Barst RJ et al (2002) Bosentan therapy for pulmonary arterial hypertension. N Engl J Med 346:896–903

Russell FD, Davenport AP (1999) Secretory pathways in endothelin synthesis. Br J Pharmacol 126(2):391–398

Sakai S, Miyauchi T, Hara J, Goto K, Yamaguchi I (2000) Hypotensive effect of endothelin-1 via endothelin-B-receptor pathway on pulmonary circulation is enhanced in rats with pulmonary hypertension. J Cardiovasc Pharmacol 36(5 Suppl 1):S95–S98

Sato K, Oka M, Hasunuma K, Ohnishi M, Kira S (1995) Effects of separate and combined ETA and ETB blockade on ET-1-induced constriction in perfused rat lungs. Am J Physiol 269(5 Pt 1):L668–L672

Sato K, Rodman DM, McMurtry IF (1999) Hypoxia inhibits increased ETB receptor-mediated NO synthesis in hypertensive rat lungs. Am J Physiol 276(4 Pt 1):L571–L581

Sauvageau S, Thorin E, Caron A, Dupuis J (2006) Evaluation of endothelin-1-induced pulmonary vasoconstriction following myocardial infarction. Exp Biol Med (Maywood) 231(6):840–846

Sauvageau S, Thorin E, Caron A, Dupuis J (2007) Endothelin-1-induced pulmonary vasoreactivity is regulated by ET(A) and ET(B) receptor interactions. J Vasc Res 44(5):375–381

Sauvageau S, Thorin E, Villeneuve L, Dupuis J (2009) Change in pharmacological effect of endothelin receptor antagonists in rats with pulmonary hypertension: role of ETB-receptor expression levels. Pulm Pharmacol Ther 22(4):311–317

Schroll S, Arzt M, Sebah D, Stoelcker B, Luchner A, Budweiser S et al (2008) Effects of selective and unselective endothelin-receptor antagonists on prostacyclin synthase gene expression in experimental pulmonary hypertension. Scand J Clin Lab Invest 68(4):270–276

Schroll S, Arzt M, Sebah D, Nuchterlein M, Blumberg F, Pfeifer M (2010) Improvement of bleomycin-induced pulmonary hypertension and pulmonary fibrosis by the endothelin receptor antagonist Bosentan. Respir Physiol Neurobiol 170(1):32–36

Sessa WC, Kaw S, Hecker M, Vane JR (1991) The biosynthesis of endothelin-1 by human polymorphonuclear leukocytes. Biochem Biophys Res Commun 174(2):613–618

Shichiri M, Kato H, Marumo F, Hirata Y (1997) Endothelin-1 as an autocrine/paracrine apoptosis survival factor for endothelial cells. Hypertension 30(5):1198–1203

Shi-Wen X, Denton CP, Dashwood MR, Holmes AM, Bou-Gharios G, Pearson JD et al (2001) Fibroblast matrix gene expression and connective tissue remodeling: role of endothelin-1. J Invest Dermatol 116(3):417–425

Shi-Wen X, Chen Y, Denton CP, Eastwood M, Renzoni EA, Bou-Gharios G et al (2004) Endothelin-1 promotes myofibroblast induction through the ETA receptor via a rac/phosphoinositide 3-kinase/Akt-dependent pathway and is essential for the enhanced contractile phenotype of fibrotic fibroblasts. Mol Biol Cell 15(6):2707–2719

Sidharta PN, Atsmon J, Dingemanse J (2010) Investigation of the effect of ketoconazole on the pharmacokinetics of macitentan in healthy male subjects. Br J Clin Pharmacol 70:930–31

Sidharta PN, van Giersbergen PL, Halabi A, Dingemanse J (2011) Macitentan: entry-into-humans study with a new endothelin receptor antagonist. Eur J Clin Pharmacol 67(10):977–984

Sidharta PN, van Giersbergen PLM, Wolzt M, Dingemanse J (2012a) Lack of clinically relevant pharmacokinetic interactions between the dual endothelin receptor antagonist macitentan and sildenafil in healthy subjects. Am J Respir Crit Care Med 185:A4802

Sidharta PN, van Giersbergen PLM, Wolzt M, Dingemanse J (2012b) Lack of relevant pharmacokinetic and pharmacodynamic interactions between the new dual endothelin receptor antagonist macitentan and warfarin in healthy subjects. Eur Respir J 40(Suppl 56):166s

Sitbon O, Gressin V, Speich R, Macdonald PS, Opravil M, Cooper DA, Fourme T, Humbert M, Delfraissy JF, Simonneau G (2004) Bosentan for the treatment of human immunodeficiency virus-associated pulmonary arterial hypertension. Am J Respir Crit Care Med 170:1212–1217

Sitbon O, McLaughlin VV, Badesch DB et al (2005) Survival in patients with class III idiopathic pulmonary arterial hypertension treated with first line oral bosentan compared with an historical cohort of patients started on intravenous epoprostenol. Thorax 60:1025–1030

Stenmark KR, Meyrick B, Galie N, Mooi WJ, McMurtry IF (2009) Animal models of pulmonary arterial hypertension: the hope for etiological discovery and pharmacological cure. Am J Physiol Lung Cell Mol Physiol 297(6):L1013–L1032

Stewart DJ, Levy RD, Cernacek P, Langleben D (1991) Increased plasma endothelin-1 in pulmonary hypertension: marker or mediator of disease? Ann Intern Med 114(6):464–469

Tilton RG, Munsch CL, Sherwood SJ, Chen SJ, Chen YF, Wu C et al (2000) Attenuation of pulmonary vascular hypertension and cardiac hypertrophy with sitaxsentan sodium, an orally active ET(A) receptor antagonist. Pulm Pharmacol Ther 13(2):87–97

Trakada G, Nikolaou E, Pouli A, Tsiamita M, Spiropoulos K (2003) Endothelin-1 levels in interstitial lung disease patients during sleep. Sleep Breath 7(3):111–118

Treiber A, Schneiter R, Hausler S, Stieger B (2007) Bosentan is a substrate of human OATP1B1 and OATP1B3: inhibition of hepatic uptake as the common mechanism of its interactions with cyclosporin A, rifampicin, and sildenafil. Drug Metab Dispos 35(8):1400–1407

Trow TK, Taichman DB (2009) Endothelin receptor blockade in the management of pulmonary arterial hypertension: selective and dual antagonism. Respir Med 103(7):951–962

Tuder RM, Cool CD, Yeager M, Taraseviciene-Stewart L, Bull TM, Voelkel NF (2001) The pathobiology of pulmonary hypertension. Endothelium. Clin Chest Med 22(3):405–418

Underwood DC, Bochnowicz S, Osborn RR, Louden CS, Hart TK, Ohlstein EH et al (1998) Chronic hypoxia-induced cardiopulmonary changes in three rat strains: inhibition by the endothelin receptor antagonist SB 217242. J Cardiovasc Pharmacol 31(Suppl 1):S453–S455

Vancheeswaran R, Magoulas T, Efrat G, Wheeler-Jones C, Olsen I, Penny R et al (1994) Circulating endothelin-1 levels in systemic sclerosis subsets–a marker of fibrosis or vascular dysfunction? J Rheumatol 21(10):1838–1844

Wagner OF, Christ G, Wojta J, Vierhapper H, Parzer S, Nowotny PJ et al (1992) Polar secretion of endothelin-1 by cultured endothelial cells. J Biol Chem 267(23):16066–16068

Weitzberg E, Hemsen A, Rudehill A, Modin A, Wanecek M, Lundberg JM (1996) Bosentan-improved cardiopulmonary vascular performance and increased plasma levels of endothelin-1 in porcine endotoxin shock. Br J Pharmacol 118(3):617–626

Willette RN, Ohlstein EH, Mitchell MP, Sauermelch CF, Beck GR, Luttmann MA et al (1997) Nonpeptide endothelin receptor antagonists. VIII: attenuation of acute hypoxia-induced pulmonary hypertension in the dog. J Pharmacol Exp Ther 280(2):695–701

Yanagisawa MH, Kurihara H, Kimura S, Tomobe Y, Kobayashi M, Mitsui Y et al (1988) A novel potent vasoconstrictor peptide produced by vascular endothelial cells. Nature 332:411–415

Yoshibayashi M, Nishioka K, Nakao K, Saito Y, Matsumura M, Ueda T et al (1991) Plasma endothelin concentrations in patients with pulmonary hypertension associated with congenital heart defects. Evidence for increased production of endothelin in pulmonary circulation. Circulation 84(6):2280–2285

Yoshizumi M, Kurihara H, Sugiyama T, Takaku F, Yanagisawa M, Masaki T et al (1989) Hemodynamic shear stress stimulates endothelin production by cultured endothelial cells. Biochem Biophys Res Commun 161(2):859–864

Yuyama H, Fujimori A, Sanagi M, Koakutsu A, Sudoh K, Sasamata M et al (2004) The orally active nonpeptide selective endothelin ETA receptor antagonist YM598 prevents and reverses the development of pulmonary hypertension in monocrotaline-treated rats. Eur J Pharmacol 496(1–3):129–139

Zamora MR, O'Brien RF, Rutherford RB, Weil JV (1990) Serum endothelin-1 concentrations and cold provocation in primary Raynaud's phenomenon. Lancet 336(8724):1144–1147

Zhao YD, Springall DR, Hamid Q, Yacoub MH, Levene M, Polak JM (1995) Localization and characterization of endothelin-1 binding sites in the transplanted human lung. J Cardiovasc Pharmacol 26(Suppl 3):S336–S340

Phosphodiesterase-5 Inhibitors

Barbara A. Cockrill and Aaron B. Waxman

Abstract Nitric oxide (NO) signaling plays a key role in modulating vascular tone and remodeling in the pulmonary circulation. The guanylate cyclase/cyclic guanylate monophosphate-signaling pathway primarily mediates nitric oxide signaling. This pathway is critical in normal regulation of the pulmonary vasculature, and is an important target for therapy in patients with pulmonary hypertension. In the pulmonary vasculature, degradation of cGMP is primarily regulated by PDE-5, and inhibition of this enzyme has important effects on pulmonary vasculature smooth muscle tone. Large randomized placebo-controlled trials of PDE-5 inhibitors demonstrated improved exercise capacity, hemodynamics and quality of life in adult patients with PAH. This chapter will discuss the mechanisms of NO signaling in the vasculature, characteristics of the PDE5-inhibitors approved for treatment of PH, and review available data on the use of phosphodiesterase inhibitors in PH.

Keywords cGMP • Phosphodiesterase 5 • Sildenafil • Tadalafil • Guanyl cyclase

Contents

1 Introduction	230
2 Signaling Pathways	231
2.1 Nitric Oxide	231
2.2 cGMP-Dependent Protein Kinases (PKG)	231
2.3 Phosphodiesterases and Phosphodiesterase-5	233
2.4 PDE-5 Inhibitors	235
2.5 Clinical Pharmacology	236
3 Clinical Efficacy	238
3.1 Group-1 PAH	239

B.A. Cockrill • A.B. Waxman (✉)
Pulmonary Vascular Disease Program, Brigham and Women's Hospital, Harvard Medical School, 75 Francis Street, Boston, MA 02115, USA
e-mail: bcockrill@partners.org; abwaxman@partners.org

3.2 Pediatric Pulmonary Arterial Hypertension 243
3.3 Pulmonary Hypertension and Left Ventricular Dysfunction 245
3.4 Group-3 Pulmonary Hypertension 247
3.5 Adverse Effects of PDE-5 Inhibitors 249
4 Conclusion 249
References 250

List of Abbreviations

6MWD	6-min walk distance
ANP	Atrial natriuretic peptide
BNP	Brain-type natriuretic peptide
cAMP	Cyclic adenosine monophosphate
cGK	cGMP-dependent protein kinases
cGMP	Cyclic guanosine monophosphate
CNP	C-type natriuretic peptide
CrCL	Creatinine clearance
EMA	European Medicines Agency
eNOS	Endothelial nitric oxide synthase
FDA	United States Food and Drug Administration
GTP	Guanosine triphosphate
HFpEF	Heart failure preserved ejection fraction
HFrEF	Heart failure reduced ejection fraction
NO	Nitric oxide
PAH	Pulmonary arterial hypertension
PAP	Pulmonary artery pressure
PDE	Phosphodiesterase
PH	Pulmonary hypertension
PKG	Protein kinase-G
QOL	Quality of life
RHC	Right heart catheterization
RV	Right ventricle
sGC	Soluble guanylate cyclase

1 Introduction

Abnormalities in nitric oxide (NO) pathway signaling pathways are present in patients with pulmonary hypertension (PH) (Giaid and Saleh 1995; Ghofrani et al. 2004a) leading to defective vasorelaxation and abnormal vascular remodeling. The recognition that inhaled NO caused pulmonary vasodilation via increased cyclic guanosine monophosphate (cGMP) led to investigation of phosphodiesterase inhibitors, which block the breakdown of cGMP, as treatment for PH. This chapter discusses the mechanisms of NO signaling in the vasculature and compound characteristics of sildenafil, vardenafil, and tadalafil and reviews available data on the use of phosphodiesterase inhibitors in PH.

2 Signaling Pathways

2.1 Nitric Oxide

Nitric Oxide (NO) is an important player in mechanisms of cardiovascular homeostasis and is involved in vasorelaxation and platelet aggregation. NO is produced and released from endothelial cells and acts in a paracrine fashion. Defects of NO production and/or bioavailability have been implicated in varied conditions including hypertension, hypercholesterolemia, aging, diabetes, erectile dysfunction, heart failure, and pulmonary vascular disease. Endothelial cells are the major source of plasma NO that is synthesized via endothelial nitric oxide synthase (eNOS). NO synthesis and release from endothelial cells are increased in response to the mechanical shear stress of blood flow. NO crosses the plasma membrane of smooth muscle cells and acts as a signaling molecule and induces changes in target protein function. NO binds to a prosthetic heme on the β-subunit of soluble guanylate cyclase (sGC), activating the enzyme (Derbyshire and Marletta 2009). Activation of sGC drives the conversion of guanosine triphosphate (GTP) to cyclic-guanosine monophosphate (cGMP), resulting in increased concentration of cGMP (Waldman and Murad 1988). cGMP can also be synthesized by membrane-bound guanylyl cyclases that are activated by specific peptides including atrial natriuretic peptide (ANP) (Chinkers 1994), brain-type natriuretic peptide (BNP) (Schulz et al. 1989), and c-type natriuretic peptide (CNP) (Koller et al. 1991). However, NO-induced increase in cGMP is crucial in a number of physiologic processes including relaxation of vascular smooth muscle, inhibition of platelet aggregation, blunting of cardiac hypertrophy, and protection against ischemia reperfusion injury of the heart. NO-induced increase in cGMP and the downstream effects on contraction in smooth muscle appear to be mediated specifically by protein kinase-G (PKG) (Hofmann et al. 2000).

2.2 cGMP-Dependent Protein Kinases (PKG)

PKG represents the principal intracellular mediator of cGMP signals. There are two PKG families, or cGMP-dependent protein kinases (cGK). They are both serine/threonine kinases that are widely distributed in eukaryotes and are encoded by two genes, *prkg1* and *prkg2* that code for PKG1 and PKG2, respectively. Two isozymes, PKG1α and PKG1β, are generated from the *prkg1* gene and differ only in the N-terminal 100-amino acids (Hofmann et al. 2009). The PKG1 family is more commonly involved when NO mediates cGMP signaling. cGMP binds to allosteric sites in the PKG1 regulatory domain and increases phosphotransferase activity three- to tenfold. The N-terminal binding site in PKG1 isozymes has higher affinity for cGMP than does the more C-terminal site and is found in particular subcellular membrane fractions, in complex with certain cytosolic proteins, and as free cytosolic proteins (Klein et al. 2007; Schlossmann and Desch 2009).

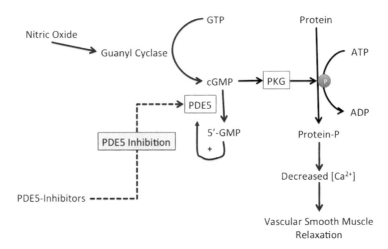

Fig. 1 Schematic diagram of the nitric oxide (NO)/cyclic guanosine monophosphate (cGMP)/ cyclic nucleotide phosphodiesterase-5 (PDE-5) pathway and the site of action of PDE-5 inhibitors

Ligand-induced elevation of cGMP triggers activation of PKG1 leading to the catalytic transfer of the γ-phosphate from adenosine triphosphate (ATP) to a serine or threonine residue on the target protein. This phosphorylated protein then mediates the translation of the extracellular stimulus (NO) into a specific biological function. There are specific physiological substrates for PKG1 in smooth muscle (Fig. 1).

Smooth muscle tone is regulated by the rise and fall of the intracellular Ca^{2+} concentration. In most cells, contraction is initiated by receptor-mediated generation of inositol 1,4,5-triphosphate (IP_3). IP_3 induces Ca^{2+} release from intracellular stores followed by an influx of extracellular Ca^{2+} through voltage-dependent and independent Ca^{2+} channels. The contractile state of smooth muscle is largely determined by the phosphorylation state and activity of the myosin light chain. The rise in intracellular calcium concentration initiates contraction by activation of the Ca^{2+}/calmodulin-dependent myosin light chain kinase, which phosphorylates myosin light chain and, subsequently, activates myosin ATPase. Smooth muscle contraction can be modulated when the intracellular Ca^{2+} concentration is held constant by changing the Ca^{2+} sensitivity of the contractile apparatus (Ogut and Brozovich 2003).

Myosin light-chain phosphatase reverses smooth muscle contraction by catalyzing the dephosphorylation of the myosin light chain. NO-induced activation of cGMP-dependent PKG1 increases the open probability of large conductance Ca^{2+}-activated K^+ channels (Fukao et al. 1999) and negatively regulates Ca^{2+} release from IP_3 sensitive stores (Schlossmann et al. 2000), resulting in a marked lowering of intracellular free calcium, desensitization of contractile proteins to Ca^{2+}, and smooth muscle relaxation. NO/cGMP/PKG1 signaling and phosphorylation of multiple intracellular proteins contributes to a reduction of intracellular Ca^{2+} concentration through decreased mobilization from intracellular stores and decreased entry from the extracellular space, or reduction in sensitivity to Ca^{2+} leading to decreased smooth muscle tone (Schlossmann et al. 2003; Vanderheyden et al. 2009).

2.3 Phosphodiesterases and Phosphodiesterase-5

The level of intracellular cGMP is regulated by both synthesis and degradation. Decreased synthesis of cGMP occurs as the NO signal declines and NO dissociates from NO-sGC, which rapidly returns sGC to the inactive state. Degradation of cGMP is regulated by cyclic nucleotide phosphodiesterases (PDE), a superfamily of metallophosphohydrolases, which are involved with catabolism of cAMP and GMP signaling (Rybalkin et al. 2003a). PDEs fall into 11 families according to their substrate specificity, mechanism of action, and subcellular location; PDE-5 is specific for cGMP and abundantly present in the lung vascular smooth muscle (Soderling and Beavo 2000). PDE-5, as a major cGMP-hydrolyzing PDE expressed in smooth muscle cells, can effectively control the NO/cGMP/PKG1 signaling pathway, especially under conditions of low calcium. In adult rat lungs, PDE-5 is highly expressed in smooth muscle cells in the medial layer of pulmonary arteries and veins (Sebkhi et al. 2003) and increased in hypoxia-induced pulmonary hypertension (Maclean et al. 1997). PDE-5 plays a key role in vascular smooth muscle tone particularly in the venous system of the corpus cavernosum and the pulmonary vascular bed. PDE-5 has been found in all types of vascular and visceral smooth muscle cells as well as cardiac myocytes.

Phosphodiesterase-5 contains two major functional domains: the catalytic domain (C domain) located in the more C-terminal portion of the protein and a regulatory domain (R domain) located in the more N-terminal portion (Corbin and Francis 1999; Francis and Corbin 1999). Via the catalytic C-domain, intracellular cGMP is rapidly degraded by PDE-5. PDE-5 limits the activity of cGMP on its substrate PKG1 by specifically cleaving the $3'$, $5'$-cyclic phosphate moiety of cGMP to produce $5'$-GMP, an inactive molecule that has no second messenger activity.

The R domain contains specific allosteric cGMP-binding sites that regulate enzyme function. Regulatory allosteric cGMP-binding sites are located in a GAF subdomain in the N-terminal portion of the respective proteins (Francis et al. 2006). The allosteric cGMP-binding sites and phosphorylation of a single serine near the N-terminus by PKG activate the PDE-5 catalytic activity and provide negative feedback regulation of cGMP (Francis et al. 2001; Mullershausen et al. 2003; Rybalkin et al. 2003b). When cGMP binds to the allosteric site a conformational change occurs that exposes the serine allowing for phosphorylation and activation of the PDE-5 catalytic activity (Francis et al. 2001). When cGMP occupies the allosteric site, the affinity of the catalytic site for cGMP is enhanced, which further enhances the PDE-5 catalytic site function (Mullershausen et al. 2003).

In addition to the upregulation caused by cGMP feedback, a second feedback mechanism upregulates PDE-5 activity in the NO/cGMP/PKG1 signaling pathway. Once activated, PKG1 phosphorylates and activates the cGMP-specific PDE-5 in smooth muscle cells causing increased activity (Corbin et al. 2000; Wyatt et al. 1998). In humans, phosphorylation of PDE-5 occurs at serine (Ser-102) in

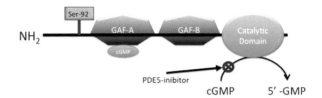

Fig. 2 Domain structure of PDE-5. Binding of cGMP to GAF-A via a high-affinity binding site results in activation of the catalytic domain. Within the catalytic domain, cGMP can interact with PDE-5 to undergo hydrolysis to 5′-GMP. GAF-A binding also enhances phosphorylation of serine-92, inducing conformational changes and increasing the affinity for cGMP

the regulatory domain and increases both the affinity of the enzyme for the allosteric sites of cGMP substrate, and the V_{max} of the enzyme (Corbin et al. 2000; Rybalkin et al. 2002; Thomas et al. 1992), thus increasing cGMP hydrolysis. The high affinity cGMP binding sites on PDE-5 are on the N-terminal regulatory GAF-A domain of the enzyme (Zoraghi et al. 2005) (Fig. 2). The principal result of PDE-5 activity is to decrease cGMP levels leading to decreased stimulation of PKG1. Binding of cGMP to a GAF domain induces a conformational change in the PDE to activate its catabolic activity for cGMP. GAF binding enhances phosphorylation at Ser-92 of PKG1 (Turko et al. 1998). PKG1 phosphorylation induces conformational changes increasing cGMP-binding affinity in the regulatory GAF domain and enhancing cGMP catalytic activity (Corbin et al. 2000). Allosteric binding of cGMP to PDE-5 could sequester the nucleotide as an intracellular storage site, providing protection from catalysis and possibly releasing the pool when cGMP levels decline as a result of other signaling events (Kotera et al. 2004). Both the binding of cGMP to allosteric sites in PDE-5 and its phosphorylation by PKG1 provide a potent negative feedback on NO/cGMP/PKG1 signaling.

PDEs can be differentially regulated in chronic conditions and may contribute to pathophysiology of specific diseases. PDE-5 upregulation in both expression and activity levels has been observed in disease conditions such as pulmonary hypertension (Corbin et al. 2005), congestive heart failure (Forfia et al. 2007), and right ventricular hypertrophy (Nagendran et al. 2007). The upregulation of PDE-5 underlies reduced responsiveness to NO as well as an enhanced responsiveness to PDE-5 inhibitors; a potential mechanism is the feedback activation of the PDE-5 promoter by increased levels of cGMP (Lin et al. 2001). The physiological importance of PDE-5 in regulation of smooth muscle tone has been most effectively demonstrated by the successful clinical use of its specific inhibitors, sildenafil, vardenafil, and tadalafil in the treatment of erectile dysfunction (Ravipati et al. 2007) and pulmonary hypertension (Galie et al. 2005). By inhibiting PDE-5 hydrolytic activity, sildenafil causes a higher rate of accumulation of cGMP in response to NO, thus enhancing NO's vasodilatory effect.

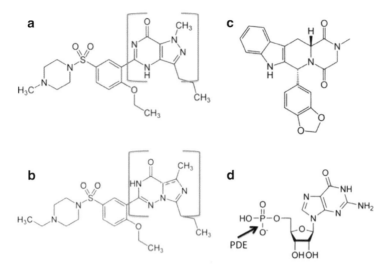

Fig. 3 Comparison of the molecular structures of (**a**) sildenafil, (**b**) vardenafil, (**c**) tadalafil, and (**d**) cyclic guanosine monophosphate (cGMP). The *arrow* indicates the site of the phosphodiester bond that undergoes PDE hydrolysis. The area between the brackets is the ringed structure that mimics the purine moiety of cGMP

2.4 PDE-5 Inhibitors

A number of studies confirm that basal activity of sGC, driven by NO, generates cGMP concentrations sufficient to induce strong vasorelaxation when PDE-5-dependent cGMP breakdown is reduced in rat aorta (Teixeira et al. 2006), intrapulmonary arteries (Andersen et al. 2005), and guinea pig basilar artery (Kruuse et al. 2001). The selective PDE-5 inhibitors sildenafil, vardenafil, and tadalafil evoked sustained relaxation of endothelium-intact aortic rings through mechanisms involving both NO-dependent and -independent pathways. The relaxation effect evoked by PDE-5 inhibitors was markedly reduced after endothelial cell removal, following inhibition of NO synthesis, inhibition of sGC, or trapping of endothelium-derived NO (Teixeira et al. 2006).

Sildenafil, tadalafil, and vardenafil are competitive and reversible inhibitors of cGMP hydrolysis by the catalytic site of PDE-5. The structures of sildenafil and vardenafil are similar in that both contain a structural mimic of the purine ring of cGMP (Fig. 3) that contributes to their capacities to act as competitive inhibitors of PDE-5. However, there are differences in the molecular structure of sildenafil and vardenafil that are understood to convey greater potency (Teixeira et al. 2006). The distinguishing components of the molecular structures are what provide for interactions with the catalytic site of PDE-5 and dramatically improves the affinity of these compounds for PDE-5 compared with cGMP, and the selectivity of these compounds for PDE-5 compared with other PDEs (Gupta et al. 2005).

2.5 Clinical Pharmacology

In the USA and Europe, there are currently three PDE-5 inhibitors approved for the treatment of PAH, sildenafil, vardenafil, and tadalafil. All three drugs were initially approved for treating erectile dysfunction. There are clear differences between the compounds as to their selectivity and specificity for PDE inhibition. Sildenafil and vardenafil are very similar in chemical structure, while tadalafil is quite different based on a methyldione structure (Fig. 3). These differences are reflected in the different pharmacokinetics.

Sildenafil citrate was the first selective PDE-5 inhibitor approved. It is a potent inhibitor of PDE-5 (IC_{50} of 3.9 nM), with a high selectivity (>1,000-fold) for human PDE-5, moderate selectivity (>80-fold) over PDE-1, and approximately tenfold selectivity over PDE-6. The overlap with PDE-6, which is found in the photoreceptors of the retina, is thought to account for color vision abnormalities observed with high plasma levels of sildenafil (Center for Drug Evaluation and Research 2000). Vardenafil has a high potency in vitro ($IC_{50} = 0.1$–0.7 nM) and a high selectivity for PDE-5 compared with the other known phosphodiesterase inhibitors. Tadalafil is a selective and potent inhibitor of PDE-5 with an IC_{50} of 0.94 nm. It exhibits high selectivity for PDE-5 compared to other PDEs (Rosen and Kostis 2003). Tadalafil is structurally different from both sildenafil and vardenafil and has distinct differences in the clinical pharmacology.

Sildenafil is relatively lipophilic with a weakly basic center in the piperazine tertiary amine ($pK_a = 6.5$), resulting in only partial ionization at physiological pH (Walker et al. 1999). Following oral administration, sildenafil is rapidly absorbed and reaches peak plasma concentrations within 1 h (range, 0.5–2 h). The mean absolute oral bioavailability of sildenafil is approximately 38–41 %. However, 92 % of the administered dose is absorbed, and the relatively low oral bioavailability of sildenafil is a result of extensive gut wall and hepatic first-pass metabolism (Walker et al. 1999). Bioavailability of sildenafil is further reduced by 11 % after a high-fat meal (Nichols et al. 2002).

Both sildenafil and its major active metabolite (*N*-desmethyl sildenafil) are highly bound to plasma proteins ($\approx$96 %). Sildenafil is extensively metabolized, with no unchanged sildenafil being detected in either urine or feces. Metabolites are predominantly excreted into the feces (73–88 %) and to a lesser extent into the urine (6–15 %) (Muirhead et al. 2002a). Plasma concentrations of sildenafil and its *N*-desmethyl metabolite were reported to decline biexponentially, with a mean terminal half-life of 3–5 h for both of them, independent of the route of administration (Nichols et al. 2002). Vardenafil is a class 2 low-solubility/high-permeability drug that is rapidly absorbed, with plasma concentrations being detected in all subjects within 8–15 min after oral administration. Peak plasma concentrations were observed 0.25–3 h after administration, with a median of 0.7 h for the 20- and 40-mg dose level, and slightly later, with 0.9 h for the 10-mg dose level (Center for Drug Evaluation and Research 2003a).

Sildenafil is primarily metabolized by the cytochrome P-450 (CYP) isoenzyme CYP3A4 and to a lesser extent CYP2C9. Since CYP3A4 is the major metabolic pathway for sildenafil, all inhibitors of CYP3A4 have the potential to interfere with sildenafil elimination (Fukao et al. 1999). The principal routes of metabolism are N-demethylation, oxidation, and aliphatic hydroxylation (Schlossmann et al. 2003). Grapefruit juice, an inhibitor of intestinal CYP3A4, alters the pharmacokinetics of sildenafil and its metabolite, N-desmethyl sildenafil. Grapefruit juice increases sildenafil bioavailability, tends to delay sildenafil absorption (by 0.25 h), and sincreases sildenafil AUC (1.23-fold) and the AUC of N-desmethyl sildenafil (1.24-fold), which, based on the average change, may be judged not clinically significant.

The absolute bioavailability of vardenafil was described as approximately 15 % (range, 8–25 %; 95 % confidence interval, 12–18 %). Based on in vitro investigations in human plasma, approximately 93–95 % of the drug is bound to plasma proteins, approximately 80 % to albumin, and 11 % to α_1-acid glycoprotein (Center for Drug Evaluation and Research 2003a).

Vardenafil is extensively metabolized and predominantly excreted as metabolites in the feces ($\approx$91–95 % of administered oral dose) and to a small extent in urine ($\approx$2–6 % of administered oral dose). Only 1 % of the administered dose is excreted into urine in unchanged form (Corbin et al. 2005). Metabolism is predominantly mediated by CYP3A4 and to a smaller extent by CYP3A5 and CYP2C isoforms. The systemic exposure of vardenafil was shown to be largely increased when coadministered with potent CY3A4 inhibitors such as erythromycin, ketoconazole, indinavir, and ritonavir. Grapefruit juice as a weak inhibitor of CYP3A gut wall metabolism may give rise to modest increases in vardenafil systemic exposure when coadministered (Center for Drug Evaluation and Research 2003a).

Severe renal impairment can lead to an increase in systemic exposure for all three drugs that merits dose reductions for sildenafil and tadalafil in affected patients. Patients with severe renal impairment (CrCL <30 mL/min) have a reduced oral clearance of sildenafil compared to healthy subjects, resulting in an approximately twofold higher AUC and C_{max} of the drug and its metabolite. There was no significant effect on sildenafil systemic exposure in patients with mild (CrCL 50–80 mL/min) or moderate (CrCL 30–49 mL/min) renal dysfunction (Muirhead et al. 2002b). In subjects with hepatic dysfunction (Child-Pugh A and B), sildenafil oral clearance was reduced by 46 %, resulting in an 84 % increase in AUC, a 47 % increase in C_{max}, and a 34 % prolongation of $t_{1/2}$. N-desmethyl sildenafil AUC and C_{max} increased by 87 % and 154 %, respectively. The larger effect of hepatic impairment on N-desmethyl sildenafil compared to sildenafil increased the AUC ratio from metabolite to parent drug from 52 to 71 %, suggesting that hepatic dysfunction affects elimination of the metabolite more than the parent drug (Muirhead et al. 2002b).

In the case of vardenafil, patients with moderate (CrCL 30–50 mL/min) and severe renal impairment (CrCL < 30 mL/min) had a slightly higher vardenafil exposure as compared to healthy subjects, with increases of 31 % and 21 %, respectively. Plasma concentrations were similar, however, between healthy subjects and patients with mild renal impairment (Center for Drug Evaluation and Research 2003a).

Tadalafil is a class 2 low-solubility/high-permeability drug that is rapidly absorbed after oral administration with a median time to reach peak plasma concentration of 2 h (range, 0.5–6 h) (Center for Drug Evaluation and Research 2003b). While the absolute bioavailability of tadalafil following oral dosing has not been reported, at least 36 % of the dose is absorbed from an oral solution. The absorption and pharmacodynamic properties of tadalafil are not affected by either food or alcohol, and thus the drug can be administered without regard for food or alcohol consumption (Abdel-Aziz et al. 2011).

Tadalafil is excreted primarily as inactive metabolites, largely in the feces ($\approx$61 % of the administered oral dose) and to a lesser extent in urine ($\approx$36 % of the administered oral dose). The mean elimination half-life for tadalafil was 17.5 h, and the mean apparent oral clearance was 2.5 L/h in healthy subjects. Tadalafil is primarily metabolized by CYP3A4 to a catechol metabolite, which further undergoes extensive methylation and glucuronidation to form methylcatechol and methylcatechol glucuronide metabolites (Center for Drug Evaluation and Research 2003b). Renal impairment increased the systemic exposure to tadalafil but had a greater effect on the levels of its renally cleared methylcatechol metabolite and its conjugated form than the parent compound. In mild (CrCL 51–80 mL/min) to moderate (31–50 mL/min) renal insufficiency, tadalafil systemic exposure was twice as high compared to that of healthy subjects, and in end-stage renal disease (ESRD) patients, it was 3.7- to 4.1-fold higher (Gupta et al. 2005).

3 Clinical Efficacy

Although treatment of pulmonary hypertension (PH) has progressed dramatically in the last two decades, definitive randomized clinical trials are particularly difficult to perform in this population. Because PH is a relatively rare condition and large numbers of patients are not available for clinical trials, most clinical studies are small. The diagnosis of PH requires a right heart catheterization (RHC) (Galie et al. 2009a). However, because this is an invasive and expensive procedure, both clinicians and investigators have attempted to use alternative means to identify patients with PH. Cardiac echocardiography, while a good screening test, commonly misclassifies patients (Taleb et al. 2013). Unfortunately, making the diagnosis of PH by means other than hemodynamic data measured invasively is fraught with error and may misidentify patients and invalidate conclusions of therapeutic investigations. The remainder of this chapter reviews existing data on the use of PDE-5 inhibitors, and when available will note the criteria used to identify the diagnosis of PH.

3.1 Group-1 PAH

In 2000, the first case report describing the use of sildenafil in a single patient with PAH was published (Prasad et al. 2000). The first randomized trial of sildenafil treatment in PAH was a single center placebo-controlled double-blind cross-over study that investigated the use of sildenafil (dose range 25–100 three times per day) vs. placebo for 6 weeks. The diagnosis of PAH was based on echocardiography. Twenty-two patients completed the trial. After 6 weeks, improvements in exercise time (475 ± 168 s prior to treatment vs. 686 ± 224 s after sildenafil, $p < 0.0001$)), cardiac index as measured by echocardiography (2.80 ± 0.9 to 3.45 ± 1.1 L/m^2, $p < 0.0001$) and quality of life (QOL) were noted (25 % improvement in dyspnea, $p = 0.009$; 8 % improvement in fatigue, $p = 0.04$) (Sastry et al. 2004).

The first multicenter double-blind placebo-controlled trial to demonstrate efficacy of a PDE-5 inhibitor to treat PH (SUPER-1) randomized 278 patients to receive placebo or sildenafil (20, 40 or 80 mg) three times daily for 12 weeks (Galie et al. 2005). Patients with PAH, PH related to connective tissue diseases, or repaired congenital systemic to pulmonary shunts as determined by RHC were eligible. The primary outcome measurement was 6-min walk test distance (6MWD). Secondary endpoints included hemodynamic measurements, symptoms, and time to clinical worsening (defined as death, transplantation, hospitalization, or the need for additional PAH therapy). Idiopathic PAH was the diagnosis in 2/3 of patients and 96 % were WHO functional class II and III. Baseline characteristics were well matched between study groups. In this study, sildenafil treatment was associated with improvement in exercise capacity at 4, 8, and 12 weeks in all study groups compared with placebo, with a trend toward greater improvement at 12 weeks in the group receiving 80 mg three times per day. At 12 weeks, the mean placebo-corrected treatment effects among 266 patients were 45 m among those receiving 20 mg of sildenafil (99 % confidence interval, 21–70; $p < 0.001$), 46 m for those receiving 40 mg (99 % confidence interval, 20–72; $p < 0.001$), and 50 m for those receiving 80 mg (99 % confidence interval, 23–77; $p < 0.001$). At 12 weeks compared with placebo, all treatment groups demonstrated an improvement in mean pulmonary artery pressure (PAP) (mean change PAP vs. placebo in mmHg with confidence intervals from baseline to week 12: 20 mg −2.1 (−4.3 to 0.0), $p = 0.04$; 40 mg −2.6 (−4.4 to −0.9), $p = 0.01$; 80 mg −4.7 (−6.7 to −2.8) $p < 0.001$), cardiac index, pulmonary vascular resistance and WHO functional class. No differences were found in clinical worsening or survival.

Patients who completed the first 12 weeks of the SUPER-1 study were eligible to enter a long-term open-label observational extension study of sildenafil treatment (SUPER-2) (Rubin et al. 2011). Primary endpoints were 6MWD and WHO functional class. The dose was titrated to 80 mg three times per day as tolerated. Two hundred fifty-nine of the 277 patients initially randomized in SUPER-1 entered the extension study; the study was extended until the last patient received 3 years of sildenafil treatment. At 3 years, 183 patients remained on sildenafil treatment and 87 % were receiving 80 mg three times per day. For patients who completed the

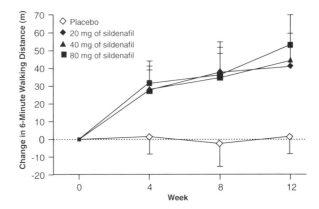

Fig. 4 Comparison of changes in 6-min walk distance with sildenafil treatment compared with placebo. The figure shows mean changes from baseline, with 95 % confidence intervals, in the 6-min walking distance at week 12 in the placebo and sildenafil groups. With the use of a two-sample t-test stratified according to baseline walking distance and cause of pulmonary arterial hypertension, $p < 0.001$ for the comparison of sildenafil in doses of 20, 40, and 80 mg with placebo. In this intention-to-treat analysis, 266 patients for whom outcome data were available were included. The dosing schedule for all study medication was three times daily (Galie et al. 2005)

study, compared to the baseline at the time of randomization in SUPER-1, 127 (69 %) increased 6MWD (81 patients with >60 m improvement, 22 patients with >30–60 m improvement, 24 patients with 0–30 m improvement) (Fig. 4), 49 (22 %) decreased 6MWD (18 patients with ≥0–30 m worsening, 12 patients with >30–60 m worsening, 19 patients with >60 m worsening), and WHO functional class was improved in 81 (44 %) (71 patients improved one class, 10 patients improved two classes, 86 patients had no change, and 15 patients worsened by one class), maintained in 86 (47 %), and declined in 15 (8 %). At 3 years from post-SUPER-1 baseline, 187 patients were alive, 53 had died, and 37 did not have data available. Of patients with known outcomes the survival rate was 79 %: a conservative estimate assuming all censored patients had died was 68 %.

Unfortunately, the SUPER-2 study design was based on dosing strategies from SUPER-1, and 87 % of patients who completed the extension study received sildenafil 80 mg three times per day. While SUPER-2 was ongoing, the United States Food and Drug Administration (FDA) and the European Medicines Agency (EMA) approved a maximum dose of sildenafil 20 mg three times per day and thus the findings of SUPER-2 may not be applicable.

Tadalafil, a longer acting PDE-5 inhibitor that can be taken once per day, became available shortly after sildenafil. A small case series investigated the transition from sildenafil to tadalafil in 12 patients with PAH (4 iPAH, 4 congenital heart disease, and 2 connective tissue disease) (Tay et al. 2008). Sildenafil had resulted in significant improvement in 6MWD, New York Heart Association score (NYHA), and dyspnea; improvements were maintained after transition to tadalafil. Compared with baseline, the mean increase in 6MWD after sildenafil was 56.45 m

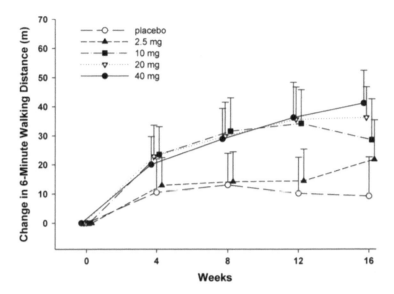

Fig. 5 Tadalafil mean changes from baseline, with 95 % CIs, in the 6-min walking distance at week 16 in the tadalafil and placebo groups. With the use of a permutation test on rank stratified according to baseline walking distance, cause of PAH, and bosentan use, $p = 0.402$, $p = 0.047$, $p = 0.028$, and $p < 0.001$ for the comparison of tadalafil in doses of 2.5, 10, 20, and 40 mg, respectively, with placebo. In this intention-to-treat analysis, 392 patients for whom outcome data were available were included in descriptive statistics. The dosing schedule for all study medications was once daily (Galie et al. 2009b)

and the improvement was unchanged after transition to tadalafil (56.73 m, $p = 0.99$.) Compared with baseline, 4/12 patients improved NYHA score after sildenafil and 5/12 were improved after tadalafil. The mean improvement in Borg Dyspnea Index after sildenafil was 2.25 and after tadalafil was 3.00 ($p = 0.486$).

Tadalafil was studied in the double-blind placebo-controlled, multicenter Pulmonary Arterial Hypertension and Response to Tadalafil (PHIRST-1) study (Galie et al. 2009b). Four hundred five patients with iPAH or associated PH were randomized to receive placebo or tadalafil at doses of 2.5, 10, 20, or 40 mg once daily. Patients who had PH confirmed by RHC and were treatment naïve or on background therapy were eligible: 53 % of subjects were already taking bosentan at the time of study entry. The primary endpoint was 6MWD. Similar to the SUPER-1 trial of sildenafil, most patients had iPAH (61 %) and were WHO functional class II (32 %) or III (65 %). At doses of 10 mg and above, tadalafil treatment was associated with a dose-dependent increase in 6MWD; however only the 40 mg dose met predefined criteria for a significant change ($p < 0.01$) (Fig. 5). The mean placebo-corrected improvements in 6MWD were 20 m for patients receiving tadalafil 10 mg ($n = 78$; 95 % CI 1–39); 27 m for patients receiving 20 mg ($n = 82$; 95 % CI 11–44); and 33 m for patients receiving 40 mg ($n = 70$, 95 % CI, 15–50). Tadalafil 40 mg/day was also associated with an improvement in time to clinical worsening ($p = 0.041$) and QOL. Two measures of QOL were used and both showed improvement in the tadalafil 40 group; at week 16 compared with baseline, statistically significant improvements were found in 6/8 domains of the Medical Outcomes Study 36-item short form version 2 health survey (all $p < 0.01$) and for all sections of the EuroQol-5D questionnaire (all $p < 0.02$). Unlike the SUPER-1 study, there was no change in WHO functional class.

The PHIRST-1 study also had an extension protocol (PHIRST-2) (Oudiz et al. 2012). Patients who completed PHIRST-1 were eligible to continue treatment with tadalafil at either 20 or 40 mg. Because the results of PHIRST-1 were not known at the beginning of the extension study, the 20 mg dose was included. Subjects were discontinued from the study if they initiated prostacyclin analogs, PDE-5 inhibitors, and/or endothelin receptor antagonists.

As discussed previously, vardenafil appears to be a more potent inhibitor of PDE-5 than either sildenafil or tadalafil (Rosen and Kostis 2003; Corbin and Francis 2002). As with sildenafil and tadalafil, early reports indicated favorable treatment effects in patients with PH (Jing et al. 2009; Ghofrani et al. 2004b). Vardenafil was recently studied in a randomized placebo-controlled trial in China (Jing et al. 2011). Sixty-six patients with Group-1 PAH on no other specific PAH treatment were randomized 2:1 to receive vardenafil or placebo. Subjects received 5 mg once daily for 4 weeks, increasing to 5 mg twice daily for a total of 12 weeks. The primary endpoint was 6MWD; secondary endpoints were cardiopulmonary hemodynamics, the Borg dyspnea index, WHO functional class and time to clinical worsening. After 12 weeks, compared with placebo, there was significant improvement in the treatment group in all of the endpoints including the occurrence of clinical worsening. With vardenafil treatment 6MWD increased by 59 m compared with a decrease of 10 m in the placebo group yielding a mean placebo-corrected treatment effect of 59 m (95 % CI 41–98 m, $p = 0.001$). Hemodynamic variables were also significantly improved with treatment: placebo-corrected mean PAP was 5.3 mmHg lower (95 % CI -10.6 to -0.1, $p = 0.047$), cardiac index was 0.4 L/min/m^2 higher (95 % CI 0.1–0.7, $p = 0.005$) and pulmonary vascular resistance (PVR) was 4.7 Wood Units lower (95 % CI -7.8 to -1.7, $p = 0.003$). More patients in the treatment group improved WHO function class compared with placebo. After 12 weeks of vardenafil treatment, 10 (22.7 %) of 44 patients improved at least one WHO functional class, 32 (72.8 %) remained stable, and two (4.5 %) deteriorated to class III and IV ($p < 0.032$). In the placebo group, only 1 (5 %) of 20 patients improved from class III to class II, 15 (75 %) remained stable, two (10 %) deteriorated from class II to class III, and two deteriorated from class III to class IV (the two patients who died were classified as class IV) ($p \leq 0.0585$). Notably, vardenafil is the only PDE-5 inhibitor to demonstrate a delay in clinical worsening as a monotherapy at approved dosing. During the 12-week study period, five patients had clinical worsening events: four patients in the placebo group (20 %) vs. one in the vardenafil group (2.3 %, $p = 0.044$). Patients who completed 12 weeks of treatment were eligible for an open-label study of vardenafil 5 mg twice daily for 12 more weeks. Improvements in 6MWD were maintained in the treatment group, and subjects initially on placebo showed significant improvement during the extension phase.

Currently, there are no long-term investigations to guide the decision regarding which PDE-5 inhibitor is the best choice for patients. The acute effects of sildenafil, tadalafil, and vardenafil on hemodynamics and oxygenation were studied in 60 patients undergoing RHC (Ghofrani et al. 2004b). Patients received short-term NO inhalation at 20–40 ppm. After the hemodynamic status had returned to baseline, patients received one oral dose of a PDE-5 inhibitor. All three drugs caused significant pulmonary vasodilation. Compared with baseline sildenafil 50 mg caused a decrease of mPAP by -16.2 % (95 % CI -11.6 to -21.5, $p < 0.001$). Tadalafil, given at doses of 20 and 40 mg caused a significant decrease of mPAP of -12.6 % (CI -3.6 to -24.4, $p = 0.01$), -18.3 % (CI -13.9 to -21.8, $p < 0.001$). The decrease in mPAP with tadalafil 60 mg did not reach statistical significance (-10.0 % (CI -22.2 to 1.1). Oral vardenafil at doses of 10 and 20 mg caused a significant decrease of mPAP of -14.3 % (CI -5.6 to -23.1, $p < 0.001$) and -12.1 % (CI -7.3 to -15.8, $p = 0.001$), respectively. A reduction in pulmonary to systemic vascular resistance ratio was seen with both sildenafil 50 mg (15.5 % decrease) and tadalafil only at 40 mg (16.0 % decrease), but not vardenafil at either 10 or 20 mg dose. Only sildenafil was associated with an improvement in arterial oxygenation that was equivalent to that of inhaled NO.

Survival benefit with PDE-5 inhibitor treatment has not been confirmed in a randomized trial. However, given the proven benefit on other clinical parameters, it is unlikely that further placebo-controlled trials will be done. Recently, a small retrospective observational study from Japan indicated that treatment with sildenafil was associated with improved survival (Zeng et al. 2012). Patients were not randomized, and the criteria for the treatment decision cannot be determined from the report. However, survival rates in PAH patients treated with sildenafil after 1, 2, and 3 years were 88 %, 72 %, and 68 % compared with 61 %, 36 %, and 27 % for patients treated with previously conventional therapy.

3.2 Pediatric Pulmonary Arterial Hypertension

Currently, no specific PAH drugs are approved for treatment of PAH in children. However, because of similar clinical characteristics and histopathology, the data for use of treatment in adults has been extrapolated to the pediatric population. Evaluating the efficacy of PAH treatment in children is fraught with difficulty. Most of the standard trial endpoints—6MWD, functional class, and QOL surveys—are not applicable or validated in children. In addition, lung vasculature continues to develop after birth and epigenetic influences may be more prominent. These issues are reinforced by the recent Pulmonary Vascular Research Institute Taskforce recommendation to develop a classification of pulmonary vascular hypertensive disease that is specific to children (Cerro et al. 2011). Because sildenafil is available in an oral liquid from, it is especially suited for treatment of children. However, as discussed below, recent data raise serious questions regarding the safety of PDE-5 use in the pediatric population.

Sildenafil has been studied in a randomized, double-blind, placebo-controlled trial (Sildenafil in Treatment-Naïve Children, Aged 1–17 Years with Pulmonary Arterial Hypertension—STARTS-1.) (Barst et al. 2012). In this multicenter international trial, 235 treatment-naïve children were randomized to receive sildenafil at low-, medium-, or high dose or placebo for 16 weeks. The primary endpoint, peak oxygen consumption as measured by cycle ergometry, was determined only for the 115 patients who could exercise reliably. Secondary endpoints, cardiopulmonary hemodynamics and functional class, were measured in all patients. After 16 weeks, the primary endpoint showed a placebo-corrected increase from baseline of 7.7 % in treated subjects, which was not statistically significant ($p = 0.056$). However, in post hoc analysis, the low-dose treatment was not different from placebo: when only medium- and high-dose groups were included, peak oxygen consumption did show a significant increase from placebo-controlled baseline (9.7 %, $p = 0.023$). Compared with placebo, high-dose sildenafil treatment was associated with improvement in hemodynamic parameters: mPAP decreased by 7.3 mmHg (CI -12.4 to -2.1, $p = 0.006$) and PVRI ratio decreased to 0.727 (CI 0.612–0.863, $p < 0.001$); medium dose was also associated with a decrease in PVRI ratio (0.819, CI 0.684–0.981, $p = 0.031$), but the decrease in mPAP (-3.9 mmHg) was not statistically significant. Low-dose sildenafil subjects did not show hemodynamic improvement. Major safety concerns surfaced regarding the use of sildenafil in children after an interim analysis of the extension trial, STARTS-2, prompting the FDA to issue a recommendation that sildenafil not be used in children (Administration. USFDA 2012). All patients entered into the STARTS-1 trial were eligible for entry into an extension study of low, medium and high dose. There were 35 deaths during the extension phase: 26 patients died while still on therapy, and 9 after completing treatment. The interim analysis of patients treated for ≥ 3 years revealed an apparent dose-related higher mortality rate in patients started on treatment with high-dose sildenafil compared with medium- or low-dose sildenafil (20 % vs. 14 % vs. 9 %, respectively). Patients who died had more severe disease at baseline: Of patients who died, 74 %, 69 %, and 71 % had baseline values above median values for PVRI (15.1 Wood units m^2), mPAP (62 mmHg), and right atrial pressure (7 mmHg), respectively; 40 % were classified as FC III or IV at baseline (versus 15 % of the overall study population). In contrast to the FDA warning, the authors concluded that the overall efficacy and safety profile favors the medium-dose regimen. Currently, the EMA endorses the use of sildenafil in pediatric pulmonary hypertension but recommends against high-dose treatment. All agree that in patients currently treated with sildenafil, the drug should be discontinued only under careful medical supervision.

The safety and efficacy of tadalafil in pediatric pulmonary hypertension was investigated in a retrospective study of 33 children (Takatsuki et al. 2012). Four patients were started on tadalafil as initial therapy and 29 were switched from sildenafil to tadalafil primarily due to the ease of once/day dosing. Fourteen of the 29 patients who switched medication underwent RHC while on sildenafil and again after tadalafil treatment: in this small group statistically significant improvements were found in mean PA (53.2 ± 18.3 mmHg vs. 47.4 ± 13.7; $p < 0.005$) and PVR index (12.2 ± 7.0 vs. 10.6 ± 7.2 Units/m^2; $p < 0.05$.) Clinical improvement was also noted in the treatment-naïve patients after starting tadalafil. The authors speculated that the improvement in hemodynamics after transition from sildenafil to the longer-acting tadalafil may have been due to improved compliance with once/day dosing and/or more consistent serum levels.

3.3 Pulmonary Hypertension and Left Ventricular Dysfunction

Patients with longstanding left ventricular dysfunction and pulmonary venous hypertension frequently develop an increase in pulmonary vascular resistance and secondary pulmonary hypertension. In fact, Group-2 patients account for the largest number of patients with pulmonary hypertension (Oudiz 2007). Mechanistically, pulmonary vasodilation without a concomitant decrease in left heart pressures would lead to increased pulmonary capillary pressures and pulmonary edema. Importantly, targeting the PH in patients with heart failure due to reduced ejection fraction (HFrEF) with epoprostenol and endothelin antagonists has been shown to be ineffective or harmful (Califf et al. 1997; Anand et al. 2004). Conversely, initial reports on the use of PDE-5 inhibitors to treat patients with Group-2 pulmonary hypertension are promising. In small studies of short duration, PDE-5 inhibitors have been shown to have a number of beneficial effects in patients with Group-2 PH including increased oxygen uptake, improved pulmonary vascular resistance, and cardiac output (Behling et al. 2008; Melenovsky et al. 2009; Lewis et al. 2007). There are a number of reasons that PDE-5 inhibitors may be effective in heart failure. PDE-5 expression is increased in hypertrophied myocardium and may be a marker of oxidative stress (Lu et al. 2010). Cyclic-GMP modulates contractility in the hypertrophied RV and inhibition of breakdown of cGMP acutely increases RV contractility (Nagendran et al. 2007).

3.3.1 PDE-5 Inhibitors in HFrEF

In short-term studies, sildenafil has been shown to increase exercise capacity and quality of life in patients with HFrEF. In a randomized trial of 34 patients with PH and systolic dysfunction, treatment with sildenafil for 12 weeks resulted in a 20 % reduction in resting and peak exercise PVR, a 38 % increase in peak exercise cardiac output, a 15 % increase in $VO_{2\ max}$ and improved QOL (Lewis et al. 2007). Another study treated patients with PH and HFrEF for 24 weeks and found a 25 % reduction in systolic PA, a 25 % increase in $VO_{2\ max}$ and improved ventilator efficiency (Guazzi et al. 2010).

In a recent 1-year study, 45 patients with HFrEF and PH were randomized to receive placebo or sildenafil 50 mg three times daily (Guazzi et al. 2011a). Primary endpoints were left ventricular diastolic function, chamber dimensions, and mass; secondary endpoints were functional capacity and QOL. At 6 months, sildenafil increased LVEF (6.8 %), improved echocardiographic indices of diastolic function, reduced PASP (13 mmHg), and improved QOL ($p < 0.01$ for all measures). Improvements were maintained at 1 year.

The favorable early studies of sildenafil on hemodynamics, cardiac function, and quality of life suggest an important role for PDE-5 inhibitors in treatment of PH-HFrEF and prevention of adverse cardiac and vascular remodeling. The available data has led to extensive off-label use of PDE-5 inhibitors in HFrEF. Recently, a large multinational randomized trial, Sildenafil in Heart Failure (SiHF) was initiated by investigators to confirm earlier findings of safety and efficacy (Cooper et al. 2013).

3.3.2 PDE-5 Inhibitors in HFpEF

In a placebo-controlled trial, Guazzi et al. (2011b) randomized 44 patients with HFPEF and PH to receive placebo or sildenafil 50 mg three times per day for 1 year. The primary endpoints were improvement in pulmonary hemodynamics and right ventricular function. The patient groups were well matched at baseline. After 6 months, in comparison with baseline and the placebo group, the treatment group showed significant improvements in hemodynamic variables and right ventricular function. The pulmonary vascular resistance decreased from 3.88 ± 1.38 Wood units to 1.18 ± 2.1 Wood units ($p < 0.01$) and was associated with a small but statistically significant decrease in pulmonary wedge pressure. Pulmonary function as measured by spirometry, lung volumes and diffusing capacity show significant improvement and likely indicates a decrease in pulmonary congestion. Secondary endpoints of breathlessness, fatigue emotional function also improved with treatment and were unchanged with placebo. Importantly, the positive changes associated with sildenafil treatment were maintained at 1 year. Although not specified as entry criteria, patients in this study had markedly elevated right atrial pressures at baseline, and thus represent a particularly ill subset of HFPEF. This study also found improvements in measurements of left ventricular function. The mechanism of the decrease in improved LV function may be related to interventricular interaction or direct effects of PDE-5 inhibition on LV mechanics.

However, in a randomized controlled trial of sildenafil (20 mg three times daily) in 113 patients with HFpEF (Phosphodiesterase-5 Inhibition to Improve Clinical Status and Exercise Capacity in Diastolic Heart Failure—the RELAX trial) there were no significant differences in the primary endpoint of change in peak oxygen consumption at 24 weeks, cardiovascular or renal hospitalization, echocardiographic cardiac magnetic resonance indices of ventricular remodeling and diastolic function, pulmonary artery pressures, or QOL (Redfield et al. 2012).

3.4 Group-3 Pulmonary Hypertension

Pulmonary hypertension is a common complication of severe pulmonary parenchymal lung disease. Currently, there are no adequate trials to determine whether specific treatment of PH in this setting is beneficial. Advanced therapy is not approved by the FDA in this setting, and according to published guidelines its use is recommended only in the context of a clinical trial. The most important consideration is to treat the underlying lung disease and provide supplemental oxygen in patients who are hypoxemic. Furthermore, the use of pulmonary vasodilators in patients with parenchymal lung disease carries the added risk of worsening oxygenation by inhibiting hypoxic pulmonary vasoconstriction and increasing blood flow to poorly ventilated lung units. However, there are data to support the concept that pulmonary hypertension contributes to symptoms and prognosis in patients with parenchymal lung disease: the presence of PH is clearly associated with a worse prognosis and the degree of PH is not clearly correlated with the severity of lung function abnormalities. Thus, the use of advanced PH therapy is an area of great interest for patients with Group-3 PH.

There are a few small studies that have investigated the use of PDE-5 inhibitors in patients with parenchymal lung disease. Importantly, rather than specifically targeting patients with PH associated with parenchymal lung disease, many studies include a mix of subjects with both normal and elevated pulmonary artery pressure. In patients with moderate-to-severe COPD, the acute effects of sildenafil were investigated in 18 patients: only 5 had mPA > 25 mmHg at rest, but all developed increased PA pressure with exertion (Holverda et al. 2008). A single dose of sildenafil attenuated the rise in PA pressure with exertion and was not associated with changes in stroke volume or cardiac output. This same group studied 15 COPD patients after 3 months of sildenafil therapy and found no change in exercise capacity or stroke volume as assessed by magnetic resonance imaging (MRI) (Rietema et al. 2008). A recent study of sildenafil treatment in COPD patients *without* PH showed worsened gas exchange and QOL with no benefit in exercise capacity (Lederer et al. 2012).

A small study, which specifically enrolled COPD patients with documented PH, randomized 20 to sildenafil at low dose (20 mg three times per day) or high dose (40 mg three times per day) (Blanco et al. 2010). Baseline hemodynamic and gas exchange data were collected, and patients were treated for 3 months. Sildenafil treatment was associated with a decrease in mean PAP both at rest and during exercise (21 ± 9 % at rest, 19 ± 10 % with exercise, $p < 0.001$ for both). A decrease in PaO_2 at rest was found with treatment thought due to the inhibition of hypoxic vasoconstriction. The authors caution that treatment of patients with COPD-PH with sildenafil should be closely monitored to assess for changes in gas exchange.

Patients with diffuse parenchymal lung disease (DPLD), such as idiopathic pulmonary fibrosis (IPF) or pulmonary fibrosis associated with collagen vascular diseases, may also develop severe PH, which contributes to morbidity and mortality

(Corte et al. 2009; Helman et al. 2007). Unfortunately, as with COPD, convincing data regarding the efficacy of advanced treatment of the PH are lacking. In a randomized controlled open-label trial of 16 patients with documented PH and pulmonary fibrosis, a single dose of sildenafil acutely caused preferential pulmonary vasodilation and, in contrast to IV epoprostenol, actually improved PaO_2. This study found an improvement in ventilation–perfusion matching with sildenafil administration (Ghofrani et al. 2002).

Small-randomized studies of PDE-5 inhibition in DPLD have had variable results. Of 14 patients with PH-ILD treated for 3 months with sildenafil, 57 % had an improvement in 6MWD (Collard et al. 2007). Another study of 29 IPF patients, which targeted patients with PH, did not document any improvement in dyspnea or 6MWD. This study, however, used only echocardiography to define PH; thus, the number of subjects who actually had PH is uncertain. A third study retrospectively identified 15 patients with PH-ILD who were treated with sildenafil (Corte et al. 2010). Patients (11/15) had PH documented by right heart catheterization (mean PA $= 41.3 \pm 11.0$ mmHg) or had markedly elevated right ventricular systolic pressure on echocardiography (RVSP $= 73.8 \pm 17$ mmHg). After 6 months, there was an improvement in 6MWD and a decreased in BNP levels.

A prospective observational study of a mixed group of patients with nongroup-1 PAH (chronic inoperable thromboembolic disease $n = 11$, COPD $n = 6$, DPLD $n = 5$, and valvular heart disease $n = 3$) followed patients who were treated with sildenafil (Chapman et al. 2009). All patients had PH diagnosed by right heart catheterization (mean PA > 25 mmHg at rest) at baseline and had follow-up catheterization after treatment. Although the numbers were small, after 12 months of treatment with sildenafil 50 mg three times per day, there was a significant improvement in hemodynamics for patients with CTEPH, DPLD, and valvular heart disease. However, patients with COPD did not show improvement in any parameter.

The largest published investigation of PDE-5 inhibitors to treat advanced IPF enrolled 180 patients in a double- blind, randomized placebo-controlled trial (Zisman et al. 2010). Importantly, enrollment criteria were based on the degree of lung function abnormality; measurement or estimation of pulmonary artery pressure was not reported. Subjects received sildenafil (20 mg three times per day) or placebo for 12 weeks. After the initial study period, all patients received sildenafil in an open-label 12-week period. An improvement in 6MWD ≥ 20 % was the primary outcome; changes in oxygenation, dyspnea, and QOL were secondary outcomes. After sildenafil treatment, there was no difference in primary outcome. There were small, but statistically significant differences in arterial oxygenation, dyspnea, carbon monoxide diffusing capacity, and QOL all favoring sildenafil treatment. Because confirmed PH was *not* required for study entry, the number of patients who actually had PH is not known, and greatly confounds the interpretation of the results.

From the currently available data, it is not known whether treatment of Group-3 PH with PDE-5 inhibitors or any advanced PH therapy is beneficial. It is clear that treatment of "all comers" with COPD or IPF in the absence of documented PH is not useful. It is the opinion of the authors that some early results are promising in patients with known PH and further study is needed. Critically important to the design of randomized, placebo-controlled trials is to study the subjects most likely to benefit: only patients who have verified PH should be enrolled.

3.5 Adverse Effects of PDE-5 Inhibitors

The two largest studies of PDE-5 inhibitors in patients with PAH confirmed that these agents are generally well tolerated in patients with PH. The SUPER-1 trial found the most common adverse effects attributable to the highest dose of sildenafil (80 mg three times daily) to be headache (49 % vs. 39 % with placebo), flushing (15 % vs. 4 %), dyspepsia (13 % vs. 7 %), myalgia (14 % vs. 3 %), and pyrexia (10 % vs. 2 %) (Galie et al. 2005). The PHIRST trial had similar results with tadalafil 40 mg once per day: headache (42 % vs. 15 % with placebo), myalgia (14 % vs. 3 %), and flushing (13 % vs. 2 %) being most common (Galie et al. 2009b).

Serious side effects, although not seen in the trials, are also recognized. Systemic hypotension may occur with PDE-5 inhibitor treatment, especially in patients taking nitrates. Therefore, patients should not take nitrates with PDE-5 inhibitors and should be clearly warned about this interaction. Post-marketing studies have reported sudden vision loss due to nonarteritic anterior ischemic optic neuropathy in patients using sildenafil for erectile dysfunction. However, it appears this correlation was due to patients with diabetes were at increased risk for this condition and were also more likely to use medication for erectile dysfunction. Recently, a large, multicenter trial and an in-depth case review found no increase in ocular risk in patients taking sildenafil (Wirostko et al. 2012; Azzouni and Abu samra 2011). Sensorineural hearing loss, however, does appear to be a true risk associated with the use of PDE-5 inhibitors; it has been reported with all three available drugs and may be related to elevated cGMP level effects on cochlear hair cells (Snodgrass et al. 2010; Khan et al. 2011; Maddox et al. 2009).

4 Conclusion

The NO/cGMP/PKG1 signaling pathway is critical in normal regulation of the pulmonary vasculature, and patients with PH have documented abnormalities in this system. In the pulmonary vasculature, degradation of cGMP is primarily regulated by PDE-5, and inhibition of this enzyme has salutary effects on pulmonary vasculature smooth muscle tone. Large randomized placebo-controlled trials of PDE-5 inhibitors demonstrated improved exercise capacity, hemodynamics, and quality of life in adult patients with PAH. Studies of these agents in patients with other forms of PH are promising but inconclusive. Currently, there is no definitive evidence that PDE-5 treatment prolongs survival.

References

Abdel-Aziz AA, Asiri YA, El-Azab AS, Al-Omar MA, Kunieda T (2011) Tadalafil. Profiles Drug Subst Excip Relat Methodol 36:287–329

Administration. USFDA (2012) Revatio (sildenafil): drug safety communication – recommendation against use in children

Anand I, McMurray J, Cohn JN, Konstam MA, Notter T, Quitzau K et al (2004) Long-term effects of darusentan on left-ventricular remodelling and clinical outcomes in the EndothelinA Receptor Antagonist Trial in Heart Failure (EARTH): randomised, double-blind, placebo-controlled trial. Lancet 364(9431):347–354

Andersen CU, Mulvany MJ, Simonsen U (2005) Lack of synergistic effect of molsidomine and sildenafil on development of pulmonary hypertension in chronic hypoxic rats. Eur J Pharmacol 510(1–2):87–96

Azzouni F, Abu samra K (2011) Are phosphodiesterase type 5 inhibitors associated with vision-threatening adverse events? A critical analysis and review of the literature. J Sex Med 8(10):2894–2903

Barst RJ, Ivy DD, Gaitan G, Szatmari A, Rudzinski A, Garcia AE et al (2012) A randomized, double-blind, placebo-controlled, dose-ranging study of oral sildenafil citrate in treatment-naive children with pulmonary arterial hypertension. Circulation 125(2):324–334

Behling A, Rohde LE, Colombo FC, Goldraich LA, Stein R, Clausell N (2008) Effects of 5′-phosphodiesterase four-week long inhibition with sildenafil in patients with chronic heart failure: a double-blind, placebo-controlled clinical trial. J Card Fail 14(3):189–197

Blanco I, Gimeno E, Munoz PA, Pizarro S, Gistau C, Rodriguez-Roisin R et al (2010) Hemodynamic and gas exchange effects of sildenafil in patients with chronic obstructive pulmonary disease and pulmonary hypertension. Am J Respir Crit Care Med 181(3):270–278

Califf RM, Adams KF, McKenna WJ, Gheorghiade M, Uretsky BF, McNulty SE et al (1997) A randomized controlled trial of epoprostenol therapy for severe congestive heart failure: the Flolan International Randomized Survival Trial (FIRST). Am Heart J 134(1):44–54

Center for Drug Evaluation and Research (2000) Guidance document: Waiver of in vivo bioavailability and bioequivalence studies for immediate release solid oral dosage forms based on biopharmaceutics classification system. In: Department of Health and Human Services UFaDA (ed) Rockville, MD

Center for Drug Evaluation and Research (2003a) NDA 021400 Levitra (Vardenafil Hydrochloride) Tablets: Clinical Pharmacology/Biopharmaceutics Review. In: Department of Health and Human Services UFaDA (ed) Rockville, MD

Center for Drug Evaluation and Research (2003b) NDA 021368 Cialis (Tadalafil) Tablets: Clinical Pharmacology/Biopharmaceutics Review. In: Department of Health and Human Services UFaDA (ed) Rockville, MD

Cerro MJ, Abman S, Diaz G, Freudenthal AH, Freudenthal F, Harikrishnan S et al (2011) A consensus approach to the classification of pediatric pulmonary hypertensive vascular disease: report from the PVRI Pediatric Taskforce, Panama 2011. Pulm Circ 1(2):286–298

Chapman TH, Wilde M, Sheth A, Madden BP (2009) Sildenafil therapy in secondary pulmonary hypertension: is there benefit in prolonged use? Vascul Pharmacol 51(2–3):90–95

Chinkers M (1994) Targeting of a distinctive protein-serine phosphatase to the protein kinase-like domain of the atrial natriuretic peptide receptor. Proc Natl Acad Sci USA 91(23):11075–11079

Collard HR, Anstrom KJ, Schwarz MI, Zisman DA (2007) Sildenafil improves walk distance in idiopathic pulmonary fibrosis. Chest 131(3):897–899

Cooper TJ, Guazzi M, Al-Mohammad A, Amir O, Bengal T, Cleland JG et al (2013) Sildenafil in Heart failure (SilHF). An investigator-initiated multinational randomized controlled clinical trial: rationale and design. Eur J Heart Fail 15(1):119–122

Corbin JD, Francis SH (1999) Cyclic GMP phosphodiesterase-5: target of sildenafil. J Biol Chem 274(20):13729–13732

Corbin JD, Francis SH (2002) Pharmacology of phosphodiesterase-5 inhibitors. Int J Clin Pract 56 (6):453–459

Corbin JD, Turko IV, Beasley A, Francis SH (2000) Phosphorylation of phosphodiesterase-5 by cyclic nucleotide-dependent protein kinase alters its catalytic and allosteric cGMP-binding activities. Eur J Biochem 267(9):2760–2767

Corbin JD, Beasley A, Blount MA, Francis SH (2005) High lung PDE5: a strong basis for treating pulmonary hypertension with PDE5 inhibitors. Biochem Biophys Res Commun 334 (3):930–938

Corte TJ, Wort SJ, Gatzoulis MA, Macdonald P, Hansell DM, Wells AU (2009) Pulmonary vascular resistance predicts early mortality in patients with diffuse fibrotic lung disease and suspected pulmonary hypertension. Thorax 64(10):883–888

Corte TJ, Gatzoulis MA, Parfitt L, Harries C, Wells AU, Wort SJ (2010) The use of sildenafil to treat pulmonary hypertension associated with interstitial lung disease. Respirology 15 (8):1226–1232

Derbyshire ER, Marletta MA (2009) Biochemistry of soluble guanylate cyclase. Handb Exp Pharmacol (191):17–31

Forfia PR, Lee M, Tunin RS, Mahmud M, Champion HC, Kass DA (2007) Acute phosphodiesterase 5 inhibition mimics hemodynamic effects of B-type natriuretic peptide and potentiates B-type natriuretic peptide effects in failing but not normal canine heart. J Am Coll Cardiol 49 (10):1079–1088

Francis SH, Corbin JD (1999) Cyclic nucleotide-dependent protein kinases: intracellular receptors for cAMP and cGMP action. Crit Rev Clin Lab Sci 36(4):275–328

Francis SH, Turko IV, Corbin JD (2001) Cyclic nucleotide phosphodiesterases: relating structure and function. Prog Nucleic Acid Res Mol Biol 65:1–52

Francis S, Zoraghi R, Kotera J, Ke H, Bessay E, Blount M et al (2006) Phosphodiesterase-5: molecular characteristics relating to structure, function, and regulation. In: Beavo JA, Francis SH, Houslay MD (eds) Cyclic nucleotide phosphodiesterases in health and disease. CRC, Boca Raton, FL

Fukao M, Mason HS, Britton FC, Kenyon JL, Horowitz B, Keef KD (1999) Cyclic GMP-dependent protein kinase activates cloned BKCa channels expressed in mammalian cells by direct phosphorylation at serine 1072. J Biol Chem 274(16):10927–10935

Galie N, Ghofrani HA, Torbicki A, Barst RJ, Rubin LJ, Badesch D et al (2005) Sildenafil citrate therapy for pulmonary arterial hypertension. N Engl J Med 353(20):2148–2157

Galie N, Hoeper MM, Humbert M, Torbicki A, Vachiery JL, Barbera JA et al (2009a) Guidelines for the diagnosis and treatment of pulmonary hypertension. Eur Respir J 34(6):1219–1263

Galie N, Brundage BH, Ghofrani HA, Oudiz RJ, Simonneau G, Safdar Z et al (2009b) Tadalafil therapy for pulmonary arterial hypertension. Circulation 119(22):2894–2903

Ghofrani HA, Wiedemann R, Rose F, Schermuly RT, Olschewski H, Weissmann N et al (2002) Sildenafil for treatment of lung fibrosis and pulmonary hypertension: a randomised controlled trial. Lancet 360(9337):895–900

Ghofrani HA, Pepke-Zaba J, Barbera JA, Channick R, Keogh AM, Gomez-Sanchez MA et al (2004a) Nitric oxide pathway and phosphodiesterase inhibitors in pulmonary arterial hypertension. J Am Coll Cardiol 43(12 Suppl S):68S–72S

Ghofrani HA, Voswinckel R, Reichenberger F, Olschewski H, Haredza P, Karadas B et al (2004b) Differences in hemodynamic and oxygenation responses to three different phosphodiesterase-5 inhibitors in patients with pulmonary arterial hypertension: a randomized prospective study. J Am Coll Cardiol 44(7):1488–1496

Giaid A, Saleh D (1995) Reduced expression of endothelial nitric oxide synthase in the lungs of patients with pulmonary hypertension. N Engl J Med 333(4):214–221

Guazzi M, Myers J, Peberdy MA, Bensimhon D, Chase P, Arena R (2010) Ventilatory efficiency and dyspnea on exertion improvements are related to reduced pulmonary pressure in heart failure patients receiving Sildenafil. Int J Cardiol 144(3):410–412

Guazzi M, Vicenzi M, Arena R, Guazzi MD (2011a) PDE5 inhibition with sildenafil improves left ventricular diastolic function, cardiac geometry, and clinical status in patients with stable systolic heart failure: results of a 1-year, prospective, randomized, placebo-controlled study. Circ Heart Fail 4(1):8–17

Guazzi M, Vicenzi M, Arena R, Guazzi MD (2011b) Pulmonary hypertension in heart failure with preserved ejection fraction: a target of phosphodiesterase-5 inhibition in a 1-year study. Circulation 124(2):164–174

Gupta M, Kovar A, Meibohm B (2005) The clinical pharmacokinetics of phosphodiesterase-5 inhibitors for erectile dysfunction. J Clin Pharmacol 45(9):987–1003

Helman DL Jr, Brown AW, Jackson JL, Shorr AF (2007) Analyzing the short-term effect of placebo therapy in pulmonary arterial hypertension: potential implications for the design of future clinical trials. Chest 132(3):764–772

Hofmann F, Ammendola A, Schlossmann J (2000) Rising behind NO: cGMP-dependent protein kinases. J Cell Sci 113(10):1671–1676

Hofmann F, Bernhard D, Lukowski R, Weinmeister P (2009) cGMP regulated protein kinases (cGK). In: Schmidt HHW, Hofmann F, Stasch J-P (eds) cGMP: generators, effectors and therapeutic implications. Springer, Heidelberg, pp 137–162

Holverda S, Rietema H, Bogaard HJ, Westerhof N, Postmus PE, Boonstra A et al (2008) Acute effects of sildenafil on exercise pulmonary hemodynamics and capacity in patients with COPD. Pulm Pharmacol Ther 21(3):558–564

Jing ZC, Jiang X, Wu BX, Xu XQ, Wu Y, Ma CR et al (2009) Vardenafil treatment for patients with pulmonary arterial hypertension: a multicentre, open-label study. Heart 95(18):1531–1536

Jing ZC, Yu ZX, Shen JY, Wu BX, Xu KF, Zhu XY et al (2011) Vardenafil in pulmonary arterial hypertension: a randomized, double-blind, placebo-controlled study. Am J Respir Crit Care Med 183(12):1723–1729

Khan AS, Sheikh Z, Khan S, Dwivedi R, Benjamin E (2011) Viagra deafness – sensorineural hearing loss and phosphodiesterase-5 inhibitors. Laryngoscope 121(5):1049–1054

Klein T, Eltze M, Grebe T, Hatzelmann A, Komhoff M (2007) Celecoxib dilates guinea-pig coronaries and rat aortic rings and amplifies NO/cGMP signaling by PDE5 inhibition. Cardiovasc Res 75(2):390–397

Koller K, Lowe D, Bennett G, Minamino N, Kangawa K, Matsuo H et al (1991) Selective activation of the B natriuretic peptide receptor by C-type natriuretic peptide (CNP). Science 252(5002):120–123

Kotera J, Francis SH, Grimes KA, Rouse A, Blount MA, Corbin JD (2004) Allosteric sites of phosphodiesterase-5 sequester cyclic GMP. Front Biosci 9:378–386

Kruuse C, Rybalkin SD, Khurana TS, Jansen-Olesen I, Olesen J, Edvinsson L (2001) The role of cGMP hydrolysing phosphodiesterases 1 and 5 in cerebral artery dilatation. Eur J Pharmacol 420(1):55–65

Lederer DJ, Bartels MN, Schluger NW, Brogan F, Jellen P, Thomashow BM et al (2012) Sildenafil for chronic obstructive pulmonary disease: a randomized crossover trial. COPD 9(3):268–275

Lewis GD, Shah R, Shahzad K, Camuso JM, Pappagianopoulos PP, Hung J et al (2007) Sildenafil improves exercise capacity and quality of life in patients with systolic heart failure and secondary pulmonary hypertension. Circulation 116(14):1555–1562

Lin CS, Chow S, Lau A, Tu R, Lue TF (2001) Regulation of human PDE5A2 intronic promoter by cAMP and cGMP: identification of a critical Sp1-binding site. Biochem Biophys Res Commun 280(3):693–699

Lu Z, Xu X, Hu X, Lee S, Traverse JH, Zhu G et al (2010) Oxidative stress regulates left ventricular PDE5 expression in the failing heart. Circulation 121(13):1474–1483

Maclean MR, Johnston ED, McCulloch KM, Pooley L, Houslay MD, Sweeney G (1997) Phosphodiesterase isoforms in the pulmonary arterial circulation of the rat: changes in pulmonary hypertension. J Pharmacol Exp Ther 283(2):619–624

Maddox PT, Saunders J, Chandrasekhar SS (2009) Sudden hearing loss from PDE-5 inhibitors: a possible cellular stress etiology. Laryngoscope 119(8):1586–1589

Melenovsky V, Al-Hiti H, Kazdova L, Jabor A, Syrovatka P, Malek I et al (2009) Transpulmonary B-type natriuretic peptide uptake and cyclic guanosine monophosphate release in heart failure and pulmonary hypertension: the effects of sildenafil. J Am Coll Cardiol 54(7):595–600

Muirhead GJ, Rance DJ, Walker DK, Wastall P (2002a) Comparative human pharmacokinetics and metabolism of single-dose oral and intravenous sildenafil. Br J Clin Pharmacol 53(Suppl 1):13S–20S

Muirhead GJ, Wilner K, Colburn W, Haug-Pihale G, Rouviex B (2002b) The effects of age and renal and hepatic impairment on the pharmacokinetics of sildenafil. Br J Clin Pharmacol 53 (Suppl 1):21S–30S

Mullershausen F, Friebe A, Feil R, Thompson WJ, Hofmann F, Koesling D (2003) Direct activation of PDE5 by cGMP: long-term effects within NO/cGMP signaling. J Cell Biol 160 (5):719–727

Nagendran J, Archer SL, Soliman D, Gurtu V, Moudgil R, Haromy A et al (2007) Phosphodiesterase type 5 is highly expressed in the hypertrophied human right ventricle, and acute inhibition of phosphodiesterase type 5 improves contractility. Circulation 116(3):238–248

Nichols DJ, Muirhead GJ, Harness JA (2002) Pharmacokinetics of sildenafil after single oral doses in healthy male subjects: absolute bioavailability, food effects and dose proportionality. Br J Clin Pharmacol 53(Suppl 1):5S–12S

Ogut O, Brozovich FV (2003) Regulation of force in vascular smooth muscle. J Mol Cell Cardiol 35(4):347–355

Oudiz RJ (2007) Pulmonary hypertension associated with left-sided heart disease. Clin Chest Med 28(1):233–241, x

Oudiz RJ, Brundage BH, Galie N, Ghofrani HA, Simonneau G, Botros FT et al (2012) Tadalafil for the treatment of pulmonary arterial hypertension: a double-blind 52-week uncontrolled extension study. J Am Coll Cardiol 60(8):768–774

Prasad S, Wilkinson J, Gatzoulis MA (2000) Sildenafil in primary pulmonary hypertension. N Engl J Med 343(18):1342

Ravipati G, McClung JA, Aronow WS, Peterson SJ, Frishman WH (2007) Type 5 phosphodiesterase inhibitors in the treatment of erectile dysfunction and cardiovascular disease. Cardiol Rev 15(2):76–86

Redfield MM, Borlaug BA, Lewis GD, Mohammed SF, Semigran MJ, Lewinter MM et al (2012) PhosphdiesteRasE-5 Inhibition to Improve CLinical Status and EXercise Capacity in Diastolic Heart Failure (RELAX) trial: rationale and design. Circ Heart Fail 5(5):653–659

Rietema H, Holverda S, Bogaard HJ, Marcus JT, Smit HJ, Westerhof N et al (2008) Sildenafil treatment in COPD does not affect stroke volume or exercise capacity. Eur Respir J 31 (4):759–764

Rosen RC, Kostis JB (2003) Overview of phosphodiesterase 5 inhibition in erectile dysfunction. Am J Cardiol 92(9):9–18

Rubin LJ, Badesch DB, Fleming TR, Galie N, Simonneau G, Ghofrani HA et al (2011) Long-term treatment with sildenafil citrate in pulmonary arterial hypertension: the SUPER-2 study. Chest 140(5):1274–1283

Rybalkin SD, Rybalkina IG, Feil R, Hofmann F, Beavo JA (2002) Regulation of cGMP-specific phosphodiesterase (PDE5) phosphorylation in smooth muscle cells. J Biol Chem 277 (5):3310–3317

Rybalkin SD, Yan C, Bornfeldt KE, Beavo JA (2003a) Cyclic GMP phosphodiesterases and regulation of smooth muscle function. Circ Res 93(4):280–291

Rybalkin SD, Rybalkina IG, Shimizu-Albergine M, Tang XB, Beavo JA (2003b) PDE5 is converted to an activated state upon cGMP binding to the GAF A domain. EMBO J 22 (3):469–478

Sastry BK, Narasimhan C, Reddy NK, Raju BS (2004) Clinical efficacy of sildenafil in primary pulmonary hypertension: a randomized, placebo-controlled, double-blind, crossover study. J Am Coll Cardiol 43(7):1149–1153

Schlossmann J, Desch M (2009) cGK substrates. In: Schmidt HHW, Hofmann F, Stasch J-P (eds) cGMP: generators, effectors and therapeutic implications. Springer, Heidelberg, pp 163–193

Schlossmann J, Ammendola A, Ashman K, Zong X, Huber A, Neubauer G et al (2000) Regulation of intracellular calcium by a signalling complex of IRAG, IP3 receptor and cGMP kinase I [beta]. Nature 404(6774):197–201

Schlossmann J, Feil R, Hofmann F (2003) Signaling through NO and cGMP-dependent protein kinases. Ann Med 35(1):21–27

Schulz S, Singh S, Bellet RA, Singh G, Tubb DJ, Chin H et al (1989) The primary structure of a plasma membrane guanylate cyclase demonstrates diversity within this new receptor family. Cell 58(6):1155–1162

Sebkhi A, Strange JW, Phillips SC, Wharton J, Wilkins MR (2003) Phosphodiesterase type 5 as a target for the treatment of hypoxia-induced pulmonary hypertension. Circulation 107 (25):3230–3235

Snodgrass AJ, Campbell HM, Mace DL, Faria VL, Swanson KM, Holodniy M (2010) Sudden sensorineural hearing loss associated with vardenafil. Pharmacotherapy 30(1):112

Soderling SH, Beavo JA (2000) Regulation of cAMP and cGMP signaling: new phosphodiesterases and new functions. Curr Opin Cell Biol 12(2):174–179

Takatsuki S, Calderbank M, Ivy DD (2012) Initial experience with tadalafil in pediatric pulmonary arterial hypertension. Pediatr Cardiol 33(5):683–688

Taleb M, Khuder S, Tinkel J, Khouri SJ (2013) The diagnostic accuracy of Doppler echocardiography in assessment of pulmonary artery systolic pressure: a meta-analysis. Echocardiography 30(3):258–265

Tay EL, Geok-Mui MK, Poh-Hoon MC, Yip J (2008) Sustained benefit of tadalafil in patients with pulmonary arterial hypertension with prior response to sildenafil: a case series of 12 patients. Int J Cardiol 125(3):416–417

Teixeira CE, Priviero FBM, Webb RC (2006) Differential effects of the phosphodiesterase type 5 inhibitors sildenafil, vardenafil, and tadalafil in rat aorta. J Pharmacol Exp Ther 316 (2):654–661

Thomas MK, Francis SH, Beebe SJ, Gettys TW, Corbin JD (1992) Partial mapping of cyclic nucleotide sites and studies of regulatory mechanisms of phosphodiesterases using cyclic nucleotide analogues. Adv Second Messenger Phosphoprotein Res 25:45–53

Turko IV, Francis SH, Corbin JD (1998) Binding of cGMP to both allosteric sites of cGMP-binding cGMP-specific phosphodiesterase (PDE5) is required for its phosphorylation. Biochem J 329(Pt 3):505–510

Vanderheyden V, Devogelaere B, Missiaen L, De Smedt H, Bultynck G, Parys JB (2009) Regulation of inositol 1,4,5-trisphosphate-induced Ca2+ release by reversible phosphorylation and dephosphorylation. Biochim Biophys Acta 1793(6):959–970

Waldman SA, Murad F (1988) Biochemical mechanisms underlying vascular smooth muscle relaxation: the guanylate cyclase-cyclic GMP system. J Cardiovasc Pharmacol 12(Suppl 5): S115–S118

Walker DK, Ackland MJ, James GC, Muirhead GJ, Rance DJ, Wastall P et al (1999) Pharmacokinetics and metabolism of sildenafil in mouse, rat, rabbit, dog and man. Xenobiotica 29 (3):297–310

Wirostko BM, Tressler C, Hwang LJ, Burgess G, Laties AM (2012) Ocular safety of sildenafil citrate when administered chronically for pulmonary arterial hypertension: results from phase III, randomised, double masked, placebo controlled trial and open label extension. BMJ 344:e554

Wyatt TA, Naftilan AJ, Francis SH, Corbin JD (1998) ANF elicits phosphorylation of the cGMP phosphodiesterase in vascular smooth muscle cells. Am J Physiol 274(2 Pt 2):H448–H455

Zeng WJ, Sun YJ, Gu Q, Xiong CM, Li JJ, He JG (2012) Impact of sildenafil on survival of patients with idiopathic pulmonary arterial hypertension. J Clin Pharmacol 52(9):1357–1364

Zisman DA, Schwarz M, Anstrom KJ, Collard HR, Flaherty KR, Idiopathic Pulmonary Fibrosis Clinical Research Network et al (2010) A controlled trial of sildenafil in advanced idiopathic pulmonary fibrosis. N Engl J Med 363(7):620–628

Zoraghi R, Bessay EP, Corbin JD, Francis SH (2005) Structural and functional features in human PDE5A1 regulatory domain that provide for allosteric cGMP binding, dimerization, and regulation. J Biol Chem 280(12):12051–12063

Inhaled Nitric Oxide for the Treatment of Pulmonary Arterial Hypertension

Steven H. Abman

Abstract Following the recognition of nitric oxide (NO) as the "endothelium-derived relaxing factor," an explosion of laboratory and clinical research led to the development of inhaled NO as a potential therapy for patients with pulmonary arterial hypertension (PAH). Despite clear demonstration of its selective and potent pulmonary vasodilator properties, inhaled NO therapy has only been formally approved by the US Food and Drug Administration and European Medicine Evaluation Agency for clinical use in the treatment of term and near-term infants with severe persistent pulmonary hypertension of the newborn (PPHN) with acute hypoxemic respiratory failure. Over the past decades, inhaled NO remains the central therapy for PPHN and is commonly used for acute pulmonary vasoreactivity testing during right heart catheterization and for treating pediatric and adult patients with PAH associated with postoperative cardiac surgery, severe respiratory failure, pulmonary hypertension crises, and other disorders. This review will describe the current use of inhaled NO in clinical practice and briefly discuss its potential role for the treatment of chronic PAH.

Keywords Persistent pulmonary hypertension of the newborn • Acute respiratory distress syndrome • Cyclic GMP • Soluble guanylate cyclase • Phosphodiesterase inhibitors • Pediatric pulmonary hypertension

---

S.H. Abman (✉)
University of Colorado School of Medicine, Aurora, CO, USA

Pulmonary Medicine, Children's Hospital Colorado, Mail Stop B395, 13123 East 16th Avenue, Aurora, CO 80045, USA
e-mail: Steven.Abman@ucdenver.edu

Contents

1	Introduction	258
2	Physiologic Basis for Inhaled NO Therapy	259
	2.1 Mechanisms of NO–cGMP Signaling	259
	2.2 Physiologic Effects of Inhaled NO	260
3	Inhaled NO and Neonatal Pulmonary Hypertension	261
	3.1 NO and the Developing Lung	261
	3.2 Inhaled NO in Experimental PPHN	262
	3.3 Inhaled NO for Clinical PPHN	264
	3.4 Inhaled NO and Preterm Infants	265
4	Inhaled NO and Acute Vasoreactivity Testing in PAH	267
5	Inhaled NO and Postoperative Cardiac Patients	268
6	Inhaled NO in Acute Respiratory Distress Syndrome	269
7	Potential Role for Inhaled NO for Chronic Treatment of PAH	269
8	Conclusions	271
References		271

1 Introduction

Nearly all pharmacologic therapies that are currently available for the treatment of pulmonary arterial hypertension (PAH) are based on studies of adult patients and are exclusively approved by the Food and Drug Administration (FDA) for use in adults. Children with PAH are generally treated with the same drug therapies as adults, representing off-label use of these medications. In striking contrast, inhaled NO has been most extensively studied and is FDA approved exclusively for the treatment of term and near-term neonates with pulmonary hypertension and acute hypoxemic respiratory failure. Interestingly, this represents only 25 % of the current use of inhaled NO in most centers, including hospitals in the USA and in Europe. That is, most of the clinical use of inhaled NO is off-label, including its use for testing acute pulmonary vasoreactivity during right heart catheterization, postoperative cardiac patients, acute respiratory failure, and other settings. This chapter provides a brief overview of the physiologic basis and roles for inhaled NO therapy in the treatment of sick newborns, including term and preterm infants. Additional studies regarding its potential application in the care of pediatric and adult patients with PAH in settings such as cardiac catheterization, postoperative care, and acute and chronic lung disease will be reviewed. Finally, the potential role for inhaled NO for the chronic treatment of PAH is also discussed.

2 Physiologic Basis for Inhaled NO Therapy

2.1 Mechanisms of NO–cGMP Signaling

The Nobel Prize-winning work of Furchgott, Murad, and Ignarro led to the initial recognition and identification of the "endothelium-derived relaxing factor" as NO and set the stage for extensive work regarding its complex biology. NO, a highly diffusible gas with a very short half-life (seconds), is synthesized from the terminal nitrogen of L-arginine by the enzyme NO synthase (NOS). These enzymes are complex oxidoreductases that function as dimers and exist as three distinct isoforms, including the endothelial (or type III) NOS (eNOS). NO production is partly dependent upon substrate availability (L-arginine), as well as the presence of essential cofactors, including heat shock protein 90 (Hsp 90), an intracellular chaperone, and tetrahydrobiopterin (BH4), a bioactive form of folic acid. Depletion of these cofactors or inhibition of Hsp90–eNOS interactions leads to "uncoupled" eNOS activity, which promotes generation of the toxic reactive oxygen species, superoxide (O_2^-), rather than NO. Multiple factors contribute to the regulation of NO production and activity, including oxygen tension, shear stress, diverse hormones and growth factors, and endogenous inhibitors, such as asymmetric dimethyl-arginine. NO also interacts with O_2^-, thereby limiting NO bioavailability and resulting in the formation of the potent oxidant peroxynitrite ($ONOO^-$).

As a gas, NO diffuses freely from the endothelium to subjacent vascular smooth muscle cells (SMC). The biologic effects of NO in vascular SMC are largely dependent upon direct activation of soluble guanylate cyclase (sGC), which converts GTP to cGMP and relaxes SMC through stimulation of cGMP-gated ion channels and activation of cGMP-dependent protein kinase type I (Francis et al. 2010; Gao 2010; Hofmann et al. 2009; Somlyo and Solyo 2003). The actions of cGMP are limited through catabolism by phosphodiesterases (PDE), especially by type 5 cGMP-dependent PDE (PDE5), which are key determinants of the magnitude and duration of the vasodilator response to endogenous and exogenous NO.

NO has multiple effects on cell signaling, inflammation, growth and differentiation, and metabolism. NO based signaling via soluble GC explains much of its actions; however, exciting new studies suggest that the effects of NO on vasodilation are far more complex than provided by this simple paradigm. NO also elicits physiologic effects via cGMP-independent mechanisms, which include direct activation of K+ channels and through interactions with other heme-containing molecules and proteins containing reactive thiol groups (reviewed in Hare and Stamler 2005). Alternative NO signaling pathways likely exist through the reaction of NO with protein thiols to form S-nitrosothiols (SNO), which may induce vasodilaton or protein modification. That is, NO regulates protein activities through S-nitrosylation, in which addition or removal of NO modulates protein function. SNO homeostasis is a balance between S-nitrosylation and de-nitrosylation, which can either up- or downregulate protein functions (Torko et al. 2012). Furthermore, the view that NO gas directly diffuses to airway and vascular cells to mediate its effects has been challenged. Extracellular NO may increase intracellular SNO accumulation in vascular endothelium and SMC through nitrosation of extracellular L-cysteine to form SNO-L-cysteine, which is then imported via type L amino acid transporters.

2.2 Physiologic Effects of Inhaled NO

Inhaled NO is a selective pulmonary vasodilator that can acutely lower pulmonary artery pressure (PAP) and pulmonary vascular resistance (PVR) without altering systemic arterial pressure (Frostell et al. 1991; Pepke-Zaba et al. 1991). The local effects of NO inhalation on the lung are primarily due to the rapid absorption and scavenging through avid binding to hemoglobin, followed by rapid oxidation to form nitrite, which interacts with oxyhemoglobin, leading to the formation of nitrate and met-hemoglobin (metHgb). In the presence of oxygenated hemoglobin (Hgb), NO is rapidly metabolized to nitrate with formation of met-Hgb, which is rapidly reduced to ferrous-Hgb by met-Hb reductase in red blood cells. In contrast with inhaled NO, systemically administered vasodilators, including prostacyclin, sildenafil, and pharmacologic NO donors, can dilate the pulmonary vasculature, but their efficacy is often limited by systemic hypotension. Additionally, in the setting of acute lung injury or with chronic lung disease, systemic vasodilator drugs can impair matching of ventilation with perfusion leading to systemic arterial hypoxemia.

In addition to its lung-selective effects, inhaled NO has the unique ability to regulate the distribution of pulmonary blood flow in the setting of parenchymal lung disease, including acute respiratory distress syndrome (ARDS) (Rossaint et al. 1993). This "micro-selective" effect of inhaled NO accounts for its ability to increase pulmonary blood flow to well-aerated and ventilated lung regions in preference to poorly inflated lung regions. NO-induced improvement in oxygenation appears largely dose dependent; although higher doses of inhaled NO (20–80 ppm) can cause progressive pulmonary vasodilation in many settings, low doses of inhaled NO (1–10 ppm) achieve better improvements in oxygenation than that observed with higher doses. In fact, higher doses of inhaled NO may paradoxically worsen oxygenation in some patients with acute respiratory failure due to the loss of this "micro-selective" effect (Gerlach et al. 1993).

Since the effects of NO are short-lived, inhaled NO is generally administered continuously or with a pulsing device that is rapidly triggered with the onset of inspiration. Inhaled NO has been administered via pulsed nasal delivery, which may provide an effective mode of delivery for ambulatory therapy in the setting of chronic cardiopulmonary disease (as discussed below). NO is a colorless and odorless gas that readily reacts with oxygen to form nitrogen dioxide (NO_2), which can cause lung injury. As a result, NO gas must be stored in nitrogen or another inert gas and administered to the patient in a manner designed to minimize the duration of exposure to oxygen. Because the pulmonary vasodilator effects of

NO are transient when the gas is discontinued, NO must be administered continuously with careful monitoring of NO and NO_2 concentrations.

Routinely available commercial equipment permits the safe delivery of NO gas in intubated and non-intubated patients. Met-hemoglobin is a source of potential toxicity with exposures to high doses of NO ($\geq$80 ppm) especially for prolonged periods. Met-hemoglobinemia typically responds to a reduction of the inhaled concentration or to discontinuation of NO therapy. Abrupt discontinuation of NO inhalation can result in "rebound" pulmonary hypertension leading to a decreased cardiac output and systemic hypotension. Rebound pulmonary hypertension after NO withdrawal may be due to suppression of endogenous NOS activity by exogenous NO (Assruey et al. 1993). The risk of rebound pulmonary hypertension can be minimized by gradually weaning the concentration of inhaled NO gas delivered to the patient. In the presence of severe left ventricular (LV) failure, administration of NO to relieve pulmonary vasoconstriction may augment LV filling and increase left atrial and pulmonary capillary wedge pressure, causing pulmonary edema. Thus, caution should be exercised when patients with LV failure or diastolic dysfunction are treated with inhaled NO. Similarly, inhaled NO in the setting of pulmonary veno-occlusive disease can induce "flash pulmonary edema," causing marked hypoxemia and acute respiratory distress, which has been observed during acute vasoreactivity testing during evaluation of patients with PAH.

3 Inhaled NO and Neonatal Pulmonary Hypertension

3.1 NO and the Developing Lung

PVR is high throughout fetal life, especially in comparison with the low resistance of the systemic circulation. As a result, the fetal lung receives less than 8 % of combined ventricular output, with most of the right ventricular output crossing the ductus arteriosus (DA) to the aorta. Mechanisms that contribute to high basal PVR in the fetus include low oxygen tension, relatively low basal production of vasodilator products, such as NO and prostacyclin (PgI_2), increased production of vasoconstrictors, including endothelin-1 (ET-1) or leukotrienes, and altered smooth muscle cell reactivity including high myogenic tone (Gao and Raj 2010a; Rudolph et al. 1977). The fetal pulmonary circulation is characterized by a progressive increase in responsiveness to vasoactive stimuli during development. In the ovine fetus, the pulmonary circulation is largely unresponsive to increased oxygen tension during the early canalicular period, but the vasodilator response to high and low fetal PO_2 increases with gestation (Morin et al. 1988; Rasanen et al. 1998). These observations parallel clinical findings that maternal hyperoxia does not increase pulmonary blood flow between 20 and 26 weeks gestation, but increased PO_2 causes pulmonary vasodilation in the 31–36-week fetus (Rasanen et al. 1998). Thus, in addition to structural maturation and growth of the developing lung circulation, the vessel wall also undergoes functional maturation, leading to enhanced vasoreactivity during fetal life.

Several mechanisms contribute to progressive changes in pulmonary vasoreactivity during development, including maturational changes in endothelial cell function, especially with regard to NO production (Abman et al. 1990; 1991; North et al. 1994; Halbower et al. 1994; Parker et al. 2000). Lung eNOS mRNA and protein are present in the early fetus and increase with advancing gestation in utero and during the early postnatal period in rats and sheep (Parker et al. 2000). Endothelial NOS expression and activity are regulated by several factors, including oxygen tension, hemodynamic forces, hormonal stimuli (e.g., estradiol), paracrine factors (including vascular endothelial growth factor, VEGF), substrate and cofactor availability, superoxide production (which inactivates NO), and others. NO causes vasodilation by stimulating sGC in vascular SMC, increasing SMC cGMP content. The ability to respond to endogenous or exogenous NO is operative very early during fetal life, suggesting critical roles for NO–cGMP throughout lung development (Kinsella et al. 1994a). Fetal lung PDE5 expression and activity are high in comparison with the normal postnatal lung, contribute significantly to the regulation of pulmonary vascular tone and reactivity in utero as well as after birth, and are further increased in PPHN (Hanson et al. 1998; Farrow et al. 2008a). The role of NO–cGMP signaling in fetal pulmonary vasoregulation has been recently reviewed (Gao and Raj 2010b).

At birth, a rapid and dramatic decrease in PVR causes an eight- to tenfold increase in pulmonary blood flow. Systemic vascular resistance increases at birth, partly due to removal of the low-resistance vascular bed of the placenta. The largest drop in PVR occurs immediately after birth, but PVR continues to progressively fall throughout infancy. As pulmonary artery pressure falls below systemic levels, blood flow through the ductus arteriosus (DA) and foramen ovale becomes left to right prior to functional closure, thereby establishing the normal postnatal circulatory pattern. Mechanisms contributing to the fall in PVR at birth include establishment of an air–liquid interface, rhythmic lung distension, increased oxygen tension, and altered production of vasoactive substances. Inhibition of NOS activity attenuates the decline in PVR after delivery of fetal lambs, suggesting that about 50 % of the rise in pulmonary blood flow at birth may be directly related to the acute release of NO (Abman et al. 1990). Failure to achieve or sustain this drop in PVR at birth leads to profound hypoxemia and constitutes the syndrome known as *persistent pulmonary hypertension of the newborn (PPHN)*.

3.2 Inhaled NO in Experimental PPHN

PPHN is associated with diverse cardiorespiratory diseases in newborns or can be idiopathic, often accompanies hypoxemic respiratory failure, and contributes to significant mortality and morbidity. Newborns with PPHN are at risk for severe asphyxia and its complications, including death, chronic lung disease, neuro-developmental sequelae, and other problems. Successful transition of the pulmonary circulation at birth requires the precise orchestration of multiple growth factors and signaling pathways to ensure normal functional and structural maturation of the lung circulation.

Mechanisms that lead to the failure of PVR to fall at birth have been pursued in diverse animal models, including exposure to acute or chronic hypoxia after birth, chronic hypoxia in utero, placement of meconium into the airways of neonatal animals, sepsis, and others. Each model demonstrates important physiologic changes that may be especially relevant to particular clinical settings. However, most studies examined only brief changes in the pulmonary circulation, while the mechanisms underlying altered lung vascular structure, growth, and function in PPHN remain poorly understood. Clinical observations that neonates with severe PPHN who die during the first days after birth already have pathologic signs of chronic pulmonary vascular disease suggest that intrauterine events may play an important role in this syndrome (Geggel and Reid 1984; Murphy et al. 1981). Adverse intrauterine stimuli during late gestation, such as abnormal hemodynamics, changes in substrate or hormone delivery to the lung, hypoxia, inflammation, or others, may potentially alter lung vascular function and structure, thereby contributing to abnormalities of postnatal adaptation.

Pulmonary hypertension induced by early closure of the DA in fetal lambs alters lung vascular reactivity and structure, causing the failure of postnatal adaptation at delivery and providing an experimental model of PPHN (Abman et al. 1989; Morin 1989). Over days, pulmonary artery pressure and PVR progressively increased in this model, but flow remained low and PaO_2 unchanged (Abman et al. 1989). Marked right ventricular hypertrophy and structural remodeling of small pulmonary arteries develop after 8 days of hypertension. After delivery, these lambs have persistent elevation of PVR despite mechanical ventilation with high oxygen concentrations. Studies with this model show that chronic hypertension without high flow can alter fetal lung vascular structure and function. This model is further characterized by endothelial cell dysfunction and altered smooth muscle cell reactivity and growth. Other hallmarks include findings of impaired NO production and activity and downregulation of lung endothelial NO synthase mRNA and protein expression (Villamor et al. 1997; Shaul et al. 1997). Fetal pulmonary hypertension is also associated with decreased cGMP concentrations, which are also partly due to decreased sGC and increased PDE5 activities, suggesting further impairments in downstream signaling (Steinhorn et al. 1995). Thus, multiple alterations in the NO–cGMP cascade appear to play an essential role in the pathogenesis and pathophysiology of experimental PPHN by contributing to altered structure and function of the developing lung circulation, and leading to failure of postnatal cardiorespiratory adaptation. High production of reactive oxygen species, such as superoxide in the pulmonary vasculature, may further contribute to the disruption in NO–cGMP signaling in this model (Farrow et al. 2008b).

3.3 Inhaled NO for Clinical PPHN

Based on strong preclinical data, early pilot studies demonstrated that inhaled NO can increase oxygenation with brief exposure (Roberts et al. 1992) and cause sustained improvements with prolonged therapy, obviating the need for ECMO therapy in severe PPHN (Kinsella et al. 1992). The primary goal of PPHN therapy is selective pulmonary vasodilation. Inhaled NO therapy (5–20 ppm) improves oxygenation and decreases the need for ECMO therapy in patients with diverse causes of PPHN (Roberts et al. 1997; No Authors 1997; Clark et al. 2000; Davidson et al. 1998). Inhaled NO is well suited for the treatment of PPHN: It is a rapid and potent vasodilator, and because nitric oxide is a small gas molecule, it can be delivered as inhalation therapy to airspaces approximating the pulmonary vascular bed. In contrast, intravenous dilators such as prostacyclin, tolazoline, and sodium nitroprusside may produce nonselective effects on the systemic circulation leading to hypotension as well as increased right-to-left shunting and impaired oxygenation. Large placebo-controlled trials that enrolled infants with an oxygenation index (OI) over 25 provided clear evidence that inhaled NO significantly decreases the need for extracorporeal life support in newborns with diverse causes of hypoxemic respiratory failure and PPHN. These studies also showed that doses of 5–20 ppm were effective and that increasing the dose beyond 20 ppm in nonresponders did not improve outcomes (No Authors 1997). Further, sustained treatment with 80 ppm NO increased the risk of methemoglobinemia (Davidson et al. 1998). Weaning can generally be accomplished in 4–5 days; any prolonged need for inhaled NO therapy without resolution of disease should lead to a more extensive evaluation to determine whether other underlying anatomic lung or cardiovascular disease is present, such as pulmonary venous stenosis, alveolar capillary dysplasia, severe lung hypoplasia, and others.

These clinical trials suggest that up to 40 % of infants will not respond or sustain a response to inhaled NO (Roberts et al. 1997; No Authors 1997; Clark et al. 2000). The reasons for an inadequate response are diverse and require the clinician to carefully analyze the relative roles of parenchymal lung disease, pulmonary vascular disease, and cardiac dysfunction for each infant. For instance, if severe airspace disease is associated with PPHN, strategies such as high-frequency ventilation or more aggressive lung recruitment strategies that optimize lung expansion are likely to be effective. The two therapies used together are more effective than either used individually (Kinsella et al. 1997a). Cardiac function should also be carefully assessed longitudinally. In particular, infants with severe left ventricular dysfunction are likely to have pulmonary venous hypertension and are unlikely to respond to pulmonary vasodilation unless there is also optimization of cardiac performance.

Recent studies suggest that infants with PPHN may respond well to the administration of sildenafil, a selective PDE5 inhibitor, which has been approved by the FDA for the treatment of adult pulmonary hypertension (Ichinose et al. 2001; Weimann et al. 2000; Shekerdemian et al. 2002, 2004). Sildenafil is a potent and highly specific PDE5 inhibitor that has been used in several preclinical studies on animal models of pulmonary hypertension. One study of a small cohort of human infants with PPHN demonstrated that enteral sildenafil improved oxygenation and survival compared to placebo in centers where inhaled NO was not available (Baquero et al. 2006). Animal studies suggest that sildenafil augments pulmonary vasodilator response to inhaled NO in experimental PPHN (Deruelle et al. 2005). A recent study of intravenous sildenafil given to infants with severe pulmonary hypertension indicates that the drug was generally well tolerated and notably improved oxygenation (Steinhorn et al. 2009). While systemic administration of sildenafil improved oxygenation, sildenafil had little effect on systemic blood pressure except with high-dose bolus infusions. Sildenafil has the potential to independently decrease PVR and improve oxygenation in human infants with PPHN; ongoing studies are exploring the relative roles for sildenafil in infants with severe PPHN who have poor or partial responses to inhaled NO.

In addition to high PDE5 activity, recent studies suggest that oxidized sGC may contribute to high PVR and the failure to respond to inhaled NO in experimental models of PPHN (Deruelle et al. 2006; Chester et al. 2009). In these studies, sGC activators and stimulators induced greater pulmonary vasodilation than inhaled NO after delivery of lambs with severe intrauterine pulmonary hypertension (Deruelle et al. 2006; Chester et al. 2009). These findings support the need for clinical studies to explore the potential role for sGC activators as a novel intervention in infants with severe PPPHN who remain hypoxic despite inhaled NO therapy.

3.4 Inhaled NO and Preterm Infants

Surfactant therapy, the use of prenatal steroids, and safer and more effective modes and strategies of mechanical ventilation and airway care have markedly improved outcomes for extremely low birth weight, preterm newborns. However, some preterm infants present with severe respiratory failure associated with severe PAH. As described above, inhaled NO therapy has been proven as safe and effective for near-term and term newborns with PPHN, but its role in the treatment of the premature newborn has been far more controversial. Many early concerns included uncertainty regarding the role of PAH in preterm infants with or without respiratory distress syndrome (RDS), as well as potential risks for lung injury, inducing or extending problems with intracranial hemorrhage (ICH) and related issues associated with extreme immaturity.

The rationale and potential role for inhaled NO treatment of PAH in preterm infants has been well established through laboratory studies. Preclinical studies suggested that inhaled NO causes potent and selective pulmonary vasodilation in very immature animals and that inhaled NO lowers PVR and improves oxygenation in premature lambs with severe RDS (Kinsella et al. 1994b). Importantly, these studies and others demonstrated improved pulmonary hemodynamics without the development of pulmonary edema or hemorrhage despite the presence of a patent ductus arteriosus. Early clinical reports further supported a potential role for inhaled

NO therapy in premature newborns, as reflected by marked improvement in oxygenation caused by effective treatment of severe pulmonary hypertension and resolution of extra-pulmonary right-to-left shunting, as well as other preterm infants with severe respiratory failure.

Multiple clinical reports subsequently demonstrated that inhaled NO can acutely improve oxygenation in preterm infants with moderate or severe respiratory failure, especially in the settings of sepsis or prolonged premature rupture of membranes with oligohydramnios (Abman et al. 1993; Peliowski et al. 1995). Initial concerns regarding high risks for adverse effects, including ICH due to potential impairment of platelet function and prolonged bleeding time, were not confirmed by subsequent trials of low doses of inhaled NO in preterm infants (Kinsella et al. 1999). In fact, early studies suggested that low-dose inhaled NO therapy may actually attenuate the severity of ICH in preterm infants with high mortality (Kinsella et al. 1999).

Thus, strong evidence suggests that inhaled NO can effectively lower PVR and improve oxygenation in preterm infants with severe hypoxemia and PPHN-type physiology. A recent NIH workshop suggested that further placebo-controlled studies would be helpful to better ascertain the role of inhaled NO in these profoundly ill infants (Cole et al. 2011a). Currently, however, this subpopulation of preterm newborns with marked respiratory failure and PAH is relatively small and due to the risk for high mortality in this group, most centers routinely employ inhaled NO therapy in this setting, limiting the likelihood of successfully performing a randomized controlled trial.

In addition to its potential role in the treatment of PAH, marked interest has grown regarding the potential role for inhaled NO therapy in preterm infants for the prevention of bronchopulmonary dysplasia (BPD). BPD is the chronic lung disease that results from the inhibition or disruption of normal pulmonary alveolar and vascular development due to premature birth and oxygen- and ventilator-induced lung injury (Kinsella et al. 2006a). Inhaled NO has been shown to lower PAP and PVR and improve gas exchange in preterm infants with established BPD (Mourani et al. 2004; Banks et al. 1999), but whether early treatment with inhaled NO could enhance long-term cardiopulmonary outcomes is uncertain. Laboratory studies have demonstrated that in addition to its effects on pulmonary vasodilation, inhaled NO also reduces lung inflammation in preterm lambs with RDS and attenuates acute lung injury in other animal models (Kinsella et al. 1997b). Inhaled NO has been shown to improve lung vascular and alveolar growth after hyperoxia in neonatal rats and improves lung structure in preterm lambs or baboons after prolonged mechanical ventilation.

Whether inhaled NO can prevent BPD in human preterm infants remains unclear. Several multicenter randomized trials have been performed to directly test this hypothesis, but the results have been variable and conclusions regarding its potential roles remain incomplete (Steinhorn and Porta 2007). Inhaled NO was shown to decrease the incidence of BPD and death in prematurely born infants with moderate RDS at a single-center study (Schreiber et al. 2003). This observation was partly supported by data from a subsequent multicenter trial in which inhaled NO decreased the rate of BPD and death in premature infants with a birth weight >1,000 g, but not for the entire study population (Kinsella et al. 2006b). Another multicenter study, in which inhaled NO was provided at 7–21 days after birth, reported that NO inhalation improved survival without BPD (Ballard et al. 2006). However, a more recent large-scale MRCT failed to demonstrate any difference in the incidence of BPD with early therapy (Mercier et al. 2010).

Although these results are encouraging when considered together, additional studies need to be performed before the precise role of inhaled NO for the prevention of BPD can be defined. The study designs for the large MRCTs involved different doses and duration of treatment with inhaled NO, and there were striking differences in patients regarding postnatal age and respiratory status at enrollment. In fact, a recent NIH workshop concluded that based on extensive meta-analyses, there are currently insufficient data to recommend the routine use of inhaled NO therapy for the prevention of BPD in preterm infants (Cole et al. 2011b). Mechanisms that contribute to the failure of inhaled NO to more consistently improve long-term outcomes of preterm infants are uncertain, but may be due to impaired sGC expression or activation by exogenous NO, as suggested in an experimental model of BPD in sheep (Bland et al. 2003).

4 Inhaled NO and Acute Vasoreactivity Testing in PAH

Inhaled NO has been used extensively to perform acute vascular reactivity testing (AVT) in the cardiac catheterization laboratory in patients with PAH. In patients with PAH, demonstration of a positive response to vasodilator agents correlates with a favorable long-term clinical outcome (Rich et al. 1992). Inhaled NO has largely replaced other agents, such as prostacyclin or adenosine, in most centers due to its ease of administration, selectivity for the pulmonary vascular bed, and rapid "on and off" nature of its effects. While a number of vasodilators including intravenous prostacyclin and calcium channel blockers have been utilized for diagnostic testing in PH patients, systemic administration of these agents can produce severe hypotension, increased intrapulmonary right-to-left shunting, and death. In contrast, a number of studies have indicated that inhaled NO can be safely and effectively used to assess the capacity for pulmonary vasodilation in pediatric (Atz et al. 1999) and adult patients (Ricciardi et al. 1998) with PH, without causing systemic hypotension. The ability of inhaled NO to decrease PAP or PVR can be used to predict the subsequent response to therapy with oral vasodilators, such as nifedipine (Ricciardi et al. 1998), and a better midterm survival in adult patients with PH due to congenital heart disease (Post et al. 2004). Risks of inhaled NO or other vasodilator use during short exposures include the rapid development of pulmonary edema in patients with pulmonary veno-occlusive disease (PVOD), which is often considered diagnostic of PVOD in unsuspected cases. AVT with or without inhaled NO must also be performed cautiously in patients with left ventricular dysfunction. A positive pulmonary vasodilator response to NO inhalation has been used as a criterion to select patients for cardiac transplantation (Fojon et al. 2005).

5 Inhaled NO and Postoperative Cardiac Patients

Pulmonary hypertension leading to acute right heart failure can complicate the management of patients during and after cardiac surgery employing cardiopulmonary bypass (CPB). A number of uncontrolled studies have demonstrated that inhalation of inhaled NO (20–40 ppm) effectively decreases PAP in cases of coronary artery bypass grafting or surgery for valvular heart disease that are complicated by perioperative PAH (Fullerton et al. 1996). Pulmonary hypertension in cardiac transplant recipients is a major cause of right heart failure and early death. Inhaled NO has been reported to selectively reduce right ventricular (RV) afterload and enhance RV stroke work after cardiac transplantation (Ardehali et al. 2001). Although there is physiologic data and clinical experience, suggesting that NO inhalation may be beneficial in patients during or after cardiac surgery complicated by PAH and RV dysfunction, whether or not inhaled NO can improve clinically important outcomes remains to be determined by randomized controlled trials.

PAH contributes to significant morbidity and mortality in cardiac surgery patients with congenital heart disease (CHD) (Hopkins et al. 1991; Hoskote et al. 2010). PAH associated with CHD may be due to increased pulmonary blood flow, high PVR, or both. In some settings (Fontan patients), even modest elevation of PVR can markedly reduce cardiac output. In many children with CHD, elevated PAP is due to high PBF (due to left-to-right shunts) and PAH returns to normal after surgical correction of the anatomic lesion. In some cases, however, patients develop high PVR due to significant vascular injury, which may not resolve after surgical correction, which can be exacerbated by endothelial injury after cardiopulmonary bypass (CPB). In addition, PAH contributes to poor outcomes following cardiac transplantation (Costard-Jackle and Fowler 1992; Giglia and Humpl 2010). In these instances, PAH may be a significant clinical problem and requires treatment. The exact incidence of PAH in children undergoing surgical interventions remains unclear and has changed due to earlier surgical repairs, but it has been estimated between 10 and 30 % (Hopkins et al. 1991). PAH also affects morbidity as well as mortality (Hoskote et al. 2010). Among infants undergoing CPB, postoperative PAH is strongly associated with increased time on mechanical ventilation and longer durations of ICU admission (Brown et al. 2003; Schulze-Neick et al. 2001).

Studies of inhaled NO have been consistently shown to lower PAP and PVR without affecting systemic arterial pressure in postoperative PAH associated with CHD (Day et al. 2000; Journois et al. 1994), Fontan procedures (Miller et al. 2000), and orthotopic heart transplantation. Inhaled NO has also been shown to be effective in preventing transient episodes of postoperative PH crisis and to stabilize patients during PH crisis (Goldman et al. 1996). A retrospective study suggested significant decreases in mortality in a patient group receiving inhaled NO compared with a control group receiving conventional care (24 % vs. 56 %) (Journois et al. 2005). In a randomized study, inhaled NO use was associated with fewer pulmonary hypertensive crises and shorter time to extubation (Miller et al. 2000). A postoperative approach that includes inhaled NO may allow for shorter stays in the pediatric intensive care unit after CPB surgery (Gothberg and Edberg 2000). These studies show that in pediatric patients with PH after surgery for CHD, inhaled NO effectively improved hemodynamics and postoperative outcomes.

Abrupt withdrawal from inhaled NO may lead to rebound PH, which may be decreased by gradually reducing inhaled NO dose to 1 ppm before withdrawal, or by the addition of a second PAH drug treatment, such as sildenafil, dipyridamole, or prostacyclin (Vonbank et al. 2003; Ivy et al. 1998a). In pediatric patients who previously failed to wean from inhaled NO after surgery for CHD, sildenafil has been shown to facilitate weaning (Lee et al. 2008).

6 Inhaled NO in Acute Respiratory Distress Syndrome

Inhaled NO is a potent vasodilator that is delivered directly to areas of the ventilated lung to improve ventilation–perfusion mismatch, resulting in improved oxygenation and lowering PAH. Early observational studies of patients with severe acute respiratory distress syndrome (ARDS) consistently demonstrated that inhaled NO can cause selective pulmonary vasodilation and improve systemic oxygenation in adults (Rossaint et al. 1993). Similar studies in children further demonstrated that inhaled NO lowers PAP and PVR, increases cardiac index, and improves oxygenation in children with ARDS (Abman et al. 1994). Despite acute improvement in pulmonary hemodynamics and gas exchange, however, subsequent randomized trials have failed to demonstrate benefits with regard to mortality, duration of mechanical ventilation, or the number of days alive and off mechanical ventilation (Dobyns et al. 1999; Dellinger et al. 1998; Michael et al. 1998; Lundin et al. 1999). Whether or not inhaled NO can improve clinical outcome in a subgroup of severely hypoxemic patients with respiratory failure and how to better identify candidates remain unknown. For example, greater benefits were observed with inhaled NO in children with ARDS who were treated with combined high-frequency oscillatory ventilation than conventional ventilation (Dobyns et al. 1999).

Past studies used various doses and duration of inhaled NO therapy, which could affect their outcomes based on the dynamic dose responsiveness over time. Therefore, the current evidence suggests that inhaled NO should not be routinely used in patients with ARDS, but may be considered as adjunctive therapy in selected patients (e.g., those with coexisting PAH) to transiently improve oxygenation in patients with severe ARDS while other therapies are considered.

7 Potential Role for Inhaled NO for Chronic Treatment of PAH

Although traditionally administered through an endotracheal tube during mechanical ventilation, inhaled NO can also be safely and effectively delivered noninvasively by face mask or nasal cannula delivery devices. Ivy et al. first described the use of noninvasive NO delivery in a young infant with severe idiopathic PAH (IPAH), who responded well to therapy until transitioned to chronic intravenous prostacyclin therapy (Ivy et al. 1994). Nasal inhaled NO provided prolonged and effective vasodilator therapy in infants with severe CDH with persistent PAH after extubation (Kinsella et al. 2003). Several reports have described the successful ambulatory use of inhaled NO. Channick et al. reported that ambulatory NO by nasal cannula provided sustained reductions in PAP and PVR in eight adult patients with IPAH (Channick et al. 1996). Importantly, no adverse signs, such as syncope, cyanosis, or respiratory distress, were noted with brief discontinuation of therapy, although the potential for "rebound PH" with sudden withdrawal remains a concern. Additionally, Ivy et al. reported successful management in 26 children and young adults with PAH through pulsed NO delivery (Ivy et al. 2003). Another study demonstrated improvements in WHO functional class, 6 min walk distance, and brain natriuretic peptide (BNP) levels after 1 month of ambulatory inhaled NO treatment (Ivy et al. 1998b). Patients demonstrated improved pulmonary hemodynamics and cardiac index after 1 year, although additional PAH-specific therapies were often required for ongoing care. The combination of inhaled NO with PDE5 inhibition may provide an additional strategy for long-term therapy of PAH (Perez-Penate et al. 2008).

The dose of inhaled NO actually delivered to the distal lung is uncertain, but NO from nasopharyngeal aspirates was measured at roughly half of the administered dose (Kinsella et al. 2003; Ivy et al. 1998b, 2003). A potential advantage of inhaled NO, in contrast with intravenous vasodilators, is its pulmonary selectivity and the ability to avoid the need for a central line and such associated risks as line infections or thrombosis. Potential advantages of using a pulsed delivery device include greater efficiency of delivery during early inspiration, thereby allowing for more consistent alveolar delivery of gas, efficacy at lower concentrations of gas, and with less wasted delivery to the environment. Other potential candidates for prolonged ambulatory therapy could potentially include patients with COPD or other chronic lung diseases, PAH associated with sickle cell disease, as a "bridge to transplantation," or for temporary use with disease exacerbations. Thus, despite availability of numerous other PAH-specific therapies, inhaled NO may provide an additional strategy for chronic management of severe PAH. Much still needs to be achieved to improve the applicability of ambulatory NO therapy, including the need for high concentration tanks for greater portability, studies that determine whether continuous 24/7 therapy is necessary or if shorter durations of treatment are sufficient, and other questions. Further studies are needed to determine the effect of long-term ambulatory breathing of pulsed NO and oxygen on the quality of life and mortality rate of patients with severe COPD.

8 Conclusions

Over the past 25 years, extensive basic and clinical research has increased our understanding of the pathobiology of NO–cGMP signaling in pulmonary hypertension and has led to the safe and rapid translation of inhaled NO therapy for patients with severe PAH. Based on the strength of extensive multicenter randomized controlled trials (MRCTs) in term and near-term newborns with PPHN, inhaled NO continues to provide an effective and safe therapy in this critical care setting. Although clinical observations have supported the use of inhaled NO in preterm infants with PPHN-type physiology and profound hypoxemia, MRCTs are lacking in this population to provide more clear evidence for use in this population. Although low doses of inhaled NO are safe in preterm infants, whether inhaled NO therapy prevents the development of BPD or brain injury in preterm newborns remains highly controversial, and current recommendations do not support its routine use for the prevention of chronic lung disease. Inhaled NO is commonly used for acute vasoreactivity testing of PAH in the cardiac catheterization lab, in stabilization of postoperative cardiac patients, and in selective cases of severe ARDS with PAH, but further work is needed in these settings to better prove efficacy. Finally, chronic ambulatory therapy for the long-term treatment of PAH has been suggested from case studies, and with continued improvements in technology, inhaled NO may provide an alternate strategy for chronic PAH as well as in the acute setting.

References

Abman SH, Shanley PF, Accurso FJ (1989) Failure of postnatal adaptation of the pulmonary circulation after chronic intrauterine pulmonary hypertension in fetal lambs. J Clin Invest 83:1849–1858

Abman SH, Chatfield BA, Hall SL et al (1990) Role of endothelium-derived relaxing factor during transition of pulmonary circulation at birth. Am J Physiol 259:H1921–H1927

Abman SH, Chatfield BA, Rodman DM, Hall SL, McMurtry IF (1991) Maturation-related changes in endothelium-dependent relaxation of ovine pulmonary arteries. Am J Physiol 260: L280–L285

Abman SH, Kinsella JP, Schaffer MS, Wilkening RB (1993) Inhaled nitric oxide therapy in the management of a premature newborn with severe respiratory distress and pulmonary hypertension. Pediatrics 92:606–609

Abman SH, Griebel J, Schmidt J, Parker D, Swanton D, Kinsella JP (1994) Acute effects of inhaled nitric oxide in severe hypoxemic respiratory failure in pediatrics. J Pediatr 174:681–688

Ardehali A, Hughes K, Sadeghi A, Esmailian F, Marelli D, Moriguchi J et al (2001) Inhaled nitric oxide for pulmonary hypertension after heart transplantation. Transplantation 72:638–641

Assruey J, Cunha FQ, Liew FY, Moncada S (1993) Feedback inhibition of nitric oxide synthase activity by NO. Br J Pharmacol 108:833–837

Atz AM, Adatia I, Lock JE, Wessel DL (1999) Combined effects of nitric oxide and oxygen during acute pulmonary vasodilator testing. J Am Coll Cardiol 33:813–819

Ballard RA, Truog WE, Cnaan A, Martin RJ, Ballard PL, Merrill JD et al (2006) Inhaled nitric oxide in preterm infants undergoing mechanical ventilation. N Engl J Med 355:343–353

Banks BA, Seri I, Ischiropoulos H, Merrill J, Rychik J, Ballard RA (1999) Changes in oxygenation with inhaled NO in severe BPD. Pediatrics 103:870–874

Baquero H, Soliz A, Neira F et al (2006) Oral sildenafil in infants with persistent pulmonary hypertension of the newborn: a pilot randomized blinded study. Pediatrics 117:1077–1083

Bland RD, Ling CY, Albertine KH, Carlton DP, MacRitchie AJ, Day RW, Dahl MJ (2003) Pulmonary vascular dysfunction in preterm lambs with chronic lung disease. Am J Physiol Lung Cell Mol Physiol 285:L76–L85

Brown KL et al (2003) Risk factors for long intensive care unit stay after cardiopulmonary bypass in children. Crit Care Med 31:28–33

Channick RN, Newhart JW, Johnson FW, Williams PJ, Auger WR, Fedullo PF et al (1996) Pulsed delivery of inhaled nitric oxide to patients with primary pulmonary hypertension: an ambulatory delivery system and initial clinical tests. Chest 109:1545–1549

Chester M, Tourneux P, Seedorf G, Grover TR, Abman SH (2009) Cinaciguat, a soluble guanylate cyclase activator, causes potent and sustained pulmonary vasodilation in the ovine fetus. Am J Physiol 297:L318–L325

Clark RH, Kueser TJ, Walker MW, Southgate WM, Huckaby JL, Perez JA et al (2000) Low-dose nitric oxide therapy for persistent pulmonary hypertension of the newborn. Clinical Inhaled Nitric Oxide Research Group. N Engl J Med 342:469–474

Cole FS, Alleyne C, Barks JD et al (2011) NIH consensus development conference statement: inhaled NO therapy for premature infants. Pediatrics 127:363–369

Costard-Jackle A, Fowler MB (1992) Influence of preoperative pulmonary artery pressure on mortality after heart transplantation: testing of potential reversibility of pulmonary hypertension with nitroprusside is useful in defining a high risk group. J Am Coll Cardiol 19:48–54

Davidson D, Barefield ES, Kattwinkel J, Dudell G, Damask M, Straube R et al (1998) Inhaled nitric oxide for the early treatment of persistent pulmonary hypertension of the term newborn: a randomized, double masked, placebo-controlled, dose-response, multicenter study. Pediatrics 101:325–334

Day RW et al (2000) Randomized controlled study of inhaled nitric oxide after operation for congenital heart disease. Ann Thorac Surg 69:1907–1912

Dellinger RP, Zimmerman JL, Taylor RW, Straube RC (1998) Placebo and inhaled nitric oxide mortality the same in ARDS clinical trial. Crit Care Med 26:619

Deruelle P, Grover TR, Abman SH (2005) Pulmonary vascular effects of nitric oxide-cGMP augmentation in a model of chronic pulmonary hypertension in fetal and neonatal sheep. Am J Physiol Lung Cell Mol Physiol 289:L788–L806

Deruelle P, Balasubramanuam V, Kunig AM, Seedorf G, Markham NE, Abman SH (2006) Bay 41-2272, a direct activator of soluble guanylate cyclase, reduces right ventricular hypertrophy and improves pulmonary vascular structure during chronic hypoxia in neonatal rats. Biol Neonate 90:135–144

Dobyns EL, Cornfield DN, Anas NG, Fortenberry JD, Tasker RC, Lynch A, Liu P, Eells PL, Griebel J, Baier M, Kinsella JP, Abman SH (1999) Multicenter randomized trial of the effects of inhaled NO therapy on gas exchange in children with acute hypoxemic respiratory failure. J Pediatr 134:406–412

Farrow KN, Groh BS, Schumacker PT et al (2008a) Hyperoxia increases phosphodiesterase 5 expression and activity in ovine fetal pulmonary artery smooth muscle cells. Circ Res 102:226–233

Farrow KN, Lakshminrusimha S, Reda WJ et al (2008b) Superoxide dismutase restores eNOS expression and function in resistance pulmonary arteries from neonatal lambs with persistent pulmonary hypertension. Am J Physiol Lung Cell Mol Physiol 295:L979–L987

Fojon S, Fernandez-Gonzalez C, Sanchez-Andrade J, Lopez-Perez JM, Hermida LF, Rodriguez JA et al (2005) Inhaled nitric oxide through a noninvasive ventilation device to assess reversibility of pulmonary hypertension in selecting recipients for heart transplant. Transplant Proc 37:4028–4030

Francis SH, Bsuch JL, Corbin JD, Sibley D (2010) cGMP dependent protein kinases and cGMP phosphodiesterases in NO and cGMP action. Pharmacol Rev 62:525–563

Frostell C, Fratacci MD, Wain JC, Jones R, Zapol WM (1991) Inhaled nitric oxide. A selective pulmonary vasodilator reversing hypoxic pulmonary vasoconstriction. Circulation 83:2038–2047

Fullerton DA, Jones SD, Jaggers J, Piedalue F, Grover FL, McIntyre RC Jr (1996) Effective control of pulmonary vascular resistance with inhaled nitric oxide after cardiac operation. J Thorac Cardiovasc Surg 111:753–762

Gao Y (2010) Multiple actions of NO. Pflugers Archiv Eur J Phyisol 459:829–839

Gao Y, Raj JU (2010) Regulation of the pulmonary circulation in the fetus and newborn. Physiol Rev 90:1291–1335

Geggel R, Reid LM (1984) The structural basis for PPHN. Clin Perinatol 11:525–549

Gerlach H, Roissant R, Pappert D, Falke KJ (1993) Time-course and dose-response of NO inhalation for systemic oxygenation and pulmonary hypertension in patients with ARDS. Eur J Clin Invest 23:499–502

Giglia TM, Humpl T (2010) Preoperative pulmonary hemodynamics and assessment of operability: is there a pulmonary vascular resistance that precludes cardiac operation? Pediatr Crit Care Med 11:S57–S69

Goldman AP et al (1996) Pharmacological control of pulmonary blood flow with inhaled nitric oxide after the fenestrated Fontan operation. Circulation 94:II44–II48

Gothberg S, Edberg KE (2000) Inhaled nitric oxide to newborns and infants after congenital heart surgery on cardiopulmonary bypass: a dose-response study. Scand Cardiovasc J 34:154–158

Halbower AC, Tuder RM, Franklin WA, Pollock JS, Forstermann U, Abman SH (1994) Maturation-related changes in endothelial NO synthase immunolocalization in the developing ovine lung. Am J Physiol 267:L585–L591

Hanson KA, Beavo JA, Abman SH, Clarke WR (1998) Chronic pulmonary hypertension increases fetal lung cGMP activity. Am J Physiol 275:L931–L941

Hare JM, Stamler JS (2005) NO/redox disequilibrium in the failing heart and cardiovascular system. J Clin Invest 115:509–517

Hofmann F, Bernhard D, Lukowski R, Weinmeister P (2009) cGMP regulated protein kinases (cGK). Handb Exp Pharmacol 191:137–162

Hopkins RA et al (1991) Pulmonary hypertensive crises following surgery for congenital heart defects in young children. Eur J Cardiothorac Surg 5:628–634

Hoskote A et al (2010) Acute right ventricular failure after pediatric cardiac transplant: predictors and long-term outcome in current era of transplantation medicine. J Thorac Cardiovasc Surg 139:146–153

Ichinose F, Erana-Garcia J, Hromi J et al (2001) Nebulized sildenafil is a selective pulmonary vasodilator in lambs with acute pulmonary hypertension. Crit Care Med 29:1000–1005

Ivy DD, Wiggins JW, Badesch D, Kinsella JP, Kelminson LL, Abman SH (1994) Treatment of an infant with severe primary pulmonary hypertension using inhaled nitric oxide and prostacyclin. Am J Cardiol 74:414–416

Ivy DD et al (1998a) Dipyridamole attenuates rebound pulmonary hypertension after inhaled nitric oxide withdrawal in postoperative congenital heart disease. J Thorac Cardiovasc Surg 115:875–882

Ivy DD, Griebel JL, Kinsella JP, Abman SH (1998b) Acute hemodynamic effects of pulsed delivery of low flow nasal nitric oxide in children with pulmonary hypertension. J Pediatr 133:453–456

Ivy DD, Parker D, Doran A, Parker D, Kinsella JP, Abman SH (2003) Acute hemodynamic effects and home therapy using a novel pulsed nasal nitric oxide delivery system in children and young adults with pulmonary hypertension. Am J Cardiol 92:886–890

Journois D et al (1994) Inhaled nitric oxide as a therapy for pulmonary hypertension after operations for congenital heart defects. J Thorac Cardiovasc Surg 107:1129–1135

Journois D et al (2005) Effects of inhaled nitric oxide administration on early postoperative mortality in patients operated for correction of atrioventricular canal defects. Chest 128:3537–3544

Kinsella JP, Neish SR, Shaffer E, Abman SH (1992) Low-dose inhalational nitric oxide in persistent pulmonary hypertension of the newborn. Lancet 340:819–820

Kinsella JP, Ivy DD, Abman SH (1994a) Ontogeny of NO activity and response to inhaled NO in the developing ovine pulmonary circulation. Am J Physiol 267:H1955–H1961

Kinsella JP, Ivy DD, Abman SH (1994b) Inhaled nitric oxide improves gas exchange and lowers pulmonary vascular resistance in severe experimental hyaline membrane disease. Pediatr Res 36:402–408

Kinsella JP, Truog WE, Walsh WF et al (1997a) Randomized, multicenter trial of inhaled nitric oxide and high-frequency oscillatory ventilation in severe, persistent pulmonary hypertension of the newborn. J Pediatr 131:55–62

Kinsella JP, Parker TA, Galan H, Sheridan BC, Halbower AC, Abman SH (1997b) Effects of inhaled NO on pulmonary edema and lung neutrophil accumulation in severe experimental HMD. Pediatr Res 41:457–463

Kinsella JP, Walsh WF, Bose CL, Gerstmann DR, Labella JJ, Sardesai S et al (1999) Inhaled nitric oxide in premature neonates with severe hypoxaemic respiratory failure: a randomised controlled trial. Lancet 354:1061–1065

Kinsella JP, Parker TA, Ivy DD, Abman SH (2003) Non-invasive delivery of inhaled NO therapy for late pulmonary hypertension in newborns with congenital diaphragmatic hernia. J Pediatr 142:397–401

Kinsella JP, Greenough A, Abman SH (2006a) Bronchopulmonary dysplasia. Lancet 367:1421–1431

Kinsella JP, Cutter GR, Walsh WF, Gerstmann DR, Bose CL, Hart C et al (2006b) Early inhaled nitric oxide therapy in premature newborns with respiratory failure. N Engl J Med 355:354–364

Lee JE, Hillier SC, Knoderer CA (2008) Use of sildenafil to facilitate weaning from inhaled nitric oxide in children with pulmonary hypertension following surgery for congenital heart disease. J Intensive Care Med 23:329–334

Lundin S, Mang H, Smithies M, Stenqvist O, Frostell C (1999) Inhalation of nitric oxide in acute lung injury: results of a European multicentre study. The European Study Group of Inhaled Nitric Oxide. Intensive Care Med 25:911–919

Mercier JC, Hummler H, Durrmeyer X et al (2010) Inhaled NO for prevention of BPD in premature babies (EUNO): a randomized controlled trial. Lancet 376:346–354

Michael JR, Barton RG, Saffle JR, Mone M, Markewitz BA, Hillier K et al (1998) Inhaled nitric oxide versus conventional therapy: effect on oxygenation in ARDS. Am J Respir Crit Care Med 157:1372–1380

Miller OI et al (2000) Inhaled nitric oxide and prevention of pulmonary hypertension after congenital heart surgery: a randomized double-blind study. Lancet 356:1464–1469

Morin FC (1989) Ligating the ductus arteriosus before birth causes persistent pulmonary hypertension in the newborn lamb. Pediatr Res 25:245–250

Morin FC, Egan EA, Ferguson W, Lundgren CEG (1988) Development of pulmonary vascular response to oxygen. Am J Physiol 254:H542–H546

Mourani P, Ivy DD, Gao D, Abman SH (2004) Pulmonary vascular effects of inhaled NO and oxygen tension in older children and adolescents with bronchopulmonary dysplasia. Am J Respir Crit Care Med 170:1006–1013

Murphy JD, Rabinovitch M, Goldstein JD et al (1981) The structural basis of persistent pulmonary hypertension of the newborn infant. J Pediatr 98:962–967

No Authors (1997) Inhaled nitric oxide in full-term and nearly full-term infants with hypoxic respiratory failure. The Neonatal Inhaled Nitric Oxide Study Group. N Engl J Med 336:597–604
North AJ, Star RA, Brannon TS, Ujiie K, Wells LB, Lowenstien CJ, Snyder SH, Shaul PW (1994) NO synthase type I and type III gene expression are developmentally regulated in rat lung. Am J Physiol 266:L635–L641
Parker TA, Le Cras TD, Kinsella JP, Abman SH (2000) Developmental changes in endothelial NO synthase expression in the ovine fetal lung. Am J Physiol 278:L202–L208
Peliowski A, Finer NN, Etches PC, Tierney AJ, Ryan CA (1995) Inahled NO for premature infants after prolonged rupture of the membranes. J Pediatr 126:450–453
Pepke-Zaba J, Higenbottam TW, Dinh-Xuan AT, Stone D, Wallwork J (1991) Inhaled NO as a cause of selective pulmonary vasodilation in pulmonary hypertension. Lancet 338:1173–1174
Perez-Penate GM, Julia-Serda G, Ojeda-Betancort N, Garcia-Quintana A, Pulido-Duque J, Rodriguez-Perez A et al (2008) Long-term inhaled nitric oxide plus phosphodiesterase 5 inhibitors for severe pulmonary hypertension. J Heart Lung Transplant 27:1326–1332
Post MC, Janssens S, Van de Werf F, Budts W (2004) Responsiveness to inhaled nitric oxide is a predictor for mid-term survival in adult patients with congenital heart defects and pulmonary arterial hypertension. Eur Heart J 25:1651–1656
Rasanen J, Wood DC, Debbs RH, Cohen J, Weiner S, Huhta JC (1998) Reactivity of the human fetal pulmonary circulation to maternal hyperoxygenation increases during the second half of pregnancy. A randomized study. Circulation 97:257–262
Ricciardi MJ, Knight BP, Martinez FJ, Rubenfire M (1998) Inhaled nitric oxide in primary pulmonary hypertension: a safe and effective agent for predicting response to nifedipine. J Am Coll Cardiol 32:1068–1073
Rich S, Kaufmann E, Levy PS (1992) The effect of high doses of calcium channel blockers on survival in primary pulmonary hypertension. N Engl J Med 327:76–81
Roberts JD Jr, Polaner DM, Lang P, Zapol WM (1992) Inhaled nitric oxide in persistent pulmonary hypertension of the newborn. Lancet 340:818–819
Roberts JD Jr, Fineman JR, Morin FC III, Shaul PW, Rimar S, Schreiber MD et al (1997) Inhaled nitric oxide and persistent pulmonary hypertension of the newborn. N Engl J Med 336:605–610
Rossaint R, Falke KJ, Lopez F, Slama K, Pison U, Zapol WM (1993) Inhaled nitric oxide for the adult respiratory distress syndrome. N Engl J Med 328:399–405
Rudolph AM, Heymann MA, Lewis AB (1977) Physiology and pharmacology of the pulmonary circulation in the fetus and newborn. In: Hodson W (ed) Development of the lung. Marcel Dekker, New York, pp 497–523
Schreiber MD, Gin-Mestan K, Marks JD, Huo D, Lee G, Srisuparp P (2003) Inhaled nitric oxide in premature infants with the respiratory distress syndrome. N Engl J Med 349:2099–2107
Schulze-Neick I et al (2001) Pulmonary vascular resistance after cardiopulmonary bypass in infants: effect on postoperative recovery. J Thorac Cardiovasc Surg 121:1033–1039
Shaul PW, Yuhanna IS, German Z et al (1997) Pulmonary endothelial NO synthase gene expression is decreased in fetal lambs with pulmonary hypertension. Am J Physiol Lung Cell Mol Physiol 272:L1005–L1012
Shekerdemian L, Ravn H, Penny D (2002) Intravenous sildenafil lowers pulmonary vascular resistance in a model of neonatal pulmonary hypertension. Am J Respir Crit Care Med 165:1098–2002
Shekerdemian LS, Ravn HB, Penny DJ (2004) Interaction between inhaled nitric oxide and intravenous sildenafil in a porcine model of meconium aspiration syndrome. Pediatr Res 55:413–418
Somlyo AP, Solyo AV (2003) Ca2+ sensitivity of smooth muscle and non-smooth muscle myosin II: modulated be G proteins, kinases and myosin phosphatase. Physiol Rev 83:1325–1358
Steinhorn R, Porta N (2007) Use of inhaled nitric oxide in the preterm infant. Curr Opin Pediatr 19:137–141

Steinhorn RH, Russell JA, Morin FC (1995) Disruption of cGMP production in pulmonary arteries isolated from fetal lambs with pulmonary hypertension. Am J Physiol 268:H1483–H1489

Steinhorn RH, Kinsella JP, Pierce C et al (2009) Intravenous sildenafil in the treatment of neonates with persistent pulmonary hypertension. J Pediatr 155:841–847.e1

Torko JA, Brahmajohi MV, Zhu H, Tinch BT, Auten RL, McMahon TJ (2012) Transpulmonary flux of S-nitrosothiols and pulmonary vasodilation during NO inhalation. Role of Transport. Am J Respir Cell Mol Biol 47:37–43

Villamor E, LeCras TD, Horan MP et al (1997) Chronic intrauterine pulmonary hypertension impairs endothelial nitric oxide synthase in the ovine fetus. Am J Physiol Lung Cell Mol Physiol 272:L1013–L1020

Vonbank K, Ziesche R, Higenbottam TW, Stiebellehner L, Petkov V, Schenk P et al (2003) Controlled prospective randomised trial on the effects on pulmonary haemodynamics of the ambulatory long term use of nitric oxide and oxygen in patients with severe COPD. Thorax 58:289–293

Weimann J, Ullrich R, Hromi J et al (2000) Sildenafil is a pulmonary vasodilator in awake lambs with acute pulmonary hypertension. Anesthesiology 92:1702–1712

Part III
Pulmonary Hypertension: Novel Pathways and Emerging Therapies

Soluble Guanylate Cyclase Stimulators in Pulmonary Hypertension

Johannes-Peter Stasch and Oleg V. Evgenov

Abstract Soluble guanylate cyclase (sGC) is a key enzyme in the nitric oxide (NO) signalling pathway. On binding of NO to its prosthetic haem group, sGC catalyses the synthesis of the second messenger cyclic guanosine monophosphate (cGMP), which promotes vasodilation and inhibits smooth muscle proliferation, leukocyte recruitment, platelet aggregation and vascular remodelling through a number of downstream mechanisms. The central role of the NO–sGC–cGMP pathway in regulating pulmonary vascular tone is demonstrated by the dysregulation of NO production, sGC activity and cGMP degradation in pulmonary hypertension (PH). The sGC stimulators are novel pharmacological agents that directly stimulate sGC, both independently of NO and in synergy with NO. Optimisation of the first sGC stimulator, YC-1, led to the development of the more potent and more specific sGC stimulators, BAY 41-2272, BAY 41-8543 and riociguat (BAY 63-2521). Other sGC stimulators include CFM-1571, BAY 60-4552, vericiguat (BAY 1021189), the acrylamide analogue A-350619 and the aminopyrimidine analogues. BAY 41-2272, BAY 41-8543 and riociguat induced marked dose-dependent reductions in mean pulmonary arterial pressure and vascular resistance with a concomitant increase in cardiac output, and they also reversed vascular remodelling and right heart hypertrophy in several experimental models of PH. Riociguat is the first sGC stimulator that has entered clinical development.

J.-P. Stasch (✉)
Cardiology Research, Bayer Pharma AG, Aprather Weg 18a, 42096 Wuppertal, Germany

Institute of Pharmacy, University Halle-Wittenberg, Halle (Saale), Germany
e-mail: johannes-peter.stasch@bayer.com

O.V. Evgenov
Department of Anesthesia, Critical Care and Pain Medicine, Massachusetts General Hospital, Harvard Medical School, 55 Fruit St, Gray/Jackson 424, Boston, MA 02114, USA
e-mail: oevgenov@partners.org

Clinical trials have shown that it significantly improves pulmonary vascular haemodynamics and increases exercise ability in patients with pulmonary arterial hypertension (PAH), chronic thromboembolic PH and PH associated with interstitial lung disease. Furthermore, riociguat reduces mean pulmonary arterial pressure in patients with PH associated with chronic obstructive pulmonary disease and improves cardiac index and pulmonary vascular resistance in patients with PH associated with left ventricular systolic dysfunction. These promising results suggest that sGC stimulators may constitute a valuable new therapy for PH. Other trials of riociguat are in progress, including a study of the haemodynamic effects and safety of riociguat in patients with PH associated with left ventricular diastolic dysfunction, and long-term extensions of the phase 3 trials investigating the efficacy and safety of riociguat in patients with PAH and chronic thromboembolic PH. Finally, sGC stimulators may also have potential therapeutic applications in other diseases, including heart failure, lung fibrosis, scleroderma and sickle cell disease.

Keywords Pulmonary hypertension • Nitric oxide • Soluble guanylate cyclase • sGC stimulator • Riociguat

Contents

1 Soluble Guanylate Cyclase Signalling in Pulmonary Endothelium and Vascular Smooth Muscle	281
2 Dysregulation of the NO–sGC–cGMP Pathway in Pulmonary Hypertension	281
2.1 Nitric Oxide Production	281
2.2 Activity of Soluble Guanylate Cyclase	283
2.3 Cyclic Guanosine Monophosphate Degradation	284
2.4 Genetic Factors	284
3 Mechanisms of sGC Stimulation	284
3.1 Activation of sGC by NO	284
3.2 Pharmacological Stimulation of sGC	286
4 Discovery of sGC Stimulators	287
4.1 YC-1	287
4.2 BAY 41-2272	287
4.3 BAY 41-8543	294
4.4 Other sGC Stimulators	295
5 Riociguat	296
5.1 Discovery	296
5.2 Preclinical Evidence	296
5.3 Clinical Trials	297
5.4 Ongoing Clinical Trials	303
5.5 Future Directions	303
References	304

1 Soluble Guanylate Cyclase Signalling in Pulmonary Endothelium and Vascular Smooth Muscle

Nitric oxide (NO), a diffusible, short-lived, highly reactive gaseous molecule produced from L-arginine by NO synthases, plays a major role in controlling nearly every cellular and organ function in the body and performs diverse functions in human physiology and disease (Coggins and Bloch 2007; Derbyshire and Marletta 2012; Hirst and Robson 2011; Tonelli et al. 2013). The haemoprotein soluble guanylate cyclase (sGC), a key enzyme in the cardiopulmonary system, is the physiological receptor for endogenous NO. It catalyses the generation of the signalling molecule cyclic guanosine monophosphate (cGMP). This cyclic nucleotide is an important second messenger that is involved in the regulation of many vascular processes. It promotes vasodilation (i.e. vascular smooth muscle relaxation) via cGMP-dependent protein kinase G (PKG), inhibits proliferation of vascular smooth muscle cells in cooperation with bone morphogenetic protein (BMP) and Smad signalling, inhibits fibrosis and platelet aggregation, and also plays a role in protecting against vascular inflammation (Garg and Hassid 1989; Mellion et al. 1981; Pfeifer et al. 1998; Rizzo et al. 2010; Sawada et al. 2009; Schlossmann and Schinner 2012; Schwappacher et al. 2013).

The central role of sGC in the regulation of vascular tone is demonstrated by the phenotype of sGC-knockout mice, in which NO-dependent aortic relaxation is lost (Bryan et al. 2009; Friebe et al. 2007). In addition, smooth muscle relaxation and conventional haemodynamic response following NO administration are attenuated in mice that express a mutant form of sGC with only basal activity (Thoonen et al. 2013a). Further studies using these mouse models promise to help elucidate how sGC contributes to cardiovascular homeostasis (Thoonen et al. 2013b).

2 Dysregulation of the NO–sGC–cGMP Pathway in Pulmonary Hypertension

The pathophysiology of pulmonary hypertension (PH) includes endothelial dysfunction that results in insufficient stimulation of the NO–sGC–cGMP pathway. This pathway may be affected at a number of different steps in many types of PH (Fig. 1).

2.1 Nitric Oxide Production

Endogenous NO levels are reduced in patients with pulmonary arterial hypertension (PAH), PH associated with connective tissue disease, chronic obstructive pulmonary disease (COPD) and interstitial lung disease (Clini et al. 2000; Cremona et al. 1994; Girgis et al. 2005; Kaneko et al. 1998; Kawaguchi et al. 2006; Malerba et al. 2007; Riley et al. 1997). Moreover, impaired vasodilatory responsiveness to NO has been found in the pulmonary arteries of rats with experimentally induced

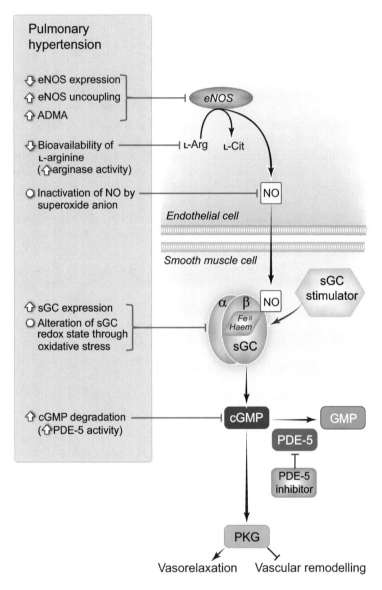

Fig. 1 Steps in the nitric oxide-soluble guanylate cyclase–cyclic guanosine monophosphate signalling pathway that may be disrupted in pulmonary hypertension. *ADMA* asymmetrical dimethylarginine, *cGMP* cyclic guanosine monophosphate, *eNOS* endothelial nitric oxide synthase, Fe^{II} ferrous iron, *GMP* guanosine monophosphate, *L-arg* L-arginine, *L-cit* L-citrulline, *NO* nitric oxide, *PDE-5* phosphodiesterase-5, *PKG* protein kinase G, *sGC* soluble guanylate cyclase

PH (Mam et al. 2010). The mechanisms leading to low bioavailability of NO in PH include changes in the expression and activity of endothelial nitric oxide synthase (eNOS), changes in the bioavailability of L-arginine, and oxidative stress.

Expression of eNOS is downregulated in the vascular endothelium of the pulmonary arteries in several forms of PH (i.e. PAH, PH associated with COPD, interstitial lung disease and bronchiectasis) (Giaid and Saleh 1995). Interestingly, in patients with plexiform lesions associated with PAH, eNOS expression correlates inversely with pulmonary vascular resistance (Giaid and Saleh 1995). The expression of eNOS is also reduced in platelets from patients with idiopathic PAH, which exhibit defective activation and aggregation (Aytekin et al. 2012).

In addition to the downregulation of eNOS expression, eNOS activity may be suppressed by endogenous inhibitors such as asymmetrical dimethylarginine (ADMA). Plasma levels of ADMA are elevated in patients with idiopathic PAH and chronic thromboembolic pulmonary hypertension (CTEPH) compared with healthy individuals (Kielstein et al. 2005; Skoro-Sajer et al. 2007). Plasma ADMA levels have also been found to be higher in patients with congenital heart disease accompanied by PAH than in those who do not have PAH (Sanli et al. 2012).

NO levels may also be reduced due to decreased bioavailability of its precursor L-arginine. Increased serum arginase activity and consequently reduced circulating levels of L-arginine have been observed in patients with PAH compared with healthy individuals (Xu et al. 2004). In addition, dysregulated arginine metabolism, characterised by both diminished systemic arginine bioavailability and raised ADMA levels, is directly associated with elevated pulmonary artery pressure in patients with advanced decompensated heart failure (Shao et al. 2012).

Furthermore, NO levels are reduced in conditions of oxidative stress, and biomarkers of oxidative stress are increased in the lungs of patients with idiopathic PAH and PH associated with connective tissue disease, left heart disease or hypoxia (Bowers et al. 2004; Cracowski et al. 2001; Hoshikawa et al. 2001). Under conditions of oxidative stress, NO levels are depleted by a reaction with superoxide (Laursen et al. 2001). This reaction generates the strong oxidant peroxynitrite, which, in concert with other oxidants, induces cell damage via lipid peroxidation, inactivation of enzymes and other proteins by oxidation and nitration, and activation of matrix metalloproteinases. Importantly, it also oxidises eNOS, resulting in its uncoupling. The uncoupling of eNOS ultimately abrogates NO synthesis and enhances the production of superoxide, thus amplifying the oxidative stress, which leads to further cellular dysfunction and finally cell death.

2.2 Activity of Soluble Guanylate Cyclase

Oxidative stress also impairs sGC activity. A study by Perez et al. demonstrated a link between oxidative stress and the regulation of sGC activity in newborn lambs with persistent PH. The increased levels of reactive oxygen species and decreased sGC activity that resulted from ventilating these lambs with oxygen were both normalised by a high-dose hydrocortisone treatment (Perez et al. 2012). Oxidation of the haem group on sGC causes it to dissociate from the enzyme and thus decreases the latter's activity and NO responsiveness (Roy et al. 2008; Thoonen et al. 2013a).

Although total sGC expression is increased in the vascular smooth muscle cells of small pulmonary arteries of patients with idiopathic PAH, this increase may be attributable to the increase in haem-free, NO-insensitive sGC that has been described in a range of animal models involving oxidative stress and in human cardiovascular diseases (Ahrens et al. 2011; Schermuly et al. 2008; Stasch et al. 2006).

2.3 Cyclic Guanosine Monophosphate Degradation

Increased cGMP degradation may also affect sGC signalling. Elevated expression and activity of phosphodiesterase (PDE)-5, resulting in reduced cGMP levels, have been found in two animal models of PH: sheep with PH associated with left heart disease and rats with hypoxia-induced PH (Black et al. 2001; Maclean et al. 1997). Furthermore, PDE-5 activity increased in response to oxidative stress in the pulmonary vascular smooth muscle cells of mice exposed to short-term hyperoxia (Farrow et al. 2012).

2.4 Genetic Factors

Advances in our understanding of the genetic factors associated with PH are important in further elucidating its underlying pathological mechanisms. Variants in many genes are known to be associated with systemic hypertension, indicating that they may play a central role in the control of blood pressure (Ehret et al. 2011). These include variants in three genes encoding components of pathways that regulate cGMP levels (i.e. the α and β subunits of sGC, and the natriuretic peptide clearance receptor, which sequesters the ligands of particulate guanylate cyclase) (Ehret et al. 2011). In addition, a hypertension sensitivity locus has been identified in the promoter region of the gene encoding eNOS (Salvi et al. 2012). The impact of these genetic variations on PH is yet to be determined.

3 Mechanisms of sGC Stimulation

3.1 Activation of sGC by NO

The heterodimeric α/β haemoprotein sGC is the physiological receptor for the endogenous gaseous second messenger NO. The prosthetic haem group of sGC is noncovalently bound to its β subunit via a haem-binding motif and the axial haem ligand histidine-105. Binding of NO to the reduced central iron atom of the haem moiety results in an up to 200-fold increase in the conversion rate of guanosine triphosphate (GTP) into cGMP (Fig. 2) (Arnold et al. 1977; Evgenov et al. 2006; Humbert et al. 1990; Lee et al. 2000; Murad 2006; Thoonen et al. 2013a; Wolin et al. 1982). However, this simple model of sGC activation is not consistent with

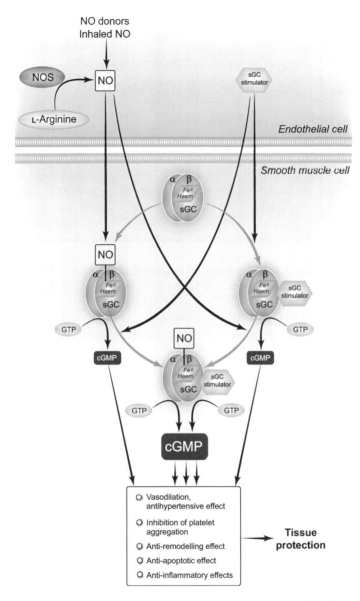

Fig. 2 Dual mode of action of soluble guanylate cyclase (sGC) stimulators. sGC is a key enzyme in the cardiopulmonary system and is the receptor for nitric oxide (NO). It catalyses the generation of the signalling molecule cyclic guanosine monophosphate (cGMP), which plays a pivotal role in regulating cellular and organ functions, such as vascular tone, proliferation, fibrosis, platelet aggregation and inflammation. Stimulators of sGC increase cGMP production by directly targeting sGC to induce its full activation via a dual mode of action. They stimulate sGC independently of NO, due to their distinct binding site on sGC, and they are also able to sensitise sGC to low levels of NO by stabilising NO–sGC binding. *cGMP* cyclic guanosine monophosphate, Fe^{II} ferrous iron, *GTP* guanosine triphosphate, *NO* nitric oxide, *NOS* nitric oxide synthase, *sGC* soluble guanylate cyclase

some subsequently observed activation characteristics of the enzyme. For example, some research groups have demonstrated that the conversion of the hexa-coordinated intermediate state into the penta-coordinated active nitrosyl–haem complex depends on the concentration of free NO, suggesting that there is a second binding site for NO (Cary et al. 2005; Derbyshire and Marletta 2012; Friebe and Koesling 1998).

3.2 Pharmacological Stimulation of sGC

Stimulators of sGC are a novel class of compounds that directly stimulate sGC with a dual mode of action (Fig. 2) (Stasch et al. 2001, 2002a; Stasch and Hobbs 2009). First, they bind at a different site than NO and can increase sGC activity in the absence of NO (Friebe et al. 1996; Mülsch et al. 1997; Stasch et al. 2001). Secondly, they can act synergistically with NO by stabilising the nitrosyl–haem complex to maintain sGC in its active conformation (Russwurm et al. 2002; Schmidt et al. 2003). Therefore, the combined effect of an sGC stimulator and NO on the catalysis of cGMP synthesis is more pronounced than the effects of each compound alone. The signalling molecule cGMP plays an important role in regulating cellular and organ functions such as vascular tone, proliferation, fibrosis and inflammation (Schlossmann and Schinner 2012; Stasch et al. 2011).

The non-haem-binding site of sGC is yet to be identified, but it has been postulated that sGC stimulators bind at the amino-terminus of the α1 subunit of sGC or the pseudosymmetric substrate site of its catalytic domain (Friebe et al. 1999; Lamothe et al. 2004; Stasch et al. 2001; Yazawa et al. 2006). The elucidation of the full-length structure of sGC and co-crystallisation studies will provide further insight into the mechanism of sGC activation and its potentiation by sGC stimulators.

Stimulation of sGC has several potential advantages compared with other therapies targeting the NO–sGC–cGMP pathway. The disadvantages of inhaled NO and NO donor treatment include the resistance of a significant proportion of patients with PH to NO, the development of tolerance to NO, the potential cytotoxic effects of NO and rebound PH after discontinuation of inhaled NO administration (Burney et al. 1999; Erusalimsky and Moncada 2007; Krasuski et al. 2000; Münzel et al. 2005; Radi 2004).

Stimulation of sGC may also have significant advantages over PDE-5 inhibition. Owing to their NO-independent mode of action, sGC stimulators could be effective even when NO production is severely compromised or absent; by contrast, PDE-5 inhibitors merely prevent the degradation of cGMP. When NO signalling is disrupted in PH, the production of cGMP may be severely limited, and the preservation of low cGMP levels by PDE-5 inhibitors may be ineffective. Moreover, when PDE-5 is inhibited, the activity of other PDEs may compensate for it (Stasch et al. 2011).

Additionally, sGC expression is upregulated in vascular smooth muscle cells of small pulmonary arteries in patients with PH compared with those of healthy lung donors, presenting an opportunity for sGC stimulation (Schermuly et al. 2008). This contrasts with the upregulation of PDE-5 in pulmonary arteries (Black et al. 2001; Maclean et al. 1997).

Another potential advantage of sGC stimulators is their synergistic action with the low levels of endogenous NO produced in the lungs of patients with PH. This synergistic activity could ensure optimal ventilation–perfusion matching to maintain or even enhance arterial haemoglobin oxygen saturation (Becker et al. 2011).

4 Discovery of sGC Stimulators

The search for drugs targeting sGC began at Bayer in 1994 with the testing of nearly 20,000 compounds for sGC-inducing activity in primary cultures of endothelial cells. This led to the discovery of a class of compounds, the 5-substituted-2-furaldehyde-hydrazone derivatives, that are direct NO-independent sGC stimulators (Evgenov et al. 2006). However, it was found that exposure to light increased the potency of these substances in stimulating sGC and relaxing isolated blood vessels, which made them unsuitable for further drug development (Evgenov et al. 2006). It was serendipitous that, in 1994, the synthetic bezylindazole YC-1 was discovered to be an inhibitor of platelet aggregation mediated by intracellular cGMP increase. It had obvious structural similarity to the 5-substituted-2-furaldehyde-hydrazone derivatives and it was found to directly stimulate isolated sGC in the same NO-independent, but haem-dependent, manner (Friebe et al. 1996; Ko et al. 1994; Mülsch et al. 1997). This was the starting point for a chemical optimisation programme and the development of sGC stimulators (Stasch et al. 2001, 2002a).

4.1 YC-1

YC-1 demonstrates both NO-independent and NO-dependent sGC stimulation (Ko et al. 1994; Mülsch et al. 1997). YC-1 increases the activity of purified haem-bound sGC up to 12-fold, and synergy with NO increases this effect by an order of magnitude compared with YC-1 alone (Friebe et al. 1996). YC-1 has weak vasodilatory properties, and inhibits vascular smooth muscle remodelling and platelet aggregation (Cetin et al. 2004; Ko et al. 1994; Mülsch et al. 1997; O'Reilly et al. 2001; Pan et al. 2004; Tulis et al. 2002). YC-1 attenuates hypoxia-induced PAH, right ventricular (RV) hypertrophy and pulmonary vascular remodelling in mice (Huh et al. 2011). However, the clinical utility of YC-1 is limited by its lack of specificity: it inhibits PDEs and has many cGMP-independent effects (Friebe et al. 1998; Galle et al. 1999; Hwang et al. 2003a, b).

4.2 BAY 41-2272

BAY 41-2272 was developed during an optimisation programme that generated more than 2,000 new compounds using YC-1 as a lead structure (Stasch et al. 2001). BAY 41-2272 has a greater specificity than YC-1; up to a concentration of 10 µM it has no

PDE-5 inhibitory activity, and it does not inhibit other cGMP-metabolising PDEs such as PDE-1, PDE-2 or PDE-9 (Bischoff and Stasch 2004; Stasch et al. 2001).

BAY 41-2272 has greater potency than YC-1; in the absence of NO, it stimulates sGC approximately 20-fold above baseline, and in synergy with the NO donor diethylamine NONOate (DEA/NO), it stimulates sGC activity up to 400-fold above baseline (Stasch et al. 2001). BAY 41-2272 is also over 30 times more potent than YC-1 in causing rabbit aortic ring relaxation (Straub et al. 2001).

4.2.1 In Vitro Results

BAY 41-2272 inhibits the conversion of lung fibroblasts to myofibroblasts, which is a feature of pulmonary fibrosis (Dunkern et al. 2007). This suggests that sGC stimulation may be useful in mitigating the progression of pulmonary fibrosis.

BAY 41-2272 also has beneficial effects on other pathological processes. The compound elicits cardioprotective effects in vitro, inhibiting hypertrophy of cultured neonatal rat cardiomyocytes (Irvine et al. 2012). In addition, BAY 41-2272 has been found to protect against ischaemia–reperfusion-induced lung injury in isolated, intact rabbit lungs (Egemnazarov et al. 2009). Moreover, its administration at high doses inhibits platelet aggregation ex vivo (Roger et al. 2010; Stasch et al. 2001).

4.2.2 Studies in Experimental Models of Pulmonary Hypertension

An overview of preclinical studies of sGC stimulators is given in Table 1. The potent vasodilatory activity of BAY 41-2272 has been demonstrated in many animal models of PH. Intravenous infusion of BAY 41-2272 in sheep with acute PH induced by a thromboxane A_2 analogue caused dose-dependent pulmonary and systemic vasodilation, and enhanced and prolonged the pulmonary vasodilatory response to inhaled NO (Evgenov et al. 2004). Inhibition of nitric oxide synthase (NOS) by N-nitro-L-arginine methyl ester (L-NAME) in these sheep abolished the systemic but not the pulmonary vasodilatory effects, suggesting that, in the pulmonary vasculature, BAY 41-2272 acts independently of endogenous NO production (Evgenov et al. 2004). In the same model, inhalation of BAY 41-2272 produced selective pulmonary vasodilation (Evgenov et al. 2007). This was also observed following BAY 41-2272 infusion combined with NO inhalation in rabbits with acute lung injury (induced by oleic acid injection) and PH. In addition, the combination of BAY 41-2272 and inhaled NO improved arterial oxygenation and reduced intrapulmonary shunting in these rabbits (Weidenbach et al. 2005).

Intravenous infusion of BAY 41-2272 markedly attenuated heparin–protamine-induced PH in dogs. BAY 41-2272 administration also led to both pulmonary and systemic vasodilation in these dogs and improved arterial oxygen saturation compared with vehicle infusion alone (Freitas et al. 2007). In addition, intravenous infusion of low doses of BAY 41-2272 reduced pulmonary arterial pressure and pulmonary vascular resistance in a canine model of acute pulmonary thromboembolism (Cau et al. 2008).

Table 1 Preclinical studies of soluble guanylate cyclase (sGC) stimulators in pulmonary hypertension

Study design	Treatment	Effects of sGC stimulators	References
Ovine model of acute PH – Pharmacologically induced by the thromboxane A_2 analogue U-46619	Intravenous infusion of BAY 41-2272	• Dose-dependent pulmonary and systemic vasodilation • Augmentation and prolongation of the pulmonary vasodilatory response to inhaled NO • Pharmacological inhibition of eNOS abolished systemic but not pulmonary vasodilator effects	Evgenov et al. (2004)
Ovine model of severe persistent PH of the newborn – Induced by partial ligation of the ductus arteriosus	BAY 41-2272 infusion into the pulmonary artery	• Potent pulmonary vasodilation	Deruelle et al. (2005a)
Ovine fetal model	Infusion of BAY 41-2272 into the left pulmonary artery	• Sustained pulmonary vasodilation was not attenuated by an eNOS inhibitor • More prolonged pulmonary vasodilatory response than following sildenafil administration	Deruelle et al. (2005b)
Rabbit model of acute lung injury and PH – Pharmacologically induced by oleic acid injection	Combined administration of inhaled NO and intravenously infused BAY 41-2272	• Selective pulmonary vasodilation • Improved arterial oxygenation • Reduced intrapulmonary shunting	Weidenbach et al. (2005)
Rat model of PH – Pharmacologically induced by monocrotaline	Daily oral administration of BAY 41-2272 after fully established PH	• Reduced RV systolic pressure	Dumitrascu et al. (2006)
Mouse model of PH – Induced by chronic hypoxia		• Reversal of RV hypertrophy and pulmonary vascular remodelling	
Neonatal rat model of PH – Induced by hypoxia	Daily intramuscular treatment with BAY 41-2272	• Reduced RV hypertrophy • Attenuation of pulmonary arterial wall thickening	Deruelle et al. (2006)

(continued)

Table 1 (continued)

Study design	Treatment	Effects of sGC stimulators	References
Canine model of PH – Induced by heparin-protamine reaction	Intravenous infusion of BAY 41-2272	• Pulmonary and systemic vasodilation • Improved arterial oxygen saturation	Freitas et al. (2007)
Ovine model of acute PH – Pharmacologically induced by the thromboxane A_2 analogue U-46619	Inhalation of BAY 41-2272 or BAY 41-8543	• Selective pulmonary vasodilation	Evgenov et al. (2007)
	BAY 41-8543 inhalation	• Improved systemic arterial oxygenation • Augmentation of the magnitude and duration of the pulmonary vasodilatory response to inhaled NO	
	Concurrent administration of the phosphodiesterase inhibitor zaprinast and BAY 41-8543	• Augmentation and prolongation of the pulmonary vasodilator effect of BAY 41-8543	
Canine model of acute pulmonary embolism – Induced by microsphere injection	Intravenous infusion of BAY 41-2272	• Pulmonary vasodilation • Systemic vasodilation with higher doses • No adverse effects on arterial oxygen saturation	Cau et al. (2008)
Rat model of PH – Pharmacologically induced by monocrotaline Mouse model of PH – Induced by chronic hypoxia	Daily oral administration of riociguat after fully established PH	• Partial reversal of PH • Partial reversal of RV hypertrophy and pulmonary vascular remodelling • No significant effect on systemic arterial pressure	Schermuly et al. (2008)
Neonatal rat model of PH – Induced by hypoxia	Daily intraperitoneal administration of BAY 41-2272	• Prevention of hypoxia-induced increase in RV systolic pressure and RV hypertrophy comparable to sildenafil treatment • Acute pulmonary and systemic vasodilation	Thorsen et al. (2010)
Rat model of PH – Pharmacologically induced by monocrotaline	Inhaled and oral BAY 41-8543	• Decreased pulmonary vascular remodelling • Improved cardiac function	Egemnazarov et al. (2010)
	Inhaled BAY 41-8543	• Selective pulmonary vasodilation	

(continued)

Table 1 (continued)

Study design	Treatment	Effects of sGC stimulators	References
Rat model of acute PH – Pharmacologically induced by the thromboxane A_2 analogue U-46619	Intravenous injection of BAY 41-8543	• Pulmonary and systemic vasodilation that was blunted by pharmacological inhibition of eNOS	Badejo et al. (2010)
Pig model of PH – Induced by hypoxia	Right atrial infusion of BAY 41-8543	• Reversal of pulmonary vasoconstriction • Systemic vasodilation	Lundgren et al. (2012)
Rat model of moderate systemic hypertension – Renin transgenic rats treated with the eNOS inhibitor L-NAME Rat model of malignant systemic hypertension – Induced by 5/6 nephrectomy	Daily oral administration of riociguat	• Normalisation of blood pressure • Reduced heart and kidney damage • Improved survival	Sharkovska et al. (2010)
Rat carotid artery balloon injury	Perivascular administration of BAY 41-2272	• Reduction in neointimal growth after arterial balloon injury	Joshi et al. (2011)
Rat model of acute pulmonary embolism – Induced by microsphere injection	BAY 41-8543 infusion	• Normalisation of pulmonary haemodynamics • Improved arterial oxygenation	Watts et al. (2011)
Pig model of septic shock – Induced by endotoxin infusion	Inhalation or intravenous administration of BAY 41-8543	• Reduced pulmonary vascular resistance • Increase in cardiac output	Kronas et al. (2011)
Mouse model of pulmonary fibrosis and PH – Pharmacologically induced by bleomycin	Daily oral administration of riociguat	• Attenuation of PH and RV hypertrophy to a greater extent than by sildenafil treatment • Amelioration of pulmonary inflammation and fibrosis • Improved survival	Evgenov et al. (2011)
Rat model of malignant systemic hypertension – Dahl salt-sensitive rats fed a high salt diet	Daily oral administration of riociguat	• Attenuation of systemic hypertension and systolic dysfunction • Attenuation of fibrotic tissue remodelling and degenerative changes in the myocardium and renal cortex • Improved survival	Geschka et al. (2011)

(continued)

Table 1 (continued)

Study design	Treatment	Effects of sGC stimulators	References
Pig model of PH – Induced by hypoxia	Right atrial infusion of BAY 41-8543	• Augmentation of the pulmonary vasodilatory effect of the endothelin receptor blocker tezosentan	Lundgren et al. (2012)
Rat model of acute pulmonary embolism – Induced by microsphere injection	BAY 41-8543 infusion	• Amelioration of RV dysfunction • Reduction of pulmonary vascular resistance	Watts et al. (2013)
Rat model of PH – Pharmacologically induced by monocrotaline	Intravenous administration of BAY 41-8543	• Decreased pulmonary and systemic arterial pressures • Slightly greater decrease in pulmonary arterial pressure in rats that also received chronic sodium nitrite therapy than in those that received BAY 41-8543 alone	Pankey et al. (2012)
Rat model of angioproliferative PAH – Induced by hypoxia and the vascular endothelial growth factor receptor antagonist SU5416	Oral administration of riociguat	• Increased the open-to-occluded artery ratio and decreased the neointima-to-media ratio • Decrease in RV systolic pressure • Decrease in pulmonary vascular resistance • Increase in cardiac output • Greater decrease in RV hypertrophy and improvement in RV function than following sildenafil administration	Lang et al. (2012)
Mouse model of COPD – Induced by tobacco smoke exposure	Daily oral administration of riociguat	• Prevention of PH development • Prevention of RV hypertrophy • Prevention of vascular remodelling	Seimetz et al. (2012)

(continued)

Table 1 (continued)

Study design	Treatment	Effects of sGC stimulators	References
Diabetic eNOS-knockout mice	Co-administration of riociguat and angiotensin II receptor blocker	• Reduced blood pressure • Attenuation of nephropathy	Ott et al. (2012)
Porcine model of CTEPH – Induced by left pulmonary artery ligation and weekly histoacryl pulmonary artery embolisation	Administration of BAY 41-8543 or sildenafil	• BAY 41-8543 led to a marked decrease in pulmonary vascular resistance and concomitant increase in cardiac output to a greater extent than sildenafil	Mercier et al. (2012)
Guinea pigs chronically exposed to cigarette smoke	Daily administration of BAY 41-2272	• BAY 41-2272 prevented tobacco smoke-induced pulmonary vascular remodelling and emphysema	Lobo et al. (2013)
Rat model of angioproliferative PAH – Induced by hypoxia and the vascular endothelial growth factor receptor antagonist SU5416	Oral administration of riociguat	• Increased Smad 1/5 phosphorylation in PAH vascular lesions • Decreased medial thickness of PAH vascular lesions	Schwappacher et al. (2013)

COPD chronic obstructive pulmonary disease, *CTEPH* chronic thromboembolic pulmonary hypertension, *eNOS* endothelial nitric oxide synthase, L-*NAME* N-nitro-L-arginine methyl ester, *NO* nitric oxide, *PAH* pulmonary arterial hypertension, *PH* pulmonary hypertension, *RV* right ventricular, *sGC* soluble guanylate cyclase

In newborn sheep with persistent PH (induced by partial ligation of the ductus arteriosus), a pulmonary artery infusion of BAY 41-2272 resulted in marked pulmonary vasodilation (Deruelle et al. 2005a). BAY 41-2272 infusion in sheep fetuses also induced a more prolonged pulmonary vasodilator response than sildenafil administration, and this effect was not attenuated by administration of a NOS inhibitor (Deruelle et al. 2005b).

BAY 41-2272 has been shown to have beneficial effects on other cardiopulmonary parameters in several animal models. Intramuscular injections of BAY 41-2272 into newborn rats with hypoxic PH reduced RV hypertrophy and attenuated pulmonary arterial wall thickening (Deruelle et al. 2006). Intraperitoneal administration of BAY 41-2272 in the same rat model prevented a hypoxia-induced increase in RV systolic pressure and RV hypertrophy, and resulted in acute pulmonary and systemic vasodilatation compared with vehicle administration alone (Thorsen et al. 2010).

Oral administration of BAY 41-2272 in rats with monocrotaline-induced PH and in wild-type mice with hypoxia-induced PH reduced RV systolic pressure, and reversed

RV hypertrophy and pulmonary vascular remodelling (Dumitrascu et al. 2006). BAY 41-2272 reduced vascular smooth muscle growth by PKG-specific antimigratory and protein kinase A-specific antiproliferative mechanisms (Joshi et al. 2011; Wharton et al. 2005). Its anti-remodelling activity was demonstrated by a marked reduction in neointimal growth after arterial balloon injury in animals that received perivascular BAY 41-2272 compared with vehicle-treated controls (Joshi et al. 2011).

The clinical use of BAY 41-2272 was, however, not possible because of its strong inhibition and induction of cytochrome P450 isoenzymes (Mittendorf et al. 2009).

4.3 BAY 41-8543

BAY 41-8543 was the sGC stimulator with the most promising potency, specificity and oral activity of the 2,000 compounds derived from the optimisation of YC-1 (Mittendorf et al. 2009; Stasch et al. 2002a, b). BAY 41-8543 has a greater potency than both YC-1 and BAY 41-2272; it stimulates sGC activity up to 92-fold over baseline in the absence of NO and 362-fold over baseline in synergy with the NO donor sodium nitroprusside (Stasch et al. 2002a). Its vasodilatory potency is 500-fold higher than that of YC-1 and three times higher than that of BAY 41-2272 following intravenous administration in normotensive rats (Stasch et al. 2002a; Straub et al. 2001). BAY 41-8543 also has greater specificity than YC-1; it does not inhibit PDE-1, PDE-2, PDE-5 or PDE-9 (Stasch et al. 2002a).

4.3.1 Studies in Experimental Models of Pulmonary Hypertension

BAY 41-8543 has demonstrated beneficial effects in animal models of PH, most notably potent pulmonary vasodilation (Table 1). In sheep with acute PH induced by a thromboxane A_2 analogue, BAY 41-8543 inhalation caused selective dose-dependent reductions in pulmonary artery pressure and pulmonary vascular resistance, improved systemic arterial oxygenation and augmented the magnitude and duration of the pulmonary vasodilatory response to NO inhalation (Evgenov et al. 2007). In a rat model of acute PH induced in the same way, intravenous injection of BAY 41-8543 also resulted in dose-dependent pulmonary and systemic vasodilation, which was blunted by NOS inhibition (Badejo et al. 2010).

Intravenous injection of BAY 41-8543 in rats with monocrotaline-induced PH decreased pulmonary and systemic arterial pressures. Interestingly, the decrease in pulmonary arterial pressure induced by BAY 41-8543 was slightly greater in rats that also received chronic sodium nitrite therapy than in those that received BAY 41-8543 alone (Pankey et al. 2012). Right atrial infusion of BAY 41-8543 in pigs with hypoxic PH reversed pulmonary vasoconstriction and produced systemic vasodilation (Lundgren et al. 2012). It also reversed hypoxia-induced pulmonary vasoconstriction when administered alone or in combination with the endothelin receptor blocker tezosentan in a porcine model (Lundgren et al. 2012).

Moreover, BAY 41-8543 attenuates vascular remodelling. In rats with monocrotaline-induced PH, intratracheal administration of BAY 41-8543 decreased pulmonary vascular remodelling, improved cardiac function and induced selective pulmonary vasodilation (Egemnazarov et al. 2010).

BAY 41-8543 administration has been shown to produce other favourable effects in animal models, for example inhibition of thrombosis (Stasch et al. 2002a, b). Infusion of BAY 41-8543 also normalised pulmonary haemodynamics, improved arterial oxygenation and ameliorated RV dysfunction after experimental pulmonary embolism in a rat model (Watts et al. 2011, 2013). Inhalation or intravenous administration of BAY 41-8543 in pigs with endotoxin-induced septic shock reduced pulmonary vascular resistance and increased cardiac output (Kronas et al. 2011). In addition, BAY 41-8543 and, to a lesser extent, sildenafil produced a marked decrease in pulmonary vascular resistance with a concurrent increase in cardiac output in a CTEPH model in piglets (Mercier et al. 2012).

Despite promising results in animal models, the clinical application of BAY 41-8543 is precluded by unfavourable drug metabolism and pharmacokinetic properties, including high clearance and dose non-linearity of plasma concentrations (Mittendorf et al. 2009).

4.4 Other sGC Stimulators

CFM-1571 is another sGC stimulator that was developed using YC-1 as a lead structure (Selwood et al. 2001). CFM-1571 synergises with NO to stimulate sGC and has no PDE inhibitory activity. However, it has low oral bioavailability and low potency (Selwood et al. 2001; Stasch and Hobbs 2009).

Acrylamide analogues such as A-350619 are direct haem-dependent stimulators of sGC with a similar mode of action to YC-1; they act independently of and in synergy with NO, although they have no structural similarity with YC-1 (Miller et al. 2003; Zhang et al. 2003). Aminopyrimidine analogues are another class of sGC stimulators currently in clinical development (Brockunier et al. 2009).

BAY 60-4552, a close chemical analogue of riociguat (see Sect. 5), has been shown to attenuate cardiac hypertrophy and improve survival in spontaneously hypertensive stroke-prone rats (SHR-SP) maintained on a high salt/fat diet (Costell et al. 2012). In patients with biventricular heart failure and PH, BAY 60-4552 induces potent vasodilation and a significant increase in cardiac index (Gheorghiade et al. 2013; Mitrovic et al. 2009). Another sGC stimulator, vericiguat (BAY 1021189), is currently in clinical phase 1 trials for heart failure (Bayer 2013).

5 Riociguat

5.1 Discovery

Extended exploration of the substituent-active region of the pyrazolopyridines (the compound class to which BAY 41-8543 and BAY 41-2272 belong) was performed to address their specificity and pharmacokinetic limitations (Mittendorf et al. 2009). Pharmacokinetic optimisation of more than 800 pyrimidine derivatives led to the identification of the potent sGC stimulator riociguat (BAY 63-2521), which exhibits an improved drug metabolism/pharmacokinetic profile compared with its predecessors, and favourable effects on pulmonary haemodynamics and exercise capacity in patients with PH (Mittendorf et al. 2009; Stasch et al. 2011).

5.2 Preclinical Evidence

5.2.1 In Vitro Results

Riociguat stimulates recombinant sGC in a dose-dependent manner, up to 73-fold above baseline at 100 µM (Schermuly et al. 2008). Riociguat synergises with the NO donor DEA/NO to increase sGC activity up to 112-fold above baseline (Schermuly et al. 2008). Riociguat is highly specific and does not inhibit PDEs, including PDE-1, PDE-2, PDE-5 and PDE-9 (Schermuly et al. 2008).

The vasodilatory properties of riociguat have been demonstrated in a number of ex vivo experiments. Riociguat reduced coronary perfusion pressure in rat heart preparations without affecting left ventricular pressure or heart rate, inhibited contraction of rabbit aortic rings and porcine coronary artery rings and promoted vasorelaxation in arteries isolated from nitrate-tolerant rabbits (Stasch and Hobbs 2009). Riociguat also decreased the acute pulmonary vasoconstriction induced by hypoxia in isolated mouse lungs (Schermuly et al. 2008).

5.2.2 Studies in Experimental Models of Cardiopulmonary Diseases

The vasodilatory potency of riociguat has also been demonstrated in animal models of PH (Table 1). Daily oral administration of riociguat in rats with monocrotaline-induced PH and mice with hypoxia-induced PH attenuated PH, RV hypertrophy and pulmonary vascular remodelling without a significant effect on systemic arterial pressure (Schermuly et al. 2008). In addition, riociguat effectively suppressed pulmonary vascular remodelling in other animal models; in rats with severe PH, it increased the open-to-occluded artery ratio and decreased the neointima-to-media ratio, and RV systolic pressure was reduced compared with placebo (Lang et al. 2012). A potential mechanism for the suppression of pulmonary vascular remodelling by riociguat is via its activation of PKG isotype I, which significantly increased Smad1/5 phosphorylation and decreased the medial thickness of pulmonary vascular lesions in a rat model of PAH (Schwappacher et al. 2013). Riociguat decreased RV hypertrophy and improved RV function to a greater extent than sildenafil in rats with severe PH, and decreased pulmonary vascular resistance and increased cardiac output compared with placebo (Lang et al. 2012). Riociguat was also more effective than sildenafil in attenuating PH and RV hypertrophy in mice with bleomycin-induced PH, and reduced their mortality compared with vehicle-treated mice (Evgenov et al. 2011). Moreover, riociguat treatment prevented tobacco smoke-induced PH, RV hypertrophy and vascular remodelling in a mouse model of COPD (Seimetz et al. 2012).

Riociguat has beneficial effects on other pathological processes. For example, a daily oral administration of riociguat improved survival, normalised blood pressure and reduced heart and kidney damage in rat models of moderate and malignant hypertension (Sharkovska et al. 2010). Riociguat also improved survival and attenuated systemic hypertension, systolic dysfunction and fibrotic tissue remodelling and degeneration in the myocardium and renal cortex in another rat model of malignant systemic hypertension (Geschka et al. 2011). Administration of riociguat in combination with an angiotensin II receptor blocker reduced blood pressure in diabetic eNOS-knockout mice and attenuated nephropathy, as evidenced by reduced urinary albumin excretion (Ott et al. 2012). Finally, riociguat treatment ameliorated pulmonary inflammation and fibrosis in bleomycin-exposed mice (Evgenov et al. 2011).

5.3 Clinical Trials

Riociguat is the first sGC stimulator that has entered clinical development. Its dual mode of action (i.e. synergy with endogenous NO and NO-independent stimulation of sGC) restores NO–sGC–cGMP signalling, resulting in significant improvements in pulmonary vascular haemodynamics and increased exercise ability in patients with PH.

A summary of clinical studies of riociguat is given in Table 2. In a phase 1 clinical trial, a single dose of riociguat decreased mean arterial and diastolic pressures and increased heart rate compared with placebo in healthy male volunteers (Frey et al. 2008). Systolic pressure was not affected and no serious adverse events were reported. Riociguat was readily absorbed and demonstrated dose-proportional pharmacokinetics in this study. In another phase 1 study, the drug showed complete oral absorption in healthy men and a dose-proportional increase in systemic exposure (Becker et al. 2012). Riociguat was well tolerated and exhibited no clinically relevant food effects, or pharmacodynamic or pharmacokinetic interactions with aspirin or warfarin in healthy volunteers (Becker et al. 2012; Frey et al. 2011, 2013b). In addition, two exploratory studies have concluded that the safety profile of riociguat is similar in individuals with and without hepatic impairment. These studies also showed that riociguat exposure was significantly increased in patients with moderate, but not mild, hepatic impairment, compared with healthy individuals (Frey et al. 2013a).

Table 2 Clinical studies of riociguat in pulmonary hypertension

Study design	Treatment	Effects of sGC stimulators	References
Randomised, placebo-controlled, single-blinded, parallel-group, single-dose trial – 58 healthy male volunteers	Single dose of riociguat 0.25–5 mg	• Well tolerated with a favourable safety profile • Slight but statistically significant decreases in MAP and DBP, but not SBP	Frey et al. (2008)
Uncontrolled, open-label, single-dose trial – 12 patients with PAH – 6 patients with CTEPH – 1 patient with PH-ILD	Single dose of riociguat, 1 or 2.5 mg	• Favourable safety profile • Improvements in all major pulmonary haemodynamic parameters • Mean SBP remained above 110 mmHg	Grimminger et al. (2009)
Uncontrolled, open-label, 12-week trial – 42 patients with CTEPH – 33 patients with PAH	Dose titration of riociguat from 1 mg t.i.d. to a maximum of 2.5 mg t.i.d. according to SBP and tolerability	• Improved pulmonary haemodynamics • Improved exercise capacity	Ghofrani et al. (2010b)
Uncontrolled, open-label, long-term extension ($\leq$24 months) of 12-week trial – 41 patients with CTEPH – 27 patients with PAH	Riociguat 0.5–2.5 mg t.i.d. for at least 15 months	• Sustained improvements in exercise capacity and functional class for at least 15 months (interim analysis)	Ghofrani et al. (2010a)
Single-centre, randomised, double-blind, placebo-controlled, crossover warfarin interaction study – 22 healthy male volunteers	The addition of a single dose of warfarin sodium 25 mg to riociguat 2.5 mg t.i.d. (10 days)	• Favourable safety profile • No pharmacodynamic interactions and no clinically relevant pharmacokinetic interactions with warfarin	Frey et al. (2011)

Three open-label, randomised, cross-over studies
– Healthy male volunteers
  (a) 22 for bioavailability study
  (b) 24 for food effect and dose-proportionality studies
	Riociguat 0.5–2.5 mg	• Complete oral absorption
		• Dose-proportional increase in systemic exposure
		• No clinically relevant food effect
		• Well tolerated
			Becker et al. (2012)

(continued)

Table 2 (continued)

Study design	Treatment	Effects of sGC stimulators	References
Randomised, open-label, crossover aspirin interaction study – 15 healthy male volunteers	Single dose of riociguat 2.5 mg and two doses of aspirin 500 mg/day	• No clinically relevant pharmacodynamic or pharmacokinetic interactions with aspirin	Frey et al. (2013b)
Two non-randomised, non-blinded exploratory studies with group stratification – 32 patients with liver cirrhosis (16, Child-Pugh A and 16, Child-Pugh B)	Single oral dose of 1 mg riociguat	• Safety profile of riociguat similar in individuals with and without hepatic impairment • Exposure to riociguat unchanged in patients with Child-Pugh A (mild) hepatic impairment • Exposure to riociguat significantly increased in patients with Child-Pugh B (moderate) hepatic impairment	Frey et al. (2013a)
Uncontrolled, open-label, 12-week trial – 22 patients with PH-ILD	Dose titration of riociguat 1.0–2.5 mg t.i.d. continued for 12 weeks	• Substantial reduction in pulmonary vascular resistance • Increases in cardiac output and 6-min walk distance	Hoeper et al. (2013)
Uncontrolled, open-label, two-dose trial – 22 patients with PH-COPD	Single dose of riociguat, 1 or 2.5 mg	• Decreased mean pulmonary arterial pressure • Decreased pulmonary vascular resistance	Ghofrani et al. (2013c)
PATENT-1 trial Phase 3, randomised, placebo-controlled, double-blind 12-week trial – 443 patients with PAH	Dose titration of riociguat 0.5–2.5 mg t.i.d., continued for 12 weeks	• Increased 6-min walk distance • Improved pulmonary vascular resistance, functional class, time to clinical worsening and Borg dyspnoea score • Well tolerated	Ghofrani et al. (2013a)
PATENT-2 Phase 3, open-label, long-term extension study – 363 patients with PAH who had completed PATENT-1	Dose titration of riociguat 0.5–2.5 mg t.i.d., continued for 1 year	• Improved 6-min walk distance and functional class • Well tolerated	Rubin et al. (2013)

(continued)

Table 2 (continued)

Study design	Treatment	Effects of sGC stimulators	References
PATENT PLUS Phase 2, placebo-controlled, double-blind interaction 12-week and long-term extension study – 18 patients with symptomatic PAH receiving stable sildenafil (i.e. ≥90 days)	Dose titration of riociguat 0.5–2.5 mg t.i.d., continued for 12 weeks and long-term extension	• Similar changes in blood pressure after 12 weeks of riociguat or placebo • Combined riociguat and sildenafil therapy associated with high rate of discontinuation	Galiè et al. (2013)
CHEST-1 trial Phase 3, randomised, placebo-controlled, double-blind 16-week trial – 261 patients with inoperable or persistent CTEPH	Dose titration of riociguat 0.5–2.5 mg t.i.d., continued for 16 weeks	• Increased 6-min walk distance and cardiac index • Improved pulmonary vascular resistance and functional class • Reduced right atrial pressure and mean pulmonary arterial pressure • Positive trend in time to clinical worsening, Borg dyspnoea score and quality of life • Well tolerated	Ghofrani et al. (2013b), Tse (2012), Bayer (2012), Kim et al. (2013)
LEPHT trial Randomised, placebo-controlled, double-blind dose-finding 16-week trial – 210 patients with PH associated with left ventricular systolic dysfunction	Dose titration of riociguat 0.5–2 mg t.i.d., continued for 16 weeks	• Improved cardiac index, pulmonary and systemic vascular resistance and quality of life • No significant changes in mean pulmonary arterial pressure and systemic blood pressure • Well tolerated	Bonderman et al. (2013)

COPD chronic obstructive pulmonary disease, *CTEPH* chronic thromboembolic pulmonary hypertension, *DBP* diastolic blood pressure, *MAP* mean arterial pressure, *PAH* pulmonary arterial hypertension, *PH-COPD* pulmonary hypertension due to chronic obstructive pulmonary disease, *PH-ILD* pulmonary hypertension due to interstitial lung disease, *SBP* systolic blood pressure, *sGC* soluble guanylate cyclase, *t.i.d.* three times daily

Several phase 2 clinical trials assessing the safety and efficacy of riociguat in the treatment of different types of PH have been completed. The results of two large phase 3 placebo-controlled trials, PATENT-1 and CHEST-1, investigating the efficacy and safety of oral riociguat in patients with PAH and CTEPH, respectively, have also been presented (Ghofrani et al. 2013a, b).

5.3.1 Pulmonary Arterial Hypertension

In a first study of 19 patients with moderate-to-severe PH, including 12 patients with PAH, single doses of riociguat (2.5 or 1 mg) had a favourable safety profile (Grimminger et al. 2009). Riociguat significantly reduced mean pulmonary arterial pressure and pulmonary vascular resistance and increased cardiac index in a dose-dependent manner compared with baseline, and to a greater extent than inhaled NO. The drug had no effect on gas exchange or ventilation–perfusion matching. Riociguat also exerted dose-dependent systemic effects in these patients, but mean systolic blood pressure remained above 110 mmHg.

In a follow-up study of 33 patients with PAH, treatment with riociguat for 12 weeks had a favourable safety profile and improved their symptoms, pulmonary haemodynamics (e.g. reduced pulmonary vascular resistance) and exercise capacity [as demonstrated by an increased 6-min walk distance (6MWD)], compared with baseline (Ghofrani et al. 2010b). The improvements from baseline in 6MWD and functional class in patients with PAH after 12 weeks' treatment with riociguat were sustained for at least a further 15 months. Long-term riociguat treatment was also well tolerated by these patients (Ghofrani et al. 2010a).

In the phase 3 PATENT-1 trial, riociguat significantly improved pulmonary haemodynamics, dyspnoea score and functional class, and increased exercise ability in patients with PAH (Ghofrani et al. 2013a). In this trial, 443 patients with symptomatic PAH were randomised to receive placebo or riociguat for 12 weeks. Riociguat significantly increased 6MWD from baseline compared with placebo both in patients receiving endothelin receptor antagonists (ERAs) or inhaled prostanoids and in treatment-naïve patients (Ghofrani et al. 2013a). A number of other clinically relevant measures of efficacy were also consistently improved by riociguat compared with placebo, including pulmonary vascular resistance, N-terminal prohormone of brain natriuretic peptide (NT-proBNP) levels, functional class, time to clinical worsening and Borg dyspnoea score. Riociguat was well tolerated and had a favourable safety profile in this study, both as a monotherapy and in combination with ERAs or inhaled prostanoids (Ghofrani et al. 2013a). In the PATENT-2 long-term extension study, riociguat treatment for an additional year was also well tolerated and improved 6MWD and functional class compared with baseline (Rubin et al. 2013).

In the PATENT PLUS interaction study, administration of riociguat to patients with symptomatic PAH receiving stable sildenafil therapy resulted in similar changes in blood pressure as placebo after 12 weeks. However, in a long-term extension of this study, the combined administration of riociguat and sildenafil was associated with a high rate of discontinuation due to hypotension and there was no evidence that the combination exerted any favourable clinical effect (Galiè et al. 2013).

5.3.2 Chronic Thromboembolic Pulmonary Hypertension

The first study of riociguat in patients with PH included six patients with CTEPH (Grimminger et al. 2009). Results from the overall study population are described in the previous section. A sub-analysis of the patients with CTEPH showed a significant improvement in cardiac index following administration of riociguat compared with NO inhalation.

The outcome of 12 weeks' treatment with riociguat in a group of 42 patients with CTEPH was similar to that observed in patients with PAH; their symptoms and pulmonary haemodynamics (including pulmonary vascular resistance) improved and their exercise capacity increased compared with baseline (Ghofrani et al. 2010b). These beneficial effects persisted for at least 15 months and were associated with a favourable safety profile (Ghofrani et al. 2010a).

Preliminary results of the phase 3 CHEST-1 trial have demonstrated that riociguat benefits patients with CTEPH by improving pulmonary haemodynamics, dyspnoea score and functional class and increasing 6MWD and cardiac index (Bayer 2012; Ghofrani et al. 2013b; Kim et al. 2013). A total of 261 patients with inoperable, persistent or recurrent CTEPH were randomised to receive placebo or riociguat for 16 weeks (Ghofrani et al. 2013b). Riociguat treatment resulted in a significant increase in mean 6MWD of 46 m from baseline compared with placebo (Ghofrani et al. 2013b; Tse 2012).

5.3.3 Pulmonary Hypertension Associated with Interstitial Lung Disease

After 12 weeks' treatment with riociguat, a slight improvement in 6MWD, a reduction in pulmonary vascular resistance and an increase in cardiac output were observed in patients with pulmonary hypertension associated with interstitial lung disease (PH-ILD) compared with baseline. Riociguat was well tolerated by most of these patients (Hoeper et al. 2013).

5.3.4 Pulmonary Hypertension Due to Chronic Obstructive Pulmonary Disease

Riociguat administered to patients with PH due to COPD decreased mean pulmonary arterial pressure and pulmonary vascular resistance compared with baseline. Selected parameters of lung function were slightly improved compared with baseline, whereas gas exchange was not significantly affected in these patients (Ghofrani et al. 2013c). The maintenance of gas exchange suggests that the vasodilation induced by riociguat did not cause ventilation–perfusion mismatch in these patients.

5.3.5 Pulmonary Hypertension Associated with Left Ventricular Systolic Dysfunction

Riociguat has also been shown to provide significant benefits to patients with PH associated with left ventricular systolic dysfunction already receiving optimised therapy for heart failure (Semigran et al. 2012). In the LEPHT trial, 210 patients with chronic systolic heart failure and PH underwent right heart catheterisation before and after receiving placebo or riociguat for 16 weeks (Bonderman et al. 2013). Riociguat treatment was well tolerated and improved cardiac index, pulmonary and systemic vascular resistance and quality of life in these patients, without significantly changing mean pulmonary arterial pressure or systemic blood pressure compared with placebo (Ghio et al. 2012; Semigran et al. 2012).

5.4 Ongoing Clinical Trials

The results of several clinical trials of riociguat are awaited. CHEST-2 (NCT00910429) is a long-term, open-label extension of the phase 3 trial CHEST-1 trial. Other trials include a phase 2 study of riociguat in patients with PH-ILD (NCT00694850) and a phase 2 DILATE trial (NCT01172756) assessing the haemodynamic effects and safety of riociguat in patients with PH associated with left ventricular diastolic dysfunction.

5.5 Future Directions

Stimulators of sGC may have therapeutic applications in diseases other than PH. For example the restoration of NO–sGC–cGMP signalling and the favourable haemodynamic effects and cardiac protective properties of sGC stimulators may be beneficial for patients with heart failure (Gheorghiade et al. 2013; Mitrovic et al. 2009; Stasch and Hobbs 2009). In particular, the antifibrotic properties of sGC stimulators may benefit these patients. This possibility is supported by the results of experiments in a mouse model of chronic RV pressure overload in which riociguat treatment reduced the collagen content of the right ventricle and improved the RV ejection fraction (Schymura et al. 2012).

New evidence indicates that sGC stimulators may constitute a potential new therapy for lung fibrosis. This includes the attenuation of pulmonary fibrosis by orally administered riociguat in a mouse model, and the inhibition of the conversion of lung fibroblasts to myofibroblasts by BAY 41-2272 in combination with sildenafil (Dunkern et al. 2007; Evgenov et al. 2011). In addition, the antifibrotic activity and vasoactive properties of sGC stimulators make them promising candidates for the treatment of scleroderma. It has recently been discovered that sGC stimulators exert potent antifibrotic effects in experimental models of systemic sclerosis (Beyer et al. 2012).

NO production in the airways is reduced in patients with COPD (Clini et al. 2000). The use of sGC stimulators may therefore have beneficial effects beyond the alleviation of PH in these patients. Support for this hypothesis is provided by experiments in mice that showed that riociguat prevented the development of tobacco smoke-induced emphysema (Seimetz et al. 2011; Stasch et al. 2011). In similar experiments in guinea pigs, BAY 41-2272 prevented both tobacco smoke-induced emphysema and pulmonary vascular remodelling (Lobo et al. 2013).

Sickle cell disease is another condition that may be amenable to sGC stimulator therapy. NO bioavailability is reduced in this disease due to oxidative stress and the sequestration of NO by free haemoglobin released into the plasma following erythrocyte haemolysis. It has been suggested that sGC stimulators may be able to bypass NO scavenging by free haemoglobin in sickle cell disease (as well as in other haemolytic diseases and during the transfusion of aged blood or haemoglobin-based oxygen carriers) to reverse haemoglobin-mediated vasoconstriction (Raat et al. 2013). It has also been shown that the increased adhesion of neutrophils, which contributes to veno-occlusion in sickle cell disease, is reduced by BAY 41-2272 (Canalli et al. 2008; Miguel et al. 2011). Furthermore, in a murine model of sickle cell disease, priapism associated with dysregulation of NO–sGC–cGMP signalling may be ameliorated by treatment with sGC stimulators (Claudino et al. 2009).

Disrupted NO–sGC–cGMP signalling is a common pathogenic feature of many cardiovascular diseases. Although clinical studies of sGC stimulators are currently focusing on the treatment of PH, their therapeutic potential has been, and continues to be, explored in a wide range of experimental disease models. In addition, research is in progress to identify and optimise new compounds in this drug class for the treatment of a number of specific diseases.

References

Ahrens I, Habersberger J, Baumlin N, Qian H, Smith BK, Stasch JP, Bode C, Schmidt HH, Peter K (2011) Measuring oxidative burden and predicting pharmacological response in coronary artery disease patients with a novel direct activator of haem-free/oxidised sGC. Atherosclerosis 218:431–434

Arnold WP, Mittal CK, Katsuki S, Murad F (1977) Nitric oxide activates guanylate cyclase and increases guanosine $3':5'$-cyclic monophosphate levels in various tissue preparations. Proc Natl Acad Sci USA 74:3203–3207

Aytekin M, Aulak KS, Haserodt S, Chakravarti R, Cody J, Minai OA, Dweik RA (2012) Abnormal platelet aggregation in idiopathic pulmonary arterial hypertension: role of nitric oxide. Am J Physiol Lung Cell Mol Physiol 302:L512–L520

Badejo AM Jr, Nossaman VE, Pankey EA, Bhartiya M, Kannadka CB, Murthy SN, Nossaman BD, Kadowitz PJ (2010) Pulmonary and systemic vasodilator responses to the soluble guanylyl cyclase stimulator, BAY 41-8543, are modulated by nitric oxide. Am J Physiol Heart Circ Physiol 299:H1153–H1159

Bayer (2012) Bayer's riociguat first drug to demonstrate efficacy in patients with chronic thromboembolic pulmonary hypertension. Bayer Investor News, Leverkusen, Germany, 23 Oct 2012

Bayer (2013) Bayer HealthCare Development Pipeline. http://www.bayerpharma.com/en/research-and-development/development-pipeline/index.php?phase=1. Accessed 4 June 2013

Becker EM, Stasch JP, Bechem M, Truebel H (2011) Comparison of different vasodilators, endothelin antagonist, PDE5 inhibitor and sGC stimulators in an animal model of secondary pulmonary hypertension: effects on "desaturation". BMC Pharmacol 11(Suppl 1):P5 (abstract)

Becker C, Frey R, Hesse C, Unger S, Reber M, Mück W (2012) Absorption behavior of riociguat (BAY 632521): bioavailability, food effects, and dose-proportionality. Eur Respir Soc P951 (abstract)

Beyer C, Reich N, Schindler SC, Akhmetshina A, Dees C, Tomcik M, Hirth-Dietrich C, von Degenfeld G, Sandner P, Distler O et al (2012) Stimulation of soluble guanylate cyclase reduces experimental dermal fibrosis. Ann Rheum Dis 71:1019–1026

Bischoff E, Stasch JP (2004) Effects of the sGC stimulator BAY 41-2272 are not mediated by phosphodiesterase 5 inhibition. Circulation 110:e320–e321

Black SM, Sanchez LS, Mata-Greenwood E, Bekker JM, Steinhorn RH, Fineman JR (2001) sGC and PDE5 are elevated in lambs with increased pulmonary blood flow and pulmonary hypertension. Am J Physiol Lung Cell Mol Physiol 281:L1051–L1057

Bonderman D, Ghio S, Felix SB, Ghofrani HA, Michelakis E, Mitrovic V, Oudiz RJ, Boateng F, Scalise AV, Roessig L, Semigran MJ (2013) Left ventricular systolic dysfunction associated with pulmonary hypertension riociguat trial (LEPHT) study group. Circulation 128(5):502–511

Bowers R, Cool C, Murphy RC, Tuder RM, Hopken MW, Flores SC, Voelkel NF (2004) Oxidative stress in severe pulmonary hypertension. Am J Respir Crit Care Med 169:764–769

Brockunier LL, Guo J, Parmee ER, Raghavan S, Rosauer K, Stelmach JE, Schmidt DR (2009) Soluble guanylate cyclase activators. Patent application number PCT/US2009/064570, Merck Sharp & Dohme Corporation

Bryan NS, Bian K, Murad F (2009) Discovery of the nitric oxide signaling pathway and targets for drug development. Front Biosci 14:1–18

Burney S, Caulfield JL, Niles JC, Wishnok JS, Tannenbaum SR (1999) The chemistry of DNA damage from nitric oxide and peroxynitrite. Mutat Res 424:37–49

Canalli AA, Franco-Penteado CF, Saad ST, Conran N, Costa FF (2008) Increased adhesive properties of neutrophils in sickle cell disease may be reversed by pharmacological nitric oxide donation. Haematologica 93:605–609

Cary SP, Winger JA, Marletta MA (2005) Tonic and acute nitric oxide signaling through soluble guanylate cyclase is mediated by nonheme nitric oxide, ATP, and GTP. Proc Natl Acad Sci USA 102:13064–13069

Cau SB, Dias-Junior CA, Montenegro MF, de Nucci G, Antunes E, Tanus-Santos JE (2008) Dose-dependent beneficial hemodynamic effects of BAY 41-2272 in a canine model of acute pulmonary thromboembolism. Eur J Pharmacol 581:132–137

Cetin A, Kaya T, Demirkoprulu N, Karadas B, Duran B, Cetin M (2004) YC-1, a nitric oxide-independent activator of soluble guanylate cyclase, inhibits the spontaneous contractions of isolated pregnant rat myometrium. J Pharmacol Sci 94:19–24

Claudino MA, Franco-Penteado CF, Corat MA, Gimenes AP, Passos LA, Antunes E, Costa FF (2009) Increased cavernosal relaxations in sickle cell mice priapism are associated with alterations in the NO-cGMP signaling pathway. J Sex Med 6:2187–2196

Clini E, Cremona G, Campana M, Scotti C, Pagani M, Bianchi L, Giordano A, Ambrosino N (2000) Production of endogenous nitric oxide in chronic obstructive pulmonary disease and patients with cor pulmonale. Correlates with echo-Doppler assessment. Am J Respir Crit Care Med 162:446–450

Coggins MP, Bloch KD (2007) Nitric oxide in the pulmonary vasculature. Arterioscler Thromb Vasc Biol 27:1877–1885

Costell MH, Ancellin N, Bernard RE, Zhao S, Upson JJ, Morgan LA, Maniscalco K, Olzinski AR, Ballard VL, Herry K et al (2012) Comparison of soluble guanylate cyclase stimulators and activators in models of cardiovascular disease associated with oxidative stress. Front Pharmacol 3:128

Cracowski JL, Cracowski C, Bessard G, Pepin JL, Bessard J, Schwebel C, Stanke-Labesque F, Pison C (2001) Increased lipid peroxidation in patients with pulmonary hypertension. Am J Respir Crit Care Med 164:1038–1042

Cremona G, Higenbottam T, Borland C, Mist B (1994) Mixed expired nitric oxide in primary pulmonary hypertension in relation to lung diffusion capacity. QJM 87:547–551

Derbyshire ER, Marletta MA (2012) Structure and regulation of soluble guanylate cyclase. Annu Rev Biochem 81:533–559

Deruelle P, Grover TR, Abman SH (2005a) Pulmonary vascular effects of nitric oxide-cGMP augmentation in a model of chronic pulmonary hypertension in fetal and neonatal sheep. Am J Physiol Lung Cell Mol Physiol 289:L798–L806

Deruelle P, Grover TR, Storme L, Abman SH (2005b) Effects of BAY 41-2272, a soluble guanylate cyclase activator, on pulmonary vascular reactivity in the ovine fetus. Am J Physiol Lung Cell Mol Physiol 288:L727–L733

Deruelle P, Balasubramaniam V, Kunig AM, Seedorf GJ, Markham NE, Abman SH (2006) BAY 41-2272, a direct activator of soluble guanylate cyclase, reduces right ventricular hypertrophy and prevents pulmonary vascular remodeling during chronic hypoxia in neonatal rats. Biol Neonate 90:135–144

Dumitrascu R, Weissmann N, Ghofrani HA, Dony E, Beuerlein K, Schmidt H, Stasch JP, Gnoth MJ, Seeger W, Grimminger F et al (2006) Activation of soluble guanylate cyclase reverses experimental pulmonary hypertension and vascular remodeling. Circulation 113:286–295

Dunkern TR, Feurstein D, Rossi GA, Sabatini F, Hatzelmann A (2007) Inhibition of TGF-beta induced lung fibroblast to myofibroblast conversion by phosphodiesterase inhibiting drugs and activators of soluble guanylyl cyclase. Eur J Pharmacol 572:12–22

Egemnazarov B, Sydykov A, Schermuly RT, Weissmann N, Stasch JP, Sarybaev AS, Seeger W, Grimminger F, Ghofrani HA (2009) Novel soluble guanylyl cyclase stimulator BAY 41-2272 attenuates ischemia-reperfusion-induced lung injury. Am J Physiol Lung Cell Mol Physiol 296:L462–L469

Egemnazarov B, Amirjanians V, Kojonazarov B, Sydykov A, Stasch JP, Weissmann N, Grimminger F, Seeger W, Schermuly RT, Ghofrani HA (2010) Inhalative application of soluble guanylyl cyclase stimulator BAY 41-8543 for treatment of pulmonary arterial hypertension. Am J Respir Crit Care Med 181:A6307 (abstract)

Ehret GB, Munroe PB, Rice KM, Bochud M, Johnson AD, Chasman DI, Smith AV, Tobin MD, Verwoert GC, Hwang SJ et al (2011) Genetic variants in novel pathways influence blood pressure and cardiovascular disease risk. Nature 478:103–109

Erusalimsky JD, Moncada S (2007) Nitric oxide and mitochondrial signaling: from physiology to pathophysiology. Arterioscler Thromb Vasc Biol 27:2524–2531

Evgenov OV, Ichinose F, Evgenov NV, Gnoth MJ, Falkowski GE, Chang Y, Bloch KD, Zapol WM (2004) Soluble guanylate cyclase activator reverses acute pulmonary hypertension and augments the pulmonary vasodilator response to inhaled nitric oxide in awake lambs. Circulation 110:2253–2259

Evgenov OV, Pacher P, Schmidt PM, Hasko G, Schmidt HH, Stasch JP (2006) NO-independent stimulators and activators of soluble guanylate cyclase: discovery and therapeutic potential. Nat Rev Drug Discov 5:755–768

Evgenov OV, Kohane DS, Bloch KD, Stasch JP, Volpato GP, Bellas E, Evgenov NV, Buys ES, Gnoth MJ, Graveline AR et al (2007) Inhaled agonists of soluble guanylate cyclase induce selective pulmonary vasodilation. Am J Respir Crit Care Med 176:1138–1145

Evgenov OV, Zou L, Zhang M, Mino-Kenudson M, Mark EJ, Buys ES, Raher MJ, Li Y, Feng Y, Jones RC et al (2011) Nitric oxide-independent stimulation of soluble guanylate cyclase attenuates pulmonary fibrosis. BMC Pharmacol 11(Suppl 1):O9 (abstract)

Farrow KN, Lee KJ, Perez M, Schriewer JM, Wedgwood S, Lakshminrusimha S, Smith CL, Steinhorn RH, Schumacker PT (2012) Brief hyperoxia increases mitochondrial oxidation and increases phosphodiesterase 5 activity in fetal pulmonary artery smooth muscle cells. Antioxid Redox Signal 17:460–470

Freitas CF, Morganti RP, Annichino-Bizzacchi JM, De Nucci G, Antunes E (2007) Effect of BAY 41-2272 in the pulmonary hypertension induced by heparin-protamine complex in anaesthetized dogs. Clin Exp Pharmacol Physiol 34:10–14

Frey R, Mück W, Unger S, Artmeier-Brandt U, Weimann G, Wensing G (2008) Single-dose pharmacokinetics, pharmacodynamics, tolerability, and safety of the soluble guanylate cyclase stimulator BAY 63-2521: an ascending-dose study in healthy male volunteers. J Clin Pharmacol 48:926–934

Frey R, Mück W, Kirschbaum N, Krätzschmar J, Weimann G, Wensing G (2011) Riociguat (BAY 63-2521) and warfarin: a pharmacodynamic and pharmacokinetic interaction study. J Clin Pharmacol 51:1051–1060

Frey R, Becker C, Unger S, Schmidt A, Wensing G, Mück W (2013a) Pharmacokinetics of the soluble guanylate cyclase stimulator riociguat in individuals with hepatic impairment. Am J Respir Crit Care Med 187:A3310 (abstract)

Frey R, Mück W, Unger S, Reber M, Krätzschmar J, Wensing G (2013b) Riociguat (BAY 63-2521) and aspirin: a pharmacodynamic and pharmacokinetic interaction study. J Clin Pharmacol (in submission)

Friebe A, Koesling D (1998) Mechanism of YC-1-induced activation of soluble guanylyl cyclase. Mol Pharmacol 53:123–127

Friebe A, Schultz G, Koesling D (1996) Sensitizing soluble guanylyl cyclase to become a highly CO-sensitive enzyme. EMBO J 15:6863–6868

Friebe A, Mullershausen F, Smolenski A, Walter U, Schultz G, Koesling D (1998) YC-1 potentiates nitric oxide- and carbon monoxide-induced cyclic GMP effects in human platelets. Mol Pharmacol 54:962–967

Friebe A, Russwurm M, Mergia E, Koesling D (1999) A point-mutated guanylyl cyclase with features of the YC-1-stimulated enzyme: implications for the YC-1 binding site? Biochemistry 38:15253–15257

Friebe A, Mergia E, Dangel O, Lange A, Koesling D (2007) Fatal gastrointestinal obstruction and hypertension in mice lacking nitric oxide-sensitive guanylyl cyclase. Proc Natl Acad Sci USA 104:7699–7704

Galiè N, Neuser D, Muller K, Scalise A, Grunig E (2013) A placebo-controlled, double-blind phase II interaction study to evaluate blood pressure following addition of riociguat to patients with symptomatic pulmonary arterial hypertension (PAH) receiving sildenafil (PATENT PLUS). Am J Respir Crit Care Med 187:A3530 (abstract)

Galle J, Zabel U, Hubner U, Hatzelmann A, Wagner B, Wanner C, Schmidt HH (1999) Effects of the soluble guanylyl cyclase activator, YC-1, on vascular tone, cyclic GMP levels and phosphodiesterase activity. Br J Pharmacol 127:195–203

Garg UC, Hassid A (1989) Nitric oxide-generating vasodilators and 8-bromo-cyclic guanosine monophosphate inhibit mitogenesis and proliferation of cultured rat vascular smooth muscle cells. J Clin Invest 83:1774–1777

Geschka S, Kretschmer A, Sharkovska Y, Evgenov OV, Lawrenz B, Hucke A, Hocher B, Stasch JP (2011) Soluble guanylate cyclase stimulation prevents fibrotic tissue remodeling and improves survival in salt-sensitive Dahl rats. PLoS One 6:e21853

Gheorghiade M, Marti CN, Sabbah HN, Roessig L, Greene SJ, Bohm M, Burnett JC, Campia U, Cleland JG, Collins SP et al (2013) Soluble guanylate cyclase: a potential therapeutic target for heart failure. Heart Fail Rev 18:123–134

Ghio S, Bonderman D, Felix SB, Ghofrani HA, Michelakis ED, Mitrovic V, Oudiz RJ, Frey R, Roessig L, Semigran MJ (2012) Left ventricular systolic dysfunction associated with pulmonary hypertension riociguat trial (LEPHT): rationale and design. Eur J Heart Fail 14:946–953

Ghofrani HA, Hoeper MM, Halank M, Meyer FJ, Staehler G, Behr J, Ewert R, Binnen T, Weimann G, Grimminger F (2010a) Riociguat for chronic thromboembolic pulmonary hypertension and pulmonary arterial hypertension: first long-term extension data from a phase II study. Am J Respir Crit Care Med 181:A6770 (abstract)

Ghofrani HA, Hoeper MM, Halank M, Meyer FJ, Staehler G, Behr J, Ewert R, Weimann G, Grimminger F (2010b) Riociguat for chronic thromboembolic pulmonary hypertension and pulmonary arterial hypertension: a phase II study. Eur Respir J 36:792–799

Ghofrani HA, Galiè N, Grimminger F, Grüning E, Humbert M, Jing ZC, Keogh A, Langleben D, Ochan Kilama M, Fritsch A, Neuser D, Rubin L (2013a) Riociguat for the treatment of pulmonary arterial hypertension. N Engl J Med 369:330–340

Ghofrani HA, D'Armini AM, Grimminger F, Hoeper M, Jansa P, Kim NH, Mayer E, Simonneau G, Wilkins M, Fritsch A, Neuser D, Weimann G, Wang C (2013b) Riociguat for the treatment of chronic thromboembolic pulmonary hypertension. N Engl J Med 369:319–329

Ghofrani HA, Stähler G, Grünig E, Halank M, Mitrovic V, Unger S, Mück W, Frey R, Grimminger F, Schermuly RT et al (2013c) Riociguat in pulmonary hypertension associated with chronic obstructive pulmonary disease. Eur Resp J (in submission)

Giaid A, Saleh D (1995) Reduced expression of endothelial nitric oxide synthase in the lungs of patients with pulmonary hypertension. N Engl J Med 333:214–221

Girgis RE, Champion HC, Diette GB, Johns RA, Permutt S, Sylvester JT (2005) Decreased exhaled nitric oxide in pulmonary arterial hypertension: response to bosentan therapy. Am J Respir Crit Care Med 172:352–357

Grimminger F, Weimann G, Frey R, Voswinckel R, Thamm M, Bolkow D, Weissmann N, Mück W, Unger S, Wensing G et al (2009) First acute haemodynamic study of soluble guanylate cyclase stimulator riociguat in pulmonary hypertension. Eur Respir J 33:785–792

Hirst DG, Robson T (2011) Nitric oxide physiology and pathology. Methods Mol Biol 704:1–13

Hoeper MM, Halank M, Wilkens H, Gunther A, Weimann G, Gebert I, Leuchte HH, Behr J (2013) Riociguat for interstitial lung disease and pulmonary hypertension: a pilot trial. Eur Respir J 41:853–860

Hoshikawa Y, Ono S, Suzuki S, Tanita T, Chida M, Song C, Noda M, Tabata T, Voelkel NF, Fujimura S (2001) Generation of oxidative stress contributes to the development of pulmonary hypertension induced by hypoxia. J Appl Physiol 90:1299–1306

Huh JW, Kim SY, Lee JH, Lee YS (2011) YC-1 attenuates hypoxia-induced pulmonary arterial hypertension in mice. Pulm Pharmacol Ther 24:638–646

Humbert P, Niroomand F, Fischer G, Mayer B, Koesling D, Hinsch KD, Gausepohl H, Frank R, Schultz G, Bohme E (1990) Purification of soluble guanylyl cyclase from bovine lung by a new immunoaffinity chromatographic method. Eur J Biochem 190:273–278

Hwang TL, Hung HW, Kao SH, Teng CM, Wu CC, Cheng SJ (2003a) Soluble guanylyl cyclase activator YC-1 inhibits human neutrophil functions through a cGMP-independent but cAMP-dependent pathway. Mol Pharmacol 64:1419–1427

Hwang TL, Wu CC, Guh JH, Teng CM (2003b) Potentiation of tumor necrosis factor-alpha expression by YC-1 in alveolar macrophages through a cyclic GMP-independent pathway. Biochem Pharmacol 66:149–156

Irvine JC, Ganthavee V, Love JE, Alexander AE, Horowitz JD, Stasch JP, Kemp-Harper BK, Ritchie RH (2012) The soluble guanylyl cyclase activator bay 58-2667 selectively limits cardiomyocyte hypertrophy. PLoS One 7:e44481

Joshi CN, Martin DN, Fox JC, Mendelev NN, Brown TA, Tulis DA (2011) The soluble guanylate cyclase stimulator BAY 41-2272 inhibits vascular smooth muscle growth through the cAMP-dependent protein kinase and cGMP-dependent protein kinase pathways. J Pharmacol Exp Ther 339:394–402

Kaneko FT, Arroliga AC, Dweik RA, Comhair SA, Laskowski D, Oppedisano R, Thomassen MJ, Erzurum SC (1998) Biochemical reaction products of nitric oxide as quantitative markers of primary pulmonary hypertension. Am J Respir Crit Care Med 158:917–923

Kawaguchi Y, Tochimoto A, Hara M, Kawamoto M, Sugiura T, Katsumata Y, Okada J, Kondo H, Okubo M, Kamatani N (2006) NOS2 polymorphisms associated with the susceptibility to pulmonary arterial hypertension with systemic sclerosis: contribution to the transcriptional activity. Arthritis Res Ther 8:R104

Kielstein JT, Bode-Boger SM, Hesse G, Martens-Lobenhoffer J, Takacs A, Fliser D, Hoeper MM (2005) Asymmetrical dimethylarginine in idiopathic pulmonary arterial hypertension. Arterioscler Thromb Vasc Biol 25:1414–1418

Kim NH, d'Armini A, Grunig E, Hoeper MM, Jansa P, Mayer E, Simonneau G, Torbicki A, Want C, Wilkins MR et al (2013) Hemodynamic assessment of patients with inoperable chronic thromboembolic pulmonary hypertension (CTEPH) in the phase III CHEST-1 study. Am J Respir Crit Care Med 187:A3529 (abstract)

Ko FN, Wu CC, Kuo SC, Lee FY, Teng CM (1994) YC-1, a novel activator of platelet guanylate cyclase. Blood 84:4226–4233

Krasuski RA, Warner JJ, Wang A, Harrison JK, Tapson VF, Bashore TM (2000) Inhaled nitric oxide selectively dilates pulmonary vasculature in adult patients with pulmonary hypertension, irrespective of etiology. J Am Coll Cardiol 36:2204–2211

Kronas N, Peters B, Goetz AE, Kubitz JC (2011) Inhaled and intravenous application of a stimulator of the soluble guanylate cyclase (BAY 41-8543) reduces pulmonary vascular resistance in a model of septic shock. BMC Pharmacol 11(Suppl 1):P41 (abstract)

Lamothe M, Chang FJ, Balashova N, Shirokov R, Beuve A (2004) Functional characterization of nitric oxide and YC-1 activation of soluble guanylyl cyclase: structural implication for the YC-1 binding site? Biochemistry 43:3039–3048

Lang M, Kojonazarov B, Tian X, Kalymbetov A, Weissmann N, Grimminger F, Kretschmer A, Stasch JP, Seeger W, Ghofrani HA et al (2012) The soluble guanylate cyclase stimulator riociguat ameliorates pulmonary hypertension induced by hypoxia and SU5416 in rats. PLoS One 7:e43433

Laursen JB, Somers M, Kurz S, McCann L, Warnholtz A, Freeman BA, Tarpey M, Fukai T, Harrison DG (2001) Endothelial regulation of vasomotion in apoE-deficient mice: implications for interactions between peroxynitrite and tetrahydrobiopterin. Circulation 103:1282–1288

Lee YC, Martin E, Murad F (2000) Human recombinant soluble guanylyl cyclase: expression, purification, and regulation. Proc Natl Acad Sci USA 97:10763–10768

Lobo B, Puig-Pey R, Ferrer E, Dominguez-Fandos D, Coll N, Garcia J, Musri MM, Peinado VI, Barberà JA (2013) Stimulation of soluble guanylate cyclase in guinea pigs chronically exposed to cigarette smoke reduces intrapulmonary vascular remodeling and prevents emphysema. Am J Respir Crit Care Med 187:A4666

Lundgren J, Kylhammar D, Hedelin P, Radegran G (2012) sGC stimulation totally reverses hypoxia-induced pulmonary vasoconstriction alone and combined with dual endothelin-receptor blockade in a porcine model. Acta Physiol (Oxf) 206:178–194

Maclean MR, Johnston ED, McCulloch KM, Pooley L, Houslay MD, Sweeney G (1997) Phosphodiesterase isoforms in the pulmonary arterial circulation of the rat: changes in pulmonary hypertension. J Pharmacol Exp Ther 283:619–624

Malerba M, Radaeli A, Ragnoli B, Airo P, Corradi M, Ponticiello A, Zambruni A, Grassi V (2007) Exhaled nitric oxide levels in systemic sclerosis with and without pulmonary involvement. Chest 132:575–580

Mam V, Tanbe AF, Vitali SH, Arons E, Christou HA, Khalil RA (2010) Impaired vasoconstriction and nitric oxide-mediated relaxation in pulmonary arteries of hypoxia- and monocrotaline-induced pulmonary hypertensive rats. J Pharmacol Exp Ther 332:455–462

Mellion BT, Ignarro LJ, Ohlstein EH, Pontecorvo EG, Hyman AL, Kadowitz PJ (1981) Evidence for the inhibitory role of guanosine 3′, 5′-monophosphate in ADP-induced human platelet aggregation in the presence of nitric oxide and related vasodilators. Blood 57:946–955

Mercier O, Guihaire J, Boulate D, Nickl W, Truebel H (2012) sGC-stimulation vs. PDE5-inhibition in a model of chronic thromboembolic pulmonary hypertension (CTEPH). Am J Respir Crit Care Med 185:A4775 (abstract)

Miguel LI, Almeida CB, Traina F, Canalli AA, Dominical VM, Saad ST, Costa FF, Conran N (2011) Inhibition of phosphodiesterase 9A reduces cytokine-stimulated in vitro adhesion of neutrophils from sickle cell anemia individuals. Inflamm Res 60:633–642

Miller LN, Nakane M, Hsieh GC, Chang R, Kolasa T, Moreland RB, Brioni JD (2003) A-350619: a novel activator of soluble guanylyl cyclase. Life Sci 72:1015–1025

Mitrovic V, Swidnicki B, Ghofrani HA, Mück W, Kirschbaum N, Mittendorf J, Stasch JP, Wensing G, Frey R, Lentini S (2009) Acute hemodynamic response to single oral doses of BAY 60-4552, a soluble guanylate cyclase stimulator, in patients with biventricular heart failure. BMC Pharmacol 9(Suppl 1):P51 (abstract)

Mittendorf J, Weigand S, Alonso-Alija C, Bischoff E, Feurer A, Gerisch M, Kern A, Knorr A, Lang D, Muenter K et al (2009) Discovery of riociguat (BAY 63-2521): a potent, oral stimulator of soluble guanylate cyclase for the treatment of pulmonary hypertension. ChemMedChem 4:853–865

Mülsch A, Bauersachs J, Schafer A, Stasch JP, Kast R, Busse R (1997) Effect of YC-1, an NO-independent, superoxide-sensitive stimulator of soluble guanylyl cyclase, on smooth muscle responsiveness to nitrovasodilators. Br J Pharmacol 120:681–689

Münzel T, Daiber A, Mülsch A (2005) Explaining the phenomenon of nitrate tolerance. Circ Res 97:618–628

Murad F (2006) Shattuck Lecture. Nitric oxide and cyclic GMP in cell signaling and drug development. N Engl J Med 355:2003–2011

O'Reilly DA, McLaughlin BE, Marks GS, Brien JF, Nakatsu K (2001) YC-1 enhances the responsiveness of tolerant vascular smooth muscle to glyceryl trinitrate. Can J Physiol Pharmacol 79:43–48

Ott IM, Alter ML, Von Websky K, Kretschmer A, Tsuprykov O, Sharkovska Y, Krause-Relle K, Raila J, Henze A, Stasch JP et al (2012) Effects of stimulation of soluble guanylate cyclase on diabetic nephropathy in diabetic eNOS knockout mice on top of angiotensin II receptor blockade. PLoS One 7(8):e42623

Pan SL, Guh JH, Chang YL, Kuo SC, Lee FY, Teng CM (2004) YC-1 prevents sodium nitroprusside-mediated apoptosis in vascular smooth muscle cells. Cardiovasc Res 61:152–158

Pankey EA, Badejo AM, Casey DB, Lasker GF, Riehl RA, Murthy SN, Nossaman BD, Kadowitz PJ (2012) Effect of chronic sodium nitrite therapy on monocrotaline-induced pulmonary hypertension. Nitric Oxide 27:1–8

Perez M, Lakshminrusimha S, Wedgwood S, Czech L, Gugino SF, Russell JA, Farrow KN, Steinhorn RH (2012) Hydrocortisone normalizes oxygenation and cGMP regulation in lambs with persistent pulmonary hypertension of the newborn. Am J Physiol Lung Cell Mol Physiol 302:L595–L603

Pfeifer A, Klatt P, Massberg S, Ny L, Sausbier M, Hirneiss C, Wang GX, Korth M, Aszodi A, Andersson KE et al (1998) Defective smooth muscle regulation in cGMP kinase I-deficient mice. EMBO J 17:3045–3051

Raat NJ, Tabima DM, Specht PA, Tejero J, Champion HC, Kim-Shapiro D, Baust J, Mik EG, Hildesheim M, Stasch JP et al (2013) Direct sGC activation bypasses NO scavenging reactions of intravascular free oxy-hemoglobin and limits vasoconstriction. Antioxid Redox Signal. doi:10.1089/ars.2013.5181

Radi R (2004) Nitric oxide, oxidants, and protein tyrosine nitration. Proc Natl Acad Sci USA 101:4003–4008

Riley MS, Porszasz J, Miranda J, Engelen MP, Brundage B, Wasserman K (1997) Exhaled nitric oxide during exercise in primary pulmonary hypertension and pulmonary fibrosis. Chest 111:44–50

Rizzo NO, Maloney E, Pham M, Luttrell I, Wessells H, Tateya S, Daum G, Handa P, Schwartz MW, Kim F (2010) Reduced NO-cGMP signaling contributes to vascular inflammation and insulin resistance induced by high-fat feeding. Arterioscler Thromb Vasc Biol 30:758–765

Roger S, Badier-Commander C, Paysant J, Cordi A, Verbeuren TJ, Feletou M (2010) The antiaggregating effect of BAY 41-2272, a stimulator of soluble guanylyl cyclase, requires the presence of nitric oxide. Br J Pharmacol 161:1044–1058

Roy B, Mo E, Vernon J, Garthwaite J (2008) Probing the presence of the ligand-binding haem in cellular nitric oxide receptors. Br J Pharmacol 153:1495–1504

Rubin LJ, Galiè N, Grimminger F, Grunig E, Humbert MJC, Jing ZC, Keogh AM, Langleben D, Fritsch A, Ochan Kilama M et al (2013) Riociguat for the treatment of pulmonary arterial hypertension (PAH): a phase III long-term extension study (PATENT-2). Am J Respir Crit Care Med 187:A3531 (abstract)

Russwurm M, Mergia E, Mullershausen F, Koesling D (2002) Inhibition of deactivation of NO-sensitive guanylyl cyclase accounts for the sensitizing effect of YC-1. J Biol Chem 277:24883–24888

Salvi E, Kutalik Z, Glorioso N, Benaglio P, Frau F, Kuznetsova T, Arima H, Hoggart C, Tichet J, Nikitin YP et al (2012) Genomewide association study using a high-density single nucleotide polymorphism array and case-control design identifies a novel essential hypertension susceptibility locus in the promoter region of endothelial NO synthase. Hypertension 59:248–255

Sanli C, Oguz D, Olgunturk R, Tunaoglu FS, Kula S, Pasaoglu H, Gulbahar O, Cevik A (2012) Elevated homocysteine and asymmetric dimethyl arginine levels in pulmonary hypertension associated with congenital heart disease. Pediatr Cardiol 33:1323–1331

Sawada N, Itoh H, Miyashita K, Tsujimoto H, Sone M, Yamahara K, Arany ZP, Hofmann F, Nakao K (2009) Cyclic GMP kinase and RhoA Ser188 phosphorylation integrate pro- and antifibrotic signals in blood vessels. Mol Cell Biol 29:6018–6032

Schermuly RT, Stasch JP, Pullamsetti SS, Middendorff R, Muller D, Schluter KD, Dingendorf A, Hackemack S, Kolosionek E, Kaulen C et al (2008) Expression and function of soluble guanylate cyclase in pulmonary arterial hypertension. Eur Respir J 32:881–891

Schlossmann J, Schinner E (2012) cGMP becomes a drug target. Naunyn Schmiedebergs Arch Pharmacol 385:243–252

Schmidt P, Schramm M, Schroder H, Stasch JP (2003) Mechanisms of nitric oxide independent activation of soluble guanylyl cyclase. Eur J Pharmacol 468:167–174

Schwappacher R, Kilic A, Kojonazarov B, Lang M, Diep T, Zhuang S, Gawlowski T, Schermuly RT, Pfeifer A, Boss GR et al (2013) A molecular mechanism for therapeutic effects of cGMP-elevating agents in pulmonary arterial hypertension. J Biol Chem. doi:10.1074/jbc.M113.458729

Schymura Y, Janssen W, Eule U, Stasch JP, Weissmann N, Ghofrani HA, Grimminger F, Seeger W, Schermuly RT (2012) Antifibrotic effects of riociguat in a murine model of chronic right ventricular pressure overload. Am J Respir Crit Care Med 185:A6850

Seimetz M, Parajuli N, Pichl A, Stasch JP, Frey R, Schermuly RT, Ghofrani HA, Seeger W, Grimminger F, Weissmann N (2011) Effects of the soluble guanylate cyclase stimulator riociguat on emphysema development in tobacco-smoke exposed mice. Am J Respir Crit Care Med 183:A3107 (abstract)

Seimetz M, Parajuli N, Pichl A, Stasch JP, Frey R, Schermuly RT, Seeger W, Grimminger F, Ghofrani HA, Weissmann N (2012) Prevention of cigarette smoke-induced pulmonary hypertension by the soluble guanylate cyclase stimulator riociguat. Am J Respir Crit Care Med 185:A3416

Selwood DL, Brummell DG, Budworth J, Burtin GE, Campbell RO, Chana SS, Charles IG, Fernandez PA, Glen RC, Goggin MC et al (2001) Synthesis and biological evaluation of novel pyrazoles and indazoles as activators of the nitric oxide receptor, soluble guanylate cyclase. J Med Chem 44:78–93

Semigran M, Bonderman D, Ghio S, Felix S, Ghofrani HA, Michelakis ED, Mitrovic V, Oudiz RJ, Roessig L, Scalise AV (2012) Left ventricular systolic dysfunction associated with pulmonary hypertension riociguat trial (LEPHT). Circulation 126:2789–2790

Shao Z, Wang Z, Shrestha K, Thakur A, Borowski AG, Sweet W, Thomas JD, Moravec CS, Hazen SL, Tang WHW (2012) Pulmonary hypertension associated with advanced systolic heart failure. J Am Coll Cardiol 59:1150–1158

Sharkovska Y, Kalk P, Lawrenz B, Godes M, Hoffmann LS, Wellkisch K, Geschka S, Relle K, Hocher B, Stasch JP (2010) Nitric oxide-independent stimulation of soluble guanylate cyclase reduces organ damage in experimental low-renin and high-renin models. J Hypertens 28:1666–1675

Skoro-Sajer N, Mittermayer F, Panzenboeck A, Bonderman D, Sadushi R, Hitsch R, Jakowitsch J, Klepetko W, Kneussl MP, Wolzt M et al (2007) Asymmetric dimethylarginine is increased in chronic thromboembolic pulmonary hypertension. Am J Respir Crit Care Med 176:1154–1160

Stasch JP, Hobbs AJ (2009) NO-independent, haem-dependent soluble guanylate cyclase stimulators. Handb Exp Pharmacol 191:277–308

Stasch JP, Becker EM, Alonso-Alija C, Apeler H, Dembowsky K, Feurer A, Gerzer R, Minuth T, Perzborn E, Pleiss U et al (2001) NO-independent regulatory site on soluble guanylate cyclase. Nature 410:212–215

Stasch JP, Alonso-Alija C, Apeler H, Dembowsky K, Feurer A, Minuth T, Perzborn E, Schramm M, Straub A (2002a) Pharmacological actions of a novel NO-independent guanylyl cyclase stimulator, BAY 41-8543: in vitro studies. Br J Pharmacol 135:333–343

Stasch JP, Dembowsky K, Perzborn E, Stahl E, Schramm M (2002b) Cardiovascular actions of a novel NO-independent guanylyl cyclase stimulator, BAY 41-8543: in vivo studies. Br J Pharmacol 135:344–355

Stasch JP, Schmidt PM, Nedvetsky PI, Nedvetskaya TY, Kumar HSA, Meurer S, Deile M, Taye A, Knorr A, Lapp H et al (2006) Targeting the heme-oxidized nitric oxide receptor for selective vasodilatation of diseased blood vessels. J Clin Invest 116:2552–2561

Stasch JP, Pacher P, Evgenov OV (2011) Soluble guanylate cyclase as an emerging therapeutic target in cardiopulmonary disease. Circulation 123:2263–2273

Straub A, Stasch JP, Alonso-Alija C, Benet-Buchholz J, Ducke B, Feurer A, Furstner C (2001) NO-independent stimulators of soluble guanylate cyclase. Bioorg Med Chem Lett 11:781–784

Thoonen R, Buys ES, Cauwels A, Nimmegeers S, Geschka S, Delanghe J, Hochepied T, De Cauwer L, Rogge E, Sips P et al (2013a) A critical role for heme-free soluble guanylate cyclase in cardiovascular disease. Nat Med (under review)

Thoonen R, Sips PY, Bloch KD, Buys ES (2013b) Pathophysiology of hypertension in the absence of nitric oxide/cyclic GMP signaling. Curr Hypertens Rep 15:47–58

Thorsen LB, Eskildsen-Helmond Y, Zibrandtsen H, Stasch JP, Simonsen U, Laursen BE (2010) BAY 41-2272 inhibits the development of chronic hypoxic pulmonary hypertension in rats. Eur J Pharmacol 647:147–154

Tonelli AR, Haserodt S, Aytekin M, Dweik RA (2013) Nitric oxide deficiency in pulmonary hypertension: pathobiology and implications for therapy. Pulm Circ 3:20–30

Tse MT (2012) Trial watch: Phase III success for first-in-class pulmonary hypertension drug. Nat Rev Drug Discov 11:896

Tulis DA, Bohl Masters KS, Lipke EA, Schiesser RL, Evans AJ, Peyton KJ, Durante W, West JL, Schafer AI (2002) YC-1-mediated vascular protection through inhibition of smooth muscle cell proliferation and platelet function. Biochem Biophys Res Commun 291:1014–1021

Watts JA, Gellar MA, Fulkerson MB, Kline JA (2011) Pulmonary vascular reserve during experimental pulmonary embolism: effects of a soluble guanylate cyclase stimulator, BAY 41-8543. Crit Care Med 39:2700–2704

Watts JA, Gellar MA, Fulkerson MB, Kline JA (2013) A soluble guanylate cyclase stimulator, BAY 41-8543, preserves right ventricular function in experimental pulmonary embolism. Pulm Pharmacol Ther 26:205–211

Weidenbach A, Stasch JP, Ghofrani HA, Weissmann N, Grimminger F, Seeger W, Schermuly RT (2005) Inhaled NO and the guanylate cyclase stimulator BAY 41-2272 in oleic acid induced acute lung injury in rabbits. BMC Pharmacol 5:P61

Wharton J, Strange JW, Moller GM, Growcott EJ, Ren X, Franklyn AP, Phillips SC, Wilkins MR (2005) Antiproliferative effects of phosphodiesterase type 5 inhibition in human pulmonary artery cells. Am J Respir Crit Care Med 172:105–113

Wolin MS, Wood KS, Ignarro LJ (1982) Guanylate cyclase from bovine lung. A kinetic analysis of the regulation of the purified soluble enzyme by protoporphyrin IX, heme, and nitrosyl-heme. J Biol Chem 257:13312–13320

Xu W, Kaneko FT, Zheng S, Comhair SA, Janocha AJ, Goggans T, Thunnissen FB, Farver C, Hazen SL, Jennings C et al (2004) Increased arginase II and decreased NO synthesis in endothelial cells of patients with pulmonary arterial hypertension. FASEB J 18:1746–1748

Yazawa S, Tsuchiya H, Hori H, Makino R (2006) Functional characterization of two nucleotide-binding sites in soluble guanylate cyclase. J Biol Chem 281:21763–21770

Zhang HQ, Zhiren X, Teodozyj K, Dinges J (2003) A concise synthesis of ortho-substituted aryl-acrylamides – potent activators of soluble guanylate cyclase. Tetrahedron Lett 44:8661–8663

Therapeutics Targeting of Dysregulated Redox Equilibrium and Endothelial Dysfunction

Michael G. Risbano and Mark T. Gladwin

Abstract All forms of WHO Group 1 PAH share a progressive and complex vasculopathy. At the center of this derangement lies the pulmonary vascular endothelium, which plays a crucial role in maintaining a delicate and precise balance of opposing vasoconstricting and vasodilating forces. In PAH, endothelial cell damage and dysfunction alter vascular homeostasis in favor of vasoconstriction. There is evidence of increased expression and activity in the vasoconstrictor and mitogen endothelin-1 signaling system and a decreased production of the potent vasodilator prostacyclin. These pathways have been a major focus of FDA approved PAH-specific therapies. Beyond these pathways, there is the dysfunction within the endothelial nitric oxide (NO) synthase signaling pathway and dysregulation of reactive oxygen and nitrogen species (ROS) that contribute to the pathogenesis of PAH. The dysregulation of vasodilator systems in PAH in large part involves the NO pathway, with almost every step subject to impairments. This includes a reduction in endothelial NO synthase function (eNOS), the enzymatic "uncoupling" of eNOS, increased scavenging of NO by superoxide and cell-free hemoglobin, the elaboration of endogenous competitive inhibitors of eNOS (ADMA), and the oxidation of the NO target, soluble guanylyl cyclase. The dysregulation of NO signaling pathways occurs in the setting of parallel upregulation of vascular oxidases that generate ROS. Enzymatic sources of ROS in PH that have been identified include the NAPDPH oxidases 1, 2, and 4, xanthine oxidase, uncoupled eNOS, and complex III of the mitochondrial electron transport chain. Superoxide produced from these sources reacts with NO to form the reactive nitrogen species peroxynitrate, further diverting bioavailable NO to more injuries species. In PAH, this upstream dysregulation of ROS/NO redox homeostasis severely impairs vascular tone and contributes to the pathological activation of

M.G. Risbano • M.T. Gladwin (✉)
Vascular Medicine Institute, University of Pittsburgh, Pittsburgh, PA, USA

Division of Pulmonary Allergy and Critical Care Medicine, University of Pittsburgh Medical Center, Pittsburgh, PA, USA
e-mail: gladwinmt@upmc.edu

mitogenic pathways, leading to cellular proliferation and obliteration of the pulmonary vasculature. Therapeutic strategies are being evaluated that target the associated dysregulated redox equilibrium and endothelial dysfunction in PAH. Therapeutic interventions reviewed in this chapter include NO donor or NO generating drugs, therapies that recouple eNOS or directly increase cGMP levels via inhibition of phosphodiesterase 5 or stimulation of soluble guanylyl cyclase, and therapies that inhibit vascular oxidases or scavenge ROS.

Keywords Endothelial dysfunction • Nitric oxide • Redox equilibrium

Contents

1	Introduction	316
	1.1 Pulmonary Arterial Hypertension, Defined	316
	1.2 Pulmonary Arterial Hypertension: A Disorder of Vasoconstriction and Vasodilation	317
2	Mechanisms for Decreased NO Bioavailability in PAH	317
	2.1 NO Signaling and Endothelial Dysfunction	317
	2.2 eNOS Uncoupling	319
	2.3 Arginase (I and II)	320
	2.4 Hemolysis as a Mechanism of Endothelial Dysfunction and Vasculopathy	320
	2.5 Oxidative Stress, ROS, and RNS	323
3	Therapies	328
	3.1 Therapies Using NO or NO-Generating Drugs	328
	3.2 Therapies That Increase cGMP Levels	333
	3.3 Therapies Targeting ROS Signaling	336
	3.4 Therapies Targeting *S*-nitrosoglutathione Reductase Inhibitors	338
4	Summary and Future Directions	338
References		339

List of Abbreviations

eNOS   Endothelial nitric oxide synthase
PAH   Pulmonary arterial hypertension
PH   Pulmonary hypertension
RNS   Reactive nitrogen species
ROS   Reactive oxygen species

1 Introduction

1.1 Pulmonary Arterial Hypertension, Defined

Pulmonary arterial hypertension (PAH) is characterized by obstructive cellular proliferation and vasoconstriction of the pulmonary arterial vasculature (Tuder et al. 1994). A decrease in pulmonary artery lumen is a consequence of a derangement in multiple, complex pathways and includes (a) endothelial cell dysfunction and increased contractility of small pulmonary arteries, (b) proliferation and remodeling of endothelial and smooth muscle cells, and (c) in situ thrombosis. This pathologic remodeling ultimately leads to an increase in pulmonary vascular resistance with an increased afterload, causing a reduction in cardiac output as the right heart fails. The diagnosis of pulmonary arterial hypertension must be demonstrated hemodynamically by right heart catheterization with a resting mean pulmonary artery pressure (mPAP) $\geq$ 25 mmHg without an elevation in pulmonary capillary wedge pressure (<15 mmHg) (McLaughlin et al. 2009a, b).

PAH is a rare disease with an estimated prevalence of 15 cases per million persons in France (Humbert et al. 2006) with a historical survival of a median of 2.8 years after diagnosis (McLaughlin et al. 2009a, b). Even in the current era of PAH-specific therapy prognosis remains poor with a 1-, 2-, and 3-year survival of 87 %, 76 %, and 67 % respectively (Humbert et al. 2010).

1.2 Pulmonary Arterial Hypertension: A Disorder of Vasoconstriction and Vasodilation

All forms of WHO Group 1 PAH share a progressive and complex vasculopathy. At the center of this derangement lies the pulmonary vascular endothelium, which plays a crucial role in maintaining a delicate and precise balance of opposing vasoconstricting and vasodilating forces. In PAH, endothelial cell damage and dysfunction alter vascular homeostasis in favor of vasoconstriction. There is evidence of increased expression and activity in the vasoconstrictor and mitogen endothelin-1 signaling system (Stewart et al. 1991; Giaid et al. 1993), a decreased production of the potent vasodilator prostacyclin with increased release of the vasoconstrictor thromboxane A_2 (Christman et al. 1992). These pathways have been a major focus of FDA approved PAH-specific therapies. Beyond these pathways, there is the dysfunction within the endothelial nitric oxide synthase signaling pathway and dysregulation of reactive oxygen and nitrogen species that contribute to the pathogenesis of pulmonary arterial hypertension. Investigation into these alternative pathways has produced potential PAH-specific treatments aimed at increasing nitric oxide production or generation, increasing cGMP and GSNO or reducing ROS/RNS. These therapeutic interventions may serve to compliment current existing treatment modalities.

2 Mechanisms for Decreased NO Bioavailability in PAH

2.1 NO Signaling and Endothelial Dysfunction

Nitric oxide is a diatomic gas molecule that is produced by endothelial NO synthase enzyme via the 5-electron oxidation of L-arginine to form L-citrulline and NO. NO then diffuses as a paracrine messenger from endothelium to smooth muscle, where it

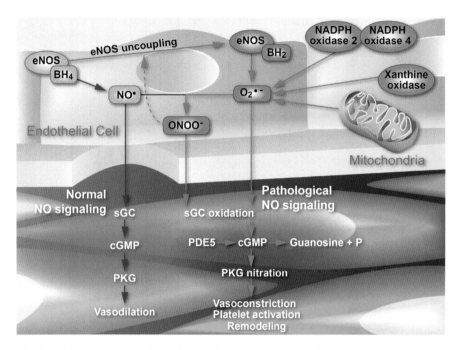

Fig. 1 Schematic representation of the various forms of free radical production from eNOS uncoupling, xanthine oxidase, NADPH, and downstream effects within the pulmonary endothelial cell. Reproduced with permission from Tabima et al. (2012)

binds to the heme group of soluble guanylyl cyclase (sGC) (see Fig. 1). This activates sGC, which converts guanosine triphosphate (GTP) to the cyclic guanosine nucleotide monophosphate (cGMP). cGMP in turn activates protein kinases to produce smooth muscle relaxation, vasodilation and increases in blood flow (Arnold et al. 1977; Ignarro et al. 1982, 1987; Palmer et al. 1987, 1988). As an endothelium-derived relaxing factor, nitric oxide (NO) is an important signaling molecule that regulates basal vasomotor tone, blood pressure, and inhibition of vascular cell growth. NO also regulates vascular smooth muscle proliferation and migration (Kibbe et al. 2000; Zuckerbraun et al. 2007). The normal production of NO by the endothelium is impaired in patients with cardiovascular disease, associated risk factors of increasing age, smoking, diabetes, obesity, hypercholesterolemia, and in PAH.

This NO–sGC–cGMP pathway regulates approximately 25 % of resting blood flow and blood pressure (Quyyumi et al. 1995; Cannon et al. 2001). With endothelial dysfunction, the NO synthase enzyme fails to produce NO. This can be measured in humans (and animals) by infusing acetylcholine into the brachial artery. Acetylcholine activates muscarinic receptors to increase intracellular levels of calcium and activate eNOS, which increases blood flow. Alternatively infusions of N-monomethyl-L-arginine (L-NMMA) - a nonselective competitive inhibitor of eNOs, will decrease blood flow. In patients with endothelial dysfunction the normal vasoconstriction to L-NMMA and vasodilation to acetylcholine are blunted, indicating a failure of endothelial eNOS to produce NO under basal and stimulated conditions, respectively. In studies of endothelial dysfunction, the direct NO donor sodium nitroprusside is infused to test eNOS and endothelium-independent NO signaling, since NO produced by nitroprusside will directly activate sGC.

The dysregulation of vasodilator systems in PAH in large part involves the NO pathway, with almost every step subject to impairments. This includes a reduction in endothelial NO synthase function (eNOS), the enzymatic "uncoupling" of eNOS (see Fig. 1), increased arginase activity that metabolizes L-arginine (Zuckerbraun et al. 2011), increased scavenging of NO by superoxide and cell-free hemoglobin, the elaboration of endogenous competitive inhibitors of eNOS (ADMA), and the oxidation of the NO target, soluble guanylyl cyclase. The endothelial dysfunction in PAH is characterized by early impaired endothelium-dependent vasodilation, as evidenced by a lack of vasodilator response to acetylcholine, bradykinin, or calcium ionophore and an increased superoxide production in the vessel wall with subsequent decreased NO bioavailability (Rabinovitch 2008).

In the following section, we summarize key aspects of the current status of NO signaling in PAH related to (a) reduction, dysfunction, or uncoupling of eNOS, (b) downstream effects on NO–cGMP signaling, (c) Arginase I and II, and (d) catabolism of NO by ROS or hemoglobin.

2.2 eNOS Uncoupling

Despite reports of varying levels of eNOS measured in the lung it is the functional reduction in NO signaling that is consistently reported in the animal models and in patients with PAH. The reduction in endogenous NO formation, as measured by reduced basal or stimulated NO enzymatic synthesis, has been demonstrated in patients with primary PAH (Demoncheaux et al. 2005) and in the hypoxic rat model of PH. A decrease in NO production results in a reduction of vasodilation regardless of normal or increased eNOS protein levels (Weerackody et al. 2009). The paradox of a normal or increased eNOS expression and reduced NO signaling is most likely explained by a pathologic mechanism called "eNOS uncoupling" (see Fig. 1).

Under normal circumstances eNOS forms a homodimer, and each monomer contains an oxygenase domain at the N-terminus. This oxygenase domain contains binding sites for tetrahydrobiopterin (BH4), the heme iron, and L-arginine (Gielis et al. 2011), while there is a reductase domain at the C-terminus (List et al. 1997; Venema et al. 1997; Forstermann and Munzel 2006). In the dysfunctional state of the uncoupled eNOS enzyme, electrons that normally transfer from the NOS reductase domain to the oxygenase domain are diverted to molecular oxygen (O_2) thus forming superoxide ($O_2^{\cdot-}$) rather than transferring electrons to L-arginine (Beckman and Koppenol 1996) to form NO. eNOS therefore exhibits oxidase activity rather than NO synthase activity. NO levels are reduced by primary diversion of electrons to oxygen and via secondary NO scavenging reaction by NOS-generated superoxide, which can react to form peroxynitrite ($ONOO^-$). Triggers for eNOS uncoupling include the depletion of necessary cofactors L-arginine and BH4 (Kunuthur et al. 2011). The oxidation of BH4 to dihydrobiopterin (BH2) is thought to be central to enzyme coupling (Cunnington and Channon 2010), with a number of models suggesting that the eNOS-bound BH2 may be more important than low levels of BH4. This is measured as either a decrease in the BH4:BH2 ratio or increases in BH2. A key observation is that eNOS generates superoxide when either BH4 levels decline or oxidized BH2 increases (Khoo et al. 2005; Crabtree et al. 2009). Consistent with this pathobiology, animal models of PH have low BH4:BH2 ratios. (Khoo et al. 2005; Grobe et al. 2006).

2.3 Arginase (I and II)

NO production is the result of the oxidation of L-arginine to citrulline. As expected, a decrease in the substrate L-arginine leads to a decrease in NO bioavailability. In chronic hemolytic diseases, such as sickle cell anemia, a chronic elevation in circulating arginase I has been noted, which decreases NO bioavailability (Morris et al. 2005, 2008). Arginases are increased in sickle cell disease as a result of liver dysfunction, inflammation, and most importantly the release from red blood cells during hemolysis. The levels of arginase II are also elevated in idiopathic PAH (Xu et al. 2004). Arginase catalyzes arginine to ornithine and urea, and its activity can be measured in patients as a decreasing ratio of arginine/ornithine (Morris et al. 2003, 2005; Xu et al. 2004). This reaction occurs with arginase I (the final step of the urea cycle, which occurs mainly in the cytoplasm of the liver, and also in human red blood cells) and arginase II (which occurs within mitochondrion intracellularly). Overexpression of arginases affects NO synthase activity and decreases the NO-dependent smooth muscle relaxation by depleting the L-arginine substrate that would otherwise be available to eNOS. In fact, Langle et al. (1995) has demonstrated that an acute release of hepatic arginase into the pulmonary circulation in the setting of liver transplant and reperfusion resulted in acute pulmonary vasoconstriction.

Sickle cell patients with increased plasma arginase activity have reduced L-arginine:ornithine ratios and in turn have more severe PAH and higher mortality (Morris et al. 2005). In IPAH patients plasma levels of L-arginine correlate with right atrial pressure, cardiac index and 6-minute walk distance (Beyer et al. 2008). Additionally arginase activity alone may promote the development and vasculopathy of PAH (Sharma et al. 2009) through NO-independent pathways, such as production of ornithine which may promote vascular remodeling and proliferation (Xu et al. 2004; Chen et al. 2009).

2.4 Hemolysis as a Mechanism of Endothelial Dysfunction and Vasculopathy

A number of disease states culminate in the hemolysis of erythrocytes; most notably sickle cell anemia, thalassemia, hemodialysis-associated hemolysis, malaria, and paroxysmal nocturnal hemoglobinuria (PNH). Hemolysis causes a decompartmentalization of cellular contents into the intravascular space. The release of materials including hemoglobin, arginase I, and asymmetric dimethylarginine (ADMA) produces a net effect of resistance to NO signaling and NO donor infusions (Reiter et al. 2002; Gladwin et al. 2004; Morris et al. 2005; Schnog et al. 2005; Landburg et al. 2008; Kato et al. 2009; D'Alecy and Billecke 2010; Landburg et al. 2010).

During steady-state sickle cell disease, there is a high rate of intravascular hemolysis. This is reflected by a depletion of haptoglobin in steady state in patients with homozygous HbSS disease and measurable cell-free hemoglobin in the plasma (Reiter et al. 2002; Rother et al. 2005). Hemoglobin, when decompartmentalized from the red cell, will react with NO at the near diffusion limit, to oxidize the NO to nitrate (Doherty et al. 1998; Dou et al. 2002). This reaction is so fast and irreversible that even levels of cell-free plasma hemoglobin of only 6–10 µM are sufficient to inhibit all NO signaling and produce vasoconstriction (Reiter et al. 2002). In addition to cell-free hemoglobin, hemolysis releases other red cell enzymes that have the potential to inhibit NO signaling. Arginase I is present in abundance in red cells and metabolizes arginine to ornithine, reducing arginine availability for NO synthesis (Morris et al. 2005). Hemolysis is also associated with ADMA accumulation, an endogenous competitive NO synthase inhibitor (Schnog et al. 2005; Kato et al. 2009). Hemoglobin also drives oxidative reactions via fenton and peroxidase chemistry and possibly via the induction of vascular oxidases, such as xanthine oxidase and NADPH oxidase, which are both upregulated in mouse models of SCD (Aslan et al. 2001; Wood et al. 2005, 2008). The net effect of hemolysis is a resistance to NO signaling and NO donor infusions (Kaul et al. 2000; Reiter et al. 2002; Kaul et al. 2004). For example, in humans and in transgenic mouse models that express human hemoglobin S, there is a reduction in the vasodilator effect of infusions of NO donors. In both humans with SCD and in mouse models of SCD, the magnitude of the blunted responses to NO donors is proportional to the plasma hemoglobin level (Dasgupta et al. 2010; Reiter et al. 2002; Kaul et al. 2004). In addition to direct inhibition of NO by reaction with hemoglobin, NO resistance can also occur secondary to inactivation reactions with superoxide, as well as direct oxidation and inactivation of soluble guanylate cyclase, the target for NO (Gladwin 2006).

Most studies have supported this hypothesis over the last decade using animal models, human vascular studies, and large epidemiological cohort studies. For example, humans with SCD and transgenic mice with SCD have impaired vasodilator responses to NO that are proportional to the levels of plasma hemoglobin (Reiter et al. 2002; Kaul et al. 2004; Hsu et al. 2007). Multiple mouse models of hemolysis develop spontaneous PH and right heart failure including the Berkeley sickle cell mouse (Hsu et al. 2007), the spherocytosis mouse (Frei et al. 2008), and the alloimmune hemolysis mouse (Hsu et al. 2007). In transgenic sickle mice, the blunted vasodilator response to NO directly relates to plasma hemoglobin levels (Kaul et al. 2004) and NO metabolite levels are inversely correlated with plasma hemoglobin levels (Dasgupta et al. 2010). NO resistance, PH, and systemic hypertension can be produced experimentally in animals by induced intravascular hemolysis or by infusions of hemoglobin or hemolysate (Minneci et al. 2005; Tofovic et al. 2009).

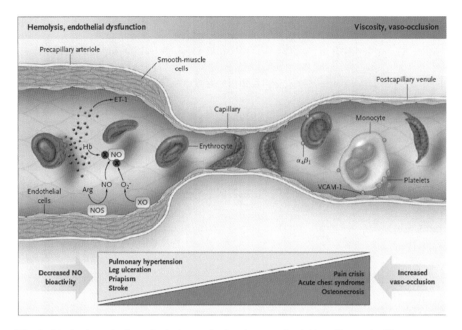

Fig. 2 Mechanisms for hypothetical vascular involvement in sickle cell anemia. There are two overlapping mechanisms proposed for vascular related disease in sickle cell anemia, (a) on the *left*, a low steady-state hemoglobin levels and high rates of intravascular hemolysis are a result of endothelial dysfunction, mediated by the inactivation of NO by plasma-free hemoglobin, increased ROS production and arginase activity; (b) on the *right*, increased steady-state leukocyte counts and increased hemoglobin levels may cause hemolysis-associated endothelial dysfunction through hemostatic activation and intimal/smooth muscle proliferation and therefore obstruction of capillaries and post-capillary venules. The relative frequency of associated clinical phenotypes with the degree of hemolysis/endothelial dysfunction and increased viscosity/vaso-occlusion are depicted by the two complementary triangles. *Arg* arginine, *ET-1* endothelin-1, *NO* nitric oxide, *NOS* nitric oxide synthase, O_2^- superoxide, *VCAM-1* vascular cell adhesion molecule-1, *XO* xanthine oxidase. Reproduced with permission Gladwin and Vichinsky (2008)

In humans, impairment in NO signaling, directly measured via venous occlusion strain-gauge plethysmography, correlates with high levels of plasma hemoglobin or its surrogate lactate dehydrogenase (Reiter et al. 2002). The relationship in humans between increasing plasma hemoglobin levels and direct measures of decreasing NO-dependent blood flow or low NO bioavailability have now been reproduced in numerous studies in disparate diseases, such as hemodialysis-associated hemolysis (Meyer et al. 2010), malaria (Yeo et al. 2007, 2009), and PNH (Hill et al. 2010).

From an epidemiological standpoint, in patients with sickle cell disease their clinical vasculopathic complications are significantly associated with markers of hemolytic anemia, including PH, priapism, and leg ulceration (Gladwin et al. 2004; Nolan et al. 2005; Kato et al. 2006; Nolan et al. 2006) (see Fig. 2). Many cohort studies have consistently associated the severity of hemolytic anemia with increasing Doppler-estimated pulmonary artery systolic hypertension and high risk of death, including the NIH-PH cohort (Gladwin et al. 2004), the Duke cohort

(De Castro et al. 2008), the UNC cohort (Ataga et al. 2006), the Muti-Centers Study of Hydroxyurea cohort (Machado et al. 2006), the Pediatric Hypoxic Response (PUSH) cohort (Onyekwere et al. 2008; Minniti et al. 2009; Naoman et al. 2010), and a cohort study from Greece (Voskaridou et al. 2010). An analysis of banked plasma samples from the classic cooperative study of sickle cell disease (CSSCD) cohort revealed that an abnormally high NT-proBNP level $\geq$ 160 pg/ml, a biomarker for pulmonary hypertension in patients with sickle cell disease (Machado et al. 2006), was present in 27.6 % of adult SCD patients, and high levels were associated with markers of hemolytic anemia such as a low hemoglobin level ($P < 0.001$), high lactate dehydrogenase ($P < 0.001$), and high total bilirubin levels ($P < 0.007$) (Machado et al. 2009). Most importantly, an NT-proBNP level $\geq$ 160 pg/ml was a major and independent predictor of mortality (RR 6.24, 95 % CI 2.9–13.3, $P < 0.0001$). The recently published analysis of >600 screening patients in the Walk-PHASST cohort confirms similar strong associations between indices of hemolytic anemia, high NT-proBNP, low walk distance and increased Doppler-echocardiographic estimates of pulmonary artery systolic pressures (Sachdev et al. 2011).

Three additional recently published studies in patients with sickle cell disease confirmed the associations between the severity of hemolytic anemia and pulmonary hypertension defined by right heart catheterization (Parent et al. 2011; Fonseca et al. 2012; Mehari et al. 2012).

2.5 Oxidative Stress, ROS, and RNS

Free radicals are oxygen-derived reactive oxygen species (ROS) that have one or more unpaired electrons, superoxide ($O_2^{\cdot-}$), hydroperoxyl ($HO_2^{\cdot}$), peroxyl ($RO_2^{\cdot}$), and hydroxyl ($OH^{\cdot-}$) that along with non-radical peroxides (ROOH) like hydrogen peroxide ($H_2O_2^{\cdot}$) are regarded as ROS. These ROS oxidize various targets. Both human subjects (Joppa et al. 2007) with PAH and animal models of PH have demonstrated increased levels of oxidative stress (DeMarco et al. 2008; Jankov et al. 2008).

Reactive nitrogen species or RNS ($NO^{\cdot}$, $NO_2^{\cdot}$, nitrogen dioxide; N_2O_3, nitrogen trioxide; $ONOO^-$, peroxynitrite) are chemically derived from NO and result from the oxidation of the terminal nitrogen of L-arginine. RNS are produced during the urea cycle in the presence of O_2 and NOS (Brown and Borutaite 2006). RNS reacts with reactive sulfur species creating sulfenic acid and other thiol or thyl radicals (RSOH; RS– and $RS^{\cdot}$). N_2O_3 acts as a nitrosating agent to add a NO^+ group to secondary amines or thiols to produce N-nitrosamines and S-nitrosothiols (RSNO), respectively. Such S-nitrosation of the cysteine thiol of glutathione by NO^+ or by transnitrosation reactions with other S-nitrosothiols results in the formation of S-nitrosoglutathione (GSNO). The production of RSNO affects the downstream signaling of the NO pathway both physiologically and pathologically (Moya et al. 2002; Foster et al. 2003; Gaston et al. 2006).

It is increasingly apparent that enhanced production and dysregulation of ROS/RNS signaling contributes to the pathogenesis of PAH (DeMarco et al. 2008). The generation of ROS/RNS has been described in antimicrobial defense as well as physiological responses in redox signaling ranging from cellular differentiation and antioxidant gene expression to proliferation and apoptosis (Fialkow et al. 2007). Oxidative stress is the result of the imbalance of excessive formation of ROS and RNS and/or a decrease in antioxidant defenses. Cellular homeostasis is affected through oxidative damage and posttranslational redox modifications of proteins, lipids, and nucleic acids with resultant defective cellular function, apoptosis, and vasculopathy.

The role of reactive oxygen and nitrogen species has been recently reviewed in the pathogenesis of pulmonary hypertension (Tabima et al. 2012). For example, peroxynitrite ($ONOO^-$) can have variable effects that have been proposed to contribute to the pathogenesis of pulmonary hypertension. Not only can it can cause toxicity in smooth muscle and endothelial cells to cause cell death but also cellular proliferation through ERK and protein kinase C activation (Agbani et al. 2011). Peroxynitrite is derived from NO and superoxide and can form from uncoupled eNOS or other oxidases that produce superoxide in the presence of NO. As an oxidant $ONOO^-$ in turn modifies eNOS activity, reduces NO production and enhanced superoxide production, which reacts with remaining NO to produce additional $ONOO^-$. Peroxynitrite at low concentrations can inactivate prostacyclin synthase, reducing the vasodilatory effects of prostacyclin and increasing vasoconstrictive mediators thromboxane A2 and prostaglandin H2 (Agbani et al. 2011).

2.5.1 ROS/RNS and eNOS Uncoupling

As previously described eNOS is essential in vascular homeostasis through the production of NO. Uncoupled eNOS shifts away from the production NO to the generation of $O_2^{\bullet-}$ with subsequent reductions in vascular dilation. There is evidence that eNOS uncoupling is involved in the pathogenesis of PH which comes from studies using the Caveolin-1 (Cav-1) knockout mice. Cav-1 regulates eNOS signaling as caveolin binds to eNOS (Garcia-Cardena et al. 1997; Bernatchez et al. 2005) as a negative regulator of its activity. Released activated uncoupled eNOS produces oxidative stress and contributes to the development of cardiopulmonary disease (Wunderlich et al. 2008; Karuppiah et al. 2011). Although not all mice with uncoupled eNOS develop PH (Ozaki et al. 2002), Cav-1-deficient mice develop spontaneous PH which is reversed with treatment with L-NAME, which inhibits eNOS and reduces superoxide production (Wunderlich et al. 2008). This further supports that eNOS uncoupling can produce a PH phenotype. Additionally eNOS-dependent peroxynitrite formation from the reaction of NO and superoxide contributes to pulmonary vasculopathy (Zhao et al. 2009).

2.5.2 NOXs

Recent research shows that the oxidase systems nicotinamide adenine dinucleotide dehydrogenase (NADH) oxidase (Nox) and xanthine oxidoreductase (XOR) are involved in the development of pulmonary vascular disease and PH. Members of the Nox family that are found in the human vasculature include Nox1, Nox4, Nox5, and, the phagocyte NADPH oxidase gp91^{phox}, known as Nox2. It is notable that Nox5 is not expressed in rodents.

The primary role of Nox is ROS generation, in particular superoxide production (Takeya and Sumimoto 2003). All Nox have transmembrane helices that regulate electron transport across the membranes to reduce O_2 to $O_2^{\cdot-}$. Within the six membrane domains are two highly conserved heme binding histidine sites (heme domains) and a binding site in the cytoplasmic tail for NADPH and flavin adenine dinucleotide (FAD) (Vignais 2002; Bedard and Krause 2007). The FAD is constitutively bound to the flavin domain and the NADPH acts a substrate. The NADPH oxidases use NADPH as an electron donor to reduce molecular oxygen to $O_2^{\cdot-}$ and/or H_2O_2 while maintaining their own discrete biochemical function for activity, regulation, and expression. Two electrons are transferred from NADPH to FAD, which is reduced to $FADH_2$ on the inner C-terminal tail (Nisimoto et al. 1999). The electrons transfer from the inner heme to an outer heme and then bind to an O_2 molecule to produce $O_2^{\cdot-}$ on the external membrane (Cross and Segal 2004). Different Nox families produce different ROS, as Nox1–3 and 5 produce $O_2^{\cdot-}$ and Nox4, Duox 1, 2 produce H_2O_2. The production of H_2O_2 by Nox is controversial and may result from the rapid dismutation by the Nox catalytic site.

Noxs have been shown to be a major source for the generation of ROS in vascular smooth muscle cells (Wedgwood et al. 2001), endothelial cells (Chatterjee et al. 2008), and fibroblasts (Pagano et al. 1997). The ROS derived from the NADPH oxidases have been implicated in the pathogenesis of pulmonary hypertension (Brennan et al. 2003; Sanders and Hoidal 2007) and in particular the development of PH related to chronic hypoxia (Mittal et al. 2007; Sanders and Hoidal 2007). Within the pulmonary vasculature, Nox1, Nox2, and Nox4 produce ROS, which blunts vascular relaxation to exogenous NO resulting in impaired tone (Brennan et al. 2003) and contractile response in hypoxic PH models (Fresquet et al. 2006; Liu et al. 2006; Mittal et al. 2007; Fike et al. 2008). Elevated superoxide levels produced within the pulmonary arterioles in the persistent PH of the newborn model (Deruelle et al. 2005; Farrow et al. 2008b) have been defined as the primary oxidant causing oxidative stress mainly derived from Nox and also secondarily from uncoupled eNOS (Wedgwood and Black 2003; Deruelle et al. 2005; Wedgwood et al. 2005; Farrow et al. 2008a, b).

Chronic hypoxia models of PH result in clear changes in NO and ROS production. Hypoxia results in increased HIF-1 gene expression, which increases TGF-β1 levels (Sturrock et al. 2006) and increases Nox-4 expression and ROS production in PASMC (Sanders and Hoidal 2007; Nisbet et al. 2010). Studies have shown that inhibition or knock-down silencing of Nox4 reduces ROS formation and reverses

PH in cultured human pulmonary artery smooth muscle cells through a reduction of smooth muscle proliferation (Sturrock et al. 2006; Mittal et al. 2007). The Nox2 isoform plays a role in the NO-dependent endothelial dysfunction in pulmonary arteries by inhibiting NO-mediated activation of sGC and cGMP production during chronic hypoxia (Li et al. 1999; Li and Forstermann 2000). Accordingly, the Nox2 knockout mouse is protected from hypoxic PH, with decreases in right ventricular pressures and hypertrophy, pulmonary artery wall thickening, and superoxide production (Liu et al. 2006). Increased ROS from Nox4 can alter the balance of $NADP^+$:NADPH and tetrahydrobiopterin to dihydrobiopterin (BH4:BH2) and contribute to an uncoupled eNOS state. Not only are the NADPH oxidase enzymes responsible for superoxide production but also they serve as regulators of xanthine oxidase and eNOS (Wolin et al. 2005; Jankov et al. 2008).

2.5.3 Xanthine Oxidoreductase

Xanthine oxidoreductase (XOR) primarily exists in two forms, in cells as the dehydrogenase form (XDH), and in the circulation as the oxidase form (XO). Xanthine oxidoreductase catalyzes the final two steps in purine degradation hypoxanthine → xanthine → uric acid. Substrate-derived electrons from the oxidation of hypoxanthine to xanthine reduce NAD^+ to NADH. This process has been well characterized in gout; however, XOR has also been examined as a contributor of intracellular ROS in the setting of inflammatory conditions and vascular injury. Inflammation has been recognized as a significant contribution to vascular remodeling in human and experimental models of PH (Hassoun et al. 2009). During inflammatory conditions, the cysteine residue of XOR is oxidized and converts XDH to XO (Amaya et al. 1990), which transfers substrate-derived electrons to O_2 to generate intracellular H_2O_2 and $O_2^{\cdot-}$. Both XO and XDH are responsible for the generation of ROS (Harris and Massey 1997) and moderate hypoxia is a potent stimulus for XOR activity in arterial endothelial cells (Kelley et al. 2006).

It is important to note three crucial distinctions in the activity of XOR (1) the conversion to XO is not essential for ROS production, as XDH has partial oxidase activity during conditions where $NADH/NAD^+$ ratios are increased (i.e., hypoxemia) (Williamson et al. 1993; Harris and Massey 1997); (2) about 80–95 % of the ROS produced by XOR under normal and pathophysiological conditions is H_2O_2 (Kelley et al. 2010); and (3) hypoxia induces XOR expression and activity (Kelley et al. 2006). XOR can be either produced within the vasculature or distributed to critical sites in the vasculature. When this occurs XOR serves as a significant source of ROS thereby modulating vascular function. Studies have shown enhanced xanthine oxidoreductase activity within the arteries of PAH patients (Spiekermann et al. 2009), in cultured endothelial cells (Cote et al. 1996; Houston et al. 1999), and in a hypoxia exposed rat model of PH (Hoshikawa et al. 2001). It is therefore postulated that xanthine oxidoreductase can serve as a critical regulator of ROS formation within the vasculature and drive pathological processes involved in the pathogenesis of PAH.

In chronic hypoxic models of PH, an increase in vascular $O_2^{\cdot-}$ production results in impaired NO signaling and endothelial dysfunction (Houston et al. 1999; Zulueta et al. 2002). This directly contributes to vascular remodeling and has been partly attributed to enhanced xanthine oxidoreductase activity (Hoshikawa et al. 2001). Treatment of chronic hypoxia-induced PAH rats with XOR-specific inhibitors, such as allopurinol attenuates hypoxia, normalizes blood pressure, reduces RV hypertrophy as well as pulmonary vascular thickening (Hoshikawa et al. 2001; Jankov et al. 2008). It is unclear, however, if XOR's role in PAH is either contributory or as a nitrite reductase that prevents/reverses experimental PH (Zuckerbraun et al. 2010). One potential explanation for this unresolved paradox and discordant role of XOR may be explained by the presence of nitrite. Electrons may be diverted away from oxygen and pathological superoxide formation to nitrite and protective NO.

2.5.4 Mitochondria as a Source of ROS

Under normal aerobic conditions, mitochondrion in the endothelial and smooth muscle cells of pulmonary arteries converts a small amount (1–2 %) of O_2 into $O_2^{\cdot-}$ via the electron transport chain (Boveris et al. 1972; Wolin 2009). The production of superoxide can be affected by the modulation of mitochondrial redox status, biogenesis, fusion and fission processes, as well as changes in mitochondrial bioenergetics (Erusalimsky and Moncada 2007). Superoxide dismutase 2 (SOD2) converts mitochondrial superoxide to H_2O_2, which play a variable role in regulation of vascular tone and cellular proliferation and apoptosis. There are beneficial effects when AMPK, JNK, and nuclear NF-κB are activated and result in the upregulation of cytoprotective genes involved in glucose transportation, antiapoptosis, glycolysis, antioxidant defenses, and repair mechanisms (Biswas et al. 2003). Additionally, H_2O_2 can have a variable effect on the pulmonary artery depending on the anatomical aspect of the vessel, where H_2O_2 will either act to increase contractility or vasodilation. In the absence of PAH, under normal physiological conditions, the contribution of mitochondrial $O_2^{\cdot-}$ to $ONOO^-$ is not clear (Erusalimsky and Moncada 2007). In the setting of PAH, the ability to dismutase superoxide are either saturated or impaired (Farrow et al. 2008b; Nozik-Grayck et al. 2008) and the electron transport chain production of $ONOO^-$ (from the reaction of NO and superoxide) and $O_2^{\cdot-}$ is enhanced (Burwell et al. 2006; Dahm et al. 2006).

The reason for mitochondrial dysfunction in PAH may be explained by two distinct theories. The first suggests that the mitochondrial electron transport chain serves as a source of ROS promoting senescence, necrosis, or apoptosis (Wolin et al. 2005) thus resulting in pulmonary arterial vasculopathy. The second possibility considers that mitochondrial dysfunction and a reduction in physiologic ROS levels (H_2O_2) drive pulmonary hypertension through a net decrease in cellular redox state, which inhibits redox-sensitive voltage-gated potassium channels Kv1.512 activity, causing cellular depolarization, increased intracellular Ca^{2+}, and then vasoconstriction of pulmonary artery smooth muscle cells (Archer et al. 2010). A possible reconciliation of these two theories can be postulated by considering the enzymatic source and the identities and quantities of the measured ROS. In the setting of hypertension and inflammation, high levels of NADPH and XOR-derived superoxide and $ONOO^-$ are likely pathologic, while the potassium-gated channels Kv1.512 may be tonically activated by the physiologic flux of H_2O_2 thus limiting intracellular calcium levels.

3 Therapies

The therapeutic strategies discussed target the dysregulated redox equilibrium and endothelial dysfunction that occurs in PAH. Many of the therapies discussed are experimental, and they are being tested in preclinical animal models, with some currently being tested in human clinical trials in human PAH (BH4 and sodium nitrite).

3.1 Therapies Using NO or NO-Generating Drugs

3.1.1 Inhaled Nitric Oxide Gas

In both animal and human studies, inhaled NO (5–80 ppm) is a potent, short-acting and selective pulmonary vasodilator that reduces pulmonary vascular resistance and right ventricular afterload (Rossaint et al. 1993; Atz and Wessel 1997; Roberts et al. 1997; Markewitz and Michael 2000). iNO is delivered directly to the pulmonary vascular resistance vessels where it is immediately inactivated by hemoglobin (Hb), limiting the systemic vasodilatory effects of the drug. This selectivity of NO to Hb exemplifies the beneficial attributes of iNO as a pulmonary-specific vasodilator compared to other intravenous vasodilators, which result in a decrease in systemic vascular resistance. In patients with acute lung injury (ALI) and acute respiratory distress syndrome (ARDS), iNO is delivered to well-ventilated lung units and can improve V/Q mismatching in injured lung to improve oxygenation; however, despite significant improvements in hypoxemia, therapeutic trials with iNO have not shown a mortality benefit or shorter duration of mechanical ventilation. As a therapeutic agent in PH patients, iNO is mostly utilized as adjuvant therapy to improve gas exchange in newborns with severe PH (Ivy et al. 1997). However, its usefulness as a therapeutic treatment for other forms of PH is currently limited (Coggins and Bloch 2007). It remains a commonly used agent in the cardiac catheterization laboratory to test pulmonary vasodilator responsiveness.

iNOs therapeutic value is limited by the dose and duration of exposure, with doses higher of 80 ppm leading to methemoglobin levels above 5 % (Weinberger et al. 2001). Other safety concerns regard the withdrawal of iNO and rebound pulmonary hypertension. Several studies have noted a potentially life-threatening increase in pulmonary vascular resistance upon acute withdrawal of iNO (Miller et al. 1995; Atz et al. 1996; Lavoie et al. 1996; Cueto et al. 1997). This form of "rebound pulmonary hypertension" is manifested by an acute increase in pulmonary vascular resistance and compromised cardiac output associated with severe hypoxemia. The mechanism behind this rebound phenomena may be in part explained by a finding that exogenous NO exposure inhibits endogenous eNOS activity (Black et al. 1999). This suggests that a transient decrease in endogenous eNOS activity (rather than endothelial NOS) likely results in a decrease in of plasma cGMP and nitrate concentrations during inhaled NO therapy resulting in the rebound pressure increase. The coadministration of a phosphodiesterase-5 inhibitor (PDE5i) (sildenafil) in children has been shown to ameliorate this rebound effect (Namachivayam et al. 2006), suggesting that inhaled NO may work synergistically with current treatment options such as phosphodiesterase-5 inhibitors (see below). New delivery systems that allow for pulsed NO delivery combined with pulsed oxygen from liquid oxygen sources may allow for small portable devices of inhaled NO. Large clinical trials of inhaled NO in patients with PAH on background phosphodiesterase-5 inhibitor therapies are required to address these questions.

In the adult PAH population, however, inhaled NO (iNO) has been primarily utilized as a diagnostic agent to assess acute pulmonary vasoreactivity to guide therapeutic options, i.e., calcium channel blockers versus PAH-specific agents (Sitbon et al. 2005). The ACCF/AHA 2009 Expert Consensus Document on Pulmonary Hypertension (McLaughlin et al. 2009a, b) favored the use of iNO over other vasodilators largely due to ease of administration compared to other agents and limited side effects when used correctly (Sitbon et al. 1998). Typical dose range is 20–80 ppm for 5–10 min, with repeated hemodynamic measurements while on iNO to measure the degree of response (Barst et al. 2004; Sitbon et al. 2005). Response to iNO may also portend improved long-term outcomes independent of treatment with calcium channel blockers (Krasuski et al. 2011; Malhotra et al. 2011).

3.1.2 NO Donors

Organic nitrates such as nitroglycerin are commonly utilized agents that relax capacitance vessels and decrease pulmonary and systemic vascular resistance at higher doses. The underlying mechanisms of action are unclear and are likely multifactorial (Elkayam 1991) and may include the metabolism to NO and nitrite with secondary activation of sGC (Artz et al. 2002; Ignarro et al. 2002). The development of nitrate tolerance with treatment can limit the class utility of these drugs.

Nitrate tolerance appears to be related to both increases in ROS formation, which inactivates the NO formed by organic nitrate metabolism, and by the oxidative inhibition of aldehyde dehydrogenase, the enzyme that converts the organic nitrates to NO. Munzel et al. (1995b) demonstrated an increase in vascular superoxide production after nitroglycerin treatment. Tolerance was prevented by the coadministration of superoxide dismutase (SOD), thus highlighting the role of ROS in this inhibition (Munzel et al. 1995b). A later study by the same group identified a membrane-bound Nox1 as a likely source of superoxide production (Munzel et al. 1996), and more recent studies demonstrated a role for endothelin-1 in the pathway (Kurz et al. 1999). It is likely that the increase in superoxide inactivates the NO released from nitroglycerin, increasing the formation of peroxynitrite. This may in turn promotes a supersensitivity to vasoconstrictors due to a tonic activation of protein kinase C, leading to nitrate tolerance (Munzel et al. 1995a). Proposed superoxide sources include NADPH oxidase, uncoupled eNOS, and the mitochondria. Superoxide and NO rapidly combine to form peroxynitrite. This aggravates tolerance by promoting the uncoupling of NO synthase and the inhibition of sGC and prostacyclin synthase (Elkayam 1991). More recent studies have defined a new tolerance mechanism through the inhibition of mitochondrial aldehyde dehydrogenase, an enzyme that metabolizes nitroglycerin to form NO and nitrite (Sydow et al. 2004; Zhang et al. 2004).

3.1.3 Nitrite

Nitrite (NO_2^-), an endogenous anion salt, has historically been considered an inert oxidation product of NO and oxygen (see Fig. 3). There is increasing evidence over the past 10 years that has supported the use of nitrite as a potent physiological vasodilator (Gladwin et al. 2000; Modin et al. 2001; Cosby et al. 2003; Dejam et al. 2004; Lundberg et al. 2008) and reservoir for NO (Lundberg et al. 2008). Nitrite is reduced to NO by a variety of enzyme systems as oxygen tension drops along the physiological and pathological hypoxic gradient. Enzymes that have been identified with nitrite reductase activity include hemoglobin, myoglobin, neuroglobin, eNOS, xanthine oxidoreductase, and acidic disproportionation (Flogel et al. 2004; Huang et al. 2005; Rassaf et al. 2007; Shiva et al. 2007; Hendgen-Cotta et al. 2008; Shiva and Gladwin 2009; Tota et al. 2010; Tiso et al. 2011). Nitrite has been shown to mediate hypoxic pulmonary vasodilation in a deoxyhemoglobin and pH-dependent fashion (Hunter et al. 2004).

Nitrite has been proposed as a therapy for PH based on inhaled studies in newborn sheep with hypoxia- and thromboxane-induced PH (Hunter et al. 2004). Recent studies have shown that inhaled, low-dose nebulized sodium nitrite can prevent and reverse experimental PAH and heart failure in both the chronic hypoxia mouse model and monocrotaline rat model (Zuckerbraun et al. 2010). Zuckerbraun et al. (2010) showed that a low dose of sodium nitrite has a potent antiproliferative effect on smooth muscle that is dependent on nitrite reduction by the enzyme xanthine oxidoreductase. Nitrite in this context signals through cGMP to increase the levels of the cell cycle checkpoint inhibitor P21.

Recently, a study demonstrated that a hypoxia-induced PH mouse model treated with oral dietary nitrate had a reduction in right ventricular pressures and hypertrophy and with decreased pulmonary vascular remodeling (Baliga et al. 2012). Plasma and

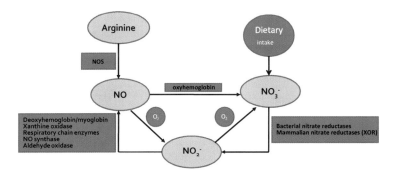

Fig. 3 The production of nitrite from nitric oxide and dietary nitrate. Within cells nitric oxide synthase (NOS) produces nitric oxide (NO) from L-arginine. The enterosalivary circulation produces NO through dietary intake of nitrate rich foods (spinach, lettuce, or beetroot juice). Ingested nitrate is taken from the circulation into the salivary glands and concentrated in the saliva. Salivary nitrate is reduced to nitrite by oral cavity bacteria, which when swallowed generates nitrogen oxide in the acidic stomach. Adapted from Lundberg et al. (2009)

lung tissue levels of nitrite and cGMP were found to be elevated in mice supplemented with nitrate. In comparison, mice that either lacked eNOS or were treated with allopurinol, a xanthine oxidoreductase inhibitor, reduced the beneficial effects of oral nitrite/nitrate supplementation.

Animal toxicology studies and phase Ia and Ib human studies have been completed with plans for a proof of concept phase II study over the next 2 years.

3.1.4 Recoupling of eNOS

Tetrahydrobiopterin

As a critical cofactor for endothelial NO synthase, suboptimal concentrations of tetrahydrobiopterin (BH4) result in decreased synthesis of NO, increased superoxide production, which contributes to eNOS uncoupling and endothelial dysfunction. BH4 is a potential therapeutic target for patients with pulmonary hypertension. Mice with congenital low tissue levels of BH4 tend to develop PH with RVH due to distal vessel muscularization (Khoo et al. 2005; Nandi et al. 2005). After hypoxic challenge, treatment with BH4 reduced muscularization of the distal pulmonary arteries (Francis et al. 2010). A BH4 analogue, 6-acetyl-7,7-dimethyl-7, 8-dihydropterin (ADDP), studied in a hypoxia-induced rat model of PH demonstrated enhanced eNOS expression in the pulmonary vascular endothelium with improved NO-mediated pulmonary artery dilation (Kunuthur et al. 2011).

Caveolin-1 (Cav1)-deficient mice, prone to developing PAH and a hypertrophied right ventricle, have enhanced eNOS activity but do not increase BH4 levels. This causes a discordance between eNOS activity and BH4 levels, resulting in dysfunctional eNOS signaling. Chronic supplementation with BH4 in these mice results in a normalized BH4/BH2 ratio, decreased superoxide production and a reduction in pulmonary artery pressures and right ventricular congestion manifest by a lower central venous pressure. These findings were not reproduced with a similar antioxidant tetrahydrobiopterin (H4N), which has no NOS activity (Wunderlich et al. 2008).

Supplementation with sepiapterin, an oxidized BH4 analogue, increased cellular BH4 content and NO production, recoupled eNOS, and decreased superoxide formation in the pulmonary artery endothelial cells isolated from fetal lambs with PPHN (Teng et al. 2011).

Exogenous treatment with BH4 or therapies to increase endogenous BH4 to modulate eNOS activity in humans are of great clinical interest and are currently under investigation. An open-label dose escalation safety study of the optically active form of BH4, sapropterin (6r-BH4) in 18 Group 1 PAH or inoperable CTEPH patients on concurrent oral PH therapy with either ERAs or PDE5i demonstrated tolerability without significant systemic hypotension (Robbins et al. 2011). Additionally, 6-min walk distance improved in the treatment group as well as a small but significant decrease in the inflammatory marker MCP-1. There were no significant changes in measures of NO synthesis or oxidative stress.

Other therapies that increase BH4 levels include statins (Hattori et al. 2003), certain angiotensin converting enzyme inhibitors (Oak and Cai 2007; Imanishi et al. 2008), telmisartan (Wenzel et al. 2008), and insulin (Ishii et al. 2001). Larger, placebo-controlled human studies in the use of oral BH4 or increases endogenous BH4 in the setting of PH are needed.

L-Arginine

L-arginine is the nitrogen donor to nitric oxide production. As previously stated, the enzyme arginase hydrolyzes arginine to urea and ornithine, therefore competing with eNOS for arginine. In certain forms of PH, such as sickle cell anemia, arginase I levels are increased which result in impaired pulmonary vasorelaxation due to alterations in the NO pathway (Morris et al. 2005, 2008). Additionally, decreased plasma levels of arginine have been reported in infants with PPHN (Vosatka et al. 1994), which correlates with decreased NO levels (Pearson et al. 2001). Increased bioavailability of L-arginine through exogenous administration is a possible therapeutic target in the treatment of some types of pulmonary hypertension.

In chronically hypoxic rats, L-arginine restored endothelial-dependent vasorelaxation to acetylcholine within the pulmonary circulation (Eddahibi et al. 1992). Dietary L-arginine supplementation in hypoxia-induced PH maintained ameliorated PH by preventing hypoxia-induced increase in resistance of the precapillary pulmonary vessels, possibly be inducing angiogenesis (Howell et al. 2009).

Mehta et al. (1995) evaluated the short-term vasodilatory effects of an infusion of L-arginine in ten subjects with various forms of pulmonary hypertension. Compared to hypertonic saline L-arginine infusion resulted in decreases in pulmonary and systemic pressures, pulmonary vascular resistance, and an increase in cardiac output. Compared to an infusion of the potent vasodilator, prostacyclin, L-arginine produced a similar reduction in pulmonary artery pressures; however, prostacyclin infusion resulted in a significant reduction in pulmonary vascular resistance. The first study to evaluate the short-term oral administration of L-arginine in 19 subjects primarily with Group 1 PAH and CTEPH resulted in a decrease in pulmonary hemodynamics (16 % reduction in pulmonary vascular resistance) and demonstrated improvement in exercise capacity (Nagaya et al. 2001). Similar reductions in pulmonary artery pressures were shown in a sickle cell population with pulmonary hypertension (Morris et al. 2003). Despite the hemodynamic responses that the above clinical trials demonstrated, L-arginine therapy has not been in the forefront of treatment for pulmonary hypertension.

3.2 Therapies That Increase cGMP Levels

3.2.1 Phosphodiesterase-5 Inhibitors

Therapies that increase cGMP levels, such as the phosphodiesterase-5 inhibitors, sildenafil and tadalafil, have been reviewed in detail in a previous chapter. We briefly review the available data pertaining to the selective cGMP-specific phosphodiesterase type 5 inhibitors.

Phosphodiesterase type 5 enzyme blockade prevents the normal hydrolysis of cGMP and increases the effects of NO including pulmonary vasodilation and inhibition of smooth muscle cell growth (Friebe and Koesling 2003). Sildenafil and tadalafil are the only FDA approved PDE-5 inhibitors (PDE5i) for the treatment of moderate–severity PAH. Sildenafil, which was approved in 1998, is available as a thrice daily oral formulation for iPAH and connective tissue disease-associated PAH and has recently been made available as an IV formulation (Vachiery et al. 2011). A 12-week, double-blind trial of moderate Group 1 PAH (Galie et al. 2005) compared sildenafil and placebo and met the primary endpoint of improved 6MWD in the sildenafil group as well as hemodynamic endpoints of decreased mPAP and PVR.

Tadalafil approved 5 years after sildenafil is available as once a day oral therapy for iPAH, heritable PAH, and connective tissue disease-associated PAH. PHIRST 1, a 16-week, randomized, double-blind, double-dummy, placebo-controlled multicenter trial studying the efficacy and tolerability of tadalafil on Group 1 PAH subjects with and without background bosentan therapy (Galie et al. 2009). Tadalafil significantly decreased time to clinical worsening when compared to placebo and improved exercise capacity. An evaluation of a hemodynamic subgroup demonstrated improved cardiopulmonary hemodynamics, in particular decreased mPAP, PVR and improved CI in the tadalafil group. A double-blind uncontrolled 52-week extension study of PHIRST-1 demonstrated that prolonged tadalafil treatment was well tolerated and initial improvements noted in 6MWD noted at 16 weeks were preserved at week 52 (Oudiz et al. 2012).

There has been great interest in investigating novel treatment options for patients with interstitial lung disease (ILD)-associated pulmonary hypertension as there are limited therapies available for this population. A randomized, open-label study of 16 subjects with severe pulmonary hypertension (mPAP >35 mmHg) and pulmonary fibrosis compared the acute cardiopulmonary response to IV epoprostenol to oral sildenafil (Ghofrani et al. 2002). Both epoprostenol and sildenafil reduced pulmonary vascular resistance; however, sildenafil had a reduction in pulmonary to systemic vascular resistance; additionally, epoprostenol increased ventilation perfusion (V/Q) mismatch of the lung and decreased arterial O_2 saturations. The sildenafil group maintained V/Q matching and partial pressure of arterial O_2. The likely explanation for this is that eNOS requires molecular oxygen as a substrate for the L-arginine oxidation to NO, so more NO and cGMP are produced in well-oxygenated regions of the lung. Therefore, the inhibition of PDE-5 will increase cGMP levels more in oxygenated lung vasculature.

Based upon the finding that sildenafil preferentially induced vasodilation in well-ventilated regions of lung tissue, the investigators of Sildenafil Trial of Exercise Performance in Idiopathic Pulmonary Fibrosis (STEP-IPF) evaluated the use of sildenafil in 180 subjects in a double-blind, randomized, placebo-controlled trial of 20 mg sildenafil thrice daily (Zisman et al. 2010). The study did not meet the primary endpoint of a 20 % increase in 6-min walk distance from baseline. However, compared to the placebo group treatment, subjects had a significant improvement in the diffusing capacity for carbon monoxide, arterial O_2 saturations, and partial pressure of arterial O_2 and maintained a stable degree of shortness of breath at week 12. This study did not evaluate subjects for the presence of pulmonary vascular disease. It is unclear if the beneficial effects can be attributed to that of a particular subgroup. Additional large clinical trials targeting patients with ILD and pulmonary hypertension are needed.

3.2.2 sGC Activators and Stimulators

As previously reviewed, guanylyl cyclase (GC) is a heterodimeric enzyme that catalyzes the conversion of cGTP to cGMP following stimulation with NO in vascular smooth muscle. The impaired signaling of the NO–sGC–cGMP pathway has been implicated in the pathogenesis of pulmonary hypertension. The activators of sGC bind to and activate the oxidized or heme-free enzyme and do not require NO (Evgenov et al. 2006). The stimulators directly sensitize sGC to low levels of nitric oxide in the presence of a reduced or ferrous synthetic heme (Stasch et al. 2001). These agents are potent vasodilators in the pulmonary and systemic circulation (see Fig. 4).

sGC activators such as BAY 58-2667 and HMR 1766 have been evaluated in animal models and preclinical studies of PH. The sGC activators have been shown to have potent and selective pulmonary vasodilator properties without significant

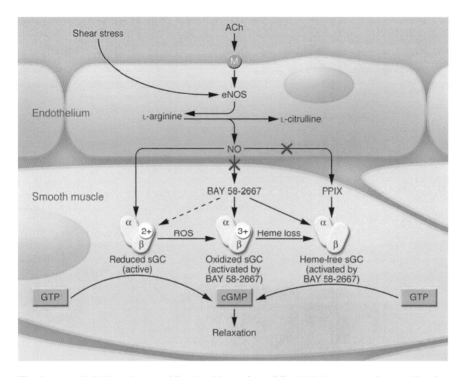

Fig. 4 BAY 58-2667 activates oxidized and heme-free sGC, which is unresponsive to NO. After activation either by shear stress or acetylcholine (ACh), endothelial NOS produces NO, which diffuses to smooth muscle and binds to the reduced hemes (Fe^{2+} ferrous heme) on the α/β heterodimer sGC. This converts GTP to cGMP and leads to smooth muscle relaxation with resultant vasodilation. Endothelial dysfunction can be associated with the accumulation of oxidized (Fe3+ ferric heme) and heme-free sGC that is not activated by NO. *M* muscarinic receptor. Reproduced with permission Gladwin (2006)

abnormalities in pulmonary gas exchange (Evgenov et al. 2006). Stimulators of sGC such as BAY 41-2272 and BAY 63-2521 have been shown to both potently vasodilate the pulmonary vascular bed and decrease the workload of the right ventricle (Mittendorf et al. 2009). The evaluation of BAY 41-2272 in a high renin, low-NO rat model of hypertension (transgenic renin rats with additional renin gene, TG(mren2)27) showed an improvement in oxygenation, potent vasodilator effects, augmented response to inhaled nitric oxide, decreased blood pressure, and improved survival in drug treated subjects (Stasch et al. 2001).

BAY 63-2521, also known as riociguat, is one of the first sGC stimulators in clinical development to be evaluated for the treatment of pulmonary hypertension (Grimminger et al. 2009). The safety profile of a single dose of riociguat was evaluated the in Group 1, 3, and 4 pulmonary hypertension. Despite the lack of pulmonary vasodilator selectivity, riociguat significantly improved pulmonary hemodynamics, increased cardiac index better than inhaled NO, and did not cause clinically significant systemic hypotension (Grimminger et al. 2009) As a follow-up, Ghofrani et al. (2010) completed an open-label phase II study in patients with chronic thromboembolic PH as well as Class I PAH patients and found that riociguat reduced pulmonary vascular resistance and increased the 6-min walk distance. Side effects related to therapy with riociguat included hypotension, shortness of breath, and headache. Larger randomized trials are currently underway and include evaluation of combination therapy of riociguat with the PDE-5 inhibitor sildenafil and the use of riociguat in non-Group 1 PAH, such as COPD.

3.3 Therapies Targeting ROS Signaling

As ROS has been implicated in the pathogenesis of PAH, there is an increasing interest in diminishing ROS levels as potential therapeutics for PAH. Scavengers of ROS, such as recombinant human MnSOD, increased eNOS expression levels and function in the ductal ligation model of the pulmonary hypertensive lamb through superoxide metabolism and restoration of BH4 levels (Farrow et al. 2008b). The downregulation of SOD2 precedes the development of PAH in fawn-hooded rats. SOD2 regulates the production of H_2O_2 within the mitochondrion during respiration, which is less toxic than superoxide. Treatment with the SOD mimetic Manganese(III)tetrakis(4-benzoic acid) porphyrin chloride in fawn-hooded rats (a mutant strain that spontaneously develops PAH) reduces mean pulmonary artery pressure, improves exercise capacity, reduces right ventricle free wall thickness, and decreased medial wall thickness in the precapillary resistance arteries (Archer et al. 2010).

Endothelin-1 stimulates the production of H_2O_2 in pulmonary artery smooth muscle cells (PASMC) isolated from the fetal lamb model of pulmonary hypertension. Exposure of the PASMC to H_2O_2 decreases sGC levels and NO-dependent GMP expression. The addition of the ROS scavenger polyethylene glycol catalase, which is specific to H_2O_2, normalizes the vasodilator response to NO, a finding that was not observed in the control arm (Wedgwood et al. 2005)

Apart from ROS scavengers, other strategies have been aimed at the inhibition of ROS generation. Cav1-deficient mice show evidence of heart failure and pulmonary hypertension demonstrate increase in superoxide production related to uncoupled eNOS. The acute inhibition of uncoupled eNOS with L-nitro-arginine methyl ester (L-NAME) reduced vascular superoxide production (Wunderlich et al. 2008). Phosphodiesterase-5 inhibitors, a current standard treatment for PAH, have also been reported to lower superoxide levels in the mouse model of bleomycin-induced pulmonary fibrosis and pulmonary hypertension (Hemnes et al. 2008).

Increased superoxide production in the ductus arteriosus ligation model of persistent pulmonary hypertension of the newborn (PPHN) derived from NADPH oxidase impairs vascular tone and contractility of the pulmonary arteries. Pretreatment of the isolated pulmonary arteries with superoxide scavengers tiron and PEG-SOD and with NADPH oxidase inhibitor diphenyliodonium enhanced the relaxation of pulmonary arteries to the NO donor s-nitrosyl-acetyl-penicillamine (SNAP) (Brennan et al. 2003).

Transgenic rat (mRen2)27 models characterized by the overexpression of renin in extrarenal tissue have increased synthesis of angiotensin II, resulting in increased angiotensin II ROS formation with systemic and pulmonary hypertension. This occurs as a result of Nox2-induced oxidative stress through the upstream activation of the renin–angiotensin system in the pulmonary vascular tissue and parenchyma (DeMarco et al. 2008, 2009). Human iPAH subjects have increased renin–angiotensin–aldosterone (RAAS) activity and shown to have increased systemic levels of circulating renin and angiotensin I and II, which are associated with disease progression and mortality and pulmonary (de Man et al. 2012). Angiotensin-converting enzyme inhibitors (ACEi) have been shown to reduce pulmonary artery pressures and thickening of pulmonary arteries (Nong et al. 1996). The effectiveness of the ACEi in PH models likely lies in the upregulation of NO production and a decrease in oxidative stress related to angiotensin II-dependent Nox2 activation (Kanno et al. 2001).

An inhibitor of fatty acid oxidation, the anti-anginal drug trimetazidine, attenuated cardiac fibrosis in a transverse aortic constriction rat model. Trimetazidine acted as an antioxidant reducing ROS generation and directly affected NADPH oxidase in the fibroblasts of cardiac tissue, rather than serving as a free radical scavenger (Liu et al. 2010).

There has also been recent interest in the role of PPAR gamma in pulmonary hypertension (Hansmann and Zamanian 2009; Tian et al. 2009; Sutliff et al. 2010). Patients with idiopathic pulmonary arterial hypertension have been shown to have reduced mRNA expression levels of PPAR-gamma (Ameshima et al. 2003), a ligand-activated nuclear receptor and transcription factor that regulates glucose metabolism and adipogenesis (He et al. 2003). Rosiglitazone is a synthetic PPAR-gamma agonist that reverses pulmonary hypertension by attenuating Nox4 induction and ROS generation in the lungs of chronic hypoxic mice (Nisbet et al. 2010). Furthermore, activation of PPAR-gamma with rosiglitazone attenuated H_2O_2 production, Nox4 expression and cellular proliferation in human pulmonary artery cells exposed to hypoxia (Lu et al. 2010). Together, the inhibition of fatty acid oxidation and the stimulus of PPAR-gamma support other findings that metabolic modulators can be selective for the pulmonary circulation (Sutendra et al. 2010). Since fatty acid oxidation inhibitors and agonists of PPAR-gamma are readily available, clinical trials to evaluate their effect on PAH are needed. Metabolic modulators are new and exciting avenues to pursue in the therapy of pulmonary hypertension.

3.4 Therapies Targeting S-nitrosoglutathione Reductase Inhibitors

Decreased *S*-nitrosothiols (RSNO) have been initially demonstrated in pulmonary hypertension (Foster et al. 2003) in particular hypoxic newborns (Gaston et al. 1993) or infants with presumed PPHN (Gaston et al. 1998). As previously mentioned RSNO are produced by the *S*-nitrosylation of the cysteine thiol of proteins by nitrosonium (NO^+), with *S*-nitrosoglutathione (GSNO) representing one of these intracellular *S*-nitrosothiols (Liu et al. 2004). Through *S*-nitrosylation of proteins, NO has been shown to posttranslationally regulate protein function (Benhar and Stamler 2005; Kelleher et al. 2007; Whalen et al. 2007). Denitrosylation pathways within cells can terminate the *S*-nitrosothiols (SNOs) pathways in vivo, in particular the enzyme GSNO reductase (GSNOR), previously characterized as formaldehyde dehydrogenase (Liu et al. 2001, 2004). To date, only a few studies have evaluated the therapeutic benefit of *S*-nitrosoglutathione reductase inhibitors (GSNORi) in pulmonary hypertension. The role of GSNOR in the regulation of pulmonary vascular tone and remodeling has been evaluated in mice models of hypoxic pulmonary hypertension (Wu et al. 2010). Wu et al. (2010) showed that hypoxic mice with PH significantly increased GSNOR mRNA and protein expression in lung tissue compared to normoxic mice with resultant interference of RSNO metabolism

Currently, there are no clinical trials evaluating the use of GSNORi in pulmonary hypertension. However, a small number of basic and preclinical studies evaluating GSNOR inhibitors have been conducted including a study by Sanghani et al. (2009), which identified three novel GSNOR inhibitors through a high-throughput screening approach. One of the novel GSNOR inhibitors, C3, demonstrated properties that suppressed NF-κB activation (a master regulator of the inflammatory response), increased cGMP formation, and promoted the relaxation of the murine aorta. The first-in-class drug N6022, a potent inhibitor of GSNOR has shown safety and efficacy in animal models of asthma and COPD and is currently being evaluated in early phase human clinical studies. The results of these human safety studies have not been published. Further evaluation of GSNOR inhibitors such as N6022 in the pulmonary hypertensive models and clinical PH trials is required.

4 Summary and Future Directions

As demonstrated in this chapter, the pathogenesis of pulmonary arterial hypertension of various clinical and experimental forms have been associated with endothelial dysfunction, reduced NO signaling, and increased oxidative stress. Therapies outlined in this chapter that enhance NO signaling, recouple eNOS, scavenge ROS, directly inhibit Nox2 and 4 and xanthine oxidoreductase, and restore mitochondrial function share a promising role as future therapeutics for pulmonary arterial hypertension.

References

Agbani EO, Coats P et al (2011) Peroxynitrite stimulates pulmonary artery endothelial and smooth muscle cell proliferation: involvement of ERK and PKC. Pulm Pharmacol Ther 24(1):100–109

Amaya Y, Yamazaki K et al (1990) Proteolytic conversion of xanthine dehydrogenase from the NAD-dependent type to the O2-dependent type. Amino acid sequence of rat liver xanthine dehydrogenase and identification of the cleavage sites of the enzyme protein during irreversible conversion by trypsin. J Biol Chem 265(24):14170–14175

Ameshima S, Golpon H et al (2003) Peroxisome proliferator-activated receptor gamma (PPARgamma) expression is decreased in pulmonary hypertension and affects endothelial cell growth. Circ Res 92(10):1162–1169

Archer SL, Marsboom G et al (2010) Epigenetic attenuation of mitochondrial superoxide dismutase 2 in pulmonary arterial hypertension: a basis for excessive cell proliferation and a new therapeutic target. Circulation 121(24):2661–2671

Arnold WP, Mittal CK et al (1977) Nitric oxide activates guanylate cyclase and increases guanosine 3′:5′-cyclic monophosphate levels in various tissue preparations. Proc Natl Acad Sci USA 74(8):3203–3207

Artz JD, Schmidt B et al (2002) Effects of nitroglycerin on soluble guanylate cyclase: implications for nitrate tolerance. J Biol Chem 277(21):18253–18256

Aslan M, Ryan TM et al (2001) Oxygen radical inhibition of nitric oxide-dependent vascular function in sickle cell disease. Proc Natl Acad Sci USA 98(26):15215–15220

Ataga KI, Moore CG et al (2006) Pulmonary hypertension in patients with sickle cell disease: a longitudinal study. Br J Haematol 134(1):109–115

Atz AM, Wessel DL (1997) Inhaled nitric oxide in the neonate with cardiac disease. Semin Perinatol 21(5):441–455

Atz AM, Adatia I et al (1996) Rebound pulmonary hypertension after inhalation of nitric oxide. Ann Thorac Surg 62(6):1759–1764

Baliga RS, Milsom AB et al (2012) Dietary nitrate ameliorates pulmonary hypertension: cytoprotective role for endothelial nitric oxide synthase and xanthine oxidoreductase. Circulation 125(23):2922–2932

Barst RJ, McGoon M et al (2004) Diagnosis and differential assessment of pulmonary arterial hypertension. J Am Coll Cardiol 43(12 Suppl S):40S–47S

Beckman JS, Koppenol WH (1996) Nitric oxide, superoxide, and peroxynitrite: the good, the bad, and ugly. Am J Physiol 271(5 Pt 1):C1424–C1437

Bedard K, Krause KH (2007) The NOX family of ROS-generating NADPH oxidases: physiology and pathophysiology. Physiol Rev 87(1):245–313

Benhar M, Stamler JS (2005) A central role for S-nitrosylation in apoptosis. Nat Cell Biol 7(7):645–646

Bernatchez PN, Bauer PM et al (2005) Dissecting the molecular control of endothelial NO synthase by caveolin-1 using cell-permeable peptides. Proc Natl Acad Sci USA 102(3):761–766

Beyer J, Kolditz M et al (2008) L-arginine plasma levels and severity of idiopathic pulmonary arterial hypertension. Vasa 37(1):61–67

Biswas G, Anandatheerthavarada HK et al (2003) Mitochondria to nucleus stress signaling: a distinctive mechanism of NFkappaB/Rel activation through calcineurin-mediated inactivation of IkappaBbeta. J Cell Biol 161(3):507–519

Black SM, Heidersbach RS et al (1999) Inhaled nitric oxide inhibits NOS activity in lambs: potential mechanism for rebound pulmonary hypertension. Am J Physiol 277(5 Pt 2): H1849–H1856

Boveris A, Oshino N et al (1972) The cellular production of hydrogen peroxide. Biochem J 128(3):617–630

Brennan LA, Steinhorn RH et al (2003) Increased superoxide generation is associated with pulmonary hypertension in fetal lambs: a role for NADPH oxidase. Circ Res 92(6):683–691

Brown GC, Borutaite V (2006) Interactions between nitric oxide, oxygen, reactive oxygen species and reactive nitrogen species. Biochem Soc Trans 34(Pt 5):953–956

Burwell LS, Nadtochiy SM et al (2006) Direct evidence for S-nitrosation of mitochondrial complex I. Biochem J 394(Pt 3):627–634

Cannon RO 3rd, Schechter AN et al (2001) Effects of inhaled nitric oxide on regional blood flow are consistent with intravascular nitric oxide delivery. J Clin Invest 108(2):279–287

Chatterjee A, Black SM et al (2008) Endothelial nitric oxide (NO) and its pathophysiologic regulation. Vascul Pharmacol 49(4–6):134–140

Chen B, Calvert AE et al (2009) Hypoxia promotes human pulmonary artery smooth muscle cell proliferation through induction of arginase. Am J Physiol Lung Cell Mol Physiol 297(6): L1151–L1159

Christman BW, McPherson CD et al (1992) An imbalance between the excretion of thromboxane and prostacyclin metabolites in pulmonary hypertension. N Engl J Med 327(2):70–75

Coggins MP, Bloch KD (2007) Nitric oxide in the pulmonary vasculature. Arterioscler Thromb Vasc Biol 27(9):1877–1885

Cosby K, Partovi KS et al (2003) Nitrite reduction to nitric oxide by deoxyhemoglobin vasodilates the human circulation. Nat Med 9(12):1498–1505

Cote CG, Yu FS et al (1996) Regulation of intracellular xanthine oxidase by endothelial-derived nitric oxide. Am J Physiol 271(5 Pt 1):L869–L874

Crabtree MJ, Tatham AL et al (2009) Quantitative regulation of intracellular endothelial nitric-oxide synthase (eNOS) coupling by both tetrahydrobiopterin-eNOS stoichiometry and biopterin redox status: insights from cells with tet-regulated GTP cyclohydrolase I expression. J Biol Chem 284(2):1136–1144

Cross AR, Segal AW (2004) The NADPH oxidase of professional phagocytes – prototype of the NOX electron transport chain systems. Biochim Biophys Acta 1657(1):1–22

Cueto E, Lopez-Herce J et al (1997) Life-threatening effects of discontinuing inhaled nitric oxide in children. Acta Paediatr 86(12):1337–1339

Cunnington C, Channon KM (2010) Tetrahydrobiopterin: pleiotropic roles in cardiovascular pathophysiology. Heart 96(23):1872–1877

D'Alecy LG, Billecke SS (2010) Massive quantities of asymmetric dimethylarginine (ADMA) are incorporated in red blood cell proteins and may be released by proteolysis following hemolytic stress. Blood Cells Mol Dis 45(1):40

Dahm CC, Moore K et al (2006) Persistent S-nitrosation of complex I and other mitochondrial membrane proteins by S-nitrosothiols but not nitric oxide or peroxynitrite: implications for the interaction of nitric oxide with mitochondria. J Biol Chem 281(15):10056–10065

Dasgupta T, Fabry ME et al (2010) Antisickling property of fetal hemoglobin enhances nitric oxide bioavailability and ameliorates organ oxidative stress in transgenic-knockout sickle mice. Am J Physiol Regul Integr Comp Physiol 298(2):R394–R402

De Castro LM, Jonassaint JC et al (2008) Pulmonary hypertension associated with sickle cell disease: clinical and laboratory endpoints and disease outcomes. Am J Hematol 83(1):19–25

de Man FS, Tu L et al (2012) Dysregulated renin-angiotensin-aldosterone system contributes to pulmonary arterial hypertension. Am J Respir Crit Care Med 186(8):780–789

Dejam A, Hunter CJ et al (2004) Emerging role of nitrite in human biology. Blood Cells Mol Dis 32(3):423–429

DeMarco VG, Habibi J et al (2008) Oxidative stress contributes to pulmonary hypertension in the transgenic (mRen2)27 rat. Am J Physiol Heart Circ Physiol 294(6):H2659–H2668

DeMarco VG, Habibi J et al (2009) Rosuvastatin ameliorates the development of pulmonary arterial hypertension in the transgenic (mRen2)27 rat. Am J Physiol Heart Circ Physiol 297(3): H1128–H1139

Demoncheaux EA, Higenbottam TW et al (2005) Decreased whole body endogenous nitric oxide production in patients with primary pulmonary hypertension. J Vasc Res 42(2):133–136

Deruelle P, Grover TR et al (2005) Pulmonary vascular effects of nitric oxide-cGMP augmentation in a model of chronic pulmonary hypertension in fetal and neonatal sheep. Am J Physiol Lung Cell Mol Physiol 289(5):L798–L806

Doherty DH, Doyle MP et al (1998) Rate of reaction with nitric oxide determines the hypertensive effect of cell-free hemoglobin. Nat Biotechnol 16(7):672–676

Dou Y, Maillett DH et al (2002) Myoglobin as a model system for designing heme protein based blood substitutes. Biophys Chem 98(1–2):127–148

Eddahibi S, Adnot S et al (1992) L-arginine restores endothelium-dependent relaxation in pulmonary circulation of chronically hypoxic rats. Am J Physiol 263(2 Pt 1):L194–L200

Elkayam U (1991) Tolerance to organic nitrates: evidence, mechanisms, clinical relevance, and strategies for prevention. Ann Intern Med 114(8):667–677

Erusalimsky JD, Moncada S (2007) Nitric oxide and mitochondrial signaling: from physiology to pathophysiology. Arterioscler Thromb Vasc Biol 27(12):2524–2531

Evgenov OV, Pacher P et al (2006) NO-independent stimulators and activators of soluble guanylate cyclase: discovery and therapeutic potential. Nat Rev Drug Discov 5(9):755–768

Farrow KN, Groh BS et al (2008a) Hyperoxia increases phosphodiesterase 5 expression and activity in ovine fetal pulmonary artery smooth muscle cells. Circ Res 102(2):226–233

Farrow KN, Lakshminrusimha S et al (2008b) Superoxide dismutase restores eNOS expression and function in resistance pulmonary arteries from neonatal lambs with persistent pulmonary hypertension. Am J Physiol Lung Cell Mol Physiol 295(6):L979–L987

Fialkow L, Wang Y et al (2007) Reactive oxygen and nitrogen species as signaling molecules regulating neutrophil function. Free Radic Biol Med 42(2):153–164

Fike CD, Slaughter JC et al (2008) Reactive oxygen species from NADPH oxidase contribute to altered pulmonary vascular responses in piglets with chronic hypoxia-induced pulmonary hypertension. Am J Physiol Lung Cell Mol Physiol 295(5):L881–L888

Flogel U, Godecke A et al (2004) Role of myoglobin in the antioxidant defense of the heart. FASEB J 18(10):1156–1158

Fonseca GH, Souza R et al (2012) Pulmonary hypertension diagnosed by right heart catheterisation in sickle cell disease. Eur Respir J 39(1):112–118

Forstermann U, Munzel T (2006) Endothelial nitric oxide synthase in vascular disease: from marvel to menace. Circulation 113(13):1708–1714

Foster MW, McMahon TJ et al (2003) S-nitrosylation in health and disease. Trends Mol Med 9(4):160–168

Francis BN, Wilkins MR et al (2010) Tetrahydrobiopterin and the regulation of hypoxic pulmonary vasoconstriction. Eur Respir J 36(2):323–330

Frei AC, Guo Y et al (2008) Vascular dysfunction in a murine model of severe hemolysis. Blood 112(2):398–405

Fresquet F, Pourageaud F et al (2006) Role of reactive oxygen species and gp91phox in endothelial dysfunction of pulmonary arteries induced by chronic hypoxia. Br J Pharmacol 148(5):714–723

Friebe A, Koesling D (2003) Regulation of nitric oxide-sensitive guanylyl cyclase. Circ Res 93(2):96–105

Galie N, Ghofrani HA et al (2005) Sildenafil citrate therapy for pulmonary arterial hypertension. N Engl J Med 353(20):2148–2157

Galie N, Brundage BH et al (2009) Tadalafil therapy for pulmonary arterial hypertension. Circulation 119(22):2894–2903

Garcia-Cardena G, Martasek P et al (1997) Dissecting the interaction between nitric oxide synthase (NOS) and caveolin. Functional significance of the nos caveolin binding domain in vivo. J Biol Chem 272(41):25437–25440

Gaston B, Reilly J et al (1993) Endogenous nitrogen oxides and bronchodilator S-nitrosothiols in human airways. Proc Natl Acad Sci USA 90(23):10957–10961

Gaston B, Fry E et al (1998) Umbilical arterial S-nitrosothiols in stressed newborns: role in perinatal circulatory transition. Biochem Biophys Res Commun 253(3):899–901

Gaston B, Singel D et al (2006) S-nitrosothiol signaling in respiratory biology. Am J Respir Crit Care Med 173(11):1186–1193

Ghofrani HA, Wiedemann R et al (2002) Sildenafil for treatment of lung fibrosis and pulmonary hypertension: a randomised controlled trial. Lancet 360(9337):895–900

Ghofrani HA, Hoeper MM et al (2010) Riociguat for chronic thromboembolic pulmonary hypertension and pulmonary arterial hypertension: a phase II study. Eur Respir J 36(4):792–799

Giaid A, Yanagisawa M et al (1993) Expression of endothelin-1 in the lungs of patients with pulmonary hypertension. N Engl J Med 328(24):1732–1739

Gielis JF, Lin JY et al (2011) Pathogenetic role of eNOS uncoupling in cardiopulmonary disorders. Free Radic Biol Med 50(7):765–776

Gladwin MT (2006) Deconstructing endothelial dysfunction: soluble guanylyl cyclase oxidation and the NO resistance syndrome. J Clin Invest 116(9):2330–2332

Gladwin MT, Vichinsky E (2008) Pulmonary complications of sickle cell disease. N Engl J Med 359(21):2254–2265

Gladwin MT, Shelhamer JH et al (2000) Role of circulating nitrite and S-nitrosohemoglobin in the regulation of regional blood flow in humans. Proc Natl Acad Sci USA 97(21):11482–11487

Gladwin MT, Sachdev V et al (2004) Pulmonary hypertension as a risk factor for death in patients with sickle cell disease. N Engl J Med 350(9):886–895

Grimminger F, Weimann G et al (2009) First acute haemodynamic study of soluble guanylate cyclase stimulator riociguat in pulmonary hypertension. Eur Respir J 33(4):785–792

Grobe AC, Wells SM et al (2006) Increased oxidative stress in lambs with increased pulmonary blood flow and pulmonary hypertension: role of NADPH oxidase and endothelial NO synthase. Am J Physiol Lung Cell Mol Physiol 290(6):L1069–L1077

Hansmann G, Zamanian RT (2009) PPARgamma activation: a potential treatment for pulmonary hypertension. Sci Transl Med 1(12):12ps14

Harris CM, Massey V (1997) The oxidative half-reaction of xanthine dehydrogenase with NAD; reaction kinetics and steady-state mechanism. J Biol Chem 272(45):28335–28341

Hassoun PM, Mouthon L et al (2009) Inflammation, growth factors, and pulmonary vascular remodeling. J Am Coll Cardiol 54(1 Suppl):S10–S19

Hattori Y, Nakanishi N et al (2003) HMG-CoA reductase inhibitor increases GTP cyclohydrolase I mRNA and tetrahydrobiopterin in vascular endothelial cells. Arterioscler Thromb Vasc Biol 23(2):176–182

He W, Barak Y et al (2003) Adipose-specific peroxisome proliferator-activated receptor gamma knockout causes insulin resistance in fat and liver but not in muscle. Proc Natl Acad Sci USA 100(26):15712–15717

Hemnes AR, Zaiman A et al (2008) PDE5A inhibition attenuates bleomycin-induced pulmonary fibrosis and pulmonary hypertension through inhibition of ROS generation and RhoA/Rho kinase activation. Am J Physiol Lung Cell Mol Physiol 294(1):L24–L33

Hendgen-Cotta UB, Merx MW et al (2008) Nitrite reductase activity of myoglobin regulates respiration and cellular viability in myocardial ischemia-reperfusion injury. Proc Natl Acad Sci USA 105(29):10256–10261

Hill A, Rother RP et al (2010) Effect of eculizumab on haemolysis-associated nitric oxide depletion, dyspnoea, and measures of pulmonary hypertension in patients with paroxysmal nocturnal haemoglobinuria. Br J Haematol 149(3):414–425

Hoshikawa Y, Ono S et al (2001) Generation of oxidative stress contributes to the development of pulmonary hypertension induced by hypoxia. J Appl Physiol 90(4):1299–1306

Houston M, Estevez A et al (1999) Binding of xanthine oxidase to vascular endothelium. Kinetic characterization and oxidative impairment of nitric oxide-dependent signaling. J Biol Chem 274(8):4985–4994

Howell K, Costello CM et al (2009) L-Arginine promotes angiogenesis in the chronically hypoxic lung: a novel mechanism ameliorating pulmonary hypertension. Am J Physiol Lung Cell Mol Physiol 296(6):L1042–L1050

Hsu LL, Champion HC et al (2007) Hemolysis in sickle cell mice causes pulmonary hypertension due to global impairment in nitric oxide bioavailability. Blood 109(7):3088–3098

Huang KT, Keszler A et al (2005) The reaction between nitrite and deoxyhemoglobin. Reassessment of reaction kinetics and stoichiometry. J Biol Chem 280(35):31126–31131
Humbert M, Sitbon O et al (2006) Pulmonary arterial hypertension in France: results from a national registry. Am J Respir Crit Care Med 173(9):1023–1030
Humbert M, Sitbon O, Yaïci A, Montani D, O'Callaghan DS, Jaïs X, Parent F, Savale L, Natali D, Günther S, Chaouat A, Chabot F, Cordier JF, Habib G, Gressin V, Jing ZC, Souza R, Simonneau G, French Pulmonary Arterial Hypertension Network (2010) Survival in incident and prevalent cohorts of patients with pulmonary arterial hypertension. Eur Respir J 36(3): 549–55. doi:10.1183/09031936.00057010, Epub 2010 Jun 18
Hunter CJ, Dejam A et al (2004) Inhaled nebulized nitrite is a hypoxia-sensitive NO-dependent selective pulmonary vasodilator. Nat Med 10(10):1122–1127
Ignarro LJ, Degnan JN et al (1982) Activation of purified guanylate cyclase by nitric oxide requires heme. Comparison of heme-deficient, heme-reconstituted and heme-containing forms of soluble enzyme from bovine lung. Biochim Biophys Acta 718(1):49–59
Ignarro LJ, Buga GM et al (1987) Endothelium-derived relaxing factor produced and released from artery and vein is nitric oxide. Proc Natl Acad Sci USA 84(24):9265–9269
Ignarro LJ, Napoli C et al (2002) Nitric oxide donors and cardiovascular agents modulating the bioactivity of nitric oxide: an overview. Circ Res 90(1):21–28
Imanishi T, Ikejima H et al (2008) Addition of eplerenone to an angiotensin-converting enzyme inhibitor effectively improves nitric oxide bioavailability. Hypertension 51(3):734–741
Ishii M, Shimizu S et al (2001) Stimulation of tetrahydrobiopterin synthesis induced by insulin: possible involvement of phosphatidylinositol 3-kinase. Int J Biochem Cell Biol 33(1):65–73
Ivy DD, Parker TA et al (1997) Prolonged endothelin A receptor blockade attenuates chronic pulmonary hypertension in the ovine fetus. J Clin Invest 99(6):1179–1186
Jankov RP, Kantores C et al (2008) Contribution of xanthine oxidase-derived superoxide to chronic hypoxic pulmonary hypertension in neonatal rats. Am J Physiol Lung Cell Mol Physiol 294(2):L233–L245
Joppa P, Petrasova D et al (2007) Oxidative stress in patients with COPD and pulmonary hypertension. Wien Klin Wochenschr 119(13–14):428–434
Kanno S, Wu YJ et al (2001) Angiotensin-converting enzyme inhibitor preserves p21 and endothelial nitric oxide synthase expression in monocrotaline-induced pulmonary arterial hypertension in rats. Circulation 104(8):945–950
Karuppiah K, Druhan LJ et al (2011) Suppression of eNOS-derived superoxide by caveolin-1: a biopterin-dependent mechanism. Am J Physiol Heart Circ Physiol 301(3):H903–H911
Kato GJ, McGowan V et al (2006) Lactate dehydrogenase as a biomarker of hemolysis-associated nitric oxide resistance, priapism, leg ulceration, pulmonary hypertension, and death in patients with sickle cell disease. Blood 107(6):2279–2285
Kato GJ, Wang Z et al (2009) Endogenous nitric oxide synthase inhibitors in sickle cell disease: abnormal levels and correlations with pulmonary hypertension, desaturation, haemolysis, organ dysfunction and death. Br J Haematol 145(4):506–513
Kaul DK, Liu XD et al (2000) Impaired nitric oxide-mediated vasodilation in transgenic sickle mouse. Am J Physiol Heart Circ Physiol 278(6):H1799–H1806
Kaul DK, Liu XD et al (2004) Effect of fetal hemoglobin on microvascular regulation in sickle transgenic-knockout mice. J Clin Invest 114(8):1136–1145
Kelleher ZT, Matsumoto A et al (2007) NOS2 regulation of NF-kappaB by S-nitrosylation of p65. J Biol Chem 282(42):30667–30672
Kelley EE, Hock T et al (2006) Moderate hypoxia induces xanthine oxidoreductase activity in arterial endothelial cells. Free Radic Biol Med 40(6):952–959
Kelley EE, Khoo NK et al (2010) Hydrogen peroxide is the major oxidant product of xanthine oxidase. Free Radic Biol Med 48(4):493–498
Khoo JP, Zhao L et al (2005) Pivotal role for endothelial tetrahydrobiopterin in pulmonary hypertension. Circulation 111(16):2126–2133

Kibbe MR, Li J et al (2000) Inducible nitric oxide synthase (iNOS) expression upregulates p21 and inhibits vascular smooth muscle cell proliferation through p42/44 mitogen-activated protein kinase activation and independent of p53 and cyclic guanosine monophosphate. J Vasc Surg 31(6):1214–1228

Krasuski RA, Devendra GP et al (2011) Response to inhaled nitric oxide predicts survival in patients with pulmonary hypertension. J Card Fail 17(4):265–271

Kunuthur SP, Milliken PH et al (2011) Tetrahydrobiopterin analogues with NO-dependent pulmonary vasodilator properties. Eur J Pharmacol 650(1):371–377

Kurz S, Hink U et al (1999) Evidence for a causal role of the renin-angiotensin system in nitrate tolerance. Circulation 99(24):3181–3187

Landburg PP, Teerlink T et al (2008) Plasma concentrations of asymmetric dimethylarginine, an endogenous nitric oxide synthase inhibitor, are elevated in sickle cell patients but do not increase further during painful crisis. Am J Hematol 83(7):577–579

Landburg PP, Teerlink T et al (2010) Plasma asymmetric dimethylarginine concentrations in sickle cell disease are related to the hemolytic phenotype. Blood Cells Mol Dis 44(4):229–232

Langle F, Roth E et al (1995) Arginase release following liver reperfusion. Evidence of hemodynamic action of arginase infusions. Transplantation 59(11):1542–1549

Lavoie A, Hall JB et al (1996) Life-threatening effects of discontinuing inhaled nitric oxide in severe respiratory failure. Am J Respir Crit Care Med 153(6 Pt 1):1985–1987

Li H, Forstermann U (2000) Nitric oxide in the pathogenesis of vascular disease. J Pathol 190(3):244–254

Li D, Zhou N et al (1999) Soluble guanylate cyclase gene expression and localization in rat lung after exposure to hypoxia. Am J Physiol 277(4 Pt 1):L841–L847

List BM, Klosch B et al (1997) Characterization of bovine endothelial nitric oxide synthase as a homodimer with down-regulated uncoupled NADPH oxidase activity: tetrahydrobiopterin binding kinetics and role of haem in dimerization. Biochem J 323(Pt 1):159–165

Liu L, Hausladen A et al (2001) A metabolic enzyme for S-nitrosothiol conserved from bacteria to humans. Nature 410(6827):490–494

Liu L, Yan Y et al (2004) Essential roles of S-nitrosothiols in vascular homeostasis and endotoxic shock. Cell 116(4):617–628

Liu JQ, Zelko IN et al (2006) Hypoxic pulmonary hypertension: role of superoxide and NADPH oxidase (gp91phox). Am J Physiol Lung Cell Mol Physiol 290(1):L2–L10

Liu X, Gai Y et al (2010) Trimetazidine inhibits pressure overload-induced cardiac fibrosis through NADPH oxidase-ROS-CTGF pathway. Cardiovasc Res 88(1):150–158

Lu X, Murphy TC et al (2010) PPAR{gamma} regulates hypoxia-induced Nox4 expression in human pulmonary artery smooth muscle cells through NF-{kappa}B. Am J Physiol Lung Cell Mol Physiol 299(4):L559–L566

Lundberg JO, Weitzberg E et al (2008) The nitrate-nitrite-nitric oxide pathway in physiology and therapeutics. Nat Rev Drug Discov 7(2):156–167

Lundberg JO, Gladwin MT et al (2009) Nitrate and nitrite in biology, nutrition and therapeutics. Nat Chem Biol 5(12):865–869

Machado RF, Anthi A et al (2006) N-terminal pro-brain natriuretic peptide levels and risk of death in sickle cell disease. JAMA 296(3):310–318

Machado RF, Hildescheim M et al (2009) NT-pro brain natriuretic peptide levels and the risk of stroke and death in the cooperative study of sickle cell disease. Blood 114:1541

Malhotra R, Hess D et al (2011) Vasoreactivity to inhaled nitric oxide with oxygen predicts long-term survival in pulmonary arterial hypertension. Pulm Circ 1(2):250–258

Markewitz BA, Michael JR (2000) Inhaled nitric oxide in adults with the acute respiratory distress syndrome. Respir Med 94(11):1023–1028

McLaughlin VV, Archer SL et al (2009a) ACCF/AHA 2009 expert consensus document on pulmonary hypertension a report of the American College of Cardiology Foundation Task Force on Expert Consensus Documents and the American Heart Association developed in collaboration with the American College of Chest Physicians; American Thoracic Society, Inc.; and the Pulmonary Hypertension Association. J Am Coll Cardiol 53(17):1573–1619

McLaughlin VV, Archer SL et al (2009b) ACCF/AHA 2009 expert consensus document on pulmonary hypertension: a report of the American College of Cardiology Foundation Task Force on Expert Consensus Documents and the American Heart Association: developed in collaboration with the American College of Chest Physicians, American Thoracic Society, Inc., and the Pulmonary Hypertension Association. Circulation 119(16):2250–2294

Mehari A, Gladwin MT et al (2012) Mortality in adults with sickle cell disease and pulmonary hypertension. JAMA 307(12):1254–1256

Mehta S, Stewart DJ et al (1995) Short-term pulmonary vasodilation with L-arginine in pulmonary hypertension. Circulation 92(6):1539–1545

Meyer C, Heiss C et al (2010) Hemodialysis-induced release of hemoglobin limits nitric oxide bioavailability and impairs vascular function. J Am Coll Cardiol 55(5):454–459

Miller OI, Tang SF et al (1995) Rebound pulmonary hypertension on withdrawal from inhaled nitric oxide. Lancet 346(8966):51–52

Minneci PC, Deans KJ et al (2005) Hemolysis-associated endothelial dysfunction mediated by accelerated NO inactivation by decompartmentalized oxyhemoglobin. J Clin Invest 115(12):3409–3417

Minniti CP, Sable C et al (2009) Elevated tricuspid regurgitant jet velocity in children and adolescents with sickle cell disease: association with hemolysis and hemoglobin oxygen desaturation. Haematologica 94(3):340–347

Mittal M, Roth M et al (2007) Hypoxia-dependent regulation of nonphagocytic NADPH oxidase subunit NOX4 in the pulmonary vasculature. Circ Res 101(3):258–267

Mittendorf J, Weigand S et al (2009) Discovery of riociguat (BAY 63-2521): a potent, oral stimulator of soluble guanylate cyclase for the treatment of pulmonary hypertension. ChemMedChem 4(5):853–865

Modin A, Bjorne H et al (2001) Nitrite-derived nitric oxide: a possible mediator of 'acidic-metabolic' vasodilation. Acta Physiol Scand 171(1):9–16

Morris CR, Morris SM Jr et al (2003) Arginine therapy: a new treatment for pulmonary hypertension in sickle cell disease? Am J Respir Crit Care Med 168(1):63–69

Morris CR, Kato GJ et al (2005) Dysregulated arginine metabolism, hemolysis-associated pulmonary hypertension, and mortality in sickle cell disease. JAMA 294(1):81–90

Morris CR, Gladwin MT et al (2008) Nitric oxide and arginine dysregulation: a novel pathway to pulmonary hypertension in hemolytic disorders. Curr Mol Med 8(7):620–632

Moya MP, Gow AJ et al (2002) Inhaled ethyl nitrite gas for persistent pulmonary hypertension of the newborn. Lancet 360(9327):141–143

Munzel T, Giaid A et al (1995a) Evidence for a role of endothelin 1 and protein kinase C in nitroglycerin tolerance. Proc Natl Acad Sci USA 92(11):5244–5248

Munzel T, Sayegh H et al (1995b) Evidence for enhanced vascular superoxide anion production in nitrate tolerance. A novel mechanism underlying tolerance and cross-tolerance. J Clin Invest 95(1):187–194

Munzel T, Kurz S et al (1996) Hydralazine prevents nitroglycerin tolerance by inhibiting activation of a membrane-bound NADH oxidase. A new action for an old drug. J Clin Invest 98 (6):1465–1470

Nagaya N, Uematsu M et al (2001) Short-term oral administration of L-arginine improves hemodynamics and exercise capacity in patients with precapillary pulmonary hypertension. Am J Respir Crit Care Med 163(4):887–891

Namachivayam P, Theilen U et al (2006) Sildenafil prevents rebound pulmonary hypertension after withdrawal of nitric oxide in children. Am J Respir Crit Care Med 174(9):1042–1047

Nandi M, Miller A et al (2005) Pulmonary hypertension in a GTP-cyclohydrolase 1-deficient mouse. Circulation 111(16):2086–2090

Naoman SG, Nouraie M et al (2010) Echocardiographic findings in patients with sickle cell disease. Ann Hematol 89(1):61–66

Nisbet RE, Bland JM et al (2010) Rosiglitazone attenuates chronic hypoxia-induced pulmonary hypertension in a mouse model. Am J Respir Cell Mol Biol 42(4):482–490

Nisimoto Y, Motalebi S et al (1999) The p67(phox) activation domain regulates electron flow from NADPH to flavin in flavocytochrome b(558). J Biol Chem 274(33):22999–23005

Nolan VG, Wyszynski DF et al (2005) Hemolysis associated priapism in sickle cell disease. Blood 106(9):3264–3267

Nolan VG, Adewoye A et al (2006) Sickle cell leg ulcers: associations with haemolysis and SNPs in Klotho, TEK and genes of the TGF-beta/BMP pathway. Br J Haematol 133(5):570–578

Nong Z, Stassen JM et al (1996) Inhibition of tissue angiotensin-converting enzyme with quinapril reduces hypoxic pulmonary hypertension and pulmonary vascular remodeling. Circulation 94(8):1941–1947

Nozik-Grayck E, Suliman HB et al (2008) Lung EC-SOD overexpression attenuates hypoxic induction of Egr-1 and chronic hypoxic pulmonary vascular remodeling. Am J Physiol Lung Cell Mol Physiol 295(3):L422–L430

Oak JH, Cai H (2007) Attenuation of angiotensin II signaling recouples eNOS and inhibits nonendothelial NOX activity in diabetic mice. Diabetes 56(1):118–126

Onyekwere OC, Campbell A et al (2008) Pulmonary hypertension in children and adolescents with sickle cell disease. Pediatr Cardiol 29(2):309–312

Oudiz RJ, Brundage BH et al (2012) Tadalafil for the treatment of pulmonary arterial hypertension: a double-blind 52-week uncontrolled extension study. J Am Coll Cardiol 60(8):768–774

Ozaki M, Kawashima S et al (2002) Overexpression of endothelial nitric oxide synthase accelerates atherosclerotic lesion formation in apoE-deficient mice. J Clin Invest 110(3):331–340

Pagano PJ, Clark JK et al (1997) Localization of a constitutively active, phagocyte-like NADPH oxidase in rabbit aortic adventitia: enhancement by angiotensin II. Proc Natl Acad Sci USA 94(26):14483–14488

Palmer RM, Ferrige AG et al (1987) Nitric oxide release accounts for the biological activity of endothelium-derived relaxing factor. Nature 327(6122):524–526

Palmer RM, Ashton DS et al (1988) Vascular endothelial cells synthesize nitric oxide from L-arginine. Nature 333(6174):664–666

Parent F, Bachir D et al (2011) A hemodynamic study of pulmonary hypertension in sickle cell disease. N Engl J Med 365(1):44–53

Pearson DL, Dawling S et al (2001) Neonatal pulmonary hypertension – urea-cycle intermediates, nitric oxide production, and carbamoyl-phosphate synthetase function. N Engl J Med 344(24):1832–1838

Quyyumi AA, Dakak N et al (1995) Nitric oxide activity in the human coronary circulation. Impact of risk factors for coronary atherosclerosis. J Clin Invest 95(4):1747–1755

Rabinovitch M (2008) Molecular pathogenesis of pulmonary arterial hypertension. J Clin Invest 118(7):2372–2379

Rassaf T, Flogel U et al (2007) Nitrite reductase function of deoxymyoglobin: oxygen sensor and regulator of cardiac energetics and function. Circ Res 100(12):1749–1754

Reiter CD, Wang X et al (2002) Cell-free hemoglobin limits nitric oxide bioavailability in sickle-cell disease. Nat Med 8(12):1383–1389

Robbins IM, Hemnes AR et al (2011) Safety of sapropterin dihydrochloride (6r-bh4) in patients with pulmonary hypertension. Exp Lung Res 37(1):26–34

Roberts JD Jr, Fineman JR et al (1997) Inhaled nitric oxide and persistent pulmonary hypertension of the newborn. The Inhaled Nitric Oxide Study Group. N Engl J Med 336(9):605–610

Rossaint R, Falke KJ et al (1993) Inhaled nitric oxide for the adult respiratory distress syndrome. N Engl J Med 328(6):399–405

Rother RP, Bell L et al (2005) The clinical sequelae of intravascular hemolysis and extracellular plasma hemoglobin: a novel mechanism of human disease. JAMA 293(13):1653–1662

Sachdev V, Kato GJ et al (2011) Echocardiographic markers of elevated pulmonary pressure and left ventricular diastolic dysfunction are associated with exercise intolerance in adults and adolescents with homozygous sickle cell anemia in the United States and United Kingdom. Circulation 124(13):1452–1460

Sanders KA, Hoidal JR (2007) The NOX on pulmonary hypertension. Circ Res 101(3):224–226

Sanghani PC, Davis WI et al (2009) Kinetic and cellular characterization of novel inhibitors of S-nitrosoglutathione reductase. J Biol Chem 284(36):24354–24362

Schnog JB, Teerlink T et al (2005) Plasma levels of asymmetric dimethylarginine (ADMA), an endogenous nitric oxide synthase inhibitor, are elevated in sickle cell disease. Ann Hematol 84(5):282–286

Sharma S, Kumar S et al (2009) Alterations in lung arginine metabolism in lambs with pulmonary hypertension associated with increased pulmonary blood flow. Vascul Pharmacol 51(5–6):359–364

Shiva S, Gladwin MT (2009) Shining a light on tissue NO stores: near infrared release of NO from nitrite and nitrosylated hemes. J Mol Cell Cardiol 46(1):1–3

Shiva S, Huang Z et al (2007) Deoxymyoglobin is a nitrite reductase that generates nitric oxide and regulates mitochondrial respiration. Circ Res 100(5):654–661

Sitbon O, Humbert M et al (1998) Inhaled nitric oxide as a screening agent for safely identifying responders to oral calcium-channel blockers in primary pulmonary hypertension. Eur Respir J 12(2):265–270

Sitbon O, Humbert M et al (2005) Long-term response to calcium channel blockers in idiopathic pulmonary arterial hypertension. Circulation 111(23):3105–3111

Spiekermann S, Schenk K et al (2009) Increased xanthine oxidase activity in idiopathic pulmonary arterial hypertension. Eur Respir J 34(1):276

Stasch JP, Becker EM et al (2001) NO-independent regulatory site on soluble guanylate cyclase. Nature 410(6825):212–215

Stewart DJ, Levy RD et al (1991) Increased plasma endothelin-1 in pulmonary hypertension: marker or mediator of disease? Ann Intern Med 114(6):464–469

Sturrock A, Cahill B et al (2006) Transforming growth factor-beta1 induces Nox4 NAD(P)H oxidase and reactive oxygen species-dependent proliferation in human pulmonary artery smooth muscle cells. Am J Physiol Lung Cell Mol Physiol 290(4):L661–L673

Sutendra G, Bonnet S et al (2010) Fatty acid oxidation and malonyl-CoA decarboxylase in the vascular remodeling of pulmonary hypertension. Sci Transl Med 2(44):44ra58

Sutliff RL, Kang BY et al (2010) PPARgamma as a potential therapeutic target in pulmonary hypertension. Ther Adv Respir Dis 4(3):143–160

Sydow K, Daiber A et al (2004) Central role of mitochondrial aldehyde dehydrogenase and reactive oxygen species in nitroglycerin tolerance and cross-tolerance. J Clin Invest 113(3):482–489

Tabima DM, Frizzell S et al (2012) Reactive oxygen and nitrogen species in pulmonary hypertension. Free Radic Biol Med 52(9):1970–1986

Takeya R, Sumimoto H (2003) Molecular mechanism for activation of superoxide-producing NADPH oxidases. Mol Cells 16(3):271–277

Teng RJ, Du J et al (2011) Sepiapterin improves angiogenesis of pulmonary artery endothelial cells with in utero pulmonary hypertension by recoupling endothelial nitric oxide synthase. Am J Physiol Lung Cell Mol Physiol 301(3):L334–L345

Tian J, Smith A et al (2009) Effect of PPARgamma inhibition on pulmonary endothelial cell gene expression: gene profiling in pulmonary hypertension. Physiol Genomics 40(1):48–60

Tiso M, Tejero J et al (2011) Human neuroglobin functions as a redox-regulated nitrite reductase. J Biol Chem 286(20):18277–18289

Tofovic SP, Jackson EK et al (2009) Adenosine deaminase-adenosine pathway in hemolysis-associated pulmonary hypertension. Med Hypotheses 72(6):713–719

Tota B, Quintieri AM et al (2010) The emerging role of nitrite as an endogenous modulator and therapeutic agent of cardiovascular function. Curr Med Chem 17(18):1915–1925

Tuder RM, Groves B et al (1994) Exuberant endothelial cell growth and elements of inflammation are present in plexiform lesions of pulmonary hypertension. Am J Pathol 144(2):275–285

Vachiery JL, Huez S et al (2011) Safety, tolerability and pharmacokinetics of an intravenous bolus of sildenafil in patients with pulmonary arterial hypertension. Br J Clin Pharmacol 71(2):289–292

Venema VJ, Zou R et al (1997) Caveolin-1 detergent solubility and association with endothelial nitric oxide synthase is modulated by tyrosine phosphorylation. Biochem Biophys Res Commun 236(1):155–161

Vignais PV (2002) The superoxide-generating NADPH oxidase: structural aspects and activation mechanism. Cell Mol Life Sci 59(9):1428–1459

Vosatka RJ, Kashyap S et al (1994) Arginine deficiency accompanies persistent pulmonary hypertension of the newborn. Biol Neonate 66(2–3):65–70

Voskaridou E, Christoulas D et al (2010) The effect of prolonged administration of hydroxyurea on morbidity and mortality in adult patients with sickle cell syndromes: results of a 17-year, single-center trial (LaSHS). Blood 115(12):2354–2363

Wedgwood S, Black SM (2003) Role of reactive oxygen species in vascular remodeling associated with pulmonary hypertension. Antioxid Redox Signal 5(6):759–769

Wedgwood S, Bekker JM et al (2001) Shear stress regulation of endothelial NOS in fetal pulmonary arterial endothelial cells involves PKC. Am J Physiol Lung Cell Mol Physiol 281(2):L490–L498

Wedgwood S, Steinhorn RH et al (2005) Increased hydrogen peroxide downregulates soluble guanylate cyclase in the lungs of lambs with persistent pulmonary hypertension of the newborn. Am J Physiol Lung Cell Mol Physiol 289(4):L660–L666

Weerackody RP, Welsh DJ et al (2009) Inhibition of p38 MAPK reverses hypoxia-induced pulmonary artery endothelial dysfunction. Am J Physiol Heart Circ Physiol 296(5):H1312–H1320

Weinberger B, Laskin DL et al (2001) The toxicology of inhaled nitric oxide. Toxicol Sci 59(1):5–16

Wenzel P, Schulz E et al (2008) AT1-receptor blockade by telmisartan upregulates GTP-cyclohydrolase I and protects eNOS in diabetic rats. Free Radic Biol Med 45(5):619–626

Whalen EJ, Foster MW et al (2007) Regulation of beta-adrenergic receptor signaling by S-nitrosylation of G-protein-coupled receptor kinase 2. Cell 129(3):511–522

Williamson JR, Chang K et al (1993) Hyperglycemic pseudohypoxia and diabetic complications. Diabetes 42(6):801–813

Wolin MS (2009) Reactive oxygen species and the control of vascular function. Am J Physiol Heart Circ Physiol 296(3):H539–H549

Wolin MS, Ahmad M et al (2005) The sources of oxidative stress in the vessel wall. Kidney Int 67(5):1659–1661

Wood KC, Hebbel RP et al (2005) Endothelial cell NADPH oxidase mediates the cerebral microvascular dysfunction in sickle cell transgenic mice. FASEB J 19(8):989–991

Wood KC, Hsu LL et al (2008) Sickle cell disease vasculopathy: a state of nitric oxide resistance. Free Radic Biol Med 44(8):1506–1528

Wu X, Du L et al (2010) Increased nitrosoglutathione reductase activity in hypoxic pulmonary hypertension in mice. J Pharmacol Sci 113(1):32–40

Wunderlich C, Schober K et al (2008) The adverse cardiopulmonary phenotype of caveolin-1 deficient mice is mediated by a dysfunctional endothelium. J Mol Cell Cardiol 44(5):938–947

Xu W, Kaneko FT et al (2004) Increased arginase II and decreased NO synthesis in endothelial cells of patients with pulmonary arterial hypertension. FASEB J 18(14):1746–1748

Yeo TW, Lampah DA et al (2007) Impaired nitric oxide bioavailability and L-arginine reversible endothelial dysfunction in adults with falciparum malaria. J Exp Med 204(11):2693–2704

Yeo TW, Lampah DA et al (2009) Relationship of cell-free hemoglobin to impaired endothelial nitric oxide bioavailability and perfusion in severe falciparum malaria. J Infect Dis 200(10):1522–1529

Zhang J, Chen Z et al (2004) Role of mitochondrial aldehyde dehydrogenase in nitroglycerin-induced vasodilation of coronary and systemic vessels: an intact canine model. Circulation 110(6):750–755

Zhao YY, Zhao YD et al (2009) Persistent eNOS activation secondary to caveolin-1 deficiency induces pulmonary hypertension in mice and humans through PKG nitration. J Clin Invest 119(7):2009–2018

Zisman DA, Schwarz M et al (2010) A controlled trial of sildenafil in advanced idiopathic pulmonary fibrosis. N Engl J Med 363(7):620–628

Zuckerbraun BS, Stoyanovsky DA et al (2007) Nitric oxide-induced inhibition of smooth muscle cell proliferation involves S-nitrosation and inactivation of RhoA. Am J Physiol Cell Physiol 292(2):C824–C831

Zuckerbraun BS, Shiva S et al (2010) Nitrite potently inhibits hypoxic and inflammatory pulmonary arterial hypertension and smooth muscle proliferation via xanthine oxidoreductase-dependent nitric oxide generation. Circulation 121(1):98–109

Zuckerbraun BS, George P et al (2011) Nitrite in pulmonary arterial hypertension: therapeutic avenues in the setting of dysregulated arginine/nitric oxide synthase signalling. Cardiovasc Res 89(3):542–552

Zulueta JJ, Sawhney R et al (2002) Modulation of inducible nitric oxide synthase by hypoxia in pulmonary artery endothelial cells. Am J Respir Cell Mol Biol 26(1):22–30

Rho-Kinase Inhibitors

Yoshihiro Fukumoto and Hiroaki Shimokawa

Abstract Pulmonary arterial hypertension (PAH) is a fatal disease with poor prognosis characterized by progressive elevation of pulmonary arterial pressure and vascular resistance due to pulmonary artery hyperconstriction and remodeling; however, the precise mechanism of PAH still remains to be elucidated. Although anticoagulant agents, pulmonary vasodilators, and lung transplantation are currently used for the treatment of PAH, more effective treatment needs to be developed. Rho-kinase causes vascular smooth muscle hyperconstriction and vascular remodeling through inhibition of myosin phosphatase and activation of its downstream effectors. In a series of experimental and clinical studies, it has been demonstrated that Rho-kinase-mediated pathway plays an important role in various cellular functions not only in vascular smooth muscle hyperconstriction but also in actin cytoskeleton organization, cell adhesion and motility, cytokinesis, and gene expressions, all of which may be involved in the pathogenesis of arteriosclerosis. Rho-kinase is activated in animal models of PAH (monocrotaline and chronic hypoxia) associated with enhanced pulmonary vasoconstriction and proliferation, impaired endothelial vasodilator functions, and pulmonary remodeling. Therapeutic application of Rho-kinase inhibitors reverses established experimental pulmonary hypertension. Further, administration or inhalation of Rho-kinase inhibitors exerts acute pulmonary vasodilation in patients with PAH who were refractory to conventional therapies. Taken together, Rho-kinase is a novel and important therapeutic target of PAH, and Rho-kinase inhibitors are a promising new class of drugs for this fatal disorder.

Y. Fukumoto • H. Shimokawa (✉)
Department of Cardiovascular Medicine, Tohoku University Graduate School of Medicine, 1-1 Seiryo-machi, Aoba-ku, Sendai 980-8575, Japan
e-mail: shimo@cardio.med.tohoku.ac.jp

Keywords Rho-kinase • Pulmonary arterial hypertension • Pulmonary arteriopathy • Pulmonary arterial hyperconstriction

Contents

1 Introduction	352
2 Novel Pathophysiological Pathways of PAH	353
2.1 Important Role of Rho-Kinase in the Cardiovascular Fields	353
2.2 PAH and Rho-Kinase Pathway	355
3 Novel Therapeutic Target as Rho-Kinase Pathway in PAH	357
3.1 Potential Importance of Rho-Kinase Inhibitors for the Treatment of PAH	357
3.2 Enhanced Rho-Kinase Expression and Activity in Patients with PAH	358
4 New Developments of Novel Rho Kinase Inhibitors	360
5 Future Perspectives and Conclusions	360
References	360

1 Introduction

In mid 1990s, two Japanese groups and one Singapore group independently identified Rho-kinase/ROK/ROCK as an effecter of the small GTP-binding protein Rho (Leung et al. 1995; Ishizaki et al. 1996; Amano et al. 1997), which plays an important role in various cellular functions, including smooth muscle contraction, actin cytoskeleton organization, cell adhesion and motility, cytokinesis, and gene expressions (Narumiya 1996; Shimokawa and Takeshita 2005; Loirand et al. 2006). The Rho/Rho-kinase pathway has recently attracted much attention in the cardiovascular research field for several reasons (Fig. 1). First, the Rho/Rho-kinase pathway plays an important role in various cellular functions that are involved in the pathogenesis of a variety of cardiovascular diseases (Shimokawa and Takeshita 2005; Shimokawa 2000). Second, this intracellular signaling pathway is substantially involved in the effects of many vasoactive substances that are implicated in the pathogenesis of cardiovascular diseases (Shimokawa and Takeshita 2005; Shimokawa 2000). Third, the so-called pleiotropic effects of statins, especially those of high-doses of statins, may be mediated, at least in part, by their inhibitory effects on Rho with a resultant inhibition on Rho-kinase (Shimokawa and Takeshita 2005; Shimokawa 2000). Fourth, the important roles of the Rho-kinase pathway have been recently demonstrated in the pathogenesis of pulmonary arterial hypertension (PAH) (Fig. 2) (Fukumoto et al. 2005, 2007; Shimokawa 2002; Abe et al. 2004, 2006; Do e et al. 2009; Fukumoto and Shimokawa 2011).

In this article, we will briefly review the role of Rho-kinase pathway on PAH, in terms of pathophysiology and treatment.

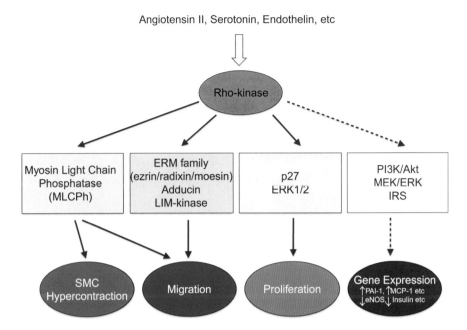

Fig. 1 Pathophysiology of Rho-kinase pathway. Rho-kinase pathway plays an important role in various cellular functions that are involved in the pathogenesis of a variety of cardiovascular diseases

2 Novel Pathophysiological Pathways of PAH

The pathological changes of the pulmonary arteries in PH include endothelial injury, proliferation and hyperconstriction of vascular smooth muscle cells (VSMC), and migration of inflammatory cells (Fig. 2) (Fukumoto and Shimokawa 2011; Humbert et al. 2004; McLaughlin and McGoon 2006).

2.1 Important Role of Rho-Kinase in the Cardiovascular Fields

Recent advances in molecular biology have elucidated the substantial involvement of intracellular signaling pathways mediated by small GTP-binding proteins (G proteins), such as Rho, Ras, Rab, Sarl/Arf, and Ran families (Fukata et al. 2001; Takai et al. 2001). The Rho/Rho-kinase pathway has recently attracted much attention in various research fields, especially in the cardiovascular research field (Shimokawa and Takeshita 2005; Shimokawa 2000, 2002; Shimokawa and Rashid 2007a).

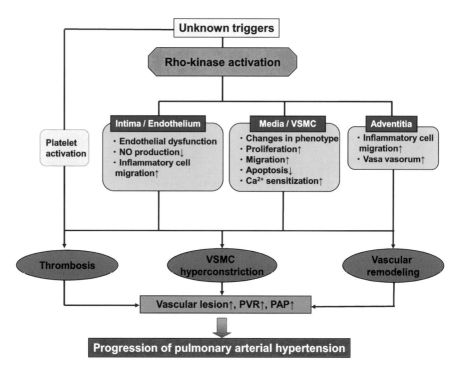

Fig. 2 Roles of Rho-kinase pathway in the pathogenesis of pulmonary hypertension. Rho-kinase activation has been confirmed to be substantially involved in the pathological changes of the pulmonary arteries. *NO* nitric oxide, *PAP* pulmonary arterial pressure, *PVR* pulmonary vascular resistance, *VSMC* vascular smooth muscle cell [from Fukumoto and Shimokawa 2011 with permission]

It has been previously demonstrated that Rho-kinase is a novel therapeutic target in ischemic heart disease (Shimokawa and Takeshita 2005; Shimokawa 2002). Rho-kinase suppresses myosin phosphatase activity by phosphorylating the myosin-binding subunit of the enzyme and thus augments VSMC contraction at a given intracellular calcium concentration (Fig. 3) (Uehata et al. 1997; Somlyo and Somlyo 2000). It has been also demonstrated that the Rho-kinase pathway is associated with enhanced myosin light chain (MLC) phosphorylations at the hyperconstrictive artery segments in animals (Shimokawa and Takeshita 2005; Shimokawa 2002). The activity and the expression of Rho-kinase are enhanced at the hyperconstrictive coronary segments, thereby suppressing myosin phosphatase through phosphorylation of its myosin-binding subunit with a resultant increase in MLC phosphorylations and hyperconstriction (Shimokawa and Takeshita 2005; Shimokawa 2002). Thus, VSMC hyperconstriction mediated by activated Rho-kinase plays a key role in patients with coronary artery spasm (Shimokawa and Takeshita 2005; Shimokawa 2002; Masumoto et al. 2001, 2002), suggesting that Rho-kinase inhibition is an important therapeutic strategy for vasospastic angina (Shimokawa 2002; Masumoto et al. 2002). Moreover, it has recently been

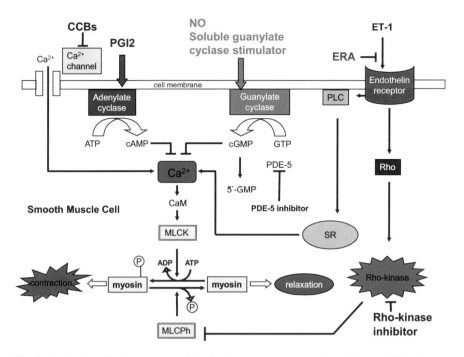

Fig. 3 Mechanism of pulmonary vasodilatation in response to conventional drugs and Rho-kinase inhibitors. *5-HT* serotonin, *CaM* calmodulin, *CCBs* calcium channel blockers, *ERA* endothelin receptor antagonist, *ET-1* endothelin-1, *MLCK* myosin light chain kinase, *MLCPh* myosin light chain phosphatase, *NO* nitric oxide, *PDE-5* phosphodiesterase-5, *PLC* phospholipase C, *VSMC* vascular smooth muscle cell, PGI_2 prostacyclin, *PLC* phospholipase C, *SR* sarcoplasmic reticulum

demonstrated that Rho-kinase inhibition increases eNOS expression and decreases inflammatory cell migration and angiotensin II-induced upregulation of atherogenic molecules [e.g., monocyte chemoattractant protein (MCP)-1, plasminogen activator inhibitor (PAI)-1, and NADPH oxidase] and cardiovascular hypertrophy both in vitro and in vivo (Shimokawa and Takeshita 2005; Shimokawa 2002; Higashi et al. 2003).

2.2 PAH and Rho-Kinase Pathway

Increased pulmonary vascular resistance in PAH is caused by both pulmonary vascular remodeling and sustained pulmonary vasoconstriction (Giaid and Saleh 1995; Higenbottam and Laude 1998; Yuan et al. 1998), in which endothelial dysfunction and VSMC hyperconstriction may be involved through Rho-kinase activation (Fig. 2) (Higenbottam and Laude 1998; Yuan et al. 1998). Indeed, in patients with PAH, eNOS expression is reduced and pulmonary VSMC are hyperreactive (Giaid and Saleh 1995; Yuan et al. 1998). It is thus conceivable that the Rho-kinase pathway plays an important role in the pathogenesis of PAH (Figs. 2 and 3).

Rho-kinase suppresses myosin phosphatase activity by phosphorylating the myosin-binding subunit of the enzyme, thus augmenting VSMC contraction at a given intracellular calcium concentration (Uehata et al. 1997; Somlyo and Somlyo 2000). VSMC hypercontraction mediated by activated Rho-kinase plays a key role not only in coronary artery spasm but also in PAH (Fukumoto et al. 2005, 2007; Shimokawa 2002; Fukumoto and Shimokawa 2011; Masumoto et al. 2002; Mohri et al. 2003; Fujita et al. 2010). Rho-kinase inhibition may be preferable to calcium channel blockers because of its selective spasmolytic effect on vascular hyperconstrictive segments (Fukumoto et al. 2007; Shimokawa 2002; Masumoto et al. 2002).

2.2.1 Rho-Kinase and Inflammation

A number of studies have suggested that inflammation may be involved in the pathogenesis of PAH (Fukumoto et al. 2007; Dorfmuller et al. 2003). Some patients with idiopathic PAH have immunological disturbances (e.g., circulating autoantibodies, such as antinuclear antibodies) and elevated circulating levels of pro-inflammatory cytokines (e.g., interleukin-1 and -6) (Dorfmuller et al. 2003). It has been demonstrated that Rho-kinase is upregulated by inflammatory stimuli (Shimokawa and Takeshita 2005; Hiroki et al. 2004; Liao et al. 2007) and that Rho-kinase inhibition increases endothelial nitric oxide synthase (eNOS) expression and inhibits inflammatory cell migration and angiotensin II-induced upregulation of monocyte chemoattractant protein-1 and plasminogen activator inhibitor-1 in vivo or in vitro (Shimokawa 2002), in which the Rho-kinase pathway may play an important role in the development of PAH.

2.2.2 Rho-Kinase and PAH

Indeed, it has been demonstrated that the long-term inhibition of Rho-kinase ameliorates monocrotaline (MCT)-induced PAH and hypoxia-induced PAH in animal models (Fukumoto et al. 2007; Abe et al. 2004, 2006; Fukumoto and Shimokawa 2011; Oka et al. 2008; Dahal et al. 2010). In those studies, Rho-kinase activity of pulmonary arteries was enhanced irrespective of the different etiologies and long-term treatment with Rho-kinase inhibitors ameliorated endothelial dysfunction and suppressed hypercontraction and proliferation of VSMC and migration of inflammatory cells (Abe et al. 2004, 2006; Oka et al. 2008). In clinical studies, it also has been demonstrated that a Rho-kinase inhibitor, fasudil, acutely improves pulmonary hemodynamics in patients with PAH (Fukumoto et al. 2005; Fujita et al. 2010; Ishikura et al. 2006).

3 Novel Therapeutic Target as Rho-Kinase Pathway in PAH

3.1 Potential Importance of Rho-Kinase Inhibitors for the Treatment of PAH

We have It has recently been demonstrated that Rho-kinase is a novel therapeutic target not only in ischemic heart disease and essential hypertension but also in PAH (Shimokawa 2002; Abe et al. 2004, 2006; Fukumoto et al. 2005; Jiang et al. 2007; Tawara et al. 2007; Yasuda et al. 2011). Potent and selective inhibitors of Rho-kinase like fasudil and azaindole-1 are available possessing an inhibitory effect on Rho-kinase being 100 times and 1,000 times more potent than on other protein kinases C and myosin light chain kinase, respectively (Shimokawa 2002; Shimokawa et al. 1999; Shimokawa and Rashid 2007b; Kast et al. 2007). Rho-kinase inhibition ameliorates VSMC hyperconstriction, increases eNOS expression, and decreases inflammatory cell migration and angiotensin II-induced upregulation of MCP-1 and PAI-1 (Shimokawa 2002; Morishige et al. 2001).

In the rat model of monocrotaline (MCT)-induced PAH, fasudil markedly suppressed the development of PAH when started simultaneously with monocrotaline and even induced a regression of the disorder when started after the establishment of PAH (Abe et al. 2004). Further, the highly selective Rho-kinase inhibitor azaindole-1 has shown to reverse MCT-induced pulmonary hypertension (Dahal et al. 2010; Lohn et al. 2009). Oral treatment with fasudil and azaindole-1 is also effective to inhibit the development of PAH induced by chronic hypoxia in mice, in which both eNOS-dependent and -independent mechanisms were involved (Abe et al. 2006; Dahal et al. 2010; Lohn et al. 2009; Pankey et al. 2012; Peng et al. 2012). Inhalation of fasudil is effective to reduce pulmonary vascular resistance in animal models of PAH with various etiologies (Nagaoka et al. 2005). Rho-kinase inhibitors appear to have pulmonary vasodilator effects through mechanism different from conventional drugs (Fig. 3). Therefore, it is possible that the combination of a Rho-kinase inhibitor and conventional drugs exerts additive or synergistic effects. Indeed, prostacyclin lacks direct inhibitory effects on Rho-kinase and the combination of oral prostacyclin analogue, beraprost, and a Rho-kinase inhibitor, fasudil, is more effective than each monotherapy for ameliorating monocrotaline-induced PH in rats (Tawara et al. 2007; Abe et al. 2005). Finally, the acute beneficial effects of intravenous fasudil have been shown in patients with PAH (Fukumoto et al. 2005; Fujita et al. 2010).

We have performed a clinical trial to examin the mid-term effects of Rho-kinase inhibitor, addressing the efficacy and safety in patients with PAH. (Fukumoto et al. 2013). Through those trials, we expect to learn whether the long-term inhibition of Rho-kinase is a novel therapeutic strategy for the treatment in patients with PAH.

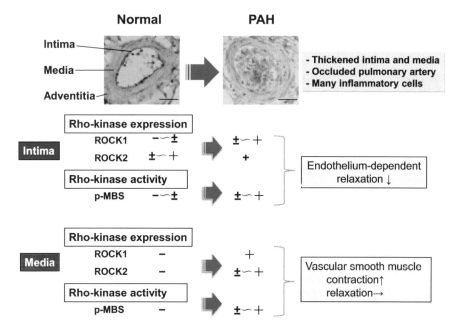

Fig. 4 Enhanced Rho-kinase expression and activity of pulmonary arteries from patients with pulmonary arterial hypertension. Immunohistological findings of Rho-kinase expression (both isoforms, ROCK 1 and 2) and activity in pulmonary arteries from patients with pulmonary arterial hypertension. Activated Rho-kinase expression and activity causes impairment of endothelium-dependent relaxation and enhances vascular smooth muscle contraction. *Scale bar*, 50 μm (from Fukumoto and Shimokawa 2011 with permission)

3.2 Enhanced Rho-Kinase Expression and Activity in Patients with PAH

Recently, the direct evidence for Rho-kinase activation has been demonstrated in patients with PAH (Do e et al. 2009), where Rho-kinase activity is enhanced in circulating neutrophils and the pulmonary arteries from patients with PAH, resulting in hypercontractions of the artery (Fig. 4). These findings support the previous findings in animal models of PAH and during right heart cardiac catheterization in patients with PAH (Shimokawa and Takeshita 2005; Abe et al. 2004, 2006; Fukumoto et al. 2005; Fujita et al. 2010; Oka et al. 2008; Ishikura et al. 2006). Thus, increased pulmonary vascular resistance may be caused, at least in part, by activated Rho-kinase pathway (Do e et al. 2009). In addition, in patients with PAH, eNOS expression is reduced and pulmonary VSMC are hyperreactive (Do e et al. 2009; Giaid and Saleh 1995; Xu et al. 2004). Indeed, activated Rho-kinase causes several important abnormalities, including eNOS downregulation in endothelial cells, VSMC hypercontraction through inhibition of myosin phosphatase, VSMC proliferation and migration, and inhibition of VSMC apoptosis (Figs. 1 and 2)

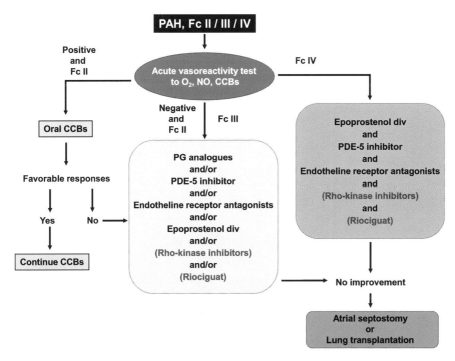

Fig. 5 Potential treatment algorithm for pulmonary arterial hypertension. In Fc II/III, the combination therapy can be beneficial compared with the mono-therapy. In Fc IV, the maximum use of medicines can be required. *CCBs* calcium channel blockers, *Fc* functional class, *NO* nitric oxide, *PAH* pulmonary arterial hypertension, *PG* prostaglandin, *PDE-5* phosphodiesterase-5 (from Fukumoto and Shimokawa 2011 with permission)

(Shimokawa and Takeshita 2005; Shimokawa 2002; Liao et al. 2007; Shimokawa and Rashid 2007b; Takemoto et al. 2002). Also, the direct evidence has been recently demonstrated that endothelial vasodilator function is impaired and VSMC contraction is enhanced in pulmonary arteries from patients with PAH (Fig. 4) (Do e et al. 2009). These findings are consistent with the previous studies with MCT-induced PH in rats and hypoxia-induced PH in mice and previous clinical studies with PAH patients (Abe et al. 2004, 2006; Fukumoto et al. 2005; Fujita et al. 2010; Oka et al. 2008; Ishikura et al. 2006). Furthermore, the inhibition of Rho-kinase abolishes VSMC hypercontraction of pulmonary arteries from idiopathic PAH patients (Do e et al. 2009), which is also consistent with the previous clinical study that acute inhibition of Rho-kinase improves pulmonary hemodynamics in PAH patients (Fukumoto et al. 2005; Fujita et al. 2010; Ishikura et al. 2006). However, it still remains to be examined whether those functional abnormalities of pulmonary arteries could be ameliorated by the long-term treatment with a Rho-kinase inhibitor in patients with PAH. For this purpose, the effects of long-acting oral form of fasudil in PAH patients are examined in a clinical trial.

4 New Developments of Novel Rho Kinase Inhibitors

Although a clinical trial with long-acting oral form of fasudil was done for the mid-term treatment, further clinical trials are required. Ahead of cardiovascular fields including PAH (Fukumoto et al. 2013), clinical use of Rho-kinase inhibitors has begun for glaucoma and ocular hypertension in ophthalmology (K-115 from D Western Therapeutics Institute/Kowa, AR-12286 from Aerie Pharmaceuticals Inc, AMA-0076 from Amakem NV) (Williams et al. 2011). As Rho-kinase pathway can be a new therapeutic target in various fields, new developments are required to establish safe and effective Rho-kinase inhibitors.

5 Future Perspectives and Conclusions

PAH still remains a fatal disease, leading to right ventricular failure and premature death. Although significant research progress has been made on the pathogenesis of PAH, especially with regard to Rho-kinase, the detailed mechanisms of the disorder still remain to be elucidated. In the clinical practice, significant progress has also been made for new treatment (e.g., endothelin receptor antagonists, Rho-kinase inhibitors, and riociguat, Fig. 5). The usefulness of those new drugs remains to be fully examined in future studies.

It is expected that clinical trials with long-term oral treatment with Rho-kinase inhibitors will elucidate their effectiveness and safety for the treatment of PAH in humans.

Acknowledgments The authors' works presented in this article were supported in part by the grants-in-aid from the Japanese Ministry of Education, Culture, Sports, Science, and Technology, Tokyo, Japan, grants-in-aid for Scientific Research on Innovative Areas (Research in a Proposed Research Area), MEXT, Japan, and Tohoku University Global COE for Conquest of Signal Transduction Diseases with Network Medicine, Sendai, Japan.

References

Abe K, Shimokawa H, Morikawa K, Uwatoku T, Oi K, Matsumoto Y, Hattori T, Nakashima Y, Kaibuchi K, Sueishi K, Takeshit A (2004) Long-term treatment with a Rho-kinase inhibitor improves monocrotaline-induced fatal pulmonary hypertension in rats. Circ Res 94:385–393

Abe K, Morikawa K, Hizume T, Uwatoku T, Oi K, Seto M, Ikegaki I, Asano T, Kaibuchi K, Shimokawa H (2005) Prostacyclin does not inhibit Rho-kinase: an implication for the treatment of pulmonary hypertension. J Cardiovasc Pharmacol 45:120–124

Abe K, Tawara S, Oi K, Hizume T, Uwatoku T, Fukumoto Y, Kaibuchi K, Shimokawa H (2006) Long-term inhibition of rho-kinase ameliorates hypoxia-induced pulmonary hypertension in mice. J Cardiovasc Pharmacol 48:280–285

Amano M, Chihara K, Kimura K, Fukata Y, Nakamura N, Matsuura Y, Kaibuchi K (1997) Formation of actin stress fibers and focal adhesions enhanced by Rho-kinase. Science 275:1308–1311

Dahal BK, Kosanovic D, Pamarthi PK, Sydykov A, Lai YJ, Kast R, Schirok H, Stasch JP, Ghofrani HA, Weissmann N, Grimminger F, Seeger W, Schermuly RT (2010) Therapeutic efficacy of azaindole-1 in experimental pulmonary hypertension. Eur Respir J 36:808–818

Do e Z, Fukumoto Y, Takaki A, Tawara S, Ohashi J, Nakano M, Tada T, Saji K, Sugimura K, Fujita H, Hoshikawa Y, Nawata J, Kondo T, Shimokawa H (2009) Evidence for rho-kinase activation in patients with pulmonary arterial hypertension. Circ J 73:1731–1739

Dorfmuller P, Perros F, Balabanian K, Humbert M (2003) Inflammation in pulmonary arterial hypertension. Eur Respir J 22:358–363

Fujita H, Fukumoto Y, Saji K, Sugimura K, Demachi J, Nawata J, Shimokawa H (2010) Acute vasodilator effects of inhaled fasudil, a specific Rho-kinase inhibitor, in patients with pulmonary arterial hypertension. Heart Vessels 25:144–149

Fukata Y, Amano M, Kaibuchi K (2001) Rho-Rho-kinase pathway in smooth muscle contraction and cytoskeletal reorganization of non-muscle cells. Trends Pharmacol Sci 22:32–39

Fukumoto Y, Shimokawa H (2011) Recent progress in the management of pulmonary hypertension. Circ J 75:1801–1810

Fukumoto Y, Matoba T, Ito A, Tanaka H, Kishi T, Hayashidani S, Abe K, Takeshita A, Shimokawa H (2005) Acute vasodilator effects of a Rho-kinase inhibitor, fasudil, in patients with severe pulmonary hypertension. Heart 91:391–392

Fukumoto Y, Tawara S, Shimokawa H (2007) Recent progress in the treatment of pulmonary arterial hypertension: expectation for Rho-kinase inhibitors. Tohoku J Exp Med 211:309–320

Fukumoto Y, Yamada N, Matsubara H, Mizoguchi M, Uchino K, Yao A, Kihara Y, Kawano M, Watanabe H, Takeda Y, Adachi T, Osanai S, Tanabe N, Inoue T, Nakano T, Shimokawa H (2013) A double-blind, placebo-controlled clinical trial with a Rho-kinase inhibitor in pulmonary arterial hypertension; a pilot efficacy trial. http://dx.doi.org/10.1253/circj.CJ-13-0443. Circ J 77:2691–2625

Giaid A, Saleh D (1995) Reduced expression of endothelial nitric oxide synthase in the lungs of patients with pulmonary hypertension. N Engl J Med 333:214–221

Higashi M, Shimokawa H, Hattori T, Hiroki J, Mukai Y, Morikawa K, Ichiki T, Takahashi S, Takeshita A (2003) Long-term inhibition of Rho-kinase suppresses angiotensin II-induced cardiovascular hypertrophy in rats in vivo: effect on endothelial NAD(P)H oxidase system. Circ Res 93:767–775

Higenbottam TW, Laude EA (1998) Endothelial dysfunction providing the basis for the treatment of pulmonary hypertension: Giles f. Filley lecture. Chest 114:72S–79S

Hiroki J, Shimokawa H, Higashi M, Morikawa K, Kandabashi T, Kawamura N, Kubota T, Ichiki T, Amano M, Kaibuchi K, Takeshita A (2004) Inflammatory stimuli upregulate rho-kinase in human coronary vascular smooth muscle cells. J Mol Cell Cardiol 37:537–546

Humbert M, Morrell NW, Archer SL, Stenmark KR, MacLean MR, Lang IM, Christman BW, Weir EK, Eickelberg O, Voelkel NF, Rabinovitch M (2004) Cellular and molecular pathobiology of pulmonary arterial hypertension. J Am Coll Cardiol 43:13S–24S

Ishikura K, Yamada N, Ito M, Ota S, Nakamura M, Isaka N, Nakano T (2006) Beneficial acute effects of Rho-kinase inhibitor in patients with pulmonary arterial hypertension. Circ J 70:174–178

Ishizaki T, Maekawa M, Fujisawa K, Okawa K, Iwamatsu A, Fujita A, Watanabe N, Saito Y, Kakizuka A, Morii N, Narumiya S (1996) The small GTP-binding protein Rho binds to and activates a 160 kDa Ser/Thr protein kinase homologous to myotonic dystrophy kinase. EMBO J 15:1885–1893

Jiang BH, Tawara S, Abe K, Takaki A, Fukumoto Y, Shimokawa H (2007) Acute vasodilator effect of fasudil, a Rho-kinase inhibitor, in monocrotaline-induced pulmonary hypertension in rats. J Cardiovasc Pharmacol 49:85–89

Kast R, Schirok H, Figueroa-Perez S, Mittendorf J, Gnoth MJ, Apeler H, Lenz J, Franz JK, Knorr A, Hutter J, Lobell M, Zimmermann K, Munter K, Augstein KH, Ehmke H, Stasch JP (2007) Cardiovascular effects of a novel potent and highly selective azaindole-based inhibitor of Rho-kinase. Br J Pharmacol 152:1070–1080

Leung T, Manser E, Tan L, Lim L (1995) A novel serine/threonine kinase binding the Ras-related Rhoa GPTase which translocates the kinase to peripheral membranes. J Biol Chem 270:29051–29054

Liao JK, Seto M, Noma K (2007) Rho kinase (ROCK) inhibitors. J Cardiovasc Pharmacol 50:17–24

Lohn M, Plettenburg O, Ivashchenko Y, Kannt A, Hofmeister A, Kadereit D, Schaefer M, Linz W, Kohlmann M, Herbert JM, Janiak P, O'Connor SE, Ruetten H (2009) Pharmacological characterization of sar407899, a novel Rho-kinase inhibitor. Hypertension 54:676–683

Loirand G, Guerin P, Pacaud P (2006) Rho kinases in cardiovascular physiology and pathophysiology. Circ Res 98:322–334

Masumoto A, Hirooka Y, Shimokawa H, Hironaga K, Setoguchi S, Takeshita A (2001) Possible involvement of Rho-kinase in the pathogenesis of hypertension in humans. Hypertension 38:1307–1310

Masumoto A, Mohri M, Shimokawa H, Urakami L, Usui M, Takeshita A (2002) Suppression of coronary artery spasm by the rho-kinase inhibitor fasudil in patients with vasospastic angina. Circulation 105:1545–1547

McLaughlin VV, McGoon MD (2006) Pulmonary arterial hypertension. Circulation 114:1417–1431

Mohri M, Shimokawa H, Hirakawa Y, Masumoto A, Takeshita A (2003) Rho-kinase inhibition with intracoronary fasudil prevents myocardial ischemia in patients with coronary microvascular spasm. J Am Coll Cardiol 41:15–19

Morishige K, Shimokawa H, Eto Y, Kandabashi T, Miyata K, Matsumoto Y, Hoshijima M, Kaibuchi K, Takeshita A (2001) Adenovirus-mediated transfer of dominant-negative Rho-kinase induces a regression of coronary arteriosclerosis in pigs in vivo. Arterioscler Thromb Vasc Biol 21:548–554

Nagaoka T, Fagan KA, Gebb SA, Morris KG, Suzuki T, Shimokawa H, McMurtry IF, Oka M (2005) Inhaled Rho kinase inhibitors are potent and selective vasodilators in rat pulmonary hypertension. Am J Respir Crit Care Med 171:494–499

Narumiya S (1996) The small GTPase Rho: cellular functions and signal transduction. J Biochem 120:215–228

Oka M, Fagan KA, Jones PL, McMurtry IF (2008) Therapeutic potential of Rhoa/Rho kinase inhibitors in pulmonary hypertension. Br J Pharmacol 155:444–454

Pankey EA, Byun RJ, Smith WB 2nd, Bhartiya M, Bueno FR, Badejo AM, Stasch JP, Murthy SN, Nossaman BD, Kadowitz PJ (2012) The Rho kinase inhibitor azaindole-1 has long-acting vasodilator activity in the pulmonary vascular bed of the intact chest rat. Can J Physiol Pharmacol 90:825–835

Peng G, Ivanovska J, Kantores C, Van Vliet T, Engelberts D, Kavanagh BP, Enomoto M, Belik J, Jain A, McNamara PJ, Jankov RP (2012) Sustained therapeutic hypercapnia attenuates pulmonary arterial Rho-kinase activity and ameliorates chronic hypoxic pulmonary hypertension in juvenile rats. Am J Physiol Heart Circ Physiol 302:H2599–H2611

Shimokawa H (2000) Cellular and molecular mechanisms of coronary artery spasm: lessons from animal models. Jpn Circ J 64:1–12

Shimokawa H (2002) Rho-kinase as a novel therapeutic target in treatment of cardiovascular diseases. J Cardiovasc Pharmacol 39:319–327

Shimokawa H, Rashid M (2007a) Development of Rho-kinase inhibitors for cardiovascular medicine. Trends Pharmacol Sci 28:296–302

Shimokawa H, Rashid M (2007b) Rho-kinase inhibitors for cardiovascular medicine. Its rationale and current status. Trends Pharmacol Sci 28:296–302

Shimokawa H, Takeshita A (2005) Rho-kinase is an important therapeutic target in cardiovascular medicine. Arterioscler Thromb Vasc Biol 25:1767–1775

Shimokawa H, Seto M, Katsumata N, Amano M, Kozai T, Yamawaki T, Kuwata K, Kandabashi T, Egashira K, Ikegaki I, Asano T, Kaibuchi K, Takeshita A (1999) Rho-kinase-mediated pathway induces enhanced myosin light chain phosphorylations in a swine model of coronary artery spasm. Cardiovasc Res 43:1029–1039

Somlyo AP, Somlyo AV (2000) Signal transduction by G-proteins, Rho-kinase and protein phosphatase to smooth muscle and non-muscle myosin II. J Physiol 522(Pt 2):177–185

Takai Y, Sasaki T, Matozaki T (2001) Small GTP-binding proteins. Physiol Rev 81:153–208

Takemoto M, Sun J, Hiroki J, Shimokawa H, Liao JK (2002) Rho-kinase mediates hypoxia-induced downregulation of endothelial nitric oxide synthase. Circulation 106:57–62

Tawara S, Fukumoto Y, Shimokawa H (2007) Effects of combined therapy with a Rho-kinase inhibitor and prostacyclin on monocrotaline-induced pulmonary hypertension in rats. J Cardiovasc Pharmacol 50:195–200

Uehata M, Ishizaki T, Satoh H, Ono T, Kawahara T, Morishita T, Tamakawa H, Yamagami K, Inui J, Maekawa M, Narumiya S (1997) Calcium sensitization of smooth muscle mediated by a Rho-associated protein kinase in hypertension. Nature 389:990–994

Williams RD, Novack GD, van Haarlem T, Kopczynski C (2011) Ocular hypotensive effect of the rho kinase inhibitor ar-12286 in patients with glaucoma and ocular hypertension. Am J Ophthalmol 152:834–841

Xu W, Kaneko FT, Zheng S, Comhair SA, Janocha AJ, Goggans T, Thunnissen FB, Farver C, Hazen SL, Jennings C, Dweik RA, Arroliga AC, Erzurum SC (2004) Increased arginase II and decreased no synthesis in endothelial cells of patients with pulmonary arterial hypertension. FASEB J 18:1746–1748

Yasuda T, Tada Y, Tanabe N, Tatsumi K, West J (2011) Rho-kinase inhibition alleviates pulmonary hypertension in transgenic mice expressing a dominant-negative type II bone morphogenetic protein receptor gene. Am J Physiol Lung Cell Mol Physiol 301:L667–L674

Yuan JX, Aldinger AM, Juhaszova M, Wang J, Conte JV Jr, Gaine SP, Orens JB, Rubin LJ (1998) Dysfunctional voltage-gated K+ channels in pulmonary artery smooth muscle cells of patients with primary pulmonary hypertension. Circulation 98:1400–1406

Serotonin Transporter and Serotonin Receptors

Serge Adnot, Amal Houssaini, Shariq Abid, Elisabeth Marcos, and Valérie Amsellem

Abstract The nature of the primary defect responsible for triggering and maintaining pulmonary artery smooth muscle (PA-SMC) proliferation in pulmonary artery hypertension (PAH) is poorly understood but may be either an inherent characteristic of PA-SMCs or a secondary response to an external abnormality, such as upregulation of growth factors. The serotonin hypothesis of PAH originated in the 1960s after an outbreak of the disease was reported among patients taking the anorexigenic drugs aminorex. The anorexiant dexfenfluramine which inhibits 5-HT neuronal uptake, causes 5-HT platelet depletion, and increases plasma levels of 5-HT, was then shown to increase the relative risk of developing PAH in the adults. More recently, the incidence of persistent pulmonary hypertension of the newborn was shown to be increased by the use of selective 5-HT reuptake inhibitors taken in late pregnancy. Serotonin is a vasoconstrictor and a potent mitogen for pulmonary smooth muscle cells (PA-SMC), an effect which depends upon activity of both the 5-HT transporter (5-HTT) and the 5-HT receptors. Expression analysis of lung tissues from PAH patients undergoing lung transplantation revealed an increased expression of the 5-HT transporter (5-HTT) and an enhanced proliferative growth response of isolated pulmonary arterial smooth muscle cells (PASMC) to 5-HT. Serotonin is contained in platelets but is also synthesized by pulmonary endothelial cells which express tryptophan hydroxylase 1, the rate-limiting enzyme of 5-HT synthesis. While inhibitors of 5-HTT and of 5-HT2B receptors can reverse experimental PH, 5-HTT-overexpressing mice spontaneously develop PH. In patients with chronic lung disease, a close association has been found between a 5-HTT gene polymorphism and the severity of pulmonary hypertension. Agents capable of selectively inhibiting 5-HTT-mediated PA-SMC proliferation deserve to be investigated as potential treatments for pulmonary hypertension. However, the

S. Adnot (✉) • A. Houssaini • S. Abid • E. Marcos • V. Amsellem
INSERM U955 and Département de Physiologie, Hôpital Henri Mondor, AP-HP, 94010 Créteil, France
e-mail: serge.adnot@inserm.fr

5-HT pathway is still studied only on a preclinical level and the usefulness of drugs interacting with the 5-HT pathway remains to be established in human PAH.

Keywords Serotonin • Pulmonary hypertension • Smooth muscle

Contents

1 Introduction 366
2 Serotonin and Serotonin Synthesis in Pulmonary Hypertension 368
   2.1 Serotonin 368
   2.2 Serotonin Synthesis in Human and Experimental PH: Serotonin Synthesis as a Pharmacological Target in Pulmonary Hypertension? 368
3 Serotonin-Mediated Effects on Target Cells: The Serotonin Transporter (5-HTT) as a Potential Therapeutic Target in Pulmonary Hypertension 370
   3.1 The Serotonin Transporter (5-HTT) 370
   3.2 Role of the Serotonin Transporter (5-HTT) in Pulmonary Hypertension 371
   3.3 Overexpression of 5-HTT and PA-SMC Hyperplasia in Idiopathic PH 371
   3.4 5-HTT Gene Polymorphism and Increased 5-HTT Expression in Pulmonary Arteries from Patients with Idiopathic PH 372
   3.5 Potential Usefulness of 5-HTT Inhibitors in the Treatment of Adult Patients with PH 373
   3.6 Serotonin and Lung Development: Persistent Pulmonary Hypertension in the Newborn 373
4 Role of the Serotonin Receptors in Pulmonary Hypertension 374
   4.1 $5HT_{2A}$ Receptor 374
   4.2 $5HT_{2B}$ Receptor 375
   4.3 $5HT_{1B}$ Receptor 375
5 Serotonin Signaling in PH 376
6 Conclusions 377
References 377

1 Introduction

The effects of serotonin (5-hydroxytryptamine, 5-HT) on the pulmonary circulation attracted the interest of researchers when an increased risk of idiopathic pulmonary hypertension (IPH) was reported in patients who had used appetite suppressants that interact with 5-HT (Abenhaim et al. 1996). An association between the anorexigen aminorex and IPH was described in the 1960s. In the 1980s, fenfluramine use was shown to be associated with an epidemic of IPH in France and Belgium (Abenhaim et al. 1996). The serotonin hypothesis received support from the observation that pulmonary hypertension (PH) develops spontaneously in Fawn-hooded rats, which have an inherited platelet-storage defect (MacLean et al. 2000).

Early studies focused on circulating 5-HT and its potential effects on the pulmonary vascular bed. They showed that patients with PAH had increased circulating serotonin levels even after heart–lung transplantation (Herve

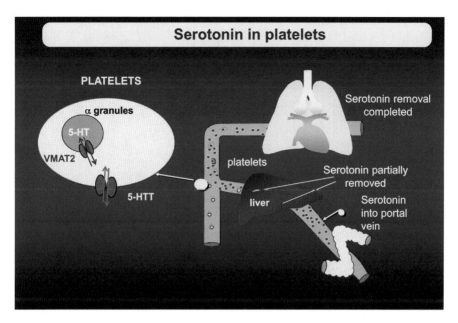

Fig. 1 Besides 5-HT, platelets contain three serotonergic components: the 5-HT2A receptor (5-HT2A), the plasma membrane serotonin reuptake transporter (5-HTT), and the vesicular monoamine transporter 2 (VMAT2). Intact platelets take up 5-HT by the 5-HTT and store the monoamine in vesicles by the VMAT2. Fenfluramine (Fen) is SERT-dependently transported into the cytoplasm and reverses the VMAT2 transport, increasing the cytoplasmic 5-HT concentration [5-HT]i to levels that reverse the SERT activity, resulting in a continuous slow release of 5-HT. In contrast, selective inhibitors of the 5-HT transporter (5-HTT) block the uptake of 5-HTT activity without affecting VMAT2 activity and therefore, do not favor the release of 5-HT from platelet stores. Activation of 5-HT2A receptors potentiates the aggregation response to various platelet activators

et al. 1995). In addition to its vasoactive effects, 5-HT exerts mitogenic and co-mitogenic effects on pulmonary artery smooth muscle cells (PA-SMCs). Convincing evidence indicates that 5-HT-related stimulation of PA-SMC proliferation requires 5-HT internalization through the 5-HT transporter (5-HTT) (Chin et al. 1999; Dennery et al. 1998; Eddahibi et al. 2001; Lee et al. 1991) and activation of 5-HT_{2B} receptors (Launay et al. 2002) in PA-SMCs. In contrast, the constricting action of 5-HT on smooth muscle cells is mainly mediated by 5-HT receptors (5-HT 1B/D, 2A, and 2B) (MacLean et al. 2000). Accordingly, drugs that competitively inhibit 5-HTT also block the mitogenic effects of 5-HT on smooth muscle cells (Eddahibi and Adnot 2002). The appetite suppressants fenfluramine, d-fenfluramine, and aminorex differ from selective serotonin transporter inhibitors in that they not only inhibit serotonin reuptake but also trigger indoleamine release and interact with 5-HTT and 5-HT receptors in a specific manner (Eddahibi and Adnot 2002; Rothman et al. 1999) (Fig. 1).

Over the last few years, the role of the 5-HT pathway in the pathogenesis of PH has been extensively investigated, and severe alterations in several critical steps of this pathway have been found during the progression of PH. These alterations include changes in 5-HT synthesis and bioavailability, changes in 5-HT-mediating effects on target cells (PA-SMCs and fibroblasts), and alterations in intracellular signaling. These critical steps represent targets for future therapies and will be discussed individually.

2 Serotonin and Serotonin Synthesis in Pulmonary Hypertension

2.1 Serotonin

Serotonin (5-hydroxytryptamine, 5-HT) is an endogenous vasoactive indoleamine substance found primarily in enterochromaffin tissue, brain, and blood platelets. The neutral amino acid L-tryptophan is the primary precursor of serotonin, the synthesis of which involving first hydroxylation of the benzene ring of tryptophan by the rate-limiting monooxygenase, tryptophan hydroxylase, to form 5-hydroxytryptophan, and second, decarboxylation of its amino-carboxy terminal group by an aromatic L-amino acid decarboxylase, to form 5-HT. The synthesis and degradation of serotonin are very active processes, the total serotonin pool being replaced every 24 h. In humans, approximately 90 % of the body's total serotonin is located in the intestines where it plays a role in mechano and chemotransduction. The most part of 5-HT that is released from enterochromaffin cells reaches the portal circulation where it is avidly taken up and stored in small electron-dense granules within platelets, the half-life of serotonin in platelets being 33–48 h. Accumulation of serotonin in platelets which are incapable of synthesizing the amine, is accomplished mostly via a sodium-dependent active transport process which is also present in endothelial, smooth muscle, and neural cells. The remaining circulating amine is either catabolized by the monoamine oxidase A (MAO-A) in the liver or taken up by active transport in the lungs and degraded inside endothelial and smooth muscle cells successively by monoamino oxidase B and an aldehyde deshydrogenase to form 5-hydroxyindole acetic acid (5-HIAA). These processes result in very low levels of free 5-HT in plasma under normal physiological conditions.

2.2 Serotonin Synthesis in Human and Experimental PH: Serotonin Synthesis as a Pharmacological Target in Pulmonary Hypertension?

Serotonin production outside the central nervous system occurs chiefly in the enterochromaffin cells, from which indoleamine is released into the bloodstream before being taken up by circulating platelets. Thus, one hypothesis is that only 5-HT released from platelets acts on pulmonary vessels and that lung 5-HT bioavailability is independent from the rate of 5-HT synthesis (Fig. 1). However, in recent studies, we showed that 5-HT was also produced in the human lung by pulmonary endothelial cells (Eddahibi et al. 2006). The 5-HT synthesis rate was increased in endothelial cells from patients with idiopathic PH, supporting a contribution to pulmonary vascular remodeling of increased 5-HT availability in the immediate vicinity of PA-SMCs.

The functional impact of variations in 5-HT synthesis on PH development was recently examined in animal studies (Izikki et al. 2007; Morecroft et al. 2007). The critical step in 5-HT biosynthesis is catalyzed by the rate-limiting enzyme tryptophan hydroxylase (Tph). Tph activity therefore serves as a marker for 5-HT synthesis. Tph, which was initially thought to derive from a single gene, exists as two isoforms, Tph1 and Tph2, encoded by separate genes. Tph1 is expressed mainly in the gut and pineal gland (Darmon et al. 1988). The recently identified *Tph2* gene appears exclusively expressed in neurons and is known as the neuronal isoform (Walther et al. 2003a). Thus, 5-HT synthesis is thought to be controlled chiefly by Tph2 in the central nervous system and Tph1 in peripheral organs. As mentioned above, peripherally produced 5-HT is synthesized chiefly by enterochromaffin cells in the gut, but small amounts of 5-HT may also be formed at other sites including endothelial cells in the systemic and pulmonary arteries (Eddahibi et al. 2006; Ni et al. 2008). Human pulmonary microvascular endothelial cells produce 5-HT and express Tph1, and both 5-HT synthesis and Tph1 expression are increased in cells from patients with IPH compared to cells from controls (Eddahibi et al. 2006).

In recent studies, we also found that the severity of hypoxic PH in mice was directly linked to the rate of 5-HT synthesis catalyzed by the Tph enzyme isoforms Tph1 and Tph2 (Izikki et al. 2007; Morecroft et al. 2007). In mice lacking the *Tph1* gene and therefore exhibiting marked reductions in 5-HT synthesis rates and contents in peripheral organs, hypoxic PH was far less severe than in wild-type mice (Izikki et al. 2007; Morecroft et al. 2007). Treatment with the Tph inhibitor PCPA further reduced PH severity in *Tph1*−/− mice, normalizing right ventricular systolic pressure and pulmonary artery muscularization, in keeping with persistence of some peripheral Tph activity in *Tph1*−/− mice (Izikki et al. 2007). Evidence for involvement of Tph2 was obtained in mice harboring a polymorphism of the *Tph2* gene and showing differences in brain 5-HT synthesis rates according to their genotype. These animals also showed differences in the severity of hypoxic PH, which were abolished by PCPA treatment, indicating a contribution for neuronal Tph2 activity to the severity of hypoxic PH. In addition, hypoxia was found to increase 5-HT synthesis in the forebrain and lung, although no changes were noted in either Tph1 or Tph2 mRNA levels. Although mRNA levels are not necessarily reflective of protein expression, this finding suggests that the hypoxia-induced increase in 5-HT synthesis may result from enhanced enzyme activity as opposed to increased gene expression. Hypoxia not only increases 5-HT synthesis but also markedly diminishes lung 5-HT contents and blood 5-HT levels, a combination suggesting a sharp increase in the indoleamine turnover rate during hypoxia exposure and PH development (Izikki et al. 2007).

Dysregulation of serotonin synthesis in vivo therefore appears closely linked to the PH phenotype in mice. Moreover, variations in the rate of 5-HT synthesis seem to directly affect the extent of pulmonary vascular remodeling, and, therefore, the severity of PH. Taken together with the observation that both 5-HT synthesis and Tph1 expression are increased in endothelial cells from patients with IPH compared to cells from controls, these results suggest that *Tph1*, and possibly *Tph2*, may deserve consideration as candidate genes for human PH. However, targeting Tph activity does not seem to hold promise for treating human PH, as unacceptable side effects might occur. Recent results, however, provided proof of concept that selectively inhibiting Tph1 may represent a therapeutic option for PAH.

3 Serotonin-Mediated Effects on Target Cells: The Serotonin Transporter (5-HTT) as a Potential Therapeutic Target in Pulmonary Hypertension

3.1 The Serotonin Transporter (5-HTT)

The serotonin transporter (5-HTT) belongs to a large family of integral membrane proteins responsible for terminating the action of neurotransmitters released from presynaptic neurons. This family includes GABA transporter, norepinephrine transporter, and dopamine transporter. In addition to the reuptake of 5-HT into neurons, the 5-HTT is also responsible for the uptake of serotonin by platelets, endothelial, and vascular smooth muscle cells. The 5-HT uptake process is driven by transmembrane gradients of Na^+ which is maintained by the activity of the Na^+/K^+-ATPase. Transport of serotonin is sensitive to nanomolar concentrations of tricyclic and heterocyclic antidepressants, such as imipramine, fluoxetine, paroxetine, and citalopram which bind to the carrier protein. The identity of the neuronal and non-neuronal 5-HT transporter indicates that these proteins are encoded by the same single-copy gene. The human 5-HT gene mapped to chromosome 17q11.1–17q12 is organized in 14 exons spanning 35 kb. It has been initially cloned from the human midbrain raphe complex and platelets. The coding sequence for the 5-HTT covers about 2 kb which corresponds to approximately 617–630 amino acids. Analysis of the amino acid sequences predicted by the transporter clones suggests the presence of 12 putative transmembrane domains with a large extracellular loop between transmembrane domains 3 and 4 with both the N- and C-termini residing within the cytoplasm. There are multiple sites for NH_2-linked glycosylation and multiple consensus sequences in the intracellular domains for phosphorylation by protein kinases such as protein kinase A and protein kinase C. The promoter activity also appears to be regulated by an interplay of several positive and negative regulatory elements including AP1, CRE, NF-kB, and SP1.

3.2 Role of the Serotonin Transporter (5-HTT) in Pulmonary Hypertension

Accumulating evidence suggests that the 5-HT transporter (5-HTT) in the lung may be a key determinant of pulmonary vessel remodeling because of its effects on PA-SMC growth. 5-HTT is abundantly expressed in the lung, where it is predominantly located on PA-SMCs (Eddahibi et al. 2001; Fanburg and Lee 1997). It is encoded by a single gene expressed in several cell types such as neurons, platelets, and pulmonary vascular endothelial and smooth muscle cells. The level of 5-HTT expression in humans seems considerably higher in the lung than in the brain, suggesting that altered 5-HTT expression may have direct consequences on PA-SMC function. The requirement for 5-HTT as a mediator for the mitogenic activity of 5-HT appears specific of PA-SMCs, since no such effect has been reported with other SMC types. Direct evidence that 5-HTT plays a key role in pulmonary vascular remodeling was recently provided by studies showing that mice with targeted 5-HTT gene disruption developed less severe hypoxic PH than did wild-type controls and that selective 5-HTT inhibitors attenuated hypoxia- and monocrotaline-induced PH (Eddahibi et al. 2000; Guignabert et al. 2005). Conversely, increased 5-HTT expression was associated with increased severity of hypoxic PH (MacLean et al. 2004). To investigate whether 5-HTT overexpression in PA-SMCs was sufficient to produce PH, we generated transgenic mice overexpressing 5-HTT under the control of the SM22 promoter. We found that these transgenic mice with selective 5-HTT overexpression in smooth muscle cells spontaneously developed PH (Guignabert et al. 2005). Importantly, PH in these SM22-5-HTT+ mice developed without any alterations in 5-HT bioavailability and, therefore, as a consequence solely of 5-HTT protein overexpression in smooth muscle cells. Interestingly, only female mice that ubiquitously overexpress the human 5-HTT gene develop PH, suggesting that some interactions between estrogens and the 5-HTT gene influence the PAH phenotype (White et al. 2011). Taken together, these observations suggest a close correlation between 5-HTT expression and/or activity and the extent of pulmonary vascular remodeling during experimental PH.

3.3 Overexpression of 5-HTT and PA-SMC Hyperplasia in Idiopathic PH

Evidence that 5-HTT plays an important role in the pathogenesis of human IPH is now clearly established. 5-HTT expression was shown to be increased in platelets and lungs from patients with IPH, where it predominated in the media of thickened pulmonary arteries and in onion-bulb lesions (Eddahibi et al. 2001). Interestingly, the higher levels of 5-HTT protein and activity persisted in cultured smooth muscle cells from pulmonary arteries of patients with IPH, compared to cells from controls

(Eddahibi et al. 2001). Moreover, PA-SMCs from patients with IPH grew faster than PA-SMCs from controls when stimulated by serotonin or serum, as a consequence of increased expression of the serotonin transporter (Eddahibi et al. 2001). In the presence of 5-HTT inhibitors, the growth-stimulating effects of serum and serotonin were markedly reduced, and the difference between growth of PA-SMCs from patients and controls was abolished. The proliferative response of PA-SMCs to various growth factors such as PDGF, EGF, TGFβ, FGFa, and IGF did not differ between patients with IPH and controls (Eddahibi et al. 2001). It follows that 5-HTT overexpression and/or activity in PA-SMCs from patients with IPH is responsible for the increased mitogenic response to serotonin and to serum (which contains micromolar concentrations of serotonin).

3.4 5-HTT Gene Polymorphism and Increased 5-HTT Expression in Pulmonary Arteries from Patients with Idiopathic PH

That 5-HTT expression is genetically controlled has been convincingly demonstrated: a polymorphism in the promoter region of the human 5-HTT gene alters the level of transcription (Lesch et al. 1993). This polymorphism consists of two common alleles, a 44-bp insertion or deletion, designated the L and S allele, respectively. The L allele drives a two- to threefold higher level of *5-HTT* gene transcription than does the S allele. In studies of PA-SMCs from controls, we found that cells from LL individuals had a twofold increase in 5-HTT mRNA expression, compared to cells from SS subjects, and that LS individuals had an intermediate level of expression (Eddahibi et al. 2001). Accordingly, the growth-stimulating effects of 5-HT or serum were more marked in LL cells than in LS or SS cells, indicating that the ability of PA-SMCs to proliferate in response to serotonin or serum was directly linked to the functional polymorphism of the 5-HTT gene promoter. While studying a population of 89 patients with IPH, we found that the L allelic variant of the 5-HTT gene promoter, associated with 5-HTT overexpression and increased PA-SMC growth, was present homozygously in 65 % of patients and 27 % of controls (Eddahibi et al. 2001). This finding suggested that the 5-HTT gene polymorphism might either convey susceptibility to PH or act as an important modifier of the PH phenotype. However, recent studies investigating larger populations of patients with IPH, associated forms of PH (APH), or familial PH (FPH) did not confirm our previous results (Machado et al. 2006; Willers et al. 2006). This apparent discrepancy might be related to specific characteristics of our cohort, which included a relatively high percentage of lung-transplant recipients. Because patients referred to lung transplantation are selected on the basis of faster disease progression or poor responsiveness to conventional treatment, this may have introduced a bias in our cohort. Support for this possibility comes from our unpublished observation that among 50 patients with PH who underwent lung transplantation between 1998 and 2005 at the Marie Lannelongue hospital (Le Plessis Robinson, France), 57 % had the LL genotype (Dewachter et al. 2006). Moreover, we previously reported a high frequency of the LL genotype among lung-transplant recipients with APH (Marcos et al. 2004). Thus, one possible explanation for the differences between these studies may be related to differences in the disease progression profiles of the study populations. Therefore, the LL genotype may not deserve to be considered a susceptibility factor for the development of IPH. The possibility that it may affect disease progression remains to be evaluated in a larger population in which transplantation and death are censored.

3.5 Potential Usefulness of 5-HTT Inhibitors in the Treatment of Adult Patients with PH

5-HTT inhibitors in the dosages commonly used in depression may induce therapeutic effects in patients with IPH. Inhibition of in vitro SMC proliferation occurs with micromolar concentrations, which are at the lower end of the plasma level range obtained with the usual dosages of these drugs. Increasing the dosage within the authorized range would produce effective levels in the pulmonary circulation. 5-HTT inhibitors are considered safe, despite recent evidence of an increased risk of persistent pulmonary hypertension (PPHN) in neonates born to mothers treated during pregnancy (Chambers et al. 2006).

At present, the effects of 5-HTT inhibitors on the progression of human PH have not been assessed in a controlled study using appropriate drug dosages. These drugs, however, have occasionally been given to patients with IPH, in the usual recommended dosages, with no major effects detectable in individual patients.

3.6 Serotonin and Lung Development: Persistent Pulmonary Hypertension in the Newborn

Persistent pulmonary hypertension of the newborn (PPHN) is a life-threatening condition that occurs in up to 2 infants per 1,000 live births and that is caused by a failure of pulmonary vascular resistance to decrease after birth. Recent epidemiological studies have linked the use of selective serotonin reuptake inhibitors (SSRIs) for the treatment of maternal depression to a two- to sixfold increase in the incidence of PPHN when taken in late pregnancy (Chambers et al. 2006; Kieler et al. 2012; Delaney and Cornfield 2012). This increased risk seems to be a drug class effect since it appears to be of similar magnitude for each of the specific SSRIs used, including sertraline, citalopram, paroxetine, and fluoxetine (Kieler et al. 2012). The mechanism by which SSRIs may affect the pulmonary vasculature and cause PPHN remains unknown. Possible contributing factors include the accumulation of SSRIs in the lungs, with some consequences on lung development, as well as some direct effects of serotonin on the pulmonary vasculature. Recently, the hemodynamic effects of 5-HT, 5-HT receptor antagonists, and SSRIs were studied in fetal sheep (Delaney et al. 2012). Infusion of 5-HT was associated with marked pulmonary vasoconstriction which was abolished by ketanserin, a 5-HT2A receptor antagonist, but not by 5-HT1B or 5-HT2B receptor antagonists. Brief infusions of the SSRIs, sertraline, and fluoxetine, caused potent and sustained elevations of pulmonary vascular resistance. Moreover, in the lamb with experimental PPHN, SSRIs and 5-HT caused further elevation of pulmonary vascular resistance. Thus, 5-HT, by causing pulmonary vasoconstriction, may contribute to maintenance of high pulmonary vascular resistance in the normal fetus through stimulation of 5-HT2A receptors. By preventing 5-HT internalization through inhibition of the 5-HT transporter, SSRIs may contribute to increase 5-HT bioavailability in the fetal pulmonary circulation, thereby increasing 5-HT binding to 5-HT receptors. The effects of SSRIs on the developing pulmonary circulation in the neonate may therefore differ to those exerted in adult pulmonary hypertension. Whether it is due to differential effects of SSRIs on 5-HT bioavailability or whether 5-HTT signaling may differ in the neonate and adult pulmonary circulations remain opened questions.

4 Role of the Serotonin Receptors in Pulmonary Hypertension

On release from platelets or nerve endings, serotonin can activate up to 14 structurally distinct 5HT receptors, which are classified into seven families ($5HT_{1-7}$). Of these, the $5HT_{2A}$, $5HT_{2B}$, and $5HT_{1B}$ receptors have been investigated with respect to PH.

4.1 *$5HT_{2A}$ Receptor*

In most nonhuman mammals, the $5HT_{2A}$ receptor mediates vasoconstriction in both the systemic and the pulmonary circulation. For example, $5HT_{2A}$ receptor-mediated vasoconstriction occurs in the mouse and rat lung under control conditions (Keegan et al. 2001; MacLean et al. 1996a). The $5HT_{2A}$ receptor antagonist ketanserin has proved clinically effective in the treatment of systemic hypertension, especially in the elderly (Frishman et al. 1995). Thus, ketanserin does not act specifically on the pulmonary circulation and its systemic effects limit its use in both IPH and secondary PH, where it fails to significantly improve pulmonary hemodynamics (Frishman et al. 1995). Recent studies performed in fetal animals, however, reported that 5-HT behaved as a strong pulmonary vasoconstrictor and that this effect was mediated primarily by 5-HT2A receptors. 5-HT2A mediated 5-HT induced pulmonary vascular constriction may therefore contribute to the mechanism by which maternal use of SSRIs increase the risk of PPHN.

4.2 5HT$_{2B}$ Receptor

Strong evidence for the 5-HT2B receptor as a therapeutic target in PAH has emerged. 5-HT2B knockout animals are resistant to hypoxia-induced PH and administration of the specific 5-HTR2B antagonist RS-127445 prevented the increase in pulmonary arterial pressure (Ppa) in mice that were challenged to hypoxia (Launay et al. 2002). Recently, terguride, a potent 5-HTR2A/2B antagonist, was reported to exert anti-remodeling effects upon long-term use in rats with monocrotaline-induced PH (Dumitrascu et al. 2011). Immunohistochemistry against 5-HTR2A/B on human lungs revealed their localisation to the vascular smooth muscle layer and quantitative RT-PCR showed 5-HTR2B upregulation in pulmonary artery smooth muscle cells (PASMC) isolated from PAH patients. Proliferation and migration of cultured primary human PASMC were dose-dependently blocked by terguride (Dumitrascu et al. 2011). Terguride is approved for ovulation disorders due to hyperprolactinaemia by acting as a partial dopamine receptor (DR) D2 agonist in the pituitary gland (Ciccarelli and Camanni 1996). In addition, it is a strong antagonist of 5-HTR2A and 5-HTR2B (Millan et al. 2002), and, therefore, seems well suited for treatment of PAH. However, terguride doses needed to achieve 5-HT2B receptor antagonistic properties are higher than those required for its dopaminergic agonistic properties. Terguride is well tolerated at dopaminergic doses. Whether terguride or other drugs with 5-HT2B receptor antagonistic properties may be safely used in patients with PH deserve further studies.

4.3 5HT$_{1B}$ Receptor

$5HT_{1B}$ receptors mediate constriction of human pulmonary arteries (MacLean et al. 1996a; Morecroft et al. 1999). There is also evidence supporting a role for the $5HT_{1B}$ receptor in PH development (Keegan et al. 2001; MacLean et al. 1996b). (1) Inhibition of $5HT_{1B}$ activity, either by genetic knockout or by antagonism, reduces hypoxia-induced pulmonary vascular remodeling (Keegan et al. 2001). (2) Serotonin is required for the proliferative effect of the calcium-binding protein S100A4/Mts1, and the $5HT_{1B}$ receptor and 5-HTT are codependent in regulating S100A4/Mts1-induced human PA-SMC proliferation (Lawrie et al. 2005). (3) Cooperation occurs between the 5-HT$_{1B}$ receptor and 5-HTT in mediating pulmonary vascular contraction (Morecroft et al. 2005). (4) The $5HT_{1B}$ receptor is increased in mice overexpressing human 5-HTT and in the Fawn-hooded rat, which also exhibits increased 5-HTT expression (Morecroft et al. 2005); both models are particularly susceptible to hypoxia-induced pulmonary vascular remodeling. (5) The $5HT_{1B}$ receptor and 5-HTT act cooperatively to facilitate serotonin-induced proliferation and rho-kinase-induced nuclear translocation of extracellular regulated kinase (ERK)1/ERK2 in PA-SMCs (Liu et al. 2004). (6) The $5HT_{1B}$ receptor plays a role in PH development in a pig model (Rondelet et al. 2003). (7) Finally, remodeled PAs from patients with PH overexpress the $5HT_{1B}$ receptor (Launay et al. 2002). These data suggest that serotonin-induced activation of the $5HT_{1B}$ receptor may be involved in pulmonary vascular remodeling and that cooperation exists between the $5-HT_{1B}$ receptor and 5-HTT. As $5HT_{1B}$ receptor-mediated changes are specific of the pulmonary circulation, this receptor is an attractive therapeutic target for PH.

5 Serotonin Signaling in PH

As mentioned above, 5-HT transport within the cell via 5-HTT is involved in the proliferation of PA-SMCs and fibroblasts. How 5-HT internalization is linked to cell proliferation signaling remains unknown, although several intracellular signaling pathways are known to be involved. ROS generation and ERK1/2 activation have been suggested as potential mechanisms (Lee et al. 1998, 2001). ROS may be generated either via 5-HT breakdown by monoamine-oxidase or via activation of NADPH oxidase. In bovine PA-SMCs, ROS may mediate the phosphorylation of ERK1/2 (Lee et al. 2001). Phosphorylated ERK1/2 may then be translocated to the nucleus, where it may increase the DNA binding of transcription factors such as GATA-4, Egr-1, and Elk-1, leading to the expression of proteins such as S1004/Mts1 that play a major role in both human and experimental PH (Lawrie et al. 2005). Most of these studies, however, did not specify the role for 5-HTT or 5-HT receptors in mediating the activation of these pathways.

More recently, evidence has been provided that activation of the small G protein RhoA and its target Rho kinase may represent a major pathway involved in 5-HT-induced PA-SMC proliferation via 5-HTT (Walther et al. 2003b). First, it was shown that 5-HT internalized in platelets via 5-HTT was covalently linked to RhoA by intracellular type 2 transglutaminase (TG2), which led to constitutive RhoA activation (Walther et al. 2003b). This link was then demonstrated in vivo in pulmonary arteries of rats under hypoxic conditions, providing a connection between 5-HTT and RhoA activation (Guilluy et al. 2007). Moreover, we recently provided evidence that RhoA activation by 5-HT occurred both in platelets and in PA-SMCs during PH progression and that this event contributed to PA-SMC proliferation (Guilluy et al. 2009). In mice characterized by 5-HTT overexpression in smooth muscle cells and spontaneous development of PH, treatment with 5-HTT inhibitors limited both PH progression and RhoA/Rho kinase activation. Only partial limitation of PH progression, however, is observed after treatment with the

Rho-kinase inhibitor fasudil, suggesting that activation of the RhoA pathway may generate only part of the intracellular signals originating from internalization of 5-HT via 5-HTT (Guilluy et al. 2009).

In conclusion, direct involvement of the 5-HTT/RhoA/Rho kinase signaling pathway in 5-HT-mediated PA-SMC proliferation and platelet activation during PH progression identifies RhoA/Rho kinase signaling as a promising target for new treatments against PH.

6 Conclusions

In conclusion, the data reviewed here support a crucial role for serotonin synthesis and functional activities of serotonin receptors and transporter in the pathogenesis of pulmonary vascular remodeling. Agents targeting these molecular effectors deserve to be investigated as potential treatments for PH.

References

Abenhaim L, Moride Y, Brenot F, Rich S, Benichou J, Kurz X, Higenbottam T, Oakley C, Wouters E, Aubier M, Simonneau G, Begaud B (1996) Appetite-suppressant drugs and the risk of primary pulmonary hypertension. International Primary Pulmonary Hypertension Study Group. N Engl J Med 335:609–616

Chambers CD, Hernandez-Diaz S, Van Marter LJ, Werler MM, Louik C, Jones KL, Mitchell AA (2006) Selective serotonin-reuptake inhibitors and risk of persistent pulmonary hypertension of the newborn. N Engl J Med 354:579–587

Chin L, Artandi SE, Shen Q, Tam A, Lee SL, Gottlieb GJ, Greider CW, DePinho RA (1999) p53 deficiency rescues the adverse effects of telomere loss and cooperates with telomere dysfunction to accelerate carcinogenesis. Cell 97:527–538

Ciccarelli E, Camanni F (1996) Diagnosis and drug therapy of prolactinoma. Drugs 51:954–965

Darmon MC, Guibert B, Leviel V, Ehret M, Maitre M, Mallet J (1988) Sequence of two mRNAs encoding active rat tryptophan hydroxylase. J Neurochem 51:312–316

Delaney C, Cornfield DN (2012) Risk factors for persistent pulmonary hypertension of the newborn. Pulm Circ 2:15–20

Delaney C, Gien J, Grover TR, Roe G, Abman SH (2012) Pulmonary vascular effects of serotonin and selective serotonin reuptake inhibitors in the late-gestation ovine fetus. Am J Physiol Lung Cell Mol Physiol 301:L937–L944

Dennery PA, Spitz DR, Yang G, Tatarov A, Lee CS, Shegog ML (1998) Oxygen toxicity and iron accumulation in the lungs of mice lacking heme oxygenase-2. J Clin Invest 101:1001–1011

Dewachter L, Adnot S, Fadel E, Humbert M, Maitre B, Barlier-Mur AM, Simonneau G, Hamon M, Naeije R, Eddahibi S (2006) Angiopoietin/Tie2 pathway influences smooth muscle hyperplasia in idiopathic pulmonary hypertension. Am J Respir Crit Care Med 174:1025–1033

Dumitrascu R, Kulcke C, Konigshoff M, Kouri F, Yang X, Morrell N, Ghofrani HA, Weissmann N, Reiter R, Seeger W, Grimminger F, Eickelberg O, Schermuly RT, Pullamsetti SS (2011) Terguride ameliorates monocrotaline-induced pulmonary hypertension in rats. Eur Respir J 37:1104–1118

Eddahibi S, Adnot S (2002) Anorexigen-induced pulmonary hypertension and the serotonin (5-HT) hypothesis: lessons for the future in pathogenesis. Respir Res 3:9–13

Eddahibi S, Hanoun N, Lanfumey L, Lesch KP, Raffestin B, Hamon M, Adnot S (2000) Attenuated hypoxic pulmonary hypertension in mice lacking the 5-hydroxytryptamine transporter gene. J Clin Invest 105:1555–1562

Eddahibi S, Humbert M, Fadel E, Raffestin B, Darmon M, Capron F, Simonneau G, Dartevelle P, Hamon M, Adnot S (2001) Serotonin transporter overexpression is responsible for pulmonary artery smooth muscle hyperplasia in primary pulmonary hypertension. J Clin Invest 108:1141–1150

Eddahibi S, Guignabert C, Barlier-Mur AM, Dewachter L, Fadel E, Dartevelle P, Humbert M, Simonneau G, Hanoun N, Saurini F, Hamon M, Adnot S (2006) Cross talk between endothelial and smooth muscle cells in pulmonary hypertension: critical role for serotonin-induced smooth muscle hyperplasia. Circulation 113:1857–1864

Fanburg B, Lee S-L (1997) A new role for an old molecule: serotonin as a mitogen. Am J Physiol 272:L795–L806

Frishman WH, Huberfeld S, Okin S, Wang YH, Kumar A, Shareef B (1995) Serotonin and serotonin antagonism in cardiovascular and non-cardiovascular disease. J Clin Pharmacol 35:541–572

Guignabert C, Raffestin B, Benferhat R, Raoul W, Zadigue P, Rideau D, Hamon M, Adnot S, Eddahibi S (2005) Serotonin transporter inhibition prevents and reverses monocrotaline-induced pulmonary hypertension in rats. Circulation 111:2812–2819

Guilluy C, Rolli-Derkinderen M, Tharaux PL, Melino G, Pacaud P, Loirand G (2007) Transglutaminase-dependent RhoA activation and depletion by serotonin in vascular smooth muscle cells. J Biol Chem 282:2918–2928

Guilluy C, Eddahibi S, Agard C, Guignabert C, Izikki M, Tu L, Savale L, Humbert M, Fadel E, Adnot S, Loirand G, Pacaud P (2009) RhoA and Rho kinase activation in human pulmonary hypertension: role of 5-HT signaling. Am J Respir Crit Care Med 179:1151–1158

Herve P, Launay JM, Scrobohaci ML, Brenot F, Simonneau G, Petitpretz P, Poubeau P, Cerrina J, Duroux P, Drouet L (1995) Increased plasma serotonin in primary pulmonary hypertension. Am J Med 99:249–254

Izikki M, Hanoun N, Marcos E, Savale L, Barlier-Mur AM, Saurini F, Eddahibi S, Hamon M, Adnot S (2007) Tryptophan hydroxylase 1 knockout and tryptophan hydroxylase 2 polymorphism: effects on hypoxic pulmonary hypertension in mice. Am J Physiol Lung Cell Mol Physiol 293:L1045–L1052

Keegan A, Morecroft I, Smillie D, Hicks M, MacLean M (2001) Contribution of the 5-HT1B receptor to hypoxia-induced pulmonary hypertension. Circ Res 89:1231–1239

Kieler H, Artama M, Engeland A, Ericsson O, Furu K, Gissler M, Nielsen RB, Norgaard M, Stephansson O, Valdimarsdottir U, Zoega H, Haglund B (2012) Selective serotonin reuptake inhibitors during pregnancy and risk of persistent pulmonary hypertension in the newborn: population based cohort study from the five Nordic countries. BMJ 344:d8012

Launay JM, Herve P, Peoc'h K, Tournois C, Callebert J, Nebigil CG, Etienne N, Drouet L, Humbert M, Simonneau G, Maroteaux L (2002) Function of the serotonin 5-hydroxytryptamine 2B receptor in pulmonary hypertension. Nat Med 8:1129–1135

Lawrie A, Spiekerkoetter E, Martinez EC, Ambartsumian N, Sheward WJ, MacLean MR, Harmar AJ, Schmidt AM, Lukanidin E, Rabinovitch M (2005) Interdependent serotonin transporter and receptor pathways regulate S100A4/Mts1, a gene associated with pulmonary vascular disease. Circ Res 97:227–235

Lee SL, Wang WW, Moore BJ, Fanburg BL (1991) Dual effect of serotonin on growth of bovine pulmonary artery smooth muscle cells in culture. Circ Res 68:1362–1368

Lee SL, Wang WW, Fanburg BL (1998) Superoxide as an intermediate signal for serotonin-induced mitogenesis. Free Radic Biol Med 24:855–858

Lee SL, Simon AR, Wang WW, Fanburg BL (2001) H(2)O(2) signals 5-HT-induced ERK MAP kinase activation and mitogenesis of smooth muscle cells. Am J Physiol Lung Cell Mol Physiol 281:L646–L652

Lesch KP, Wolozin BL, Estler HC, Murphy DL, Rieder P (1993) Isolation of a cDNA encoding the human brain serotonin transporter. J Neural Transm Gen Sect 91:67–72

Liu Y, Suzuki YJ, Day RM, Fanburg BL (2004) Rho kinase-induced nuclear translocation of ERK1/ERK2 in smooth muscle cell mitogenesis caused by serotonin. Circ Res 95:579–586

Machado RD, Koehler R, Glissmeyer E, Veal C, Suntharalingam J, Kim M, Carlquist J, Town M, Elliott CG, Hoeper M, Fijalkowska A, Kurzyna M, Thomson JR, Gibbs SR, Wilkins MR, Seeger W, Morrell NW, Gruenig E, Trembath RC, Janssen B (2006) Genetic association of the serotonin transporter in pulmonary arterial hypertension. Am J Respir Crit Care Med 173:793–797

MacLean MR, Clayton RA, Templeton AG, Morecroft I (1996a) Evidence for 5-HT1-like receptor-mediated vasoconstriction in human pulmonary artery. Br J Pharmacol 119:277–282

MacLean MR, Sweeney G, Baird M, McCulloch KM, Houslay M, Morecroft I (1996b) 5-Hydroxytryptamine receptors mediating vasoconstriction in pulmonary arteries from control and pulmonary hypertensive rats. Br J Pharmacol 119:917–930

MacLean M, Herve P, Eddahibi S, Adnot S (2000) 5-hydroxytryptamine and the pulmonary circulation: receptors, transporters and relevance to pulmonary arterial hypertension. Br J Pharmacol 131:161–168

MacLean MR, Deuchar GA, Hicks MN, Morecroft I, Shen S, Sheward J, Colston J, Loughlin L, Nilsen M, Dempsie Y, Harmar A (2004) Overexpression of the 5-hydroxytryptamine transporter gene: effect on pulmonary hemodynamics and hypoxia-induced pulmonary hypertension. Circulation 109:2150–2155

Marcos E, Fadel E, Sanchez O, Humbert M, Dartevelle P, Simonneau G, Hamon M, Adnot S, Eddahibi S (2004) Serotonin-induced smooth muscle hyperplasia in various forms of human pulmonary hypertension. Circ Res 94:1263–1270

Millan MJ, Maiofiss L, Cussac D, Audinot V, Boutin JA, Newman-Tancredi A (2002) Differential actions of antiparkinson agents at multiple classes of monoaminergic receptor. I. A multivariate analysis of the binding profiles of 14 drugs at 21 native and cloned human receptor subtypes. J Pharmacol Exp Ther 303:791–804

Morecroft I, Heeley RP, Prentice HM, Kirk A, MacLean MR (1999) 5-hydroxytryptamine receptors mediating contraction in human small muscular pulmonary arteries: importance of the 5-HT1B receptor. Br J Pharmacol 128:730–734

Morecroft I, Loughlin L, Nilsen M, Colston J, Dempsie Y, Sheward J, Harmar A, MacLean MR (2005) Functional interactions between 5-hydroxytryptamine receptors and the serotonin transporter in pulmonary arteries. J Pharmacol Exp Ther 313:539–548

Morecroft I, Dempsie Y, Bader M, Walther DJ, Kotnik K, Loughlin L, Nilsen M, MacLean MR (2007) Effect of tryptophan hydroxylase 1 deficiency on the development of hypoxia-induced pulmonary hypertension. Hypertension 49:232–236

Ni W, Geddes TJ, Priestley JR, Szasz T, Kuhn DM, Watts SW (2008) The existence of a local 5-hydroxytryptaminergic system in peripheral arteries. Br J Pharmacol 154:663–674

Rondelet B, Van Beneden R, Kerbaul F, Motte S, Fesler P, McEntee K, Brimioulle S, Ketelslegers JM, Naeije R (2003) Expression of the serotonin 1b receptor in experimental pulmonary hypertension. Eur Respir J 22:408–412

Rothman RB, Ayestas MA, Dersch CM, Baumann MH (1999) Aminorex, fenfluramine, and chlorphentermine are serotonin transporter substrates. Implications for primary pulmonary hypertension. Circulation 100:869–875

Walther DJ, Peter JU, Bashammakh S, Hortnagl H, Voits M, Fink H, Bader M (2003a) Synthesis of serotonin by a second tryptophan hydroxylase isoform. Science 299:76

Walther DJ, Peter JU, Winter S, Holtje M, Paulmann N, Grohmann M, Vowinckel J, Alamo-Bethencourt V, Wilhelm CS, Ahnert-Hilger G, Bader M (2003b) Serotonylation of small GTPases is a signal transduction pathway that triggers platelet alpha-granule release. Cell 115:851–862

White K, Dempsie Y, Nilsen M, Wright AF, Loughlin L, MacLean MR (2011) The serotonin transporter, gender, and 17beta oestradiol in the development of pulmonary arterial hypertension. Cardiovasc Res 90:373–382

Willers ED, Newman JH, Loyd JE, Robbins IM, Wheeler LA, Prince MA, Stanton KC, Cogan JA, Runo JR, Byrne D, Humbert M, Simonneau G, Sztrymf B, Morse JA, Knowles JA, Roberts KE, McElroy JJ, Barst RJ, Phillips JA 3rd (2006) Serotonin transporter polymorphisms in familial and idiopathic pulmonary arterial hypertension. Am J Respir Crit Care Med 173:798–802

Targeting of Platelet-Derived Growth Factor Signaling in Pulmonary Arterial Hypertension

Eva Berghausen, Henrik ten Freyhaus, and Stephan Rosenkranz

Abstract Despite recent advances in the management of patients with pulmonary arterial hypertension (PAH), this disease remains a devastating condition with limited survival. While the current therapies primarily target the vasoconstrictor/vasodilator imbalance in the pulmonary circulation, there is currently no cure for PAH, and pulmonary vascular remodeling—representing the underlying cause of the disease—is only modestly affected. Hence, novel therapeutic approaches directly targeting the vascular remodeling process are warranted. Recent studies provided compelling evidence that peptide growth factors, which elicit their signals via receptor tyrosine kinases, are important contributors to the development and progression of PAH. In particular, platelet-derived growth factor (PDGF) is a strong mitogen for pulmonary vascular smooth muscle cells and protects these cells from apoptosis, thus representing an important mediator of pulmonary vascular remodeling. PDGF ligand and receptors are upregulated in PAH, and experimental studies have shown that inhibition of PDGF receptor signaling by pharmacological or genetic approaches prevents the development of PAH in animal models and is even able to reverse pulmonary vascular remodeling once it has been established. Consistently, results from phase II and phase III clinical trials indicate that the tyrosine kinase inhibitor imatinib mesylate, which potently inhibits the PDGF receptor, is effective in improving exercise capacity and pulmonary hemodynamics as add-on therapy in patients with severe PAH (i.e., pulmonary vascular resistance >800 dynes s cm^{-5}). Future studies will evaluate the long-term clinical efficacy and safety of imatinib, including patients with less impaired hemodynamics. Based on the current knowledge, targeting of PDGFR signaling is likely to become an anti-proliferative treatment option for patients with PAH and has the potential to at least partially correct the pathology of the disease.

E. Berghausen • H. ten Freyhaus • S. Rosenkranz (✉)
Klinik III für Innere Medizin, Herzzentrum der Universität zu Köln, Cologne, Germany

Center for Molecular Medicine Cologne (CMMC), Universität zu Köln, Cologne, Germany
e-mail: stephan.rosenkranz@uk-koeln.de

Keywords Pulmonary arterial hypertension • Receptor tyrosine kinase • Platelet-derived growth factor • Tyrosine kinase inhibitor • Imatinib

Contents

1 Introduction	382
2 Pathobiology of Pulmonary Arterial Hypertension	383
3 Targeted PAH Therapies and Therapeutic Limitations	384
4 Pathogenic Role of Platelet-Derived Growth Factor in PAH	386
5 PDGF Ligands and Receptors	387
6 Signal Relay Downstream of PDGF Receptors	387
7 Control of PDGF Signaling	391
8 Therapeutic Potential of PDGF Receptor Inhibition in PAH	392
8.1 Tyrosine Kinase Inhibitors: Imatinib Mesylate	392
8.2 Preclinical Studies	393
8.3 Clinical Studies	394
9 Limitations and Future Aspects	397
10 Safety and Tolerability of Tyrosine Kinase Inhibitors in PAH	398
10.1 Cardiotoxicity	399
10.2 Induction of PAH by Tyrosine Kinase Inhibitors	400
10.3 Subdural Hematomas	400
11 Conclusions	401
References	401

1 Introduction

Pulmonary hypertension (PH) is characterized by a marked and sustained elevation of pulmonary arterial pressure (PAP) and pulmonary vascular resistance (PVR) (Badesch et al. 2009; McLaughlin et al. 2009; Galiè et al. 2009b). It is defined by an elevation of the mean PAP to ≥ 25 mmHg at rest. The hemodynamic definition also distinguishes precapillary PH from postcapillary PH, in which the elevation of the PAP occurs as a consequence of elevated filling pressures in the left heart. Furthermore, pulmonary *arterial* hypertension (PAH) is distinguished from other more prevalent forms of PH that may develop secondary to left heart disease, chronic lung disease, and chronic thromboembolic disease (Galiè et al. 2009a–c; Rosenkranz 2007; Simonneau et al. 2009). The current clinical classification of PH (Dana Point 2008) defines five subgroups (Simonneau et al. 2009): group 1 comprises patients with PAH including idiopathic PAH (IPAH), heritable PAH (HPAH), PAH associated with connective tissue disease, drug- and toxin-induced PAH, portopulmonary hypertension, PAH due to HIV infection, PAH associated with congenital heart disease, and other rare causes of PAH. The other four groups comprise patients with PH owing to left heart disease (group 2), chronic lung disease/hypoxia (group 3), chronic thromboembolic PH (CTEPH; group 4), and PH with unclear or multifactorial etiologies (group 5).

PAH is a severe disorder which is characterized by incremental pulmonary vasculopathy and—without sufficient treatment—a poor life expectancy (Farber and Loscalzo 2004; Rosenkranz 2007). Chronically elevated PAP and PVR represent an increased afterload of the right heart, which over time leads to right ventricular hypertrophy and failure, eventually causing premature death (van Wolferen et al. 2007; Haworth 2007). Data from the US registry that were collected during the 1980s—when there was no efficient treatment available—showed that the median survival was limited to only 2.8 years (D'Alonzo et al. 1991). In the last decade, a number of drugs have been approved for the targeted treatment of PAH, and these compounds have proven effective in improving outcome and quality of life. However, these therapies are not able to cure the disease, as they primarily target the diminished pulmonary vascular relaxation response, while proliferative remodeling of the small pulmonary resistance vessels underlying the disease is only modestly affected. Although the current treatments improve exercise capacity and delay disease progression, their efficacy with regard to improvement of pulmonary hemodynamics is limited, and mortality remains unacceptably high (Benza et al. 2010; Humbert et al. 2010). Given the lack of curative therapeutic options at present, it is essential to develop innovative additional concepts that further improve pulmonary hemodynamics, clinical status, and survival. To this end, a key aim is to reverse or at least to halt pulmonary vascular remodeling, representing the underlying cause of PAH. Recent studies have indicated that peptide growth factors, which elicit their signals through receptor tyrosine kinases (RTK), play a crucial role in the development and progression of pulmonary vascular remodeling. In particular, several studies have indicated that platelet-derived growth factor (PDGF) plays a prominent role in abnormal remodeling of pulmonary resistance vessels (Schermuly et al. 2005; Perros et al. 2008). Hence, inhibition of PDGF signaling by tyrosine kinase inhibitors (TKI) appears as a promising concept for the treatment of PAH.

2 Pathobiology of Pulmonary Arterial Hypertension

PAH is a complex and multifactorial disease, and three key mechanisms are involved in its development and progression: (1) inappropriate pulmonary vasoconstriction, (2) vascular remodeling, and (3) in situ thrombosis. Various cell types residing in all layers of the vessel wall such as endothelial cells, vascular smooth muscle cells (VSMC), fibroblasts, platelets, and inflammatory cells contribute to the pathobiology of PAH (Hassoun et al. 2009; Morrell et al. 2009; Galiè et al. 2010). While a genetic background may predispose individuals for PAH, associated diseases and/or trigger factors are necessary for manifestation of the disease. Genetic alterations include modifications in the genes that comprise the signaling pathways of transforming growth factor-β (bone morphogenic protein receptor-2, BMPR2; activin receptor-like kinase-1, ALK1) and serotonin (5-hydroxitryptamine transporter; HTT) (Newman et al. 2004; Machado et al. 2009). Since only 15–20 %

of individuals with such a genetic condition develop PAH, the genetic variants are recognized as predisposing factors, and associated diseases, environmental factors, and/or gene–gene interactions are required for the manifestation of PAH (Farber and Loscalzo 2004; Machado et al. 2009).

A key feature in the pathobiology of PAH is a vasoconstrictor/vasodilator imbalance, which results in abnormal pulmonary vasoconstriction. While the bioavailability of pulmonary vasodilators such as nitric oxide (NO) and prostacyclin is decreased, the expression and release of potent vasoconstrictors such as endothelin (ET)-1 and serotonin are enhanced. These abnormalities are accompanied by the dysregulation of further modulators of vascular tone, including overexpression of the serotonin transporter and phosphodiesterase type 5 (PDE5) in the media, reduced expression and activity of voltage gated potassium channels (Kv), and alterations in the cross talk of vascular cells (Hassoun et al. 2009; Morrell et al. 2009). The current targeted PAH therapies all target on the imbalance between pulmonary vasodilators and constrictors.

A second key feature of the vascular pathology in PAH is abnormal remodeling of the small pulmonary resistance vessels. In fact, these structural alterations occurring in all layers of the vessel wall are believed to represent the main underlying cause of the disease. Disorganized endothelial cell proliferation may lead to the formation of plexiform lesions in the vicinity of obstructed arterioles, which contain monoclonal endothelial cell conglomerates and inflammatory cells (Morrell et al. 2009; Tuder et al. 2009). Additional pathological characteristics are owing to pulmonary arterial smooth muscle cell proliferation and migration, which leads to medial hypertrophy of the muscular resistance vessels, and an extension of pulmonary arterial smooth muscle cells into the distal, normally non-muscular peripheral pulmonary arterioles (Hassoun et al. 2009; Morrell et al. 2009). Further changes include intimal fibrosis, chronic inflammatory responses, increased connective tissue deposition in the medial layer, hypertrophy/thickening of the adventitia, and prothrombotic abnormalities causing in situ thrombosis (Hassoun et al. 2009; Morrell et al. 2009; Tuder et al. 2009; Stenmark et al. 2002; Tuder and Voelkel 2010). Both concentric and eccentric pulmonary vascular remodeling lead to progressive vascular narrowing and thus a marked loss of the cross-sectional luminal area. This results in an increase of PVR and right ventricular pressure load, which in turn causes right heart hypertrophy and failure and eventually premature death.

3 Targeted PAH Therapies and Therapeutic Limitations

The current treatment for PAH consists of supportive therapies (e.g., long-term oxygen therapy, oral anticoagulants, diuretics) and targeted drug treatment (Badesch et al. 2007; Barst et al. 2009; Galiè et al. 2009a–c). In few patients who demonstrate a favorable response in acute vasoreactivity testing, high-dose calcium channel blockers may be effective. However, most patients fail to meet the vasoresponder criteria and do not benefit from such treatment (Sitbon et al. 2005). Hence, the vast majority of patients requires targeted PAH therapy.

In recent years, there has been considerable progress in our understanding of the pathobiology and the targeted treatment of PAH. A number of pharmacological interventions have been developed which have provided symptomatic relief and improved outcome. All currently approved therapies for PAH belong to three classes of drugs: Prostacyclin analogues (prostanoids; epoprostenol, iloprost, treprostinil), endothelin receptor antagonists (ERA; bosentan, ambrisentan), and phosphodiesterase type 5 inhibitors (PDE5i; sildenafil, tadalafil). The efficacy and safety of each of these drugs were demonstrated in randomized, placebo-controlled trials (Barst et al. 1996; Galiè et al. 2005, 2009a; Olschewski et al. 2002; Oudiz et al. 2009; Rubin et al. 2002, Simonneau et al. 2002), and all above compounds have been approved by the European Medicines Agency (EMA) and/or the Food and Drug Administration (FDA). According to current guidelines, members of these drug classes may be prescribed for the treatment of PAH alone or in combination, depending on the patient's functional status, hemodynamic parameters, and the individual response to therapy (Barst et al. 2009; Galiè et al. 2009b).

While the advent of targeted PAH therapies has significantly improved the clinical situation and outcome of affected patients, the current treatment of PAH is not satisfactory. Prostanoids require subcutaneous/intravenous administration or frequent inhalations, and adverse reactions include flushing, headache, diarrhea, injection site pain, and catheter-related complications such as sepsis and pump failure (Galiè et al. 2009b; Simonneau et al. 2002). ERA are associated with liver toxicity requiring periodic monitoring of liver enzymes (Humbert et al. 2007), and PDE5i are not effective in all patients while dosing remains uncertain (Chockalingam et al. 2005; Galiè et al. 2005; Simonneau et al. 2008). Although the current treatment options have markedly improved quality of life and survival (Galiè et al. 2009c, 2010), their efficacy is still limited. Data from randomized, controlled trials have shown that the improvements in exercise capacity and mean PAP and PVR are only moderate. Many patients remain in functional class III, and the mortality of treated PAH patients remains unacceptably high. Recent data from large registries demonstrate that the annual mortality rate in the modern treatment era remains at approximately 10 % (Benza et al. 2010; Humbert et al. 2010).

Since the current PAH therapies primarily act as pulmonary vasodilators and do not markedly affect the underlying morphological changes of the small pulmonary arterioles, it is apparent that these treatments are only partially effective in providing symptomatic relief and in delaying disease progression. Hence, it is essential to establish additional therapies that more effectively target pulmonary vascular remodeling. While some data suggest that prostanoids, PDE5i, and ERA may elicit modest anti-proliferative effects (Chen et al. 1995; Falcetti et al. 2010; Wharton et al. 2000, 2005), this is most likely secondary to their vasodilatory properties and decreased shear stress, which itself represents a mitogenic stimulus (Haga et al. 2003; Sterpetti et al. 1992). Direct targeting of vascular remodeling by anti-proliferative approaches provides a promising strategy which holds the potential to reverse the remodeling process and to take a step closer to more efficiently affect or even cure PAH.

4 Pathogenic Role of Platelet-Derived Growth Factor in PAH

Compelling evidence reveals that PDGF is a major contributor to the pathobiology of vascular disorders including PAH (Raines 2004; Grimminger and Schermuly 2010). PDGF is expressed in multiple cell types including endothelial cells, smooth muscle cells, and macrophages and acts as a strong mitogen and chemokine. In fact, PDGF represents the most potent mitogenic factor for VSMC (Heldin and Westermark 1999; Kazlauskas 1994; Rosenkranz and Kazlauskas 1999; Schwartz 1997). It exerts its actions via the activation of two PDGF receptor (PDGFR) subtypes, PDGFRα and PDGFRβ. Both are expressed in numerous cell types of mesenchymal origin and promote various cellular responses including proliferation, migration, cellular survival, and transformation (Rosenkranz and Kazlauskas 1999; ten Freyhaus et al. 2006).

In the normal vessel wall, both PDGF ligand and the PDGFRα and PDGFRβ subtypes are expressed at very low levels. However, increased expression of PDGF ligands and receptors was found in atherosclerotic plaques and in neointimal lesions following balloon injury (Abe et al. 1998; Raines 2004; Tanizawa et al. 1996). Interventional studies in animals have shown that inhibition of PDGF signaling by TKIs, neutralizing antibodies, or oligonucleotides against the βPDGFR was able to reduce neointima formation and the accumulation of VSMCs in intimal lesions (Raines 2004; Levitzki 2005). Furthermore, genetic modification of the βPDGFR in F3/F3 mice also resulted in reduced neointima formation following balloon injury (Caglayan et al. 2011).

A role for PDGF in pulmonary vascular remodeling was first suggested by the observation that PDGF ligands (A and B chains) are upregulated in pulmonary cells and lung tissue in monocrotaline (MCT)- and hypoxia-induced experimental PH (Arcot et al. 1993; Cai et al. 1996; Huang and Sun 1997; Katayose et al. 1993). Upregulation of PDGF-B was also observed in a microarray-based expression study in laser-microdissected intrapulmonary arteries in hypoxia-induced PH in the mouse (Kwapiszewska et al. 2005), and a marked shift in the expression of PDGF ligands (A/B) and receptors was found in progenitor smooth muscle cells developing in microvessels of experimental PH (Jones et al. 2006). In humans, similar alterations with increased expression of PDGF ligand and the βPDGFR were found in lung tissue of patients with PAH (Humbert et al. 1998; Lanner et al. 2005). In small pulmonary arteries showing signs of vascular remodeling, upregulation of PDGF ligands (A/B) was mainly found in VSMC and endothelial cells, whereas increased expression of the βPDGFR was restricted to VSMC (Schermuly et al. 2005; Perros et al. 2008).

Several interventional experimental studies suggest that PDGF is causally involved in the development of PH. While chronically hypoxic mice carrying a constitutively active PDGFR-β (D849N) developed exaggerated pulmonary vascular remodeling and PH as compared to wild-type mice (Dahal et al. 2011a), inhibition of PDGF signaling in animal models of PH was effective in perinatal PH in the fetal lamb (Balasubramaniam et al. 2003), as well as in MCT- or hypoxia-induced PH (Schermuly et al. 2005), and ablation of PDGF-dependent downstream signaling in a genetic model prevented hypoxia-induced PH in the mouse (ten Freyhaus et al. 2012). Importantly, inhibition of PDGF signaling by imatinib reversed pulmonary vascular remodeling once it had been established (Schermuly et al. 2005). In summary, the above studies collectively indicate that PDGF is an important contributor to the pathobiology of PAH, and targeting of PDGFR signaling is therefore considered an important therapeutic target.

5 PDGF Ligands and Receptors

PDGF is a dimeric protein consisting of two of the four known homologous chains (A–D), which are linked by disulfide bonds, and exists as homo- (PDGF-AA, -BB, -CC, -DD) or heterodimers (PCGF-AB) (Bergsten et al. 2001; Gilbertson et al. 2001; Heldin and Westermark 1999; Kazlauskas 1994; LaRochelle et al. 2001; Li et al. 2000; Rosenkranz and Kazlauskas 1999). Whereas PDGF-A and -B are secreted as active ligands, PDGF-C and -D need to undergo cleavage of the N-terminal CUB domain in order to become activated (Bergsten et al. 2001; Gilbertson et al. 2001; LaRochelle et al. 2001; Li et al. 2000).

The biological responses of the various PDGF isoforms are transmitted via activation of two highly specific transmembrane RTKs, termed PDGFR α and β (Heldin and Westermark 1999; Kazlauskas 1994; Rosenkranz and Kazlauskas 1999). Both receptor subunits contain five extracellular immunoglobin-like domains, a juxtamembrane domain, a kinase domain which is disrupted by a kinase insert, and an N-terminal tail region. Upon binding of the dimeric ligand, two PDGFR subunits merge to form PDGFR homo- ($\alpha\alpha$, $\beta\beta$) or heterodimers ($\alpha\beta$). The two receptor subtypes differ in their binding affinity for the various PDGF isoforms. Whereas the PDGF B-chain associates with both receptor subtypes, the A-chain binds exclusively to the αPDGFR. Consequently, PDGF-AA induces only PDGFR-$\alpha\alpha$ homodimers, PDGF-AB may induce the assembly of PDGFR-$\alpha\alpha$ and -$\alpha\beta$, and PDGF-BB leads to the formation of all possible receptor dimers ($\alpha\alpha$, $\alpha\beta$, $\beta\beta$) (Fig. 1a). The affinity of the PDGFR subtypes for PDGF-C and -D is less well established. Apparently, PDGF-CC assembles both $\alpha\alpha$ and $\alpha\beta$ receptor dimers, whereas PDGF-DD mainly binds to $\beta\beta$, although it may also induce the formation of $\alpha\beta$ heterodimers under certain conditions. This is of relevance, as the two PDGFR subtypes differ from one another in both their signaling properties and biological responses (Rosenkranz and Kazlauskas 1999).

6 Signal Relay Downstream of PDGF Receptors

Binding of PDGF ligands leads to dimerization of two receptor subunits and a subsequent increase of the receptor's intrinsic tyrosine kinase activity, resulting in trans-/autophosphorylation of the receptor. Once PDGFRs are activated,

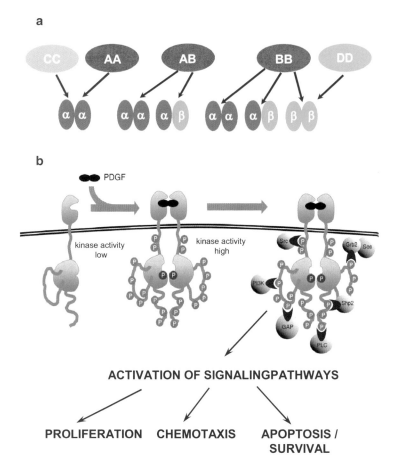

Fig. 1 Signal transduction by PDGF receptors. (a) Binding specificity of the PDGF ligand chains (A, B, C, D) to the α- and βPDGFR subtypes. (b) Receptor dimerization, trans-/autophosphorylation, and cellular effects induced by activated PDGFR

phosphorylated tyrosine residues in the intracellular domain serve as specific docking sites for intracellular signaling molecules, which selectively associate with the receptors at specific sites. Once bound to the receptor, each of the receptor-associated signaling molecules is activated and triggers downstream signal relay cascades which mediate PDGF-dependent cellular responses such as protein synthesis, proliferation, migration, protection against apoptosis, and cellular transformation (Fig. 1b).

Signaling via the βPDGFR subtype is required for the proper development of the vessel wall during embryogenesis and is also essential for maintaining the integrity of vascular structures (Leveen et al. 1994; Lindahl et al. 1997; Mellgren et al. 2008; Soriano 1994; Tallquist et al. 1999, 2000). The activated βPDGFR associates with SH2 (Src homology 2) domain containing signaling molecules such as Src family

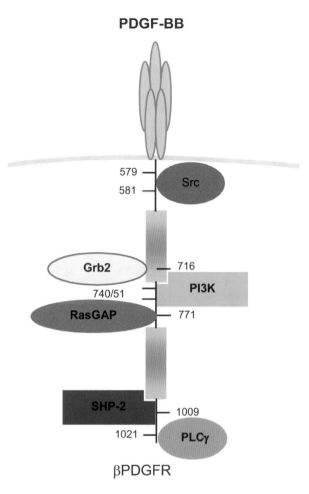

Fig. 2 Signaling molecules that interact with the activated βPDGFR. *Numbers* indicate the amino acid position in the intact human βPDGFR

kinases (Src), phosphatidylinositol 3-kinase (PI3K), the GTPase-activating protein of Ras (RasGAP), the tyrosine phosphatase SHP-2, and phospholipase Cγ1 (PLCγ), which specifically bind to phosphorylated tyrosine residues (Fig. 2) (Rosenkranz and Kazlauskas 1999). In addition, SH2-SH3 domain containing adaptor proteins without enzymatic activity such as Grb2, Grb7, Grb10, Nck, and Shc also associate with the activated receptor, but their role in signal transduction is yet less clear. Each of the receptor-associated signaling molecules has a unique role in the activation of downstream targets such as Erk1/2, Akt, JNK, and FAK, and the induction of immediate early response genes such as *c-myc*, *c-fos*, *egr-1*, and *JunB*.

The mechanisms of signal transduction by PDGF receptors have been extensively studied in cellular systems. In particular, the contribution of

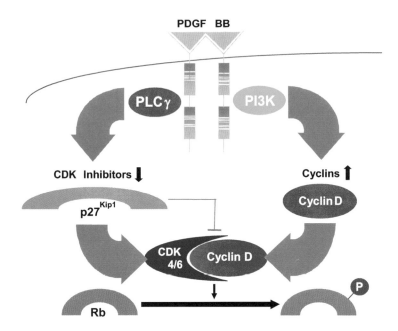

Fig. 3 Schematic diagram illustrating the differential roles of PDGF-dependent PI3K and PLCγ activity for cellular responses. The activated βPDGFR recruits multiple signaling molecules including PI3K and PLCγ. While PI3K is required for efficient upregulation of cyclin D1, PLCγ is mainly responsible for downregulation of $p27^{kip1}$. Both events are required for Rb phosphorylation which is crucial for cellular events such as cell cycle progression and chemotaxis (modified from Caglayan et al. 2011)

PDGFR-associated signaling molecules to downstream pathways and cellular responses was investigated by utilizing mutated PDGFRs, in which the selective tyrosine binding sites for intracellular signaling molecules were individually deleted, and that were thus unable to bind and activate single receptor-associated signal relay enzymes. PDGFR mutants have been expressed in PDGFR-deficient cell lines including human HepG2 cells, A431 cells, murine *Ph* cells (3T3-like), and pig aortic endothelial cells (PAE) (Bäumer et al. 2008; Chung et al. 1994; DeMali et al. 1997; Fambrough et al. 1999; Joly et al. 1994; Klinghoffer et al. 1996; Kundra et al. 1994; Montmayeur et al. 1997; Rosenkranz et al. 1999; Valius and Kazlauskas 1993; Valius et al. 1995). These studies have identified relevant signaling pathways and have also shown cell type-specific differences with regard to signal relay of PDGFR-mediated responses. In VSMCs, a critical role of PI3K and PLCγ for PDGF-dependent VSMC proliferation and migration was demonstrated by use of a system of chimeric CSF1R/βPDGFR mutants (Caglayan et al. 2011). These investigations have shown that PI3K is mainly required for efficient upregulation of cyclin D1, whereas PLCγ is responsible for downregulation of the cdk-inhibitor $p27^{kip1}$. Both events are necessary for the activation of cyclin-dependent kinases (cdks) and Rb phosphorylation which are required for both cell cycle progression and chemotaxis (Fig. 3). Disruption of PDGF-dependent PI3K and PLCγ signaling in mice was sufficient to attenuate neointima formation following balloon injury in vivo. Recent evidence indicates that PI3K and PLCγ are also critically involved in PDGF-dependent pulmonary vascular remodeling (ten Freyhaus et al. 2011), and ablation of PI3K and PLCγ signaling downstream of the βPDGFR in F3/F3 mice markedly attenuated the development of hypoxia-induced PH (ten Freyhaus et al. 2012).

7 Control of PDGF Signaling

The control of PDGF signaling is facilitated by several mechanisms. RasGAP directly associates with the activated βPDGFR and serves as a negative regulator of PDGF signaling by inactivating Ras (Rosenkranz and Kazlauskas 1999; Klinghoffer et al. 1996; Valius et al. 1995). A recent study demonstrated that downstream signaling by the PDGFR and other RTKs is amplified by the adaptor protein p130Cas which is upregulated and over-activated in serum and pulmonary vascular cells (PASMC, PEC) in PAH (Tu et al. 2012). The role of Src family kinases, which directly interact with PDGFR, is somewhat controversial. While Src is considered an oncogene, and some studies indicate that it is involved in pulmonary VSMC proliferation and the development of PH (Pullamsetti et al. 2012), Src family members appear to be involved in the termination of PDGF-induced signals, as they promote the ubiquitination and degradation of the active ligand–receptor complex via the phosphorylation of c-Cbl (Miyake et al. 1999; Rosenkranz et al. 2000). PDGFR activity is also negatively regulated by a set of antagonizing protein tyrosine phosphatases (PTPs) including DEP-1, SHP-2, PTP-1B, and TC-PTP, which dephosphorylate and thus inactivate RTKs (Kappert et al. 2005, 2007).

Hypoxia, which is frequently present in lung tissue of patients with pulmonary hypertension, was recently identified as an important modulator of RTK signaling. Hypoxia enhances the proliferative and migratory response of VSMCs to various growth factors including PDGF, basic fibroblast growth factor (bFGF), and epidermal growth factor (EGF) (ten Freyhaus et al. 2011; Humar et al. 2002; Schultz et al. 2006). While the expression of the respective receptors is not affected by hypoxia, it was shown that hypoxia enhances PDGF signaling in the pulmonary vasculature by hypoxia-inducible factor-1α (HIF1α)-dependent downregulation of PTP expression and activity, thus resulting in enhanced PDGFR phosphorylation and activity (ten Freyhaus et al. 2011) (Fig. 4). Further studies revealed that PDGF also induces pulmonary vascular remodeling through a 15-LO/15-HETE pathway under hypoxic conditions (Zhang et al. 2012) and that hypoxia induces unique proliferative response in adventitial fibroblasts by engaging PDGFRβ/c-Jun N-terminal kinase1 (JNK1) signaling and thus leads to adventitial remodeling in PH (Panzhinskiy et al. 2012).

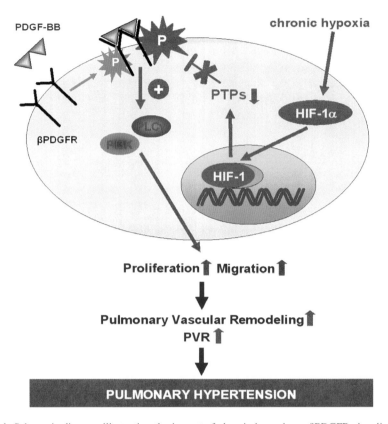

Fig. 4 Schematic diagram illustrating the impact of chronic hypoxia on βPDGFR signaling. In PASMCs, chronic hypoxia induces HIF-1α, leading to decreased expression and activity of βPDGFR-antagonizing PTPs. This leads to enhanced ligand-induced βPDGFR phosphorylation and cellular proliferation and migration. In vivo, these changes critically impact on pulmonary vascular remodeling, pulmonary vascular resistance (PVR), and the progression of PH

8 Therapeutic Potential of PDGF Receptor Inhibition in PAH

8.1 Tyrosine Kinase Inhibitors: Imatinib Mesylate

Pharmacological inhibition of PDGFRs may be achieved by inhibiting their intrinsic tyrosine kinase activity. Imatinib mesylate (STI571, Glivec) is a small molecule TKI that has been approved for the treatment of various malignant disorders including Philadelphia (Ph) chromosome-positive chronic myeloid leukemia (CML), acute lymphoblastic leukemia (ALL), and gastrointestinal stromal tumors (GIST) (Cohen et al. 2005). It is a nonspecific tyrosine kinase inhibitor, interacting with the bcr–abl fusion protein (responsible for efficacy against Ph-positive CML), both α and βPDGFR, DDR, and the c-kit receptor (Baselga 2006). The inhibitory effect of imatinib is due to binding and thereby blocking of the active site of the phosphate transferring tyrosine kinase. It is a so-called type 2 inhibitor, which binds not only to the ATP binding pocket but also to a site adjacent to the pocket, which enhances specificity and allows binding to the kinase even in an inactive conformation. Imatinib is readily absorbed with maximum blood levels achieved after 1–4 h (Gschwind et al. 2005; Le Coutre et al. 2004; Peng et al. 2005) and is metabolized mainly by the cytochrome P450 enzyme CYP3A4, leading to the formation of biologically active metabolites. Both imatinib and its metabolites are eliminated mostly via the biliary–fecal route and to a lesser extent via the urine (Gschwind et al. 2005). While the invention of imatinib has revolutionized the treatment of the described cancers, recent evidence indicates that it is also effective in the treatment of PAH.

8.2 Preclinical Studies

The observation that PDGF ligand and PDGFR are substantially upregulated in lung tissue of experimental PH and in humans with PAH (Schermuly et al. 2005; Perros et al. 2008) identified PDGF as a potential therapeutic target. On the cellular level, PDGF-dependent proliferation and migration of murine, rat and human PASMCs were potently inhibited by imatinib. Furthermore, imatinib exerts pro-apoptotic effects in PDGF-stimulated PASMCs from patients with IPAH (Nakamura et al. 2011), most likely by antagonizing the anti-apoptotic properties of PDGFR signaling (Vantler et al. 2005, 2010).

In vivo studies have shown that PDGFR inhibition by imatinib ameliorated PH in two animal models by improving pulmonary hemodynamics, reducing RV hypertrophy, and reversing pulmonary vascular remodeling (Schermuly et al. 2005). Furthermore, imatinib was shown to exert immunomodulatory effects on dendritic cells (Sato et al. 2003). In the MCT rat model and in hypoxia-induced PH in mice, the changes in RV pressure and right heart hypertrophy were reversed to near-normal levels, and this was associated with improved survival (Schermuly et al. 2005).

Although some studies suggest that TKIs such as imatinib may also modulate pulmonary vascular tone (Abe et al. 2011), such effects were observed at rather high concentrations (50 mg kg^{-1} in vivo) which by far exceed the dosages applied in humans. The primary mode of action of imatinib at relevant concentrations is to inhibit the proliferation and migration of vascular cells and to promote apoptosis (Schermuly et al. 2005; Perros et al. 2008). In addition to inhibition of PDGFR, imatinib also inhibits other kinases such as c-kit and Abl. Recent studies have indicated that c-kit may also play a role in the pathobiology of PH (Young et al. 2009), and imatinib was shown to attenuate bone marrow-derived c-kit+ cell mobilization in hypoxic PH (Gambaryan et al. 2010). While specific inhibition of c-kit was recently shown to prevent the development of PH in MCT-treated rats, it failed to reverse established PH (Dahal et al. 2011b). Although c-kit may contribute to the development of PH to some extent, the main therapeutic effect of imatinib in PH is believed to be via inhibition of PDGFR signaling. This is supported by the large body of data from in vitro findings demonstrating that the expression of PDGF ligands and both receptor subtypes is substantially enhanced in PAH, that PDGF exerts pronounced cellular effects in PASMCs, and that the prevention of PDGFR phosphorylation by TKIs such as imatinib was associated with suppression of PDGF-dependent downstream signaling pathways and cellular responses. Hence, reversal of pulmonary arterial remodeling by imatinib in experimental PH, which is even effective when pulmonary vasculopathy and the impairment of hemodynamics are fully established, may be attributed to inhibition of PDGF-induced proliferation of VSMC and increased apoptosis. The fact that other approaches such as application of the PDGF-B-antagonizing aptamer NX1975 and ablation of PDGF-dependent PI3K and PLCγ activity were also able to prevent PH in animal models further supports the important role of PDGF (Balasubramaniam et al. 2003; ten Freyhaus et al. 2012). In summary, the above studies indicate that inhibition of PDGFR signaling represents a potential causal treatment for pulmonary vascular remodeling and PAH.

8.3 Clinical Studies

In addition to experimental data, several case reports indicated that imatinib is effective as add-on treatment in humans with refractory PAH, and beneficial effects have also been reported during long-term treatment. The first published case describes a 61-year-old patient with rapidly progressing PAH who was insufficiently managed on triple therapy (bosentan, iloprost, sildenafil) and was therefore treated with add-on imatinib. Subsequently, clinical symptoms (WHO class IV–II), exercise capacity, and pulmonary hemodynamics improved dramatically (Ghofrani et al. 2005). Significant improvements after initiation of imatinib were documented from several other groups in a 52-year-old male with refractory idiopathic PAH (Patterson et al. 2006), in two patients with PAH who had received imatinib for the treatment of leukemia (Souza et al. 2006), in a 58-year-old patient with severe scleroderma-associated PAH (ten Freyhaus et al. 2009), and as additional treatment in portopulmonary hypertension (Tapper et al. 2009). Nevertheless, imatinib was reported to be not successful in other cases (Garcia-Hernandez et al. 2008).

Phase II Trial. Based on the promising data from preclinical studies and the encouraging observations from clinical case studies, a double-blind, placebo-controlled, multicenter phase II trial (NCT00477269) was implemented to evaluate the safety, tolerability, and efficacy of imatinib in PAH patients (Ghofrani et al. 2010a). Fifty-nine patients with inadequate response to established therapy with IPAH or HPAH were randomized to receive either imatinib (200–400 mg day^{-1}) or placebo for 24 weeks. Forty-two patients completed the trial. While the primary endpoint (6 min walk distance, 6MWD) was not met (+22 ± 63 vs. −1 ± 53 m, n.s.), hemodynamic measurements were in favor of imatinib (Ghofrani et al. 2010a). There was a significant decrease in PVR (−300 ± 347 vs. −78 ± 269 dynes s cm^{-5}; $p < 0.01$) and an increase in cardiac output (+0.6 ± 1.2 vs. −0.1 ± 0.9 l min^{-1}; $p = 0.02$) (Fig. 5a). A non-prescribed, post hoc subgroup analysis indicated that patients with substantial elevation of PVR ($\geq 1{,}000$ dynes s cm^{-5}) may in particular benefit from such treatment, as such patients showed a significantly increased 6MWD and a greater hemodynamic improvement on imatinib when compared to placebo than patients with lower PVR (Fig. 5b). Patients who completed the 24-week core study were invited to enter a long-term extension phase. Preliminary results indicate that the beneficial effects of imatinib on exercise capacity and functional class persist for at least 12–24 months (Ghofrani et al. 2010b).

The most common side effects included nausea, headache, and peripheral edema. More severe side effects were equally distributed between groups. While arterial and mixed venous oxygen saturation increased with imatinib treatment, there was a decrease in hemoglobin levels in patients on imatinib which may be relevant. Although there were several limitations (trend towards higher functional class and more patients on triple therapy in the placebo group) and the primary endpoint was not met (most likely due to the small sample size), the study demonstrated that imatinib is safe in the population of PAH patients and that a pure anti-proliferative approach can improve pulmonary hemodynamics (decrease of PVR, increase in cardiac output) in addition to conventional treatment.

Phase III Trial (IMPRES). In consideration of the results of the phase II trial, a randomized, double-blind, placebo-controlled multicenter phase III trial (NCT00902174) was carried out to further assess the efficacy and safety of imatinib in severe PAH. The primary endpoint of the study was the 6MWD, secondary endpoints included pulmonary hemodynamics, pharmakokinetics, and time to clinical worsening. Two hundred and two patients with severe PAH (PVR > 800 dynes s cm^{-5}) despite treatment with at least two targeted PAH therapies were randomized to receive imatinib (200–400 mg day^{-1}) or placebo for 24 weeks. At study entry, 41 % of patients in the imatinib group and 42 % of patients in the placebo group were on a triple combination therapy (ERA, PDE5i, prostanoid), and all remaining patients received dual combination therapy. At 24 weeks, there was significant improvement of the 6MWD (primary endpoint) by 32 ± 10 m in patients who received imatinib as compared to patients exposed to placebo ($p = 0.002$) (Hoeper et al. 2011a). This was accompanied by substantial hemodynamic improvement, including significant decreases in the mean PAP and PVR, and a rather impressive increase in cardiac output (1.2 vs. 0.3 l min^{-1}, $p < 0.001$) (Hoeper et al. 2011b, 2012). This was consistent with improvement in a number of echocardiographic measures of right ventricular function (Shah et al. 2012).

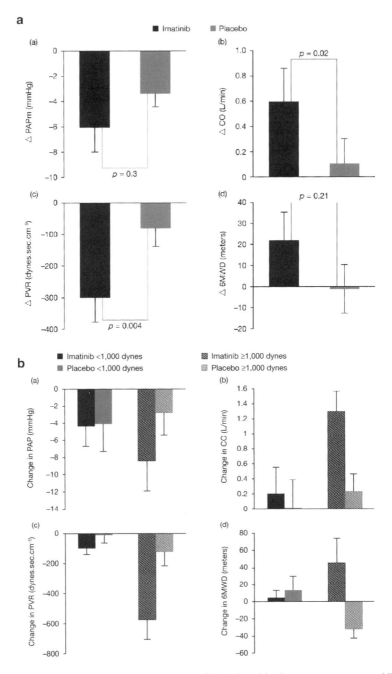

Fig. 5 Results from a phase II study in patients with PAH and inadequate response to established therapy. Effects of imatinib (200–400 mg day^{-1}) on mean pulmonary artery pressure (mPAP), cardiac output (CO), pulmonary vascular resistance (PVR), and 6 min walking distance (6MWD). Shown is the mean change from baseline to study end (24 weeks). (**a**) All patients ($n = 42$). (**b**) Post hoc analyses showed that patients with greater hemodynamic impairment (PVR $\geq$ 1,000 dynes; $n = 20$) may respond better than patients with less impairment (PVR $<$ 1,000 dynes; $n = 21$) (Ghofrani et al. 2010a)

9 Limitations and Future Aspects

Although the clinical trials with imatinib in PAH have provided encouraging results, this treatment has not been entirely successful thus far and there are a number of findings that deserve critical interpretation. In IMRES, a significant number of patients receiving imatinib (33 %) discontinued the study drug due to side effects (vs. 18 % in the placebo group), and there was no difference in mortality and the time to clinical worsening (TTCW; defined as death, hospitalization for PAH, decrease in the 6MWD by >15 %, or worsening of WHO functional class by ≥ 1 level) between the treatment and placebo groups (HR 1.16; $p = 0.56$). Given the definition of TTCW and the known imatinib side effect of fluid retention, it may be difficult to distinguish between worsening of PAH and imatinib-induced fluid retention in individual cases of peripheral edema. Of note, most of the events defining TTCW occurred during the first 8 weeks of treatment and resolved in the majority of cases despite ongoing drug treatment. Nevertheless, tolerability and the management of side effects have to be improved in PAH patients.

The results from the phase II trial have indicated that only patients with a PVR > 1,000 dynes s cm^{-5} benefit from imatinib (Ghofrani et al. 2010a), and the IMPRES trial (phase III) included only patients with a PVR > 800 dynes s cm^{-5} (Hoeper et al. 2011a, b). Thus, current data indicate that imatinib may only be effective in patients with advanced stages of PAH and severe hemodynamic impairment. This is in contrast to the current concept to diagnose and treat PAH as early as possible (Sitbon and Galiè 2010). There is no reason to believe that PDGF-dependent signals are only important at advanced stages of PAH, and it does not make sense to wait until a disease has progressed to an advanced stage before anti-proliferative treatment is initiated. Nevertheless, the current data suggest that imatinib may not be effective in patients with moderate PAH (PVR < 800 dynes s cm^{-5}). In this context, time may be a critical issue. In contrast to previous randomized, controlled PAH trials that followed up patients for 12 weeks, the observation period in the imatinib studies was extended to 24 weeks, because "reverse remodeling" strategies are believed to require more time than vasodilative approaches. However, even an observation time of 24 weeks may not be enough to capture the full therapeutic effect. At least some observations indicate that continued improvements of clinical status, hemodynamics, and other prognostically relevant parameters persist far beyond 24 weeks (ten Freyhaus et al. 2009; Ghofrani et al. 2010b), which may actually be expected from an effective anti-proliferative approach. Hence, one may speculate that the efficacy of a TKI in patients with moderately impaired hemodynamics and less severely remodeled vessels may require more time and become evident with longer treatment. It will therefore be critical to investigate the efficacy of imatinib in patients with a PVR < 800 dynes s cm^{-5}, and to monitor patients for a longer period of time.

Besides PAH, the efficacy, tolerability and safety of imatinib may also be evaluated in other forms of PH. Pulmonary vascular remodeling was shown at least in some patients with PH owing to chronic left heart disease (Delgado et al. 2005), and immunohistochemical analysis indicated high deposition of PDGF and high levels of PDGFR expression in tissue sections from pulmonary endartherectomy (PEA) samples from patients with CTEPH (Ogawa et al. 2009). While surgical PEA represents the first choice of treatment in patients with CTEPH, imatinib may therefore be effective in patients with non-operable CTEPH and those with persistent PH after PEA. However, this has to be assessed with great caution, as subdural hematomas have been observed in some patients with PAH, in whom imatinib was given in addition to oral anticoagulation (Hoeper et al. 2012). In patients with PAH, the subset of patients with scleroderma-associated PAH may be viewed separately. Stimulatory autoantibodies to the PDGFR have been described in scleroderma patients (Baroni et al. 2006). In scleroderma-associated PAH, the therapeutic effect of imatinib may therefore be attributable to inhibition of PDGFR signaling and reversal of vascular remodeling in the pulmonary vasculature, and/or to interference with stimulatory PDGFR autoantibodies and thus modulation of scleroderma per se.

From the current clinical experience with imatinib, we have learned that it appears not to be equally effective in all patients and that pulmonary hemodynamics are frequently improved but not normalized. A possible explanation for this may be that in addition to PDGF, other growth factors such as EGF and bFGF also contribute to the pathology in pulmonary arterial remodeling (Dahal et al. 2010; Izikki et al. 2009; Merklinger et al. 2005), and that the contribution of each growth factor may differ among individual patients. Therefore, inhibition of central signaling pathways that affect signal relay by all relevant RTKs may represent an even more effective strategy. In this context, PI3K was recently identified as a central mediator of pathogenic responses transmitted by the PDGFR and other RTKs. Selective inhibition of the catalytic class IA PI3K isoform p110α was shown to potently inhibit PASMC proliferation induced by multiple growth factors and to attenuate PH in mice (Berghausen et al. 2011). However, toxicity certainly remains an issue with TKIs, and—depending on the kinase profile—one has to be aware of unexpected adverse reactions.

10 Safety and Tolerability of Tyrosine Kinase Inhibitors in PAH

Imatinib is a generally safe and well-tolerated drug, particularly when considering that it profoundly interacts with basal cellular processes such as PDGF signaling. Adverse events are usually mild and mostly occur during the first month of treatment. Usual adverse reactions that have been recorded in the initial clinical trials for CML and in long-term observational studies include nausea, headache, skin rash, pruritus, anemia, and fluid retention (Druker et al. 2006). Similarly, adverse drug reactions in the PAH phase II and phase III trials included nausea (approximately 50 % of patients), headache (36 %), and peripheral edema

(25–30 %) (Ghofrani et al. 2010a; Hoeper et al. 2012). In most cases, these adverse reactions did not lead to discontinuation of the drug, and nausea was frequently controlled by taking the drug with food. In the phase II trial, serious adverse events occurred in 39 % of patients receiving imatinib and in 23 % of those receiving placebo (Ghofrani et al. 2010a, b). A potential concern in PAH patients may be liver toxicity, especially in patients on combination therapy with ERAs. However, in patients with CML, imatinib appears to be a safe drug in this respect (Tong et al. 2010). Nevertheless, one has to be aware of potential additive toxicity, and frequent monitoring of LFTs might be advisable. Particular concerns with TKI may be cardiotoxicity, the recent observation of PAH induction by at least one TKI, and the potential occurrence of subdural hematomas when imatinib is given in addition to oral anticoagulants.

10.1 Cardiotoxicity

A harmful effect of imatinib on cardiac integrity and function was shown in mice and a small number of patients who had received the drug chronically for CML (Kerkelä et al. 2006). This cardiotoxic effect was not attributed to PDGFR inhibition, but to the inhibitory effect on the non-receptor Abelson tyrosine kinase (c-Abl), leading to mitochondrial dysfunction and an ER stress response via activation of the PKR-like ER kinase (PERK) and IRE1, and to increased expression of PKCδ in cardiac myocytes (Kerkelä et al. 2006; Cheng and Force 2010). The involvement of c-Abl was confirmed by the fact that a redesigned imatinib that no longer inhibited c-Abl did not exert cardiotoxic effects (Fernandez et al. 2007). Despite these experimental data, further analyses of clinical studies revealed that congestive heart failure is a rare event in patients receiving imatinib for CML (Druker et al. 2006), but occurs more frequently in patients with preexisting cardiac conditions (Atallah et al. 2007). The latter fact may be of particular significance in patients with PAH and impaired right ventricular function since the right ventricle responds particularly sensitive to hemodynamic and/or cardiotoxic impairment. However, add-on imatinib was associated with improved right ventricular function as assessed by MRI in a case of scleroderma-associated PAH (ten Freyhaus et al. 2009), cardiac output significantly improved in the phase II and phase III trials of imatinib for PAH (Ghofrani et al. 2010a, b; Hoeper et al. 2011a, b), and in the IMPRES trial, this was associated with a marked improvement of echocardiographic measures of right ventricular function (Shah et al. 2012). Thus, the implemented clinical investigations of imatinib in patients with PAH did not provide any signals that would indicate detrimental effects on cardiac function. Nevertheless, cardiac side effects remain a concern of TKIs that requires close follow-up and regular monitoring of contractile function by echocardiography.

10.2 Induction of PAH by Tyrosine Kinase Inhibitors

Recent observations indicate that—depending on the kinase profile of a TKI—some compounds may actually have the potential to induce PAH. Although dasatinib was recently shown to be therapeutic in experimental PH (Pullamsetti et al. 2012), chronic treatment with dasatinib in several cases of imatinib-refractory CML was associated with an onset of severe PAH that mostly occurred in conjunction with pleural effusions (Dumitrescu et al. 2011; Hennigs et al. 2011; Mattei et al. 2009; Quinta's-Cardama et al. 2007; Rasheed et al. 2009). Consistently, Montani et al. recently published a series of nine cases with dasatinib-induced PAH that had been identified in the French PH registry (Montani et al. 2012). Interestingly, in many of the cases of dasatinib-associated PAH, pulmonary hypertension appeared to be reversible after termination of dasatinib. This is remarkable because complete remission and normalization of pulmonary hemodynamics is usually not achieved by medical treatment in PAH. Therefore, it is assumed that the pathophysiological mechanism leading to increased PVR differs from the vascular alterations typically seen in PAH. While the mechanism remains elusive, it may be related to the additional targeting of Src family kinases by this compound. Since Src is required for degradation of activated PDGFRs and the termination of PDGF signals (Miyake et al. 1999; Rosenkranz et al. 2000), its inhibition may result in increased signaling by PDGF and possibly other growth factors. Alternatively, Src kinase inhibition may directly alter the pulmonary vascular tone through a functional mechanism. Src was recently shown to control the potassium channel TASK-1, thus acting as a cofactor for setting a negative resting membrane potential in human PASMCs and a low resting pulmonary vascular tone. Consequently, inhibition of Src decreased the tyrosine-phosphorylation state of TASK-1 and potassium current density, resulting in considerable depolarization and an increase in PAP in experimental settings (Nagaraj et al. 2013). An additional explanation for the induction of PAH by dasatinib may be related to off-kinase inhibition which may affect pulmonary vascular wall integrity, possibly leading to (peri)vascular edema and subsequent PVR increase. Although such effects have not been reported for imatinib, these observations remind us that one should be aware of potential unexpected side effects whenever a drug of this class is used.

10.3 Subdural Hematomas

A potential concern with the use of imatinib in PAH is the unexpected occurrence of subdural hematomas (SDH), which has recently been reported in some patients who were on concomitant oral anticoagulation with vitamin K antagonists (Hoeper et al. 2012). While such events have not been reported at increased incidence in patients on imatinib for CML or GIST, the indication for anticoagulation in PAH patients as suggested by current guidelines (Galiè et al. 2009a–c) may represent an important difference between the populations of CML/GIST vs. PAH patients. A detailed analysis of all cases with SDH and the potential underlying mechanism(s) are warranted.

11 Conclusions

In summary, a series of preclinical studies has identified PDGF as a major contributor to pulmonary vascular remodeling, increased PVR, and subsequent right ventricular failure. Indeed, experimental studies demonstrating that inhibition of PDGF signaling by the TKI imatinib or by genetic approaches is able to prevent or even reverse pulmonary vascular remodeling in vivo are complemented by the observation that PDGF ligands and receptors are upregulated and activated in experimental and human PAH. It is therefore not surprising that two randomized clinical trials that have investigated the safety, tolerability, and efficacy of imatinib in human PAH have shown that this anti-proliferative compound is well tolerated in most patients and produces significant improvements of exercise capacity and pulmonary hemodynamics in patients with advanced PAH. Therefore, inhibition of PDGFR signaling, e.g., by imatinib mesylate represents an encouraging therapeutic concept and is likely to become an additional targeted therapy for PAH. In addition to existing therapies, this anti-proliferative approach has the potential to establish a sustained reduction of the underlying pulmonary vascular remodeling, which in turn allows hemodynamic improvement. The possibility to reverse pulmonary vascular remodeling in human disease would indeed represent an important step forward in the endeavor to establish a causal treatment for PAH that is able to correct the pathology of the disease. While the current data provide evidence for therapeutic efficacy only in patients with severe hemodynamic impairment, the role of PDGFR inhibition in less severe PAH and possibly in other forms of PH remains to be investigated. Furthermore, one should be alert for unexpected adverse events including induction of PAH by TKIs, cardiotoxicity, and SDH, and the long-term benefit remains to be documented.

References

Abe J, Deguchi J, Takuwa Y et al (1998) Tyrosine phosphorylation of platelet derived growth factor beta receptors in coronary artery lesions: implications for vascular remodelling after directional coronary atherectomy and unstable angina pectoris. Heart 79:400–406

Abe K, Toba M, Alzoubi A, Koubsky K, Ito M, Ota H, Gairhe S, Gerthoffer WT, Fagan KA, McMurtry IF, Oka M (2011) Tyrosine kinase inhibitors are potent acute pulmonary vasodilators in rats. Am J Respir Cell Mol Biol 45:804–808

Arcot SS, Lipke DW, Gillespie MN, Olson JW (1993) Alterations of growth factor transcripts in rat lungs during development of monocrotaline-induced pulmonary hypertension. Biochem Pharmacol 46:1086–1091

Atallah E, Durand JB, Kantarijan H, Cortes J (2007) Congestive heart failure is a rare event in patients receiving imatinib therapy. Blood 110:1233–1237

Badesch DB, Abman SH, Simonneau G et al (2007) Medical therapy for pulmonary arterial hypertension. Updated ACCP evidence-based clinical practice guidelines. Chest 131:1917–1928

Badesch DB, Champion HC, Sanchez MA et al (2009) Diagnosis and assessment of pulmonary arterial hypertension. J Am Coll Cardiol 54(Suppl S):55–66

Balasubramaniam V, Le Cras TD, Ivy DD, Grover TR, Kinsella JP, Abman SH (2003) Role of platelet-derived growth factor in vascular remodeling during pulmonary hypertension in the ovine fetus. Am J Physiol Lung Cell Mol Physiol 284:L826–L833

Baroni SS, Santillo M, Bevilacqua F, Luchetti M, Spadoni T, Mancini M, Fraticelli P, Sambo P, Funaro A, Kazlauskas A, Avvedimento EV, Gabrielli A (2006) Stimulatory autoantibodies to the PDGF receptor in systemic sclerosis. N Engl J Med 354:2667–2676

Barst RJ, Rubin LJ et al (1996) A comparison of continuous intravenous epoprostenol (prostacyclin) with conventional therapy for primary pulmonary hypertension. The Primary Pulmonary Hypertension Study Group. N Engl J Med 334:296–302

Barst RJ, Gibbs SR, Ghofrani HA et al (2009) Updated evidence-based treatment algorithm in pulmonary arterial hypertension. J Am Coll Cardiol 54(Suppl S):S78–S84

Baselga J (2006) Targeting tyrosine kinases in cancer: the second wave. Science 312:1175–1178

Bäumer AT, Ten Freyhaus H, Sauer H et al (2008) PI3 kinase-dependent membrane recruitment of rac-1 and p47phox is critical for alpha PDGF receptor-induced production of reactive oxygen species. J Biol Chem 283:7864–7876

Benza RL, Miller DP, Gomberg-Maitland M et al (2010) Predicting survival in pulmonary arterial hypertension: insights from the Registry to Evaluate Early and Long-Term Pulmonary Arterial Hypertension Disease Management (REVEAL). Circulation 122:164–172

Berghausen E, Janssen W, Vantler M, Zimmermann T, ten Freyhaus H, Zhao JJ, Schermuly RT, Rosenkranz S (2011) The PI 3-kinase isoform p110alpha is essential for growth factor-induced vascular remodeling in pulmonary hypertension. Eur Heart J 32(Suppl):3072 (Abstract)

Bergsten E, Uutela M, Li X et al (2001) PDGF-D is a specific, protease-activated ligand for the PDGF β-receptor. Nat Cell Biol 3:512–516

Caglayan E, Vantler M, Leppänen O et al (2011) Disruption of PDGF-dependent PI 3-kinase and PLCγ activity abolishes vascular smooth muscle cell proliferation and migration and attenuates neointima formation in vivo. J Am Coll Cardiol 57:2527–2538

Cai Y, Han M, Luo L, Song W, Zhou X (1996) Increased expression of *pdgf* and *c-myc* genes in lungs and pulmonary arteries of pulmonary hypertensive rats induced by hypoxia. Chin Med Sci J 11:152–156

Chen SJ, Chen YF et al (1995) Endothelin-receptor antagonist bosentan prevents and reverses hypoxic pulmonary hypertension in rats. J Appl Physiol 79:2122–2131

Cheng H, Force T (2010) Molecular mechanisms of cardiovascular toxicity of targeted cancer therapeutics. Circ Res 106:21–34

Chockalingam A, Gnanavelu G, Venkatesan S et al (2005) Efficacy and optimal dose of sildenafil in primary pulmonary hypertension. Int J Cardiol 99:91–95

Chung J, Grammer TC, Lemon KP, Kazlauskas A, Blenis J (1994) PDGF- and insulin-dependent pp70^{S6K} activation mediated by phosphatidylinositol-3-OH-kinase. Nature 370:71–75

Cohen MH, Johnson JR, Pazdur R (2005) US Food and Drug Administration drug approval summary: conversion of imatinib mesylate (sti571; gleevec) tablets from accelerated approval to full approval. Clin Cancer Res 11:12–19

D'Alonzo GE, Barst RJ, Ayres SM et al (1991) Survival in patients with primary pulmonary hypertension. Ann Intern Med 115:343–349

Dahal BK, Cornitescu T, Tretyn A et al (2010) Role of epidermal growth factor inhibition in experimental pulmonary hypertension. Am J Respir Crit Care Med 181:158–167

Dahal BK, Heuchel R, Pullamsetti SS, Wilhelm J, Ghofrani HA, Weissmann N, Seeger W, Grimminger F, Schermuly RT (2011a) Hypoxic pulmonary hypertension in mice with constitutively active platelet-derived growth factor receptor-β. Pulm Circ 1:259–268

Dahal BK, Kosanovic D, Kaulen C, Cornitescu T, Savai R, Hoffmann J, Reiss I, Ghofrani HA, Weissmann N, Kübler WM, Seeger W, Grimminger F, Schermuly RT (2011b) Involvement of mast cells in monocrotaline-induced pulmonary hypertension in rats. Respir Res 12:60

Delgado JF, Conde E, Sanchez V et al (2005) Pulmonary vascular remodeling in pulmonary hypertension due to chronic heart failure. Eur J Heart Fail 7:1011–1016

DeMali KA, Whiteford CC, Ulug ET, Kazlauskas A (1997) Platelet-derived growth factor-dependent cellular transformation requires either phospholipase Cγ or phosphatidylinositol 3 kinase. J Biol Chem 272:9011–9018

Druker BJ, Guilhot F, O'Brian SG, IRIS Investigators et al (2006) Five-year follow-up of patients receiving imatinib for chronic myeloid leukemia. N Engl J Med 355:2408–2417

Dumitrescu D, Seck C, ten Freyhaus H, Gerhardt F, Erdmann E, Rosenkranz S (2011) Fully reversible pulmonary arterial hypertension associated with dasatinib treatment for chronic myeloid leukemia. Eur Respir J 38:218–220

Falcetti E, Hall SM, Phillips PG et al (2010) Smooth muscle proliferation and role of the prostacyclin (IP) receptor in idiopathic pulmonary arterial hypertension. Am J Respir Crit Care Med 182:1161–1170

Fambrough D, McClure K, Kazlauskas A, Lander ES (1999) Diverse signaling pathways activated by growth factor receptors induce broadly overlapping, rather than independent, sets of genes. Cell 97:727–741

Farber HW, Loscalzo J (2004) Pulmonary arterial hypertension. N Engl J Med 351:1655–1665

Fernandez A, Sanguino A et al (2007) An anticancer c-Kit kinase inhibitor is reengineered to make it more active and less cardiotoxic. J Clin Invest 117:4044–4054

Galiè N, Ghofrani HA et al (2005) Sildenafil citrate therapy for pulmonary arterial hypertension. N Engl J Med 353:2148–2157

Galiè N, Brundage BH et al (2009a) Tadalafil therapy for pulmonary arterial hypertension. Circulation 119:2894–2903

Galiè N, Hoeper MM, Humbert M et al (2009b) Guidelines for the diagnosis and treatment of pulmonary hypertension: The Task Force for the Diagnosis and Treatment of Pulmonary Hypertension of the European Society of Cardiology (ESC) and the European Respiratory Society (ERS), endorsed by the International Society of Heart and Lung Transplantation (ISHLT). Eur Heart J 30:2493–2537

Galiè N, Manes A, Negro L et al (2009c) A meta-analysis of randomized controlled trials in pulmonary arterial hypertension. Eur Heart J 30:394–403

Galiè N, Palazzini M, Manes A (2010) Pulmonary arterial hypertension: from the kingdom of the near-dead to multiple clinical trial meta-analyses. Eur Heart J 31:2080–2086

Gambaryan N, Perros F, Montani D, Cohen-Kaminski S, Mazmanian GM, Humbert M (2010) Imatinib inhibits bone marrow-derived c-kit+ cell mobilization in hypoxic pulmonary hypertension. Eur Respir J 35:1209–1211

Garcia-Hernandez FJ, Castillo-Palma MJ, Gonzalez-Leon R et al (2008) Experience with imatinib to treat pulmonary arterial hypertension. Arch Bronconeumol 44:689–691

Ghofrani HA, Seeger W, Grimminger F (2005) Imatinib for the treatment of pulmonary arterial hypertension. N Engl J Med 353:1412–1413

Ghofrani HA, Morrell NW, Hoeper MM et al (2010a) Imatinib in pulmonary arterial hypertension patients with inadequate response to established therapy. Am J Respir Crit Care Med 182:1171–1177

Ghofrani HA, Morrell NW, Hoeper MM et al (2010b) Long term use of imatinib in patients with severe pulmonary arterial hypertension. Am J Respir Crit Care Med 181(Suppl):A2513 (Abstract)

Gilbertson DG, Duff ME, West JW et al (2001) Platelet-derived growth factor C (PDGF-C), a novel growth factor that binds to PDGF α and β receptor. J Biol Chem 276:27406–27414

Grimminger F, Schermuly RT (2010) PDGF receptor and its antagonists: role in treatment of PAH. Adv Exp Med Biol 661:435–446

Gschwind HP, Pfaar U et al (2005) Metabolism and disposition of imatinib mesylate in healthy volunteers. Drug Metab Dispos 33:1503–1512

Haga M, Yamashita A et al (2003) Oscillatory shear stress increases smooth muscle cell proliferation and Akt phosphorylation. J Vasc Surg 37:1277–1284

Hassoun PM, Mouthon L, Barbera JA et al (2009) Inflammation, growth factors, and pulmonary vascular remodeling. J Am Coll Cardiol 54(Suppl S):S10–S19

Haworth SG (2007) The cell and molecular biology of right ventricular dysfunction in pulmonary hypertension. Eur Heart J 28(Suppl H):H10–H16

Heldin CH, Westermark B (1999) Mechanism of action and in vivo role of platelet-derived growth factor. Physiol Rev 79:1283–1316

Hennigs JK, Keller G, Baumann HJ et al (2011) Multi tyrosine kinase inhibitor dasatinib as novel cause of severe pre-capillary pulmonary hypertension? BMC Pulm Med 11:30

Hoeper MM, Barst RJ, Bourge R et al (2011a) Imatinib improves exercise capacity and hemodynamics at 24 weeks as add-on therapy in symptomatic pulmonary arterial hypertension patients: the IMPRES Study. Chest 140(Suppl):1045A (Abstract)

Hoeper M, Barst RJ, Galiè N et al (2011b) Imatinib in pulmonary arterial hypertension, a randomized efficacy study (IMPRES). Eur Respir J 38(Suppl):413 (Abstract)

Hoeper MM, Barst RJ, Chang HJ et al (2012) Imatinib safety and efficacy: interim analysis of IMPRES extension study in patients with pulmonary arterial hypertension. Am J Respir Crit Care Med 184(Suppl):A2495 (Abstract)

Huang Q, Sun R (1997) Changes of PDGF-α and β receptor gene expression in hypoxic rat pulmonary vessels. Zhongguo Yi Xue Ke Xue Yuan Xue Bao 19:470–473

Humar R, Kiefer FN et al (2002) Hypoxia enhances vascular cell proliferation and angiogenesis in vitro via rapamycin (mTOR)-dependent signaling. FASEB J 16:771–780

Humbert M, Monti G, Fartoukh M et al (1998) Platelet-derived growth factor expression in primary pulmonary hypertension: comparison of HIV seropositive and HIV seronegative patients. Eur Respir J 11:554–559

Humbert M, Segal ES, Kiely DG et al (2007) Results of European post-marketing surveillance of bosentan in pulmonary hypertension. Eur Respir J 30:338–340

Humbert M, Sitbon O, Chaouat A et al (2010) Survival in patients with idiopathic, familial, and anorexigen-associated pulmonary arterial hypertension in the modern management era. Circulation 122:156–163

Izikki M, Guignabert C, Fadel E et al (2009) Endothelial-derived FGF2 contributes to the progression of pulmonary hypertension in humans and rodents. J Clin Invest 119:512–523

Joly M, Kazlauskas A, Fay FS, Corvera S (1994) Disruption of PDGF receptor trafficking by mutation of its PI-3 kinase sites. Science 263:684–687

Jones R, Capen D, Jacobson M, Munn L (2006) PDGF and microvessel wall remodeling in adult rat lung: imaging PDGF-aa and PDGF-rα molecules in progenitor smooth muscle cells developing in experimental pulmonary hypertension. Cell Tissue Res 326:759–769

Kappert K, Peters KG, Bohmer FD, Ostman A (2005) Tyrosine phosphatases in vessel wall signaling. Cardiovasc Res 65:587–598

Kappert K, Paulsson J, Sparwel J et al (2007) Dynamic changes in the expression of DEP-1 and other PDGF receptor-antagonizing PTPs during onset and termination of neointima formation. FASEB J 21:523–534

Katayose D, Ohe M, Yamauchi K et al (1993) Increased expression of PDGF-α and β-chain genes in rat lungs with hypoxic pulmonary hypertension. Am J Physiol 264:L100–L106

Kazlauskas A (1994) Receptor tyrosine kinases and their targets. Curr Opin Genet Dev 4:5–14

Kerkelä R, Grazette L, Yacobi R et al (2006) Cardiotoxicity of the cancer therapeutic agent imatinib mesylate. Nat Med 12:908–916

Klinghoffer RA, Duckworth B, Valius M, Cantley L, Kazlauskas A (1996) Platelet-derived growth factor-dependent activation of phosphatidylinositol 3-kinase is regulated by receptor binding of SH2-domain-containing proteins which influence Ras activity. Mol Cell Biol 16:5905–5914

Kundra V, Escobedo JA, Kazlauskas A et al (1994) Regulation of chemotaxis by the platelet-derived growth factor receptor β. Nature 367:474–476

Kwapiszewska G, Wilhelm J, Wolff S et al (2005) Expression profiling of laser-microdissected intrapulmonary arteries in hypoxia-induced pulmonary hypertension. Respir Res 6:109

Lanner MC, Raper M et al (2005) Heterotrimeric G proteins and the platelet-derived growth factor receptor-beta contribute to hypoxic proliferation of smooth muscle cells. Am J Respir Cell Mol Biol 33:412–419

LaRochelle WJ, Jeffers M, McDonald WF et al (2001) PDGF-D, a new protease-activated growth factor. Nat Cell Biol 3:517–521

Le Coutre P, Kreuzer KA et al (2004) Pharmacokinetics and cellular uptake of imatinib and its main metabolite CGP74588. Cancer Chemother Pharmacol 53:313–323

Leveen P, Pekny M, Gebre-Medhin S, Swolin B, Larsson E, Betsholtz C (1994) Mice deficient for PDGF B show renal, cardiovascular, and hematological abnormalities. Genes Dev 8:1875–1887

Levitzki A (2005) PDGF receptor kinase inhibitors for the treatment of restenosis. Cardiovasc Res 65:581–586

Li X, Pontén A, Aase K et al (2000) PDGF-C is a new protease-activated ligand for the PDGF α-receptor. Nat Cell Biol 2:302–309

Lindahl P, Johansson BR, Leveen P, Betsholtz C (1997) Pericyte loss and microaneurysm formation in PDGF-B-deficient mice. Science 277:242–245

Machado R, Eickelberg O, Elliott CG et al (2009) Genetics and genomics of pulmonary arterial hypertension. J Am Coll Cardiol 54(Suppl S):S32–S42

Mattei D, Feola M, Orzan F, Mordini N, Rapezzi D, Gallamini A (2009) Reversible dasatinib-induced pulmonary arterial hypertension and right ventricle failure in a previously allografted CML patient. Bone Marrow Transplant 43:967–968

McLaughlin VV, Archer SL, Badesch DB et al (2009) ACCF/AHA 2009 expert consensus document on pulmonary hypertension: a report of the American College of Cardiology Foundation Task Force on Expert Consensus Documents and the American Heart Association developed in collaboration with the American College of Chest Physicians; American Thoracic Society, Inc.; and the Pulmonary Hypertension Association. J Am Coll Cardiol 53:1573–1619

Mellgren AM, Smith CL, Olsen GS et al (2008) Platelet-derived growth factor receptor beta signaling is required for efficient epicardial cell migration and development of two distinct coronary vascular smooth muscle cell populations. Circ Res 103:1393–1401

Merklinger SL, Jones PL, Martinez EC, Rabinovitch M (2005) Epidermal growth factor receptor blockade mediates smooth muscle cell apoptosis and improves survival in rats with pulmonary hypertension. Circulation 112:423–431

Miyake S, Mullane-Robinson KP, Lill NL, Douillard P, Band H (1999) Cbl-mediated negative regulation of platelet-derived growth factor receptor-dependent cell proliferation. A critical role for Cbl tyrosine kinase-binding domain. J Biol Chem 274:16619–16628

Montani D, Bergot E, Günther S, Savale L, Bergeron A, Bourdin A, Bouvaist H, Canuet M, Pison C, Macro M, Poubeau P, Girerd B, Natali D, Guignabert C, Perros F, O'Callaghan DS, Jaïs X, Tubert-Bitter P, Zalcman G, Sitbon O, Simonneau G, Humbert M (2012) Pulmonary arterial hypertension in patients treated by dasatinib. Circulation 125:2128–2137

Montmayeur JP, Valius M, Vandenheede J, Kazlauskas A (1997) The platelet-derived growth factor β-receptor triggers multiple cytoplasmic signaling cascades that arrive at the nucleus at distinguishable inputs. J Biol Chem 272:32670–32678

Morrell NW, Adnot S, Archer SL et al (2009) Cellular and molecular basis of pulmonary arterial hypertension. J Am Coll Cardiol 54(Suppl S):S20–S31

Nagaraj C, Tang B, Bálint Z, Wygrecka M, Hrzenjak A, Kwapiszewska G, Stacher E, Lindenmann J, Weir EK, Olschewski H, Olschewski A (2013) Src tyrosine kinase is crucial for potassium channel function in human pulmonary arteries. Eur Respir J 41:85–95

Nakamura K, Akagi S, Ogawa A et al (2011) Pro-apoptotic effects of imatinib on PDGF-stimulated pulmonary artery smooth muscle cells from patients with idiopathic pulmonary arterial hypertension. Int J Cardiol 159(2):100–106

Newman JH, Trembath RC et al (2004) Genetic basis of pulmonary arterial hypertension: current understanding and future directions. J Am Coll Cardiol 43(Suppl S):33S–39S

Ogawa A, Firth AL, Yao W et al (2009) Inhibition of mTOR attenuates store-operated Ca2+ entry in cells from endarterectomized tissues of patients with chronic thromboembolic pulmonary hypertension. Am J Physiol Lung Cell Mol Physiol 297:L666–L676

Olschewski H, Simonneau G et al (2002) Inhaled iloprost for severe pulmonary hypertension. N Engl J Med 347:322–329

Oudiz RJ, Galie N et al (2009) Long-term ambrisentan therapy for the treatment of pulmonary arterial hypertension. J Am Coll Cardiol 54:1971–1981

Panzhinskiy E, Zawada WM, Stenmark KR, Das M (2012) Hypoxia induces unique proliferative response in adventitial fibroblasts by activating PDGFβ receptor-JNK1 signalling. Cardiovasc Res 95(3):356–365

Patterson KC, Weissmann A, Ahmadi T, Farber HW (2006) Imatinib mesylate in the treatment of refractory idiopathic pulmonary arterial hypertension. Ann Intern Med 145:152–153

Peng B, Lloyd P et al (2005) Clinical pharmacokinetics of imatinib. Clin Pharmacokinet 44:879–894

Perros F, Montani D, Dorfmüller P et al (2008) Platelet-derived growth factor expression and function in idiopathic pulmonary arterial hypertension. Am J Respir Crit Care Med 178:81–88

Pullamsetti SS, Berghausen EM, Dabral S, Tretyn A, Butrous E, Savai R, Butrous G, Dahal BK, Brandes RP, Ghofrani HA, Weissmann N, Grimminger F, Seeger W, Rosenkranz S, Schermuly RT (2012) Role of Src tyrosine kinases in experimental pulmonary hypertension. Arterioscler Thromb Vasc Biol 32:1354–1365

Quinta's-Cardama A, Kantarijan H, O'Brien S et al (2007) Pleural effusion in patients with chronic myelogenous leukemia treated with dasatinib after imatinib failure. J Clin Oncol 25:3908–3914

Raines EW (2004) PDGF and cardiovascular disease. Cytokine Growth Factor Rev 15:237–254

Rasheed W, Flaim B, Seymour JF (2009) Reversible severe pulmonary hypertension secondary to dasatinib in a patient with chronic myeloid leukemia. Leuk Res 33:861–864

Rosenkranz S (2007) Pulmonary hypertension: current diagnosis and treatment. Clin Res Cardiol 96:527–541

Rosenkranz S, Kazlauskas A (1999) Evidence for distinct signaling properties and biological responses induced by the PDGF receptor α and β subtypes. Growth Factors 16:201–216

Rosenkranz S, DeMali KA, Gelderloos JA, Bazenet C, Kazlauskas A (1999) Identification of the receptor-associated signaling enzymes that are required for platelet-derived growth factor-AA-dependent chemotaxis and DNA synthesis. J Biol Chem 274:28335–28343

Rosenkranz S, Ikuno Y, Leong FL et al (2000) Src family kinases negatively regulate platelet-derived growth factor alpha receptor-dependent signaling and disease progression. J Biol Chem 275:9620–9627

Rubin LJ, Badesch DB et al (2002) Bosentan therapy for pulmonary arterial hypertension. N Engl J Med 346:896–903

Sato N, Narita M, Takahashi M et al (2003) The effects of STI571 on antigen presentation in dendritic cells generated from patients with chronic myelogenous leukemia. Hematol Oncol 2:67–75

Schermuly RT, Dony E, Ghofrani HA et al (2005) Reversal of experimental pulmonary hypertension by PDGF inhibition. J Clin Invest 115:2811–2821

Schultz K, Fanburg BL et al (2006) Hypoxia and hypoxia-inducible factor-1alpha promote growth factor-induced proliferation of human vascular smooth muscle cells. Am J Physiol 290: H2528–H2534

Schwartz SM (1997) Perspectives series: cell adhesion in vascular biology: smooth muscle migration in atherosclerosis and restenosis. J Clin Invest 99:2814–2816

Shah AM, Campbell P, Peacock A, Barst RJ, Quinn D, Salomon SD, for the IMPRES Investigators (2012) Effect of imatinib as add-on therapy on echocardiographic measures of right ventricular function in patients with advanced PAH: the Imatinib in Pulmonary arterial hypertension, a Randomized Efficacy Study (IMPRES) echocardiography sub-study. Heart Fail (Abstract)

Simonneau G, Barst RJ, Galiè N et al (2002) Continuous subcutaneous infusion of treprostinil, a prostacyclin analogue, in patients with pulmonary arterial hypertension: a double-blind, randomized, placebo-controlled trial. Am J Respir Crit Care Med 165:800–804

Simonneau G, Rubin LJ, Galiè N et al (2008) Addition of sildenafil to long-term intravenous epoprostenol therapy in patients with pulmonary arterial hypertension: a randomized trial. Ann Intern Med 149:521–530

Simonneau G, Robbins IM, Beghetti M et al (2009) Updated clinical classification of pulmonary hypertension. J Am Coll Cardiol 54(Suppl S):S43–S54

Sitbon O, Galiè N (2010) Treat-to-target strategies in pulmonary arterial hypertension: the importance of using multiple goals. Eur Respir Rev 19:272–278

Sitbon O, Humbert M, Jais X et al (2005) Long-term response to calcium channel blockers in idiopathic pulmonary arterial hypertension. Circulation 111:3105–3111

Soriano P (1994) Abnormal kidney development and haematological disorders in PDGF β-receptor mutant mice. Genes Dev 8:1888–1896

Souza R, Sitbon O, Parent G, Simonneau G, Humbert M (2006) Long term imatinib treatment in pulmonary arterial hypertension. Thorax 61:736

Stenmark KR, Gerasimovskaya E, Nemenoff RA, Das M (2002) Hypoxic activation of adventitial fibroblasts: role in vascular remodeling. Chest 122(Suppl 6):326S–334S

Sterpetti AV, Cucina A et al (1992) Modulation of arterial smooth muscle cell growth by haemodynamic forces. Eur J Vasc Surg 6:16–20

Tallquist MD, Soriano P, Klinghoffer RA (1999) Growth factor signaling pathways in vascular development. Oncogene 18:7917–7932

Tallquist MD, Klinghoffer RA, Heuchel R, Mueting-Nelsen PF, Corrin PD, Heldin C-H, Johnson RJ, Soriano P (2000) Retention of PDGFR-β function in mice in the absence of phosphatidylinositol 3′-kinase and phospholipase Cγ signalling pathways. Genes Dev 14:3179–3190

Tanizawa S, Ueda M et al (1996) Expression of platelet derived growth factor B chain and beta receptor in human coronary arteries after percutaneous transluminal coronary angioplasty: an immunohistochemical study. Heart 75:549–556

Tapper EB, Knowles D, Heffron T et al (2009) Portopulmonary hypertension: imatinib is a novel tretament and the Emory experience with this condition. Transplant Proc 41:1969–1971

Ten Freyhaus H, Huntgeburth M et al (2006) Novel Nox inhibitor VAS2870 attenuates PDGF-dependent smooth muscle cell chemotaxis, but not proliferation. Cardiovasc Res 71:331–341

Ten Freyhaus H, Dumitrescu D, Bovenschulte H, Erdmann E, Rosenkranz S (2009) Significant improvement of right ventricular function by imatinib mesylate in scleroderma-associated pulmonary arterial hypertension. Clin Res Cardiol 98:265–267

Ten Freyhaus H, Dagnell M, Leuchs M et al (2011) Hypoxia enhances platelet-derived growth factor signaling in the pulmonary vasculature by down-regulation of protein tyrosine phosphatases. Am J Respir Crit Care Med 183:1092–1102

Ten Freyhaus H, Janssen W, Leuchs M, Zierden M, Vantler M, Caglayan E, Schermuly RT, Tallquist MD, Rosenkranz S (2012) Disruption of βPDGFR-dependent PI3K and PLCγ signaling protects from hypoxia-induced pulmonary hypertension. Circulation 125(Suppl) (Abstract)

Tong WG, Kantarjian H, O'Brien S, Faderl S, Ravandi F, Borthakur G, Shan J, Pierce S, Rios MB, Cortes J (2010) Imatinib front-line therapy is safe and effective in patients with chronic myelogenous leukemia with pre-existing liver and/or renal dysfunction. Cancer 116:3152–3159

Tu L, De Man FS, Girerd B, Huertas A, Chaumais MC, Lecerf F, François C, Perros F, Dorfmüller P, Fadel E, Montani D, Eddahibi S, Humbert M, Guignabert C (2012) A critical role for p130Cas in the progression of pulmonary hypertension in humans and rodents. Am J Respir Crit Care Med 186(7):666–676

Tuder RM, Voelkel N (2010) Pulmonary hypertension and inflammation. J Lab Clin Med 132:16–24

Tuder RM, Abman SH, Braun T et al (2009) Development and pathology of pulmonary hypertension. J Am Coll Cardiol 54(Suppl S):S3–S9

Valius M, Kazlauskas A (1993) Phospholipase C-γ1 and phosphatidylinositol 3 kinase are the downstream mediators of the PDGF receptor's mitogenic signal. Cell 73:321–334

Valius M, Secrist J-P, Kazlauskas A (1995) The GTPase activating protein of Ras suppresses platelet-derived growth factor beta receptor signaling by silencing phospholipase C-γ1. Mol Cell Biol 15:3058–3071

Van Wolferen SA, Marcus JT, Boonstra A et al (2007) Prognostic value of right ventricular mass, volume, and function in idiopathic pulmonary arterial hypertension. Eur Heart J 28:1250–1257

Vantler M, Caglayan E, Zimmermann WH, Bäumer AT, Rosenkranz S (2005) Systematic evaluation of anti-apoptotic growth factor signaling in vascular smooth muscle cells. Only phosphatidylinositol 3′-kinase is important. J Biol Chem 280:14168–14176

Vantler M, Karikkineth BC, Naito H et al (2010) PDGF-BB protects cardiomyocytes from apoptosis and improves contractile function of engineered heart tissue. J Mol Cell Cardiol 48:1316–1323

Wharton J, Davie N et al (2000) Prostacyclin analogues differentially inhibit growth of distal and proximal human pulmonary artery smooth muscle cells. Circulation 102:3130–3136

Wharton J, Strange JW et al (2005) Antiproliferative effects of phosphodiesterase type 5 inhibition in human pulmonary artery cells. Am J Respir Crit Care Med 172:105–113

Young KC, Torres E, Hehre D, Suguihara C, Hare J (2009) Neonatal c-kit mutant mice exhibit decreased susceptibility to hypoxia-induced pulmonary hypertension. Circulation 120:S750–S751

Zhang L, Ma J, Shen T, Wang S, Ma C, Liu Y, Ran Y, Wang L, Liu L, Zhu D (2012) Platelet-derived growth factor (PDGF) induces pulmonary vascular remodeling through 15-LO/15-HETE pathway under hypoxic condition. Cell Signal 24:1931–1939

Emerging Molecular Targets for Anti-proliferative Strategies in Pulmonary Arterial Hypertension

Ly Tu and Christophe Guignabert

Abstract The combination of pulmonary vasoconstriction, in situ thrombosis, and pulmonary arterial wall remodeling is largely responsible for the rise in pulmonary vascular resistance (PVR) and pulmonary arterial pressure (PAP) in patients with pulmonary arterial hypertension (PAH). Even though several drugs have been developed over the past decades, at this time there is no cure for PAH. The overriding goals of the current therapeutic options seek to compensate for the defects in the relative balance of competing vasoconstrictor and vasodilator influences. Because the past decade has seen great strides in our understanding of the pathogenesis of PAH, interest has been growing in the potential use of anti-proliferative approaches in PAH. Indeed anti-proliferative strategies could offer ways not only to reinstate the homeostatic balance between cell proliferation and apoptosis but also to reverse the progressive pulmonary vascular obstruction in PAH. However, further efforts still need to be made in order to establish the long-term safety and efficacy of those anti-proliferative approaches in PAH and their potential additive benefit with other drugs.

Keywords Pulmonary hypertension · Pulmonary vascular remodeling · Anti-proliferative compounds · Signal transduction · Growth factors

---

L. Tu · C. Guignabert (✉)
INSERM UMR 999, "Pulmonary Hypertension: Physiopathology and Novel Therapies", LabEx LERMIT, Le Plessis-Robinson, France

DHU Thorax Innovation (TORINO), School of Medicine, University of Paris-Sud, Le Kremlin-Bicêtre, France
e-mail: christophe.guignabert@inserm.fr

Contents

1 Introduction 410
2 Growth Factor Signaling Pathways as Targets for Anti-remodeling Therapies in
  Pulmonary Hypertension 412
  2.1 The Platelet-Derived Growth Factor Signaling System in PAH 412
  2.2 The Fibroblast Growth Factor Signaling System in PAH 413
  2.3 The Epidermal Growth Factor Signaling System in PAH 414
  2.4 The Serotoninergic System in PAH 415
  2.5 Other Growth Factor Signaling Systems in PAH 418
3 Restitution of the Aberrant Extracellular Matrix Remodeling in PH 424
4 Restitution of the Dysfunctional BMPR-II Signaling System in PAH 424
5 Recovery of Oxidative Metabolism in PAH 425
6 Conclusions and Challenges 426
References 427

1 Introduction

Pulmonary arterial hypertension (PAH) is a rapidly progressive disease characterized by sustained elevation of pulmonary vascular resistance (PVR) and pulmonary arterial pressure (PAP) leading to right heart failure and death. Although the exact mechanisms of remodeling of pulmonary arteries, leading to the onset and progression of the disease, are still largely unclear, many disease-predisposing factors and/or contributing factors have been identified, including inflammation, endothelial dysfunction, aberrant vascular wall cell proliferation, as well as mutations in the *bone morphogenetic protein receptor type 2 (Bmpr2)* gene (Humbert et al. 2004; Rabinovitch 2005; Morrell et al. 2009) (Fig. 1).

Development of therapeutic agents that modulate abnormalities in three main pathobiologic pathways for PAH endothelin (ET)-1, prostacyclin (PGI2), and nitric oxide (NO) has revolutionized our approach to the treatment of PAH and has changed the course of this devastating disease (O'Callaghan et al. 2011). However, although the spectrum of therapeutic options for PAH has expanded in the last decade, available therapies remain essentially palliative.

Irreversible remodeling of the pulmonary vasculature is the cause of increased PAP in PAH and frequently leads to progressive functional decline in patients despite treatment with current available therapeutic options. This process is ascribed to the increased proliferation, migration, and survival of pulmonary vascular cells within the pulmonary artery wall, i.e., myofibroblasts, pulmonary vascular smooth muscle (SMCs), and endothelial cells (ECs). The increasing knowledge on PAH pathogenesis has revealed the complex nature of these structural and functional changes in the pulmonary arteries of patients with PAH and highlighted the need to elucidate the molecular mechanisms involved. Over recent years, special attention has been devoted to the disease-promoting roles of three different growth factors and their corresponding tyrosine kinase

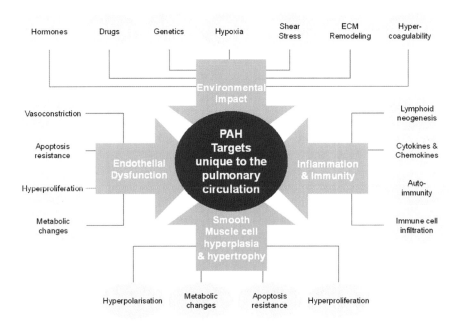

Fig. 1 Conceptual framework for the development of pulmonary arterial hypertension (PAH) therapies. *ECM* extracellular matrix

receptors: platelet-derived growth factor (PDGF), epidermal growth factor (EGF), and fibroblast growth factor (FGF) receptors. Altered expression and/or increased activity of these three signaling pathways as well as their contribution to the abnormal proliferation and migration of smooth muscle and endothelial cells have been demonstrated in human and experimental models of pulmonary hypertension. Furthermore, their inhibition by specific inhibitors, tyrosine kinase inhibitors (TKIs), has been shown to exert beneficial effects in rodent models (Merklinger et al. 2005; Schermuly et al. 2005; Perros et al. 2008; Izikki et al. 2009; Tu et al. 2011, 2012). Currently, imatinib (a TKI of PDGF receptors), which is widely used for the treatment of chronic myeloid leukemia, is being tested as a new potential therapy in PAH and ongoing studies seek to better evaluate the overall risk benefit ratio of this anti-proliferative molecule in PAH.

The last few years have seen the emergence of the concept that anti-proliferative strategies could offer a novel approach for the treatment of PAH, by downregulating the progression of the disease and reversal of pulmonary vascular remodeling. However, further studies are needed to better understand the mechanisms underlying this abnormal over-activation of some growth factor-stimulated signaling pathways in PAH and to identify novel pharmacological targets for the development of new, better-tolerated, and more powerful therapeutic tools. This chapter aims to provide an overview on current status and future perspectives of target-based anti-proliferative therapies in pulmonary hypertension.

2 Growth Factor Signaling Pathways as Targets for Anti-remodeling Therapies in Pulmonary Hypertension

Growth factors are diffusible proteins that act through activation of diverse signaling pathways to regulate a myriad of critical cellular functions for normal lung development and homeostasis. As growth factors dictate growth, proliferation, and survival, there has been an increasing interest in the cellular biology of growth factor signaling in PAH in recent years. Excessive release of growth factors that are encrypted in the extracellular matrix, and/or modification of growth factor production, and receptor expression, and/or alterations in the intracellular mitogenic signals, have been reported to have a critical role in the disease; however, it is unknown whether or not this imbalance is the cause or the consequence of PAH. Many molecular therapies that target aberrant growth factor signaling pathways are being investigated, with some agents in the late stages of clinical testing.

2.1 The Platelet-Derived Growth Factor Signaling System in PAH

The PDGF family consists of PDGF-A, -B, -C, and -D, which form either homo- or heterodimers (PDGF-AA, -AB, -BB, -CC, -DD). The four PDGFs are inactive in their monomeric forms. The PDGF-AA, -AB, and -BB dimers are processed intracellularly and secreted as active dimers. In contrast, PDGF-CC and PDGF-DD differ from the others in that they are secreted as an inactive form until their N-terminal complement C1r/C1s, Uegf, Bmp1 (CUB) domain is cleaved. There are two receptor subunits -α and -β that dimerize upon binding one PDGF dimer, leading to three possible receptor combinations, namely, PDGFR-$\alpha\alpha$, -$\beta\beta$, and -$\alpha\beta$. PDGF-AA binds exclusively to PDGFR-$\alpha\alpha$, while PDGF-BB is the only PDGF that can bind all three receptor combinations with high affinity. PDGF-AB and -CC could assemble and activate PDGFR-$\alpha\alpha$ and -$\alpha\beta$. PDGF-DD activates PDGFR-$\beta\beta$ with high affinity, and under certain conditions the PDGFR-$\alpha\beta$. The PDGFR autophosphorylation in the kinase insert region regulates interactions with different cell proteins involved in the initiation of the intracellular signaling pathway.

Since PDGF is an important autocrine and paracrine mitogen for vascular smooth muscle cells, mediating both hyperplasia and migration for pulmonary vascular remodeling, studies on animal models of PH as well as in human tissues have been undertaken (Tanabe et al. 2000; Berk 2001; Yamboliev and Gerthoffer 2001). Balasubramaniam et al. (2003) were the first to note a PDGF upregulation in a lamb model with chronic intrauterine PH and to suggest a pathogenic role of the PDGF signaling system for the disease. This hypothesis was tested and validated by a study from Schermuly et al. (2005), showing that the PDGF receptor antagonist imatinib reverses pulmonary vascular remodeling in the monocrotaline-induced PH in rats and the chronic hypoxia-induced PH in mice. They also reported a substantial increase in the protein levels of PDGFR-β in total lung homogenates of patients with PAH as compared with controls. Finally, by investigating abnormalities within human tissue of patients with idiopathic PAH, Perros et al. (2008) have reported increased levels of mRNA encoding PDGF-A, PDGF-B, PDGFR-α, and PDGFR-β in microdissected small pulmonary arteries of patients with idiopathic PAH. In addition, they demonstrated a marked increase in the protein levels of PDGFR-β in total lung homogenates of patients with PAH as compared with controls. Consistent with these observations, intense immunoreactivity of the phosphorylated form of PDGFR-β has been shown in vascular lesions in lungs of idiopathic PAH patients. In addition, it has also been demonstrated that PDGFR, specifically PDGFR-β, is transactivated by sphingosine-1 phosphate (Baudhuin et al. 2004), G protein-coupled receptors (GPCRs) stimulated by lysophosphatidic acid (LPA) (Herrlich et al. 1998), angiotensin II (Saito and Berk 2001), and serotonin transporter (5-HTT or SERT) (Ren et al. 2011). Furthermore, Tie2-mediated loss of peroxisome proliferator-activated receptor (PPAR)-γ in mice causes PDGFR-β-dependent pulmonary arterial muscularization (Guignabert et al. 2009a).

2.2 The Fibroblast Growth Factor Signaling System in PAH

FGFs are a group of at least 23 structurally related heparin-binding polypeptide mitogens that are expressed in almost all tissues and constitute a signaling system conserved throughout animal evolution. FGFs interact with a family of four distinct, high-affinity RTKs, designated FGFR 1 to 4. The FGF system has a broad range of biological activities that not only stimulate cell proliferation, migration, and differentiation but also inhibit cell death. FGFRs are protein tyrosine kinase receptors, which consist of three extracellular immunoglobulin-type domains (D1–D3), a single-span transmembrane domain and an intracellular split tyrosine kinase domain. FGFRs 1, 2, and 3 undergo an alternative splicing event in which two alternative exons (IIIb and IIIc) can be used to encode the carboxy terminal portion of the third immunoglobulin-like loop. The splicing variant IIIa of FGFRs is a secreted FGF-binding protein. In addition, other types of alternate mRNA splicing events have been described. Each alternatively spliced variants binds to a specific subset of FGFs with variable affinities and have distinct tissue-specific expression patterns. Upon binding of FGF, FGFR monomers dimerize, and subsequently the TK domains autophosphorylate, which initiates the intracellular signaling pathway.

Since FGF2 is a known potent pro-angiogenic stimulator of endothelial and smooth muscle cell proliferation, migration, and synthesis of various extracellular components (Goncalves 1998), there is a strong interest for this growth factor in PAH. In addition, FGF2 is up regulated in response to hypoxia and shear stress in pulmonary vascular cells (Quinn et al. 2002; Li et al. 2003). Lung and circulating FGF2 levels are increased in both experimental and human PH. Abnormally high levels of FGF2 were found in the blood of 51 % and in the urine of 21 % of patients with idiopathic PAH (Benisty et al. 2004). FGF2 levels are increased in two animal models, a lamb model of PH developed by inserting an aorto-pulmonary vascular bypass graft (Wedgwood et al. 2007) and the rat model of monocrotaline-induced PH (Arcot et al. 1995; Izikki et al. 2009). In PAH patients, we have shown that FGF2 is markedly overproduced by pulmonary endothelial cells in walls of distal arteries and in isolated primary cells than in controls. Furthermore, we have demonstrated that this FGF2 overexpression contributes significantly not only to smooth muscle hyperplasia by a paracrine action, but also to the abnormal endothelial phenotype by an autocrine action (Izikki et al. 2009; Tu et al. 2011, 2012). In idiopathic PAH, dysregulation of the FGF2 signaling pathway was associated not only with FGF2 overproduction, but also with altered expression of the FGF receptor. We found increased FGFR2 expression in pulmonary endothelial cells derived from idiopathic PAH patients as compared to control cells. Interestingly, Matsunaga et al. (2009) recently demonstrated that overexpression of a constitutively active FGFR2 in endothelial cells in vitro enhanced migration and survival, and augmented autocrine FGF2 production. Selection of endothelial cells that naturally overexpress FGF2 in early stages of the disease and/or deficient activity of PPARγ may also explain the augmented FGF2 production by endothelial cells of idiopathic PAH patients (Tian et al. 2009). Moreover, FGF2 can be sequestered and stored as a complex in the extracellular matrix and then released by proteolytic processes to bind and activate cell targets, thereby promoting mitogenesis (Benezra et al. 1993; Thompson and Rabinovitch 1996; Buczek-Thomas and Nugent 1999; George et al. 2001). Recently, we reported that daily treatment with the FGFR inhibitor dovitinib started 2 weeks after a subcutaneous monocrotaline injection substantially attenuated the abnormal increase in p130^{cas} and ERK1/2 activation and regressed established PH (Tu et al. 2012).

2.3 The Epidermal Growth Factor Signaling System in PAH

The EGF family consists of EGF, transforming growth factor-α (TGF-α), heparin-binding EGF-like growth factor (HB-EGF), epiregulin, amphiregulin (AR), epigen, beta-cellulin (BTC), and neuregulin 1, 2, 3, and 4. Each ligand displays overlapping but distinct binding affinities toward ErbB receptors and subsequently induces the formation of homo- and heterodimers of these receptors. There are four types of EGF receptors (EGFR) including ErbB1, also referred to as EGFR or Her1, ErbB2 (Neu/Her2), ErbB3 (Her3), and ErbB4 (Her4). All EGF receptors have a common extracellular ligand-binding region, a single membrane-spanning region, and a cytoplasmic protein tyrosine kinase domain. EGF, HB-EGF, TGF-α, AR, BTC, and epiregulin all bind ErbB1. HB-EGF, epiregulin, and BTC are known to bind to ErbB4 as well as ErbB1. The other group includes the neuregulins, which are ligands for ErbB3 and ErbB4. Upon ligand binding, EGFR monomers dimerize, and subsequently the TK domains autophosphorylate, which leads to the activation of intracellular signaling pathways.

Multiple lines of evidence suggest that the EGF signaling system contributes to the SMC proliferative response and are involved in the initiation and/or progression of the pulmonary vascular remodeling in PAH. Several studies have demonstrated that EGF co-localizes with Tenascin C, a component of the extracellular matrix (ECM) that is overabundant in obstructive lesions of patients with PAH and thus leads to an EGF-dependent proliferation and migration of pulmonary vascular cells (Jones and Rabinovitch 1996; Jones et al. 1997a, b, 1999; Cowan et al. 1999, 2000b). Consistent with this pathogenic role of the EGF signaling system, transgenic mice that over-express TGF-α under the control of the human surfactant protein (SP)-C promoter (the TGF-α mice) developed severe PH and vascular remodeling characterized by abnormally extensive muscularization of small pulmonary arteries (Le Cras et al. 2003). This phenotype was prevented in bi-transgenic mice expressing both TGF-α and a dominant-negative mutant EGF receptor under the control of the SP-C promoter. In addition, Merklinger et al. (2005) have shown that inhibition of EGFR by PKI166, a dual EGFR/HER2 inhibitor, mediated PASMC apoptosis and improved survival of monocrotaline-injected rats. However, Dahal et al. (2010) reported no changes in the levels of mRNA encoding of BTC, AR, HB-EGF, ErbB1, ErbB2, ErbB3, and ErbB4 as well as in the protein levels of ErbB1 between lung homogenates from idiopathic PAH (late stage of the disease) and normal subjects. In the same study, Dahal et al. (2010) found that three clinically approved EGFR antagonists (gefitinib, erlotinib, and lapatinib) substantially reduced the EGF-induced proliferation of PASMCs isolated from healthy or monocrotaline-injected rats. In this rat monocrotaline model of PH, an upregulation of EGF, TGF-α, ErbB1, ErbB2, and ErbB3 mRNA levels has been noted in lung homogenates from monocrotaline-injected rats as compared with control rats. However, no differences were noted between the two groups regarding the levels of mRNA encoding for HB-EGF, epiregulin, AR, and ErbB4 in this rodent PH model. They also found in rats that daily treatment with gefitinib and erlotinib but not with lapatinib started 3 weeks after a subcutaneous monocrotaline injection substantially regressed established PH. In contrast to the beneficial effects of gefitinib and erlotinib treatments in the monocrotaline model, no substantial improvements were observed with these drugs in the chronic hypoxia mouse model of PH. The discordant findings may be attributable to differences in species, disease severity, mechanisms underlying PH induction and/or may suggest a much more complex regulatory network. Indeed, cooperative and synergistic signaling exists between EGF and other factors including TGF-β1 and FGF2 (Kelvin et al. 1989; Ciccolini and Svendsen 1998; Park et al. 2000; Murillo et al. 2005; Ding et al. 2007; Grouf et al. 2007; Uttamsingh et al. 2008).

2.4 The Serotoninergic System in PAH

Serotonin (5-hydroxytryptamine or 5-HT) is an endogenous vasoactive indolamine found mainly in enterochromaffin tissue, brain, and blood platelets. 5-HT is synthesized from amino acid tryptophan in two steps: the hydroxylation of tryptophan to form the 5-hydroxytryptophan by the enzyme tryptophan hydroxylase (TPH), and then the decarboxylation of this intermediate by the aromatic L-amino acid decarboxylase. 5-HT is predominantly synthesized in the enterochromaffin cells of the intestine, representing more than 95 % of total body 5-HT. In the circulation, 5-HT is actively incorporated into platelets and stored in platelet dense storage, keeping the free circulating 5-HT in low levels (Nilsson et al. 1985; Vanhoutte 1991; Brenner et al. 2007). 5-HT is also synthesized in the raphe nuclei of the brain, pineal gland, and in pulmonary vascular endothelial cells. In addition to SERT, 5-HT concentration is regulated by the mitochondrial enzyme monoamine oxidase (MAO) and by 5-HT storage. 5-HT is metabolized by monoamine oxidase (MAO) in 5-hydroxyindole acetic acid (5-HIAA). There is one 5-HT transporter (5-HTT or SERT) encoded by the *SLC6A4* gene, and seven known families of serotonin receptors: 5-HT1A-E, P, 5-HT2A-C, 5-HT3, 5-HT4, 5-HT5, 5-HT6, and 5-HT7. The serotoninergic system is particularly important for promoting pulmonary arterial smooth muscle cell proliferation, pulmonary arterial vasoconstriction, and local microthrombosis.

The serotoninergic system has long been suspected of playing important roles in the pathogenesis of idiopathic PAH. Plasma 5-HT levels are elevated in patients with PAH and remain high even after lung transplantation, indicating that this condition is not secondary to the disease (Herve et al. 1995). 5-HTT belongs to a large family of integral membrane proteins and is responsible for 5-HT uptake (e.g., by platelets, endothelial and vascular SMCs). Analysis of distal pulmonary arteries of patients with PAH and their cultured PA-SMCs indicates that 5-HTT is overexpressed and that the level of expression correlates with PAH severity (Eddahibi et al. 2001, 2002; Marcos et al. 2004, 2005). Tryptophan hydroxylase (TPH), the rate-limiting enzyme in 5-HT biosynthesis, is also expressed at abnormally high levels in pulmonary endothelial cells from patients with idiopathic PAH and therefore raises 5-HT levels locally (Eddahibi et al. 2006). There is evidence that alterations in platelet 5-HT storage and/or increased platelet consumption by the lung may trigger the development of PAH (Herve et al. 1990, 1995; Breuer et al. 1996; Eddahibi et al. 2000b; Kereveur et al. 2000; Morecroft et al. 2005). Furthermore, serotoninergic appetite suppressant drugs have been associated with an increased risk of developing PAH (Douglas et al. 1981; Gurtner 1985; Loogen et al. 1985; Brenot et al. 1993; Abenhaim et al. 1996; Perros et al. 2008). Serotonylation of RhoA was also proposed as a possible risk factor of pulmonary vascular remodeling in PAH (Guilluy et al. 2007, 2009). Similarly, findings from another recent study from Wei et al. (2012) indicate increased serotonylation of fibronectin in human and experimental PH. During the last few years, direct evidence for a molecular interplay between 5-HTT signaling and Kv1.5 expression/activity has emerged (Guignabert et al. 2006, 2009b; Guignabert 2011).

Additionally, studies on animal models of PH consolidate all these observations obtained from human subjects. Plasma 5-HT levels are elevated not only in rodents treated with the anorectic agent dexfenfluramine (Eddahibi et al. 1998), but also in the progression of monocrotaline- and chronic hypoxia-induced PH. The chronic infusion of exogenous 5-HT via osmotic pumps can potentiate the development of PH in rats exposed to chronic hypoxia (Eddahibi et al. 1997). A BMPR-II deficiency increases susceptibility to PH induced by 5-HT in mice (Long et al. 2006). In the fawn-hooded rat, a strain with a genetic deficit in platelet 5-HT storage that causes elevated plasma 5-HT concentrations, PH develops when the animals are exposed to mild hypoxia but not in control rats (Sato et al. 1992). An abnormally high level of 5-HTT in the lungs was reported for fawn-hooded rats (Sato et al. 1992; Morecroft et al. 2005). Furthermore, rodents engineered to constitutively express angiopoietin 1 in the lung develop PH. This effect was found to be directly related to the elevated production and secretion of 5-HT by stimulated pulmonary endothelial cells (Sullivan et al. 2003). It has also been shown in the monocrotaline model that 5-HTT expression levels increased prior to the onset of PH, which strongly supports a role for 5-HTT overexpression in disease development (Guignabert et al. 2005). Treatment with selective serotonin reuptake inhibitors (e.g., fluoxetine) abrogates the disease in chronically hypoxic mice and rats with monocrotaline-induced PH (Wang et al. 2011; Marcos et al. 2003; Guignabert et al. 2005, 2009b; Jiang et al. 2007; Zhai et al. 2009; Zhu et al. 2009). Furthermore, mice carrying null mutations at the 5-HTT locus are protected from developing PH induced by prolonged hypoxia (Eddahibi et al. 2000a). Similarly, hypoxia-induced PH in mice lacking the *tph1* gene, which exhibit marked reductions in 5-HT synthesis rates and contents in their peripheral organs, was less severe than in wild-type mice (Izikki et al. 2007). More recently, direct evidence that elevated levels of *5-HTT* gene expression can promote pulmonary vascular remodeling and spontaneous PH was obtained with the creation of two different types of transgenic mice: (1) SM22 5-HTT+ mice that selectively express the human *5-HTT* gene in smooth muscle at levels close to that found in human idiopathic PAH; (2) SERT+ mice that ubiquitously express high levels of the human *5-HTT* gene from a yeast artificial chromosome (YAC) construct. SM22 5-HTT+ mice undergo pulmonary vascular remodeling, develop PH, and exhibit marked increases in right ventricular systolic pressures (RVSPs), right ventricular hypertrophy (RVH), and muscularization of pulmonary arterioles. One major point is that PH in these mice developed without any alterations in 5-HT bioavailability and therefore occurred as a sole consequence of the increased 5-HTT protein levels in SMCs. Compared to wild-type mice, SM22 5-HTT+ mice exhibited increases of three- to fourfold in lung 5-HTT mRNA and protein, together with increased lung 5-HT uptake activity. However, there were no changes in platelet 5-HTT activity or blood 5-HT levels. PH worsened as the SM22 5-HTT+ mice grew older (Guignabert et al. 2006). Consistent with these observations, female SERT+ mice housed in normoxic conditions developed a threefold increase in RVSP values compared to those of their wild-type controls (MacLean et al. 2004).

Of the 14 distinct 5-HT receptors, the 5-HT-2A, -2B, and -1B receptors are particularly relevant to the pathogenesis of PAH. High levels of 5-HT-1B, -2A, and -2B receptor immunoreactivity were reported in remodeled pulmonary arteries from patients with various forms of pulmonary hypertension, but only the 5-HTT was found to be overexpressed in pulmonary artery smooth muscle cells (Marcos et al. 2005). Several lines of evidence support the notion that functional interactions exist between some of these 5-HT receptors and 5-HTT and thus have encouraged studies to better understand these complex relationships (Lawrie et al. 2005; Launay et al. 2006). Antagonism of the 5-HT-2A receptor inhibits not only monocrotaline-induced pulmonary hypertension in mice (Hironaka et al. 2003) but also the 5-HT-induced pulmonary vasoconstriction in vessels from normoxic and hypoxic rats (Morecroft et al. 2005; Cogolludo et al. 2006). However, the 5HT-2A receptor antagonist ketanserin is not specific for pulmonary circulation, and systemic effects have limited its use in PAH (Frishman et al. 1995). 5-HT-2B knockout mice are resistant to hypoxia-induced pulmonary hypertension and administration of the specific 5-HT-2B receptor antagonist RS-127445 prevented an increase in pulmonary arterial pressure in mice challenged with hypoxia (Launay et al. 2002). Furthermore, the 5-HT-2B receptor may control 5-HT plasma levels in vivo (Callebert et al. 2006), and its functional loss may predispose humans to fenfluramine-associated PAH (Blanpain et al. 2003). A very recent study showed that terguride, a potent 5-HT-2A/5-HT-2B receptor antagonist, inhibits the proliferative effects of 5-HT on PA-SMCs and prevents the development and progression of monocrotaline-induced PH in rats (Dumitrascu et al. 2011). The 5-HT-1B receptor mediates 5-HT-induced constriction in human pulmonary arteries (Morecroft et al. 1999) and has been shown to be involved in the development of PH in rodents exposed to chronic hypoxia (Keegan et al. 2001). Recently, Morecroft et al. have reported that co-inhibition of the 5-HT-1B receptor and 5-HTT with a combined 5-HT-1B receptor/5-HTT antagonist (LY393558) is effective at preventing and reversing experimental PH in animal models and 5-HT-induced proliferation in PA-SMCs derived from idiopathic PAH patients.

2.5 Other Growth Factor Signaling Systems in PAH

Many other growth factor signaling pathways are suspected to be involved in the pathogenesis of the disease and may serve to identify potential target candidates for anti-proliferative strategies in PH; however, further studies on their pathogenic roles are important.

2.5.1 The Vascular Endothelial Growth Factor Signaling System

The vascular endothelial growth factor (VEGF) signaling system is a potent and selective endothelial cell mitogen implicated in vascularization and angiogenesis and has been shown to be abnormal in PAH. VEGF family consists of five members, VEGFA, B, C, D, and placenta growth factor (PlGF) that show different binding specificities for two tyrosine kinase receptors, VEGFR1 (also called Flt-1) and VEGFR2 (also called KDR or Flk-1). The binding of VEGF to its receptors induces receptor dimerization and autophosphorylation, which activates several downstream kinases, including protein kinases C and D (PKC and PKD), phosphatidylinositol 3-kinase (PI3K), and MAPK. (Hirose et al. 2000; Voelkel et al. 2006). The VEGF signaling system plays a crucial role for endothelial cell proliferation and promotes the release of endothelial mediators including nitric oxide and prostacyclin (Wheeler-Jones et al. 1997; Dimmeler and Zeiher 1999; He et al. 1999; Nagy et al. 2012). In the lungs of PH patients, VEGF and its receptors are upregulated with a close correlation to the plexiform lesions, suggesting a potential association with the severity of the remodeling process (Cool et al. 1997; Geiger et al. 2000; Hirose et al. 2000; Tuder et al. 2001; Voelkel et al. 2006). Furthermore, patients with PH have an increase in platelet VEGF content (Eddahibi et al. 2000b). The results from the animal models of PH are less clear with differences with the human data and also between the different models studied. On the one hand, Partovian et al. (1998) have reported reduced VEGF mRNA levels in lungs and right ventricles of monocrotaline-injected rats. On the other hand, although no changes in VEGF mRNA levels were noted in lungs from chronically hypoxic rats, an increase in VEGF mRNA levels in right ventricles was observed. Increased VEGF production in the lungs of rats with hypoxia-induced PH was also noted by other investigators (Christou et al. 1998; Laudi et al. 2007). Finally, it has been reported that the *VEGF* gene transfer in the monocrotaline model (Campbell et al. 2001) and in the chronic hypoxia model (Partovian et al. 2000; Louzier et al. 2003) is beneficial. The VEGFR2 is also increased in two other models of PH: in a canine experimental heart failure (Ray et al. 2008) and in overcirculation-induced PH in piglets (Rondelet et al. 2003). However, it is well known that VEGF signaling is required for maintenance of the alveolar structures (Kasahara et al. 2000) and VEGFR-2 blockade with SU5416 in combination with chronic hypoxia causes PH in rats (Taraseviciene-Stewart et al. 2001) and in mice (Ciuclan et al. 2011). In conclusion, further studies are required to identify the precise function of the pathogenic role of VEGF in early stages and during disease progression.

2.5.2 The Insulin-Like Growth Factor Signaling System

The insulin-like growth factor (IGF) signaling system consists of three different ligands (insulin, IGF1 and IGF2), two receptors (IGF1R, IGF2R), and from at least six high-affinity IGF-binding proteins (IGFBP 1-6). IGF1 is mainly secreted by the liver and binds with high affinity to IGF1R and with low affinity to the insulin receptor and IGF2R. IGF2 binds with high affinity to IGF2R and with low affinity to IGF1R and insulin receptor. IGFR1 and IGFR2 share 70 % homology in their protein sequences, but they present differences in signaling and functions. The binding of a ligand to IGF1R leads to activation of distinct downstream signaling pathways such as the MAPK signaling and the phosphatidylinositol 3-kinase-Akt (PI3K-Akt) pathway. Since IGFR2 does not have an intra-cytoplasmic signaling domain, the binding of a ligand to IGFR2 doesn't initiate downstream signaling events. Different subtype of IGFBPs can result in enhancement or inhibition of IGF-mediated cellular effects. The IGF signaling system plays a pleiotropic role in normal cell metabolism, growth, proliferation, differentiation, cell–cell and cell–matrix adhesion, and survival. Recently, Guo et al. (2012) have found a substantial increase in IGFR1 in pulmonary arteries from rats exposed 9 days to hypoxic environments, suggesting a potential involvement of the IGF signaling system in the disease.

2.5.3 The Protein Kinase B/Mammalian Target of Rapamycin Signaling Pathway

Mammalian target of rapamycin (mTOR) is a 289-kDa serine/threonine protein kinase and a member of the phosphatidylinositol 3-kinase-related kinase (PIKK) family. mTOR consists of several conserved functional domains: a catalytic kinase domain, a FKBP12-rapamycin-binding (FRB) domain, a putative autoinhibitory domain (repressor domain) near the C-terminus and up to 20 tandemly repeated HEAT (Huntington, EF3, A subunit of PP2A, TOR1) motifs at the amino terminus, and FAT (FRAP-ATM-TRRAP) and FATC (FAT C-terminus) domains. In fact, mTOR is found to associate with different cofactors and form two distinct multiprotein complexes, mTORC1 and mTORC2 (Wullschleger et al. 2006; Dann et al. 2007). mTORC1 is sensitive to rapamycin and regulates ribosome biogenesis, autophagy, translation, transcription, and mitochondrial metabolism. mTORC1 is activated by mitogens via the PI3K-Akt and ERK1/2 signaling pathways that promote activation of S6 kinase 1 (S6K1), cell growth, and proliferation (Krymskaya and Goncharova 2009). In contrast, mTORC2 is insensitive to rapamycin and functions as an important regulator of actin cytoskeleton through stimulation of F-actin stress fibers. mTORC2 is activated by insulin in a PI3K-dependent manner and phosphorylates protein kinase B (Akt) at Ser-473, which promotes cell cycle progression and increases cell survival (Laplante and Sabatini 2009). The development of monocrotaline- and hypoxia-induced PH in rats has been found to be associated with marked activation of the Akt/glycogen synthase kinase (GSK)-3 axis (Gary-Bobo et al. 2010). In addition to being a central regulator of cell growth, proliferation, apoptosis, and metabolism, mTOR is also linked to the phosphatidylinositol 3 kinase (PI3K)/phosphatase and tensin homolog (PTEN)/Akt/tumor suppressor complex (TSC) signaling pathway, where genetic mutations of many components in this pathway result in the development of a wide variety of cancers (Fig. 2).

Recent evidence demonstrated that mTOR activation in both mTORC1 and mTORC2 due to chronic hypoxia exposure is required for pulmonary arterial smooth muscle cell proliferation (Krymskaya et al. 2011). Similarly, Gerasimovskaya et al. (2005) showed that hypoxia-induced adventitial fibroblast proliferation requires activation and interaction of PI3K, Akt, mTOR, p70S6K, and ERK1/2 and provided evidence for hypoxic regulation of protein translational

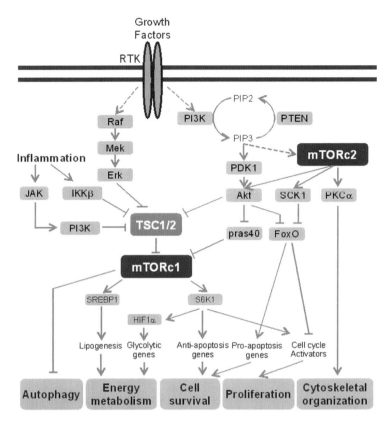

Fig. 2 Schematic representation of the mTOR-signaling pathway. *Akt* protein kinase B; *Erk* extracellular signal regulated kinase; *FoxO* Forkhead box, class O; *HIF1α* hypoxia-inducible factor 1 α; *IKKβ* inhibitor of nuclear factor kappa β; *JAK* the Janus kinase; *Mek* mitogen-activated protein kinase; *PDK1* phosphoinositide-dependent kinase-1; *PI3K* phosphoinositide 3-kinase; *PIP2* phosphatidylinositol 4,5-bisphosphate; *PIP3* phosphatidylinositol 3,4,5-trisphosphate; *PKCα* protein kinase C α; *pras40* proline-rich AKT substrate of 40 kDa; *PTEN* phosphatase and tensin homolog; *RTK* receptor tyrosine kinase; *SCK1* Sck1/2, suppressor of loss of cAMP-dependent protein kinase 1/2; *S6K1* S6 kinase 1; *SREBP1* sterol regulatory element-binding protein 1; *TSC1/2* tuberous sclerosis complex 1/2

pathways in cells exhibiting the capability to proliferate under hypoxic conditions. Furthermore, Ogawa et al. (2012) reported that PDGF-induced phosphorylation of Akt/mTOR signaling pathway in normal pulmonary arterial smooth muscle cells and that inhibition of Akt/mTOR signaling pathway by either rapamycin or Akt inhibitor attenuates PDGF-induced increase in store-operated calcium entry in these cells (Ogawa et al. 2012). Further studies into the role of Akt signaling are important, and the Akt/mTOR pathway may be a potential target for the anti-proliferative strategy in PH.

2.5.4 The Janus Kinase/Signal Transducers and Activators of Transcription Signaling System

Four JAK family kinases, including JAK1, JAK2, JAK3, and TYK2, and seven STAT family members, including STAT1, STAT2, STAT3, STAT4, STAT5a, STAT5b, and STAT6, have been identified. JAK1, JAK2, and TYK2 appear to be ubiquitously expressed, while JAK3 expression is normally limited to lymphoid cells. In addition, different isoforms of several STATs have been identified. STATs are latent cytoplasmic transcription factors that become activated after recruitment to an activated receptor complex. The Janus kinase/signal transducers and activators of transcription (JAK/STAT) pathway is activated by a wide variety of cytokines, growth factors, interferons, and some hormones. Following the binding of cytokines to their cognate receptor, STATs are activated by members of the JAK family of tyrosine kinases. Once activated, they dimerize and translocate to the nucleus and modulate the expression of target genes. STATs are not only activated by cytokine receptors that associate with JAKs but also by RTKs and G protein-coupled receptors.

Over the last few years, increased evidences have supported the role of inflammation as well as the involvement of immunologic disorders in idiopathic PAH (Humbert et al. 1995; Dorfmuller et al. 2003; Tamby et al. 2005; Terrier et al. 2008; Kherbeck et al. 2013; Tamosiuniene et al. 2011; Huertas et al. 2012; Perros et al. 2012; Price et al. 2012). An abnormal activation of the JAK/STAT signaling system was firstly noted by Mathew et al. (2004) in the monocrotaline-induced PH and by Masri et al. (2007) in human tissues of patients with idiopathic PAH. Besides its central role in chronic inflammation, dysregulated activation of the JAK/STAT3 signaling pathway was shown to play an important role in pulmonary arterial endothelial cell proliferation and survival in response to growth factors as demonstrated by pharmacological inhibition of STAT3 phosphorylation by AG-490 (Masri et al. 2007). More recent evidences of the pathogenic role of the JAK/STAT signaling system were documented by the group of Bonnet et al. (Courboulin et al. 2011, 2012; Paulin et al. 2011a, b) showing close interrelationships with Pim1 (proviral integration site for Moloney murine leukemia virus) kinase and NFATc2 (nuclear factor of activated T-cells). It is now well established that among inflammatory cytokines interleukin-6 (IL-6) plays an important role in the development of PH. Overexpression of IL-6 promotes PH in mice (Steiner et al. 2009), while IL-6-deficient mice are protected from hypoxia-induced PH (Savale et al. 2009). In addition, elevated serum IL-6 levels have been reported in patients with idiopathic PAH or PH associated with inflammatory diseases such as scleroderma and lupus (Nishimaki et al. 1999; Pendergrass et al. 2010; Soon et al. 2010). All known members of the IL-6 cytokine family (IL-6, IL-11, ciliary neurotrophic factor (CNTF), cardiotrophin-1 (CT-1), cardiotrophin-like cytokine (CLC), leukemia inhibitory factor (LIF), and oncostatin M (OSM)) through their receptors comprised of the signal transducer gp130 in combination with IL-6R, IL-11R, LIF-R, or OSM-R are known to potently activate STAT3 (Fischer and Hilfiker-Kleiner 2007). A better knowledge of the molecular processes by which JAK-STAT signaling can be turned off will certainly help to identify new targets for drug treatment in PH.

2.5.5 The RhoA/Rho-Kinase Signaling System

RhoA, a small GTPase protein, and its immediate downstream target, RhoA/Rho-kinase (ROCK), control a wide variety of signal transduction pathways. ROCKs are kinases belonging to the AGC (PKA/PKG/PKC) family of serine–threonine kinases that exist in two isoforms: ROCK1 and ROCK2 (Nakagawa et al. 1996). ROCK is comprised of an amino-terminal kinase domain, followed by a coiled-coil region that contains the ρ-binding domain. The carboxy-terminal consists of a plexstrin-homology domain, which contains an internal cysteine-rich domain. ROCK1 and ROCK2 are highly homologous, sharing an identity of 65 % in their overall amino acid sequences and 92 % in their kinase domains. In addition to their effect on actin organization, or through this effect, ROCKs have been found to regulate a wide range of fundamental cell functions such as contraction, motility, proliferation, and apoptosis.

Recent pharmacological studies suggest that activation of RhoA/ROCK signaling system is an important event in the pathogenesis of PH. In vivo, beneficial effects of treatment with Rho kinase inhibitor fasudil have been demonstrated in several animal models of PH (Nakagawa et al. 1996; Abe et al. 2004; Fagan et al. 2004; Nagaoka et al. 2004, 2005; Guilluy et al. 2009). In addition, the beneficial effect of sildenafil on PH is mediated, at least in part, by the inhibition of the RhoA/Rho kinase pathway (Guilluy et al. 2005). Serotonylation of RhoA by intracellular type 2 transglutaminase (TG2), leading to constitutive RhoA activation was also proposed as a possible risk factor of pulmonary vascular remodeling in PAH (Guilluy et al. 2007, 2009). Similarly, findings from another recent study from Wei et al. (2012) indicates increased serotonylation of fibronectin in human and experimental PH.

2.5.6 Other Growth Factor Signaling Systems

Several other studies have similarly found that other growth factor signaling might be involved in the pathogenesis of PAH such as angiotensin II (AngII) (de Man et al. 2012), connective tissue growth factor (CTGF) (Lee et al. 2005), hepatocyte growth factor (HGF) (Ono et al. 2004a, b; Hiramine et al. 2011), nerve growth factor (NGF) (Ieda et al. 2004; Kimura et al. 2007), and placenta growth factor (PlGF) (Sundaram et al. 2010; Sands et al. 2011).

3 Restitution of the Aberrant Extracellular Matrix Remodeling in PH

The proteolytic ECM remodeling not only causes qualitative and quantitative changes in the ECM, but is also actively involved in the creation of a permissive pericellular/extracellular environment for cell proliferation, survival and migration. Indeed, an aberrant ECM remodeling can (1) generate the excessive releases of growth factors and various molecules that are encrypted in the ECM; (2) expose functionally important cryptic sites in collagens, laminins, elastin, or fibronectin; (3) generate fragments of various ECM components (Giannelli et al. 1997; Rabinovitch 2001; Shang et al. 2001; Xu et al. 2001; Ma et al. 2011; Wei et al. 2012).

Various studies demonstrated an imbalance between proteases and protease inhibitors in PH which, among others, include defects in several proteolytic enzymes: elastases (Rabinovitch 1999; Kim et al. 2011), matrix metalloproteinases (Lepetit et al. 2005; George et al. 2012), chymase (Mitani et al. 1999), and tryptase (Kwapiszewska et al. 2012). In addition, impairments of both the urokinase-type plasminogen activator (uPA)—plasmin or the tissue-type plasminogen activator (tPA)—plasmin systems have also been reported in PH (Huber et al. 1994; Christ et al. 2001; Katta et al. 2008; Kouri et al. 2008). Beneficial effects of serine elastase inhibitors in several experimental models of PH have been obtained (Ilkiw et al. 1989; Maruyama et al. 1991; Cowan et al. 2000a; Zaidi et al. 2002). Similar results were found with MMP inhibitors in the monocrotaline model (Vieillard-Baron et al. 2003), but deleterious effects were found with these MMP inhibitors in the chronic hypoxia model (Vieillard-Baron et al. 2000). Differences between both animal models might partially explain the different outcome obtained with MMP inhibitors. Collectively, these findings support that restitution of the aberrant ECM remodeling in PAH may represent another strategy for inhibition of pro-migratory and pro-proliferative signaling pathways.

4 Restitution of the Dysfunctional BMPR-II Signaling System in PAH

Bone morphogenetic proteins (BMPs) are a large family of secreted molecules that belongs to the transforming growth factor (TGF) β family. To date, over 20 BMP family members and 10 antagonists have been identified and characterized. They operate with varied duration, distance, and affinity. There are two classes of transmembrane receptors, type I receptors (ACVRL-I, ACVR-I, BMPR-IA, and BMPR-IB), and the type II receptors (BMPR-II, ActR-IIA, and ActR-IIB). Following ligand binding, the kinase domain in BMPR-II phosphorylates the type 1 receptor, which then phosphorylates Smad proteins 1, 5, and 8. Following activation of Smad 1, 5, and 8, a complex with the common partner Smad4 is generated and it migrates into the nucleus and transactivates specific target genes involved in cell proliferation, survival, migration, and differentiation. *Bmpr2* gene mutations confer a reduction in the BMPR-II signaling activity (Foletta et al. 2003) resulting from a dose-dependent modulation of BMPR-II oligomerization with its co-receptor, most commonly, BMPR-IA (Gilboa et al. 2000). BMPR-II expression is also substantially reduced in patients with various form of PH without a mutation, as well as in experimental animal models (Takahashi et al. 2006; Reynolds et al. 2012). Steady-state levels of BMPR-IA are also reduced in the pulmonary vasculature of patients with pulmonary hypertension (Du et al. 2003), suggesting that disrupted BMP signaling contribute to the pathogenesis of PAH and/or represent a genetic susceptibility of developing the disease. Recently, a study by Reynolds et al. (2012) suggested a therapeutic potential for upregulation of the BMPR-II axis in PAH.

It has been shown that the BMPR-II signaling system plays pleiotropic roles, depending on the cell types: on the one hand, BMPs inhibit proliferation of smooth muscle cells; on the other hand, they promote pulmonary arterial endothelial cell survival (Teichert-Kuliszewska et al. 2006; Nasim et al. 2012). In addition, a constitutive activation of p38MAPK has been shown in primary cultured pulmonary arterial smooth muscle cells harboring a mutation in BMPR-I, a phenomenon that could contribute partly to the failure to suppress cell proliferation (Yang et al. 2005). Although further studies are required to determine the importance of these abnormalities for the initiation/progression/reversal of the disease, restitution of the dysfunctional BMPR-II signaling system may represent another anti-proliferative strategy.

5 Recovery of Oxidative Metabolism in PAH

In the presence of oxygen, normal cells completely oxidize glucose to CO_2 and H_2O, and generate ATP through aerobic oxidation. The Warburg effect is defined as an increased dependence on glycolysis for ATP synthesis, even in the presence of abundant oxygen. The Warburg effect has been found in a wide spectrum of human cancers as well as in PH (Xu et al. 2007), however the underlying mechanisms are still unclear. In cancer cells, this metabolic change has been found to be regulated by both oncogenes and tumor suppressor genes including hypoxia-inducible factors (Goda and Kanai 2012), p53 (Puzio-Kuter 2011), E2F transcription factor-1 (Puzio-Kuter 2011), and phosphatase and tensin homolog (PTEN) (Garcia-Cao et al. 2012). Interestingly, several groups have shown abnormalities in these different signaling pathways in PAH (Bonnet et al. 2006; Natali et al. 2011; Ravi et al. 2011). Restitution of oxidative metabolism with the use of dichloroacetate has been shown to be efficient in several animal models of PH (Michelakis et al. 2002; McMurtry et al. 2004; Guignabert et al. 2009b). Inhibition of pyruvate dehydrogenase kinase (PDK) by dichloroacetate frees up the mitochondrial gate-keeping enzyme pyruvate dehydrogenase (PDH), which is then able to convert pyruvate to acetyl-CoA and initiate normal oxidative phosphoryaltion via the Krebs cycle.

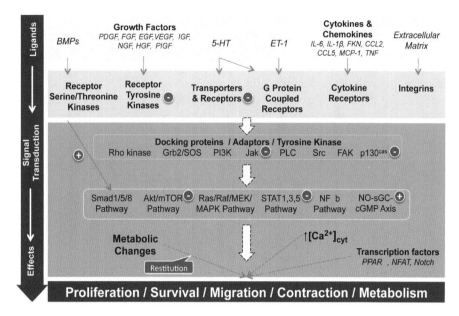

Fig. 3 Schematic representation summarizing the emerging molecular targets for the development of novel anti-proliferative therapies in pulmonary hypertension. *5-HT* serotonin; *Akt* protein kinase B; *BMPs* bone morphogenetic proteins; *CCL* CC chemokine ligands; *cGMP* cyclic guanosine monophosphate; *ET-1* endothelin-1; *FAK* focal adhesion kinase; *FKN* fractalkine; *GRB2* growth factor receptor-bound protein 2; *IL* interleukin; *Jak* the Janus kinase; *MAPK* mitogen-activated protein kinase; *MCP-1* monocyte chemotactic protein-1; *MEK* MAPK kinase; *NFκB* nuclear factor-kappaB; *NO* nitric oxide; *PI3K* phosphoinositide-3-kinase; *PLC* phospholipase C; *sGC* soluble guanylyl cyclase; *SOS* son of sevenless; *STAT* signal transducers and activators of transcription; *TNFα* tumor necrosis factor α

Since mitochondrial fatty acid oxidation (FAO) contributes to the "Randle cycle" inhibition of glucose utilization, the FAO inhibition prevents also this metabolic shift and limits the proliferative and anti-apoptotic cell phenotype observed in PH (Sutendra et al. 2010).

6 Conclusions and Challenges

In this chapter, we summarize several different emerging molecular targets for anti-proliferative strategies in pulmonary hypertension (Fig. 3).

However, growth factors, RTKs, BMPR-II, energetic and metabolic adaptation, as well as the proteolytic ECM remodeling control various aspects of normal cellular physiology, including cell growth, differentiation, motility, and death. Both the potency and selectivity of TKIs as well as of other anti-proliferative molecules are therefore important considerations, particularly as these agents are being tested as a potential therapeutic approach to PAH. Another important aspect concerns the potential impacts of these anti-proliferative molecules on the adaptive response of myocardial hypertrophy that need to be further evaluated. Therefore, a better understanding of how these factors act in the lung as well as in the heart under normal and pathologic conditions will provide a stronger rationale for their use in specific therapeutic interventions and minimize the adverse effects of less focused treatments.

Substantial work remains to be done to discover and/or develop a new, better-tolerated, and more powerful therapeutic tool for PAH that combines promotion of vasorelaxation, suppression of cellular proliferation, and activation of apoptosis within the pulmonary-artery wall. Therefore, further studies are needed to better evaluate the overall risk benefit ratio of the available and future anti-proliferative molecules in PAH as well as their efficacies in experimental models of PH.

Acknowledgments The authors thank Pr. Marc Humbert, Pr. Elie Fadel, Pr. Philippe Dartevelle and Pr. Gérald Simonneau for valuable discussions and suggestions.

References

Abe K, Shimokawa H et al (2004) Long-term treatment with a Rho-kinase inhibitor improves monocrotaline-induced fatal pulmonary hypertension in rats. Circ Res 94(3):385–393

Abenhaim L, Moride Y et al (1996) Appetite-suppressant drugs and the risk of primary pulmonary hypertension. International Primary Pulmonary Hypertension Study Group. N Engl J Med 335(9):609–616

Arcot SS, Fagerland JA et al (1995) Basic fibroblast growth factor alterations during development of monocrotaline-induced pulmonary hypertension in rats. Growth Factors 12(2):121–130

Balasubramaniam V, Le Cras TD et al (2003) Role of platelet-derived growth factor in vascular remodeling during pulmonary hypertension in the ovine fetus. Am J Physiol Lung Cell Mol Physiol 284(5):L826–L833

Baudhuin LM, Jiang Y et al (2004) S1P3-mediated Akt activation and cross-talk with platelet-derived growth factor receptor (PDGFR). FASEB J 18(2):341–343

Benezra M, Vlodavsky I et al (1993) Thrombin-induced release of active basic fibroblast growth factor-heparan sulfate complexes from subendothelial extracellular matrix. Blood 81(12):3324–3331

Benisty JI, McLaughlin VV et al (2004) Elevated basic fibroblast growth factor levels in patients with pulmonary arterial hypertension. Chest 126(4):1255–1261

Berk BC (2001) Vascular smooth muscle growth: autocrine growth mechanisms. Physiol Rev 81(3):999–1030

Blanpain C, Le Poul E et al (2003) Serotonin 5-HT(2B) receptor loss of function mutation in a patient with fenfluramine-associated primary pulmonary hypertension. Cardiovasc Res 60(3):518–528

Bonnet S, Michelakis ED et al (2006) An abnormal mitochondrial-hypoxia inducible factor-1alpha-Kv channel pathway disrupts oxygen sensing and triggers pulmonary arterial hypertension in fawn hooded rats: similarities to human pulmonary arterial hypertension. Circulation 113(22):2630–2641

Brenner B, Harney JT et al (2007) Plasma serotonin levels and the platelet serotonin transporter. J Neurochem 102(1):206–215

Brenot F, Herve P et al (1993) Primary pulmonary hypertension and fenfluramine use. Br Heart J 70(6):537–541

Breuer J, Georgaraki A et al (1996) Increased turnover of serotonin in children with pulmonary hypertension secondary to congenital heart disease. Pediatr Cardiol 17(4):214–219

Buczek-Thomas JA, Nugent MA (1999) Elastase-mediated release of heparan sulfate proteoglycans from pulmonary fibroblast cultures. A mechanism for basic fibroblast growth factor (bFGF) release and attenuation of bfgf binding following elastase-induced injury. J Biol Chem 274(35):25167–25172

Callebert J, Esteve JM et al (2006) Evidence for a control of plasma serotonin levels by 5-hydroxytryptamine(2B) receptors in mice. J Pharmacol Exp Ther 317(2):724–731

Campbell AI, Zhao Y et al (2001) Cell-based gene transfer of vascular endothelial growth factor attenuates monocrotaline-induced pulmonary hypertension. Circulation 104(18):2242–2248

Christ G, Graf S et al (2001) Impairment of the plasmin activation system in primary pulmonary hypertension: evidence for gender differences. Thromb Haemost 86(2):557–562

Christou H, Yoshida A et al (1998) Increased vascular endothelial growth factor production in the lungs of rats with hypoxia-induced pulmonary hypertension. Am J Respir Cell Mol Biol 18(6):768–776

Ciccolini F, Svendsen CN (1998) Fibroblast growth factor 2 (FGF-2) promotes acquisition of epidermal growth factor (EGF) responsiveness in mouse striatal precursor cells: identification of neural precursors responding to both EGF and FGF-2. J Neurosci 18(19):7869–7880

Ciuclan L, Bonneau O et al (2011) A novel murine model of severe pulmonary arterial hypertension. Am J Respir Crit Care Med 184(10):1171–1182

Cogolludo A, Moreno L et al (2006) Serotonin inhibits voltage-gated K+ currents in pulmonary artery smooth muscle cells: role of 5-HT2A receptors, caveolin-1, and KV1.5 channel internalization. Circ Res 98(7):931–938

Cool CD, Kennedy D et al (1997) Pathogenesis and evolution of plexiform lesions in pulmonary hypertension associated with scleroderma and human immunodeficiency virus infection. Hum Pathol 28(4):434–442

Courboulin A, Paulin R et al (2011) Role for miR-204 in human pulmonary arterial hypertension. J Exp Med 208(3):535–548

Courboulin A, Barrier M et al (2012) Plumbagin reverses proliferation and resistance to apoptosis in experimental PAH. Eur Respir J 40(3):618–629

Cowan KN, Jones PL et al (1999) Regression of hypertrophied rat pulmonary arteries in organ culture is associated with suppression of proteolytic activity, inhibition of tenascin-C, and smooth muscle cell apoptosis. Circ Res 84(10):1223–1233

Cowan KN, Heilbut A et al (2000a) Complete reversal of fatal pulmonary hypertension in rats by a serine elastase inhibitor. Nat Med 6(6):698–702

Cowan KN, Jones PL et al (2000b) Elastase and matrix metalloproteinase inhibitors induce regression, and tenascin-C antisense prevents progression, of vascular disease. J Clin Invest 105(1):21–34

Dahal BK, Cornitescu T et al (2010) Role of epidermal growth factor inhibition in experimental pulmonary hypertension. Am J Respir Crit Care Med 181(2):158–167

Dann SG, Selvaraj A et al (2007) mTOR Complex1-S6K1 signaling: at the crossroads of obesity, diabetes and cancer. Trends Mol Med 13(6):252–259

de Man FS, Tu L et al (2012) Dysregulated renin-angiotensin-aldosterone system contributes to pulmonary arterial hypertension. Am J Respir Crit Care Med 186(8):780–789

Dimmeler S, Zeiher AM (1999) Nitric oxide-an endothelial cell survival factor. Cell Death Differ 6(10):964–968

Ding W, Shi W et al (2007) Sprouty2 downregulation plays a pivotal role in mediating crosstalk between TGF-beta1 signaling and EGF as well as FGF receptor tyrosine kinase-ERK pathways in mesenchymal cells. J Cell Physiol 212(3):796–806

Dorfmuller P, Perros F et al (2003) Inflammation in pulmonary arterial hypertension. Eur Respir J 22(2):358–363

Douglas JG, Munro JF et al (1981) Pulmonary hypertension and fenfluramine. Br Med J (Clin Res Ed) 283(6296):881–883
Du L, Sullivan CC et al (2003) Signaling molecules in nonfamilial pulmonary hypertension. N Engl J Med 348(6):500–509
Dumitrascu R, Kulcke C et al (2011) Terguride ameliorates monocrotaline induced pulmonary hypertension in rats. Eur Respir J 37(5):1104–1118
Eddahibi S, Raffestin B et al (1997) Treatment with 5-HT potentiates development of pulmonary hypertension in chronically hypoxic rats. Am J Physiol 272(3 Pt 2):H1173–H1181
Eddahibi S, Raffestin B et al (1998) Effect of dexfenfluramine treatment in rats exposed to acute and chronic hypoxia. Am J Respir Crit Care Med 157(4 Pt 1):1111–1119
Eddahibi S, Hanoun N et al (2000a) Attenuated hypoxic pulmonary hypertension in mice lacking the 5-hydroxytryptamine transporter gene. J Clin Invest 105(11):1555–1562
Eddahibi S, Humbert M et al (2000b) Imbalance between platelet vascular endothelial growth factor and platelet-derived growth factor in pulmonary hypertension. Effect of prostacyclin therapy. Am J Respir Crit Care Med 162(4 Pt 1):1493–1499
Eddahibi S, Humbert M et al (2001) Serotonin transporter overexpression is responsible for pulmonary artery smooth muscle hyperplasia in primary pulmonary hypertension. J Clin Invest 108(8):1141–1150
Eddahibi S, Humbert M et al (2002) Hyperplasia of pulmonary artery smooth muscle cells is causally related to overexpression of the serotonin transporter in primary pulmonary hypertension. Chest 121(3 Suppl):97S–98S
Eddahibi S, Guignabert C et al (2006) Cross talk between endothelial and smooth muscle cells in pulmonary hypertension: critical role for serotonin-induced smooth muscle hyperplasia. Circulation 113(15):1857–1864
Fagan KA, Oka M et al (2004) Attenuation of acute hypoxic pulmonary vasoconstriction and hypoxic pulmonary hypertension in mice by inhibition of Rho-kinase. Am J Physiol Lung Cell Mol Physiol 287(4):L656–L664
Fischer P, Hilfiker-Kleiner D (2007) Survival pathways in hypertrophy and heart failure: the gp130-STAT axis. Basic Res Cardiol 102(5):393–411
Foletta VC, Lim MA et al (2003) Direct signaling by the BMP type II receptor via the cytoskeletal regulator LIMK1. J Cell Biol 162(6):1089–1098
Frishman WH, Huberfeld S et al (1995) Serotonin and serotonin antagonism in cardiovascular and non-cardiovascular disease. J Clin Pharmacol 35(6):541–572
Garcia-Cao I, Song MS et al (2012) Systemic elevation of PTEN induces a tumor-suppressive metabolic state. Cell 149(1):49–62
Gary-Bobo G, Houssaini A et al (2010) Effects of HIV protease inhibitors on progression of monocrotaline- and hypoxia-induced pulmonary hypertension in rats. Circulation 122(19):1937–1947
Geiger R, Berger RM et al (2000) Enhanced expression of vascular endothelial growth factor in pulmonary plexogenic arteriopathy due to congenital heart disease. J Pathol 191(2):202–207
George SJ, Johnson JL et al (2001) Plasmin-mediated fibroblast growth factor-2 mobilisation supports smooth muscle cell proliferation in human saphenous vein. J Vasc Res 38(5):492–501
George J, Sun J et al (2012) Transgenic expression of human matrix metalloproteinase-1 attenuates pulmonary arterial hypertension in mice. Clin Sci (Lond) 122(2):83–92
Gerasimovskaya EV, Tucker DA et al (2005) Activation of phosphatidylinositol 3-kinase, Akt, and mammalian target of rapamycin is necessary for hypoxia-induced pulmonary artery adventitial fibroblast proliferation. J Appl Physiol 98(2):722–731
Giannelli G, Falk-Marzillier J et al (1997) Induction of cell migration by matrix metalloprotease-2 cleavage of laminin-5. Science 277(5323):225–228
Gilboa L, Nohe A et al (2000) Bone morphogenetic protein receptor complexes on the surface of live cells: a new oligomerization mode for serine/threonine kinase receptors. Mol Biol Cell 11(3):1023–1035

Goda N, Kanai M (2012) Hypoxia-inducible factors and their roles in energy metabolism. Int J Hematol 95(5):457–463
Goncalves LM (1998) Fibroblast growth factor-mediated angiogenesis for the treatment of ischemia. Lessons learned from experimental models and early human experience. Rev Port Cardiol 17(Suppl 2):II11–II20
Grouf JL, Throm AM et al (2007) Differential effects of EGF and TGF-beta1 on fibroblast activity in fibrin-based tissue equivalents. Tissue Eng 13(4):799–807
Guignabert C (2011) Interplay between serotonin transporter signaling and voltage-gated potassium channel (Kv) 1.5 expression. In: Sulica R, Preston I (eds) Pulmonary hypertension – from bench research to clinical challenges. InTech Europe Rijeka, Croatia, pp 49–66
Guignabert C, Raffestin B et al (2005) Serotonin transporter inhibition prevents and reverses monocrotaline-induced pulmonary hypertension in rats. Circulation 111(21):2812–2819
Guignabert C, Izikki M et al (2006) Transgenic mice overexpressing the 5-hydroxytryptamine transporter gene in smooth muscle develop pulmonary hypertension. Circ Res 98(10):1323–1330
Guignabert C, Alvira CM et al (2009a) Tie2-mediated loss of peroxisome proliferator-activated receptor-gamma in mice causes PDGF receptor-beta-dependent pulmonary arterial muscularization. Am J Physiol Lung Cell Mol Physiol 297(6):L1082–L1090
Guignabert C, Tu L et al (2009b) Dichloroacetate treatment partially regresses established pulmonary hypertension in mice with SM22alpha-targeted overexpression of the serotonin transporter. FASEB J 23(12):4135–4147
Guilluy C, Sauzeau V et al (2005) Inhibition of RhoA/Rho kinase pathway is involved in the beneficial effect of sildenafil on pulmonary hypertension. Br J Pharmacol 146(7):1010–1018
Guilluy C, Rolli-Derkinderen M et al (2007) Transglutaminase-dependent RhoA activation and depletion by serotonin in vascular smooth muscle cells. J Biol Chem 282(5):2918–2928
Guilluy C, Eddahibi S et al (2009) RhoA and Rho kinase activation in human pulmonary hypertension: role of 5-HT signaling. Am J Respir Crit Care Med 179(12):1151–1158
Guo L, Qiu Z et al (2012) The microRNA-328 regulates hypoxic pulmonary hypertension by targeting at insulin growth factor 1 receptor and L-type calcium channel-alpha1C. Hypertension 59(5):1006–1013
Gurtner HP (1985) Aminorex and pulmonary hypertension. A review. Cor Vasa 27(2–3):160–171
He H, Venema VJ et al (1999) Vascular endothelial growth factor signals endothelial cell production of nitric oxide and prostacyclin through flk-1/KDR activation of c-Src. J Biol Chem 274(35):25130–25135
Herrlich A, Daub H et al (1998) Ligand-independent activation of platelet-derived growth factor receptor is a necessary intermediate in lysophosphatidic, acid-stimulated mitogenic activity in L cells. Proc Natl Acad Sci USA 95(15):8985–8990
Herve P, Drouet L et al (1990) Primary pulmonary hypertension in a patient with a familial platelet storage pool disease: role of serotonin. Am J Med 89(1):117–120
Herve P, Launay JM et al (1995) Increased plasma serotonin in primary pulmonary hypertension. Am J Med 99(3):249–254
Hiramine K, Sata N et al (2011) Hepatocyte growth factor improves the survival of rats with pulmonary arterial hypertension via the amelioration of pulmonary hemodynamics. Int J Mol Med 27(4):497–502
Hironaka E, Hongo M et al (2003) Serotonin receptor antagonist inhibits monocrotaline-induced pulmonary hypertension and prolongs survival in rats. Cardiovasc Res 60(3):692–699
Hirose S, Hosoda Y et al (2000) Expression of vascular endothelial growth factor and its receptors correlates closely with formation of the plexiform lesion in human pulmonary hypertension. Pathol Int 50(6):472–479
Huber K, Beckmann R et al (1994) Fibrinogen, t-PA, and PAI-1 plasma levels in patients with pulmonary hypertension. Am J Respir Crit Care Med 150(4):929–933
Huertas A, Tu L et al (2012) Leptin and regulatory T lymphocytes in idiopathic pulmonary arterial hypertension. Eur Respir J 40(4):895–904

Humbert M, Monti G et al (1995) Increased interleukin-1 and interleukin-6 serum concentrations in severe primary pulmonary hypertension. Am J Respir Crit Care Med 151(5):1628–1631

Humbert M, Morrell NW et al (2004) Cellular and molecular pathobiology of pulmonary arterial hypertension. J Am Coll Cardiol 43(12 Suppl S):13S–24S

Ieda M, Fukuda K et al (2004) Endothelin-1 regulates cardiac sympathetic innervation in the rodent heart by controlling nerve growth factor expression. J Clin Invest 113(6):876–884

Ilkiw R, Todorovich-Hunter L et al (1989) SC-39026, a serine elastase inhibitor, prevents muscularization of peripheral arteries, suggesting a mechanism of monocrotaline-induced pulmonary hypertension in rats. Circ Res 64(4):814–825

Izikki M, Hanoun N et al (2007) Tryptophan hydroxylase 1 knockout and tryptophan hydroxylase 2 polymorphism: effects on hypoxic pulmonary hypertension in mice. Am J Physiol Lung Cell Mol Physiol 293(4):L1045–L1052

Izikki M, Guignabert C et al (2009) Endothelial-derived FGF2 contributes to the progression of pulmonary hypertension in humans and rodents. J Clin Invest 119(3):512–523

Jiang GC, Tidwell K et al (2007) Neurotoxic potential of depleted uranium effects in primary cortical neuron cultures and in Caenorhabditis elegans. Toxicol Sci 99(2):553–565

Jones PL, Rabinovitch M (1996) Tenascin-C is induced with progressive pulmonary vascular disease in rats and is functionally related to increased smooth muscle cell proliferation. Circ Res 79(6):1131–1142

Jones PL, Cowan KN et al (1997a) Tenascin-C, proliferation and subendothelial fibronectin in progressive pulmonary vascular disease. Am J Pathol 150(4):1349–1360

Jones PL, Crack J et al (1997b) Regulation of tenascin-C, a vascular smooth muscle cell survival factor that interacts with the alpha v beta 3 integrin to promote epidermal growth factor receptor phosphorylation and growth. J Cell Biol 139(1):279–293

Jones PL, Jones FS et al (1999) Induction of vascular smooth muscle cell tenascin-C gene expression by denatured type I collagen is dependent upon a beta3 integrin-mediated mitogen-activated protein kinase pathway and a 122-base pair promoter element. J Cell Sci 112(Pt 4):435–445

Kasahara Y, Tuder RM et al (2000) Inhibition of VEGF receptors causes lung cell apoptosis and emphysema. J Clin Invest 106(11):1311–1319

Katta S, Vadapalli S et al (2008) t-plasminogen activator inhibitor-1 polymorphism in idiopathic pulmonary arterial hypertension. Indian J Hum Genet 14(2):37–40

Keegan A, Morecroft I et al (2001) Contribution of the 5-HT(1B) receptor to hypoxia-induced pulmonary hypertension: converging evidence using 5-HT(1B)-receptor knockout mice and the 5-HT(1B/1D)-receptor antagonist GR127935. Circ Res 89(12):1231–1239

Kelvin DJ, Simard G et al (1989) FGF and EGF act synergistically to induce proliferation in BC3H1 myoblasts. J Cell Physiol 138(2):267–272

Kereveur A, Callebert J et al (2000) High plasma serotonin levels in primary pulmonary hypertension. Effect of long-term epoprostenol (prostacyclin) therapy. Arterioscler Thromb Vasc Biol 20(10):2233–2239

Kherbeck N, Tamby MC et al (2013) The role of inflammation and autoimmunity in the pathophysiology of pulmonary arterial hypertension. Clin Rev Allergy Immunol 44(1):31–38

Kim YM, Haghighat L et al (2011) Neutrophil elastase is produced by pulmonary artery smooth muscle cells and is linked to neointimal lesions. Am J Pathol 179(3):1560–1572

Kimura K, Ieda M et al (2007) Cardiac sympathetic rejuvenation: a link between nerve function and cardiac hypertrophy. Circ Res 100(12):1755–1764

Kouri FM, Queisser MA et al (2008) Plasminogen activator inhibitor type 1 inhibits smooth muscle cell proliferation in pulmonary arterial hypertension. Int J Biochem Cell Biol 40(9):1872–1882

Krymskaya VP, Goncharova EA (2009) PI3K/mTORC1 activation in hamartoma syndromes: therapeutic prospects. Cell Cycle 8(3):403–413

Krymskaya VP, Snow J et al (2011) mTOR is required for pulmonary arterial vascular smooth muscle cell proliferation under chronic hypoxia. FASEB J 25(6):1922–1933

Kwapiszewska G, Markart P et al (2012) PAR-2 inhibition reverses experimental pulmonary hypertension. Circ Res 110(9):1179–1191

Laplante M, Sabatini DM (2009) mTOR signaling at a glance. J Cell Sci 122(Pt 20):3589–3594

Laudi S, Steudel W et al (2007) Comparison of lung proteome profiles in two rodent models of pulmonary arterial hypertension. Proteomics 7(14):2469–2478

Launay JM, Herve P et al (2002) Function of the serotonin 5-hydroxytryptamine 2B receptor in pulmonary hypertension. Nat Med 8(10):1129–1135

Launay JM, Schneider B et al (2006) Serotonin transport and serotonin transporter-mediated antidepressant recognition are controlled by 5-HT2B receptor signaling in serotonergic neuronal cells. FASEB J 20(11):1843–1854

Lawrie A, Spiekerkoetter E et al (2005) Interdependent serotonin transporter and receptor pathways regulate S100A4/Mts1, a gene associated with pulmonary vascular disease. Circ Res 97(3):227–235

Le Cras TD, Hardie WD et al (2003) Disrupted pulmonary vascular development and pulmonary hypertension in transgenic mice overexpressing transforming growth factor-alpha. Am J Physiol Lung Cell Mol Physiol 285(5):L1046–L1054

Lee YS, Byun J et al (2005) Monocrotaline-induced pulmonary hypertension correlates with upregulation of connective tissue growth factor expression in the lung. Exp Mol Med 37(1):27–35

Lepetit H, Eddahibi S et al (2005) Smooth muscle cell matrix metalloproteinases in idiopathic pulmonary arterial hypertension. Eur Respir J 25(5):834–842

Li P, Oparil S et al (2003) Fibroblast growth factor mediates hypoxia-induced endothelin – a receptor expression in lung artery smooth muscle cells. J Appl Physiol 95(2):643–651, discussion 863

Long L, MacLean MR et al (2006) Serotonin increases susceptibility to pulmonary hypertension in BMPR2-deficient mice. Circ Res 98(6):818–827

Loogen F, Worth H et al (1985) Long-term follow-up of pulmonary hypertension in patients with and without anorectic drug intake. Cor Vasa 27(2–3):111–124

Louzier V, Raffestin B et al (2003) Role of VEGF-B in the lung during development of chronic hypoxic pulmonary hypertension. Am J Physiol Lung Cell Mol Physiol 284(6):L926–L937

Ma W, Han W et al (2011) Calpain mediates pulmonary vascular remodeling in rodent models of pulmonary hypertension, and its inhibition attenuates pathologic features of disease. J Clin Invest 121(11):4548–4566

MacLean MR, Deuchar GA et al (2004) Overexpression of the 5-hydroxytryptamine transporter gene: effect on pulmonary hemodynamics and hypoxia-induced pulmonary hypertension. Circulation 109(17):2150–2155

Marcos E, Adnot S et al (2003) Serotonin transporter inhibitors protect against hypoxic pulmonary hypertension. Am J Respir Crit Care Med 168(4):487–493

Marcos E, Fadel E et al (2004) Serotonin-induced smooth muscle hyperplasia in various forms of human pulmonary hypertension. Circ Res 94(9):1263–1270

Marcos E, Fadel E et al (2005) Serotonin transporter and receptors in various forms of human pulmonary hypertension. Chest 128(6 Suppl):552S–553S

Maruyama K, Ye CL et al (1991) Chronic hypoxic pulmonary hypertension in rats and increased elastolytic activity. Am J Physiol 261(6 Pt 2):H1716–H1726

Masri FA, Xu W et al (2007) Hyperproliferative apoptosis-resistant endothelial cells in idiopathic pulmonary arterial hypertension. Am J Physiol Lung Cell Mol Physiol 293(3):L548–L554

Mathew R, Huang J et al (2004) Disruption of endothelial-cell caveolin-1alpha/raft scaffolding during development of monocrotaline-induced pulmonary hypertension. Circulation 110(11):1499–1506

Matsunaga S, Okigaki M et al (2009) Endothelium-targeted overexpression of constitutively active FGF receptor induces cardioprotection in mice myocardial infarction. J Mol Cell Cardiol 46(5):663–673

McMurtry MS, Bonnet S et al (2004) Dichloroacetate prevents and reverses pulmonary hypertension by inducing pulmonary artery smooth muscle cell apoptosis. Circ Res 95(8):830–840

Merklinger SL, Jones PL et al (2005) Epidermal growth factor receptor blockade mediates smooth muscle cell apoptosis and improves survival in rats with pulmonary hypertension. Circulation 112(3):423–431

Michelakis ED, McMurtry MS et al (2002) Dichloroacetate, a metabolic modulator, prevents and reverses chronic hypoxic pulmonary hypertension in rats: role of increased expression and activity of voltage-gated potassium channels. Circulation 105(2):244–250

Mitani Y, Ueda M et al (1999) Mast cell chymase in pulmonary hypertension. Thorax 54(1):88–90

Morecroft I, Heeley RP et al (1999) 5-hydroxytryptamine receptors mediating contraction in human small muscular pulmonary arteries: importance of the 5-HT1B receptor. Br J Pharmacol 128(3):730–734

Morecroft I, Loughlin L et al (2005) Functional interactions between 5-hydroxytryptamine receptors and the serotonin transporter in pulmonary arteries. J Pharmacol Exp Ther 313(2):539–548

Morrell NW, Adnot S et al (2009) Cellular and molecular basis of pulmonary arterial hypertension. J Am Coll Cardiol 54(1 Suppl):S20–S31

Murillo MM, del Castillo G et al (2005) Involvement of EGF receptor and c-Src in the survival signals induced by TGF-beta1 in hepatocytes. Oncogene 24(28):4580–4587

Nagaoka T, Morio Y et al (2004) Rho/Rho kinase signaling mediates increased basal pulmonary vascular tone in chronically hypoxic rats. Am J Physiol Lung Cell Mol Physiol 287(4): L665–L672

Nagaoka T, Fagan KA et al (2005) Inhaled Rho kinase inhibitors are potent and selective vasodilators in rat pulmonary hypertension. Am J Respir Crit Care Med 171(5):494–499

Nagy JA, Dvorak AM et al (2012) Vascular hyperpermeability, angiogenesis, and stroma generation. Cold Spring Harb Perspect Med 2(2):a006544

Nakagawa O, Fujisawa K et al (1996) ROCK-I and ROCK-II, two isoforms of Rho-associated coiled-coil forming protein serine/threonine kinase in mice. FEBS Lett 392(2):189–193

Nasim MT, Ogo T et al (2012) BMPR-II deficiency elicits pro-proliferative and anti-apoptotic responses through the activation of TGFbeta-TAK1-MAPK pathways in PAH. Hum Mol Genet 21(11):2548–2558

Natali D, Girerd B et al (2011) Pulmonary arterial hypertension in a patient with Cowden syndrome and anorexigen exposure. Chest 140(4):1066–1068

Nilsson O, Ericson LE et al (1985) Subcellular localization of serotonin immunoreactivity in rat enterochromaffin cells. Histochemistry 82(4):351–355

Nishimaki T, Aotsuka S et al (1999) Immunological analysis of pulmonary hypertension in connective tissue diseases. J Rheumatol 26(11):2357–2362

O'Callaghan DS, Savale L et al (2011) Treatment of pulmonary arterial hypertension with targeted therapies. Nat Rev Cardiol 8(9):526–538

Ogawa A, Firth AL et al (2012) PDGF enhances store-operated Ca2+ entry by upregulating STIM1/Orai1 via activation of Akt/mTOR in human pulmonary arterial smooth muscle cells. Am J Physiol Cell Physiol 302(2):C405–C411

Ono M, Sawa Y et al (2004a) Gene transfer of hepatocyte growth factor with prostacyclin synthase in severe pulmonary hypertension of rats. Eur J Cardiothorac Surg 26(6):1092–1097

Ono M, Sawa Y et al (2004b) Hepatocyte growth factor suppresses vascular medial hyperplasia and matrix accumulation in advanced pulmonary hypertension of rats. Circulation 110 (18):2896–2902

Park JS, Kim JY et al (2000) Epidermal growth factor (EGF) antagonizes transforming growth factor (TGF)-beta1-induced collagen lattice contraction by human skin fibroblasts. Biol Pharm Bull 23(12):1517–1520

Partovian C, Adnot S et al (1998) Heart and lung VEGF mRNA expression in rats with monocrotaline- or hypoxia-induced pulmonary hypertension. Am J Physiol 275(6 Pt 2): H1948–H1956

Partovian C, Adnot S et al (2000) Adenovirus-mediated lung vascular endothelial growth factor overexpression protects against hypoxic pulmonary hypertension in rats. Am J Respir Cell Mol Biol 23(6):762–771

Paulin R, Courboulin A et al (2011a) Signal transducers and activators of transcription-3/pim1 axis plays a critical role in the pathogenesis of human pulmonary arterial hypertension. Circulation 123(11):1205–1215

Paulin R, Meloche J et al (2011b) Dehydroepiandrosterone inhibits the Src/STAT3 constitutive activation in pulmonary arterial hypertension. Am J Physiol Heart Circ Physiol 301(5): H1798–H1809

Pendergrass SA, Hayes E et al (2010) Limited systemic sclerosis patients with pulmonary arterial hypertension show biomarkers of inflammation and vascular injury. PLoS One 5(8):e12106

Perros F, Montani D et al (2008) Platelet-derived growth factor expression and function in idiopathic pulmonary arterial hypertension. Am J Respir Crit Care Med 178(1):81–88

Perros F, Dorfmuller P et al (2012) Pulmonary lymphoid neogenesis in idiopathic pulmonary arterial hypertension. Am J Respir Crit Care Med 185(3):311–321

Price LC, Wort SJ et al (2012) Inflammation in pulmonary arterial hypertension. Chest 141 (1):210–221

Puzio-Kuter AM (2011) The role of p53 in metabolic regulation. Genes Cancer 2(4):385–391

Quinn TP, Schlueter M et al (2002) Cyclic mechanical stretch induces VEGF and FGF-2 expression in pulmonary vascular smooth muscle cells. Am J Physiol Lung Cell Mol Physiol 282(5):L897–L903

Rabinovitch M (1999) EVE and beyond, retro and prospective insights. Am J Physiol 277(1 Pt 1): L5–L12

Rabinovitch M (2001) Pathobiology of pulmonary hypertension. Extracellular matrix. Clin Chest Med 22(3):433–449, viii

Rabinovitch M (2005) Cellular and molecular pathobiology of pulmonary hypertension conference summary. Chest 128(6 Suppl):642S–646S

Ravi Y, Selvendiran K et al (2011) Dysregulation of PTEN in cardiopulmonary vascular remodeling induced by pulmonary hypertension. Cell Biochem Biophys. doi:10.1007/s12013-011-9332-z

Ray L, Mathieu M et al (2008) Early increase in pulmonary vascular reactivity with overexpression of endothelin-1 and vascular endothelial growth factor in canine experimental heart failure. Exp Physiol 93(3):434–442

Ren W, Watts SW et al (2011) Serotonin transporter interacts with the PDGFbeta receptor in PDGF-BB-induced signaling and mitogenesis in pulmonary artery smooth muscle cells. Am J Physiol Lung Cell Mol Physiol 300(3):L486–L497

Reynolds AM, Holmes MD et al (2012) Targeted gene delivery of BMPR2 attenuates pulmonary hypertension. Eur Respir J 39(2):329–343

Rondelet B, Kerbaul F et al (2003) Bosentan for the prevention of overcirculation-induced experimental pulmonary arterial hypertension. Circulation 107(9):1329–1335

Saito Y, Berk BC (2001) Transactivation: a novel signaling pathway from angiotensin II to tyrosine kinase receptors. J Mol Cell Cardiol 33(1):3–7

Sands M, Howell K et al (2011) Placenta growth factor and vascular endothelial growth factor B expression in the hypoxic lung. Respir Res 12:17

Sato K, Webb S et al (1992) Factors influencing the idiopathic development of pulmonary hypertension in the fawn hooded rat. Am Rev Respir Dis 145(4 Pt 1):793–797

Savale L, Tu L et al (2009) Impact of interleukin-6 on hypoxia-induced pulmonary hypertension and lung inflammation in mice. Respir Res 10:6

Schermuly RT, Dony E et al (2005) Reversal of experimental pulmonary hypertension by PDGF inhibition. J Clin Invest 115(10):2811–2821

Shang M, Koshikawa N et al (2001) The LG3 module of laminin-5 harbors a binding site for integrin alpha3beta1 that promotes cell adhesion, spreading, and migration. J Biol Chem 276(35):33045–33053

Soon E, Holmes AM et al (2010) Elevated levels of inflammatory cytokines predict survival in idiopathic and familial pulmonary arterial hypertension. Circulation 122(9):920–927

Steiner MK, Syrkina OL et al (2009) Interleukin-6 overexpression induces pulmonary hypertension. Circ Res 104(2):236–244, 228p following 244

Sullivan CC, Du L et al (2003) Induction of pulmonary hypertension by an angiopoietin 1/TIE2/serotonin pathway. Proc Natl Acad Sci USA 100(21):12331–12336

Sundaram N, Tailor A et al (2010) High levels of placenta growth factor in sickle cell disease promote pulmonary hypertension. Blood 116(1):109–112

Sutendra G, Bonnet S et al (2010) Fatty acid oxidation and malonyl-CoA decarboxylase in the vascular remodeling of pulmonary hypertension. Sci Transl Med 2(44):44ra58

Takahashi H, Goto N et al (2006) Downregulation of type II bone morphogenetic protein receptor in hypoxic pulmonary hypertension. Am J Physiol Lung Cell Mol Physiol 290(3):L450–L458

Tamby MC, Chanseaud Y et al (2005) Anti-endothelial cell antibodies in idiopathic and systemic sclerosis associated pulmonary arterial hypertension. Thorax 60(9):765 772

Tamosiuniene R, Tian W et al (2011) Regulatory T cells limit vascular endothelial injury and prevent pulmonary hypertension. Circ Res 109(8):867–879

Tanabe Y, Saito M et al (2000) Mechanical stretch augments PDGF receptor beta expression and protein tyrosine phosphorylation in pulmonary artery tissue and smooth muscle cells. Mol Cell Biochem 215(1–2):103–113

Taraseviciene-Stewart L, Kasahara Y et al (2001) Inhibition of the VEGF receptor 2 combined with chronic hypoxia causes cell death-dependent pulmonary endothelial cell proliferation and severe pulmonary hypertension. FASEB J 15(2):427–438

Teichert-Kuliszewska K, Kutryk MJ et al (2006) Bone morphogenetic protein receptor-2 signaling promotes pulmonary arterial endothelial cell survival: implications for loss-of-function mutations in the pathogenesis of pulmonary hypertension. Circ Res 98(2):209–217

Terrier B, Tamby MC et al (2008) Identification of target antigens of antifibroblast antibodies in pulmonary arterial hypertension. Am J Respir Crit Care Med 177(10):1128–1134

Thompson K, Rabinovitch M (1996) Exogenous leukocyte and endogenous elastases can mediate mitogenic activity in pulmonary artery smooth muscle cells by release of extracellular-matrix bound basic fibroblast growth factor. J Cell Physiol 166(3):495–505

Tian J, Smith A et al (2009) Effect of PPARgamma inhibition on pulmonary endothelial cell gene expression: gene profiling in pulmonary hypertension. Physiol Genomics 40(1):48–60

Tu L, Dewachter L et al (2011) Autocrine fibroblast growth factor-2 signaling contributes to altered endothelial phenotype in pulmonary hypertension. Am J Respir Cell Mol Biol 45(2):311–322

Tu L, de Man FS et al (2012) A critical role for p130Cas in the progression of pulmonary hypertension in humans and rodent. Am J Respir Crit Care Med 186(7):666–676

Tuder RM, Chacon M et al (2001) Expression of angiogenesis-related molecules in plexiform lesions in severe pulmonary hypertension: evidence for a process of disordered angiogenesis. J Pathol 195(3):367–374

Uttamsingh S, Bao X et al (2008) Synergistic effect between EGF and TGF-beta1 in inducing oncogenic properties of intestinal epithelial cells. Oncogene 27(18):2626–2634

Vanhoutte PM (1991) Platelet-derived serotonin, the endothelium, and cardiovascular disease. J Cardiovasc Pharmacol 17(Suppl 5):S6–S12

Vieillard-Baron A, Frisdal E et al (2000) Inhibition of matrix metalloproteinases by lung TIMP-1 gene transfer or doxycycline aggravates pulmonary hypertension in rats. Circ Res 87(5):418–425

Vieillard-Baron A, Frisdal E et al (2003) Inhibition of matrix metalloproteinases by lung TIMP-1 gene transfer limits monocrotaline-induced pulmonary vascular remodeling in rats. Hum Gene Ther 14(9):861–869

Voelkel NF, Vandivier RW et al (2006) Vascular endothelial growth factor in the lung. Am J Physiol Lung Cell Mol Physiol 290(2):L209–L221

Wang Y, Han DD et al (2011) Downregulation of osteopontin is associated with fluoxetine amelioration of monocrotaline-induced pulmonary inflammation and vascular remodelling. Clin Exp Pharmacol Physiol 38(6):365–372

Wedgwood S, Devol JM et al (2007) Fibroblast growth factor-2 expression is altered in lambs with increased pulmonary blood flow and pulmonary hypertension. Pediatr Res 61(1):32–36

Wei L, Warburton R et al (2012) Serotonylated fibronectin is elevated in pulmonary hypertension. Am J Physiol Lung Cell Mol Physiol 302(12):L1273–L1279

Wheeler-Jones C, Abu-Ghazaleh R et al (1997) Vascular endothelial growth factor stimulates prostacyclin production and activation of cytosolic phospholipase A2 in endothelial cells via p42/p44 mitogen-activated protein kinase. FEBS Lett 420(1):28–32

Wullschleger S, Loewith R et al (2006) TOR signaling in growth and metabolism. Cell 124(3):471–484

Xu J, Rodriguez D et al (2001) Proteolytic exposure of a cryptic site within collagen type IV is required for angiogenesis and tumor growth in vivo. J Cell Biol 154(5):1069–1079

Xu W, Koeck T et al (2007) Alterations of cellular bioenergetics in pulmonary artery endothelial cells. Proc Natl Acad Sci USA 104(4):1342–1347

Yamboliev IA, Gerthoffer WT (2001) Modulatory role of ERK MAPK-caldesmon pathway in PDGF-stimulated migration of cultured pulmonary artery SMCs. Am J Physiol Cell Physiol 280(6):C1680–C1688

Yang X, Long L et al (2005) Dysfunctional Smad signaling contributes to abnormal smooth muscle cell proliferation in familial pulmonary arterial hypertension. Circ Res 96(10):1053–1063

Zaidi SH, You XM et al (2002) Overexpression of the serine elastase inhibitor elafin protects transgenic mice from hypoxic pulmonary hypertension. Circulation 105(4):516–521

Zhai FG, Zhang XH et al (2009) Fluoxetine protects against monocrotaline-induced pulmonary arterial hypertension: potential roles of induction of apoptosis and upregulation of Kv1.5 channels in rats. Clin Exp Pharmacol Physiol 36(8):850–856

Zhu SP, Mao ZF et al (2009) Continuous fluoxetine administration prevents recurrence of pulmonary arterial hypertension and prolongs survival in rats. Clin Exp Pharmacol Physiol 36(8):e1–e5

Anti-inflammatory and Immunosuppressive Agents in PAH

Jolyane Meloche, Sébastien Renard, Steeve Provencher, and Sébastien Bonnet

Abstract Pulmonary arterial hypertension (PAH) pathobiology involves a remodeling process in distal pulmonary arteries, as well as vasoconstriction and in situ thrombosis, leading to enhanced pulmonary vascular resistance and pressure, to right heart failure and death. The exact mechanisms accounting for PAH development remain unknown, but growing evidence demonstrate that inflammation plays a key role in triggering and maintaining pulmonary vascular remodeling. Not surprisingly, PAH is often associated with diverse inflammatory disorders. Furthermore, pathologic specimens from PAH patients reveal an accumulation of inflammatory cells in and around vascular lesions, including macrophages, T and B cells, dendritic cells, and mast cells. Circulating levels of autoantibodies, chemokines, and cytokines are also increased in PAH patients and some of these correlate with disease severity and patients' outcome. Moreover, preclinical experiments demonstrated the key role of inflammation in PAH pathobiology. Immunosuppressive agents have also demonstrated beneficial effects in animal PAH models. In humans, observational studies suggested that immunosuppressive drugs may be effective in treating some PAH subtypes associated with marked inflammation. The present chapter reviews experimental and clinical evidence suggesting that inflammation is involved in the pathogenesis of PAH, as well the therapeutic potential of immunosuppressive agents in PAH.

Keywords PAH • Inflammatory mediators • Immunosuppressive agents

---

Jolyane Meloche and Sébastien Renard have equally contributed to this work.

J. Meloche • S. Renard • S. Provencher • S. Bonnet (✉)
Department of Medicine, Laval University, Quebec City, QC, Canada

Pulmonary Hypertension Research Group, Centre de recherche de l'Institut universitaire de cardiologie et de pneumologie de Québec, 2725 chemin Ste-Foy, Québec City, QC, Canada G1V 4G5
e-mail: jolyane.meloche@criucpq.ulaval.ca; sebastien.renard@criucpq.ulaval.ca; steve.provencher@criucpq.ulaval.ca; sebastien.bonnet@criucpq.ulaval.ca

Contents

1	Introduction	439
2	Clinical Evidence of Inflammation in Human PAH	440
	2.1 Histological and Cytological Data in Human PAH	440
	2.2 Inflammatory Mediators and Biomarkers in Human PAH	441
	2.3 Inflammatory Conditions Associated with Human PAH	441
3	Cells and Inflammatory Mediators Implicated in Pulmonary Vascular Remodeling	445
	3.1 Cell Types Implicated in Inflammatory Processes	445
	3.2 Role of Antibodies	447
	3.3 Chemical Pro-inflammatory Mediators	448
4	Inflammation in PAH: Beyond the Pulmonary Vasculature	450
5	Available Immunosuppressive Agents: Mode of Action and Use in Human Diseases Predisposing to PAH	451
	5.1 Glucocorticoids	451
	5.2 Cyclophosphamide	456
	5.3 Mycophenolate Mofetil	457
	5.4 Methotrexate	457
	5.5 Ciclosporin	457
	5.6 Anti-infectious Agents	458
6	Immunosuppressive Agents in Experimental PH	459
	6.1 Presence of Inflammation in Different PH Experimental Models	459
	6.2 Glucocorticoids	460
	6.3 Cyclophosphamide	460
	6.4 Mycophenolate Mofetil	460
	6.5 Methotrexate	461
	6.6 Ciclosporin	461
	6.7 Dehydroepiandrosterone	461
	6.8 Etanercept	462
	6.9 Sirolimus	462
7	Immunosuppressive Agents in Human PAH	462
	7.1 Updated Experiment of Immunosuppressive Agents in Human PAH	463
	7.2 Side Effects and Drug–Drug Interactions of Immunosuppressive Agents with PAH-Targeted Therapies	464
8	Conclusion	465
References		466

List of Abbreviations

BMPRII	Bone morphogenetic protein receptor 2
CCL	C-C motif ligand
CsA	Ciclosporin A
CTD	Connective tissue disease
CYC	Cyclophosphamide
DHEA	Dehydroepiandrosterone
EC	Endothelial cells
HAART	Highly active antiretroviral therapy
HHV-8	Human herpes virus 8
HIF-1α	Hypoxia inducible factor 1 alpha
HIV	Human immunodeficiency virus

IL	Interleukin (i.e., IL-6, IL-2)
iPAH	Idiopathic pulmonary arterial hypertension
MCP-1	Monocyte chemotactic protein 1
MCT	Monocrotaline (induced PAH)
MIP-1α	Macrophage inflammatory protein 1 alpha
MMF	Mycophenolate mofetil
MTX	Methotrexate
NFAT	Nuclear factor for activated T cells
NF-κB	Nuclear factor kappa-light-chain-enhancer of activated B cells
PA	Pulmonary artery
PAH	Pulmonary arterial hypertension
PASMC	Pulmonary artery smooth muscle cells
PDGF	Platelet-derived growth factor
PH	Pulmonary hypertension
RANTES	Regulated upon activation normal T cell expressed and secreted
RV	Right ventricle/ventricular
SLE	Systemic lupus erythematosus
SSc	Systemic scleroderma
Tc	Cytotoxic T lymphocytes
Th	Helper T lymphocytes
TNF-α	Tumor necrosis factor alpha
Treg	Regulatory T lymphocytes
VEGF	Vascular endothelial growth factor

1 Introduction

Pulmonary arterial hypertension (PAH) is a progressive life-threatening disease characterized by a progressive elevation of pulmonary vascular resistance, right ventricular (RV) failure, and ultimately death (Rubin 1997). According to current classification (Badesch et al. 2009), PAH constitutes the first category of pulmonary hypertension (PH) (Simonneau et al. 2009). This specific group of pre-capillary PH includes idiopathic PAH (iPAH), heritable PAH, and PAH related to drugs and toxins, congenital heart disease, connective tissue disease (CTD), human immuno-deficiency virus (HIV) infection, portal hypertension, chronic haemolytic anemia, and schistosomiasis.

Since 1994, when Tuder et al. (1994) identified for the first time inflammatory infiltrates within PAH patients' plexiform lesions, an increasing amount of experimental and human reports supported the crucial role of inflammation in PAH development (Dorfmuller et al. 2003; Hassoun et al. 2009; Price et al. 2012). Numerous inflammatory disorders and autoimmune processes are associated with the development of experimental and human PAH (Daley et al. 2008; Morse et al. 1997; Negi et al. 1998; Nicolls et al. 2005). Moreover, bone morphogenetic protein receptor type 2 (BMPRII), a pathway largely involved in heritable PAH (Deng et al. 2000; Lane et al. 2000; Machado et al. 2001), also modulates the inflammatory response (Hagen et al. 2007; Song et al. 2008). However, even if inflammation in PAH is well recognized to promote and/or perpetuate pulmonary vascular remodeling (Pullamsetti et al. 2011), its precise relationship with other components of the PAH pathogenesis like endothelial cells (EC) dysfunction, proliferation of pulmonary artery (PA) smooth muscle cells (PASMC) and fibroblast, and in situ thrombosis remains elusive. Furthermore, it is still unclear if inflammation really initiates vascular remodeling ("initial hit"), participates in its progression ("secondary hits") or represents a reactive response to remodeling only ("bystander phenomenon") (Price et al. 2012). From a therapeutic perspective, some reports have shown that immunosuppressive or anti-inflammatory drugs may improve PAH, both in animal models (Bonnet et al. 2007; Price et al. 2011; Sutendra et al. 2011; Suzuki et al. 2006; Wang et al. 2011; Zheng et al. 2010) and human cases (Jais et al. 2008; Karmochkine et al. 1996; Ogawa et al. 2011; Tanaka et al. 2002).

This chapter will review the relevance of inflammatory processes in experimental and human PAH, the implication of cells and other mediators in the inflammatory response observed in PAH, the potential targets for immunosuppressive treatment iPAH, and the updated clinical experience with available immunosuppressant agents.

2 Clinical Evidence of Inflammation in Human PAH

Over the past decades, many inflammation processes have been clearly associated with PAH development, including CTD (Asherson 1990; Cool et al. 1997; Fagan and Badesch 2002; Hachulla et al. 2005), HIV infection (Cool et al. 1997; Humbert 2008) and schistosomiasis infection (Butrous et al. 2008; Tuder 2009). PAH may also occur as an ultimate complication of exceptional autoimmune/inflammatory disorders such as POEMS syndrome (polyneuropathy, organomegaly, endocrinopathy, monoclonal gammapathy, skin changes) (Lesprit et al. 1998) or Castelman disease (Montani et al. 2005a). Also, several reports indicate an association between Hashimoto thyroiditis and PAH (Thurnheer et al. 1997). Even in iPAH, inflammatory phenomenons are present, confirming the key role of inflammation in the global pathophysiology of PA vascular remodeling (Hall et al. 2009; Humbert et al. 1995; Perros et al. 2007; Soon et al. 2010).

2.1 Histological and Cytological Data in Human PAH

Histological data were the first to support the role of inflammation in PAH pathobiology. Tuder et al. (1994) described in 1994 perivascular inflammatory cell infiltrates composed of T cells, B cells, and macrophages in seven of ten cases of iPAH with plexogenic arteriopathy. Cool et al. (1997) reported a few years later mononuclear inflammatory cells surrounding vascular sites of plexiform growth in CTD-PAH lungs. These common features between iPAH and CTD-PAH underlined a potential role of inflammation in all PAH subtypes. Since these results, many histological and fundamental reports confirmed that immune cells abnormalities are implicated in human PAH pathogenesis (Hassoun et al. 2009; Kherbeck et al. 2011; Perros et al. 2011; Price et al. 2012), including very recent data suggesting that in iPAH, the tertiary lymphoid follicle composed of B lymphocytes, T lymphocytes, and dendritic cells have connections to remodeled PA via a stromal network supplied by lymphatic channels (Perros et al. 2012).

2.2 Inflammatory Mediators and Biomarkers in Human PAH

Increased circulating levels of interleukin (IL)-1β and IL-6 were initially reported in iPAH by Humbert et al. (1995) in the early 1990s. Elevated serum cytokines have also been observed in other PAH subtypes like CTD-PAH (Gerbino et al. 2008), HIV-PAH (Humbert et al. 1995), as well as PAH associated with sickle cell disease (Niu et al. 2009) and congenital heart disease (Diller et al. 2008). Some authors even suggested inflammatory cytokines (Soon et al. 2010) and C-reactive protein (Quarck et al. 2009) levels predicted survival in PAH, although this remains controversial (Montani et al. 2011; Soon et al. 2010). Different types of chemokines are also increased in human PAH patients' serum, such as chemokine (C-C motif) ligand (CCL) 2 [known as monocyte chemotactic protein (MCP)-1] (Sanchez et al. 2007), CCL5 [known as regulated upon activation normal T cell expressed and secreted (RANTES)] (Dorfmuller et al. 2002), and CXC3CL1 (known as fractalkine) (Balabanian et al. 2002). In addition, PASMC from PAH patients demonstrate stronger migratory and proliferative response to CCL2 compared to controls (Sanchez et al. 2007). Thus, besides evidences of local inflammation in PAH, these data suggest PAH is associated with a systemic inflammatory state that may correlate with disease severity.

2.3 Inflammatory Conditions Associated with Human PAH

Clinical arguments have been accumulated supporting the role of inflammation in human PAH, especially in the spectrum of CTD and infectious diseases.

2.3.1 CTD and Inflammatory Diseases Associated with PAH

Systemic Scleroderma

The frequent association between two rare conditions is a strong argument for a link between them. Among CTD, scleroderma (SSc) is most commonly associated with PAH, occurring in 5–15 % of SSc patients (Hachulla et al. 2005; Mukerjee et al. 2003b). As in iPAH, EC activation and apoptosis, inflammatory cells recruitment, intimal proliferation, and advential fibrosis leading to vessel obliteration are observed in SSc-PAH (Cerinic et al. 2003; Dorfmuller et al. 2007; Le Pavec et al. 2011; Sgonc et al. 2000), supporting the concept that common features take part in the pathobiology of both iPAH and SSc-PAH (Terrier et al. 2008). SSc-PAH patients are also characterized as having an important number of autoantibodies (Fritzler et al. 1995; Grigolo et al. 2000; Morse et al. 1997; Mouthon et al. 2005; Nicolls et al. 2005; Okano et al. 1992; Tamby et al. 2006), including antifibroblast antibodies (Tamby et al. 2006) and anti-EC antibodies (Arends et al. 2010; Tamby et al. 2005), which activate fibroblasts and induce pro-inflammatory and pro-adhesive phenotype, promoting vascular remodeling (Chizzolini et al. 2002). Finally, the impact of the autoimmunity in PAH is further supported by the heterogeneous prevalence of PAH among SSc depending on the precise autoantibodies detected (Steen 2005).

Systemic Lupus Erythematosus

The prevalence of PAH in systemic lupus erythematosus (SLE) varies from 0.5 to 14 % depending on the diagnostic algorithm used to define PAH (Fois et al. 2010). As for other CTDs, the presence of antinuclear antibodies and rheumatoid factor, as well as abnormalities in immunoglobulin G, complement fractions, cytokines, and growth factors suggest a predominant role for immunological mechanisms in SLE-associated PAH (Quismorio et al. 1984). It is important to keep in mind that other mechanisms may lead to PH in SLE like chronic thromboembolism related to antiphopholipid syndrome, diffuse interstitial disease (Pope 2008; Torre and Harari 2011), or more rarely active pulmonary vasculitis (Asherson and Oakley 1986).

Others CTDs and Rare Inflammatory States

Other CTD inflammatory and/or autoimmune disorders such as mixed CTD (Fagan and Badesch 2002; Jais et al. 2008), polymyositis–dermatomyositis (Minai 2009), Sjogren's syndrome (Launay et al. 2007), rheumatoid arthritis (Chung et al. 2010; Hassoun 2009), Hashimoto's thyroiditis (Chu et al. 2002; Thurnheer et al. 1997), sarcoidosis (Nunes et al. 2006), and more rarely, POEMS syndrome (Jouve et al. 2007; Lesprit et al. 1998), multicentric Castleman disease (Bull et al. 2003), and systemic vasculitis (Launay et al. 2006) have been associated with PAH.

2.3.2 Infectious Diseases Associated with PAH

Evidence for inflammation in PAH infectious-related forms is also supported by histological, cytological, serological, and clinical data where the infective agent is commonly considered as a potential inflammatory trigger.

Human Immunodeficiency Virus and Human Herpes Virus 8

HIV infection is an independent risk factor for PAH (Opravil et al. 1997), occurring in 0.5 % of infected patients (Barnier et al. 2009; Sitbon et al. 2008). Histological features in HIV-PAH are similar to that in other PAH subtypes (Cool et al. 1997; Humbert 2008; Mehta et al. 2000). The mechanisms by which HIV leads to PA remodeling may involve chronic immune activation and upregulation of proinflammatory cytokines and growth factors (Humbert et al. 1998). Indeed, expression and production of platelet-derived growth factors (PDGF) and vascular endothelial growth factors (VEGF) are increased in lung tissue and HIV-infected T cells (Ascherl et al. 1999; Humbert et al. 1998). Viral proteins like Glycoproteine 120 and Tat are also associated with lung endothelial dysfunction in HIV infection mostly through endothelin-1 secretion (Ehrenreich et al. 1993; Ensoli et al. 1990). Controversial data also suggested Human herpes virus 8 (HHV-8) coinfection could account for the development of HIV-PAH (Hsue et al. 2008; Montani et al. 2005b). The effects of highly active antiretroviral therapy (HAART) on HIV-PAH are controversial as cases of PAH regression (Barnier et al. 2009; Speich et al. 2000) and worsening (Pellicelli et al. 1998) have been reported after initiation of therapy.

As described above, genes coding for the vasculotropic virus HHV-8 proteins have also been identified in plexiform lesions of iPAH patients by Cool et al. in the early 2000s (Cool et al. 2003). HHV-8 is known as Kaposi's sarcoma-associated herpes virus and is usually associated with angioproliferative disorders. HHV-8 could thus have a pathogenetic role in PAH, triggering vascular remodeling and plexiform lesions formation. Moreover, rare case reports documented an association between reversible PAH and HHV-8/HIV-associated multicentric Castelman's disease (Montani et al. 2005a). Although there are indirect evidences from in vitro and animal studies in favor of a link between HHV-8 and PAH pathophysiology, HHV-8 has not been detected in recent histological studies in human iPAH (Bendayan et al. 2008; Henke-Gendo et al. 2005; Valmary et al. 2011). Thus, the potential role of this virus on PA remodeling in PAH remains controversial.

Parasites: Schistosomiasis-Related PAH

Among the estimated 200 million people infected by schistosomiasis worldwide (King 2010), between 2 and 5 % are believed to develop PAH (Graham et al. 2010), but their survival remains, in any case, better than in iPAH (dos Santos Fernandes et al. 2010). PAH occurs almost exclusively in patients with hepatosplenic

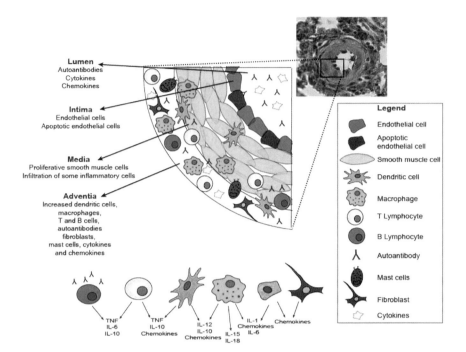

Fig. 1 Large amounts of inflammatory cell infiltrate pulmonary arteries in PAH. There is also vascular wall hypertrophy and presence of apoptotic endothelial cells. Cytokines (interleukins, TNF, MCP, etc.) and autoantibodies (anti-endothelial cells and antifibrolast for example) are present in the vascular wall and blood circulation. These autoantibodies can bind to endothelial cells and fibroblast and enhance cell modification (i.e., apoptosis and increase collagen production). Infiltration of T and B cells as well as macrophages, dendritic cells, mast cells, and fibroblast is observed in PAH. These inflammatory cells produce different cytokines as shown on the *bottom part* of this figure

infection by *Schistosoma mansoni* (Lapa et al. 2009) with subsequent portal hypertension. Histologically, plexiform lesions as well as concentric medial hypertrophy and thrombosis similar to those seen in iPAH are observed in schistosomiasis-related PAH (Tuder 2009). The mechanisms leading to PAH remain elusive, but may be related to mechanical impaction of pulmonary vessels by eggs, focal arteritis, and inflammation and increased pulmonary blood flow as consequence of portocaval shunts. Contribution of inflammation to vascular remodeling is not well understood but could be mediated by the modulation of regulatory T lymphocytes (Treg) activity resulting in the overexpression of a specific transforming growth factor superfamily (Freitas et al. 2007) or by the upregulated IL-13 signaling (Graham et al. 2010). In chronically infected animals without PH, increases in perivascular CD68-macrophages and CD45-lymphocytes have also been reported, suggesting an important switch from a Th1 to a Th2 immune response (Crosby et al. 2010). However, very few data allow extrapolating these experimental findings to human schistosomiasis-related PAH. Taken together, these non-exhaustive data suggest that increased levels of inflammatory mediators are common in human PAH (Fig. 1). Initial or latent inflammatory disorders may be an initial trigger for pulmonary vascular remodeling or act as a "second hit." Although the relationship between inflammation and PAH is most obvious for PAH related to autoimmune, inflammatory, and infectious diseases, this association is not limited to a particular type of PAH.

3 Cells and Inflammatory Mediators Implicated in Pulmonary Vascular Remodeling

Different cell types, chemokines, cytokines, and antibodies play a role in the pathogenesis of PAH. This inflammatory process is not specific to a PAH subgroup, suggesting a central role of inflammation in PAH pathology.

3.1 Cell Types Implicated in Inflammatory Processes

3.1.1 T Lymphocytes (or T cells)

T cells are part of the adaptive immune response. They can differentiate into many subtypes according to different stimuli. Major subtypes found in lungs are T helper $CD4^+$ (Th), Treg, and T cytotoxic $CD8^+$ (Tc) cells. $CD4^+$ Th cells are further divided in Th1, Th2, and Th17 according to their activation and cytokine production. Th1 play a role in cellular immunity and clearance of intracellular pathogen by producing interferon-γ. Th2 produce pro-inflammatory cytokines and are mainly implicated in humoral immunity, inflammation, and allergy. Th17 regulate tissue inflammation and autoimmunity by producing IL-17. $CD4^+$ Th cells are important regulators of adaptive immune response since they stimulate B cell differentiation and macrophage activation, which are critical in triggering the immune response. $CD8^+$ Tc are responsible of killing viral infected cells and tumor cells by binding to major histocompatibility complex class I molecules. Treg suppress autoreactive T cells and, thus, control self-tolerance and autoimmunity by balancing Th1 and Th2 responses. Different evidences support the role of T cells in PAH development. For instance, Treg are able to limit vascular endothelial injury and prevent PAH (Tamosiuniene et al. 2011). In animal models, athymic rats (no mature T cells) develop PAH more rapidly than rats with intact T cell production (Taraseviciene-Stewart et al. 2007), giving a protective role to T cells in PAH. Conversely, depletion of Th cells ameliorates the extent of PAH in other models (Sutendra et al. 2011). Thus, certain subtypes of T cells may confer beneficial effect in PAH, whereas others like Th2 may promote pulmonary vascular remodeling (Daley et al. 2008). Overall, infiltrated T cells within PA wall are increased in iPAH patients. Some studies have shown that Tc are decreased and Treg increased in PAH (Ulrich et al. 2008a) and others showed no difference in

Treg between control and PAH patients (Huertas et al. 2012). The precise role of these T cell subtypes is not yet defined. Studies in cancer demonstrated that T cells inhibit tumor growth, but their functions are often suppressed in tumor microenvironment (Koebel et al. 2007; Zitvogel et al. 2008). A similar pattern could be observed in PAH, as leptin can modulate the hyporesponsiveness of Treg (Huertas et al. 2012). Overall, it is clear that inflammation plays an important role in PAH pathogenesis and that T cells are implicated but the exact mechanism has not yet been completely elucidated.

3.1.2 Dendritic Cells

Dendritic cells are professional antigen-presenting cells, displaying antigen to activate the adaptive immune response (T cell activation). They are responsible for the initiation of the inflammatory response. Recent studies documented the role of dendritic cells in the inflammatory process in different disorders including SLE (Palucka et al. 2005). Furthermore, these cells are present in pulmonary vascular lesions and blood of PAH patients (Wang et al. 2009). Perros et al. (2007) demonstrated that immature dendritic cells are involved in pulmonary vascular remodeling and thus could be involved in PAH immunopathology. Finally, dendritic cells have the ability to differentiate into other cell phenotypes, including EC, expanding their potential role in PAH pathogenesis (Conejo-Garcia et al. 2004).

3.1.3 Mast Cells

Mast cells are derived from bone marrow precursors and reside in tissues adjacent to blood vessels. They are major effector cells of immediate hypersensitivity reactions (allergy). They contain numerous mediator-filled granules, containing histamine and heparin, and a cross-linking of their IgE to a receptor stimulates granules release as well as synthesis and secretion of other mediators leading to hypersensitivity reaction. Accumulation of mast cells is seen in different PAH types (Hamada et al. 1999; Heath and Yacoub 1991). The exact role of mast cells in PAH pathobiology is not yet well established, but they seem to be implicated in direct vasoactive effects (Heath and Yacoub 1991) and in vascular remodeling by releasing matrix metalloproteinases (Vajner et al. 2006). Conversely, mast cells can also produce IL-10, which has important anti-inflammatory and immunosuppressive effects.

3.1.4 Macrophages

Macrophages constitute the first line of immune defense and are part of the innate immune system. In normal lungs, macrophages are mainly present in alveolar airspaces to protect against inhaled pathogens. They play a crucial role in the inflammatory and immune response: they have an antigen-presenting role and display different antigens to activate T cells. Macrophages are increased in pulmonary vascular lesions and around remodeled PA in human and experimental PAH (Tuder et al. 1994). Macrophages are also responsible for producing a large spectrum of inflammatory mediators such as tumor necrosis factor (TNF), endothelin-1, different interleukins, and chemokines (Hassoun et al. 2009; Humbert et al. 2004). Recent studies demonstrated an interaction between macrophages and T cells in PAH (Gerasimovskaya et al. 2012). Indeed, macrophage migration is suppressed by activated T cells but not quiescent T cells, whereas activated macrophages partly block T cells antitumor growth effect. Activated macrophages are also able to suppress T cell activation, thus inhibiting the anti-migratory effect of T cells on macrophages, sustaining macrophage migration, and inflammation in PAH (Gerasimovskaya et al. 2012; Zitvogel et al. 2008).

3.1.5 B Lymphocytes

B cells are responsible of generating antibodies to specific antigenic epitopes which bind to antigens and tag cells for degradation by complement cascade and phagocytosis. In addition to increased antibody production observed in PAH, B cells are increased in PAH lung vasculature and in plexiform lesions and play a critical role in cell-mediated immune regulation though cytokines production (IL-6, IL-10, TNF), antigen presentation, and lymphoid organogenesis (Ulrich et al. 2008b).

3.2 Role of Antibodies

In iPAH, antinuclear antibodies are increased up to 40 % (Rich et al. 1986), whereas antoantibodies such as anti-Scl70, anticentromere, anticardiolipin, and anti-annexin C antibodies are also observed in CTD patients. The specific role of these autoantibodies in PAH pathogenesis is not yet elucidated. More recently, other autoantibodies have been detected in PAH patients.

3.2.1 Antifibroblast Antobodies

Antifibroblast antibodies are detected in 40 % of iPAH and in 30 % of SSc-PAH patients (Tamby et al. 2006). These antibodies enhance fibroblast differentiation and adhesion molecules production. Fibroblast are implicated in collagen production, which plays an important role in both CTD and PAH. Antifibroblast antibodies target different heat shock proteins, glucose-6-phosphate dehydrogenase, PI3-kinase, calumenin, and α-enolase (Terrier et al. 2008, 2009). These targets are implicated in oxidative stress-induced apoptosis resistance

(Efferth et al. 2006), cell energy metabolism, cell growth, and cytoskeleton organization (Shibasaki et al. 1994), leading to increased contractility of myofibroblasts as observed in SSc and PAH.

3.2.2 Anti-endothelial Cell Antibodies

Anti-EC antibodies are detected in over half of patients with SSc and their levels are increased in SSc-PAH compared to SSc patients without PAH (Salojin et al. 1997). Anti-EC antibodies are also present in iPAH patients' serum. A recent study identified lamins, beta tubulins, vinculin, and calumenin as anti-EC antibody targets (Dib et al. 2011). It has been suggested that these antibodies activate EC and induce apoptosis. Further studies would be needed to fully understand their role in the pathogenesis of PAH.

3.3 Chemical Pro-inflammatory Mediators

3.3.1 Cytokines and Chemokines

A large number of cytokines and chemokines (soluble cytokines acting as chemoattractants) are elevated in PAH (Balabanian et al. 2002; Soon et al. 2010). They are mainly produced not only by the innate immune system, i.e., dendritic cells and macrophages, but also by cells which form the vascular wall and adventia (Nathan 2002). Increased IL-1 and IL-6 levels are found in severe PAH (Humbert et al. 1995). Other cytokines such as IL-2, IL-4, IL-8, IL-12p70, and TNF-α are also increased in PAH patients' serum (Fig. 1). Some of these cytokines, like IL-1b, IL-8, and TNF-α, predict outcome in PAH patients (Soon et al. 2010). The implication of these cytokines in PAH development has also been shown in animal models. For example, knock-out mice for IL-6 are resistant to hypoxia-induced PAH (Savale et al. 2009) and mice overexpressing IL-6 develop severe PH (Steiner et al. 2009), suggesting a central role of IL-6 in PAH. Furthermore, circulating IL-6 levels are predictive for the presence of associated PAH among SSc patients (Gourh et al. 2009). IL-6 is known to have many effects on inflammatory and vascular cells (Steiner et al. 2009). In fact, IL-6 stimulates T lymphocytes accumulation, chemokine production (such as CXCL3) by EC (Imaizumi et al. 2004), as well as PASMC and EC proliferation (Savale et al. 2009; Steiner et al. 2009). C-reactive protein is a marker of inflammation and tissue damage and is an active player in vascular wall damage and atherosclerosis (Scirica and Morrow 2006). C-reactive protein levels predict cardiovascular events (Labarrere and Zaloga 2004). Recently, Quarck et al. (2009) demonstrated that C-reactive protein levels are also increased in PAH patients and correlate with long-term outcomes. Moreover, normalization of its levels with therapy is associated with improved functional capacity and survival in PAH (Quarck et al. 2009; Sztrymf et al. 2010).

An important chemokine potentially involved in PAH pathogenesis is fractalkine (CX3CL1 for C-X3-C motif ligand 1). Fractalkine, which can be detected as a soluble form, is anchored in EC membrane and is overexpressed in severe PAH (Balabanian et al. 2002). This chemokine may be responsible for leukocyte capture from the blood and recruitment (if they express C-X3-C motif receptor) (Balabanian et al. 2002). Circulating levels of monocyte chemoattractant protein-1 (MCP-1, also known as CCL2) are also increased in PAH. MCP-1 is secreted by EC, macrophages, and fibroblasts and stimulates leukocytes and monocytes migration. As described earlier, cytokines can also act as chemokines. Thus IL-8, which is produced by hematopoietic cell and part of the C-X-C class, is chemoattractant for T cells and neutrophils (Schall et al. 1990). Macrophage inflammatory protein-1a (MIP-1α, also known as CCL3) promotes migration of monocytes and T and B cells through the endothelial junctions and underlying tissue (Schall et al. 1993). MIP-1α mRNA expression is increased in PAH lung biopsy specimens compared to control (Fartoukh et al. 1998). RANTES is another important chemoattractant for T cells and monocytes and is increased in PAH (Dorfmuller et al. 2002; Luster 1998; Schall et al. 1990). Dorfmüller et al. (2002) demonstrated that RANTES mRNA levels are increased in EC from small remodeled PA and in plexiform lesions, thus increasing the amount of inflammatory cells in remodeled arteries of PAH patients.

3.3.2 Other Growth Factors Implicated in Inflammation

Growth factors including PDGF, epidermal growth factor, VEGF, and fibroblast growth factor 2 are implicated in the proliferative and apoptosis-resistant phenotype contributing to pulmonary vascular remodeling in PAH. There is an overlap between certain cytokines and growth factors. For example, IL-6 stimulates PASMC proliferation through an increase in VEGF and its receptor VEGFR2 (Steiner et al. 2009). PDGF-like molecules are secreted by many cell types, including PASMC, EC, and macrophages (Heldin 1992) and their expression is increased in PAH patients (Humbert et al. 1998). It has been shown that PDGF is able to induce smooth muscle cell proliferation and migration, which explains its implication in various fibroproliferative disorders such as hypoxic PH (Katayose et al. 1993; Schermuly et al. 2005).

4 Inflammation in PAH: Beyond the Pulmonary Vasculature

Right ventricular failure is an important component of the PAH pathophysiology. Indeed, RV failure is the leading cause of death in PAH. Not surprisingly, many circulating biomarkers currently used in PAH prognosis assessment are in fact cardiac-derived biomarkers (Barrier et al. 2012), such as brain natriuretic peptides (Blyth et al. 2007; Gan et al. 2006; Nagaya et al. 2000), troponins (Hoeper et al. 2004; Torbicki et al. 2003), and osteopontin (Lorenzen et al. 2011). The RV capacity to adapt to increased afterload is highly heterogeneous among PAH patients: some are adaptive remodelers with RV hypertrophy and preserved RV function, whereas others are maladaptive remodelers and rapidly develop RV failure (Sztrymf et al. 2010). Some studies have suggested the implication of inflammation in RV remodeling and hypertrophy (Overbeek et al. 2008). However, very little is known about the molecular and cellular mechanisms involved in RV adaptation.

Part of the answer could be found through the analysis of SSc-PAH patients, which have a worse prognosis than iPAH patients (Condliffe et al. 2009; Kawut et al. 2003). This difference may be partly explained by the older age of SSc patients and their comorbidities. Nevertheless, recent data suggest that their RV adapts differently compared to iPAH patients (Chung et al. 2010), including higher N-terminal BNP levels and increased neurohormonal activation found in SSc-PAH compared to iPAH despite of less severe hemodynamic abnormalities (Mathai et al. 2009; Overbeek et al. 2008). Indirect evidence of increased RV inflammation in SSc-PAH has been suggested. For instance, an increased signal intensity in T2-weighted sequence (abnormalities commonly associated with inflammatory myocarditis in the absence of coronary artery disease) is documented on cardiac magnetic resonance imaging studies in SSc-PAH patients compared to SSc-patients without PAH (Hachulla et al. 2009). Recently, Overbeek et al. (2008) confirmed that RV from SSc-PAH ($n = 5$) had more neutrophilic granulocytes, macrophages, and lymphocytes than in iPAH ($n = 9$) or controls ($n = 4$), whereas RV interstitial fibrosis was similar in all groups. However, the participation of inflammation to RV failure is also suspected in all PAH types except, perhaps, for patients with left-to-right shunt and Eisenmenger physiology. In this setting, cardiomyocyte contractions may produce a trigger for autocrine, paracrine, and neuroendocrine signaling pathways leading to a vicious circle of RV inflammation and ischemia, leading to cardiomyocytes apoptosis and RV failure (Bogaard et al. 2009a). Nonetheless, observed RV recovery after lung transplantation supports the idea that RV failure is not irreversible and may be amenable to specific interventions in PAH. Interestingly, Bogaard et al. (2009a) recently demonstrated that a mechanical model of chronic progressive RV pressure overload using PA banding did not lead to fatal RV failure. In contrast, the Sugen model, an established model of angioproliferative PAH, showed myocardial apoptosis, fibrosis, decreased RV capillary density, and VEGF expression despite increased nuclear stabilization of

Hypoxia Inducible Factor-1 (HIF-1), ultimately leading to RV dilation and failure. A more recent animal study from this group assessed a potential RV failure molecular signature (Bogaard et al. 2009b). Their results suggest that RV hypertrophy and/or failure phenotypes in Sugen rats are characterized by distinct patterns of gene expression (mRNA and microRNA) related to cell growth, angiogenesis, and energy metabolism (Drake et al. 2011). Additionally, using PA banding and monocrotaline (MCT) rat models, Piao et al. (2010a, b) demonstrated that in RV hypertrophy, there was a mitochondrial metabolic switch from glucose oxidation to glycolysis. This increase in glucose uptake by the RV can be seen on PET scan, compared to a normal RV that is usually invisible (Bokhari et al. 2011; Can et al. 2011). The increase in glycolysis may temporarily preserve the energetic balance, but ultimately becomes maladaptive (Piao et al. 2010a, b). The reversibility of mitochondrial dysfunction in animals with RV hypertrophy, by a oxidative phosphorylation restoration using a prototypic pyruvate dehydrogenase kinase inhibitor (dichloroacetate), may offer selective strategies for improving RV function; this is in addition to known positive effects of dichloroacetate on PA remodeling in multiple experimental PAH models (Guignabert et al. 2009; McMurtry et al. 2004; Michelakis et al. 2002). These data suggest that complex molecular, cellular, and hemodynamic heart–lung interactions may be involved in the transition from compensated RV hypertrophy to failure in PAH. New therapeutic approaches based, among others, on RV inflammation and abnormal glycolytic metabolism may thus be of interest in PAH.

5 Available Immunosuppressive Agents: Mode of Action and Use in Human Diseases Predisposing to PAH

The mechanisms of action of common immunosuppressive agents are summarized in Table 1 and schematized in Fig. 2.

5.1 Glucocorticoids

Glucocorticoids are small lipophilic compounds affecting B cell and T cell development, differentiation, and function. Their major mechanism for immune suppression is through NF-κB (nuclear factor kappa-light-chain-enhancer of activated B cells) inhibition. NF-κB is involved in many cytokines and/or chemokines synthesis. Inhibition of this transcription factor, therefore, blunts the immune system capacity to mount a response (Rhen and Cidlowski 2005). Glucocorticoids also suppress cell-mediated immunity by inhibiting genes coding for different cytokines such as IL-1, IL-2, IL-3, IL-4, IL-5, IL-6, IL-8, and interferon-γ. They can also stimulate T cell apoptosis by triggering the transcription of different genes (Leung and Bloom 2003). Finally, glucocorticoids not only suppress immune response but

Table 1 Immunosuppressive agents in pulmonary arterial hypertension

	Glucocorticoids (prednisone, dexamethasone...)	Cyclophosphamide (CYC)	Mycophenolate mofetil (MMF)	Methotrexate (MTX)	Ciclosporin A (CsA)
Mechanisms of action	Steroid hormones regulate function and numbers of lymphocytes. They also inhibit phospholipase A2 and cyclo-oxygenase, two important inflammation pathways (Goppelt-Struebe et al. 1989; Leung and Bloom 2003; Rhen and Cidlowski 2005)	Alkylating agent Direct cytotoxicity on bone marrow and mature lymphocytes (Brode and Cooke 2008; Manno and Boin 2010; Marder and McCune 2007; McCune et al. 1988)	Inhibitor of purine synthesis, thus inhibiting cells proliferation (Suzuki et al. 2006; Zheng et al. 2010)	Antimetabolite (dihyrdofolate reductase inhibitor), leading to antiproliferative and cytotoxic effects (Pope et al. 2001)	Calcineurin inhibitor interfering with T-cell activation and IL-2 production (Grinyo et al. 2004; Kobayashi et al. 2007; Tsuda et al. 2012)
Effects in experimental PAH	Dexamethasone (5 mg/kg) and prednisone show beneficial effects in MCT-induced PAH (Price et al. 2011; Wang et al. 2011)	Not reported	Reverses PAH in MCT-induced rats (20 and 40 mg/kg). Does not affect T cells (Suzuki et al. 2006; Zheng et al. 2010)	Not reported	Shows beneficial effect in MCT-induced and hypoxia-induced PAH (Bonnet et al. 2007; Guignabert et al. 2006)
Common dosage in inflammatory disorders	Prednisone: 0.5–1.5 mg/kg/day initially, then slowly tapered afterward. Lowest effective dose for long-term use (Jais et al. 2008)	1–2 mg/kg p.o. daily or 500–750 mg/m² i.v. monthly (Jais et al. 2008)	500–1,500 mg p.o. twice daily (Contreras et al. 2004; Derk et al. 2009; Gerbino et al. 2008; Zamora et al. 2008)	10–25 mg p.o. or im. weekly (Pope et al. 2001; van den Hoogen et al. 1996; Yildirim-Toruner and Diamond 2011)	2.5–5 mg/kg p.o. daily (Clements et al. 1993; Filaci et al. 1999)

Indications, use and clinical reports	Inflammatory manifestations of CTD and other inflammatory disorders (e.g., POEMS, sarcoidosis) (Bertsias et al. 2008; Coker 2007; Dispenzieri 2011; Kowal-Bielecka et al. 2009)	ILD related to SSc Neuropsychiatric involvement of SLE (Hoyles et al. 2006; Marder and McCune 2007; Nadashkevich et al. 2006; Nannini et al. 2008; Tashkin et al. 2006) Lupus nephritis (Barile-Fabris et al. 2005; Stojanovich et al. 2003)	ILD related to SSc (Gerbino et al. 2008; Zamora et al. 2008) SSc-skin manifestations (Derk et al. 2009) Induction therapy in lupus nephritis (Bertsias et al. 2008; Contreras et al. 2004)	Inflammatory arthritis/myositis in many CTD Unsignificant efficacy on SSc-skin disease and lung function (Pope et al. 2001; van denHoogen et al. 1996) Skin manifestations of SLE (Yildirim-Toruner and Diamond 2011)	SSc-skin involvement (Clements et al. 1993; Filaci et al. 1999) Lupus nephritis : may be effective (Moroni, G, Nephrol Dial Transplant 2009) but not recommended (Bertsias G, Ann Rheum Dis 2008)
Effects in human PAH	Observational studies suggest that 40–50 % of SLE and MCTD-PAH may respond to a combination of CYC and glucocorticoids. (Jais et al. 2008; Ribeiro et al. 2001) Rare cases of POEMS syndrome have responded to steroids (Lesprit et al. 1998)	Observational studies suggest that 40–50 % of SLE and MCTD-PAH may respond to a combination of CYC and glucocorticoids. (Jais et al. 2008; Ribeiro et al. 2001)	Not reported	One case of presumed iPAH reported clinical response to MTX coadministered with prednisone (Bellotto et al. 1999)	Not reported
Main side effects	Muscle weakness[a] (may limit physical exertion), hyperglycemia, anxiety, agitation, insomnia, hypokalemia[a] (may be aggravated by	Anemia[a] (may be poorly tolerated in patients with low cardiac output), risk of bleeding[a] in case of thrombocytopenia or hemorrhagic cystitis (may be	Gastrointestinal disturbance (may be relevant for patients treated with prostanoids whose frequently complaint of diarrhea)[a], risk of	Rare cases of pneumonitis and pulmonary fibrosis Hepatotoxicity and elevated liver function test (may be confounding in	Renal toxicity Systemic hypertension Hirsutism, tremor Gum hyperplasia Increased risk of infection[b] (Bertsias et al. 2008; Denton

(continued)

Table 1 (continued)

	Glucocorticoids (prednisone, dexamethasone...)	Cyclophosphamide (CYC)	Mycophenolate mofetil (MMF)	Methotrexate (MTX)	Ciclosporin A (CsA)
	diuretics), fluid retention[a] (may simulate RV failure signs), weight gain[a] (may exacerbate dyspnea), may increase the risk of renal crisis in SSc patients, increased risk of infection[b] (Drugs.com 2010)	aggravated by warfarin), Infertility, nausea and vomiting, alopecia, increased risk of infection[b] (Drugs.com 2010)	bleeding in case of thrombocytopenia (may be aggravated by warfarin)[a], leukopenia, increased risk of infection[b] (Derk et al. 2009; Drugs.com 2010)	patients treated with ERA)[a] Stomatitis Nausea and abdominal distress Pericardial effusion[a] Thrombocytopenia Increased risk of infection[b] (Drugs.com 2010)	et al. 1994; Kowal-Bielecka et al. 2009)
Significant interactions with specific PAH therapy and with warfarin	Not reported with PAH therapy Moderate interactions with *warfarin* (may increase or decrease the anticoagulant activity)[a] (Drugs.com 2010)	Not reported with PAH therapy Moderate interactions with *warfarin* (may increase or decrease the anticoagulant activity)[a] (Drugs.com 2010)	Not reported with PAH therapy No interactions with warfarin	The use of MTX in combination with potentially hepatotoxic agents (like *ERA*) is generally not recommended unless the potential benefit outweighs the risk[a] (Drugs.com 2010) No interactions with warfarin	*ERA*: Contraindication with bosentan[a] (Pharmaceuticals 2011; Venitz et al. 2012) (competitive inhibition of bosentan metabolism via CYP450 3A4 and/or inhibition of intestinal P-glycoprotein secretion by CsA). Ambrisentan dose should be limited to 5 mg[a] (Spence et al. 2010) (CsA inhibition of P-glycoprotein efflux

transporter and organic anion transporting polypeptide, of which ambrisentan is a substrate)

PDE5i: Coadministration with CsA (inhibitors of CYP450 3A4) may increase the plasma concentrations of PDE5i (Schwartz and Kloner 2010) (Sildenafil or Tadalafil), dosage adjustments may be appropriate[a]

Moderate interactions with *warfarin* (may increase or decrease the anticoagulant activity)[a] (Drugs.com 2010)

PAH pulmonary arterial hypertension; *MCT* monocrotaline; *CTD* connective tissue disease; *POEMS* polyneuropathy, monoclonal gammapathy, skin changes; *ILD* interstitial lung disease; *SSc* systemic scleroderma; *SLE* systemic lupus erythematosus; *MCTD* mixed CTD; *iPAH* idiopathic PAH; *RV* right ventricular; *ERA* endothelin receptor antagonist; *CYP* cytochrome; *PDE5i* phosphodiesterase type 5 inhibitors

[a]Of particular interest in PAH because of disease pathophysiology or drug interaction

[b]Particularly of concern for patients with a central venous line for epoprostenol infusion

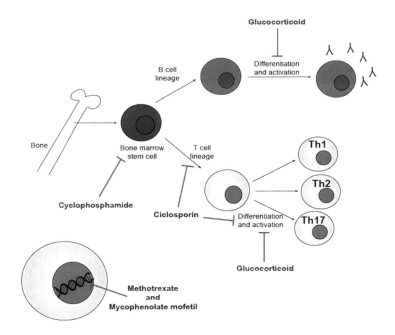

Fig. 2 Mode of action of different immunosuppressive agents as described in text. Cyclophosphamide targets bone marrow stem cells and inhibits their differentiation into B and T cells. Glucocorticoids target B and T cell differentiation and activation. Ciclosporin has a more T cell-specific mode of action as it inhibits differentiation of T cells into Th1, Th2, and Th17. Other molecules such as methotrexate and mycophenolate mofetil target DNA and RNA synthesis, thus blocking proliferation

also inhibit the two main products of inflammation: prostaglandins synthesis at the phospholipase A2 level as well as leukotrienes synthesis at the cyclooxygenase/PGE isomerase level (Goppelt-Struebe et al. 1989). Glucocorticoids are thus potent anti-inflammatory agents, regardless of the inflammation's cause. Therefore, glucocorticoids are often recommended by experts for the treatment of inflammatory manifestations of CTD (Bertsias et al. 2008; Kowal-Bielecka et al. 2009) as well as other inflammatory disorders like POEMS syndrome (Dispenzieri 2011) and sarcoidosis (Coker 2007). Some experimental data also support that glucocorticoids (prednisolone) have an antiproliferative effect on cultured PASMC from patients with iPAH (Ogawa et al. 2005).

5.2 Cyclophosphamide

Cyclophosphamide (CYC) is an alkylating agent used in chemotherapy to slow or stop cell growth. It alkylates or binds to DNA, cross-linking DNA, and RNA strands, thus inhibiting protein synthesis. These actions are neither cell cycle specific nor cell type specific. CYC also exerts its anti-inflammatory function through direct cytoxicity of bone marrow precursors and mature lymphocytes (Fig. 2), leading to a reduction of T and B cells as well as the CD4:CD8 T cell ratio (Manno and Boin 2010; Marder and McCune 2007; McCune et al. 1988), but depending on the dose used, it can also act on Treg and in this case stimulate the immune system (Brode and Cooke 2008). CYC is currently used, most commonly with concomitant corticosteroids, to treat some severe manifestations of CTD, including in interstitial lung disease related to SSc (Hoyles et al. 2006; Marder and McCune 2007; Nadashkevich et al. 2006; Nannini et al. 2008; Tashkin et al. 2006) and neuropsychiatric involvement related to lupus (Barile-Fabris et al. 2005; Stojanovich et al. 2003). Because of many side effects of CYC, however, other immunosuppressive agents are generally preferred for less severe complications of CTD or for maintaining remission.

5.3 Mycophenolate Mofetil

Mycophenolate mofetil (MMF) is an immunosuppressive drug with antiproliferative effects on inflammatory cells through the inhibition of the $5'$-monophosphate dehydrogenase. This enzyme is involved in purines synthesis, an essential step for DNA synthesis in lymphocytes (Fig. 2). This agent is mainly used following organ transplantation. MMF has also been shown to be as effective and better tolerated than CYC for the treatment of lupus nephritis (Bertsias et al. 2008; Contreras et al. 2004). In addition, small series have suggested moderate benefit for interstitial lung disease (Gerbino et al. 2008; Zamora et al. 2008) and skin score (Derk et al. 2009) in SSc.

5.4 Methotrexate

Methotrexate (MTX) is a folic acid analogue and a potent competitive inhibitor of dihydrofolate reductase. It inhibits both DNA and RNA synthesis (Fig. 2). MTX is frequently used in SSc to treat inflammatory arthritis and myositis. Its efficiency for skin disease and interstitial lung disease is however limited (Pope et al. 2001; van den Hoogen et al. 1996). Similarly, MTX has an uncontested role in the management of arthritis and skin manifestations of SLE (Yildirim-Toruner and Diamond 2011), although it has no role for major organ involvement related to SLE (Carneiro and Sato 1999).

5.5 Ciclosporin

Ciclosporin A (CsA) is frequently used for Th1-, Th2-, and Th17-related disorders. It exerts its immunosuppressive function by interfering with T-cell production as well as Th differentiation and function (Grinyo et al. 2004; Kobayashi et al. 2007) (Fig. 2). CsA inhibits calcium synergic action and suppresses the calcineurin pathway, although the precise pharmacological mechanisms of CsA have not yet been fully elucidated (Tsuda et al. 2012). Calcineurin is a key molecule for the nuclear factor for activated T cells (NFAT) activation, which is the main transcription factor for IL-2. Additional anti-fibrotic effects as well as efficiency in chronic graft-versus-host disease have prompted consideration for SSc treatment. In a small open-label study, CsA was associated with a 36 % decrease in skin scores (Clements et al. 1993). Similar improvements were demonstrated in a small randomized trial testing CsA in combination with iloprost (Filaci et al. 1999). However, side effects like systemic hypertension and renal toxicity (Denton et al. 1994) have limited the clinical development of this calcineurin inhibitor in the spectrum of SSc, and this molecule is not currently recommended as first line immunosuppressive agent in SSc (Kowal-Bielecka et al. 2009). Similarly, CsA is not recommended for the management of SLE whatever its manifestations (Bertsias et al. 2008).

5.6 Anti-infectious Agents

5.6.1 Highly Active Antiretroviral Therapy in HIV-Related PAH

Considering on one hand the absence of apparent correlation between the stage of HIV infection, the degree of immunodeficiency, the $CD4^+$ T lymphocytes counts, and the occurrence or severity of HIV-PAH (Mehta et al. 2000; Nunes et al. 2003) and, on the other hand, the heterogeneous evolution of HIV-PAH following antiretroviral therapy initiation, experts do not recommend systematic HAART initiation in case of HIV-PAH (Galie et al. 2009; Hammer et al. 2008). In fact, a large majority of patients are diagnosed for HIV-PAH while already on HAART. Nevertheless, a stable prevalence of HIV-PAH overtime despite HIV patients live much longer than in the 1980s suggest HAART may be associated with a lower incidence of PAH among HIV-infected patients (Opravil and Sereni 2008; Sitbon et al. 2008).

5.6.2 Antihelmintic Treatment in Schistosomiasis-Related PAH

Prazicantel is the most widely used antihelmintic drug to treat schistosomiasis in humans. It is a pyrazinoisoquinoline derivative only active against the adult worms (Ross et al. 2002). In mice, hepatic fibrosis and lung granuloma formation has been shown to be partly reversed by prazicantel and other antihelmintic treatment (de Almeida and Andrade 1983). In a murine model, prazicantel also reversed pulmonary vascular remodeling and prevented the development of schistosomiasis-related PAH in association with a reduction of lung mRNA expression of cytokines IL-13, IL-8, and IL-4 (Crosby et al. 2011). In humans, prazicantel is usually given to prevent further progression of the disease, even if pulmonary embolism composed of dead worms may follow this specific treatment and lead to a superimposed acute cor pulmonale (Lambertucci et al. 2000). However, antischistosomal therapy has not demonstrated any effect on pulmonary hemodynamics in schistosomiasis-related PAH, despite one case report documenting remission of PAH after prazicantel treatment (Bouree et al. 1990). New therapeutic strategies potentially targeting the inflammatory component of the disease (e.g., IL-13 antagonists) are thus needed for schistosomiasis-related PAH (Fernandes et al. 2011).

6 Immunosuppressive Agents in Experimental PH

6.1 Presence of Inflammation in Different PH Experimental Models

To study PH pathobiology, three different models are commonly used: hypoxia model, monocrotaline-induced model (MCT) and the newly developed Sugen-induced model (SU5416, a VEGF receptor inhibitor associated to chronic hypoxia) (Ryan et al. 2011). Other models [e.g., transgenic mice overexpressing 5-HTT specifically in PASMC and fawn-hooded rat are also used but at lesser extent (Guignabert et al. 2006)]. The hypoxia model develops mild to moderate PH with modest inflammation. Conversely, the MCT model develops significant inflammation with higher cytokine levels and marked inflammatory cells infiltrates in remodeled vascular wall, with inflammatory cells displaying a strong IL-6 expression (Bhargava et al. 1999; Price et al. 2011). In addition to these changes, the Sugen model shows plexiform-like lesions and, as in humans, many inflammatory cells are present in these vascular lesions. All these models show an increase in T and B cells, macrophages, mast cells, monocytes, and dendritic cells.

In these experimental models, inflammatory cells are not only innocent bystanders of the pulmonary vascular remodeling but also influence its development. For instance, absence of T cells increases vascular remodeling and worsens PH in the Sugen model (Taraseviciene-Stewart et al. 2007). Similarly, euthymic rat develop severe PH when submitted to both SU5416 injection and chronic hypoxia, whereas athymic rat develop severe vascular remodeling even in normoxia. In athymic rats, pulmonary arterioles become occluded by proliferating EC and are surrounded by mast cells, B cells, and macrophages. IL-4, proliferating cell nuclear antigen, and collagen type I levels are also markedly increased. These studies demonstrate the complex interactions between the immune system, the VEGF receptor signaling pathway, and the pulmonary vascular remodeling process. As a result, different immunosuppressive treatments have been tested in experimental PH.

6.2 Glucocorticoids

Some experimental data support that glucocorticoid treatments (with prednisolone) have an antiproliferative effect on PASMC isolated from iPAH patients (Ogawa et al. 2005). While PDGF induces iPAH-PASMC proliferation and migration, prednisolone reverses this effect in a dose-dependent manner (Ogawa et al. 2005). Very high dose of prednisolone thus caused cell cycle arrest of PASMCs with subsequent suppression of proliferation, whereas prednisolone has no effect on PASMC from controls.

Dexamethasone treatment (5 mg/kg/day) also significantly improved survival in a rat model of in MCT-induced PH (Price et al. 2011; Wang et al. 2011). Furthermore, Price et al. (2011) demonstrated that dexamethasone not only prevented but also normalized hemodynamics and right ventricular hypertrophy in a dose-dependent manner in established PH. At the vascular level, dexamethasone reversed medial and adventicial thickening, reduced MCT-induced adventitial infiltration of IL-6-expressing inflammatory cells (Price et al. 2011), inhibited pulmonary IL-6 overexpression, reduced CX3CR1 expression (Wang et al. 2011), suppressed perivascular $CD8^+$ T cells, and restored the EC integrity (Wang et al. 2011).

6.3 Cyclophosphamide

No animal data was found.

6.4 Mycophenolate Mofetil

In the past decade, different groups demonstrated the beneficial effects of MMF (20–40 mg/kg/day), a prodrug of mycophenolic acid, on MCT-induced PH (Suzuki et al. 2006; Zheng et al. 2010). MMF decreased PASMC proliferation in both MCT-induced PH and in vitro (with fibroblasts to stimulate proliferation) in a dose-dependent manner (Zheng et al. 2010). MMF was also associated with a decrease in macrophages and IL-6 production in the pulmonary vasculature (Suzuki et al. 2006), whereas T cells or mast cells were unchanged (Zheng et al. 2010). From a hemodynamic point of view, MMF decreased RV systolic pressure and hypertrophy. These beneficial effects were seen with MMF concentrations within clinical applicable range, suggesting potentiality of MMF in the treatment of human PAH (Zheng et al. 2010).

6.5 Methotrexate

No animal data was found.

6.6 Ciclosporin

Koulmann et al. (2006) suggested that CsA treatment inhibits PH and RV hypertrophy either by inhibiting HIF-1 transcriptional activity in lung, by decreasing calcineurin activity in lung and heart, by direct effects of CsA, or by a combination of these factors. The CsA dose used in this study (25 mg/kg/day) is much higher than in more recent studies. The exact mechanism by which CsA inhibits PH has been determined more recently. Indeed, Bonnet et al. (2007) demonstrated that NFAT was overexpressed both in human and experimental PH and was largely involved in the PH pathobiology. NFATc2 was in fact mostly activated in iPAH PASMC (Bonnet et al. 2007). By different techniques, they provided evidence that the inhibition of NFAT using a specific [VIVIT (Aramburu et al. 1999)] or an indirect inhibitor (CsA) had beneficial effect on experimental (in vitro and in vivo) PH (Bonnet et al. 2007). Both CsA and VIVIT decreased PASMC proliferation and restored apoptosis levels in vitro. In established MCT-induced PH, CsA (1 mg/kg, similar to doses used clinically) conferred beneficial effects on hemodynamics, RV hypertrophy, and vascular remodeling (Bonnet et al. 2007). In a PH model of mice overexpressing 5-HTT (Guignabert et al. 2006), CsA decreased the pulmonary expression of NFAT and increased Kv1.5 protein levels but did not affect PH itself (mean PA pressure for example) (Guignabert et al. 2006). These results may be explained by the fact that inflammation is not an important component of SM22-5-HTT$^+$ PH mice model (Macian 2005). Whether NFAT inhibition using indirect (e.g., CsA) or specific (e.g., VIVIT) inhibitors is effective in human PH or RV hypertrophy and failure remains unknown (McKinsey and Olson 2005).

6.7 Dehydroepiandrosterone

Our team demonstrated that dehydroepiandrosterone (DHEA) confers beneficial effects on PH not only due to its vasodilator effects but also by disrupting the IL-6/STAT3/NFAT axis known to potentiate PH (Paulin et al. 2011). In vitro data demonstrated that DHEA treatment inhibits IL-6-induced PASMC proliferation. By doing so, it confers anti-inflammatory effects. In MCT-induced PH, DHEA therapy also reverses PH, as assessed by mean pulmonary arterial pressure, right ventricular hypertrophy, and vascular wall thickness.

6.8 Etanercept

Tumor necrosis factor alpha (TNF-α) is a key proinflammatory cytokine that is increased in human and experimental PH (Stojanovich et al. 2003). Furthermore, TNF transgenic rats overexpressing TNF-α develop PH demonstrating its implication in PH pathogenesis (Fujita et al. 2001). Etanercept is a recombinant TNF-α receptor which binds to circulating TNF-α and functionally inhibits its action by blocking binding to cell membrane receptors, thus weakening its pro-inflammatory action (Zhou 2005). In MCT-induced PH, etanercept administered at clinically relevant doses (Lovell et al. 2000) at the same time as MCT injection (in prevention) or 3 weeks later of when PH was established (as a reversal therapy) decreased PH severity and RV hypertrophy (Sutendra et al. 2011). Since TNF-α expression is driven by NFAT in T cells, they confirmed that NFAT activation led to increased levels of TNF-α in an autocrine manner in PASMC. Furthermore, NFAT activation (and its downstream target Kv1.5) was similarly decreased by etanercept and VIVIT, a specific NFAT inhibitor. MCT-PH rats treated with etanercept also demonstrated lower TNF-α and IL-6 levels (another NFAT target) and decreased $CD8^+$ T cells (Sutendra et al. 2011). Etanercept also reduced PH in a late endotoxin-induced model in pigs (Mutschler et al. 2006). Conversely, etanercept at a significantly lower dose had a nonsignificant effect on RV systolic pressure in MCT-induced PH (Henriques-Coelho et al. 2008).

6.9 Sirolimus

Sirolimus (or rapamycin) is an immunosuppressive drug preventing T cells and B cells activation by blocking their response to IL-2. It also blocks the VEGF pathway. Unlike tacrolimus and CsA, sirolimus is not a calcineurin inhibitor but has similar effects on the immune system. While sirolimus may prevent the development of MCT-induced PH, it is ineffective in reversing established PH (McMurtry et al. 2007; Nishimura et al. 2001).

7 Immunosuppressive Agents in Human PAH

Even if preclinical data show some beneficial effects, therapies targeting inflammation have not been formally tested in human PAH. Nevertheless, anti-inflammatory/immunosuppressive therapies are sometimes cautionary recommended by experts in exceptional circumstances. These pharmacological agents, their reported effects on inflammatory disorders and PAH, as well as their potential adverse effects and drug interactions with current PAH therapies are summarized Table 1.

7.1 Updated Experiment of Immunosuppressive Agents in Human PAH

7.1.1 Idiopathic PAH

Rare case reports suggested immunosuppressive therapy could be effective in iPAH. In 1999, Bellotto et al. described the case of an iPAH patient with sustained clinical and hemodynamic improvement following the initiation of prednisone and MTX (Bellotto et al. 1999). Recently, Ogawa et al. (2011) reported the case of a 34-year-old female with iPAH treated with epoprostenol for 3 years, who significantly improved after prednisolone was prescribed for concomitant idiopathic thrombocytopenia. Importantly, both cases of presumed iPAH were characterized by concomitant inflammatory disorders. Apart from these rare cases, no data support the use of immunosuppressive therapy in iPAH.

7.1.2 PAH Related to CTD

According to the EULAR Scleroderma Trials and Research database, PAH is a leading cause of SSc-related deaths even in the era of modern PAH therapies (Mathai and Hassoun 2011; Tyndall et al. 2010). Current guidelines recommend SSc-PAH should follow the same treatment algorithm as in iPAH (Galie et al. 2009). To date, the efficacy of immunosuppressive therapy in SSc-PAH is not supported by observational studies. Indeed, Sanchez et al. (2006) reported that, among a retrospective cohort of 28 CTD-PAH patients, none of the six SSc-PAH responded (defined as patients remaining in New York Heart Association functional class I or II with sustained hemodynamic improvement after at least 1 year without the addition of PAH specific therapy) to monthly cyclophosphamide and glucocorticoids alone. More pronounced fibrotic vascular disease and presence of major comorbidities are proposed reasons to explain the absence of response to immunotherapy in SSc-PAH (Le Pavec et al. 2011). Conversely, cases reported before the advent of specific therapies suggested a positive impact of immunosuppressive therapy on the SLE-PAH course (Groen et al. 1993; Kawaguchi et al. 1998; Morelli et al. 1993; Pines et al. 1982). In the same retrospective study from Sanchez et al. (2006), 5 of the 12 (42 %) SLE-PAH were considered as responder under immunosuppressive therapy alone. Later and from the same group, Jais et al. (2008) reported similar findings, with half of the SLE-PAH patients responding clinically and hemodynamically to immunosuppressive therapy. Patients with less severe PAH at the time of diagnosis were more likely to benefit from immunosuppressive therapy. Normalization of hemodynamic parameters was even observed in some patients with SLE-PAH who received immunosuppressive therapy either alone or in combination with PAH-specific treatment (Heresi and Minai 2007; Jais et al. 2008; Ribeiro et al. 2001). Clinical and hemodynamic improvements have also been observed, although less frequently, in patients with PAH associated with mixed CTD (Sanchez et al. 2006). More recently, immunosuppression added to specific PAH therapy has been reported effective in a heterogeneous cohort of 13 CTD-PAH as compared to a historical group of 8 CTD-PAH receiving specific PAH therapy alone (Minai 2009). Unfortunately, repartition of CTD subtypes was different between the two groups (more SLE and less SSc in the "combination" group) limiting the interpretation of these results. Despite the limited data supporting the efficacy of immunosuppressive therapy, current guidelines suggest immunosuppression, most commonly in combination with PAH-specific agents, may be considered in SLE-PAH or mixed CTD-PAH (Jais et al. 2008; Ribeiro et al. 2001).

7.1.3 POEMS Syndrome-Related PAH

POEMS syndrome has been occasionally associated with PAH. Recently, Jouve et al. (2007) described two cases of PAH related to POEMS syndrome, including one case ascribed to PAH and one case of post-capillary/high cardiac output PAH. In any case, clinical symptoms disappear and hemodynamic study showed a significant improvement in both patients (Jouve et al. 2007). Mukerjee et al. (2003a) also reported a single case of reversible POEMS-PAH after sequential therapy including initial iloprost therapy followed by immunosuppressive therapy. Finally, among 20 cases of POEMS syndrome, 5 cases of PAH were reported during the follow-up (Lesprit et al. 1998). Of note, however, PAH was confirmed by catheterization in only two of these patients. Overproduction IL-1β, IL6, TNF-α, and VEGF was found in all cases, suggesting that cytokines may mediate the development of PAH in the POEMS syndrome. However, only two of these five patients presented a sustained improvement under immunosuppressive therapy (Lesprit et al. 1998). Results about the benefit of anti-VEGF were also conflicting (Dispenzieri 2011).

7.2 *Side Effects and Drug–Drug Interactions of Immunosuppressive Agents with PAH-Targeted Therapies*

Potential side effects and drug–drug interactions were detected using the side effect and drug interactions checker within http://www.drugs.com database (2010). Few data are currently available about the safety profile of immunosuppressive drugs in the setting of PAH. Nevertheless, among the usual side effects of immunosuppressive therapies (Table 1), some are particularly of concern in the setting of PAH. For example, glucocorticoids promote hypokalemia, which may be aggravated by diuretics and may be associated with an increased risk of arrhythmias. CYC and MMF may reduce red blood cells and platelet production. Anemia may be poorly tolerated, especially in PAH patients with low cardiac output. Patients treated with warfarin may also be at increased risk of bleeding in case of thrombocytopenia or CYC-related hemorrhagic cystitis. MMF may also worsen the gastrointestinal disturbance associated with prostanoid therapy. More widely, it is important to keep in mind that all immunosuppressive therapy increases risk of infection. This additional issue should be taken into consideration in the management of PAH patients, especially in those with a central venous line for epoprostenol infusion.

The main concern about drug–drug interaction relates to the cytochrome P450 metabolite pathway. Indeed, both bosentan and CsA are metabolized through the CYP450 3A4 isoenzyme. Coadministration of CsA and bosentan is contraindicated due to significant increased bosentan exposure and decreased CsA exposure (Pharmaceuticals 2011; Venitz et al. 2012). Clinicians should also be cautious if tacrolimus (another calcineurin inhibitor) and bosentan are used together. Conversely, coadministration of ambrisentan and CsA or tacrolimus is not contraindicated, although it is recommended to limit the ambrisentan dose to 5 mg once daily (Spence et al. 2010). Limited data are available for phosphodiesterase 5 inhibitors, which are also primarily metabolized by the CYP450 3A4 pathway (Schwartz and Kloner 2010). Therefore, coadministration with inhibitors of CYP450 3A4 may increase plasma concentrations of sildenafil or tadalafil, and possibility of prolonged and/or increased pharmacologic effects of these drugs should be considered. A Spanish renal transplant study was reassuring as sildenafil therapy was safe for the treatment of erectile dysfunction and did not modify CsA and tacrolimus blood levels (Cofan et al. 2002). MMF is not expected to compete with ambrisentan metabolism (Mandagere et al. 2010). Similarly, CYC is not expected to interact with specific PAH treatments because of its multiple metabolic pathways. Finally, no specific data are available concerning immunosuppressive therapy interactions with prostanoids agents.

8 Conclusion

Inflammation is undoubtedly part of the PAH pathobiology. Giving the poor outcomes in PAH despite recently developed treatments, there is an urgent need for new therapies targeting novel pathophysiological pathways. In many animal models of PAH, targeting the inflammation by current or new immunosuppressive agents improves PAH and pulmonary vascular remodeling. Treating the underlying inflammatory condition in human PAH associated with inflammatory disorders such as SLE, mixed CTD, and POEMS syndrome also led, in some cases, to marked clinical and hemodynamic improvements. Nevertheless, further studies are needed to characterize the elusive mechanisms by which inflammation triggers and sustains vascular remodeling and likely RV failure. More importantly, the poor safety profile of existing immunosuppressive drugs and the potential for drug–drug interactions with current PAH therapies mandate that appropriate large-scaled randomized trials are performed before immunosuppression becomes standard of care in any PAH type.

References

Actelion Pharmaceuticals (2011) Tracleer [package insert]. Actelion Pharmaceuticals US, Inc., South San Francisco, CA

Aramburu J, Yaffe MB, Lopez-Rodriguez C et al (1999) Affinity-driven peptide selection of an NFAT inhibitor more selective than cyclosporin A. Science 285:2129–2133

Arends SJ, Damoiseaux J, Duijvestijn A et al (2010) Prevalence of anti-endothelial cell antibodies in idiopathic pulmonary arterial hypertension. Eur Respir J 35:923–925

Ascherl G, Hohenadl C, Schatz O et al (1999) Infection with human immunodeficiency virus-1 increases expression of vascular endothelial cell growth factor in T cells: implications for acquired immunodeficiency syndrome-associated vasculopathy. Blood 93:4232–4241

Asherson RA (1990) Pulmonary hypertension in systemic lupus erythematosus. J Rheumatol 17:414–415

Asherson RA, Oakley CM (1986) Pulmonary hypertension and systemic lupus erythematosus. J Rheumatol 13:1–5

Badesch DB, Champion HC, Sanchez MA et al (2009) Diagnosis and assessment of pulmonary arterial hypertension. J Am Coll Cardiol 54:S55–S66

Balabanian K, Foussat A, Dorfmuller P et al (2002) CX(3)C chemokine fractalkine in pulmonary arterial hypertension. Am J Respir Crit Care Med 165:1419–1425

Barile-Fabris L, Ariza-Andraca R, Olguin-Ortega L et al (2005) Controlled clinical trial of IV cyclophosphamide versus IV methylprednisolone in severe neurological manifestations in systemic lupus erythematosus. Ann Rheum Dis 64:620–625

Barnier A, Frachon I, Dewilde J et al (2009) Improvement of HIV-related pulmonary hypertension after the introduction of an antiretroviral therapy. Eur Respir J 34:277–278

Barrier M, Meloche J, Jacob MH et al (2012) Today's and tomorrow's imaging and circulating biomarkers for pulmonary arterial hypertension. Cell Mol Life Sci 69(17):2805–2831

Bellotto F, Chiavacci P, Laveder F et al (1999) Effective immunosuppressive therapy in a patient with primary pulmonary hypertension. Thorax 54:372–374

Bendayan D, Sarid R, Cohen A et al (2008) Absence of human herpesvirus 8 DNA sequences in lung biopsies from Israeli patients with pulmonary arterial hypertension. Respiration 75:155–157

Bertsias G, Ioannidis JP, Boletis J et al (2008) EULAR recommendations for the management of systemic lupus erythematosus. Report of a Task Force of the EULAR Standing Committee for International Clinical Studies Including Therapeutics. Ann Rheum Dis 67:195–205

Bhargava A, Kumar A, Yuan N et al (1999) Monocrotaline induces interleukin-6 mRNA expression in rat lungs. Heart Dis 1:126–132

Blyth KG, Groenning BA, Mark PB et al (2007) NT-proBNP can be used to detect right ventricular systolic dysfunction in pulmonary hypertension. Eur Respir J 29:737–744

Bogaard HJ, Abe K, Vonk Noordegraaf A et al (2009a) The right ventricle under pressure: cellular and molecular mechanisms of right-heart failure in pulmonary hypertension. Chest 135:794–804

Bogaard HJ, Natarajan R, Henderson SC et al (2009b) Chronic pulmonary artery pressure elevation is insufficient to explain right heart failure. Circulation 120:1951–1960

Bokhari S, Raina A, Rosenweig EB et al (2011) PET imaging may provide a novel biomarker and understanding of right ventricular dysfunction in patients with idiopathic pulmonary arterial hypertension. Circ Cardiovasc Imaging 4:641–647

Bonnet S, Rochefort G, Sutendra G et al (2007) The nuclear factor of activated T cells in pulmonary arterial hypertension can be therapeutically targeted. Proc Natl Acad Sci USA 104:11418–11423

Bouree P, Piveteau J, Gerbal JL et al (1990) Pulmonary arterial hypertension due to bilharziasis. Apropos of a case due to Schistosoma haematobium having been cured by praziquantel. Bull Soc Pathol Exot 83:66–71

Brode S, Cooke A (2008) Immune-potentiating effects of the chemotherapeutic drug cyclophosphamide. Crit Rev Immunol 28:109–126

Bull TM, Cool CD, Serls AE et al (2003) Primary pulmonary hypertension, Castleman's disease and human herpesvirus-8. Eur Respir J 22:403–407

Butrous G, Ghofrani HA, Grimminger F (2008) Pulmonary vascular disease in the developing world. Circulation 118:1758–1766

Can MM, Kaymaz C, Tanboga IH et al (2011) Increased right ventricular glucose metabolism in patients with pulmonary arterial hypertension. Clin Nucl Med 36:743–748

Carneiro JR, Sato EI (1999) Double blind, randomized, placebo controlled clinical trial of methotrexate in systemic lupus erythematosus. J Rheumatol 26:1275–1279

Cerinic MM, Valentini G, Sorano GG et al (2003) Blood coagulation, fibrinolysis, and markers of endothelial dysfunction in systemic sclerosis. Semin Arthritis Rheum 32:285–295

Chizzolini C, Raschi E, Rezzonico R et al (2002) Autoantibodies to fibroblasts induce a proadhesive and proinflammatory fibroblast phenotype in patients with systemic sclerosis. Arthritis Rheum 46:1602–1613

Chu JW, Kao PN, Faul JL et al (2002) High prevalence of autoimmune thyroid disease in pulmonary arterial hypertension. Chest 122:1668–1673

Chung L, Liu J, Parsons L et al (2010) Characterization of connective tissue disease-associated pulmonary arterial hypertension from REVEAL: identifying systemic sclerosis as a unique phenotype. Chest 138:1383–1394

Clements PJ, Lachenbruch PA, Sterz M et al (1993) Cyclosporine in systemic sclerosis. Results of a forty-eight-week open safety study in ten patients. Arthritis Rheum 36:75–83

Cofan F, Gutierrez R, Beardo P et al (2002) Interaction between sildenafil and calcineurin inhibitors in renal transplant recipients with erectile dysfunction. Nefrologia 22:470–476

Coker RK (2007) Guidelines for the use of corticosteroids in the treatment of pulmonary sarcoidosis. Drugs 67:1139–1147

Condliffe R, Kiely DG, Peacock AJ et al (2009) Connective tissue disease-associated pulmonary arterial hypertension in the modern treatment era. Am J Respir Crit Care Med 179:151–157

Conejo-Garcia JR, Benencia F, Courreges MC et al (2004) Tumor-infiltrating dendritic cell precursors recruited by a beta-defensin contribute to vasculogenesis under the influence of Vegf-A. Nat Med 10:950–958

Contreras G, Pardo V, Leclercq B et al (2004) Sequential therapies for proliferative lupus nephritis. N Engl J Med 350:971–980

Cool CD, Kennedy D, Voelkel NF et al (1997) Pathogenesis and evolution of plexiform lesions in pulmonary hypertension associated with scleroderma and human immunodeficiency virus infection. Hum Pathol 28:434–442

Cool CD, Rai PR, Yeager ME et al (2003) Expression of human herpesvirus 8 in primary pulmonary hypertension. N Engl J Med 349:1113–1122

Crosby A, Jones FM, Southwood M et al (2010) Pulmonary vascular remodeling correlates with lung eggs and cytokines in murine schistosomiasis. Am J Respir Crit Care Med 181:279–288

Crosby A, Jones FM, Kolosionek E et al (2011) Praziquantel reverses pulmonary hypertension and vascular remodeling in murine schistosomiasis. Am J Respir Crit Care Med 184:467–473

Daley E, Emson C, Guignabert C et al (2008) Pulmonary arterial remodeling induced by a Th2 immune response. J Exp Med 205:361–372

de Almeida MA, Andrade ZA (1983) Effect of chemotherapy on experimental pulmonary schistosomiasis. Am J Trop Med Hyg 32:1049–1054

Deng Z, Morse JH, Slager SL et al (2000) Familial primary pulmonary hypertension (gene PPH1) is caused by mutations in the bone morphogenetic protein receptor-II gene. Am J Hum Genet 67:737–744

Denton CP, Sweny P, Abdulla A et al (1994) Acute renal failure occurring in scleroderma treated with cyclosporin A: a report of three cases. Br J Rheumatol 33:90–92

Derk CT, Grace E, Shenin M et al (2009) A prospective open-label study of mycophenolate mofetil for the treatment of diffuse systemic sclerosis. Rheumatology (Oxford) 48:1595–1599

Dib H, Tamby MC, Bussone G et al (2011) Targets of anti-endothelial cell antibodies in pulmonary hypertension and scleroderma. Eur Respir J 39(6):1405–1414

Diller GP, van Eijl S, Okonko DO et al (2008) Circulating endothelial progenitor cells in patients with Eisenmenger syndrome and idiopathic pulmonary arterial hypertension. Circulation 117:3020–3030

Dispenzieri A (2011) POEMS syndrome: 2011 update on diagnosis, risk-stratification, and management. Am J Hematol 86:591–601

Dorfmuller P, Zarka V, Durand-Gasselin I et al (2002) Chemokine RANTES in severe pulmonary arterial hypertension. Am J Respir Crit Care Med 165:534–539

Dorfmuller P, Perros F, Balabanian K et al (2003) Inflammation in pulmonary arterial hypertension. Eur Respir J 22:358–363

Dorfmuller P, Humbert M, Perros F et al (2007) Fibrous remodeling of the pulmonary venous system in pulmonary arterial hypertension associated with connective tissue diseases. Hum Pathol 38:893–902

dos Santos Fernandes CJ, Jardim CV, Hovnanian A et al (2010) Survival in schistosomiasis-associated pulmonary arterial hypertension. J Am Coll Cardiol 56:715–720

Drake JI, Bogaard HJ, Mizuno S et al (2011) Molecular signature of a right heart failure program in chronic severe pulmonary hypertension. Am J Respir Cell Mol Biol 45:1239–1247

Drugs.com (2010) Side effects and drug-drug interaction. Denver, CO. http://www.drugs.com, Updated: 2010

Efferth T, Schwarzl SM, Smith J et al (2006) Role of glucose-6-phosphate dehydrogenase for oxidative stress and apoptosis. Cell Death Differ 13:527–528, author reply 529–530

Ehrenreich H, Rieckmann P, Sinowatz F et al (1993) Potent stimulation of monocytic endothelin-1 production by HIV-1 glycoprotein 120. J Immunol 150:4601–4609

Ensoli B, Barillari G, Salahuddin SZ et al (1990) Tat protein of HIV-1 stimulates growth of cells derived from Kaposi's sarcoma lesions of AIDS patients. Nature 345:84–86

Fagan KA, Badesch DB (2002) Pulmonary hypertension associated with connective tissue disease. Prog Cardiovasc Dis 45:225–234

Fartoukh M, Emilie D, Le Gall C et al (1998) Chemokine macrophage inflammatory protein-1alpha mRNA expression in lung biopsy specimens of primary pulmonary hypertension. Chest 114:50S–51S

Fernandes CJ, Jardim CV, Hovnanian A et al (2011) Schistosomiasis and pulmonary hypertension. Expert Rev Respir Med 5:675–681

Filaci G, Cutolo M, Scudeletti M et al (1999) Cyclosporin A and iloprost treatment of systemic sclerosis: clinical results and interleukin-6 serum changes after 12 months of therapy. Rheumatology (Oxford) 38:992–996

Fois E, Le Guern V, Dupuy A et al (2010) Noninvasive assessment of systolic pulmonary artery pressure in systemic lupus erythematosus: retrospective analysis of 93 patients. Clin Exp Rheumatol 28:836–841

Freitas TC, Jung E, Pearce EJ (2007) TGF-beta signaling controls embryo development in the parasitic flatworm Schistosoma mansoni. PLoS Pathog 3:e52

Fritzler MJ, Hart DA, Wilson D et al (1995) Antibodies to fibrin bound tissue type plasminogen activator in systemic sclerosis. J Rheumatol 22:1688–1693

Fujita M, Shannon JM, Irvin CG et al (2001) Overexpression of tumor necrosis factor-alpha produces an increase in lung volumes and pulmonary hypertension. Am J Physiol Lung Cell Mol Physiol 280:L39–L49

Galie N, Hoeper MM, Humbert M et al (2009) Guidelines for the diagnosis and treatment of pulmonary hypertension: the Task Force for the Diagnosis and Treatment of Pulmonary Hypertension of the European Society of Cardiology (ESC) and the European Respiratory Society (ERS), endorsed by the International Society of Heart and Lung Transplantation (ISHLT). Eur Heart J 30:2493–2537

Gan CT, McCann GP, Marcus JT et al (2006) NT-proBNP reflects right ventricular structure and function in pulmonary hypertension. Eur Respir J 28:1190–1194

Gerasimovskaya EV, Kratzer A, Sidiakova A et al (2012) Interplay of macrophages and T cells in the lung vasculature. Am J Physiol Lung Cell Mol Physiol 302(10):L1014–L1022

Gerbino AJ, Goss CH, Molitor JA (2008) Effect of mycophenolate mofetil on pulmonary function in scleroderma-associated interstitial lung disease. Chest 133:455–460

Goppelt-Struebe M, Wolter D, Resch K (1989) Glucocorticoids inhibit prostaglandin synthesis not only at the level of phospholipase A2 but also at the level of cyclo-oxygenase/PGE isomerase. Br J Pharmacol 98:1287–1295

Gourh P, Arnett FC, Assassi S et al (2009) Plasma cytokine profiles in systemic sclerosis: associations with autoantibody subsets and clinical manifestations. Arthritis Res Ther 11:R147

Graham BB, Bandeira AP, Morrell NW et al 2010) Schistosomiasis-associated pulmonary hypertension: pulmonary vascular disease: the global perspective. Chest 137:20S–29S

Grigolo B, Mazzetti I, Meliconi R et al (2000) Anti-topoisomerase II alpha autoantibodies in systemic sclerosis-association with pulmonary hypertension and HLA-B35. Clin Exp Immunol 121:539–543

Grinyo JM, Cruzado JM, Millan O et al (2004) Low-dose cyclosporine with mycophenolate mofetil induces similar calcineurin activity and cytokine inhibition as does standard-dose cyclosporine in stable renal allografts. Transplantation 78:1400–1403

Groen H, Bootsma H, Postma DS et al (1993) Primary pulmonary hypertension in a patient with systemic lupus erythematosus: partial improvement with cyclophosphamide. J Rheumatol 20:1055–1057

Guignabert C, Izikki M, Tu LI et al (2006) Transgenic mice overexpressing the 5-hydroxytryptamine transporter gene in smooth muscle develop pulmonary hypertension. Circ Res 98:1323–1330

Guignabert C, Tu L, Izikki M et al (2009) Dichloroacetate treatment partially regresses established pulmonary hypertension in mice with SM22alpha-targeted overexpression of the serotonin transporter. FASEB J 23:4135–4147

Hachulla E, Gressin V, Guillevin L et al (2005) Early detection of pulmonary arterial hypertension in systemic sclerosis: a French nationwide prospective multicenter study. Arthritis Rheum 52:3792–3800

Hachulla AL, Launay D, Gaxotte V et al (2009) Cardiac magnetic resonance imaging in systemic sclerosis: a cross-sectional observational study of 52 patients. Ann Rheum Dis 68:1878–1884

Hagen M, Fagan K, Steudel W et al (2007) Interaction of interleukin-6 and the BMP pathway in pulmonary smooth muscle. Am J Physiol Lung Cell Mol Physiol 292:L1473–L1479

Hall S, Brogan P, Haworth SG et al (2009) Contribution of inflammation to the pathology of idiopathic pulmonary arterial hypertension in children. Thorax 64:778–783

Hamada H, Terai M, Kimura H et al (1999) Increased expression of mast cell chymase in the lungs of patients with congenital heart disease associated with early pulmonary vascular disease. Am J Respir Crit Care Med 160:1303–1308

Hammer SM, Eron JJ Jr, Reiss P et al (2008) Antiretroviral treatment of adult HIV infection: 2008 recommendations of the International AIDS Society-USA panel. JAMA 300:555–570

Hassoun PM (2009) Pulmonary arterial hypertension complicating connective tissue diseases. Semin Respir Crit Care Med 30:429–439

Hassoun PM, Mouthon L, Barbera JA et al (2009) Inflammation, growth factors, and pulmonary vascular remodeling. J Am Coll Cardiol 54:S10–S19

Heath D, Yacoub M (1991) Lung mast cells in plexogenic pulmonary arteriopathy. J Clin Pathol 44:1003–1006

Heldin CH (1992) Structural and functional studies on platelet-derived growth factor. EMBO J 11:4251–4259

Henke-Gendo C, Mengel M, Hoeper MM et al (2005) Absence of Kaposi's sarcoma-associated herpesvirus in patients with pulmonary arterial hypertension. Am J Respir Crit Care Med 172:1581–1585

Henriques-Coelho T, Brandao-Nogueira A, Moreira-Goncalves D et al (2008) Effects of TNF-alpha blockade in monocrotaline-induced pulmonary hypertension. Rev Port Cardiol 27:341–348

Heresi GA, Minai OA (2007) Lupus-associated pulmonary hypertension: long-term response to vasoactive therapy. Respir Med 101:2099–2107

Hoeper MM, Oudiz RJ, Peacock A et al (2004) End points and clinical trial designs in pulmonary arterial hypertension: clinical and regulatory perspectives. J Am Coll Cardiol 43:48S–55S

Hoyles RK, Ellis RW, Wellsbury J et al (2006) A multicenter, prospective, randomized, double-blind, placebo-controlled trial of corticosteroids and intravenous cyclophosphamide followed by oral azathioprine for the treatment of pulmonary fibrosis in scleroderma. Arthritis Rheum 54:3962–3970

Hsue PY, Deeks SG, Farah HH et al (2008) Role of HIV and human herpesvirus-8 infection in pulmonary arterial hypertension. AIDS 22:825–833

Huertas A, Tu L, Gambaryan N et al (2012) Leptin and regulatory T lymphocytes in idiopathic pulmonary arterial hypertension. Eur Respir J 40(4):895–904

Humbert M (2008) Mediators involved in HIV-related pulmonary arterial hypertension. AIDS 22 (Suppl 3):S41–S47

Humbert M, Monti G, Brenot F et al (1995) Increased interleukin-1 and interleukin-6 serum concentrations in severe primary pulmonary hypertension. Am J Respir Crit Care Med 151:1628–1631

Humbert M, Monti G, Fartoukh M et al (1998) Platelet-derived growth factor expression in primary pulmonary hypertension: comparison of HIV seropositive and HIV seronegative patients. Eur Respir J 11:554–559

Humbert M, Morrell NW, Archer SL et al (2004) Cellular and molecular pathobiology of pulmonary arterial hypertension. J Am Coll Cardiol 43:13S–24S

Imaizumi T, Yoshida H, Satoh K (2004) Regulation of CX3CL1/fractalkine expression in endothelial cells. J Atheroscler Thromb 11:15–21

Jais X, Launay D, Yaici A et al (2008) Immunosuppressive therapy in lupus- and mixed connective tissue disease-associated pulmonary arterial hypertension: a retrospective analysis of twenty-three cases. Arthritis Rheum 58:521–531

Jouve P, Humbert M, Chauveheid MP et al (2007) POEMS syndrome-related pulmonary hypertension is steroid-responsive. Respir Med 101:353–355

Karmochkine M, Wechsler B, Godeau P et al (1996) Improvement of severe pulmonary hypertension in a patient with SLE. Ann Rheum Dis 55:561–562

Katayose D, Ohe M, Yamauchi K et al (1993) Increased expression of PDGF A- and B-chain genes in rat lungs with hypoxic pulmonary hypertension. Am J Physiol 264:L100–L106

Kawaguchi Y, Hara M, Harigai M et al (1998) Corticosteroid pulse therapy in a patient with SLE and pulmonary hypertension. Clin Exp Rheumatol 16:510

Kawut SM, Taichman DB, Archer-Chicko CL et al (2003) Hemodynamics and survival in patients with pulmonary arterial hypertension related to systemic sclerosis. Chest 123:344–350

Kherbeck N, Tamby MC, Bussone G et al (2011) The role of inflammation and autoimmunity in the pathophysiology of pulmonary arterial hypertension. Clin Rev Allergy Immunol 44 (1):31–38

King CH (2010) Parasites and poverty: the case of schistosomiasis. Acta Trop 113:95–104

Kobayashi T, Momoi Y, Iwasaki T (2007) Cyclosporine A inhibits the mRNA expressions of IL-2, IL-4 and IFN-gamma, but not TNF-alpha, in canine mononuclear cells. J Vet Med Sci 69:887–892

Koebel CM, Vermi W, Swann JB et al (2007) Adaptive immunity maintains occult cancer in an equilibrium state. Nature 450:903–907

Koulmann N, Novel-Chate V, Peinnequin A et al (2006) Cyclosporin A inhibits hypoxia-induced pulmonary hypertension and right ventricle hypertrophy. Am J Respir Crit Care Med 174:699–705

Kowal-Bielecka O, Landewe R, Avouac J et al (2009) EULAR recommendations for the treatment of systemic sclerosis: a report from the EULAR Scleroderma Trials and Research group (EUSTAR). Ann Rheum Dis 68:620–628

Labarrere CA, Zaloga GP (2004) C-reactive protein: from innocent bystander to pivotal mediator of atherosclerosis. Am J Med 117:499–507

Lambertucci JR, Serufo JC, Gerspacher-Lara R et al (2000) Schistosoma mansoni: assessment of morbidity before and after control. Acta Trop 77:101–109

Lane KB, Machado RD, Pauciulo MW et al (2000) Heterozygous germline mutations in BMPR2, encoding a TGF-beta receptor, cause familial primary pulmonary hypertension. Nat Genet 26:81–84

Lapa M, Dias B, Jardim C et al (2009) Cardiopulmonary manifestations of hepatosplenic schistosomiasis. Circulation 119:1518–1523

Launay D, Souza R, Guillevin L et al (2006) Pulmonary arterial hypertension in ANCA-associated vasculitis. Sarcoidosis Vasc Diffuse Lung Dis 23:223–228

Launay D, Hachulla E, Hatron PY et al (2007) Pulmonary arterial hypertension: a rare complication of primary Sjogren syndrome: report of 9 new cases and review of the literature. Medicine (Baltimore) 86:299–315

Le Pavec J, Humbert M, Mouthon L et al (2011) Systemic sclerosis-associated pulmonary arterial hypertension. Am J Respir Crit Care Med 181:1285–1293

Lesprit P, Godeau B, Authier FJ et al (1998) Pulmonary hypertension in POEMS syndrome: a new feature mediated by cytokines. Am J Respir Crit Care Med 157:907–911

Leung DY, Bloom JW (2003) Update on glucocorticoid action and resistance. J Allergy Clin Immunol 111:3–22, quiz 23

Lorenzen JM, Nickel N, Kramer R et al (2011) Osteopontin in patients with idiopathic pulmonary hypertension. Chest 139:1010–1017

Lovell DJ, Giannini EH, Reiff A et al (2000) Etanercept in children with polyarticular juvenile rheumatoid arthritis. Pediatric Rheumatology Collaborative Study Group. N Engl J Med 342:763–769

Luster AD (1998) Chemokines – chemotactic cytokines that mediate inflammation. N Engl J Med 338:436–445

Machado RD, Pauciulo MW, Thomson JR et al (2001) BMPR2 haploinsufficiency as the inherited molecular mechanism for primary pulmonary hypertension. Am J Hum Genet 68:92–102

Macian F (2005) NFAT proteins: key regulators of T-cell development and function. Nat Rev Immunol 5:472–484

Mandagere A, Core B, Bird S, Bingham J, Boinpally R (2010) Absence of a clinically relevant pharmacokinetic interaction between ambrisentan and mycophenolate mofetil. In: American Journal of Respiratory and Critical Care Medicine. American Thoracic Society International Conference, New Orleans, LA, USA

Manno R, Boin F (2010) Immunotherapy of systemic sclerosis. Immunotherapy 2:863–878

Marder W, McCune WJ (2007) Advances in immunosuppressive therapy. Semin Respir Crit Care Med 28:398–417

Mathai SC, Hassoun PM (2011) Pulmonary arterial hypertension associated with systemic sclerosis. Expert Rev Respir Med 5:267–279

Mathai SC, Hummers LK, Champion HC et al (2009) Survival in pulmonary hypertension associated with the scleroderma spectrum of diseases: impact of interstitial lung disease. Arthritis Rheum 60:569–577

McCune WJ, Golbus J, Zeldes W et al (1988) Clinical and immunologic effects of monthly administration of intravenous cyclophosphamide in severe systemic lupus erythematosus. N Engl J Med 318:1423–1431

McKinsey TA, Olson EN (2005) Toward transcriptional therapies for the failing heart: chemical screens to modulate genes. J Clin Invest 115:538–546

McMurtry MS, Bonnet S, Wu X et al (2004) Dichloroacetate prevents and reverses pulmonary hypertension by inducing pulmonary artery smooth muscle cell apoptosis. Circ Res 95:830–840

McMurtry MS, Bonnet S, Michelakis ED et al (2007) Statin therapy, alone or with rapamycin, does not reverse monocrotaline pulmonary arterial hypertension: the rapamycin-atorvastatin-simvastatin study. Am J Physiol Lung Cell Mol Physiol 293:L933–L940

Mehta NJ, Khan IA, Mehta RN et al (2000) HIV-Related pulmonary hypertension: analytic review of 131 cases. Chest 118:1133–1141

Michelakis ED, McMurtry MS, Wu XC et al (2002) Dichloroacetate, a metabolic modulator, prevents and reverses chronic hypoxic pulmonary hypertension in rats: role of increased expression and activity of voltage-gated potassium channels. Circulation 105:244–250

Minai OA (2009) Pulmonary hypertension in polymyositis-dermatomyositis: clinical and hemodynamic characteristics and response to vasoactive therapy. Lupus 18:1006–1010

Montani D, Achouh L, Marcelin AG et al (2005a) Reversibility of pulmonary arterial hypertension in HIV/HHV8-associated Castleman's disease. Eur Respir J 26:969–972

Montani D, Marcelin AG, Sitbon O et al (2005b) Human herpes virus 8 in HIV and non-HIV infected patients with pulmonary arterial hypertension in France. AIDS 19:1239–1240

Montani D, Humbert M, Souza R (2011) Letter by Montani et al regarding article, "Elevated levels of inflammatory cytokines predict survival in idiopathic and familial pulmonary arterial hypertension". Circulation 123:e614, author reply e615

Morelli S, Giordano M, De Marzio P et al (1993) Pulmonary arterial hypertension responsive to immunosuppressive therapy in systemic lupus erythematosus. Lupus 2:367–369

Morse JH, Barst RJ, Fotino M et al (1997) Primary pulmonary hypertension, tissue plasminogen activator antibodies, and HLA-DQ7. Am J Respir Crit Care Med 155:274–278

Mouthon L, Guillevin L, Humbert M (2005) Pulmonary arterial hypertension: an autoimmune disease? Eur Respir J 26:986–988

Mukerjee D, Kingdon E, Vanderpump M et al (2003a) Pathophysiological insights from a case of reversible pulmonary arterial hypertension. J R Soc Med 96:403–404

Mukerjee D, St George D, Coleiro B et al (2003b) Prevalence and outcome in systemic sclerosis associated pulmonary arterial hypertension: application of a registry approach. Ann Rheum Dis 62:1088–1093

Mutschler D, Wikstrom G, Lind L et al (2006) Etanercept reduces late endotoxin-induced pulmonary hypertension in the pig. J Interferon Cytokine Res 26:661–667

Nadashkevich O, Davis P, Fritzler M et al (2006) A randomized unblinded trial of cyclophosphamide versus azathioprine in the treatment of systemic sclerosis. Clin Rheumatol 25:205–212

Nagaya N, Nishikimi T, Uematsu M et al (2000) Plasma brain natriuretic peptide as a prognostic indicator in patients with primary pulmonary hypertension. Circulation 102:865–870

Nannini C, West CP, Erwin PJ et al (2008) Effects of cyclophosphamide on pulmonary function in patients with scleroderma and interstitial lung disease: a systematic review and meta-analysis of randomized controlled trials and observational prospective cohort studies. Arthritis Res Ther 10:R124

Nathan C (2002) Points of control in inflammation. Nature 420:846–852

Negi VS, Tripathy NK, Misra R et al (1998) Antiendothelial cell antibodies in scleroderma correlate with severe digital ischemia and pulmonary arterial hypertension. J Rheumatol 25:462–466

Nicolls MR, Taraseviciene-Stewart L, Rai PR et al (2005) Autoimmunity and pulmonary hypertension: a perspective. Eur Respir J 26:1110–1118

Nishimura T, Faul JL, Berry GJ et al (2001) 40-O-(2-hydroxyethyl)-rapamycin attenuates pulmonary arterial hypertension and neointimal formation in rats. Am J Respir Crit Care Med 163:498–502

Niu X, Nouraie M, Campbell A et al (2009) Angiogenic and inflammatory markers of cardiopulmonary changes in children and adolescents with sickle cell disease. PLoS One 4:e7956

Nunes H, Humbert M, Sitbon O et al (2003) Prognostic factors for survival in human immunodeficiency virus-associated pulmonary arterial hypertension. Am J Respir Crit Care Med 167:1433–1439

Nunes H, Humbert M, Capron F et al (2006) Pulmonary hypertension associated with sarcoidosis: mechanisms, haemodynamics and prognosis. Thorax 61:68–74

Ogawa A, Nakamura K, Matsubara H et al (2005) Prednisolone inhibits proliferation of cultured pulmonary artery smooth muscle cells of patients with idiopathic pulmonary arterial hypertension. Circulation 112:1806–1812

Ogawa A, Nakamura K, Mizoguchi H et al (2011) Prednisolone ameliorates idiopathic pulmonary arterial hypertension. Am J Respir Crit Care Med 183:139–140

Okano Y, Steen VD, Medsger TA Jr (1992) Autoantibody to U3 nucleolar ribonucleoprotein (fibrillarin) in patients with systemic sclerosis. Arthritis Rheum 35:95–100

Opravil M, Sereni D (2008) Natural history of HIV-associated pulmonary arterial hypertension: trends in the HAART era. AIDS 22(Suppl 3):S35–S40

Opravil M, Pechere M, Speich R et al (1997) HIV-associated primary pulmonary hypertension. A case control study. Swiss HIV Cohort Study. Am J Respir Crit Care Med 155:990–995

Overbeek MJ, Lankhaar JW, Westerhof N et al (2008) Right ventricular contractility in systemic sclerosis-associated and idiopathic pulmonary arterial hypertension. Eur Respir J 31:1160–1166

Palucka AK, Blanck JP, Bennett L et al (2005) Cross-regulation of TNF and IFN-alpha in autoimmune diseases. Proc Natl Acad Sci USA 102:3372–3377

Paulin R, Meloche J, Jacob MH et al (2011) Dehydroepiandrosterone inhibits the Src/STAT3 constitutive activation in pulmonary arterial hypertension. Am J Physiol Heart Circ Physiol 301:H1798–H1809

Pellicelli AM, Palmieri F, D'Ambrosio C et al (1998) Role of human immunodeficiency virus in primary pulmonary hypertension – case reports. Angiology 49:1005–1011

Perros F, Dorfmuller P, Souza R et al (2007) Dendritic cell recruitment in lesions of human and experimental pulmonary hypertension. Eur Respir J 29:462–468

Perros F, Montani D, Dorfmuller P et al (2011) [Novel immunopathological approaches to pulmonary arterial hypertension]. Presse Med 40(Suppl 1):1S3–1S13

Perros F, Dorfmuller P, Montani D et al (2012) Pulmonary lymphoid neogenesis in idiopathic pulmonary arterial hypertension. Am J Respir Crit Care Med 185:311–321

Piao L, Fang YH, Cadete VJ et al (2010a) The inhibition of pyruvate dehydrogenase kinase improves impaired cardiac function and electrical remodeling in two models of right ventricular hypertrophy: resuscitating the hibernating right ventricle. J Mol Med (Berl) 88:47–60

Piao L, Marsboom G, Archer SL (2010b) Mitochondrial metabolic adaptation in right ventricular hypertrophy and failure. J Mol Med (Berl) 88:1011–1020

Pines A, Kaplinsky N, Goldhammer E et al (1982) Corticosteroid responsive pulmonary hypertension in systemic lupus erythematosus. Clin Rheumatol 1:301–304

Pope J (2008) An update in pulmonary hypertension in systemic lupus erythematosus – do we need to know about it? Lupus 17:274–277

Pope JE, Bellamy N, Seibold JR et al (2001) A randomized, controlled trial of methotrexate versus placebo in early diffuse scleroderma. Arthritis Rheum 44:1351–1358

Price LC, Montani D, Tcherakian C et al (2011) Dexamethasone reverses monocrotaline-induced pulmonary arterial hypertension in rats. Eur Respir J 37:813–822

Price LC, Wort SJ, Perros F et al (2012) Inflammation in pulmonary arterial hypertension. Chest 141:210–221

Pullamsetti SS, Savai R, Janssen W et al (2011) Inflammation, immunological reaction and role of infection in pulmonary hypertension. Clin Microbiol Infect 17:7–14

Quarck R, Nawrot T, Meyns B et al (2009) C-reactive protein: a new predictor of adverse outcome in pulmonary arterial hypertension. J Am Coll Cardiol 53:1211–1218

Quismorio FP Jr, Sharma O, Koss M et al (1984) Immunopathologic and clinical studies in pulmonary hypertension associated with systemic lupus erythematosus. Semin Arthritis Rheum 13:349–359

Rhen T, Cidlowski JA (2005) Antiinflammatory action of glucocorticoids – new mechanisms for old drugs. N Engl J Med 353:1711–1723

Ribeiro JM, Lucas M, Victorino RM (2001) Remission of precapillary pulmonary hypertension in systemic lupus erythematosus. J R Soc Med 94:32–33

Rich S, Kieras K, Hart K et al (1986) Antinuclear antibodies in primary pulmonary hypertension. J Am Coll Cardiol 8:1307–1311

Ross AG, Bartley PB, Sleigh AC et al (2002) Schistosomiasis. N Engl J Med 346:1212–1220

Rubin LJ (1997) Primary pulmonary hypertension. N Engl J Med 336:111–117

Ryan J, Bloch K, Archer SL (2011) Rodent models of pulmonary hypertension: harmonisation with the world health organisation's categorisation of human PH. Int J Clin Pract Suppl (172):15–34

Salojin KV, Bordron A, Nassonov EL et al (1997) Anti-endothelial cell antibody, thrombomodulin, and von Willebrand factor in idiopathic inflammatory myopathies. Clin Diagn Lab Immunol 4:519–521

Sanchez O, Sitbon O, Jais X et al (2006) Immunosuppressive therapy in connective tissue diseases-associated pulmonary arterial hypertension. Chest 130:182–189

Sanchez O, Marcos E, Perros F et al (2007) Role of endothelium-derived CC chemokine ligand 2 in idiopathic pulmonary arterial hypertension. Am J Respir Crit Care Med 176:1041–1047

Savale L, Tu L, Rideau D et al (2009) Impact of interleukin-6 on hypoxia-induced pulmonary hypertension and lung inflammation in mice. Respir Res 10:6

Schall TJ, Bacon K, Toy KJ et al (1990) Selective attraction of monocytes and T lymphocytes of the memory phenotype by cytokine RANTES. Nature 347:669–671

Schall TJ, Bacon K, Camp RD et al (1993) Human macrophage inflammatory protein alpha (MIP-1 alpha) and MIP-1 beta chemokines attract distinct populations of lymphocytes. J Exp Med 177:1821–1826

Schermuly RT, Dony E, Ghofrani HA et al (2005) Reversal of experimental pulmonary hypertension by PDGF inhibition. J Clin Invest 115:2811–2821

Schwartz BG, Kloner RA (2010) Drug interactions with phosphodiesterase-5 inhibitors used for the treatment of erectile dysfunction or pulmonary hypertension. Circulation 122:88–95

Scirica BM, Morrow DA (2006) Is C-reactive protein an innocent bystander or proatherogenic culprit? The verdict is still out. Circulation 113:2128–2134, discussion 2151

Sgonc R, Gruschwitz MS, Boeck G et al (2000) Endothelial cell apoptosis in systemic sclerosis is induced by antibody-dependent cell-mediated cytotoxicity via CD95. Arthritis Rheum 43:2550–2562

Shibasaki F, Fukami K, Fukui Y et al (1994) Phosphatidylinositol 3-kinase binds to alpha-actinin through the p85 subunit. Biochem J 302(Pt 2):551–557

Simonneau G, Robbins IM, Beghetti M et al (2009) Updated clinical classification of pulmonary hypertension. J Am Coll Cardiol 54:S43–S54

Sitbon O, Lascoux-Combe C, Delfraissy JF et al (2008) Prevalence of HIV-related pulmonary arterial hypertension in the current antiretroviral therapy era. Am J Respir Crit Care Med 177:108–113

Song Y, Coleman L, Shi J et al (2008) Inflammation, endothelial injury, and persistent pulmonary hypertension in heterozygous BMPR2-mutant mice. Am J Physiol Heart Circ Physiol 295: H677–H690

Soon E, Holmes AM, Treacy CM et al (2010) Elevated levels of inflammatory cytokines predict survival in idiopathic and familial pulmonary arterial hypertension. Circulation 122:920–927

Speich R, Boehler A, Rochat T et al (2000) Cystic fibrosis: current therapy. Indications for lung transplantation. Schweiz Med Wochenschr Suppl 122:57S–58S

Spence R, Mandagere A, Richards DB et al (2010) Potential for pharmacokinetic interactions between ambrisentan and cyclosporine. Clin Pharmacol Ther 88:513–520

Steen VD (2005) Autoantibodies in systemic sclerosis. Semin Arthritis Rheum 35:35–42

Steiner MK, Syrkina OL, Kolliputi N et al (2009) Interleukin-6 overexpression induces pulmonary hypertension. Circ Res 104:236–244, 228p following 244

Stojanovich L, Stojanovich R, Kostich V et al (2003) Neuropsychiatric lupus favourable response to low dose i.v. cyclophosphamide and prednisolone (pilot study). Lupus 12:3–7

Sutendra G, Dromparis P, Bonnet S et al (2011) Pyruvate dehydrogenase inhibition by the inflammatory cytokine TNFalpha contributes to the pathogenesis of pulmonary arterial hypertension. J Mol Med (Berl) 89:771–783

Suzuki C, Takahashi M, Morimoto H et al (2006) Mycophenolate mofetil attenuates pulmonary arterial hypertension in rats. Biochem Biophys Res Commun 349:781–788

Sztrymf B, Souza R, Bertoletti L et al (2010) Prognostic factors of acute heart failure in patients with pulmonary arterial hypertension. Eur Respir J 35:1286–1293

Tamby MC, Chanseaud Y, Humbert M et al (2005) Anti-endothelial cell antibodies in idiopathic and systemic sclerosis associated pulmonary arterial hypertension. Thorax 60:765–772

Tamby MC, Humbert M, Guilpain P et al (2006) Antibodies to fibroblasts in idiopathic and scleroderma-associated pulmonary hypertension. Eur Respir J 28:799–807

Tamosiuniene R, Tian W, Dhillon G et al (2011) Regulatory T cells limit vascular endothelial injury and prevent pulmonary hypertension. Circ Res 109:867–879

Tanaka E, Harigai M, Tanaka M et al (2002) Pulmonary hypertension in systemic lupus erythematosus: evaluation of clinical characteristics and response to immunosuppressive treatment. J Rheumatol 29:282–287

Taraseviciene-Stewart L, Nicolls MR, Kraskauskas D et al (2007) Absence of T cells confers increased pulmonary arterial hypertension and vascular remodeling. Am J Respir Crit Care Med 175:1280–1289

Tashkin DP, Elashoff R, Clements PJ et al (2006) Cyclophosphamide versus placebo in scleroderma lung disease. N Engl J Med 354:2655–2666

Terrier B, Tamby MC, Camoin L et al (2008) Identification of target antigens of antifibroblast antibodies in pulmonary arterial hypertension. Am J Respir Crit Care Med 177:1128–1134

Terrier B, Tamby MC, Camoin L et al (2009) Antifibroblast antibodies from systemic sclerosis patients bind to {alpha}-enolase and are associated with interstitial lung disease. Ann Rheum Dis 69:428–433

Thurnheer R, Jenni R, Russi EW et al (1997) Hyperthyroidism and pulmonary hypertension. J Intern Med 242:185–188

Torbicki A, Kurzyna M, Kuca P et al (2003) Detectable serum cardiac troponin T as a marker of poor prognosis among patients with chronic precapillary pulmonary hypertension. Circulation 108:844–848

Torre O, Harari S (2011) Pleural and pulmonary involvement in systemic lupus erythematosus. Presse Med 40:e19–e29

Tsuda K, Yamanaka K, Kitagawa H et al (2012) Calcineurin inhibitors suppress cytokine production from memory T cells and differentiation of naive T cells into cytokine-producing mature T cells. PLoS One 7:e31465

Tuder RM (2009) Pathology of pulmonary arterial hypertension. Semin Respir Crit Care Med 30:376–385

Tuder RM, Groves B, Badesch DB et al (1994) Exuberant endothelial cell growth and elements of inflammation are present in plexiform lesions of pulmonary hypertension. Am J Pathol 144:275–285

Tyndall AJ, Bannert B, Vonk M et al (2010) Causes and risk factors for death in systemic sclerosis: a study from the EULAR Scleroderma Trials and Research (EUSTAR) database. Ann Rheum Dis 69:1809–1815

Ulrich S, Nicolls MR, Taraseviciene L et al (2008a) Increased regulatory and decreased CD8+ cytotoxic T cells in the blood of patients with idiopathic pulmonary arterial hypertension. Respiration 75:272–280

Ulrich S, Taraseviciene-Stewart L, Huber LC et al (2008b) Peripheral blood B lymphocytes derived from patients with idiopathic pulmonary arterial hypertension express a different RNA pattern compared with healthy controls: a cross sectional study. Respir Res 9:20

Vajner L, Vytasek R, Lachmanova V et al (2006) Acute and chronic hypoxia as well as 7-day recovery from chronic hypoxia affects the distribution of pulmonary mast cells and their MMP-13 expression in rats. Int J Exp Pathol 87:383–391

Valmary S, Dorfmuller P, Montani D et al (2011) Human gamma-herpesviruses Epstein-Barr virus and human herpesvirus-8 are not detected in the lungs of patients with severe pulmonary arterial hypertension. Chest 139:1310–1316

van den Hoogen FH, Boerbooms AM, Swaak AJ et al (1996) Comparison of methotrexate with placebo in the treatment of systemic sclerosis: a 24 week randomized double-blind trial, followed by a 24 week observational trial. Br J Rheumatol 35:364–372

Venitz J, Zack J, Gillies H et al (2012) Clinical pharmacokinetics and drug-drug interactions of endothelin receptor antagonists in pulmonary arterial hypertension. J Clin Pharmacol 52(12):1784–1805

Wang W, Yan H, Zhu W et al (2009) Impairment of monocyte-derived dendritic cells in idiopathic pulmonary arterial hypertension. J Clin Immunol 29:705–713

Wang W, Wang YL, Chen XY et al (2011) Dexamethasone attenuates development of monocrotaline-induced pulmonary arterial hypertension. Mol Biol Rep 38:3277–3284

Yildirim-Toruner C, Diamond B (2011) Current and novel therapeutics in the treatment of systemic lupus erythematosus. J Allergy Clin Immunol 127:303–312, quiz 313–304

Zamora AC, Wolters PJ, Collard HR et al (2008) Use of mycophenolate mofetil to treat scleroderma-associated interstitial lung disease. Respir Med 102:150–155

Zheng Y, Li M, Zhang Y et al (2010) The effects and mechanisms of mycophenolate mofetil on pulmonary arterial hypertension in rats. Rheumatol Int 30:341–348

Zhou H (2005) Clinical pharmacokinetics of etanercept: a fully humanized soluble recombinant tumor necrosis factor receptor fusion protein. J Clin Pharmacol 45:490–497

Zitvogel L, Apetoh L, Ghiringhelli F et al (2008) The anticancer immune response: indispensable for therapeutic success? J Clin Invest 118:1991–2001

Vasoactive Peptides and the Pathogenesis of Pulmonary Hypertension: Role and Potential Therapeutic Application

Reshma S. Baliga, Raymond J. MacAllister, and Adrian J. Hobbs

Abstract Pulmonary hypertension (PH) is a debilitating disease with a dismal prognosis. Recent advances in therapy (e.g. prostacyclin analogues, endothelin receptor antagonists and phosphodiesterase 5 inhibitors), whilst significantly improving survival, simply delay the inexorable progression of the disease. An array of endogenous vasoconstrictors and vasodilators coordinates to maintain pulmonary vascular homeostasis and morphological integrity, and an imbalance in the expression and function of these mediators precipitates PH and related lung diseases. The vasodilator peptides, including natriuretic peptides, vasoactive intestinal peptide, calcitonin gene-related peptide and adrenomedullin, trigger the production of cyclic nucleotides (e.g. cGMP and cAMP) in many pulmonary cell types, which in tandem exert a multifaceted protection against the pathogenesis of PH, encompassing vasodilatation, inhibition of vascular smooth muscle proliferation, anti-inflammatory and anti-fibrotic effects and salutary actions on the right ventricle. This coordinated beneficial activity underpins a contemporary perception that to advance treatment of PH it is necessary to offset multiple disease mechanisms (i.e. the pulmonary vasoconstriction, pulmonary vascular remodelling, right ventricular dysfunction). Thus, there is considerable potential for harnessing the favourable activity of peptide mediators to offer a novel, efficacious therapeutic approach in PH.

R.S. Baliga • A.J. Hobbs (✉)
William Harvey Heart Centre, William Harvey Research Institute, Queen Mary University of London, Charterhouse Square, London EC1M 6BQ, UK
e-mail: a.j.hobbs@qmul.ac.uk

R.J. MacAllister
Centre for Clinical Pharmacology, University College London, The Rayne Building, 5 University Street, London WC1E 5JJ, UK

Keywords Pulmonary hypertension • Vasodilatation • Vasoconstriction • cGMP • cAMP • Endothelin-1 • Angiotensin II • Natriuretic peptides • Vasoactive intestinal peptide • Calcitonin gene-related peptide • Adrenomedullin • Somatostatin • Neutral endopeptidase • Phosphodiesterase

Contents

1	Introduction	478
2	Peptides and Pulmonary Homeostasis	479
3	Vasoconstrictor Peptides	479
	3.1 Endothelin-1	479
	3.2 Angiotensin II	480
4	Vasodilator Peptides	481
	4.1 Natriuretic Peptides	481
	4.2 Vasoactive Intestinal Peptide and Pituitary Adenylate Cyclase Activating Peptide	486
	4.3 Adrenomedullin	490
	4.4 Calcitonin Gene-Related Peptide	492
	4.5 Somatostatin	493
5	Conclusions	494
References		495

1 Introduction

Pulmonary hypertension (PH) is a multifactorial, progressive disease characterised by increased pulmonary vascular resistance (PVR), remodelling of the pulmonary small arteries, right ventricular hypertrophy (RVH) and ultimately right ventricular failure. The disease carries a substantial morbidity and mortality; even with the introduction of prostacyclin analogues (Barst et al. 1996), endothelin receptor antagonists (Channick et al. 2001) and phosphodiesterase 5 inhibitors (PDE5i) (Galie et al. 2005b), there remains no cure and clinical worsening is merely delayed, not prevented, by therapy (McLaughlin et al. 2009a, b). Advances in the understanding of the disparate aetiologies that underpin different forms of PH [for WHO classification please refer to (Simonneau et al. 2009)] have highlighted new signalling pathways and potentially novel therapeutic strategies that may improve the treatment of patients with the disease. One such area, in which considerable progress has been made in the past decade, is that of polypeptide hormones and paracrine mediators and their contribution to both pulmonary physiology and disease, particularly in the context of PH. These low molecular weight signalling peptides exert a multitude of activities that coordinate to maintain pulmonary vascular homeostasis and lung architectural integrity; moreover, dysfunctional bioactivity of these peptides also appears to underlie many aspect of PH pathogenesis. This overview provides an account of this family of pulmonary active peptides, with focus on those with vasodilator activity, and the potential for targeting these mediators therapeutically to improve management of this debilitating disorder with an extremely poor prognosis.

2 Peptides and Pulmonary Homeostasis

A plethora of systemic and lung-specific peptides are now recognised to play a role in pulmonary health and disease. PH is a disease of the small pulmonary arteries defined by an abnormal increase in pulmonary artery pressure (PAP) and PVR, which are primarily regulated locally within the lung, and influenced by many endogenous agents, both vasoconstrictors and vasodilators. Indeed, it is likely that an imbalance between the endogenous production and activity of vasoconstrictor and vasorelaxant polypeptides contributes significantly to the haemodynamic aberrations characteristic of PH. This is perhaps best exemplified by the excessive expression, release and actions of endothelin-1 (ET-1), but other endogenous vasoconstrictor peptides have been implicated in lung (patho)physiology including angiotensin II (Ang II) and somatostain-28 (SS-28). This family will be considered briefly below, but the main focus herein will be to describe the key function played by vasodilator peptides in governing pulmonary vascular and structural integrity and how aberrant activity of this cohort of peptides, including vasoactive intestinal peptide (VIP), natriuretic peptides (NP), pituitary adenylate cyclase activating peptides (PACAP), adrenomedullin (AM), calcitonin gene-related peptide (CGRP) and somatostatin-14 (SS-14), might lead to the development of PH. Moreover, consideration will be given to the possibility of therapeutic interventions targeted to these vasodilator peptides to advance therapy in this area of unmet medical need.

3 Vasoconstrictor Peptides

3.1 Endothelin-1

Endothlin-1 (ET-1) is the major cardiovascular isoform of the endothelin family, a 21 amino acid (aa) peptide with potent vasoconstrictor activity in most vascular beds (Yanagisawa et al. 1988). The gene product, prepro-ET, is cleaved by an endothelin converting enzyme (ECE) into big ET and subsequently to the biologically active ET-1 (Webb 1991). The biological activity of ET-1 is exerted via stimulation of two G-protein-coupled receptors; ET_A receptors on vascular smooth muscle cells cause vasoconstriction and proliferation, while ET_B receptors on endothelial cells stimulate NO and prostacyclin release (and clear ET-1 from the circulation), but on vascular smooth muscle cells induce vasoconstriction and mitogenesis. Plasma levels of ET-1 are significantly elevated and correlate with disease severity in experimental models and patients with PH (Rubens et al. 2001), and pulmonary expression of ET_A is also upregulated in the diseased lung (Motte et al. 2006). Experimental models of PH validated the use of endothelin receptor antagonists (ERAs), including bosentan (dual ET_A/ET_B), ambrisentan ($ET_A >$ ET_B) and sitaxsentan ($ET1_A \gg ET_B$), to treat PH (Chen et al. 1995; Iglarz et al. 2008; Ivy et al. 1997; Raja 2010; Sato et al. 1995; Underwood et al. 1998; Yuan et al. 2006). This efficacy has translated to the clinical arena where ERAs have been shown to improve pulmonary haemodynamics and exercise capacity (Barst et al. 2004; Channick et al. 2001; Galie et al. 2005a; Williamson et al. 2000). However, it remains to be determined if ERAs produce a positive effect on survival and whether selective ET_A receptor antagonists hold any clinical advantage over mixed receptor blockers (since inhibition of ET_B prevents the release of NO from the vascular endothelium and slows clearance of circulating ET-1, both potentially adverse actions).

3.2 Angiotensin II

The lung is the principal site of conversion of angiotensin I to Ang II under the influence of angiotensin converting enzyme (ACE) found in caveolae on the pulmonary vascular endothelium (Ryan et al. 1975); it is not surprising therefore that Ang II has been reported to contribute to the development of PH. For example, activation of the angiotensin receptor 1 (AT1R) leads to vasoconstriction, vascular smooth muscle proliferation, fibrosis, production of reactive oxygen species and inflammation in the lung, all processes thought to precipitate PH (Cargill and Lipworth 1995b; Morrell et al. 1995). Moreover, AT1R blockade and inhibition of ACE reduce the severity of PH in experimental hypoxia models (Morrell et al. 1995; Nong et al. 1996; Zhao et al. 1996a). Finally, single nucleotide polymorphisms (SNPs) in the angiotensinogen (precursor peptide) and ACE genes are associated with susceptibility to PH (Kanazawa et al. 2000; Solari and Puri 2004). Intriguingly, the recently identified ACE2, which hydrolyses Ang II to the biologically active fragment angiotensin$_{1-7}$ (Ang$_{1-7}$), appears to exert an opposing, cytoprotective effect in PH. Both genetic and pharmacological interventions targeted to this isoform of ACE appear to worsen the pathogenesis of experimental PH, suggesting that Ang$_{1-7}$ is an endogenous protective pathway (possibly via activation of the *Mas* receptor) (Ferreira et al. 2009; Shenoy et al. 2010; Yamazato et al. 2009).

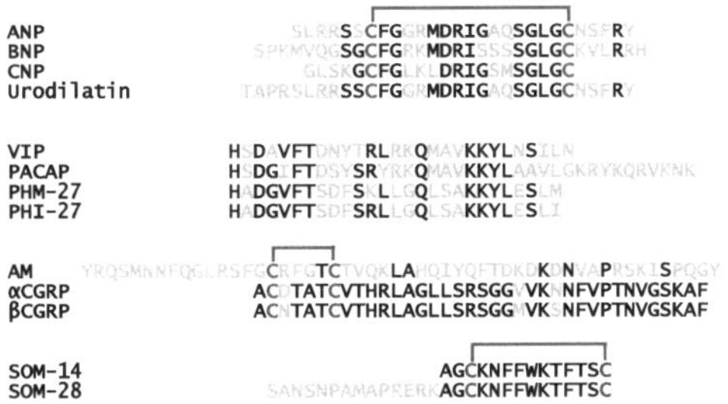

Fig. 1 Primary amino acid structure of the vasodilator peptides thought to play key roles in pulmonary vascular physiology and disease. *Red lines* indicate intramolecular disulphide bonds. Conserved residues are shown in *black*

ANP	SLRRS CFG HMDRIGA SGLGC NSFR
BNP	SPKMVQGSGCFGRKMDRI SSSGLGC VLR H
CNP	GLSKGCFGLKLDRIGSMSGLGC
Urodilatin	TAPRSLRRSSCFGGHMDRIGA SGLGCNSFR
VIP	H DVFT DNYTRLRKQMAVKKYLNS ILN
PACAP	H DGI FT DSYSRYRKQMAVKKYLAAVLGKRYKQRVKNK
PHM-27	H DGVFT SD SLQ SAKKYL ES M
PHI-27	H DGVFT SDFSRLLQS SAKKYLES I
AM	YRQSMNNFQGLRSFGCRFGTCTVQKLA HQIYQFTDKDKDNVAPRSKISPQGY
αCGRP	AC NTATCVTHRLAGLLSRSGG VK NNFVPTNVGSKAF
βCGRP	AC NTATCVTHRLAGLLSRSGG VK NNFVPTNVGSKAF
SOM-14	AGCKNFFWKTFTSC
SOM-28	SANSNPAMAPRERKAGCKNFFWKTFTSC

4 Vasodilator Peptides

4.1 Natriuretic Peptides

4.1.1 Structure and Expression

The NPs are a family of structurally similar endogenous vasoactive peptides: atrial natriuretic peptide (ANP), brain natriuretic peptide (BNP) and C-type natriuretic peptide (CNP) (Potter et al. 2006) (Fig. 1). Each peptide in the family is initially expressed as a prepro-natriuretic peptide and then enzymatically cleaved to its active form by the endopeptidases corin (ANP and BNP) and furin (CNP) (Wu et al. 2003; Yan et al. 2000). Mature human ANP is a 28 amino-acid disulphide-linked peptide that is released from atrial granules primarily in response to wall stretch resulting from increased intravascular volume, while the mature 32 amino-acid BNP is synthesised de novo and released from cardiac ventricles in response to the same stimulus (Ahluwalia et al. 2004; Potter et al. 2009; Yasue et al. 1994). In contrast, the 22 aa CNP is released from the vascular endothelium and regulates local blood flow in a paracrine fashion (Ahluwalia and Hobbs 2005; Lumsden et al. 2010).

ANP and BNP play important roles in maintaining cardiovascular and fluid homeostasis by promoting natriuresis and diuresis by the kidney (Ahluwalia et al. 2004; de Bold et al. 1981). Plasma levels of ANP are five- to tenfold higher than BNP in normal subjects (~5–10 fmol/ml versus ~1 fmol/ml); however, circulating concentrations of ANP and BNP increase rapidly and dramatically in response to haemodynamic alterations and cardiovascular disease (Potter et al. 2009). Indeed, BNP in particular is used as a diagnostic biomarker for heart failure and to gauge the effectiveness of therapy. The peptides represent an innate defence mechanism that brings about several cytoprotective effects including natriuresis, vasodilatation and anti-hypertrophic and anti-proliferative activity [particularly in the heart (Ahluwalia et al. 2004; Oliver et al. 1997; Potter et al. 2009)]. The biological actions of NPs are mediated via the activation of specific natriuretic peptide receptors (NPRs), NPR-A, NPR-B and NPR-C; NPR-A and NPR-B are activated with a rank order of potency ANP ≥ BNP ≫ CNP and CNP ≫ ANP = BNP, respectively, while NPR-C binds with high affinity to all three NPs (Bennett et al. 1991; Koller et al. 1991; Koller and Goeddel 1992).

4.1.2 Natriuretic Peptide Receptors

NPR-A and NPR-B are guanylate cyclase coupled receptors and facilitate the conversion of guanosine-5'-triphosphate to the second messenger cyclic guanosine-3',5'-monophosphate (cGMP), which governs cellular function via activation of specific ion channels, kinases and phosphodiesterases (Hobbs 1997).

The downstream effects of NPR-A stimulated cGMP formation are mediated by the same protein kinases, phosphodiesterases and channel proteins as nitric oxide (NO) and therefore exert complementary functions. NPR-A acts to reduce cardiac load, stimulating natriuresis, vasodilatation, diuresis, endothelial permeability and inhibition of the renin-angiotensin pathway (Dubois et al. 2000; Kishimoto et al. 1996; Lopez et al. 1997). The receptor also exerts a direct inhibitory effect on cardiac growth; mice lacking the NPR-A receptor are mildly hypertensive at rest, but show enhanced basal cardiac hypertrophy and fibrosis disproportionate to the systemic hypertension observed (Holtwick et al. 2003; Kishimoto et al. 2009; Kuhn et al. 2002). NPR-B is preferentially activated by CNP and is abundantly expressed in the brain, lung, bone and heart (Koller et al. 1991; Potter 2011).

NPR-B-mediated CNP signalling is essential for normal bone growth (Tamura et al. 2004). Inactivation of NPR-B causes dwarfism and early death (Chusho et al. 2001; Yasoda et al. 1998) and targeted overexpression of CNP can counteract dwarfism in a mouse model of achondroplasia (Yasoda et al. 2004). However, NPR-B activity is also critical to cardiovascular homeostasis, particularly with respect to cardiac integrity. For example, CNP produces anti-proliferative and anti-hypertrophic effects through NPR-B activation (Dickey et al. 2007; Qvigstad et al. 2010; Tokudome et al. 2004). Rats with a dominant negative NPR-BΔKC mutation display progressive, pressure-dependent cardiac hypertrophy which is exacerbated by volume overload (Langenickel et al. 2006; Pagel-Langenickel et al. 2007).

The third receptor, NPR-C, is homologous to NPR-A and NPR-B but devoid of guanylyl cyclase activity. This NPR acts as a clearance receptor as binding of NPs to NPR-C leads to internalisation and lysosomal degradation (Maack et al. 1987). However, it has also been demonstrated that NPR-C contains a 37 aa cytoplasmic domain with a Gα inhibitory protein-activating sequence and can couple to G_i/G_o-proteins to inhibit adenylate cyclase activity and activate phospholipase-Cβ3 (Anand-Srivastava et al. 1996). We have also shown CNP/NPR-C signalling as a key component of EDHF through the opening of G-protein-coupled inwardly rectifying K^+ channels (GIRK) (Chauhan et al. 2003; Hobbs et al. 2004; Villar et al. 2007). NPs are also degraded by neutral endopeptidase enzymes (and other peptidases including dipeptidyl peptidase and insulin degrading enzyme), which are found in high concentrations on the luminal surface of endothelial and epithelial cells (Roques 1998; Soleilhac et al. 1992); these are discussed further below.

4.1.3 Natriuretic Peptides and Pulmonary Hypertension

NPs and their cognate receptors have been localised to many areas of the lung, including peripheral lung tissues, airway epithelium and pulmonary circulation (Bianchi et al. 1985; Gutkowska et al. 1987; Gutkowska and Nemer 1989; Sakamoto et al. 1986; Tallerico-Melnyk et al. 1992; Toshimori et al. 1988). Accordingly, the natriuretic peptide axis has been reported to play a key role in lung physiology, regulating several processes including vasodilatation, bronchorelaxation, pulmonary permeability and surfactant production (Hulks et al. 1990; Perreault and Gutkowska 1995). ANP is a potent relaxant of pulmonary arteries (significantly greater than the systemic vasculature) (Angus et al. 1993; Baliga et al. 2008) and a potent bronchodilator in a number of species (Gutkowska et al. 1987; Gutkowska and Nemer 1989; Hulks et al. 1990). This is perhaps not surprising since we were the first to identify this pivotal role of cGMP-mediated control, i.e. the non-adrenergic, non-cholinergic (NANC) innervation of bronchial smooth muscle triggers the NO–cGMP pathway to govern airway tone (Tucker et al. 1990). Indeed, the bronchorelaxant activity of NPs in the lung may be dependent, at least in part, on NO release from the airway epithelium (Matera et al. 2011; Mohapatra et al. 2004).

Evidence for the importance of endogenous NPs in modulating the development of PH is demonstrated by the exacerbated PH and pulmonary fibrosis in numerous transgenic animals with aberrant natriuretic peptide signalling, including ANP ($Nppa^{-/-}$) and NPR-A ($Npr1^{-/-}$) mice (Chen et al. 2006; Klinger et al. 1999, 2002; Zhao et al. 1999). Similarly, chronic blockade of endogenous ANP by infusion of monoclonal antibodies stimulates the development of PH (Raffestin et al. 1992), while exogenously administered or adenovirus-mediated natriuretic peptide supplementation protects against PH and the accompanying RVH (Chen et al. 2006; Hill et al. 1994; Jin et al. 1990, 1988; Klinger et al. 1993; Louzier et al. 2001). Administration of urodilatin, an elongated ANP molecule thought to be expressed exclusively in the kidney, also reduces several indices of severity in experimental PH (Schermuly et al. 2001). Infusion of NPs has also been evaluated in small clinical studies involving healthy volunteers and patients with PH. For instance, ANP reduces the pulmonary vascular permeability index in patients with pulmonary oedema (Sakamoto et al. 2010), counteracts Ang II-mediated vasoconstriction in healthy subjects (Cargill and Lipworth 1995b) and reduces pulmonary pressure in altitude-induced PH (Cargill and Lipworth 1995a; Liu et al. 1989) and *cor pulmonale* (Cargill and Lipworth 1996). BNP administration also acutely reverses the pulmonary vasoconstriction in PH patients (Klinger et al. 2005, 2006). Natriuretic peptide receptors are also highly expressed in the lung. During hypoxia and PH, NPR-C mRNA and protein levels are significantly reduced, whereas NPR-A and NPR-B expression is relatively unaltered (Chen 2005; Sun et al. 2000, 2001). This raises the possibility that clearance of NPs from the circulation is intentionally slowed under chronic hypoxia to optimise the beneficial activity of this family of peptides in PH and other hypoxia-driven pathologies.

The short plasma half-life and negligible oral bioavailability make NPs (in common with most vasoactive peptides) poor candidates for drug therapy. An alternative strategy is to increase endogenous natriuretic peptide levels by (a) blocking clearance via NPR-C and (b) inhibiting neutral endopeptidase (NEP). Although NPR-C antagonists were developed in the 1990s, predominantly for left-sided heart failure (Veale et al. 2000), they have not been evaluated to date in PH. However, the strategy of NEP inhibition has been proven to be effective in animal models (Baliga et al. 2008; Klinger et al. 1993; Zhao et al. 1991). Indeed, our work has shown that PDE5 is pivotal in terminating the bioactivity of natriuretic peptide-induced cGMP signalling in the pulmonary vasculature, whereas this PDE isozyme is less of an important hydrolytic pathway for NPs in the systemic circulation (Baliga et al. 2008). A similar synergistic activity is observed between urodilatin and dipyridamole (Schermuly et al. 2001). Therefore, in PH patients, whose circulating natriuretic peptide concentrations are raised, it is likely that the beneficial effects of PDE5i are dependent on natriuretic peptide, rather than NO, bioactivity [since the latter pathway is deficient in PH (Humbert et al. 2004)]. In addition, this phenomenon accounts for the pulmonary selectivity of PDE5i [this thesis is supported by data from experimental models (Zhao et al. 2003)]. These observations also represent clear evidence that in PH release of NPs represents an intrinsic, host defence mechanism that slows pathogenesis. As such, the therapeutic potential of manipulating natriuretic peptide bioactivity to reverse the haemodynamic abnormalities associated with PH holds great promise. However, this is true not only for the haemodynamic dysfunction, but also for attenuating the pulmonary vascular re-modelling that also characterises the disease. This is exemplified by the fact that NPs inhibit pulmonary vascular smooth muscle proliferation and TGFβ-induced extracellular matrix expression in vitro and prevent structural changes in vivo in animal models of PH (Chen et al. 2006; Jin et al. 1990; Klinger et al. 1998, 1999; Li et al. 2007). Moreover, NPs (and cGMP per se) have been shown to play a key role in preventing the development of fibrosis in a number of organ systems, including the heart and kidney, exemplified best in mice with genetic deletions of the natriuretic peptide signalling pathway (Das et al. 2010; Ellmers et al. 2007; Knowles et al. 2001; Nishikimi et al. 2009; Oliver et al. 1997; Pandey et al. 1999; Tamura et al. 2000). Furthermore, administration of ANP or BNP halts the development of hepatic fibrosis (Ishigaki et al. 2009; Sonoyama et al. 2009), and direct activation of soluble guanylate cyclase, the NO receptor, is effective in reducing fibrosis in the heart, lung and kidney (Evgenov et al. 2011; Kalk et al. 2006; Masuyama et al. 2009). Thus, augmentation of natriuretic peptide bioactivity appears to fulfil a contemporary perception that to advance current therapy a multifaceted therapeutic approach is needed that (a) targets the lung, (b) reverses the pulmonary vasoconstriction, (c) prevents pulmonary vascular re-modelling and (d) halts the deterioration of the right ventricle.

4.1.4 Neutral Endopeptidase and Pulmonary Hypertension

Neutral endopeptidase (EC 3.4.24.11, neprilysin; CD10; common acute lymphoblastic leukaemia antigen) is an endothelial membrane-bound zinc metallopeptidase that is the major pathway for degradation of numerous vasoactive peptides, including NPs, AM, substance P, bradykinin, ET-1 and Ang II (Borson 1991; Thompson and Morice 1996; Richards et al. 1991). NEP has a wide organ (e.g. lung, gut, adrenal glands, brain and kidney) and cellular (endothelial cells, vascular smooth muscle cells, cardiac myocytes and fibroblasts) distribution (Erdos and Skidgel 1989; Graf et al. 1995; Llorens-Cortes et al. 1992; Raizada et al. 2002). NEP is also involved in processing big ET to ET-1, and so it is involved in both the activation and degradation of this peptide (Grantham et al. 2000; Murphy et al. 1994). The net effect of NEP inhibition on vascular tone therefore depends on whether the predominant substrate(s) degraded by NEP are vasodilators (e.g. NPs, AM) or vasoconstrictors (e.g. ET-1, Ang II) and the extent of NEP involvement in big ET processing. This critical balance is exemplified by studies in normotensive subjects, in whom NEP inhibitors do not lower systemic pressure, but augment the effects of ANP and lower blood pressure in hypertensive subjects (Richards et al. 1992, 1993; Ando et al. 1995; Laurent et al. 2000). NEP inhibitors have been used with limited success in hypertension and heart failure (Ando et al. 1995; Bevan et al. 1992; O'Connell et al. 1992; Ogihara et al. 1994; Rouleau et al. 2000). Tolerance to NP, altered expression of NPR receptors and activation of the renin-angiotensin–aldosterone system (RAAS) are possible explanations for this loss of chronic efficacy. Since the functional effects of natriuretic peptide augmentation can be limited by the concurrent augmentation of the RAAS system, combined ACE/NEP ('vasopeptidase') inhibitors have also been developed. Experimental studies and early clinical trials showed these drugs to have highly potent antihypertensive and anti-remodelling effects (Xu et al. 2004; Backlund et al. 2003; Blais et al. 2000; Burrell et al. 2000; Klapholz et al. 2001; McClean et al. 2001, 2002; Rouleau et al. 2000; Trippodo et al. 1999; Troughton et al. 2000). However, larger scale studies [e.g. Omapatrilat Versus Enalapril Randomized Trial of Utility in Reducing Events (OVERTURE) and Omapatrilat Cardiovascular Treatment Assessment Versus Enalapril (OCTAVE)] only revealed modest increases in efficacy (composite endpoints of CV death and hospital readmission) above placebo (ACE inhibitor alone); moreover, the vasopeptidase inhibitors were associated with an unacceptably high incidence of angioedema, likely due to the 'double hit' on bradykinin metabolism, which led to their withdrawal from clinical development (Packer et al. 2002; Solomon et al. 2005; Zanchi et al. 2003). Interestingly, recent studies suggest substituting AT_1R antagonists for ACEi can produce similar antihypertensive and anti-remodelling effects without promoting angioedema (Hegde et al. 2011).

Interestingly, though NEP inhibitors are ineffective in lowering after-load in HF, they do lower the pulmonary capillary wedge pressure (Ando et al. 1995). Further, mice lacking NEP are less susceptible to hypoxia-induced pulmonary vascular leak (Irwin et al. 2005). Experimental studies show that inhibition of NEP can decrease hypoxia-induced PH and vascular remodelling (Klinger et al. 1993; Thompson et al. 1994a, b; Winter et al. 1991) by potentiating the action of NPs (Seymour et al. 1995; Valentin et al. 1992; Yamamoto et al. 2004; Zhao et al. 1992). NEP is abundantly expressed in cardio-pulmonary tissue and can also be upregulated in injured lung and heart failure (Abassi et al. 1995; Borson 1991; Cohen et al. 1998; Hashimoto et al. 2010; Trippodo et al. 1991). Thus targeting NEP produces selective pulmonary effects in PH as a result of bolstering natriuretic peptide activity. The strategy of targeting NEP for the treatment of PH may also have the added benefit of slowing the breakdown of other protective peptides that will contribute to efficacy, including AM and VIP.

In addition to NEP, other peptidases involved in the catabolism of vasoactive peptides might be exploited in a similar strategy. ACE and ECE have been discussed briefly above. Another widely distributed peptidase with multi-potent activity is dipeptidyl peptidase 4 (DPP-4). DPP-4 inhibitors have recently been developed as novel drugs for the treatment of type 2 diabetes and might have efficacy in PH (Chrysant and Chrysant 2012).

4.2 Vasoactive Intestinal Peptide and Pituitary Adenylate Cyclase Activating Peptide

4.2.1 Structure and Expression

VIP is a 28 aa peptide originally isolated from porcine duodenum (Said and Mutt 1970a, b). VIP causes vascular and non-vascular smooth muscle relaxation via activation of adenylate cyclase and the production of cyclic adenosine-3′, 5′-monophosphate (cAMP) (Desbuquois et al. 1973). VIP is expressed in many organs and tissues, particularly in the central and peripheral nervous systems as a parasympathetic co-transmitter (Fahrenkrug and Hannibal 2004; Larsson et al. 1976; Van Geldre and Lefebvre 2004). The VIP gene is found on human Chromosome 6 (Tsukada et al. 1985) and, akin to the NPs, is expressed initially as a prepro-precursor of 170 aas that is post-transcriptionally pruned to the principal biologically active (28 aa) form, and a related 27 aa variant termed peptide histidine methionine [PHM; peptide histidine isoleucine (PHI) in rodents] (Itoh et al. 1983).

Pituitary adenylate cyclase activating protein (PACAP) is a 38 amino-acid peptide related to VIP (68 % homology; Fig. 1) that was identified following a concerted effort to search for novel hypothalamic/pituitary-derived hormones that activated adenylate cyclase (Miyata et al. 1989). A variant, consisting of 27 aas (PACAP-27), was also isolated during purification (Miyata et al. 1990) and both PACAP peptides were shown to activate adenylate cyclase with a potency far exceeding that of VIP (Miyata et al. 1990). Akin to VIP, (human) PACAP is also transcribed as a prepro-peptide of 176 amino acids (Hosoya et al. 1992, 1993) and then processed into PACAP-38, PACAP-27 and a further fragment PACAP-29 (which has lower homology with the other peptides and VIP; Fig. 1). Expression of PACAP is regulated in a parallel fashion to VIP, with both cAMP and protein kinase C playing key roles (Suzuki et al. 1994; Yamamoto et al. 1998).

4.2.2 VIP and PACAP Receptors

VIP and PACAP both bind and activate G_s-linked G-protein-coupled receptors (GPCRs) to promote adenylate cyclase activation and cAMP formation, and bring about effects on cellular function. The peptides activate the 'B' class of GPCRs, which are also stimulated by a number of other small biologically active 'secretin family' peptides, and which are almost exclusively G_s coupled. The VIP/PACAP receptor family consists of two principal transmembrane proteins: $VPAC_1R$ and $VPAC_2R$.

$VPAC_1R$ was first cloned from rat lung (Ishihara et al. 1991) and subsequently identified in human and murine tissue (Johnson et al. 1996; Sreedharan et al. 1993). VIP and PAPCAP fragments all stimulate this receptor with EC_{50} values (in terms of cAMP production) in the pM range, with a rank order of potency VIP = PACAP-27 = PACAP-38 = PHM > secretin (Dickson et al. 2006; Ishihara et al. 1991; Sreedharan et al. 1993). $VPAC_1R$ has a relatively ubiquitous distribution, found in many central and peripheral organs and tissues, including the heart, lungs, kidneys, liver, spleen, epithelial cells and lymphocytes (Hashimoto et al. 1993; Ishihara et al. 1991; Reubi 2000; Sreedharan et al. 1993).

$VPAC_2R$ was discovered shortly after the cloning of $VPAC_1R$, again initially in rat cells with human and murine homologues discovered in turn (Inagaki et al. 1994; Lutz et al. 1993; Svoboda et al. 1994). In cell-based assays, the EC_{50} values tend to be higher than for $VPAC_1R$ (in the low nM range) with a rank order of potency VIP = PACAP-27 = PACAP-38 > PHM ≫ secretin. $VPAC_2R$ also exhibits widespread heterologous tissue distribution, both centrally and peripherally, although it is most highly expressed in vascular and non-vascular smooth muscle (Harmar et al. 2004).

A third, more PACAP-selective receptor, PAC_1R, has also been identified in rat human and murine tissue (Hashimoto et al. 1993; Hosoya et al. 1993; Morrow et al. 1993; Ogi et al. 1993; Pisegna and Wank 1993; Yamamoto et al. 1998). The two PACAP fragments have highest potency at this receptor subtype (EC_{50} ~ 200 pM) whereas VIP is approximately 50-fold less potent (EC_{50} ~ 10 nM). PAC_1R is expressed in many central tissues and also in many endocrine organs peripherally (e.g. pituitary, adrenal, testes) (Hashimoto et al. 1993; Ogi et al. 1993). Splice variants of this receptor have also been identified and appear to have a more specific tissue localisation.

While VIP effects are principally mediated by the adenylate cyclase AMP pathway, VIP is often found co-localised with nNOS in the parasympathetic nervous system. There is also evidence that NO and VIP can induce each other's release, at least at the presynaptic level (Murthy and Makhlouf 1994). Based on results obtained in isolated gut smooth muscle cells, a serial post-synaptic VIP/NO model has also been proposed wherein VIP is the principal neurotransmitter acting primarily through the VPAC/cAMP pathway, but also by induction of muscular NO production, primarily through eNOS (Murthy and Makhlouf 1994; Said and Rattan 2004).

As a consequence of the ubiquitous distribution of VPAC/PAC receptors it is perhaps not surprising that they govern a plethora of cellular functions, particularly in the cardiovascular and respiratory systems. These important roles have been substantiated by the development of an array of transgenic animals with targeted alterations in different components of the signalling pathways, including both endogenous ligands and the receptors subtypes. For example, VIP overexpressing mice have augmented glucose-triggered insulin release (Kato et al. 1994), whereas $VPAC_2R$ knockout mice are significantly leaner than their WT littermates (Asnicar et al. 2002); a similar phenotype is observed in PACAP and PAC_1R KO mice (Gray et al. 2002; Hashimoto et al. 2001). In addition, the circadian rhythm is disrupted in mice deficient in VIP or $VPAC_2R$ (Colwell et al. 2003; Harmar et al. 2002), and PACAP KO mice have impaired neuropathic pain responses (Mabuchi et al. 2004). However, it is perhaps in the cardiovascular system that the physiological and pathological roles of VIP and PACAP have been best characterised. Indeed, it is in the context of PH that the importance of these vasoactive peptides, and the potential of drugs targeted to the signalling pathways they trigger, that has provided the greatest insight.

4.2.3 VIP, PACAP and Pulmonary Hypertension

VIP is found, and often co-localised with acetylcholine and neuronal NO synthase, in the parasympathetic innervation of the heart and lungs. Indeed, it had been known for many years that bronchial tone is regulated by NANC nerves innervating the airway smooth muscle (Belvisi et al. 1993; Tucker et al. 1990). VIP is also found in the myocardium, conductance tissue and vascular endothelial cells (Brum et al. 1986; Della et al. 1983; Henning and Sawmiller 2001) and heavily expressed in the lungs, including both the bronchial smooth muscle and pulmonary vascular tree (Carstairs and Barnes 1986; Dey et al. 1981; Lundberg et al. 1984). This cardiopulmonary distribution of VIP is matched by its cognate receptors. High-density VIP binding sites are found in the smooth muscle of larger airways, with density reducing with airway diameter. Similar binding sites are located in the airway epithelium, submucosal glands and alveolar walls (Carstairs and Barnes 1986; Groneberg et al. 2006).

VIP is also a strong endothelium-independent vasodilator in the pulmonary circulation (and coronary arteries) (Nandiwada et al. 1985; Smitherman et al. 1989) and exerts positive inotropic and chronotropic activity in the heart (Christophe et al. 1984; Kralios et al. 1990), as might be expected from a mediator that raises intracellular cAMP levels (akin to β-adrenoceptor stimulation). In the airways, VIP and PACAP-38 are potent dilators, causing bronchodilatation in vitro and in vivo, with a potency exceeding that brought about by isoprenaline by approximately 100-fold (Linden et al. 1999; Morice et al. 1983; Palmer et al. 1986). VIP also appears to counteract airway inflammation, characteristic of PH and other airways disorders including asthma and COPD. For example, VIP dampens the activation of mast cells and lymphocytes and inhibits the production of pro-inflammatory mediators such as IL-16, IL-12 and TNFα. Conversely, VIP enhances the production of IL-10. These anti-inflammatory effects appear to be mediated primarily via modulation of NFκB activity and a cAMP-responsive element (CRE) that prevents the expression of pro-inflammatory genes (Leceta et al. 2000; O'Dorisio et al. 1989; Undem et al. 1983). VIP also activates VPAC1R and VPAC2R to elicit anti-proliferative effects on pulmonary vascular smooth muscle (St Hilaire et al. 2010).

In the context of PH, analogous to the biological actions of NPs, VIP and PACAP peptides exert a wide range of beneficial effects, including vasodilatation, bronchodilatation, modulation of airway secretion, anti-inflammatory and anti-proliferative activity. In accord, VIP has been shown to alleviate the pathogenesis in a number of models of PH. For instance, in monocrotaline-induced PH, VIP dose dependently reduces pulmonary artery pressure and pulmonary vascular resistance and synergises with endothelin receptor antagonists in preventing pathogenesis and mortality (Gunaydin et al. 2002; Hamidi et al. 2011). Moreover, in VIP KO mice there is pulmonary hypertension accompanied by right ventricular hypertrophy, pulmonary vascular remodelling and immune cell infiltration in the lung; a phenotype that can be reversed by a chronic (4 week) infusion of VIP (Said et al. 2007). VIP infusion in neonatal piglets with PH also produces decreases in the PVR/SVR ratio, thereby executing a relatively pulmonary specific beneficial effect (Haydar et al. 2007). Moreover, expression of VIP but not PACAP receptors is markedly altered in an experimental model of PH, suggesting that abnormal VIP signalling contributes to the pathogenesis of the disease (Vuckovic et al. 2009). PAC1 receptor KO mice also develop a dramatic increase in PAP and RVH shortly after birth, entailing a significant mortality, intimating that lack of this receptor results in a severe PH phenotype (Otto et al. 2004). There is also the rationale to predict that VIP administration may be effective in PH associated with lung diseases such as COPD and IPF. This has again been substantiated with the use of VIP KO mice, which exhibit airway hyper-responsiveness and inflammation (Said 2008). VIP also prevents cytokine-elicited neutrophil accumulation in the lung and promotes epithelial healing and in response to cigarette-induced damage (Guan et al. 2006; Onoue et al. 2004), and a stable VIP analogue attenuates ovalbumin-induced airway hyper-responsiveness (Misaka et al. 2010).

In line with this positive preclinical experimental data, a single-centre (albeit not placebo-controlled) trial in patients with PAH confirmed that VIP decreases pulmonary artery pressure and augments cardiac output and venous oxygen saturation (Petkov et al. 2003). This patient cohort also had lower circulating and lung VIP levels and increased expression of VIP receptors (the latter perhaps as a compensatory mechanism). A second clinical study also reported that inhaled VIP exerts a short-lived decrease in PAP, increases stroke volume and improves mixed venous oxygen saturation in PH patients (Leuchte et al. 2008). However, a more recent multi-centre, randomised, placebo-controlled trial of inhaled VIP in PH patients with differing aetiologies has dampened enthusiasm for this approach. Here, VIP did not produce any beneficial therapeutic effect on a number of indices of disease severity including PVR, 6MWD and WHO Functional Class (although some evidence of improved 6MWD at 6 months with the higher doses of VIP was apparent)(Galie et al. 2010). Notwithstanding, it is the problem of delivery that likely holds the key to therapeutic potential of VIP and PACAP peptides in treating PH and other lung disease. In the inflammatory environment of lung disease, VIP is rapidly metabolised by degrading enzymes such as NEP. Indeed, this may explain, at least in part, the ineffectiveness of inhaled VIP in the multi-centre trial; inhalation only produced a minor increase in plasma VIP levels and this may not be the ideal route of administration. To surmount this problem, current thinking advocates the development of more stable VIP analogues, or VIP 'protected' from hydrolysis by incorporation into chaperones such as phospholipids. Indeed, the latter approach has been shown to prevent the development of PH in experimental models (Rubinstein 2005). Peptidase-resistant VIP analogues (i.e. receptor agonists) are also being evaluated in vitro and in patients with lung disease, with promising signs of efficacy (Linden et al. 2003; Schmidt et al. 2001). Time will tell if these offer a significant advance in therapy, since the inhalational route will probably be required due to the potent effects of VIP on cardiac function and the peripheral vasculature. Yet, genetic variation also provides proof-of-concept evidence that VIP acts as an innate defence mechanism in the pathogenesis of PH and that drugs targeting VIP signalling may be of therapeutic benefit. For instance, the T8129C variant in the VIP gene has been linked to the development of the disease (Zhang et al. 2009), since individuals with this genotype have lowered plasma VIP levels and higher PAP. Moreover, VIP also appears to have direct salutary effects on the right heart, a therapeutic benefit often overlooked. For example, administration of VIP dilates the coronary arteries, reduces coronary vascular resistance and, as a result, can almost double coronary artery blood flow (Brum et al. 1986; Popma et al. 1990). In addition, VIP increases fractional shortening and left ventricular contractility in human volunteers (Brum et al. 1986; Frase et al. 1987).

4.3 Adrenomedullin

4.3.1 Structure and Expression

AM is a potent, long-lasting vasodilator peptide (52 aa) originally isolated from human pheochromocytoma (Kitamura et al. 1993). The peptide shares homology with calcitonin gene-related peptide (CGRP) and amylin, and analogous to NPs and VIP, possesses intramolecular disulphide bonds (Fig. 1). AM is localised to blood vessels, kidneys and lungs and circulates in the plasma with a half-life in the order of 15 min (Ichiki et al. 1994; Sakata et al. 1994). Indeed, circulating AM levels are increased in patients with hypertension, renal failure, heart failure and MI (Ishimitsu et al. 1994; Nishikimi et al. 1995). AM has a number of cardioprotective effects, and its potent vasodilator and natriuretic/diuretic actions, akin to those exerted by NPs, suggest that the peptide has a key role to play in regulating blood volume and pressure (Kitamura et al. 1993; Majid et al. 1996). The biological actions of AM are mediated by a calcitonin receptor-like receptor (CRLR) which appears to function as a selective AM receptor when associated with receptor activity modifying protein (RAMP) subtypes 2 and 3 (McLatchie et al. 1998). Interestingly, AM appears to be able to activate a number of second messenger systems to bring about a wide array of biological functions, including cyclic nucleotides (e.g. cAMP, cGMP) and PI3K. These mechanisms underpin the ability of AM to vasodilate, inhibit vascular smooth muscle proliferation, promote diuresis and natriuresis whilst inhibiting aldosterone production and facilitate angiogenesis. Intravenous administration of AM causes vasodilatation, diuresis and a positive inotropic action in experimental models (Rademaker et al. 1997) and in patients with congestive heart failure (Nagaya et al. 2000b). AM levels are also regulated by the action of NEP and, additionally, by receptor internalisation, again paralleling the natriuretic peptide system.

4.3.2 Adrenomedullin and Pulmonary Hypertension

AM is one of the most potent dilators of the pulmonary circulation and thought to be mediated by both cAMP- and NO/cGMP-dependent pathways (Heaton et al. 1995; Lippton et al. 1994; Nakamura et al. 1997; Nossaman et al. 1996). The dilator effect of AM appears to be more pronounced in isolated pulmonary vessels (compared to systemic aorta), intimating that it may bring about a somewhat pulmonary selective dilatation (Yang et al. 1996). Circulating AM levels are elevated (two- to threefold) in patients with PH and increase with the severity of disease, mimicking the natriuretic peptide family (Kakishita et al. 1999; Nishikimi et al. 1997; Shimokubo et al. 1995; Yoshibayashi et al. 1997). Circulating AM levels are also elevated in patients with PH secondary to connective tissue disease (Nanke et al. 2000). Heterozygous $AM^{+/-}$ mice [homozygous $AM^{-/-}$ animals are embryonic lethal (Ando and Fujita 2003)] are more susceptible to hypoxia-induced PH and the consequent pulmonary vascular remodelling, suggesting that this peptide presents another intrinsic defence mechanism that prevents development of the disease (Matsui et al. 2004). Moreover, upregulated expression of RAMP2 and AM by hypoxia facilitates inhibition of ET-1 signalling by AM in PH (Dschietzig et al. 2007). Since intravenous administration of AM can decrease systemic blood pressure, alternative routes of administration have been investigated to achieve greater pulmonary selectivity. Chronic inhalation of AM reverses many indices of disease severity in monocrotaline-induced PH and increases survival (Nagaya et al. 2003c). Aerosolized AM has also been shown to reduce PAP without exerting significant systemic hypotensive effects (perhaps via inhibition of ET-1 bioactivity) (Kandler et al. 2003; von der Hardt et al. 2004) and has the added benefit of reducing pulmonary TGFβ and IL-1β gene expression (von der Hardt et al. 2002). Inhaled AM also seems to synergise with NO or prostacyclin in preventing PH in experimental models (Dani et al. 2007). AM gene transfer has also shown some evidence of efficacy in experimental models of PH (Nagaya et al. 2003a). AM infusion produces an acute reduction in pulmonary artery pressure in hypoxic rats and the binding of radiolabelled AM and CGRP is increased in the lungs of these animals (Zhao et al. 1996b). AM also lowers pulmonary artery pressure in patients with PH (Nagaya et al. 2000a), whilst concomitantly increasing cardiac index; the latter response is likely due to an AM-induced increase in myocardial cAMP (Nishikimi et al. 1998), although cAMP-independent mechanisms have been reported to underlie this phenomenon (Szokodi et al. 1998). Infusion of AM has beneficial haemodynamic effects in several PH patient cohorts (Nagaya et al. 2000a, 2003b), and inhalation of AM causes a significant reduction of pulmonary vascular resistance without affecting systemic blood pressure and increases exercise capacity in PH patients (Nagaya et al. 2004).

4.4 Calcitonin Gene-Related Peptide

4.4.1 Structure and Expression

Calcitonin gene-related peptide (CGRP) is a 33 aa peptide which is arguably the most potent endogenous vasodilator identified to date (Brain et al. 1985). The peptide exists as two forms, α and β, which differ by 3 aa in humans (two in rats) (Amara et al. 1982). αCGRP is expressed abundantly in the lung, whereas βCGRP is more prevalent in the GI tract (Mulderry et al. 1988). CGRP belongs to a superfamily of structurally and functionally related peptides (Fig. 1) including AM and amylin (a 37 aa peptide co-synthesised and secreted with insulin from islet β-cells and which reduces hypoxia-induced PH) (Keith 2000; Wimalawansa 1996). The biological effects of CGRP are mediated via interaction with cognate G_s-coupled receptors (Aiyar et al. 1997). Initially, two subtypes of CGRP receptor were identified, CGRP1 and CGRP2, defined by binding of selective CGRP-derived agonists (Aiyar et al. 1996). However, subsequent efforts have identified a calcitonin receptor-like receptor (CRLR) which appears to transduce the signals conveyed by CGRP and AM. The specificity of signalling is achieved by association of CRLR with receptor activity modifying proteins (RAMPs) whereby RAMP1 promotes CGRP binding and signalling, whereas interaction with RAMP2 or RAMP3 facilitates AM binding and triggering of downstream pathways (Fraser et al. 1999; McLatchie et al. 1998). CGRP dilates most arteries in vitro by acting on CGRP1 receptors (Aiyar et al. 1996; McCormack et al. 1989) which promote release of NO from the vascular endothelium, although some NO-independent actions of CGRP on vascular reactivity have also been reported (Brain et al. 1985; Chen and Guth 1995).

4.4.2 CGRP and Pulmonary Hypertension

Endogenous CGRP appears to maintain a cytoprotective function in PH since circulating levels of CGRP are reduced in the disease and inversely correlate with PAP (Keith and Ekman 1992; Tjen et al. 1992). Moreover, the development of experimental PH can be prevented and reversed by infusion of αCGRP (Qing et al. 2003; Tjen et al. 1992). Further still, the intrinsic protective mechanism proffered by CGRP is exemplified by exacerbated PH following infusion of the CGRP1 receptor antagonist $CGRP_{8-37}$ (Tjen et al. 1992). The beneficial activity of CGRP in PH is substantiated by the observation that adenoviral delivery of the prepro-CGRP to murine lungs prior to exposure to hypoxia significantly increases lung CGRP and cAMP levels and reduces the severity of the resulting PH (Champion et al. 2000). Administration of CGRP also alleviates hypoxia-induced PH and the antagonist $CGRP_{8-37}$ augments the pulmonary hypertensive response to Ang II (Janssen and Tucker 1994). The endogenous source of CGRP in the lung in the context of PH appears to be the C fibres, since depletion of this pool with capsaicin augments the increased PAP and RVH in response to hypoxia (Tjen et al. 1998). Moreover, 3 weeks exposure to an hypoxic environment after birth causes a drop in circulating CGRP levels and persistent PH in newborn rats (Keith et al. 2000). Moreover, in vivo gene transfer of CGRP to the lung prevents the increase in PAP, RVH and pulmonary vascular remodelling in chronically hypoxic mice (Bivalacqua et al. 2002). RAMP1 overexpression also occurs in hypoxia which, in combination with RDC-1, appears to form a functional CGRP receptor that mediates the beneficial effects of the peptide against hypoxia-induced PH (Qing and Keith 2003). CGRP also produces a potent anti-proliferative effect on pulmonary vascular smooth muscle cells (Chattergoon et al. 2005), possibly via inhibition of ERK1/2, *c-fos* and *c-myc* signalling (Li et al. 2012). The potential for a CGRP-based therapy in PH has been highlighted by the delivery of CGRP-producing endothelial progenitor cells to rats with left-to-right shunt-induced PH (Zhao et al. 2007), the administered cells lodged in the lung vasculature (visualised by immunofluorescence) reducing the PAP and vascular remodelling.

4.5 Somatostatin

Somatostatin fragments have markedly different effects in the pulmonary circulation and PH. SS-28 has been reported to augment hypoxia-induced PH whereas SS-14 appears to do the exact opposite (Tjen et al. 1992). However, this differential activity is perhaps not surprising based on the receptor pharmacology of the two related peptides, which act at different dimeric GPCRs, SST_{1-5} (Velez-Roa et al. 2004), and cause opposing effects of potassium currents, leading to dilatation and contraction (Wang et al. 1989). Interestingly, infusion of SS-14 in hypoxic rats causes release of CGRP and SS-14 itself whereas administration of SS-28 in the same fashion reduces CGRP levels. Moreover, CGRP infusion increases the

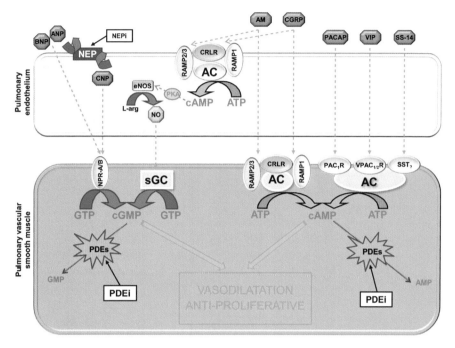

Fig. 2 Schematic representation of the important peptide vasodilator signalling pathways highlighted in this review

production of SS-14 (but not SS-28) levels, raising the possibility of a cytoprotective, positive feedback pathway (Tjen et al. 1992).

5 Conclusions

Advances in the treatment of PH over the past decade have enabled physicians to substantially improve the prognosis, yet the mortality rate remains high. Existing treatments are based predominantly on vasodilatation, whereas many emerging therapies are aimed at cell proliferation and re-modelling. It is likely that each of these processes will need to be targeted to advance therapy for PH. Ideally, a single intervention functioning in a multifaceted manner will be developed. The manipulation of vasodilator peptides holds promise in this regard. Each of those implicated in pulmonary health and disease and described herein augments levels of one or both cyclic nucleotide second messengers, which are established to be clinically effective in treating multi-aspects of PH pathogenesis (Fig. 2). This approach also raises the possibility of combination treatment with existing therapy which is underpinned by increased cAMP and/or cGMP signalling (i.e. prostacyclin analogues, PDE5i). In sum, harnessing of the cytoprotective activity of vasodilator peptides remains an encouraging strategy to reduce the morbidity and mortality associated with PH, which warrants further investigation and optimisation.

Acknowledgements The authors are supported by the British Heart Foundation, the British Lung Foundation and The Wellcome Trust.

References

Abassi ZA, Kotob S, Golomb E, Pieruzzi F, Keiser HR (1995) Pulmonary and renal neutral endopeptidase EC 3.4.24.11 in rats with experimental heart failure. Hypertension 25:1178–1184

Ahluwalia A, Hobbs AJ (2005) Endothelium-derived C-type natriuretic peptide: more than just a hyperpolarizing factor. Trends Pharmacol Sci 26:162–167

Ahluwalia A, MacAllister RJ, Hobbs AJ (2004) Vascular actions of natriuretic peptides. Cyclic GMP-dependent and -independent mechanisms. Basic Res Cardiol 99:83–89

Aiyar N, Rand K, Elshourbagy NA, Zeng Z, Adamou JE, Bergsma DJ, Li Y (1996) A cDNA encoding the calcitonin gene-related peptide type 1 receptor. J Biol Chem 271:11325–11329

Aiyar N, Disa J, Siemens IR, Nambi P (1997) Differential effects of guanine nucleotides on [125I]-hCGRP(8–37) binding to porcine lung and human neuroblastoma cell membranes. Neuropeptides 31:99–103

Amara SG, Jonas V, Rosenfeld MG, Ong ES, Evans RM (1982) Alternative RNA processing in calcitonin gene expression generates mRNAs encoding different polypeptide products. Nature 298:240–244

Anand-Srivastava MB, Sehl PD, Lowe DG (1996) Cytoplasmic domain of natriuretic peptide receptor-C inhibits adenylyl cyclase. Involvement of a pertussis toxin-sensitive G protein. J Biol Chem 271:19324–19329

Ando K, Fujita T (2003) Lessons from the adrenomedullin knockout mouse. Regul Pept 112: 185–188

Ando S, Rahman MA, Butler GC, Senn BL, Floras JS (1995) Comparison of candoxatril and atrial natriuretic factor in healthy men. Effects on hemodynamics, sympathetic activity, heart rate variability, and endothelin. Hypertension 26:1160–;1166

Angus RM, McCallum MJ, Hulks G, Thomson NC (1993) Bronchodilator, cardiovascular, and cyclic guanylyl monophosphate response to high-dose infused atrial natriuretic peptide in asthma. Am Rev Respir Dis 147:1122–1125

Asnicar MA, Koster A, Heiman ML, Tinsley F, Smith DP, Galbreath E, Fox N, Ma YL, Blum WF, Hsiung HM (2002) Vasoactive intestinal polypeptide/pituitary adenylate cyclase-activating peptide receptor 2 deficiency in mice results in growth retardation and increased basal metabolic rate. Endocrinology 143:3994–4006

Backlund T, Palojoki E, Saraste A, Gronholm T, Eriksson A, Lakkisto P, Vuolteenaho O, Nieminen MS, Voipio-Pulkki LM, Laine M, Tikkanen I (2003) Effect of vasopeptidase inhibitor omapatrilat on cardiomyocyte apoptosis and ventricular remodeling in rat myocardial infarction. Cardiovasc Res 57:727–737

Baliga RS, Zhao L, Madhani M, Lopez-Torondel B, Visintin C, Selwood D, Wilkins MR, MacAllister RJ, Hobbs AJ (2008) Synergy between natriuretic peptides and phosphodiesterase 5 inhibitors ameliorates pulmonary arterial hypertension. Am J Respir Crit Care Med 178:861–869

Barst RJ, Rubin LJ, Long WA, McGoon MD, Rich S, Badesch DB, Groves BM, Tapson VF, Bourge RC, Brundage BH (1996) A comparison of continuous intravenous epoprostenol (prostacyclin) with conventional therapy for primary pulmonary hypertension. The Primary Pulmonary Hypertension Study Group. N Engl J Med 334:296–302

Barst RJ, Langleben D, Frost A, Horn EM, Oudiz R, Shapiro S, McLaughlin V, Hill N, Tapson VF, Robbins IM, Zwicke D, Duncan B, Dixon RA, Frumkin LR (2004) Sitaxsentan therapy for pulmonary arterial hypertension. Am J Respir Crit Care Med 169:441–447

Belvisi MG, Miura M, Stretton D, Barnes PJ (1993) Endogenous vasoactive intestinal peptide and nitric oxide modulate cholinergic neurotransmission in guinea-pig trachea. Eur J Pharmacol 231:97–102

Bennett BD, Bennett GL, Vitangcol RV, Jewett JR, Burnier J, Henzel W, Lowe DG (1991) Extracellular domain-IgG fusion proteins for three human natriuretic peptide receptors. Hormone pharmacology and application to solid phase screening of synthetic peptide antisera. J Biol Chem 266:23060–23067

Bevan EG, Connell JM, Doyle J, Carmichael HA, Davies DL, Lorimer AR, McInnes GT (1992) Candoxatril, a neutral endopeptidase inhibitor: efficacy and tolerability in essential hypertension. J Hypertens 10:607–613

Bianchi C, Gutkowska J, Thibault G, Garcia R, Genest J, Cantin M (1985) Radioautographic localization of 125I-atrial natriuretic factor (ANF) in rat tissues. Histochemistry 82:441–452

Bivalacqua TJ, Hyman AL, Kadowitz PJ, Paolocci N, Kass DA, Champion HC (2002) Role of calcitonin gene-related peptide (CGRP) in chronic hypoxia-induced pulmonary hypertension in the mouse. Influence of gene transfer in vivo. Regul Pept 108:129–133

Blais C Jr, Fortin D, Rouleau JL, Molinaro G, Adam A (2000) Protective effect of omapatrilat, a vasopeptidase inhibitor, on the metabolism of bradykinin in normal and failing human hearts. J Pharmacol Exp Ther 295:621–626

Borson DB (1991) Roles of neutral endopeptidase in airways. Am J Physiol 260:L212–L225

Brain SD, Williams TJ, Tippins JR, Morris HR, MacIntyre I (1985) Calcitonin gene-related peptide is a potent vasodilator. Nature 313:54–56

Brum JM, Bove AA, Sufan Q, Reilly W, Go VL (1986) Action and localization of vasoactive intestinal peptide in the coronary circulation: evidence for nonadrenergic, noncholinergic coronary regulation. J Am Coll Cardiol 7:406–413

Burrell LM, Droogh J, Mani V, Rockell MD, Farina NK, Johnston CI (2000) Antihypertensive and antihypertrophic effects of omapatrilat in SHR. Am J Hypertens 13:1110–1116

Cargill RI, Lipworth BJ (1995a) Pulmonary vasorelaxant activity of atrial natriuretic peptide and brain natriuretic peptide in humans. Thorax 50:183–185

Cargill RI, Lipworth BJ (1995b) The role of the renin-angiotensin and natriuretic peptide systems in the pulmonary vasculature. Br J Clin Pharmacol 40:11–18

Cargill RI, Lipworth BJ (1996) Atrial natriuretic peptide and brain natriuretic peptide in cor pulmonale. Hemodynamic and endocrine effects. Chest 110:1220–1225

Carstairs JR, Barnes PJ (1986) Visualization of vasoactive intestinal peptide receptors in human and guinea pig lung. J Pharmacol Exp Ther 239:249–255

Champion HC, Bivalacqua TJ, Toyoda K, Heistad DD, Hyman AL, Kadowitz PJ (2000) In vivo gene transfer of prepro-calcitonin gene-related peptide to the lung attenuates chronic hypoxia-induced pulmonary hypertension in the mouse. Circulation 101:923–930

Channick RN, Simonneau G, Sitbon O, Robbins IM, Frost A, Tapson VF, Badesch DB, Roux S, Rainisio M, Bodin F, Rubin LJ (2001) Effects of the dual endothelin-receptor antagonist bosentan in patients with pulmonary hypertension: a randomised placebo-controlled study. Lancet 358:1119–1123

Chattergoon NN, D'Souza FM, Deng W, Chen H, Hyman AL, Kadowitz PJ, Jeter JR Jr (2005) Antiproliferative effects of calcitonin gene-related peptide in aortic and pulmonary artery smooth muscle cells. Am J Physiol Lung Cell Mol Physiol 288:L202–L211

Chauhan SD, Nilsson H, Ahluwalia A, Hobbs AJ (2003) Release of C-type natriuretic peptide accounts for the biological activity of endothelium-derived hyperpolarizing factor. Proc Natl Acad Sci USA 100:1426–1431

Chen YF (2005) Atrial natriuretic peptide in hypoxia. Peptides 26:1068–1077

Chen RY, Guth PH (1995) Interaction of endogenous nitric oxide and CGRP in sensory neuron-induced gastric vasodilation. Am J Physiol 268:G791–G796

Chen SJ, Chen YF, Meng QC, Durand J, Dicarlo VS, Oparil S (1995) Endothelin-receptor antagonist bosentan prevents and reverses hypoxic pulmonary hypertension in rats. J Appl Physiol 79:2122–2131

Chen YF, Feng JA, Li P, Xing D, Ambalavanan N, Oparil S (2006) Atrial natriuretic peptide-dependent modulation of hypoxia-induced pulmonary vascular remodeling. Life Sci 79: 1357–1365

Christophe J, Waelbroeck M, Chatelain P, Robberecht P (1984) Heart receptors for VIP, PHI and secretin are able to activate adenylate cyclase and to mediate inotropic and chronotropic effects. Species variations and physiopathology. Peptides 5:341–353

Chrysant SG, Chrysant GS (2012) Clinical implications of cardiovascular preventing pleiotropic effects of dipeptidyl peptidase-4 inhibitors. Am J Cardiol 109:1681–1685

Chusho H, Tamura N, Ogawa Y, Yasoda A, Suda M, Miyazawa T, Nakamura K, Nakao K, Kurihara T, Komatsu Y, Itoh H, Tanaka K, Saito Y, Katsuki M, Nakao K (2001) Dwarfism and early death in mice lacking C-type natriuretic peptide. Proc Natl Acad Sci USA 98:4016–4021

Cohen AJ, King TE Jr, Gilman LB, Magill-Solc C, Miller YE (1998) High expression of neutral endopeptidase in idiopathic diffuse hyperplasia of pulmonary neuroendocrine cells. Am J Respir Crit Care Med 158:1593–1599

Colwell CS, Michel S, Itri J, Rodriguez W, Tam J, Lelievre V, Hu Z, Liu X, Waschek JA (2003) Disrupted circadian rhythms in VIP- and PHI-deficient mice. Am J Physiol Regul Integr Comp Physiol 285:R939–R949

Dani C, Pavoni V, Corsini I, Longini M, Gori G, Giannesello L, Perna A, Gritti G, Paternostro F, Forestieri A, Buonocore G, Rubaltelli FF (2007) Inhaled nitric oxide combined with prostacyclin and adrenomedullin in acute respiratory failure with pulmonary hypertension in piglets. Pediatr Pulmonol 42:1048–1056

Das S, Au E, Krazit ST, Pandey KN (2010) Targeted disruption of guanylyl cyclase-A/natriuretic peptide receptor-A gene provokes renal fibrosis and remodeling in null mutant mice: role of proinflammatory cytokines. Endocrinology 151:5841–5850

de Bold AJ, Borenstein HB, Veress AT, Sonnenberg H (1981) A rapid and potent natriuretic response to intravenous injection of atrial myocardial extract in rats. Life Sci 28:89–94

Della NG, Papka RE, Furness JB, Costa M (1983) Vasoactive intestinal peptide-like immunoreactivity in nerves associated with the cardiovascular system of guinea-pigs. Neuroscience 9:605–619

Desbuquois B, Laudat MH, Laudat P (1973) Vasoactive intestinal polypeptide and glucagon: stimulation of adenylate cyclase activity via distinct receptors in liver and fat cell membranes. Biochem Biophys Res Commun 53:1187–1194

Dey RD, Shannon WA Jr, Said SI (1981) Localization of VIP-immunoreactive nerves in airways and pulmonary vessels of dogs, cat, and human subjects. Cell Tissue Res 220:231–238

Dickey DM, Flora DR, Bryan PM, Xu X, Chen Y, Potter LR (2007) Differential regulation of membrane guanylyl cyclases in congestive heart failure: NPR-B, Not NPR-A, is the predominant natriuretic peptide receptor in the failing heart. Endocrinology 148(7):3518–3522

Dickson L, Aramori I, Sharkey J, Finlayson K (2006) VIP and PACAP receptor pharmacology: a comparison of intracellular signaling pathways. Ann NY Acad Sci 1070:239–242

Dschietzig T, Richter C, Asswad L, Baumann G, Stangl K (2007) Hypoxic induction of receptor activity-modifying protein 2 alters regulation of pulmonary endothelin-1 by adrenomedullin: induction under normoxia versus inhibition under hypoxia. J Pharmacol Exp Ther 321: 409–419

Dubois SK, Kishimoto I, Lillis TO, Garbers DL (2000) A genetic model defines the importance of the atrial natriuretic peptide receptor (guanylyl cyclase-A) in the regulation of kidney function. Proc Natl Acad Sci USA 97:4369–4373

Ellmers LJ, Scott NJA, Piuhola J, Maeda N, Smithies O, Frampton CM, Richards AM, Cameron VA (2007) Npr1-regulated gene pathways contributing to cardiac hypertrophy and fibrosis. J Mol Endocrinol 38:245–257

Erdos EG, Skidgel RA (1989) Neutral endopeptidase 24.11 (enkephalinase) and related regulators of peptide hormones. FASEB J 3:145–151

Evgenov OV, Zou L, Zhang M, Mino-Kenudsen M, Mark EJ, Buys ES, Li Y, Feng Y, Raher MJ, Stasch JP, Chao W (2011) Stimulation of soluble guanylate cyclase attenuates bleomycin-induced pulmonary fibrosis in mice. Am J Respir Crit Care Med 183:A2715

Fahrenkrug J, Hannibal J (2004) Neurotransmitters co-existing with VIP or PACAP. Peptides 25:393–401

Ferreira AJ, Shenoy V, Yamazato Y, Sriramula S, Francis J, Yuan L, Castellano RK, Ostrov DA, Oh SP, Katovich MJ, Raizada MK (2009) Evidence for angiotensin-converting enzyme 2 as a therapeutic target for the prevention of pulmonary hypertension. Am J Respir Crit Care Med 179:1048–1054

Frase LL, Gaffney FA, Lane LD, Buckey JC, Said SI, Blomqvist CG, Krejs GJ (1987) Cardiovascular effects of vasoactive intestinal peptide in healthy subjects. Am J Cardiol 60:1356–1361

Fraser NJ, Wise A, Brown J, McLatchie LM, Main MJ, Foord SM (1999) The amino terminus of receptor activity modifying proteins is a critical determinant of glycosylation state and ligand binding of calcitonin receptor-like receptor. Mol Pharmacol 55:1054–1059

Galie N, Badesch D, Oudiz R, Simonneau G, McGoon MD, Keogh AM, Frost AE, Zwicke D, Naeije R, Shapiro S, Olschewski H, Rubin LJ (2005a) Ambrisentan therapy for pulmonary arterial hypertension. J Am Coll Cardiol 46:529–535

Galie N, Boonstra A, Ewert R, Gomez-Sanchez MA, Barbera JA, Torbicki A, Bremer H, Ghofrani HA, Naeije R, Gruenig E, Leuchte H, Simonneau G, Klose H, Peacock AJ, Wilkens H, Bevec D, Cavalli V, Bacher G, Rubin LJ (2010) Effects of inhaled aviptadil (vasoactive intestinal peptide) in patients with pulmonary arterial hypertension (PAH). Am J Respir Crit Care Med 181:A2516

Galie N, Ghofrani HA, Torbicki A, Barst RJ, Rubin LJ, Badesch D, Fleming T, Parpia T, Burgess G, Branzi A, Grimminger F, Kurzyna M, Simonneau G (2005b) Sildenafil citrate therapy for pulmonary arterial hypertension. N Engl J Med 353:2148–2157

Graf K, Koehne P, Grafe M, Zhang M, Auch-Schwelk W, Fleck E (1995) Regulation and differential expression of neutral endopeptidase 24.11 in human endothelial cells. Hypertension 26:230–235

Grantham JA, Schirger JA, Wennberg PW, Sandberg S, Heublein DM, Subkowski T, Burnett JC Jr (2000) Modulation of functionally active endothelin-converting enzyme by chronic neutral endopeptidase inhibition in experimental atherosclerosis. Circulation 101:1976–1981

Gray SL, Yamaguchi N, Vencova P, Sherwood NM (2002) Temperature-sensitive phenotype in mice lacking pituitary adenylate cyclase-activating polypeptide. Endocrinology 143:3946–3954

Groneberg DA, Rabe KF, Fischer A (2006) Novel concepts of neuropeptide-based drug therapy: vasoactive intestinal polypeptide and its receptors. Eur J Pharmacol 533:182–194

Guan CX, Zhang M, Qin XQ, Cui YR, Luo ZQ, Bai HB, Fang X (2006) Vasoactive intestinal peptide enhances wound healing and proliferation of human bronchial epithelial cells. Peptides 27:3107–3114

Gunaydin S, Imai Y, Takanashi Y, Seo K, Hagino I, Chang D, Shinoka T (2002) The effects of vasoactive intestinal peptide on monocrotaline induced pulmonary hypertensive rabbits following cardiopulmonary bypass: a comparative study with isoproteronol and nitroglycerine. Cardiovasc Surg 10:138–145

Gutkowska J, Nemer M (1989) Structure, expression, and function of atrial natriuretic factor in extraatrial tissues. Endocr Rev 10:519–536

Gutkowska J, Cantin M, Genest J, Sirois P (1987) Release of immunoreactive atrial natriuretic factor from the isolated perfused rat lung. FEBS Lett 214:17–20

Hamidi SA, Lin RZ, Szema AM, Lyubsky S, Jiang YP, Said SI (2011) VIP and endothelin receptor antagonist: an effective combination against experimental pulmonary arterial hypertension. Respir Res 12:141

Harmar AJ, Marston HM, Shen S, Spratt C, West KM, Sheward WJ, Morrison CF, Dorin JR, Piggins HD, Reubi JC, Kelly JS, Maywood ES, Hastings MH (2002) The VPAC(2) receptor is essential for circadian function in the mouse suprachiasmatic nuclei. Cell 109:497–508

Harmar AJ, Sheward WJ, Morrison CF, Waser B, Gugger M, Reubi JC (2004) Distribution of the VPAC2 receptor in peripheral tissues of the mouse. Endocrinology 145:1203–1210

Hashimoto H, Ishihara T, Shigemoto R, Mori K, Nagata S (1993) Molecular cloning and tissue distribution of a receptor for pituitary adenylate cyclase-activating polypeptide. Neuron 11:333–342

Hashimoto H, Shintani N, Tanaka K, Mori W, Hirose M, Matsuda T, Sakaue M, Miyazaki J, Niwa H, Tashiro F, Yamamoto K, Koga K, Tomimoto S, Kunugi A, Suetake S, Baba A (2001) Altered psychomotor behaviors in mice lacking pituitary adenylate cyclase-activating polypeptide (PACAP). Proc Natl Acad Sci USA 98:13355–13360

Hashimoto S, Amaya F, Oh-Hashi K, Kiuchi K, Hashimoto S (2010) Expression of neutral endopeptidase activity during clinical and experimental acute lung injury. Respir Res 11:164

Haydar S, Sarti JF, Grisoni ER (2007) Intravenous vasoactive intestinal polypeptide lowers pulmonary-to-systemic vascular resistance ratio in a neonatal piglet model of pulmonary arterial hypertension. J Pediatr Surg 42:758–764

Heaton J, Lin B, Chang JK, Steinberg S, Hyman A, Lippton H (1995) Pulmonary vasodilation to adrenomedullin: a novel peptide in humans. Am J Physiol 268:H2211–H2215

Hegde LG, Yu C, Renner T, Thibodeaux H, Armstrong SR, Park T, Cheruvu M, Olsufka R, Sandvik ER, Lane CE, Budman J, Hill CM, Klein U, Hegde SS (2011) Concomitant angiotensin AT1 receptor antagonism and neprilysin inhibition produces omapatrilat-like antihypertensive effects without promoting tracheal plasma extravasation in the rat. J Cardiovasc Pharmacol 57:495–504

Henning RJ, Sawmiller DR (2001) Vasoactive intestinal peptide: cardiovascular effects. Cardiovasc Res 49:27–37

Hill NS, Klinger JR, Warburton RR, Pietras L, Wrenn DS (1994) Brain natriuretic peptide: possible role in the modulation of hypoxic pulmonary hypertension. Am J Physiol 266:L308–L315

Hobbs AJ (1997) Soluble guanylate cyclase: the forgotten sibling. Trends Pharmacol Sci 18:484–491

Hobbs A, Foster P, Prescott C, Scotland R, Ahluwalia A (2004) Natriuretic peptide receptor-C regulates coronary blood flow and prevents myocardial ischemia/reperfusion injury: novel cardioprotective role for endothelium-derived C-type natriuretic peptide. Circulation 110:1231–1235

Holtwick R, van Eickels M, Skryabin BV, Baba HA, Bubikat A, Begrow F, Schneider MD, Garbers DL, Kuhn M (2003) Pressure-independent cardiac hypertrophy in mice with cardiomyocyte-restricted inactivation of the atrial natriuretic peptide receptor guanylyl cyclase-A. J Clin Invest 111:1399–1407

Hosoya M, Kimura C, Ogi K, Ohkubo S, Miyamoto Y, Kugoh H, Shimizu M, Onda H, Oshimura M, Arimura A (1992) Structure of the human pituitary adenylate cyclase activating polypeptide (PACAP) gene. Biochim Biophys Acta 1129:199–206

Hosoya M, Onda H, Ogi K, Masuda Y, Miyamoto Y, Ohtaki T, Okazaki H, Arimura A, Fujino M (1993) Molecular cloning and functional expression of rat cDNAs encoding the receptor for pituitary adenylate cyclase activating polypeptide (PACAP). Biochem Biophys Res Commun 194:133–143

Hulks G, Jardine AG, Connell JM, Thomson NC (1990) Effect of atrial natriuretic factor on bronchomotor tone in the normal human airway. Clin Sci (Lond) 79:51–55

Humbert M, Sitbon O, Simonneau G (2004) Treatment of pulmonary arterial hypertension. N Engl J Med 351:1425–1436

Ichiki Y, Kitamura K, Kangawa K, Kawamoto M, Matsuo H, Eto T (1994) Distribution and characterization of immunoreactive adrenomedullin in human tissue and plasma. FEBS Lett 338:6–10

Iglarz M, Binkert C, Morrison K, Fischli W, Gatfield J, Treiber A, Weller T, Bolli MH, Boss C, Buchmann S, Capeleto B, Hess P, Qiu C, Clozel M (2008) Pharmacology of macitentan, an orally active tissue-targeting dual endothelin receptor antagonist. J Pharmacol Exp Ther 327: 736–745

Inagaki N, Yoshida H, Mizuta M, Mizuno N, Fujii Y, Gonoi T, Miyazaki J, Seino S (1994) Cloning and functional characterization of a third pituitary adenylate cyclase-activating polypeptide receptor subtype expressed in insulin-secreting cells. Proc Natl Acad Sci USA 91:2679–2683

Irwin DC, Patot MT, Tucker A, Bowen R (2005) Neutral endopeptidase null mice are less susceptible to high altitude-induced pulmonary vascular leak. High Alt Med Biol 6:311–319

Ishigaki N, Yamamoto N, Jin H, Uchida K, Terai S, Sakaida I (2009) Continuos intravenous infusion of atrial natriuretic peptide (ANP) prevented liver fibrosis in rat. Biochem Biophys Res Commun 378:354–359

Ishihara T, Nakamura S, Kaziro Y, Takahashi T, Takahashi K, Nagata S (1991) Molecular cloning and expression of a cDNA encoding the secretin receptor. EMBO J 10:1635–1641

Ishimitsu T, Nishikimi T, Saito Y, Kitamura K, Eto T, Kangawa K, Matsuo H, Omae T, Matsuoka H (1994) Plasma levels of adrenomedullin, a newly identified hypotensive peptide, in patients with hypertension and renal failure. J Clin Invest 94:2158–2161

Itoh N, Obata K, Yanaihara N, Okamoto H (1983) Human preprovasoactive intestinal polypeptide contains a novel PHI-27-like peptide, PHM-27. Nature 304:547–549

Ivy DD, Parker TA, Ziegler JW, Galan HL, Kinsella JP, Tuder RM, Abman SH (1997) Prolonged endothelin A receptor blockade attenuates chronic pulmonary hypertension in the ovine fetus. J Clin Invest 99:1179–1186

Janssen PL, Tucker A (1994) Calcitonin gene-related peptide modulates pulmonary vascular reactivity in isolated rat lungs. J Appl Physiol 77:142–146

Jin HK, Yang RH, Chen YF, Jackson RM, Oparil S (1988) Chronic infusion of atrial natriuretic peptide prevents pulmonary hypertension in hypoxia-adapted rats. Trans Assoc Am Physicians 101:185–192

Jin H, Yang RH, Chen YF, Jackson RM, Oparil S (1990) Atrial natriuretic peptide attenuates the development of pulmonary hypertension in rats adapted to chronic hypoxia. J Clin Invest 85:115–120

Johnson MC, McCormack RJ, Delgado M, Martinez C, Ganea D (1996) Murine T-lymphocytes express vasoactive intestinal peptide receptor 1 (VIP-R1) mRNA. J Neuroimmunol 68: 109–119

Kakishita M, Nishikimi T, Okano Y, Satoh T, Kyotani S, Nagaya N, Fukushima K, Nakanishi N, Takishita S, Miyata A, Kangawa K, Matsuo H, Kunieda T (1999) Increased plasma levels of adrenomedullin in patients with pulmonary hypertension. Clin Sci (Lond) 96:33–39

Kalk P, Godes M, Relle K, Rothkegel C, Hucke A, Stasch JP, Hocher B (2006) NO-independent activation of soluble guanylate cyclase prevents disease progression in rats with 5/6 nephrectomy. Br J Pharmacol 148:853–859

Kanazawa H, Okamoto T, Hirata K, Yoshikawa J (2000) Deletion polymorphisms in the angiotensin converting enzyme gene are associated with pulmonary hypertension evoked by exercise challenge in patients with chronic obstructive pulmonary disease. Am J Respir Crit Care Med 162:1235–1238

Kandler MA, von der Hardt K, Mahfoud S, Chada M, Schoof E, Papadopoulos T, Rascher W, Dotsch J (2003) Pilot intervention: aerosolized adrenomedullin reduces pulmonary hypertension. J Pharmacol Exp Ther 306:1021–1026

Kato I, Suzuki Y, Akabane A, Yonekura H, Tanaka O, Kondo H, Takasawa S, Yoshimoto T, Okamoto H (1994) Transgenic mice overexpressing human vasoactive intestinal peptide (VIP) gene in pancreatic beta cells. Evidence for improved glucose tolerance and enhanced insulin secretion by VIP and PHM-27 in vivo. J Biol Chem 269:21223–21228

Keith IM (2000) The role of endogenous lung neuropeptides in regulation of the pulmonary circulation. Physiol Res 49:519–537

Keith IM, Ekman R (1992) Dynamic aspects of regulatory lung peptides in chronic hypoxic pulmonary hypertension. Exp Lung Res 18:205–224

Keith IM, Tjen AL, Kraiczi H, Ekman R (2000) Three-week neonatal hypoxia reduces blood CGRP and causes persistent pulmonary hypertension in rats. Am J Physiol Heart Circ Physiol 279:H1571–H1578

Kishimoto I, Dubois SK, Garbers DL (1996) The heart communicates with the kidney exclusively through the guanylyl cyclase-A receptor: acute handling of sodium and water in response to volume expansion. Proc Natl Acad Sci USA 93:6215–6219

Kishimoto I, Tokudome T, Horio T, Garbers DL, Nakao K, Kangawa K (2009) Natriuretic peptide signaling via guanylyl cyclase (GC)-A: an endogenous protective mechanism of the heart. Curr Cardiol Rev 5:45–51

Kitamura K, Kangawa K, Kawamoto M, Ichiki Y, Nakamura S, Matsuo H, Eto T (1993) Adrenomedullin: a novel hypotensive peptide isolated from human pheochromocytoma. Biochem Biophys Res Commun 192:553–560

Klapholz M, Thomas I, Eng C, Iteld BJ, Ponce GA, Niederman AL, Bilsker M, Heywood JT, Synhorst D (2001) Effects of omapatrilat on hemodynamics and safety in patients with heart failure. Am J Cardiol 88:657–661

Klinger JR, Petit RD, Warburton RR, Wrenn DS, Arnal F, Hill NS (1993) Neutral endopeptidase inhibition attenuates development of hypoxic pulmonary hypertension in rats. J Appl Physiol 75:1615–1623

Klinger JR, Warburton RR, Pietras LA, Smithies O, Swift R, Hill NS (1999) Genetic disruption of atrial natriuretic peptide causes pulmonary hypertension in normoxic and hypoxic mice. Am J Physiol 276:L868–L874

Klinger JR, Warburton RR, Pietras L, Oliver P, Fox J, Smithies O, Hill NS (2002) Targeted disruption of the gene for natriuretic peptide receptor-A worsens hypoxia-induced cardiac hypertrophy. Am J Physiol Heart Circ Physiol 282:H58–H65

Klinger JR, Houtchens J, Thaker S, Hill NS, Farber H (2005) Acute cardiopulmonary hemodynamic effects of brain natriuretic peptide in patients with pulmonary arterial hypertension. Chest 128:618S–619S

Klinger JR, Thaker S, Houtchens J, Preston IR, Hill NS, Farber HW (2006) Pulmonary hemodynamic responses to brain natriuretic peptide and sildenafil in patients with pulmonary arterial hypertension. Chest 129:417–425

Klinger JR, Warburton RR, Pietras L, Hill NS (1998) Brain natriuretic peptide inhibits hypoxic pulmonary hypertension in rats. J Appl Physiol 84:1646–1652

Knowles JW, Esposito G, Mao L, Hagaman JR, Fox JE, Smithies O, Rockman HA, Maeda N (2001) Pressure-independent enhancement of cardiac hypertrophy in natriuretic peptide receptor A-deficient mice. J Clin Invest 107:975–984

Koller KJ, Goeddel DV (1992) Molecular biology of the natriuretic peptides and their receptors. Circulation 86:1081–1088

Koller KJ, Lowe DG, Bennett GL, Minamino N, Kangawa K, Matsuo H, Goeddel DV (1991) Selective activation of the B natriuretic peptide receptor by C-type natriuretic peptide (CNP). Science 252:120–123

Kralios AC, Anderson FL, Kralios FA (1990) Myocardial electrophysiological effects of vasoactive intestinal peptide in dogs. Am J Physiol 259:H1559–H1565

Kuhn M, Holtwick R, Baba HA, Perriard JC, Schmitz W, Ehler E (2002) Progressive cardiac hypertrophy and dysfunction in atrial natriuretic peptide receptor (GC-A) deficient mice. Heart 87:368–374

Langenickel TH, Buttgereit J, Pagel-Langenickel I, Lindner M, Monti J, Beuerlein K, Al Saadi N, Plehm R, Popova E, Tank J, Dietz R, Willenbrock R, Bader M (2006) Cardiac hypertrophy in transgenic rats expressing a dominant-negative mutant of the natriuretic peptide receptor B. Proc Natl Acad Sci USA 103:4735–4740

Larsson LI, Fahrenkrug J, de Schaffalitzky MO, Sundler F, Hakanson R, Rehfeld JR (1976) Localization of vasoactive intestinal polypeptide (VIP) to central and peripheral neurons. Proc Natl Acad Sci USA 73:3197–3200

Laurent S, Boutouyrie P, Azizi M, Marie C, Gros C, Schwartz JC, Lecomte JM, Bralet J (2000) Antihypertensive effects of fasidotril, a dual inhibitor of neprilysin and angiotensin-converting enzyme, in rats and humans. Hypertension 35:1148–1153

Leceta J, Gomariz RP, Martinez C, Abad C, Ganea D, Delgado M (2000) Receptors and transcriptional factors involved in the anti-inflammatory activity of VIP and PACAP. Ann NY Acad Sci 921:92–102

Leuchte HH, Baezner C, Baumgartner RA, Bevec D, Bacher G, Neurohr C, Behr J (2008) Inhalation of vasoactive intestinal peptide in pulmonary hypertension. Eur Respir J 32:1289–1294

Li P, Oparil S, Novak L, Cao X, Shi W, Lucas J, Chen YF (2007) ANP signaling inhibits TGF-beta-induced Smad2 and Smad3 nuclear translocation and extracellular matrix expression in rat pulmonary arterial smooth muscle cells. J Appl Physiol 102:390–398

Li XW, Hu CP, Wu WH, Zhang WF, Zou XZ, Li YJ (2012) Inhibitory effect of calcitonin gene-related peptide on hypoxia-induced rat pulmonary artery smooth muscle cells proliferation: role of ERK1/2 and p27. Eur J Pharmacol 679:117–126

Linden A, Cardell LO, Yoshihara S, Nadel JA (1999) Bronchodilation by pituitary adenylate cyclase-activating peptide and related peptides. Eur Respir J 14:443–451

Linden A, Hansson L, Andersson A, Palmqvist M, Arvidsson P, Lofdahl CG, Larsson P, Lotvall J (2003) Bronchodilation by an inhaled VPAC(2) receptor agonist in patients with stable asthma. Thorax 58:217–221

Lippton H, Chang JK, Hao Q, Summer W, Hyman AL (1994) Adrenomedullin dilates the pulmonary vascular bed in vivo. J Appl Physiol 76:2154–2156

Liu LS, Cheng HY, Chin WJ, Jin HK, Oparil S (1989) Atrial natriuretic peptide lowers pulmonary arterial pressure in patients with high altitude disease. Am J Med Sci 298:397–401

Llorens-Cortes C, Huang H, Vicart P, Gasc JM, Paulin D, Corvol P (1992) Identification and characterization of neutral endopeptidase in endothelial cells from venous or arterial origins. J Biol Chem 267:14012–14018

Lopez MJ, Garbers DL, Kuhn M (1997) The guanylyl cyclase-deficient mouse defines differential pathways of natriuretic peptide signaling. J Biol Chem 272:23064–23068

Louzier V, Eddahibi S, Raffestin B, Deprez I, Adam M, Levame M, Eloit M, Adnot S (2001) Adenovirus-mediated atrial natriuretic protein expression in the lung protects rats from hypoxia-induced pulmonary hypertension. Hum Gene Ther 12:503–513

Lumsden NG, Khambata RS, Hobbs AJ (2010) C-type natriuretic peptide (CNP): cardiovascular roles and potential as a therapeutic target. Curr Pharm Des 16:4080–4088

Lundberg JM, Fahrenkrug J, Hokfelt T, Martling CR, Larsson O, Tatemoto K, Anggard A (1984) Co-existence of peptide HI (PHI) and VIP in nerves regulating blood flow and bronchial smooth muscle tone in various mammals including man. Peptides 5:593–606

Lutz EM, Sheward WJ, West KM, Morrow JA, Fink G, Harmar AJ (1993) The VIP2 receptor: molecular characterisation of a cDNA encoding a novel receptor for vasoactive intestinal peptide. FEBS Lett 334:3–8

Maack T, Suzuki M, Almeida FA, Nussenzveig D, Scarborough RM, McEnroe GA, Lewicki JA (1987) Physiological role of silent receptors of atrial natriuretic factor. Science 238:675–678

Mabuchi T, Shintani N, Matsumura S, Okuda-Ashitaka E, Hashimoto H, Muratani T, Minami T, Baba A, Ito S (2004) Pituitary adenylate cyclase-activating polypeptide is required for the development of spinal sensitization and induction of neuropathic pain. J Neurosci 24:7283–7291

Majid DS, Kadowitz PJ, Coy DH, Navar LG (1996) Renal responses to intra-arterial administration of adrenomedullin in dogs. Am J Physiol 270:F200–F205

Masuyama H, Tsuruda T, Sekita Y, Hatakeyama K, Imamura T, Kato J, Asada Y, Stasch JP, Kitamura K (2009) Pressure-independent effects of pharmacological stimulation of soluble guanylate cyclase on fibrosis in pressure-overloaded rat heart. Hypertens Res 32:597–603

Matera MG, Calzetta L, Passeri D, Facciolo F, Rendina EA, Page C, Cazzola M, Orlandi A (2011) Epithelium integrity is crucial for the relaxant activity of brain natriuretic peptide in human isolated bronchi. Br J Pharmacol 163:1740–1754

Matsui H, Shimosawa T, Itakura K, Guanqun X, Ando K, Fujita T (2004) Adrenomedullin can protect against pulmonary vascular remodeling induced by hypoxia. Circulation 109: 2246–2251

McClean DR, Ikram H, Garlick AH, Crozier IG (2001) Effects of omapatrilat on systemic arterial function in patients with chronic heart failure. Am J Cardiol 87:565–569

McClean DR, Ikram H, Mehta S, Heywood JT, Rousseau MF, Niederman AL, Sequeira RF, Fleck E, Singh SN, Coutu B, Hanrath P, Komajda M, Bryson CC, Qian C, Hanyok JJ (2002) Vasopeptidase inhibition with omapatrilat in chronic heart failure: acute and long-term hemodynamic and neurohumoral effects. J Am Coll Cardiol 39:2034–2041

McCormack DG, Mak JC, Coupe MO, Barnes PJ (1989) Calcitonin gene-related peptide vasodilation of human pulmonary vessels. J Appl Physiol 67:1265–1270

McLatchie LM, Fraser NJ, Main MJ, Wise A, Brown J, Thompson N, Solari R, Lee MG, Foord SM (1998) RAMPs regulate the transport and ligand specificity of the calcitonin-receptor-like receptor. Nature 393:333–339

McLaughlin VV, Archer SL, Badesch DB, Barst RJ, Farber HW, Lindner JR, Mathier MA, McGoon MD, Park MH, Rosenson RS, Rubin LJ, Tapson VF, Varga J (2009a) ACCF/AHA 2009 expert consensus document on pulmonary hypertension a report of the American College of Cardiology Foundation Task Force on Expert Consensus Documents and the American Heart Association developed in collaboration with the American College of Chest Physicians; American Thoracic Society, Inc.; and the Pulmonary Hypertension Association. J Am Coll Cardiol 53:1573–1619

McLaughlin VV, Archer SL, Badesch DB, Barst RJ, Farber HW, Lindner JR, Mathier MA, McGoon MD, Park MH, Rosenson RS, Rubin LJ, Tapson VF, Varga J, Harrington RA, Anderson JL, Bates ER, Bridges CR, Eisenberg MJ, Ferrari VA, Grines CL, Hlatky MA, Jacobs AK, Kaul S, Lichtenberg RC, Lindner JR, Moliterno DJ, Mukherjee D, Pohost GM, Rosenson RS, Schofield RS, Shubrooks SJ, Stein JH, Tracy CM, Weitz HH, Wesley DJ (2009b) ACCF/AHA 2009 expert consensus document on pulmonary hypertension: a report of the American College of Cardiology Foundation Task Force on Expert Consensus Documents and the American Heart Association: developed in collaboration with the American College of Chest Physicians, American Thoracic Society, Inc., and the Pulmonary Hypertension Association. Circulation 119:2250–2294

Misaka S, Aoki Y, Karaki S, Kuwahara A, Mizumoto T, Onoue S, Yamada S (2010) Inhalable powder formulation of a stabilized vasoactive intestinal peptide (VIP) derivative: anti-inflammatory effect in experimental asthmatic rats. Peptides 31:72–78

Miyata A, Arimura A, Dahl RR, Minamino N, Uehara A, Jiang L, Culler MD, Coy DH (1989) Isolation of a novel 38 residue-hypothalamic polypeptide which stimulates adenylate cyclase in pituitary cells. Biochem Biophys Res Commun 164:567–574

Miyata A, Jiang L, Dahl RD, Kitada C, Kubo K, Fujino M, Minamino N, Arimura A (1990) Isolation of a neuropeptide corresponding to the N-terminal 27 residues of the pituitary adenylate cyclase activating polypeptide with 38 residues (PACAP38). Biochem Biophys Res Commun 170:643–648

Mohapatra SS, Lockey RF, Vesely DL, Gower WR Jr (2004) Natriuretic peptides and genesis of asthma: an emerging paradigm? J Allergy Clin Immunol 114:520–526

Morice A, Unwin RJ, Sever PS (1983) Vasoactive intestinal peptide causes bronchodilatation and protects against histamine-induced bronchoconstriction in asthmatic subjects. Lancet 2: 1225–1227

Morrell NW, Morris KG, Stenmark KR (1995) Role of angiotensin-converting enzyme and angiotensin II in development of hypoxic pulmonary hypertension. Am J Physiol 269: H1186–H1194

Morrow JA, Lutz EM, West KM, Fink G, Harmar AJ (1993) Molecular cloning and expression of a cDNA encoding a receptor for pituitary adenylate cyclase activating polypeptide (PACAP). FEBS Lett 329:99–105

Motte S, McEntee K, Naeije R (2006) Endothelin receptor antagonists. Pharmacol Ther 110:386–414

Mulderry PK, Ghatei MA, Spokes RA, Jones PM, Pierson AM, Hamid QA, Kanse S, Amara SG, Burrin JM, Legon S (1988) Differential expression of alpha-CGRP and beta-CGRP by primary sensory neurons and enteric autonomic neurons of the rat. Neuroscience 25:195–205

Murphy LJ, Corder R, Mallet AI, Turner AJ (1994) Generation by the phosphoramidon-sensitive peptidases, endopeptidase-24.11 and thermolysin, of endothelin-1 and c-terminal fragment from big endothelin-1. Br J Pharmacol 113:137–142

Murthy KS, Makhlouf GM (1994) Vasoactive intestinal peptide/pituitary adenylate cyclase-activating peptide-dependent activation of membrane-bound NO synthase in smooth muscle mediated by pertussis toxin-sensitive Gi1-2. J Biol Chem 269:15977–15980

Nagaya N, Nishikimi T, Uematsu M, Satoh T, Oya H, Kyotani S, Sakamaki F, Ueno K, Nakanishi N, Miyatake K, Kangawa K (2000a) Haemodynamic and hormonal effects of adrenomedullin in patients with pulmonary hypertension. Heart 84:653–658

Nagaya N, Satoh T, Nishikimi T, Uematsu M, Furuichi S, Sakamaki F, Oya H, Kyotani S, Nakanishi N, Goto Y, Masuda Y, Miyatake K, Kangawa K (2000b) Hemodynamic, renal, and hormonal effects of adrenomedullin infusion in patients with congestive heart failure. Circulation 101:498–503

Nagaya N, Kangawa K, Kanda M, Uematsu M, Horio T, Fukuyama N, Hino J, Harada-Shiba M, Okumura H, Tabata Y, Mochizuki N, Chiba Y, Nishioka K, Miyatake K, Asahara T, Hara H, Mori H (2003a) Hybrid cell-gene therapy for pulmonary hypertension based on phagocytosing action of endothelial progenitor cells. Circulation 108:889–895

Nagaya N, Miyatake K, Kyotani S, Nishikimi T, Nakanishi N, Kangawa K (2003b) Pulmonary vasodilator response to adrenomedullin in patients with pulmonary hypertension. Hypertens Res 26(Suppl):S141–S146

Nagaya N, Okumura H, Uematsu M, Shimizu W, Ono F, Shirai M, Mori H, Miyatake K, Kangawa K (2003c) Repeated inhalation of adrenomedullin ameliorates pulmonary hypertension and survival in monocrotaline rats. Am J Physiol Heart Circ Physiol 285:H2125–H2131

Nagaya N, Kyotani S, Uematsu M, Ueno K, Oya H, Nakanishi N, Shirai M, Mori H, Miyatake K, Kangawa K (2004) Effects of adrenomedullin inhalation on hemodynamics and exercise capacity in patients with idiopathic pulmonary arterial hypertension. Circulation 109:351–356

Nakamura M, Yoshida H, Makita S, Arakawa N, Niinuma H, Hiramori K (1997) Potent and long-lasting vasodilatory effects of adrenomedullin in humans. Comparisons between normal subjects and patients with chronic heart failure. Circulation 95:1214–1221

Nandiwada PA, Kadowitz PJ, Said SI, Mojarad M, Hyman AL (1985) Pulmonary vasodilator responses to vasoactive intestinal peptide in the cat. J Appl Physiol 58:1723–1728

Nanke Y, Kotake S, Akama H, Shimamoto K, Hara M, Kamatani N (2000) Raised plasma adrenomedullin patients with systemic sclerosis complicated by pulmonary hypertension. Ann Rheum Dis 59:493–494

Nishikimi T, Saito Y, Kitamura K, Ishimitsu T, Eto T, Kangawa K, Matsuo H, Omae T, Matsuoka H (1995) Increased plasma levels of adrenomedullin in patients with heart failure. J Am Coll Cardiol 26:1424–1431

Nishikimi T, Nagata S, Sasaki T, Tomimoto S, Matsuoka H, Takishita S, Kitamura K, Miyata A, Matsuo H, Kangawa K (1997) Plasma concentrations of adrenomedullin correlate with the extent of pulmonary hypertension in patients with mitral stenosis. Heart 78:390–395

Nishikimi T, Horio T, Yoshihara F, Nagaya N, Matsuo H, Kangawa K (1998) Effect of adrenomedullin on cAMP and cGMP levels in rat cardiac myocytes and nonmyocytes. Eur J Pharmacol 353:337–344

Nishikimi T, Inaba-Iemura C, Ishimura K, Tadokoro K, Koshikawa S, Ishikawa K, Akimoto K, Hattori Y, Kasai K, Minamino N, Maeda N, Matsuoka H (2009) Natriuretic peptide/natriuretic peptide receptor-A (NPR-A) system has inhibitory effects in renal fibrosis in mice. Regul Pept 154:44–53

Nong Z, Stassen JM, Moons L, Collen D, Janssens S (1996) Inhibition of tissue angiotensin-converting enzyme with quinapril reduces hypoxic pulmonary hypertension and pulmonary vascular remodeling. Circulation 94:1941–1947

Nossaman BD, Feng CJ, Kaye AD, DeWitt B, Coy DH, Murphy WA, Kadowitz PJ (1996) Pulmonary vasodilator responses to adrenomedullin are reduced by NOS inhibitors in rats but not in cats. Am J Physiol 270:L782–L789

O'Connell JE, Jardine AG, Davidson G, Connell JM (1992) Candoxatril, an orally active neutral endopeptidase inhibitor, raises plasma atrial natriuretic factor and is natriuretic in essential hypertension. J Hypertens 10:271–277

O'Dorisio MS, Shannon BT, Fleshman DJ, Campolito LB (1989) Identification of high affinity receptors for vasoactive intestinal peptide on human lymphocytes of B cell lineage. J Immunol 142:3533–3536

Ogi K, Miyamoto Y, Masuda Y, Habata Y, Hosoya M, Ohtaki T, Masuo Y, Onda H, Fujino M (1993) Molecular cloning and functional expression of a cDNA encoding a human pituitary adenylate cyclase activating polypeptide receptor. Biochem Biophys Res Commun 196:1511–1521

Ogihara T, Rakugi H, Masuo K, Yu H, Nagano M, Mikami H (1994) Antihypertensive effects of the neutral endopeptidase inhibitor SCH 42495 in essential hypertension. Am J Hypertens 7:943–947

Oliver PM, Fox JE, Kim R, Rockman HA, Kim HS, Reddick RL, Pandey KN, Milgram SL, Smithies O, Maeda N (1997) Hypertension, cardiac hypertrophy, and sudden death in mice lacking natriuretic peptide receptor A. Proc Natl Acad Sci USA 94:14730–14735

Onoue S, Ohmori Y, Endo K, Yamada S, Kimura R, Yajima T (2004) Vasoactive intestinal peptide and pituitary adenylate cyclase-activating polypeptide attenuate the cigarette smoke extract-induced apoptotic death of rat alveolar L2 cells. Eur J Biochem 271:1757–1767

Otto C, Hein L, Brede M, Jahns R, Engelhardt S, Grone HJ, Schutz G (2004) Pulmonary hypertension and right heart failure in pituitary adenylate cyclase-activating polypeptide type I receptor-deficient mice. Circulation 110:3245–3251

Packer M, Califf RM, Konstam MA, Krum H, McMurray JJ, Rouleau JL, Swedberg K (2002) Comparison of omapatrilat and enalapril in patients with chronic heart failure: the Omapatrilat Versus Enalapril Randomized Trial of Utility in Reducing Events (OVERTURE). Circulation 106:920–926

Pagel-Langenickel I, Buttgereit J, Bader M, Langenickel TH (2007) Natriuretic peptide receptor B signaling in the cardiovascular system: protection from cardiac hypertrophy. J Mol Med (Berl) 85(8):797–810

Palmer JB, Cuss FM, Barnes PJ (1986) VIP and PHM and their role in nonadrenergic inhibitory responses in isolated human airways. J Appl Physiol 61:1322–1328

Pandey KN, Oliver PM, Maeda N, Smithies O (1999) Hypertension associated with decreased testosterone levels in natriuretic peptide receptor-A gene-knockout and gene-duplicated mutant mouse models. Endocrinology 140:5112–5119

Perreault T, Gutkowska J (1995) Role of atrial natriuretic factor in lung physiology and pathology. Am J Respir Crit Care Med 151:226–242

Petkov V, Mosgoeller W, Ziesche R, Raderer M, Stiebellehner L, Vonbank K, Funk GC, Hamilton G, Novotny C, Burian B, Block LH (2003) Vasoactive intestinal peptide as a new drug for treatment of primary pulmonary hypertension. J Clin Invest 111:1339–1346

Pisegna JR, Wank SA (1993) Molecular cloning and functional expression of the pituitary adenylate cyclase-activating polypeptide type I receptor. Proc Natl Acad Sci USA 90: 6345–6349

Popma JJ, Smitherman TC, Bedotto JB, Eichhorn EJ, Said SI, Dehmer GJ (1990) Direct coronary vasodilation induced by intracoronary vasoactive intestinal peptide. J Cardiovasc Pharmacol 16:1000–1006

Potter LR (2011) Regulation and therapeutic targeting of peptide-activated receptor guanylyl cyclases. Pharmacol Ther 130:71–82

Potter LR, Abbey-Hosch S, Dickey DM (2006) Natriuretic peptides, their receptors, and cyclic guanosine monophosphate-dependent signaling functions. Endocr Rev 27:47–72

Potter LR, Yoder AR, Flora DR, Antos LK, Dickey DM (2009) Natriuretic peptides: their structures, receptors, physiologic functions and therapeutic applications. Handb Exp Pharmacol: 341–366

Qing X, Keith IM (2003) Targeted blocking of gene expression for CGRP receptors elevates pulmonary artery pressure in hypoxic rats. Am J Physiol Lung Cell Mol Physiol 285:L86–L96

Qing X, Wimalawansa SJ, Keith IM (2003) Specific N-terminal CGRP fragments mitigate chronic hypoxic pulmonary hypertension in rats. Regul Pept 110:93–99

Qvigstad E, Moltzau LR, Aronsen JM, Nguyen CH, Hougen K, Sjaastad I, Levy FO, Skomedal T, Osnes JB (2010) Natriuretic peptides increase beta1-adrenoceptor signalling in failing hearts through phosphodiesterase 3 inhibition. Cardiovasc Res 85:763–772

Rademaker MT, Charles CJ, Lewis LK, Yandle TG, Cooper GJ, Coy DH, Richards AM, Nicholls MG (1997) Beneficial hemodynamic and renal effects of adrenomedullin in an ovine model of heart failure. Circulation 96:1983–1990

Raffestin B, Levame M, Eddahibi S, Viossat I, Braquet P, Chabrier PE, Cantin M, Adnot S (1992) Pulmonary vasodilatory action of endogenous atrial natriuretic factor in rats with hypoxic pulmonary hypertension. Effects of monoclonal atrial natriuretic factor antibody. Circ Res 70: 184–192

Raizada V, Luo W, Skipper BJ, McGuire PG (2002) Intracardiac expression of neutral endopeptidase. Mol Cell Biochem 232:129–131

Raja SG (2010) Macitentan, a tissue-targeting endothelin receptor antagonist for the potential oral treatment of pulmonary arterial hypertension and idiopathic pulmonary fibrosis. Curr Opin Investig Drugs 11:1066–1073

Reubi JC (2000) In vitro evaluation of VIP/PACAP receptors in healthy and diseased human tissues. Clinical implications. Ann NY Acad Sci 921:1–25

Richards AM, Wittert G, Espiner EA, Yandle TG, Frampton C, Ikram H (1991) Prolonged inhibition of endopeptidase 24.11 in normal man: renal, endocrine and haemodynamic effects. J Hypertens 9:955–962

Richards AM, Crozier IG, Espiner EA, Ikram H, Yandle TG, Kosoglou T, Rallings M, Frampton C (1992) Acute inhibition of endopeptidase 24.11 in essential hypertension: SCH 34826 enhances atrial natriuretic peptide and natriuresis without lowering blood pressure. J Cardiovasc Pharmacol 20:735–741

Richards AM, Wittert GA, Crozier IG, Espiner EA, Yandle TG, Ikram H, Frampton C (1993) Chronic inhibition of endopeptidase 24.11 in essential hypertension: evidence for enhanced atrial natriuretic peptide and angiotensin II. J Hypertens 11:407–416

Roques BP (1998) Cell surface metallopeptidases involved in blood pressure regulation: structure, inhibition and clinical perspectives. Pathol Biol (Paris) 46:191–200

Rouleau JL, Pfeffer MA, Stewart DJ, Isaac D, Sestier F, Kerut EK, Porter CB, Proulx G, Qian C, Block AJ (2000) Comparison of vasopeptidase inhibitor, omapatrilat, and lisinopril on exercise tolerance and morbidity in patients with heart failure: IMPRESS randomised trial. Lancet 356:615–620

Rubens C, Ewert R, Halank M, Wensel R, Orzechowski HD, Schultheiss HP, Hoeffken G (2001) Big endothelin-1 and endothelin-1 plasma levels are correlated with the severity of primary pulmonary hypertension. Chest 120:1562–1569

Rubinstein I (2005) Human VIP-alpha: an emerging biologic response modifier to treat primary pulmonary hypertension. Expert Rev Cardiovasc Ther 3:565–569

Ryan JW, Ryan US, Schultz DR, Whitaker C, Chung A (1975) Subcellular localization of pulmonary antiotensin-converting enzyme (kininase II). Biochem J 146:497–499

Said SI (2008) The vasoactive intestinal peptide gene is a key modulator of pulmonary vascular remodeling and inflammation. Ann NY Acad Sci 1144:148–153

Said SI, Mutt V (1970a) Polypeptide with broad biological activity: isolation from small intestine. Science 169:1217–1218

Said SI, Mutt V (1970b) Potent peripheral and splanchnic vasodilator peptide from normal gut. Nature 225:863–864

Said SI, Rattan S (2004) The multiple mediators of neurogenic smooth muscle relaxation. Trends Endocrinol Metab 15:189–191

Said SI, Hamidi SA, Dickman KG, Szema AM, Lyubsky S, Lin RZ, Jiang YP, Chen JJ, Waschek JA, Kort S (2007) Moderate pulmonary arterial hypertension in male mice lacking the vasoactive intestinal peptide gene. Circulation 115:1260–1268

Sakamoto M, Nakao K, Morii N, Sugawara A, Yamada T, Itoh H, Shiono S, Saito Y, Imura H (1986) The lung as a possible target organ for atrial natriuretic polypeptide secreted from the heart. Biochem Biophys Res Commun 135:515–520

Sakamoto Y, Mashiko K, Saito N, Matsumoto H, Hara Y, Kutsukata N, Yokota H (2010) Effectiveness of human atrial natriuretic peptide supplementation in pulmonary edema patients using the pulse contour cardiac output system. Yonsei Med J 51:354–359

Sakata J, Shimokubo T, Kitamura K, Nishizono M, Iehiki Y, Kangawa K, Matsuo H, Eto T (1994) Distribution and characterization of immunoreactive rat adrenomedullin in tissue and plasma. FEBS Lett 352:105–108

Sato K, Oka M, Hasunuma K, Ohnishi M, Sato K, Kira S (1995) Effects of separate and combined ETA and ETB blockade on ET-1-induced constriction in perfused rat lungs. Am J Physiol 269: L668–L672

Schermuly RT, Weissmann N, Enke B, Ghofrani HA, Forssmann WG, Grimminger F, Seeger W, Walmrath D (2001) Urodilatin, a natriuretic peptide stimulating particulate guanylate cyclase, and the phosphodiesterase 5 inhibitor dipyridamole attenuate experimental pulmonary hypertension: synergism upon coapplication. Am J Respir Cell Mol Biol 25:219–225

Schmidt DT, Ruhlmann E, Waldeck B, Branscheid D, Luts A, Sundler F, Rabe KF (2001) The effect of the vasoactive intestinal polypeptide agonist Ro 25-1553 on induced tone in isolated human airways and pulmonary artery. Naunyn Schmiedebergs Arch Pharmacol 364:314–320

Seymour AA, Abboa-Offei BE, Smith PL, Mathers PD, Asaad MM, Rogers WL (1995) Potentiation of natriuretic peptides by neutral endopeptidase inhibitors. Clin Exp Pharmacol Physiol 22:63–69

Shenoy V, Ferreira AJ, Qi Y, Fraga-Silva RA, Diez-Freire C, Dooies A, Jun JY, Sriramula S, Mariappan N, Pourang D, Venugopal CS, Francis J, Reudelhuber T, Santos RA, Patel JM, Raizada MK, Katovich MJ (2010) The angiotensin-converting enzyme 2/angiogenesis-(1–7)/Mas axis confers cardiopulmonary protection against lung fibrosis and pulmonary hypertension. Am J Respir Crit Care Med 182:1065–1072

Shimokubo T, Sakata J, Kitamura K, Kangawa K, Matsuo H, Eto T (1995) Augmented adrenomedullin concentrations in right ventricle and plasma of experimental pulmonary hypertension. Life Sci 57:1771–1779

Simonneau G, Robbins IM, Beghetti M, Channick RN, Delcroix M, Denton CP, Elliott CG, Gaine SP, Gladwin MT, Jing ZC, Krowka MJ, Langleben D, Nakanishi N, Souza R (2009) Updated clinical classification of pulmonary hypertension. J Am Coll Cardiol 54:S43–S54

Smitherman TC, Popma JJ, Said SI, Krejs GJ, Dehmer GJ (1989) Coronary hemodynamic effects of intravenous vasoactive intestinal peptide in humans. Am J Physiol 257:H1254–H1262

Solari V, Puri P (2004) Genetic polymorphisms of angiotensin system genes in congenital diaphragmatic hernia associated with persistent pulmonary hypertension. J Pediatr Surg 39: 302–306

Soleilhac JM, Lucas E, Beaumont A, Turcaud S, Michel JB, Ficheux D, Fournie-Zaluski MC, Roques BP (1992) A 94-kDa protein, identified as neutral endopeptidase-24.11, can inactivate atrial natriuretic peptide in the vascular endothelium. Mol Pharmacol 41:609–614

Solomon SD, Skali H, Bourgoun M, Fang J, Ghali JK, Martelet M, Wojciechowski D, Ansmite B, Skards J, Laks T, Henry D, Packer M, Pfeffer MA (2005) Effect of angiotensin-converting enzyme or vasopeptidase inhibition on ventricular size and function in patients with heart failure: the Omapatrilat Versus Enalapril Randomized Trial of Utility in Reducing Events (OVERTURE) echocardiographic study. Am Heart J 150:257–262

Sonoyama T, Tamura N, Miyashita K, Park K, Oyamada N, Taura D, Inuzuka M, Fukunaga Y, Sone M, Nakao K (2009) Inhibition of hepatic damage and liver fibrosis by brain natriuretic peptide. FEBS Lett 583:2067–2070

Sreedharan SP, Patel DR, Huang JX, Goetzl EJ (1993) Cloning and functional expression of a human neuroendocrine vasoactive intestinal peptide receptor. Biochem Biophys Res Commun 193:546–553

St Hilaire RC, Murthy SN, Kadowitz PJ, Jeter JR Jr (2010) Role of VPAC1 and VPAC2 in VIP mediated inhibition of rat pulmonary artery and aortic smooth muscle cell proliferation. Peptides 31:1517–1522

Sun JZ, Chen SJ, Li G, Chen YF (2000) Hypoxia reduces atrial natriuretic peptide clearance receptor gene expression in ANP knockout mice. Am J Physiol Lung Cell Mol Physiol 279: L511–L519

Sun JZ, Oparil S, Lucchesi P, Thompson JA, Chen YF (2001) Tyrosine kinase receptor activation inhibits NPR-C in lung arterial smooth muscle cells. Am J Physiol Lung Cell Mol Physiol 281: L155–L163

Suzuki N, Harada M, Hosoya M, Fujino M (1994) Enhanced production of pituitary adenylate-cyclase-activating polypeptide by 1, N6-dibutyryladenosine 3',5'-monophosphate, phorbol 12-myristate 13-acetate and by the polypeptide itself in human neuroblastoma cells, IMR-32. Eur J Biochem 223:147–153

Svoboda M, Tastenoy M, Van RJ, Goossens JF, De NP, Waelbroeck M, Robberecht P (1994) Molecular cloning and functional characterization of a human VIP receptor from SUP-T1 lymphoblasts. Biochem Biophys Res Commun 205:1617–1624

Szokodi I, Kinnunen P, Tavi P, Weckstrom M, Toth M, Ruskoaho H (1998) Evidence for cAMP-independent mechanisms mediating the effects of adrenomedullin, a new inotropic peptide. Circulation 97:1062–1070

Tallerico-Melnyk T, Yip CC, Watt VM (1992) Widespread co-localization of mRNAs encoding the guanylate cyclase-coupled natriuretic peptide receptors in rat tissues. Biochem Biophys Res Commun 189:610–616

Tamura N, Ogawa Y, Chusho H, Nakamura K, Nakao K, Suda M, Kasahara M, Hashimoto R, Katsuura G, Mukoyama M, Itoh H, Saito Y, Tanaka I, Otani H, Katsuki M (2000) Cardiac fibrosis in mice lacking brain natriuretic peptide. Proc Natl Acad Sci USA 97:4239–4244

Tamura N, Doolittle LK, Hammer RE, Shelton JM, Richardson JA, Garbers DL (2004) Critical roles of the guanylyl cyclase B receptor in endochondral ossification and development of female reproductive organs. Proc Natl Acad Sci USA 101:17300–17305

Thompson JS, Morice AH (1995) Neutral endopeptidase inhibition increases the potency of ANP in isolated rat pulmonary resistance vessels and isolated blood perfused lungs. Pulm Pharmacol 8:143–147

Thompson JS, Morice AH (1996) Neutral endopeptidase inhibitors and the pulmonary circulation. Gen Pharmacol 27:581–585

Thompson JS, Sheedy W, Morice AH (1994a) Effects of the neutral endopeptidase inhibitor, SCH 42495, on the cardiovascular remodelling secondary to chronic hypoxia in rats. Clin Sci (Lond) 87:109–114

Thompson JS, Sheedy W, Morice AH (1994b) Neutral endopeptidase (NEP) inhibition in rats with established pulmonary hypertension secondary to chronic hypoxia. Br J Pharmacol 113:1121–1126

Tjen AL, Ekman R, Lippton H, Cary J, Keith I (1992) CGRP and somatostatin modulate chronic hypoxic pulmonary hypertension. Am J Physiol 263:H681–H690

Tjen AL, Kraiczi H, Ekman R, Keith IM (1998) Sensory CGRP depletion by capsaicin exacerbates hypoxia-induced pulmonary hypertension in rats. Regul Pept 74:1–10

Tokudome T, Horio T, Soeki T, Mori K, Kishimoto I, Suga S, Yoshihara F, Kawano Y, Kohno M, Kangawa K (2004) Inhibitory effect of C-type natriuretic peptide (CNP) on cultured cardiac myocyte hypertrophy: interference between CNP and endothelin-1 signaling pathways. Endocrinology 145:2131–2140

Toshimori H, Nakazato M, Toshimori K, Asai J, Matsukura S, Oura C, Matsuo H (1988) Distribution of atrial natriuretic polypeptide (ANP)-containing cells in the rat heart and pulmonary vein. Immunohistochemical study and radioimmunoassay. Cell Tissue Res 251: 541–546

Trippodo NC, Fox M, Monticello TM, Panchal BC, Asaad MM (1999) Vasopeptidase inhibition with omapatrilat improves cardiac geometry and survival in cardiomyopathic hamsters more than does ACE inhibition with captopril. J Cardiovasc Pharmacol 34:782–790

Trippodo NC, Gabel RA, Harvey CM, Asaad MM, Rogers WL (1991) Heart failure augments the cardiovascular and renal effects of neutral endopeptidase inhibition in rats. J Cardiovasc Pharmacol 18:308–316

Troughton RW, Rademaker MT, Powell JD, Yandle TG, Espiner EA, Frampton CM, Nicholls MG, Richards AM (2000) Beneficial renal and hemodynamic effects of omapatrilat in mild and severe heart failure. Hypertension 36:523–530

Tsukada T, Horovitch SJ, Montminy MR, Mandel G, Goodman RH (1985) Structure of the human vasoactive intestinal polypeptide gene. DNA 4:293–300

Tucker JF, Brave SR, Charalambous L, Hobbs AJ, Gibson A (1990) L-NG-nitro arginine inhibits non-adrenergic, non-cholinergic relaxations of guinea-pig isolated tracheal smooth muscle. Br J Pharmacol 100:663–664

Undem BJ, Dick EC, Buckner CK (1983) Inhibition by vasoactive intestinal peptide of antigen-induced histamine release from guinea-pig minced lung. Eur J Pharmacol 88:247–250

Underwood DC, Bochnowicz S, Osborn RR, Louden CS, Hart TK, Ohlstein EH, Hay DW (1998) Chronic hypoxia-induced cardiopulmonary changes in three rat strains: inhibition by the endothelin receptor antagonist SB 217242. J Cardiovasc Pharmacol 31(Suppl 1):S453–S455

Valentin JP, Qiu CB, Wiedemann E, Gardner D, Humphreys MH (1992) Inhibition of neutral endopeptidase amplifies the effects of endogenous atrial natriuretic peptide on blood pressure and fluid partition. Am J Hypertens 5:88–91

Van Geldre LA, Lefebvre RA (2004) Interaction of NO and VIP in gastrointestinal smooth muscle relaxation. Curr Pharm Des 10:2483–2497

Veale CA, Alford VC, Aharony D, Banville DL, Bialecki RA, Brown FJ, Damewood JR Jr, Dantzman CL, Edwards PD, Jacobs RT, Mauger RC, Murphy MM, Palmer W, Pine KK, Rumsey WL, Garcia-Davenport LE, Shaw A, Steelman GB, Surian JM, Vacek EP (2000) The discovery of non-basic atrial natriuretic peptide clearance receptor antagonists. Part 1. Bioorg Med Chem Lett 10:1949–1952

Velez-Roa S, Ciarka A, Najem B, Vachiery JL, Naeije R, van de Borne P (2004) Increased sympathetic nerve activity in pulmonary artery hypertension. Circulation 110:1308–1312

Villar IC, Panayiotou CM, Sheraz A, Madhani M, Scotland RS, Nobles M, Kemp-Harper B, Ahluwalia A, Hobbs AJ (2007) Definitive role for natriuretic peptide receptor-C in mediating the vasorelaxant activity of C-type natriuretic peptide and endothelium-derived hyperpolarising factor. Cardiovasc Res 74:515–525

von der Hardt K, Kandler MA, Popp K, Schoof E, Chada M, Rascher W, Dotsch J (2002) Aerosolized adrenomedullin suppresses pulmonary transforming growth factor-beta1 and interleukin-1 beta gene expression in vivo. Eur J Pharmacol 457:71–76

von der Hardt K, Kandler MA, Chada M, Cubra A, Schoof E, Amann K, Rascher W, Dotsch J (2004) Brief adrenomedullin inhalation leads to sustained reduction of pulmonary artery pressure. Eur Respir J 24:615–623

Vuckovic A, Rondelet B, Brion JP, Naeije R (2009) Expression of vasoactive intestinal peptide and related receptors in overcirculation-induced pulmonary hypertension in piglets. Pediatr Res 66:395–399

Wang HL, Bogen C, Reisine T, Dichter M (1989) Somatostatin-14 and somatostatin-28 induce opposite effects on potassium currents in rat neocortical neurons. Proc Natl Acad Sci USA 86:9616–9620

Webb DJ (1991) Endothelin receptors cloned, endothelin converting enzyme characterized and pathophysiological roles for endothelin proposed. Trends Pharmacol Sci 12:43–46

Williamson DJ, Wallman LL, Jones R, Keogh AM, Scroope F, Penny R, Weber C, Macdonald PS (2000) Hemodynamic effects of Bosentan, an endothelin receptor antagonist, in patients with pulmonary hypertension. Circulation 102:411–418

Wimalawansa SJ (1996) Calcitonin gene-related peptide and its receptors: molecular genetics, physiology, pathophysiology, and therapeutic potentials. Endocr Rev 17:533–585

Winter RJ, Zhao L, Krausz T, Hughes JM (1991) Neutral endopeptidase 24.11 inhibition reduces pulmonary vascular remodeling in rats exposed to chronic hypoxia. Am Rev Respir Dis 144:1342–1346

Wu C, Wu F, Pan J, Morser J, Wu Q (2003) Furin-mediated processing of Pro-C-type natriuretic peptide. J Biol Chem 278:25847–25852

Xu J, Carretero OA, Liu YH, Yang F, Shesely EG, Oja-Tebbe N, Yang XP (2004) Dual inhibition of ACE and NEP provides greater cardioprotection in mice with heart failure. J Card Fail 10:83–89

Yamamoto K, Hashimoto H, Hagihara N, Nishino A, Fujita T, Matsuda T, Baba A (1998) Cloning and characterization of the mouse pituitary adenylate cyclase-activating polypeptide (PACAP) gene. Gene 211:63–69

Yamamoto T, Wada A, Tsutamoto T, Ohnishi M, Horie M (2004) Long-term treatment with a phosphodiesterase type 5 inhibitor improves pulmonary hypertension secondary to heart failure through enhancing the natriuretic peptides cGMP pathway. J Cardiovasc Pharmacol 44:596–600

Yamazato Y, Ferreira AJ, Hong KH, Sriramula S, Francis J, Yamazato M, Yuan L, Bradford CN, Shenoy V, Oh SP, Katovich MJ, Raizada MK (2009) Prevention of pulmonary hypertension by Angiotensin-converting enzyme 2 gene transfer. Hypertension 54:365–371

Yan W, Wu F, Morser J, Wu Q (2000) Corin, a transmembrane cardiac serine protease, acts as a pro-atrial natriuretic peptide-converting enzyme. Proc Natl Acad Sci USA 97:8525–8529

Yanagisawa M, Kurihara H, Kimura S, Tomobe Y, Kobayashi M, Mitsui Y, Yazaki Y, Goto K, Masaki T (1988) A novel potent vasoconstrictor peptide produced by vascular endothelial cells. Nature 332:411–415

Yang BC, Lippton H, Gumusel B, Hyman A, Mehta JL (1996) Adrenomedullin dilates rat pulmonary artery rings during hypoxia: role of nitric oxide and vasodilator prostaglandins. J Cardiovasc Pharmacol 28:458–462

Yasoda A, Ogawa Y, Suda M, Tamura N, Mori K, Sakuma Y, Chusho H, Shiota K, Tanaka K, Nakao K (1998) Natriuretic peptide regulation of endochondral ossification. Evidence for possible roles of the C-type natriuretic peptide/guanylyl cyclase-B pathway. J Biol Chem 273:11695–11700

Yasoda A, Komatsu Y, Chusho H, Miyazawa T, Ozasa A, Miura M, Kurihara T, Rogi T, Tanaka S, Suda M, Tamura N, Ogawa Y, Nakao K (2004) Overexpression of CNP in chondrocytes rescues achondroplasia through a MAPK-dependent pathway. Nat Med 10:80–86

Yasue H, Yoshimura M, Sumida H, Kikuta K, Kugiyama K, Jougasaki M, Ogawa H, Okumura K, Mukoyama M, Nakao K (1994) Localization and mechanism of secretion of B-type natriuretic peptide in comparison with those of A-type natriuretic peptide in normal subjects and patients with heart failure. Circulation 90:195–203

Yoshibayashi M, Kamiya T, Kitamura K, Saito Y, Kangawa K, Nishikimi T, Matsuoka H, Eto T, Matsuo H (1997) Plasma levels of adrenomedullin in primary and secondary pulmonary hypertension in patients <20 years of age. Am J Cardiol 79:1556–1558

Yuan SH, Dai DZ, Guan L, Dai Y, Ji M (2006) CPU0507, an endothelin receptor antagonist, improves rat hypoxic pulmonary artery hypertension and constriction in vivo and in vitro. Clin Exp Pharmacol Physiol 33:1066–1072

Zanchi A, Maillard M, Burnier M (2003) Recent clinical trials with omapatrilat: new developments. Curr Hypertens Rep 5:346–352

Zhang Y, Zhang JQ, Liu ZH, Xiong CM, Ni XH, Hui RT, He JG, Pu JL (2009) VIP gene variants related to idiopathic pulmonary arterial hypertension in Chinese population. Clin Genet 75:544–549

Zhao L, Winter RJ, Krausz T, Hughes JM (1991) Effects of continuous infusion of atrial natriuretic peptide on the pulmonary hypertension induced by chronic hypoxia in rats. Clin Sci (Lond) 81:379–385

Zhao L, al-Tubuly R, Sebkhi A, Owji AA, Nunez DJ, Wilkins MR (1996a) Angiotensin II receptor expression and inhibition in the chronically hypoxic rat lung. Br J Pharmacol 119:1217–1222

Zhao L, Brown LA, Owji AA, Nunez DJ, Smith DM, Ghatei MA, Bloom SR, Wilkins MR (1996b) Adrenomedullin activity in chronically hypoxic rat lungs. Am J Physiol 271:H622–H629

Zhao L, Hughes JM, Winter RJ (1992) Effects of natriuretic peptides and neutral endopeptidase 24.11 inhibition in isolated perfused rat lung. Am Rev Respir Dis 146:1198–1201

Zhao L, Long L, Morrell NW, Wilkins MR (1999) NPR-A-Deficient mice show increased susceptibility to hypoxia-induced pulmonary hypertension. Circulation 99:605–607

Zhao L, Mason NA, Strange JW, Walker H, Wilkins MR (2003) Beneficial effects of phosphodiesterase 5 inhibition in pulmonary hypertension are influenced by natriuretic Peptide activity. Circulation 107:234–237

Zhao Q, Liu Z, Wang Z, Yang C, Liu J, Lu J (2007) Effect of prepro-calcitonin gene-related peptide-expressing endothelial progenitor cells on pulmonary hypertension. Ann Thorac Surg 84:544–552

Pulmonary Hypertension: Novel Pathways and Emerging Therapies Inhibitors of cGMP and cAMP Metabolism

Yassine Sassi and Jean-Sébastien Hulot

Abstract Cyclic nucleotides (e.g., cAMP and cGMP) are ubiquitous second messengers that affect multiple cell functions including vascular tone and vascular cell proliferation. After production, different processes can regulate the concentration of cyclic nucleotides. Cyclic nucleotides' degradation by phosphodiesterase (PDE) enzymes has well-known roles in regulating cyclic nucleotides concentrations. Recently, recognition of ATP-binding cassette (ABC) transporter contribution to both local and global regulation of cAMP has been acknowledged. Recent data support an important role of cyclic nucleotide efflux in the pathobiology of pulmonary hypertension, thus suggesting that inhibition of cyclic nucleotide efflux proteins might be a useful strategy to prevent and treat PH.

Keywords Multidrug resistance-associated proteins • Cyclic nucleotides • Pulmonary artery • Efflux

Contents

1 Cyclic Nucleotides and the Pathogenesis of Pulmonary Hypertension 514
2 Usual and Emerging Ways to Modulate cAMP/cGMP Levels 515
   2.1 Enhancing Cyclic Nucleotide Production 516
   2.2 Blocking Cyclic Nucleotides Degradation 517

Y. Sassi
INSERM/Université Pierre et Marie Curie UMRS 956, Paris, France

Institut für Pharmakologie und Toxikologie, Technische Universität München (TUM), Munich, Germany

J.-S. Hulot (✉)
INSERM/Université Pierre et Marie Curie UMRS 956, Paris, France

Cardiovascular Research Center, Mount Sinai School of Medicine, New York, NY, USA

Faculté de Médecine Pitié-Salpêtrière, ICAN Institute of Cardiometabolism & Nutrition, UMRS-956 (INSERM/UPMC), 91 Boulevard de l'hôpital, 75013 Paris, France
e-mail: jean-sebastien.hulot@psl.aphp.fr

3 Targeting Cyclic Nucleotide Efflux: An Emerging Concept to Modulate cAMP/cGMP
  Levels 518
  3.1 ABC Transporters and Cyclic Nucleotides Efflux 518
  3.2 Influence of MRP4 on Cyclic Nucleotides Homeostasis 520
  3.3 MRP4 in the Vasculature 522
4 Discussion 523
References 524

1 Cyclic Nucleotides and the Pathogenesis of Pulmonary Hypertension

Pulmonary hypertension (PH) refers to a broad group of diseases and conditions that are characterized by persistent precapillary pulmonary hypertension (PH) (increase in mean pulmonary artery pressure (mPAP) $\geq$ 25 mmHg with a normal pulmonary artery wedge pressure $\leq$15 mmHg), leading to progressive right heart failure and premature death. The pathological mechanisms leading to this progressive and fatal cardiovascular disorder remain partially determined, although pulmonary artery endothelial cell dysfunction and structural remodeling of the pulmonary vessels are early features of PAH. Pulmonary vascular remodeling includes proliferation and migration of pulmonary artery smooth muscle cells (PASMC) leading to medial hypertrophy, increased pulmonary vascular resistance, and PAH progression. Cell hyperproliferation conducts to progressive obstruction of the pulmonary arteries leading to hypertension, progressive right heart failure, and premature death.

Pulmonary arterial hypertension can be idiopathic, hereditary, or associated with other conditions including collagen vascular disease, congenital heart disease, and HIV infection. Different studies have reported that PAH is associated with abnormalities in the homeostasis of cyclic nucleotides, namely, cyclic adenosine monophosphate (cAMP) and cyclic guanosine monophosphate (cGMP), that control critical function including vascular tone and smooth muscle proliferation.

Intracellular cGMP is increased by nitric oxide (NO) via activation of intracellular soluble guanylate cyclase (Arnold et al. 1977). Changes in nitric oxide pathways have been detected in patients with PH (Giaid and Saleh 1995; Ghofrani et al. 2004; Kharitonov et al. 1997) with a reduced expression of endothelial nitric oxide synthase, thus supporting a downregulation of cGMP synthesis. In physiological conditions, atrial and brain natriuretic peptides (ANP and BNP) increase intracellular GMP by binding to natriuretic peptide receptor-A. This receptor has a ligand-binding domain on the cell surface connected to an intracellular guanylated cyclase domain (Koeller and Goeddel 1992). Ligand binding results in the cyclization of GTP to cGMP. Overexpression of ANP prevents the development of hypoxia-induced PH, and mice lacking the natriuretic peptide receptor-A reciprocally develop exaggerated PH in response to hypoxia (Klinger et al. 1999; Zhao et al. 1999). Importantly, a downregulation of ANP-mediated cGMP synthesis has been reported in PAH patients. Taken together, these results suggest that PAH is associated with a decrease in cGMP levels and consequently a reduced activity of the cGMP-downstream targets. The cGMP-dependent kinase PKG mediates most of the cGMP biological effects. In vascular muscle, PKG activation leads to smooth muscle relaxation through a decrease in intracellular calcium (Klinger 2007). PKG might also mediate cGMP-induced inhibition of pulmonary vascular smooth muscle cell proliferation (Lincoln et al. 2001).

cAMP is also a potent regulator of pulmonary vascular tone. The activation of G-protein-coupled receptors, including beta-adrenergic receptors and prostanoid receptors, stimulates the adenylyl cyclase, leads to an increase in the production of cAMP, and eventually reduces the vascular tone. It is generally accepted that cAMP induces vascular relaxation by lowering intracellular calcium levels through inhibition of IP3 formation, inhibition of Ca^{2+} release from the SR and of Ca^{2+} entry, and stimulation of Ca^{2+} reuptake (Barnes and Liu 1995). cAMP is likely to influence other different targets including potassium channels and different kinase activities. Different studies have also shown that increasing concentrations of cAMP result in a reduction in vascular smooth muscle cell proliferation in vitro (Assender et al. 1992) and in vivo (Indolfi et al. 1997, 2000). Activation of cAMP signaling typically requires a rise in cellular cAMP concentrations and a specific diffusion in the cell. A spatial organization of cAMP signaling pathways has been proposed and might determine the effectiveness of this pathway (Fischmeister et al. 2006). The role of cAMP in the pathobiology of PH is supported by the reduction in cAMP agonist levels in PAH patients, including PGI2 in the endothelial wall (Humbert et al. 2004a, b), and by the beneficial effect of agents stimulating cAMP production including inhaled iloprost (Humbert et al. 2004a, b).

2 Usual and Emerging Ways to Modulate cAMP/cGMP Levels

The activation of cAMP and cGMP pathways is generally associated with a rise in their concentrations. These cyclic nucleotides have proved to inhibit pulmonary artery smooth muscle cells' proliferation and to be implicated in the control of pulmonary vascular tone, thus supporting strategies that will eventually sustain an increase in cyclic nucleotides levels. From a general point of view, cyclic nucleotide levels result from their rate of synthesis on the one hand and their rate of elimination on the other hand (Fig. 1). As previously exposed, cyclic adenosine monophosphate (cAMP) and cyclic guanosine monophosphate (cGMP) are generated by the action of adenylate cyclase or guanylate cyclase, respectively. On the other hand, phosphodiesterases (PDEs), by hydrolyzing the cyclic nucleotides, regulate their concentrations and localization. Recently, active export of cyclic nucleotides has become recognized as a process that can impact biological

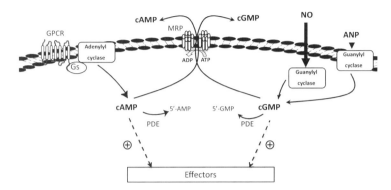

Fig. 1 Schematic representation of the synthesis, degradation, and efflux of cyclic nucleotides. *PDE* phosphodiesterase, *MRP* multidrug resistance-related protein, *GPCR* G-protein-coupled receptor, *NO* nitric oxyde, *ANP* atrial natriuretic peptide

processes. Some ABC transporters have been shown to be critical regulator of cyclic nucleotide homeostasis.

2.1 Enhancing Cyclic Nucleotide Production

2.1.1 The Prostacyclin–cAMP Pathway

cAMP is usually produced after activation of the enzyme adenylyl cyclase (AC). AC can be activated or inhibited by G proteins, which are coupled to membrane receptors. There are ten known adenylated cyclase isoforms, AC2, 5, and 8 being the most important ones for cAMP formation in rat pulmonary artery smooth muscle cells (Jourdan et al. 2001). cAMP levels are decreased during PH. Pulmonary arteries from rats with chronic hypoxia display a decrease in cAMP levels (MacLean et al. 1996). Forskolin (an AC activator) was shown to inhibit DNA synthesis in isolated pulmonary artery smooth muscle cells (Wharton et al. 2000; Davie et al. 2002). Colforsin, a water-soluble forskolin derivative, mediates a vasodilatory response in structurally remodeled pulmonary arteries from rats with PH (Yokochi et al. 2010). Many strategies have been used to increase the intracellular cAMP levels during PH; prostacyclin is one of the used targets in current therapies. Prostacyclin is the main product of arachidonic acid in the vascular endothelium (Moncada et al. 1976) and acts in the smooth muscle cells by binding to a G-protein-coupled receptor (GPCR), the prostacyclin receptor (IP) receptor. IP receptor directly stimulates the adenylyl cyclase (AC) via Gs that converts ATP to cAMP. The prostacyclin receptor agonists increase cAMP in many cell types, including smooth muscle, and the effects of prostacyclin analogs can be potentiated by agents that prevent the breakdown of cAMP. Thus, it has been proposed that cAMP is the main mediator of prostacyclin-induced effects in vascular smooth muscle cells.

Prostacyclin (Prostaglandin I_2) has proved to be a beneficial treatment for patients with severe pulmonary hypertension. In 1985, Bush et al. first reported that PGI2 might have a role in the treatment of pulmonary hypertension and associated congenital heart defects (Bush et al. 1985, 1986). In 1996, Barst and his colleagues have also shown that the continuous intravenous infusion of prostacyclin plus conventional therapy for primary pulmonary hypertension resulted in better hemodynamics, exercise endurance, quality of life, and survival than conventional therapy alone (Barst et al. 1994, 1996). Since then, many reports confirmed the beneficial effect of prostacyclin (Olschewski et al. 1996, 1999; Rosenzweig et al. 1999; McLaughlin et al. 1998). Many stable prostacyclin analogs have been developed such as beraprost, cicaprost, and iloprost (Narumiya et al. 1999).

2.1.2 The NO–cGMP Pathway

cGMP is generated by particulate or soluble guanylyl cyclases upon stimulation with natriuretic peptides or nitric oxide (NO), respectively. NO, a potent endothelium-derived vasorelaxant substance, inhibits smooth muscle cell proliferation through the cGMP pathway.

NO levels in the pulmonary circulation are decreased in patients with pulmonary hypertension (Cella et al. 2001). Pulmonary hypertension is also associated with diminished expression of endothelial nitric oxide synthase (Giaid and Saleh 1995). A consequence of defect in endothelial NO production during PH is a decrease in cGMP concentrations; for example, pulmonary arteries from rats with chronic hypoxia display a decrease in cGMP levels (MacLean et al. 1996).

NO inhalation has proved to induce vasodilatory effects in PH (Pepke-Zaba et al. 1991; Koelling et al. 1998; Hasuda et al. 2000). An alternative strategy for increasing the intracellular cGMP concentrations is the use of PDE5 inhibitors.

BAY 41-2272 (a soluble guanylate cyclase activator) augments and prolongs the pulmonary vasodilator response to inhaled NO in lambs with acute pulmonary hypertension (Evgenov et al. 2004). BAY-632521, an another soluble guanylate cyclase stimulator, partially reversed the PH, the right heart hypertrophy, and the structural remodeling of the lung vasculature in hypoxic mice and monocrotaline-injected rats (Schermuly et al. 2008). Riociguat (BAY-632521) is now in advanced clinical trials for the treatment of PH (reviewed in Schermuly et al. 2011).

2.2 Blocking Cyclic Nucleotides Degradation

The concentration of cyclic nucleotides can be regulated by processes within membrane domains (local regulation) as well as throughout the cell. The phosphodiesterase enzymes that degrade cAMP have a well-known role in both these processes.

The phosphodiesterases (PDEs) hydrolyze cAMP and cGMP to their inactive 5′ monophosphates. The predominant PDE isoforms in pulmonary arteries are PDE5 (specific for cGMP) and PDE3 (specific for cAMP) which is also inhibited by cGMP. The PDE1 family, activated by the calcium/calmodulin complex, consists of three members (PDE1A, PDE1B, and PDE1C). PDE1A and PDE1B have higher affinity for cGMP than cAMP, whereas PDE1C hydrolyzes cAMP and cGMP with similar efficiency.

Murray et al. detected the presence of PDE1A and PDE1C mRNAs and proteins in PASMC isolated from control human lungs, whereas PDE1B was absent. However, Evgenov et al. detected the presence of PDE1B mRNA in RNA extracted from the main pulmonary artery of healthy sheep (Evgenov et al. 2006). Real-time PCR and immunoblotting demonstrated an increase in the expression of PDE1A and PDE1C in PASMC from patients with idiopathic pulmonary arterial hypertension or secondary pulmonary hypertension compared with control (Murray et al. 2006). An upregulation of PDE1A and PDE1C mRNAs and proteins levels was also detected in PASMC from idiopathic PAH lungs compared with healthy donor, as well as in PASMC from chronically hypoxic mouse lungs and lungs from monocrotaline-injected rats (Schermuly et al. 2007). cGMP-PDE1 and cAMP-PDE1 activities are both increased in main pulmonary artery of chronic hypoxia-induced pulmonary hypertensive rats (Maclean et al. 1997).

Many reports proved the PDE1 inhibitor efficacy in preventing or reversing PH. An enhanced cAMP accumulation and an inhibition of cellular proliferation to a greater extent in PH PASMC than controls are observed when cells are transfected with a siRNA designed against PDE1C (Murray et al. 2006).

Inhibition of PDE1, by 8-methoxymethyl 3-isobutyl-1-methylxanthine, reverses structural lung vascular remodeling and right heart hypertrophy in hypoxic mice and monocrotaline-injected rats (Schermuly et al. 2007).

Vinpocetine, a PDE1 inhibitor, enhances pulmonary vasodilation and cGMP release induced by NO breathing without causing systemic vasodilation in lambs with acute pulmonary hypertension (Evgenov et al. 2006).

3 Targeting Cyclic Nucleotide Efflux: An Emerging Concept to Modulate cAMP/cGMP Levels

3.1 ABC Transporters and Cyclic Nucleotides Efflux

Besides the classic enzymes involved in cyclic nucleotides elimination (i.e., PDEs), other processes can also alter the intracellular cyclic nucleotide concentrations. Some studies have reported that cAMP and cGMP can be transported from the cytosol to the extracellular space in some specific cell types, including vascular smooth muscle cells and cardiac myocytes (Brunton et al. 1980; Godinho and Costa 2003). Export of monophosphorylated nucleotides was noted by Sutherland when first describing the biosynthesis of the intracellular "second messenger," 3′,5′-cyclic adenosine monophosphate (cAMP), in response to hormonal stimulation and adenylate cyclase activation (Sutherland and Rall 1958). Sutherland observed not only increased intracellular cAMP but also a concurrent rise in extracellular cAMP. cAMP export has been reported in different cell types including vascular cells (Sassi et al. 2008). This nucleotide extrusion process has general characteristics as being unidirectional, having a saturable kinetics, and being inhibited after depletion of cellular energy [i.e., adenosine triphosphate (ATP)] and after chemical inhibition by compounds such as the organic anion transport inhibitor probenecid and prostaglandins (Campbell and Taylor 1981; Doore et al. 1975; Finnegan and Carey 1998; Rindler et al. 1978).

cGMP export has similarly been reported in different cell types, sharing many of the characteristics observed with cAMP export (Billiar et al. 1992; Sager 2004).

In 1999, it was found that a member of the ATP-binding cassette transporter (ABCC4 alias MRP4) family had the ability to export an antiviral agent [i.e., a purine nucleotide monophosphate analog that resembled cAMP (Schuetz et al. 1999)]. Using membrane vesicle transport systems and altered mammalian cells engineered to stably overexpress MRP4, it was then shown that the vesicular uptake of cAMP and cGMP required ATP and that MRP4 has an affinity for cAMP and cGMP in the low micromolar range (cAMP and cGMP, with affinity constants of 45 µM and 10 µM, respectively) (Chen et al. 2001). Collectively these findings provided evidence for energy-dependent export of cyclic nucleotides from cells.

ABC transporters are transmembrane proteins that utilize the energy of ATP hydrolysis to transport various substrates across cellular membranes. In the human genome alone, ABC transporters belonging to the ABC superfamily have 48 representatives. It has progressively been shown that genetic mutations associated with failure to export a specific ligand across a lipid bilayer are the basis of inherited diseases including a neonatal surfactant deficiency (Shulenin et al. 2004), macular degeneration (Allikmets et al. 1997; Martínez-Mir et al. 1998), liver diseases (Strautnieks et al. 1998; Pauli-Magnus et al. 2005), and cystic fibrosis (Riordan et al. 1989; Cutting et al. 1990). Some human ABC transporters also appear to function in a protective capacity by exporting cytotoxic compounds (e.g., dietary cytotoxics and therapeutic drugs) out of the cell.

ABC transporters are usually composed of two distinct domains: the transmembrane domains (TMD) and the nucleotide-binding domain (NBD). The TMD recognizes the substrates and undergoes conformational changes to transport the substrate across the membrane (Dawson and Locher 2006). The TMD is embedded in the plasma membrane and its architecture is variable among ABC transporters, thus reflecting the diversity of substrates that can be translocated. The NBD is the site for ATP binding, is located in the cytoplasm, and has a highly conserved sequence.

As shown in Fig. 2, the domain organization of MRP4 is typical of ABC transporters, with a core structure composed of two TMDs each consisting of six transmembrane helices with two NBDs (Borst et al. 2007; Deeley et al. 2006; Sauna et al. 2004; Lamba et al. 2003). MRP4 is ubiquitously expressed in many tissues, including heart, lung, liver, kidney, brain, intestine, testis, and muscle. Indeed, MRP4 is present in several cell types such as cardiac myocyte (Sassi et al. 2012),

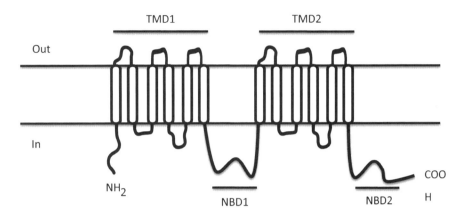

Fig. 2 Membrane topology model for MRP4. The transporter has two transmembrane domains (TMD) and two nucleotide-binding domains (NBD)

erythrocytes (Sundkvist et al. 2002), human kidney proximal tubules (Van Aubel et al. 2002), coronary artery smooth muscle cells (Sassi et al. 2008), endothelial cells (Sassi et al. 2008; Zhang et al. 2004), astrocytes and microglial cells (Hirrlinger et al. 2002), human embryonic kidney cells (Wielinga et al. 2002), hepatocytes (Assem et al. 2004; Rius et al. 2003), platelets (Jedlitschky et al. 2004), and many others cell types.

3.2 Influence of MRP4 on Cyclic Nucleotides Homeostasis

Within the large superfamily of ATP-binding cassette transporters (ABCC), three members of the subfamily of multidrug resistance-associated proteins, MRP4, MRP5, and MRP8 (also known as ABCC4, ABCC5, and ABCC11), have been shown to transport cyclic nucleotides. The affinity of MRP5 and MRP8 mediated for cyclic nucleotides remains however controversial. MRP4 was first identified as a transporter of PMEA (9-(2-phosphonylmethoxyethyl)adenine), a nucleoside monophosphate antiviral agent (Schuetz et al. 1999). MRP4 is able to transport many drugs including antiviral (adefovir, tenofovir, ganciclovir) and cytotoxic (methotrexate, 6-thioguanine, 6-mercaptopurine, topotecan) agents (reviewed in Russel et al. 2008).

In addition to the drugs transport, MRP4 mediates the cellular efflux of several endogenous metabolites. Chen and his colleagues were the first to show an ATP-dependent transport of cAMP and cGMP by MRP4 in membrane vesicles prepared from insect cells (Chen et al. 2001). Since then, many studies confirmed the cellular efflux of cyclic nucleotides by MRP4 (Lai and Tan 2002; Van Aubel et al. 2002; Wielinga et al. 2003; Reichel et al. 2007; Li et al. 2007).

Overall, the studies suggest that MRP4 (and other transporters including MRP5) are low-affinity cyclic nucleotide exporters. This assumption is also supported by the lack of substantial increase in intracellular cyclic nucleotide concentrations after inhibition of MRP4 transport. Reciprocally, overexpression of MRP4 does not lead to a substantial decrease in intracellular cAMP levels, even when the total cAMP concentrations are high (Wielinga et al. 2003). On the other hand, the changes associated with MRP4 inhibition are largely higher in the presence of adenylate cyclase activation or after phosphodiesterase inhibition (Sassi et al. 2008, 2012). These findings fostered the idea that MRP4 is a cAMP overflow pump that may exert important contribution when phosphodiesterases are limiting (Sassi et al. 2008).

Another hypothesis is that MRP4 modulates local membrane concentrations of cAMP and is thus an important contributor in the cyclic nucleotide compartmentalization process. In a recent study (Li et al. 2007), it was reported that MRP4 is physically and functionally associated with the cAMP-regulated chloride channel CFTR. In response to adenosine, the adenosine receptor (a G-protein-coupled receptor) increases intracellular cAMP and activates CFTR chloride currents by compartmentalized changes in cAMP at the apical membrane. At low adenosine concentrations (<20 µM), the MRP4 inhibitor (MK571) potentiated the cAMP-mediated activation of CFTR chloride currents, an effect that was however not observed at high adenosine concentrations (100 µM). Similarly, MRP4 inhibition resulted in moderate changes in the intracellular cyclic nucleotides levels that were however sufficient to induce a significant activation of cyclic nucleotide-mediated signaling pathways (Sassi et al. 2008, 2009). MRP4's ability to modulate local cAMP was evaluated using a FRET-based cAMP sensor, CFP-EPAC-YFP. The adenylate cyclase activation by forskolin produced local intracellular FRET changes in the CFP-EPAC-YFP sensor that reflect local changes in cAMP concentration. Stimulation with lower concentrations of forskolin led to a greater change in FRET in cells transfected with MRP4 silencing RNA as compared to controls. However, the maximal effect was similar regardless of MRP4 activity, supporting the idea that after cAMP has maximally diffused throughout the cell, MRP4 is incapable of modulating local cAMP concentrations. MRP4 inhibition was rather associated with an enhanced signaling response to a minimal cyclic nucleotide production. This supports the assumption that MRP4 physiologically acts by quenching the production of limited amounts of cyclic nucleotides at a local level. This system would be particularly suitable to avoid an inappropriate cellular response to minimal receptor stimulation and minimal cyclic nucleotide production. It is likely that the efflux is maximal in case of higher cyclic nucleotide production which however will not impact any longer the cyclic nucleotides diffusion in the cell. In this model, MRP4 inhibition (and more generally the inhibition of cyclic nucleotide efflux) will result in higher cyclic nucleotide availability after production.

Further support for a role of MRP4 in local regulation of cAMP in membrane domains is the recent finding that MRP4 localizes in caveolin-enriched fractions (Sassi et al. 2008). Because caveolins are integral membrane proteins that act as a scaffold for membrane proteins in microdomains, it is possible that caveolins physically anchor MRP4 in microdomains. Thus, like some PDEs, MRP4 might promote regulation of local membrane cAMP concentrations that are coupled to some GPCR-mediated processes to accomplish local regulation of cAMP; MRP4 can form a macromolecular complex (Li et al. 2007).

3.3 MRP4 in the Vasculature

MRP4 mRNA and protein are present in human coronary artery smooth muscle cells. MRP4 is upregulated in proliferative SMC in vitro and in vivo. Indeed, highly proliferative SMCs in the neointima rat carotid arteries after balloon injury display strong MRP4 protein expression, whereas MRP4 expression in the media is limited. MRP4 knockdown by specific siRNA designed against MRP4 mRNA induces an increase in intracellular cAMP and cGMP levels and an increase in the cAMP-responsive element binding protein (CREB) phosphorylation and activity and thus blocks SMC proliferation in vitro. MRP4 inhibition (using adenoviral vector expressing shRNA against rat Mrp4 mRNA) prevents neointimal growth in injured rat carotid arteries in vivo (Sassi et al. 2008).

MRP4 mRNA and protein were also detected in preglomerular vascular smooth muscle cells (PGVSMCs). In isoproterenol-treated PGVSMCs, MRP4 siRNA decreased the ratio of extracellular $3',5'$-cAMP to intracellular $3',5'$-cAMP and the authors conclude that MRP4 is the dominant $3',5'$-cAMP transporter in these cells (Cheng et al. 2010).

The effects of MRP4 inhibition in experimental hypoxia-induced PH were also studied (Hara et al. 2011). MRP4 was found to be upregulated in human pulmonary arteries from PAH patients compared to control patients as well as in lungs from WT mice exposed to hypoxia compared with normoxia. MRP4 silencing induces an increase in intracellular/extracellular ratios of both cAMP and cGMP in hPASMCs in vitro as well as in PKA and PKG activities. The changes in cyclic nucleotides levels after MRP4 inhibition are enhanced in the presence of the PDE5 inhibitor sildenafil. As a consequence of cAMP and cGMP regulation, PASMCs' proliferation and migration are inhibited by the MRP4 siRNA. MRP4 KO mice are protected from hypoxia-induced PH development. Indeed, exposing WT mice to hypoxia for 4 weeks resulted in an increase in distal pulmonary artery remodeling, right ventricular systolic pressure, and right ventricular hypertrophy (assessed by the Fulton index) whereas no significant changes were detected in these parameters in MRP4 KO mice (Fig. 3). The use of MK571, a pharmacological inhibitor of MRP4, reverses the PH induced by maintaining WT mice under hypoxia for 3 weeks. In fact, mice treated with MK571 (5 or 25 mg/kg/day) displayed a decrease in the medial thickening of small pulmonary arteries and arterioles and a lower right ventricular systolic pressure and Fulton index (Hara et al. 2011). Macrophage and lymphocyte cell recruitment (measured by CD68 and CD45 expression) after hypoxia was significantly reduced in MRP4 KO mice compared to WT. The amount of IL-6 was increased after hypoxia in WT mice but not in MRP4 KO mice. These results suggest a reduction of hypoxia-induced inflammatory response in MRP4 KO mice.

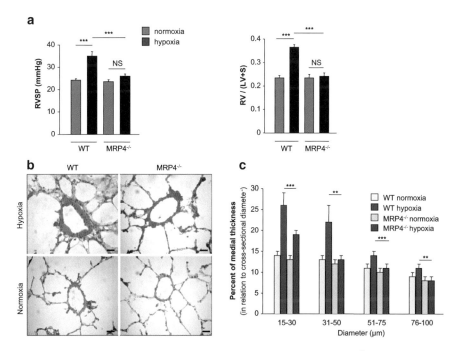

Fig. 3 Prevention of hypoxia-induced pulmonary hypertension in MRP4$^{-/-}$ mice. (**a**) Right ventricle systolic pressure (RVSP, mmHg) and right ventricle hypertrophy reflected by the right ventricle (RV) weight over left ventricle (LV) + interventricular septum (S) weights ratio (RV/[LV + S] = Fulton index) in each group ($n = 15$, ***$P < 0.001$). (**b**) Representative H&E-stained sections of small pulmonary arteries from lungs of WT and MRP4$^{-/-}$ mice in normoxia and hypoxia conditions. (**c**) Percentage of medial thickness of small arteries in relation to cross-sectional diameter (**$P < 0.01$). Scale bar: 20 μm. Reproduced with permission from Hara et al. (2011)

4 Discussion

Interestingly, changes in cyclic nucleotide levels following MRP4 silencing are further enhanced in the presence of the PDEs nonspecific inhibitor (in human coronary artery smooth muscle cells) or in the presence of the PDE5 inhibitor sildenafil (in hPASMCs). In addition, treatment with IBMX or sildenafil led to an increase of MRP4 protein in vitro and in vivo. These results suggest that MRP4 overexpression could act in a manner to compensate for PDE5 inhibition. A combination therapy with PDE5 and MRP4 inhibitors could represent an efficient therapy by attenuating the development of pulmonary hypertension compared with treatment with each drug alone (Fig. 4).

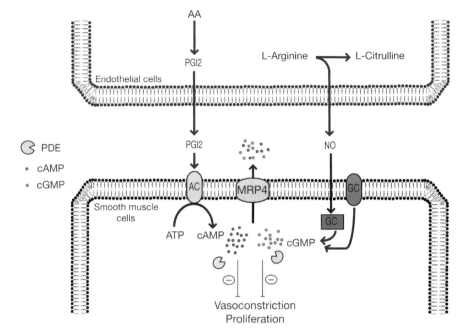

Fig. 4 Schematic representation of MRP in the cyclic nucleotide metabolism during pulmonary hypertension

References

Allikmets R, Shroyer NF, Singh N, Seddon JM, Lewis RA, Bernstein PS, Peiffer A, Zabriskie NA, Li Y, Hutchinson A, Dean M, Lupski JR, Leppert M (1997) Mutation of the Stargardt disease gene (ABCR) in age-related macular degeneration. Science 277(5333):1805–1807

Arnold WP, Mittal CK, Katsuki S, Murad F (1977) Nitric oxide activates guanylate cyclase and increases guanosine $3':5'$-cyclic monophosphate levels in various tissue preparations. Proc Natl Acad Sci USA 74(8):3203–3207

Assem M, Schuetz EG, Leggas M, Sun D, Yasuda K, Reid G, Zelcer N, Adachi M, Strom S, Evans RM, Moore DD, Borst P, Schuetz JD (2004) Interactions between hepatic Mrp4 and Sult2a as revealed by the constitutive androstane receptor and Mrp4 knockout mice. J Biol Chem 279 (21):22250–22257

Assender JW, Southgate KM, Hallett MB, Newby AC (1992) Inhibition of proliferation, but not of Ca2+ mobilization, by cyclic AMP and GMP in rabbit aortic smooth-muscle cells. Biochem J 288(Pt 2):527–532

Barnes PJ, Liu SF (1995) Regulation of pulmonary vascular tone. Pharmacol Rev 47(1):87–131

Barst RJ, Rubin LJ, McGoon MD, Caldwell EJ, Long WA, Levy PS (1994) Survival in primary pulmonary hypertension with long-term continuous intravenous prostacyclin. Ann Intern Med 121(6):409–415

Barst RJ, Rubin LJ, Long WA, McGoon MD, Rich S, Badesch DB, Groves BM, Tapson VF, Bourge RC, Brundage BH et al (1996) A comparison of continuous intravenous epoprostenol (prostacyclin) with conventional therapy for primary pulmonary hypertension. The Primary Pulmonary Hypertension Study Group. N Engl J Med 334(5):296–302

Billiar TR, Curran RD, Harbrecht BG, Stadler J, Williams DL, Ochoa JB, Di Silvio M, Simmons RL, Murray SA (1992) Association between synthesis and release of cGMP and nitric oxide biosynthesis by hepatocytes. Am J Physiol 262(4 Pt 1):C1077–C1078

Borst P, de Wolf C, van de Wetering K (2007) Multidrug resistance-associated proteins 3, 4, and 5. Pflugers Arch 453(5):661–673

Brunton LL, Buss JE (1980) Export of cyclic AMP by mammalian reticulocytes. J Cyclic Nucleotide Res 6:369–377

Bush A, Busst CM, Shinebourne EA (1985) The use of oxygen and prostacyclin as pulmonary vasodilators in congenital heart disease. Int J Cardiol 9(3):267–274

Bush A, Busst C, Booth K, Knight WB, Shinebourne EA (1986) Does prostacyclin enhance the selective pulmonary vasodilator effect of oxygen in children with congenital heart disease? Circulation 74(1):135–144

Campbell IL, Taylor KW (1981) The effect of metabolites, papaverine, and probenecid on cyclic AMP efflux from isolated rat Islets of Langerhans. Biochim Biophys Acta 676(3):357–364

Cella G, Bellotto F, Tona F, Sbarai A, Mazzaro G, Motta G, Fareed J (2001) Plasma markers of endothelial dysfunction in pulmonary hypertension. Chest 120(4):1226–1230

Chen ZS, Lee K, Kruh GD (2001) Transport of cyclic nucleotides and estradiol 17-beta-D-glucuronide by multidrug resistance protein 4. Resistance to 6-mercaptopurine and 6-thioguanine. J Biol Chem 276(36):33747–33754

Cheng D, Ren J, Jackson EK (2010) Multidrug resistance protein 4 mediates cAMP efflux from rat preglomerular vascular smooth muscle cells. Clin Exp Pharmacol Physiol 37(2):205–207

Cutting GR, Kasch LM, Rosenstein BJ, Zielenski J, Tsui LC, Antonarakis SE, Kazazian HH Jr (1990) A cluster of cystic fibrosis mutations in the first nucleotide-binding fold of the cystic fibrosis conductance regulator protein. Nature 346(6282):366–369

Davie N, Haleen SJ, Upton PD, Polak JM, Yacoub MH, Morrell NW, Wharton J (2002) ET(A) and ET(B) receptors modulate the proliferation of human pulmonary artery smooth muscle cells. Am J Respir Crit Care Med 165(3):398–405

Dawson RJ, Locher KP (2006) Structure of a bacterial multidrug ABC transporter. Nature 443(7108):180–185

Deeley RG, Westlake C, Cole SP (2006) Transmembrane transport of endo- and xenobiotics by mammalian ATP-binding cassette multidrug resistance proteins. Physiol Rev 86(3):849–899

Doore BJ, Bashor MM, Spitzer N, Mawe RC, Saier MH Jr (1975) Regulation of adenosine $3':5'$-monophosphate efflux from rat glioma cells in culture. J Biol Chem 250(11):4371–4372

Evgenov OV, Ichinose F, Evgenov NV, Gnoth MJ, Falkowski GE, Chang Y, Bloch KD, Zapol WM (2004) Soluble guanylate cyclase activator reverses acute pulmonary hypertension and augments the pulmonary vasodilator response to inhaled nitric oxide in awake lambs. Circulation 110(15):2253–2259

Evgenov OV, Busch CJ, Evgenov NV, Liu R, Petersen B, Falkowski GE, Petho B, Vas A, Bloch KD, Zapol WM, Ichinose F (2006) Inhibition of phosphodiesterase 1 augments the pulmonary vasodilator response to inhaled nitric oxide in awake lambs with acute pulmonary hypertension. Am J Physiol Lung Cell Mol Physiol 290(4):L723–L729

Finnegan RB, Carey GB (1998) Characterization of cyclic AMP efflux from swine adipocytes in vitro. Obes Res 6(4):292–298

Fischmeister R, Castro LR, Abi-Gerges A, Rochais F, Jurevicius J, Leroy J, Vandecasteele G (2006) Compartmentation of cyclic nucleotide signaling in the heart: the role of cyclic nucleotide phosphodiesterases. Circ Res 99(8):816–828

Ghofrani HA, Pepke-Zaba J, Barbera JA, Channick R, Keogh AM, Gomez-Sanchez MA, Kneussl M, Grimminger F (2004) Nitric oxide pathway and phosphodiesterase inhibitors in pulmonary arterial hypertension. J Am Coll Cardiol 43(12 Suppl S):68S–72S

Giaid A, Saleh D (1995) Reduced expression of endothelial nitric oxide synthase in the lungs of patients with pulmonary hypertension. N Engl J Med 333(4):214–221

Godinho RO, Costa VL (2003) Regulation of intracellular cyclic AMP in skeletal muscle cells involves the efflux of cyclic nucleotide to the extracellular compartment. Br J Pharmacol 138 (5):995–1003

Hara Y, Sassi Y, Guibert C, Gambaryan N, Dorfmüller P, Eddahibi S, Lompré AM, Humbert M, Hulot JS (2011) Inhibition of MRP4 prevents and reverses pulmonary hypertension in mice. J Clin Invest 121(7):2888–2897

Hasuda T, Satoh T, Shimouchi A, Sakamaki F, Kyotani S, Matsumoto T, Goto Y, Nakanishi N (2000) Improvement in exercise capacity with nitric oxide inhalation in patients with precapillary pulmonary hypertension. Circulation 101(17):2066–2070

Hirrlinger J, König J, Dringen R (2002) Expression of mRNAs of multidrug resistance proteins (Mrps) in cultured rat astrocytes, oligodendrocytes, microglial cells and neurones. J Neurochem 82(3):716–719

Humbert M, Sitbon O, Simonneau G (2004a) Treatment of pulmonary arterial hypertension. N Engl J Med 351(14):1425–1436

Humbert M, Morrell NW, Archer SL, Stenmark KR, MacLean MR, Lang IM, Christman BW, Weir EK, Eickelberg O, Voelkel NF, Rabinovitch M (2004b) Cellular and molecular pathobiology of pulmonary arterial hypertension. J Am Coll Cardiol 43(12 Suppl S):13S–24S

Indolfi C, Avvedimento EV, Di Lorenzo E, Esposito R, Rapacciuolo A, Giuliano P, Grieco D, Cavuto L, Stingone AM, Ciullo I, Condorelli G, Chiariello M (1997) Activation of cAMP-PKA signaling in vivo inhibits smooth muscle cell proliferation induced by vascular injury. Nat Med 3(7):775–779

Indolfi C, Di Lorenzo E, Rapacciuolo A, Stingone AM, Stabile E, Leccia A, Torella D, Caputo R, Ciardiello F, Tortora G, Chiariello M (2000) 8-chloro-cAMP inhibits smooth muscle cell proliferation in vitro and neointima formation induced by balloon injury in vivo. J Am Coll Cardiol 36(1):288–293

Jedlitschky G, Tirschmann K, Lubenow LE, Nieuwenhuis HK, Akkerman JW, Greinacher A, Kroemer HK (2004) The nucleotide transporter MRP4 (ABCC4) is highly expressed in human platelets and present in dense granules, indicating a role in mediator storage. Blood 104 (12):3603–3610

Jourdan KB, Mason NA, Long L, Philips PG, Wilkins MR, Morrell NW (2001) Characterization of adenylyl cyclase isoforms in rat peripheral pulmonary arteries. Am J Physiol Lung Cell Mol Physiol 280(6):L1359–L1369

Kharitonov SA, Cailes JB, Black CM, du Bois RM, Barnes PJ (1997) Decreased nitric oxide in the exhaled air of patients with systemic sclerosis with pulmonary hypertension. Thorax 52 (12):1051–1055

Klinger JR (2007) The nitric oxide/cGMP signaling pathway in pulmonary hypertension. Clin Chest Med 28(1):143–167

Klinger JR, Warburton RR, Pietras LA, Smithies O, Swift R, Hill NS (1999) Genetic disruption of atrial natriuretic peptide causes pulmonary hypertension in normoxic and hypoxic mice. Am J Physiol 276(5 Pt 1):L868–L874

Koeller KJ, Goeddel DV (1992) Molecular biology of the natriuretic peptides and their receptors. Circulation 86(4):1081–1088

Koelling TM, Kirmse M, Di Salvo TG, Dec GW, Zapol WM, Semigran MJ (1998) Inhaled nitric oxide improves exercise capacity in patients with severe heart failure and right ventricular dysfunction. Am J Cardiol 81(12):1494–1497

Lai L, Tan TM (2002) Role of glutathione in the multidrug resistance protein 4 (MRP4/ABCC4)-mediated efflux of cAMP and resistance to purine analogues. Biochem J 361(Pt 3):497–503

Lamba JK, Adachi M, Sun D, Tammur J, Schuetz EG, Allikmets R, Schuetz JD (2003) Nonsense mediated decay downregulates conserved alternatively spliced ABCC4 transcripts bearing nonsense codons. Hum Mol Genet 12(2):99–109

Li C, Krishnamurthy PC, Penmatsa H, Marrs KL, Wang XQ, Zaccolo M, Jalink K, Li M, Nelson DJ, Schuetz JD, Naren AP (2007) Spatiotemporal coupling of cAMP transporter to CFTR chloride channel function in the gut epithelia. Cell 131(5):940–951

Lincoln TM, Dey N, Sellak H (2001) cGMP-dependent protein kinase signaling mechanisms in smooth muscle: from the regulation of tone to gene expression. J Appl Physiol 91 (3):1421–1430

MacLean MR, Sweeney G, Baird M, McCulloch KM, Houslay M, Morecroft I (1996) 5-Hydroxytryptamine receptors mediating vasoconstriction in pulmonary arteries from control and pulmonary hypertensive rats. Br J Pharmacol 119(5):917–930

Maclean MR, Johnston ED, Mcculloch KM, Pooley L, Houslay MD, Sweeney G (1997) Phosphodiesterase isoforms in the pulmonary arterial circulation of the rat: changes in pulmonary hypertension. J Pharmacol Exp Ther 283(2):619–624, Nat Genet 18(1):11–12

Martínez-Mir A, Paloma E, Allikmets R, Ayuso C, del Rio T, Dean M, Vilageliu L, Gonzàlez-Duarte R, Balcells S (1998) Retinitis pigmentosa caused by a homozygous mutation in the Stargardt disease gene ABCR. Nat Genet 18(1):11–12

McLaughlin VV, Genthner DE, Panella MM, Rich S (1998) Reduction in pulmonary vascular resistance with long-term epoprostenol (prostacyclin) therapy in primary pulmonary hypertension. N Engl J Med 338(5):273–277

Moncada S, Gryglewski R, Bunting S, Vane JR (1976) An enzyme isolated from arteries transforms prostaglandin endoperoxides to an unstable substance that inhibits platelet aggregation. Nature 263(5579):663–665

Murray F, Patel HH, Suda RY, Zhang S, Thistlethwaite PA, Yuan JX, Insel PA (2006) Expression and activity of cAMP phosphodiesterase isoforms in pulmonary artery smooth muscle cells from patients with pulmonary hypertension: role for PDE1. Am J Physiol Lung Cell Mol Physiol 292(1):L294–L303

Narumiya S, Sugimoto Y, Ushikubi F (1999) Prostanoid receptors: structures, properties, and functions. Physiol Rev 79(4):1193–1226

Olschewski H, Walmrath D, Schermuly R, Ghofrani A, Grimminger F, Seeger W (1996) Aerosolized prostacyclin and iloprost in severe pulmonary hypertension. Ann Intern Med 124(9):820–824

Olschewski H, Ghofrani HA, Walmrath D, Schermuly R, Temmesfeld-Wollbruck B, Grimminger F, Seeger W (1999) Inhaled prostacyclin and iloprost in severe pulmonary hypertension secondary to lung fibrosis. Am J Respir Crit Care Med 160(2):600–607

Pauli-Magnus C, Stieger B, Meier Y, Kullak-Ublick GA, Meier PJ (2005) Enterohepatic transport of bile salts and genetics of cholestasis. J Hepatol 43(2):342–357

Pepke-Zaba J, Higenbottam TW, Dinh-Xuan AT, Stone D, Wallwork J (1991) Inhaled nitric oxide as a cause of selective pulmonary vasodilatation in pulmonary hypertension. Lancet 338 (8776):1173–1174

Reichel V, Masereeuw R, van den Heuvel JJ, Miller DS, Fricker G (2007) Transport of a fluorescent cAMP analog in teleost proximal tubules. Am J Physiol Regul Integr Comp Physiol 293(6):R2382–R2389

Rindler MJ, Bashor MM, Spitzer N, Saier MH Jr (1978) Regulation of adenosine $3':5'$-monophosphate efflux from animal cells. J Biol Chem 253(15):5431–5436

Riordan JR, Rommens JM, Kerem B, Alon N, Rozmahel R, Grzelczak Z, Zielenski J, Lok S, Plavsic N, Chou JL et al (1989) Identification of the cystic fibrosis gene: cloning and characterization of complementary DNA. Science 245(4922):1066–1073

Rius M, Nies AT, Hummel-Eisenbeiss J, Jedlitschky G, Keppler D (2003) Cotransport of reduced glutathione with bile salts by MRP4 (ABCC4) localized to the basolateral hepatocyte membrane. Hepatology 38(2):374–384

Rosenzweig EB, Kerstein D, Barst RJ (1999) Long-term prostacyclin for pulmonary hypertension with associated congenital heart defects. Circulation 99(14):1858–1865

Russel FG, Koenderink JB, Masereeuw R (2008) Multidrug resistance protein 4 (MRP4/ABCC4): a versatile efflux transporter for drugs and signalling molecules. Trends Pharmacol Sci 29 (4):200–207

Sager G (2004) Cyclic GMP transporters. Neurochem Int 45(6):865–873

Sassi Y, Lipskaia L, Vandecasteele G, Nikolaev VO, Hatem SN, Cohen Aubart F, Russel FG, Mougenot N, Vrignaud C, Lechat P, Lompré AM, Hulot JS (2008) Multidrug resistance-associated protein 4 regulates cAMP-dependent signaling pathways and controls human and rat SMC proliferation. J Clin Invest 118(8):2747–2757

Sassi Y, Hara Y, Lompré AM, Hulot JS (2009) Multi-drug resistance protein 4 (MRP4/ABCC4) and cyclic nucleotides signaling pathways. Cell Cycle 8(7):962–963

Sassi Y, Abi-Gerges A, Fauconnier J, Mougenot N, Reiken S, Haghighi K, Kranias EG, Marks AR, Lacampagne A, Engelhardt S, Hatem SN, Lompre AM, Hulot JS (2012) Regulation of cAMP homeostasis by the efflux protein MRP4 in cardiac myocytes. FASEB J 26(3):1009–1017

Sauna ZE, Nandigama K, Ambudkar SV (2004) Multidrug resistance protein 4 (ABCC4)-mediated ATP hydrolysis: effect of transport substrates and characterization of the post-hydrolysis transition state. J Biol Chem 279(47):48855–48864

Schermuly RT, Pullamsetti SS, Kwapiszewska G, Dumitrascu R, Tian X, Weissmann N, Ghofrani HA, Kaulen C, Dunkern T, Schudt C, Voswinckel R, Zhou J, Samidurai A, Klepetko W, Paddenberg R, Kummer W, Seeger W, Grimminger F (2007) Phosphodiesterase 1 upregulation in pulmonary arterial hypertension: target for reverse-remodeling therapy. Circulation 115 (17):2331–2339

Schermuly RT, Stasch JP, Pullamsetti SS, Middendorff R, Müller D, Schlüter KD, Dingendorf A, Hackemack S, Kolosionek E, Kaulen C, Dumitrascu R, Weissmann N, Mittendorf J, Klepetko W, Seeger W, Ghofrani HA, Grimminger F (2008) Expression and function of soluble guanylate cyclase in pulmonary arterial hypertension. Eur Respir J 32(4):881–891

Schermuly RT, Janssen W, Weissmann N, Stasch JP, Grimminger F, Ghofrani HA (2011) Riociguat for the treatment of pulmonary hypertension. Expert Opin Investig Drugs 20 (4):567–576

Schuetz JD, Connelly MC, Sun D, Paibir SG, Flynn PM, Srinivas RV, Kumar A, Fridland A (1999) MRP4: a previously unidentified factor in resistance to nucleoside-based antiviral drugs. Nat Med 5(9):1048–1051

Shulenin S, Nogee LM, Annilo T, Wert SE, Whitsett JA, Dean M (2004) ABCA3 gene mutations in newborns with fatal surfactant deficiency. N Engl J Med 350(13):1296–1303

Strautnieks SS, Bull LN, Knisely AS, Kocoshis SA, Dahl N, Arnell H, Sokal E, Dahan K, Childs S, Ling V, Tanner MS, Kagalwalla AF, Németh A, Pawlowska J, Baker A, Mieli-Vergani G, Freimer NB, Gardiner RM, Thompson RJ (1998) A gene encoding a liver-specific ABC transporter is mutated in progressive familial intrahepatic cholestasis. Nat Genet 20 (3):233–238

Sundkvist E, Jaeger R, Sager G (2002) Pharmacological characterization of the ATP-dependent low K(m) guanosine $3',5'$-cyclic monophosphate (cGMP) transporter in human erythrocytes. Biochem Pharmacol 63(5):945–949

Sutherland EW, Rall TW (1958) Fractionation and characterization of a cyclic adenine ribonucle-otide formed by tissue particles. J Biol Chem 232:1077–1091

Van Aubel RA, Smeets PH, Peters JG, Bindels RJ, Russel FG (2002) The MRP4/ABCC4 gene encodes a novel apical organic anion transporter in human kidney proximal tubules: putative efflux pump for urinary cAMP and cGMP. J Am Soc Nephrol 13(3):595–603

Wharton J, Davie N, Upton PD, Yacoub MH, Polak JM, Morrell NW (2000) Prostacyclin analogues differentially inhibit growth of distal and proximal human pulmonary artery smooth muscle cells. Circulation 102(25):3130–3136

Wielinga PR, Reid G, Challa EE, van der Heijden I, van Deemter L, de Haas M, Mol C, Kuil AJ, Groeneveld E, Schuetz JD, Brouwer C, De Abreu RA, Wijnholds J, Beijnen JH, Borst P (2002) Thiopurine metabolism and identification of the thiopurine metabolites transported by MRP4 and MRP5 overexpressed in human embryonic kidney cells. Mol Pharmacol 62(6):1321–1331

Wielinga PR, van der Heijden I, Reid G, Beijnen JH, Wijnholds J, Borst P (2003) Characterization of the MRP4- and MRP5-mediated transport of cyclic nucleotides from intact cells. J Biol Chem 278(20):17664–17671

Yokochi A, Itoh H, Maruyama J, Zhang E, Jiang B, Mitani Y, Hamada C, Maruyama K (2010) Colforsin-induced vasodilation in chronic hypoxic pulmonary hypertension in rats. J Anesth 24(3):432–440

Zhang Y, Schuetz JD, Elmquist WF, Miller DW (2004) Plasma membrane localization of multidrug resistance-associated protein homologs in brain capillary endothelial cells. J Pharmacol Exp Ther 311(2):449–455

Zhao L, Long L, Morrell NW, Wilkins MR (1999) NPR-A-Deficient mice show increased susceptibility to hypoxia-induced pulmonary hypertension. Circulation 99(5):605–607

Pulmonary Hypertension: Old Targets Revisited (Statins, PPARs, Beta-Blockers)

Geoffrey Watson, Eduardo Oliver, Lan Zhao, and Martin R. Wilkins

Abstract Pulmonary arterial hypertension is a therapeutic challenge. Despite progress in recent years with three drug classes—prostanoids, endothelin receptor antagonists and phosphodiesterase type 5 inhibitors—long-term patient survival remains poor. Importantly, the introduction and commercial success of these new treatments has been accompanied by growing interest in the pathology of pulmonary hypertension. This, in turn, has stimulated a re-evaluation of the molecular factors driving the structural remodelling of pulmonary arterioles and the opportunities to preserve right ventricular function in pulmonary hypertension. Academics with restricted access to new chemicals have turned to existing drugs to investigate new ideas. It is in this context that the role of statins, peroxisome proliferator-activated receptors (PPARs) and beta-blockers are of interest as potential treatments for pulmonary hypertension.

Keywords Treatment • Repurposing • Drug targets • Statins • PPARs • Betablockers

Contents

1 Introduction 532
2 Statins 532
  2.1 Molecular Action 533
  2.2 Animal Models 534
  2.3 Clinical Studies 535
  2.4 Future Prospects 536
3 Peroxisome Proliferator-Activated Receptor Agonists 537
  3.1 Molecular Action 537
  3.2 Animal Studies 538
  3.3 Clinical Potential 539

G. Watson • E. Oliver • L. Zhao • M.R. Wilkins (✉)
Imperial College London, Hammersmith Hospital, London W12 0NN, UK
e-mail: m.wilkins@imperial.ac.uk

4	Beta-Blockers	539
	4.1 Molecular Action	540
	4.2 Animal Studies	542
	4.3 Clinical Trials	543
	4.4 Future Prospects	543
References		543

1 Introduction

The costs and risks associated with the development of new drugs for new targets of therapeutic interest are well documented. The market for the drug has to be substantial to attract investment. Regulators have been prepared to recognise this as a problem for orphan diseases with an unmet clinical need and fast track drugs with therapeutic promise. The success of prostanoids, endothelin receptor antagonists and phosphodiesterase inhibitors as treatments for pulmonary arterial hypertension has stimulated interest in recognising the condition, among academics, practitioners and the pharmaceutical industry. As a result, physicians are recognising the disease more often, the demographics of the disease is changing and its prevalence increasing—but perhaps not enough to make pulmonary hypertension a large commercial market.

A valuable strategy for reducing the cost and risk of drug development is to "repurpose" drugs that have been developed for other diseases for pulmonary hypertension. Experience gained with patient exposure in another disease enables informed decisions on potential dose and safety monitoring in pulmonary hypertension. As new knowledge about the biology of pulmonary hypertension emerges, academics in particular look to the pharmacopeia for agents to investigate. Here we discuss the role of statins, peroxisome proliferator-activated receptors (PPARs) and beta-blockers as treatments for pulmonary hypertension.

2 Statins

Statins have a primary and secondary preventative role in reducing serum cholesterol levels and improving morbidity and mortality in cardiovascular disease (Pedersen et al. 2004). Statins inhibit 3-hydroxy-3-methylglutaryl-coenzyme A (HMG-CoA) reductase, which is required for cholesterol biosynthesis (Wang et al. 2008). However, there has been growing interest in the pleiotropic, cholesterol-independent effects of this drug class, particularly the role of statins in cell proliferation and apoptosis. The marked vascular remodelling that characterises pulmonary arterial hypertension (PAH) has sparked interest in statins as a treatment for this condition. An impressive body of data from animal

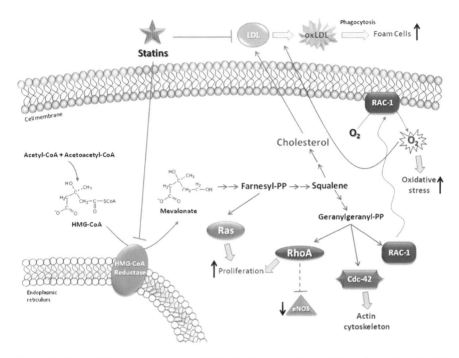

Fig. 1 Cholesterol biosynthesis pathway. Statins' pleiotropic effect on RhoA, Rac1, Ras, Cdc-42 small GTPases

models has triggered three human studies. But experience in patients has been disappointing and suggests statins are not a therapeutic option for targeting the vascular pathology.

2.1 Molecular Action

Statins inhibit the conversion of HMG-CoA to mevalonic acid, interrupting the synthesis of a number of key downstream molecules, notably the production of isoprenoids, such as farnesylpyrophosphate (FPP) and geranylgeranylpyrophosphate (GGPP) (Goldstein and Brown 1990). These are precursors of cholesterol synthesis but are essential for the post-translational isoprenylation, membrane localization and activation of Rho, Ras and Rac small GTP-binding protein families (Maurer-Stroh et al. 2007) (Fig. 1).

These small GTPases participate in the regulation of diverse cellular functions including smooth muscle contraction, cytoskeletal rearrangement, cell migration, proliferation, apoptosis and gene expression. RhoA and its downstream effector Rho-associated protein kinase (ROCK) are believed to be the most important regulators of calcium sensitivity in smooth muscle (Nagaoka et al. 2004; Nunes et al. 2010), via inhibition of myosin light chain phosphorylation (Somlyo and Somlyo 2003) and activation of downstream mediators (Sah et al. 2000). Rho/ROCK signalling has been linked with PAH; chronic blockade of ROCK by fasudil or Y-27632 reduces pulmonary vasoconstriction and vascular remodelling in monocrotaline or hypoxia-exposed rodent models (Fagan et al. 2004; Abe et al. 2004; Hyvelin et al. 2005).

Statins have been shown to enhance nitric oxide (NO) production by stabilising endothelial nitric oxide synthase (eNOS) mRNA (Laufs et al. 1997) and augmenting eNOS phosphorylation and catalytic activity (Kureishi et al. 2000). Interestingly, statins have also been reported to increase expression of bone morphogenetic protein receptor type II (BMPR2) (Hu et al. 2006) and have antithrombotic and anti-inflammatory effects. Collectively, this profile of properties makes statins an attractive drug class for PAH.

2.2 Animal Models

Most studies have used simvastatin as the probe compound, but other statins have also been investigated, particularly atorvastatin, pravastatin and rosuvastatin. These agents differ in potency and lipid solubility, with pravastatin the most hydrophilic.

Simvastatin (20 mg/kg/day) both prevents and regresses pulmonary hypertension, right ventricular hypertrophy (RVH) and vascular remodelling in chronically hypoxic rats (Girgis et al. 2003). More impressively, simvastatin 2 mg/kg/day orally both prevents and reverses pulmonary hypertension and neointima formation, and improves survival, in a challenging rat model in which pulmonary hypertension is induced by monocrotaline and pneumonectomy (Nishimura et al. 2002, 2003). McMurtry et al. found simvastatin (2 mg/kg/day, gavage) less effective in the monocrotaline model, retarding the progression of pulmonary hypertension and right ventricular hypertrophy (RVH) at 4 weeks but with no effect on survival (McMurtry et al. 2007). Simvastatin (10 mg/kg/day by gavage) led to a significant attenuation of the aggressive pulmonary vascular remodelling induced by vascular endothelial growth factor receptor blocker, Sugen 5416, in combination with chronic hypoxia (SuHx) (Taraseviciene-Stewart et al. 2006). Supporting observations have been made with pravastatin (Guerard et al. 2006), atorvastatin (Laudi et al. 2007), fluvastatin (Murata et al. 2005) and rosuvastatin (Sun and Ku 2008) in the monocrotaline and hypoxia rodent models.

In clinical practice, statins would be given in combination with other targeted treatments. Simvastatin and atorvastatin show a synergistic effect in combination with sildenafil (Satoh et al. 2009; Zhao et al. 2009). The combination of simvastatin and sildenafil augments eNOS expression, leading to a greater increase in lung cGMP levels and reduction in RhoA-GTP expression levels (Zhao et al. 2009).

Upregulation of eNOS and/or NO production has been reported with atorvastatin (Sun and Ku 2008) and pravastatin (Guerard et al. 2006). Changes in expression of the serotonin transporter (5-HTT), which has been implicated in the pathology of PAH, have been reported with atorvastatin treatment (Laudi et al. 2007).

2.3 Clinical Studies

Kao reported his experience with simvastatin in 16 patients with pulmonary hypertension of mixed aetiology. The drug was well tolerated with positive reports of an improvement in exercise capacity (Kao 2005).

The lack of a randomised control group was addressed in the simvastatin as a Treatment for Pulmonary Hypertension Trial (SiPHT) (Wilkins et al. 2010). This was a double-blind, randomised, placebo-controlled study (SiPHT) in 42 PAH patients in functional class II and III. Patients received simvastatin 80 mg/day or placebo, as an add-on to licensed oral treatment, for 6 months and thereafter offered open-labelled simvastatin. The primary outcome was RV mass assessed by cardiac magnetic resonance (CMR). This was based on the observation that a reduction in RV mass was the most consistent finding from the animal studies. Data on exercise capacity and biomarkers were also collected. At 6 months, simvastatin 80 mg in combination with sildenafil and/or endothelin antagonist was associated with a 5 % reduction in RV mass and a 13 % reduction in N-terminal pro-B-type natriuretic peptide (NT-pro-BNP). However, these effects were only transient and both markers returned to baseline in patients who continued the statin for 12 months. There were no significant changes in exercise capacity (6-minute walk test), cardiac index or circulating cytokines. A rise in muscle enzymes was seen in two patients on high-dose simvastatin and sildenafil, likely the result of competitive inhibition of simvastatin metabolism via CYP3A4, leading to a reduction in the dose of simvastatin.

The ASA-STAT trial recruited 65 patients with PAH and randomised them to aspirin or placebo and simvastatin 40 mg or placebo (Kawut et al. 2011). The primary outcome was change in 6-minute walk distance. Activated platelets produce thromboxane leading to platelet aggregation, vasoconstriction and smooth muscle hypertrophy. Aspirin was considered a potential treatment as it reduces prostaglandin–thromboxane (Tx-PGI$_2$) ratio in PAH and inhibits platelet activity (Robbins et al. 2006). After adjustment for baseline 6-minute walk distance, there was no significant difference in the 6-minute walk distance at 6 months between aspirin and placebo or between simvastatin and placebo.

The absence of haemodynamic data has been addressed by a recent large double-blind study of 220 Chinese patients with PAH or chronic thromboembolic pulmonary hypertension (CTEPH) randomised to receive atorvastatin 10 mg or placebo daily in addition to supportive care (Zeng et al. 2012). At 6 months, atorvastatin 10 mg had no significant effect on pulmonary vascular resistance (PVR), cardiac output or 6-minute walk distance in either PAH or CTEPH.

2.4 Future Prospects

The present data do not support the use of statins as a specific treatment for PAH. Statins may be used for cholesterol lowering in patients with PAH, an increasingly common occurrence as the mean age of PAH patients is rising and co-existing diseases, such as coronary artery disease, are more likely in the older age group. Care should be used with statins metabolised by CYP3A4, as sildenafil and bosentan are metabolised by the same enzyme and clinically significant drug interactions may occur.

The discrepancy between the results from animal studies and clinical studies is an important discussion point. There is concern about the failure of current animal models to predict drug response in patients. The animal models do not replicate the human pathology. Other factors include the dose of drugs used, species differences in metabolism and the design of animal experiments (such as the timing of initiating treatment with respect to the course of the disease, the lack of blinding and randomisation).

Could we have missed an important effect from statins? The drugs that make up this class vary in their physiochemical profiles. Lipophilic statins (simvastatin, atorvastatin, mevastatin and fluvastatin) exhibit direct effects on pulmonary artery smooth muscle cells (PASMCs) (Ali et al. 2011). Fluvastatin and atorvastatin have been shown to elicit more potent pleiotropic effects compared to simvastatin in other smooth muscle studies (Turner et al. 2007). But all inhibit mevalonate synthesis and so reduce FPP and GGPP levels. It is possible that specific targeting of FPP with preservation of GGPP levels might be an advantage, but this remains to be explored.

A confounding factor could be high-density lipoprotein cholesterol (HDL-C). HDL-C is linked to survival in PAH. Heresi et al. (2010) reported low plasma levels of HDL-C in patients with PAH, associated with worse clinical outcomes independent of other cardiovascular risk factors. It is acknowledged that patients with advanced PAH may have low HDL-C due to reduced exercise capacity and worsening obesity, but HDL-C was a prognostic marker when adjusted for age and body mass index (Heresi et al. 2010). It has been suggested that HDL-C may have a role in the initiation and/or progression of pulmonary vascular disease. Statins lower low-density lipoprotein (LDL) but do not lower HDL-C; indeed HDL-C increases marginally with some statins. Vickers and colleagues demonstrated HDL transports endogenous microRNAs (miRNAs) and delivers them to recipient cells with both endogenous and exogenous miRNA delivery producing direct targeting of messenger RNA reporters (Vickers et al. 2011). miRNAs are short non-coding RNAs that influence regulatory roles by targeting mRNA for cleavage and translational expression (Bartel 2004). Statin-induced changes in HDL-C could affect miRNA levels, but more extensive studies are needed to understand the role of miRNAs in PAH (Joshi et al. 2011).

3 Peroxisome Proliferator-Activated Receptor Agonists

The PPAR agonists, specifically the thiazolidinediones (rosiglitazone, pioglitazone), have achieved clinical recognition as oral hypoglycaemic agents that improve insulin sensitivity in the management of diabetes mellitus. The broader role of PPARs in regulating the transcription of genes involved in inflammation, proliferation and apoptosis in smooth muscle (de Dios et al. 2003) and the increasing appreciation of metabolic syndrome in patients with PAH have generated interest in PPARs as targets for this condition.

3.1 Molecular Action

PPARs are members of subfamily 1 of the nuclear hormone receptor superfamily (Nuclear Receptors Nomenclature Committee 1999). The three separate PPAR isoforms (α, β/δ, γ) are distributed throughout different tissues. PPARα is expressed predominantly in liver, heart, kidney, brown adipose tissue and small intestine (Mandard et al. 2004). PPAR β/δ has a broad distribution. PPARγ has four isoforms. PPARγ1 is expressed in virtually all tissues, including heart, muscle, colon, kidney, pancreas and spleen. PPARγ2 is expressed mainly in adipose tissue (30 amino acids longer). PPARγ3 is expressed in macrophages, large intestine and white adipose tissue. PPARγ4 is expressed in endothelial cells. PPARγ is activated by both natural lipid-derived substrates and synthetic ligands including the thiazolidinediones and fibrates. PPARγ has received most attention as the target isoform in PH but PPAR β/δ is also of interest

Activated PPARs form heterodimers with the retinoic acid X receptor (RXR) (Fig. 2). The PPAR–RXR heterodimeric complex binds to AGGTCA DNA sequences and regulates multiple genes, including those regulating interleukin-6, adiponectin, endothelin and nitric oxide (Jiang et al. 1998; Maeda et al. 2001; Calnek et al. 2003; Polikandriotis et al. 2005).

PPARγ is expressed in both endothelial and smooth muscle cells (SMCs) in the pulmonary vasculature of normal lungs. Ameshima et al. (2003) report reduced PPARγ gene and protein expression in the lungs from patients with severe PH and loss of PPARγ expression in their complex vascular lesions. Fluid shear stress reduced PPARγ expression in ECV304 endothelial cells in culture. ECV304 cells that stably express dominant-negative PPARγ (DN-PPARγ ECV304) form lumen-obliterating lung vascular lesions when injected into nude mice. The authors concluded that loss of PPARγ expression characterises an abnormal, proliferating, apoptosis-resistant endothelial cell phenotype in PAH.

PPARγ appears to be regulated by bone morphogenetic protein 2 (BMP2) and many of the downstream targets of PPARγ are factors that have been implicated in the pathogenesis of PAH. Hansmann and colleagues have described a BMP2–PPARγ–apolipoprotein E (apoE) pathway in human and murine vascular

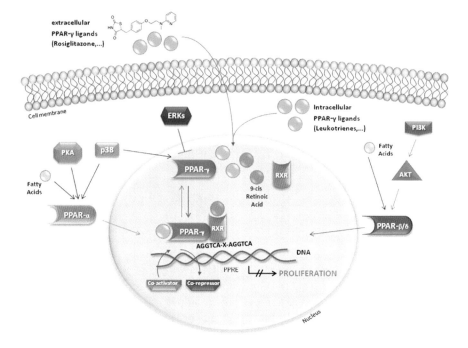

Fig. 2 Mechanism of PPAR activation. *PKA* protein Kinase A, *ERKs* extraceullar signal-regulated kinases, *RXR* retinoid X receptor, *AKT* protein kinase B

smooth muscle cells. Pharmacological activation of PPARγ inhibits PDGF-BB-induced proliferation of human and rat pulmonary artery smooth muscle cell by reducing nuclear ERK phosphorylation. Further investigation by this group revealed that mice with deletion of PPARγ in vascular smooth muscle cells spontaneously developed PAH. Very recently apelin has been identified as a transcriptional target of the PPARγ/β-catenin complex (Alastalo et al. 2011).

3.2 Animal Studies

Both pioglitazone and troglitazone have been reported to ameliorate monocrotaline-induced pulmonary hypertension in rodents (Matsuda et al. 2005).

Crossno et al. (2007) investigated the effects of rosiglitazone in the chronic hypoxic rat model of pulmonary hypertension. Rosiglitazone inhibited pulmonary vascular remodelling and, of particular note, reversed remodelling due to prior hypoxic exposure. The decreased remodelling appears to be due to repression of cell proliferation, extracellular matrix deposition and the inhibition of accumulation of c-Kit-positive cells in the pulmonary artery wall. But there was a dissociation between remodelling and pulmonary artery pressure which remained elevated.

Kim et al. (2010) found that the expression of PPARγ was reduced in the lungs of hypoxic rats, an effect that might be mediated via TGFβ signalling (Gong et al. 2011). Administration of rosiglitazone from the onset of hypoxia reduced pulmonary artery pressure, prevented the reduction in PPARγ levels and decreased the expression of ET-1 and VEGF.

Nisbet et al. (2010) have reported that rosiglitazone both prevents and reverses the development of pulmonary hypertension in the chronically hypoxic mouse. Nox4 expression and superoxide production in the lung were reduced, along with PDGFRβ activation and PTEN expression. Rosiglitazone also attenuated hypoxia-induced increases in Nox4 expression in pulmonary endothelial cells in vitro despite hypoxia-induced reductions in PPARγ expression.

Recently, PPAR β/δ agonists have been shown to relax pulmonary arteries with greater potency than PPARγ agonists. This effect is likely due to an action independent of target gene induction. Pharmacological studies suggest involvement of RhoA kinase and potassium channels (Harrington et al. 2010). Chronic administration of a PPAR β/δ agonist reduced pulmonary hypertension in the hypoxic rat.

3.3 Clinical Potential

The data from animal and in vitro models lend support to the therapeutic potential of PPAR agonists, particularly PPARγ and PPAR β/δ, in pulmonary hypertension. But clinical experience with PPAR agonists urges caution. Both rosiglitazone and pioglitazone have been under considerable scrutiny recently for their adverse cardiovascular risk profile (Tzoulaki et al. 2009; Sgarra et al. 2012). Of particular concern is fluid retention, the mechanism of which remains to be elucidated but which would not be well tolerated by patients with PAH. It may be that further refinement of agonists to develop an agent with a more favourable safety profile is possible, but this remains to be demonstrated.

4 Beta-Blockers

Catecholamines are integral to many physiological and metabolic processes. The effects of both natural and synthetic catecholamines rely on interactions with specific adrenergic receptors located on the cell membrane, resulting in either α- and/or β-mediated effects (Frishman 2008).

4.1 Molecular Action

There are three major adrenoceptors—α_1, α_2, and β (Bylund et al. 1994). Three β-adrenoceptor subtypes are recognised: β_1—comprising around 75 % of all β-adrenoceptors are found predominantly in the heart (Bylund et al. 1994); β_2— located in vascular and bronchial smooth muscle (Lopez-Sendon et al. 2004); and β_3—located in adipocytes and of interest in insulin resistance (Pott et al. 2003, 2006) but also in human myocardium (Gauthier et al. 1996). Numerous polymorphisms of beta-adrenoceptors have been described and some are associated with altered response to agonists. β_1- and β_2-adrenoceptors couple to Gs and thereby activate adenylyl cyclase and elevate cyclic AMP. β_3-adrenoceptors exert their effects through Gi coupling to cGMP–NO (Gauthier et al. 1998) (Fig. 3).

Beta-blockers vary in their cardio-selectivity and vasodilatory properties. Cardioselective agents (bisoprolol, metoprolol, nebivolol) are largely β_1-adrenoceptor selective but have some β_2 effects at higher doses. Nebivolol is a β_1-adrenoceptor antagonist but produces vasodilatation through β_3-adrenoceptor stimulation. Vasodilation can also be effected through α-adrenoceptor blockade (e.g., labetalol) (Ladage et al. 2013). Carvedilol activates both these vasodilatory mechanisms. Non-cardioselective agents, like propranolol, block β_1 and β_2 without vasodilation (McDevitt 1987).

Chronic heart failure is associated with downregulation of myocardial β_1- and β_2-adrenoceptors, perhaps attributable to sustained elevated levels of catecholamines (Bristow et al. 1986; Gaemperli et al. 2010). The reduction in β_1 signalling may be a protective mechanism causing downregulation of the PKA-independent CaMKII-mediated pro-apoptotic pathway. In contrast, β_2 adrenoceptor downregulation may be counterproductive (Bristow 2000). Unlike β_1/β_2-adrenoceptors, the β_3-adrenoceptor is upregulated in heart failure (Moniotte et al. 2001). Whether this is a protective mechanism against chronic sympathetic stimulation or a contributor to heart failure is unclear. Recently, β_3-adrenoceptor stimulation has been reported to reduce cardiac remodelling and improve cardiac function in a mouse model of aortic banding, suggesting a protective function (Niu et al. 2012).

Beta-blockade is one of the cornerstones of the treatment of left ventricular heart failure. The $\alpha_1/\beta_1/\beta_2$-adrenoceptor blocker, carvedilol, and the selective β_1-adrenoceptor blockers, bisoprolol and metoprolol, reduce mortality by about one-third. The data suggest reversal of maladaptive cardiac remodelling, improved cardiac function and prevention of arrhythmias (Reiter and Reiffel 1998; Hjalmarson et al. 2000; Packer et al. 2001; Bogaard et al. 2010).

Physicians are much more cautious about using beta-blockers in patients with PAH. Right ventricular function is a major determinant of survival in PAH (D'Alonzo et al. 1991; Sandoval et al. 1994; van de Veerdonk et al. 2011). Increased beta-adrenoceptor stimulation is seen in PAH. This may drive decompensatory changes and predispose to arrhythmias (Bogaard et al. 2009). Supraventricular tachyarrhythmias are a common problem in PAH (Tongers et al. 2007).

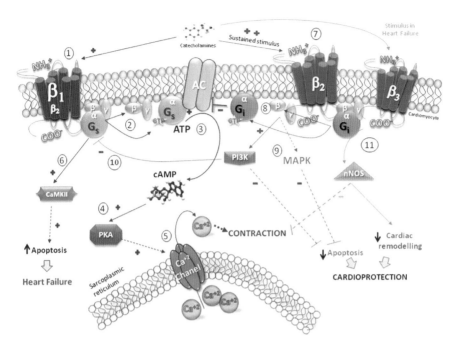

Fig. 3 Intracellular pathway mediated by β-ARs. (1) Main pathway: catecholamines bind to β-adrenoceptors inducing coupling to protein Gs; (2) uncoupling of the Gα-GTP subunit and activation of adenylyl cyclase (AC); (3) synthesis of cAMP; (4) protein kinase A (PKA) activation; (5) finally Ca^{2+} release from its stores, leading to contraction of cardiac muscle. (6) Continuous stimulation (as described in chronic heart failure) of $β_1$ induces apoptosis via Ca^{2+}/calmodulin-dependent protein kinases II (CaMKII) leading to muscular degeneration and heart damage. (7) Continued stimulation of $β_2$ (increased when using selective $β_1$-blockers) induces coupling to protein Gi; (8) AC is inhibited by the Gα-GTP subunit of protein Gi; (9) the Gβ subunit of Gi induces both the inhibition of apoptosis [via mitogen-activated protein kinases (MAPK) and phosphatidylinositol 3-kinase (PI3K)] and inhibition of protein Gs deleterious effects (10), leading to cardioprotection. (11) In heart failure, stimulation of the $β_3$-adrenoceptor leads to cardioprotection and a reduction in cardiac remodelling via neuronal nitric oxide synthase (nNOS) activation

Cardiac output in PAH is dependent upon the ability of the heart to increase its rate of contraction. Beta-blockers have been contraindicated from concern over their negative chronotropic effects as well as right ventriculo-arterial uncoupling (De Man et al. 2012). This has been supported by individual case studies of life-threatening syncope induced by beta-blockade (Peacock and Ross 2010).

Aside from effects on the myocardium, beta-blockers may also influence pulmonary vascular tone, but this is not well understood. Carvedilol inhibits proliferation of pulmonary artery smooth muscle cells isolated from patients with idiopathic PAH (Fujio et al. 2006). Recent studies in animal models suggest that further investigation of beta-blockers in PAH is warranted.

4.2 Animal Studies

The non-selective $\beta_{1,2}$-adrenoceptor blocker, propranolol, and the α_1-adrenoceptor blocker, prazosin, blunt hypoxic pulmonary vasoconstriction and hypoxia-induced pulmonary vascular remodelling (Voelkel et al. 1980; Tual et al. 2006; Bogaard et al. 2010). Recently, it was shown that pulmonary vascular remodelling in the rat monocrotaline model of pulmonary hypertension is attenuated by the α_1/β_1-adrenoceptor blocker, arotinolol (Ishikawa et al. 2009).

Bogaard and colleagues have studied the effects of carvedilol 15 mg/kg, administered for 4 weeks, in the rat SuHx model of pulmonary hypertension (Bogaard et al. 2010). The beta-blocker reduced heart rate and RV mass without changes in pulmonary vascular remodelling. Interestingly, RV systolic pressure showed no significant changes, but this may be a consequence of enhanced cardiac output with improved RV function. At the molecular level, the reversal of maladaptive RV remodelling in this model was associated with normalisation of vascular endothelial growth factor, Serca2a expression and protein kinase G (PKG) activity and enhanced expression of β-MHC, ANP and antihypertrophic MCIP-1. Exercise capacity was significantly improved in the carvedilol-treated group.

In a follow-on experiment, this group assessed whether improvements in right heart disease were due to the direct blocking of β-adrenoceptors or whether these changes were from the pleiotropic effects from the non cardio-selective properties of carvedilol. Carvedilol was compared to the more selective β_1-adrenoceptor blocker, metoprolol, in both SuHx and monocrotaline-induced PH rats. Echocardiographic assessment revealed improvement in RV function and attenuation of RV hypertrophy with both drugs, but smaller trends in the metoprolol group. Metoprolol had little effect on mRNA expression of Serca2a and MCIP-1 in the SuHx RV.

De Man et al. investigated the role of chronic low-dose bisoprolol (10 mg/kg) in the monocrotaline rat model (De Man et al. 2012). Echocardiographic measurements showed that bisoprolol delayed right heart failure by slowing down RV dilation and deterioration in cardiac output. The small dose of bisoprolol, enough to blunt heart rate response, showed minimal effects on systemic blood pressure. Bisoprolol significantly delayed time to right heart failure. RV load was not affected. Echocardiography indicated better ventriculo-arterial coupling in the bisoprolol-treated group. No changes in RV capillary density were observed, but RV fibrosis was reduced as were the number of inflammatory (CD45+) cells with bisoprolol treatment. Histological and biochemical analysis demonstrated restoration in both cMyBPC and cTnI contractile proteins compared to the controls.

Niu et al. (2012) studied the effects of β_3-adrenoceptor agonism on cardiac function and hypertrophy induced by pressure overload. C57BL/6J and neuronal NOS (nNOS) knockout mice underwent transverse aortic constriction (TAC) and were given a β_3 agonist (BRL37344) or placebo. At 3 weeks there was an increase in nitric oxide production with attenuation in left ventricular dilation and systolic function in the β_3 agonist group, with partial reduction in cardiac hypertrophy induced by TAC. This raises the therapeutic potential for targeting this receptor as part of heart failure management in the clinical setting.

4.3 Clinical Trials

Several reports have demonstrated that sympathetic activity is increased in patients with pulmonary hypertension (Velez-Roa et al. 2004; McGowan et al. 2009). Beta-blocker use is not uncommon in PAH patients. Around 12 % of subjects in the Registry to Evaluate Early and Long-term PAH Disease Management (REVEAL Registry) are recorded as receiving beta-blockers. A small study assessing ten patients with portopulmonary hypertension suggested that withdrawing propranolol or atenolol improved exercise tolerance (Provencher et al. 2006). A recent prospective cohort study of 94 consecutive patients with PAH compared outcomes in subjects who were and were not taking a beta-blocker at baseline (So et al. 2012). Beta-blocker use was common (28 %) in this cohort. After a median follow-up of 20 months, changes in pulmonary haemodynamics and right ventricular size and function were similar between groups. There were no statistically significant differences in adverse events including PAH-related hospitalisation or all-cause mortality ($p = 0.19$), presence of right heart failure ($p = 0.75$) or change in 6-minute walk distance ($p = 0.92$).

The authors conclude that beta-blocker use was not associated with detrimental effects and that a randomised study to evaluate the potential benefits and risks of this drug class in PAH was needed.

4.4 Future Prospects

There remains debate about the value and safety of beta-blockers in PAH, but there is sufficient interest to merit cautious investigation. At least two trials are recruiting patients with idiopathic PAH. In one study around 30 patients will be randomised to bisoprolol in gradually escalating doses or placebo in a crossover design and the effect on right ventricular function closely monitored by cardiac MR (clinicaltrials.gov identifier NCT01246037). In another, patients will receive carvedilol (clinicaltrials.gov identifier NCT01586156). The results of these studies will inform the future of beta-blockers as a treatment for PAH.

References

Abe K, Shimokawa H, Morikawa K, Uwatoku T, Oi K, Matsumoto Y, Hattori T, Nakashima Y, Kaibuchi K, Sueishi K, Takeshit A (2004) Long-term treatment with a Rho-kinase inhibitor improves monocrotaline-induced fatal pulmonary hypertension in rats. Circ Res 94:385–393

Alastalo TP, Li M, Perez VJ, Pham D, Sawada H, Wang JK, Koskenvuo M, Wang L, Freeman BA, Chang HY, Rabinovitch M (2011) Disruption of PPARgamma/beta-catenin-mediated regulation of apelin impairs BMP-induced mouse and human pulmonary arterial EC survival. J Clin Invest 121:3735–3746

Ali OF, Growcott EJ, Butrous GS, Wharton J (2011) Pleiotropic effects of statins in distal human pulmonary artery smooth muscle cells. Respir Res 12:137

Ameshima S, Golpon H, Cool CD, Chan D, Vandivier RW, Gardai SJ, Wick M, Nemenoff RA, Geraci MW, Voelkel NF (2003) Peroxisome proliferator-activated receptor gamma (PPARgamma) expression is decreased in pulmonary hypertension and affects endothelial cell growth. Circ Res 92:1162–1169

Bartel DP (2004) MicroRNAs: genomics, biogenesis, mechanism, and function. Cell 116:281–297

Bogaard HJ, Abe K, Vonk NA, Voelkel NF (2009) The right ventricle under pressure: cellular and molecular mechanisms of right-heart failure in pulmonary hypertension. Chest 135:794–804

Bogaard HJ, Natarajan R, Mizuno S, Abbate A, Chang PJ, Chau VQ, Hoke NN, Kraskauskas D, Kasper M, Salloum FN, Voelkel NF (2010) Adrenergic receptor blockade reverses right heart remodeling and dysfunction in pulmonary hypertensive rats. Am J Respir Crit Care Med 182:652–660

Bristow MR (2000) Beta-adrenergic receptor blockade in chronic heart failure. Circulation 101:558–569

Bristow MR, Ginsburg R, Umans V, Fowler M, Minobe W, Rasmussen R, Zera P, Menlove R, Shah P, Jamieson S et al (1986) Beta 1- and beta 2-adrenergic-receptor subpopulations in nonfailing and failing human ventricular myocardium: coupling of both receptor subtypes to muscle contraction and selective beta 1-receptor down-regulation in heart failure. Circ Res 59:297–309

Bylund DB, Eikenberg DC, Hieble JP, Langer SZ, Lefkowitz RJ, Minneman KP, Molinoff PB, Ruffolo RR Jr, Trendelenburg U (1994) International Union of Pharmacology nomenclature of adrenoceptors. Pharmacol Rev 46:121–136

Calnek DS, Mazzella L, Roser S, Roman J, Hart CM (2003) Peroxisome proliferator-activated receptor gamma ligands increase release of nitric oxide from endothelial cells. Arterioscler Thromb Vasc Biol 23:52–57

Committee NRN (1999) A unified nomenclature system for the nuclear receptor superfamily. Cell 97:161–163

Crossno JT Jr, Garat CV, Reusch JE, Morris KG, Dempsey EC, McMurtry IF, Stenmark KR, Klemm DJ (2007) Rosiglitazone attenuates hypoxia-induced pulmonary arterial remodeling. Am J Physiol Lung Cell Mol Physiol 292:L885–L897

D'Alonzo GE, Barst RJ, Ayres SM, Bergofsky EH, Brundage BH, Detre KM, Fishman AP, Goldring RM, Groves BM, Kernis JT et al (1991) Survival in patients with primary pulmonary hypertension. Results from a national prospective registry. Ann Intern Med 115:343–349

de Dios ST, Bruemmer D, Dilley RJ, Ivey ME, Jennings GL, Law RE, Little PJ (2003) Inhibitory activity of clinical thiazolidinedione peroxisome proliferator activating receptor-gamma ligands toward internal mammary artery, radial artery, and saphenous vein smooth muscle cell proliferation. Circulation 107:2548–2550

de Man FS, Handoko ML, van Ballegoij JJ, Schalij I, Bogaards SJ, Postmus PE, van der Velden J, Westerhof N, Paulus WJ, Vonk-Noordegraaf A (2012) Bisoprolol delays progression towards right heart failure in experimental pulmonary hypertension. Circ Heart Fail 5:97–105

Fagan KA, Oka M, Bauer NR, Gebb SA, Ivy DD, Morris KG, McMurtry IF (2004) Attenuation of acute hypoxic pulmonary vasoconstriction and hypoxic pulmonary hypertension in mice by inhibition of Rho-kinase. Am J Physiol Lung Cell Mol Physiol 287:L656–L664

Frishman WH (2008) beta-Adrenergic blockers: a 50-year historical perspective. Am J Ther 15:565–576

Fujio H, Nakamura K, Matsubara H, Kusano KF, Miyaji K, Nagase S, Ikeda T, Ogawa A, Ohta-Ogo K, Miura D, Miura A, Miyazaki M, Date H, Ohe T (2006) Carvedilol inhibits proliferation of cultured pulmonary artery smooth muscle cells of patients with idiopathic pulmonary arterial hypertension. J Cardiovasc Pharmacol 47:250–255

Gaemperli O, Liga R, Spyrou N, Rosen SD, Foale R, Kooner JS, Rimoldi OE, Camici PG (2010) Myocardial beta-adrenoceptor down-regulation early after infarction is associated with long-term incidence of congestive heart failure. Eur Heart J 31:1722–1729

Gauthier C, Tavernier G, Charpentier F, Langin D, Le MH (1996) Functional beta3-adrenoceptor in the human heart. J Clin Invest 98:556–562

Gauthier C, Leblais V, Kobzik L, Trochu JN, Khandoudi N, Bril A, Balligand JL, Le MH (1998) The negative inotropic effect of beta3-adrenoceptor stimulation is mediated by activation of a nitric oxide synthase pathway in human ventricle. J Clin Invest 102:1377–1384

Girgis RE, Li D, Zhan X, Garcia JG, Tuder RM, Hassoun PM, Johns RA (2003) Attenuation of chronic hypoxic pulmonary hypertension by simvastatin. Am J Physiol Heart Circ Physiol 285: H938–H945

Goldstein JL, Brown MS (1990) Regulation of the mevalonate pathway. Nature 343:425–430

Gong K, Xing D, Li P, Aksut B, Ambalavanan N, Yang Q, Nozell SE, Oparil S, Chen YF (2011) Hypoxia induces downregulation of PPAR-gamma in isolated pulmonary arterial smooth muscle cells and in rat lung via transforming growth factor-beta signaling. Am J Physiol Lung Cell Mol Physiol 301:L899–L907

Guerard P, Rakotoniaina Z, Goirand F, Rochette L, Dumas M, Lirussi F, Bardou M (2006) The HMG-CoA reductase inhibitor, pravastatin, prevents the development of monocrotaline-induced pulmonary hypertension in the rat through reduction of endothelial cell apoptosis and overexpression of eNOS. Naunyn Schmiedebergs Arch Pharmacol 373:401–414

Harrington LS, Moreno L, Reed A, Wort SJ, Desvergne B, Garland C, Zhao L, Mitchell JA (2010) The PPARbeta/delta agonist GW0742 relaxes pulmonary vessels and limits right heart hypertrophy in rats with hypoxia-induced pulmonary hypertension. PLoS One 5:e9526

Heresi GA, Aytekin M, Newman J, DiDonato J, Dweik RA (2010) Plasma levels of high-density lipoprotein cholesterol and outcomes in pulmonary arterial hypertension. Am J Respir Crit Care Med 182:661–668

Hjalmarson A, Goldstein S, Fagerberg B, Wedel H, Waagstein F, Kjekshus J, Wikstrand J, El AD, Vitovec J, Aldershvile J, Halinen M, Dietz R, Neuhaus KL, Janosi A, Thorgeirsson G, Dunselman PH, Gullestad L, Kuch J, Herlitz J, Rickenbacher P, Ball S, Gottlieb S, Deedwania P (2000) Effects of controlled-release metoprolol on total mortality, hospitalizations, and well-being in patients with heart failure: the Metoprolol CR/XL Randomized Intervention Trial in congestive heart failure (MERIT-HF). MERIT-HF Study Group. JAMA 283:1295–1302

Hu H, Sung A, Zhao G, Shi L, Qiu D, Nishimura T, Kao PN (2006) Simvastatin enhances bone morphogenetic protein receptor type II expression. Biochem Biophys Res Commun 339:59–64

Hyvelin JM, Howell K, Nichol A, Costello CM, Preston RJ, McLoughlin P (2005) Inhibition of Rho-kinase attenuates hypoxia-induced angiogenesis in the pulmonary circulation. Circ Res 97:185–191

Ishikawa M, Sato N, Asai K, Takano T, Mizuno K (2009) Effects of a pure alpha/beta-adrenergic receptor blocker on monocrotaline-induced pulmonary arterial hypertension with right ventricular hypertrophy in rats. Circ J 73:2337–2341

Jiang C, Ting AT, Seed B (1998) PPAR-gamma agonists inhibit production of monocyte inflammatory cytokines. Nature 391:82–86

Joshi SR, McLendon JM, Comer BS, Gerthoffer WT (2011) MicroRNAs-control of essential genes: implications for pulmonary vascular disease. Pulm Circ 1:357–364

Kao PN (2005) Simvastatin treatment of pulmonary hypertension: an observational case series. Chest 127:1446–1452

Kawut SM, Bagiella E, Lederer DJ, Shimbo D, Horn EM, Roberts KE, Hill NS, Barr RG, Rosenzweig EB, Post W, Tracy RP, Palevsky HI, Hassoun PM, Girgis RE (2011) Randomized clinical trial of aspirin and simvastatin for pulmonary arterial hypertension: ASA-STAT. Circulation 123:2985–2993

Kim EK, Lee JH, Oh YM, Lee YS, Lee SD (2010) Rosiglitazone attenuates hypoxia-induced pulmonary arterial hypertension in rats. Respirology 15:659–668

Kureishi Y, Luo Z, Shiojima I, Bialik A, Fulton D, Lefer DJ, Sessa WC, Walsh K (2000) The HMG-CoA reductase inhibitor simvastatin activates the protein kinase Akt and promotes angiogenesis in normocholesterolemic animals. Nat Med 6:1004–1010

Ladage D, Schwinger RH, Brixius K (2013) Cardio-selective beta-blocker: pharmacological evidence and their influence on exercise capacity. Cardiovasc Ther 31(2):76–83. doi:10.1111/j1755-5922.2011.00306x

Laudi S, Trump S, Schmitz V, West J, McMurtry IF, Mutlak H, Christians U, Weimann J, Kaisers U, Steudel W (2007) Serotonin transporter protein in pulmonary hypertensive rats treated with atorvastatin. Am J Physiol Lung Cell Mol Physiol 293:L630–L638

Laufs U, Fata VL, Liao JK (1997) Inhibition of 3-hydroxy-3-methylglutaryl (HMG)-CoA reductase blocks hypoxia-mediated down-regulation of endothelial nitric oxide synthase. J Biol Chem 272:31725–31729

Lopez-Sendon J, Swedberg K, McMurray J, Tamargo J, Maggioni AP, Dargie H, Tendera M, Waagstein F, Kjekshus J, Lechat P, Torp-Pedersen C (2004) Expert consensus document on beta-adrenergic receptor blockers. Eur Heart J 25:1341–1362

Maeda N, Takahashi M, Funahashi T, Kihara S, Nishizawa H, Kishida K, Nagaretani H, Matsuda M, Komuro R, Ouchi N, Kuriyama H, Hotta K, Nakamura T, Shimomura I, Matsuzawa Y (2001) PPARgamma ligands increase expression and plasma concentrations of adiponectin, an adipose-derived protein. Diabetes 50:2094–2099

Mandard S, Muller M, Kersten S (2004) Peroxisome proliferator-activated receptor alpha target genes. Cell Mol Life Sci 61:393–416

Matsuda Y, Hoshikawa Y, Ameshima S, Suzuki S, Okada Y, Tabata T, Sugawara T, Matsumura Y, Kondo T (2005) Effects of peroxisome proliferator-activated receptor gamma ligands on monocrotaline-induced pulmonary hypertension in rats. Nihon Kokyuki Gakkai Zasshi 43:283–288

Maurer-Stroh S, Koranda M, Benetka W, Schneider G, Sirota FL, Eisenhaber F (2007) Towards complete sets of farnesylated and geranylgeranylated proteins. PLoS Comput Biol 3:e66

McDevitt DG (1987) Pharmacologic aspects of cardioselectivity in a beta-blocking drug. Am J Cardiol 59:10F–12F

McGowan CL, Swiston JS, Notarius CF, Mak S, Morris BL, Picton PE, Granton JT, Floras JS (2009) Discordance between microneurographic and heart-rate spectral indices of sympathetic activity in pulmonary arterial hypertension. Heart 95:754–758

McMurtry MS, Bonnet S, Michelakis ED, Bonnet S, Haromy A, Archer SL (2007) Statin therapy, alone or with rapamycin, does not reverse monocrotaline pulmonary arterial hypertension: the rapamcyin-atorvastatin-simvastatin study. Am J Physiol Lung Cell Mol Physiol 293:L933–L940

Moniotte S, Kobzik L, Feron O, Trochu JN, Gauthier C, Balligand JL (2001) Upregulation of beta (3)-adrenoceptors and altered contractile response to inotropic amines in human failing myocardium. Circulation 103:1649–1655

Murata T, Kinoshita K, Hori M, Kuwahara M, Tsubone H, Karaki H, Ozaki H (2005) Statin protects endothelial nitric oxide synthase activity in hypoxia-induced pulmonary hypertension. Arterioscler Thromb Vasc Biol 25:2335–2342

Nagaoka T, Morio Y, Casanova N, Bauer N, Gebb S, McMurtry I, Oka M (2004) Rho/Rho kinase signaling mediates increased basal pulmonary vascular tone in chronically hypoxic rats. Am J Physiol Lung Cell Mol Physiol 287:L665–L672

Nisbet RE, Bland JM, Kleinhenz DJ, Mitchell PO, Walp ER, Sutliff RL, Hart CM (2010) Rosiglitazone attenuates chronic hypoxia-induced pulmonary hypertension in a mouse model. Am J Respir Cell Mol Biol 42:482–490

Nishimura T, Faul JL, Berry GJ, Vaszar LT, Qiu D, Pearl RG, Kao PN (2002) Simvastatin attenuates smooth muscle neointimal proliferation and pulmonary hypertension in rats. Am J Respir Crit Care Med 166:1403–1408

Nishimura T, Vaszar LT, Faul JL, Zhao G, Berry GJ, Shi L, Qiu D, Benson G, Pearl RG, Kao PN (2003) Simvastatin rescues rats from fatal pulmonary hypertension by inducing apoptosis of neointimal smooth muscle cells. Circulation 108:1640–1645

Niu X, Watts VL, Cingolani OH, Sivakumaran V, Leyton-Mange JS, Ellis CL, Miller KL, Vandegaer K, Bedja D, Gabrielson KL, Paolocci N, Kass DA, Barouch LA (2012) Cardioprotective effect of beta-3 adrenergic receptor agonism: role of neuronal nitric oxide synthase. J Am Coll Cardiol 59:1979–1987

Nunes KP, Rigsby CS, Webb RC (2010) RhoA/Rho-kinase and vascular diseases: what is the link? Cell Mol Life Sci 67:3823–3836

Packer M, Coats AJ, Fowler MB, Katus HA, Krum H, Mohacsi P, Rouleau JL, Tendera M, Castaigne A, Roecker EB, Schultz MK, DeMets DL (2001) Effect of carvedilol on survival in severe chronic heart failure. N Engl J Med 344:1651–1658

Peacock A, Ross K (2010) Pulmonary hypertension: a contraindication to the use of {beta}-adrenoceptor blocking agents. Thorax 65:454–455

Pedersen TR, Kjekshus J, Berg K, Haghfelt T, Faergeman O, Faergeman G, Pyorala K, Miettinen T, Wilhelmsen L, Olsson AG, Wedel H (2004) Randomised trial of cholesterol lowering in 4444 patients with coronary heart disease: the Scandinavian Simvastatin Survival Study (4S). 1994. Atheroscler Suppl 5:81–87

Polikandriotis JA, Mazzella LJ, Rupnow HL, Hart CM (2005) Peroxisome proliferator-activated receptor gamma ligands stimulate endothelial nitric oxide production through distinct peroxisome proliferator-activated receptor gamma-dependent mechanisms. Arterioscler Thromb Vasc Biol 25:1810–1816

Pott C, Brixius K, Bundkirchen A, Bolck B, Bloch W, Steinritz D, Mehlhorn U, Schwinger RH (2003) The preferential beta3-adrenoceptor agonist BRL 37344 increases force via beta1-/beta2-adrenoceptors and induces endothelial nitric oxide synthase via beta3-adrenoceptors in human atrial myocardium. Br J Pharmacol 138:521–529

Pott C, Steinritz D, Napp A, Bloch W, Schwinger RH, Brixius K (2006) On the function of beta3-adrenoceptors in the human heart: signal transduction, inotropic effect and therapeutic prospects. Wien Med Wochenschr 156:451–458

Provencher S, Herve P, Jais X, Lebrec D, Humbert M, Simonneau G, Sitbon O (2006) Deleterious effects of beta-blockers on exercise capacity and hemodynamics in patients with portopulmonary hypertension. Gastroenterology 130:120–126

Reiter MJ, Reiffel JA (1998) Importance of beta blockade in the therapy of serious ventricular arrhythmias. Am J Cardiol 82:9I–19I

Robbins IM, Kawut SM, Yung D, Reilly MP, Lloyd W, Cunningham G, Loscalzo J, Kimmel SE, Christman BW, Barst RJ (2006) A study of aspirin and clopidogrel in idiopathic pulmonary arterial hypertension. Eur Respir J 27:578–584

Sah VP, Seasholtz TM, Sagi SA, Brown JH (2000) The role of Rho in G protein-coupled receptor signal transduction. Annu Rev Pharmacol Toxicol 40:459–489

Sandoval J, Bauerle O, Palomar A, Gomez A, Martinez-Guerra ML, Beltran M, Guerrero ML (1994) Survival in primary pulmonary hypertension. Validation of a prognostic equation. Circulation 89:1733–1744

Satoh K, Fukumoto Y, Nakano M, Sugimura K, Nawata J, Demachi J, Karibe A, Kagaya Y, Ishii N, Sugamura K, Shimokawa H (2009) Statin ameliorates hypoxia-induced pulmonary hypertension associated with down-regulated stromal cell-derived factor-1. Cardiovasc Res 81:226–234

Sgarra L, Addabbo F, Potenza MA, Montagnani M (2012) Determinants of evolving metabolic and cardiovascular benefit/risk profiles of rosiglitazone therapy during the natural history of diabetes: molecular mechanisms in the context of integrated pathophysiology. Am J Physiol Endocrinol Metab 302:E1171–E1182

So PP, Davies RA, Chandy G, Stewart D, Beanlands RS, Haddad H, Pugliese C, Mielniczuk LM (2012) Usefulness of beta-blocker therapy and outcomes in patients with pulmonary arterial hypertension. Am J Cardiol 109:1504–1509

Somlyo AP, Somlyo AV (2003) Ca2+ sensitivity of smooth muscle and nonmuscle myosin II: modulated by G proteins, kinases, and myosin phosphatase. Physiol Rev 83:1325–1358

Sun X, Ku DD (2008) Rosuvastatin provides pleiotropic protection against pulmonary hypertension, right ventricular hypertrophy, and coronary endothelial dysfunction in rats. Am J Physiol Heart Circ Physiol 294:H801–H809

Taraseviciene-Stewart L, Scerbavicius R, Choe KH, Cool C, Wood K, Tuder RM, Burns N, Kasper M, Voelkel NF (2006) Simvastatin causes endothelial cell apoptosis and attenuates severe pulmonary hypertension. Am J Physiol Lung Cell Mol Physiol 291:L668–L676

Tongers J, Schwerdtfeger B, Klein G, Kempf T, Schaefer A, Knapp JM, Niehaus M, Korte T, Hoeper MM (2007) Incidence and clinical relevance of supraventricular tachyarrhythmias in pulmonary hypertension. Am Heart J 153:127–132

Tual L, Morel OE, Favret F, Fouillit M, Guernier C, Buvry A, Germain L, Dhonneur G, Bernaudin JF, Richalet JP (2006) Carvedilol inhibits right ventricular hypertrophy induced by chronic hypobaric hypoxia. Pflugers Arch 452:371–379

Turner NA, Midgley L, O'Regan DJ, Porter KE (2007) Comparison of the efficacies of five different statins on inhibition of human saphenous vein smooth muscle cell proliferation and invasion. J Cardiovasc Pharmacol 50:458–461

Tzoulaki I, Molokhia M, Curcin V, Little MP, Millett CJ, Ng A, Hughes RI, Khunti K, Wilkins MR, Majeed A, Elliott P (2009) Risk of cardiovascular disease and all cause mortality among patients with type 2 diabetes prescribed oral antidiabetes drugs: retrospective cohort study using UK general practice research database. BMJ 339:b4731

van de Veerdonk MC, Kind T, Marcus JT, Mauritz GJ, Heymans MW, Bogaard HJ, Boonstra A, Marques KM, Westerhof N, Vonk-Noordegraaf A (2011) Progressive right ventricular dysfunction in patients with pulmonary arterial hypertension responding to therapy. J Am Coll Cardiol 58:2511–2519

Velez-Roa S, Ciarka A, Najem B, Vachiery JL, Naeije R, van de Borne P (2004) Increased sympathetic nerve activity in pulmonary artery hypertension. Circulation 110:1308–1312

Vickers KC, Palmisano BT, Shoucri BM, Shamburek RD, Remaley AT (2011) MicroRNAs are transported in plasma and delivered to recipient cells by high-density lipoproteins. Nat Cell Biol 13:423–433

Voelkel NF, McMurtry IF, Reeves JT (1980) Chronic propranolol treatment blunts right ventricular hypertrophy in rats at high altitude. J Appl Physiol 48:473–478

Wang CY, Liu PY, Liao JK (2008) Pleiotropic effects of statin therapy: molecular mechanisms and clinical results. Trends Mol Med 14:37–44

Wilkins MR, Ali O, Bradlow W, Wharton J, Taegtmeyer A, Rhodes CJ, Ghofrani HA, Howard L, Nihoyannopoulos P, Mohiaddin RH, Gibbs JS (2010) Simvastatin as a treatment for pulmonary hypertension trial. Am J Respir Crit Care Med 181:1106–1113

Zeng WJ, Xiong CM, Zhao L, Shan GL, Liu ZH, Xue F, Gu Q, Ni XH, Zhao ZH, Cheng XS, Wilkins MR, He JG (2012) Atorvastatin in pulmonary hypertension (APATH) study. Eur Respir J 40(1):67–74

Zhao L, Sebkhi A, Ali O, Wojciak-Stothard B, Mamanova L, Yang Q, Wharton J, Wilkins MR (2009) Simvastatin and sildenafil combine to attenuate pulmonary hypertension. Eur Respir J 34:948–957

Part IV
Conclusions and Outlook

Pulmonary Hypertension: Current Management and Future Directions

Lewis J. Rubin

Abstract Despite major advances in our understanding of the pathobiology of pulmonary vasculopathy and in the treatment of pulmonary vascular disease, pulmonary hypertension remains a severe condition that frequently progresses to right heart failure and death. This chapter will review several novel approaches to treatment, based on recent observations in both patients and experimental models, that may lead to new and more effective treatment approaches to pulmonary hypertension in the future.

Keywords Pulmonary Artery Hypertension • Therapy • Future

Contents

1 Future Directions 553
2 Conclusions 554
References 554

The progress in elucidating the pathobiology of pulmonary vascular disease over the past two decades, and in particular over the past several years, has been both dramatic and gratifying. While all the details have not been pieced together, our basic understanding of the ways by which the pulmonary vessels can become injured, leading to vascular remodeling, has led to seminal investigations into the genetics of pulmonary vascular disease and the development of novel treatment strategies that have improved the quality of life and probably survival for many patients with pulmonary arterial hypertension (PAH).

L.J. Rubin (✉)
University of California, San Diego School of Medicine, San Diego, CA, USA
e-mail: ljrubin@ucsd.edu

It was not that long ago that therapy for PAH was empirical, based on the assumption that vasoconstriction was the predominant pathogenic factor in the development and progression of this disease. Systemic vasodilators were administered in the hope that these drugs may also have pulmonary vasodilatory properties that would provide therapeutic benefit to PAH patients; while some patients improved, particularly with calcium channel antagonists, most either had no benefit or experienced deterioration as a result of systemic vasodilation with no obvious effect on pulmonary vasomotor tone. Anticoagulants were administered based on the assumption that thrombosis, either in situ or as a result of recurrent asymptomatic embolism, may cause or complicate the disease process, and diuretics were given to maintain fluid balance once right heart failure ineroxably developed.

Given the orphan nature of PAH, most studies involved small populations and provided limited insight into optimal management. The NIH Registry on Primary Pulmonary Hypertension (D'Alonzo et al. 1991) was a landmark in the field not only by characterizing the demographic and clinical features of this condition, but also by bringing together for the first time a group of physicians representing several disciplines (cardiology, pulmonary medicine, pediatrics, pathology) to jointly study this complex and mystifying disease. Shortly after, others joined, including geneticists, clinical trial methodology experts, biostatisticians, molecular biologists, clinical pharmacologists, and surgeons, among many others. Soon multicenter and, subsequently, international clinical trials were implemented to prospectively evaluate specific drugs or treatment strategies. These trials have led to the demonstration that the pertubations in three endothelial-derived pathways (endothelin, nitric oxide, and prostacyclin) are targets for therapy, resulting in regulatory approval of drugs that specifically target these pathogenic pathways (Humbert et al. 2004).

These global collaborative networks have also resulted in the publication of guidelines for the diagnostic approach and treatment strategies for PAH (Galiè et al. 2004). While all diagnostic modalities and therapies are not universally available, these documents provide general guidance for all physicians who care for patients with pulmonary hypertension and are based on levels of evidence in the medical literature as well as consensus expert opinion. The importance of such guidelines will likely only increase in the future as novel diagnostic and therapeutic modalities make their way into the medical armamentarium for PAH. Unfortunately, it is unclear whether these widely published guidelines have gained entry into medical practice, and recent registries suggest that many physicians in community practice do not adhere to these guidances for management of patients with this complex disease process (McLaughlin et al. 2012). Recent studies have also suggested that patients are referred late in the course of their illness to experienced centers, and this delay in implementing aggressive treatment strategies impacts on outcome (Badagliacca et al. 2012).

1 Future Directions

Despite substantial improvements in the care of PAH patients, this disease remains incurable. Additionally, the optimal treatment strategy for PAH is not well defined, and many patients are unresponsive or incompletely responsive to medical therapy. Accordingly, there is still much work to be done. The following are my thoughts as to the future directions of work in this field:

1. *How can we optimize the use of currently available therapies?* While we have three classes of drugs that have shown efficacy in PAH (endothelin receptor antagonists, phosphodiesterase-type 5 inhibitors, and prostacyclin analogues), the optimal timing and utlization of these compounds remain unclear. More specifically, which drug or class of drug is the best with which to initiate therapy? Where does combination therapy fit into the scheme, early or as add-on? And which combination(s) are best, or is triple combination therapy superior to double? These questions, addressed in rigorous clinical trials, will ensure that the medications currently available are used to their maximum benefit.
2. *By what cellular mechanisms do the drugs currently available exert their beneficial effects?* Remarkably, after 15 years of availability, we still do not fully understand how prostacyclins exert their potent therapeutic effects in PAH. Some argue that it works as a pure vasodilator, but this seems unlikely since other simple vasodilators, such as calcium channel blockers, have little therapeutic benefit in most patients. Others have suggested an antiproliferative effect of both prostanoids and the other classes of drugs, but this is difficult to prove in the clinical setting. I think it is likely that prostacyclin has inotropic effects on the right ventricle that, combined with other effects on the pulmonary vessels, produces both acute and chronic benefit. Tapping into this unknown territory will lead to the development of novel therapies.
3. *What is the role of newer disease targeted therapies?* Recent trials with novel therapies designed to target specific aberrant signalling pathways in PAH have been disappointing, due to both the marginal treatment effect and the high incidence of serious side effects (Hoeper et al. 2013). Notably, despite impressive effects in animal models of pulmonary hypertension and several case reports demonstrating dramatic results with the tyrosine kinase inhibitor imatinib in a handful of patients refractory to other therapies, the results of large-scale trials failed to demonstrate a beneficial efficacy:safety ratio. Unknown at this time is whether the problem lies with the drug or with the trial design, but some experts believe that newer multikinase inhibitors may prove more effective and safe. No doubt additional potential targets will emerge in the future. Given the precious commodity of patients eligible and willing to participate in clinical trials of new agents in PAH, we must triage these targeted drugs based on the likelihood of efficacy and evidence of safety. Additionally, we need to thoughtfully develop the design of future trials to take into account currently available therapies and meaningful treatment outcomes. Finally, we

should be cautious in interpreting results of treatments in animal models of PAH, since to date many compounds that have exerted dramatic effects in animals with induced PAH, including statins, tyrosine kinase inhibitors, angiotensin-converting enzyme inhibitors, and others, have failed to demonstrate benefit in human PAH.

4. *Can we therapeutically target the right ventricle?* What kills patients with PAH is not the elevated pressures in the pulmonary vascular circuit, per se, but rather the progressive inability of the right ventricle to deal with the increased afterload, ultimately resulting in right ventricular failure. Accordingly, we have two potential targets in PAH—the vasculopathy and the failing right ventricle. The reversibility of right ventricular failure has been demonstrated after lung transplantation for PAH and pulmonary endarterectomy for chronic thromboembolic pulmonary hypertension (CTEPH), making it unnecessary to perform combined heart–lung transplantation in most PAH conditions. A greater understanding of the roles of impaired contractility and myocardial remodeling in PAH could yield novel approaches to targeted therapy of the failing right ventricle.

2 Conclusions

PAH, once a declaration of imminent death, is now a chronic condition which, for many patients, can be treated with improved quality of life and survival. However, PAH is a rare condition and its optimal management is best provided at expert centers. Nevertheless, those physicians who initially evaluate the PAH patient and participate in their ongoing collaboration with the expert centers need to be familiar with the rapid pace of developments in both early diagnosis and treatment. Finally, novel targets for therapy, such as the pathways for disordered vasoproliferation and myocardial adaptation, should be explored to enhance our treatment strategies and, ultimately, to cure this malignant disease.

References

Badagliacca R, Pezzuto B, Poscia R, Mancone M, Papa S, Marcon S, Valli G, Sardella G, Ferrante F, Iacoboni C, Parola D, Fedele F, Vizza CD (2012) Prognostic factors in severe pulmonary hypertension patients who need parenteral prostanoid therapy: the impact of late referral. J Heart Lung Transplant 31(4):364–372

D'Alonzo GE, Barst RJ, Ayres SM, Bergofsky EH, Brundage BH, Detre KM, Fishman AP, Goldring RM, Groves BM, Kernis JT et al (1991) Survival in patients with primary pulmonary hypertension. Results from a national prospective registry. Ann Intern Med 115(5):343–349

Galiè N, Torbicki A, Barst R, Dartevelle P, Haworth S, Higenbottam T, Olschewski H, Peacock A, Pietra G, Rubin LJ, Simonneau G, Priori SG, Garcia MA, Blanc JJ, Budaj A, Cowie M, Dean V, Deckers J, Burgos EF, Lekakis J, Lindahl B, Mazzotta G, McGregor K, Morais J,

Oto A, Smiseth OA, Barbera JA, Gibbs S, Hoeper M, Humbert M, Naeije R, Pepke-Zaba J (2004) Guidelines on diagnosis and treatment of pulmonary arterial hypertension. The Task Force on Diagnosis and Treatment of Pulmonary Arterial Hypertension of the European Society of Cardiology. Eur Heart J 25(24):2243–2278

Hoeper MM, Barst RJ, Bourge RC, Feldman J, Frost AE, Galié N, Gómez-Sánchez MA, Grimminger F, Grünig E, Hassoun PM, Morrell NW, Peacock AJ, Satoh T, Simonneau G, Tapson VF, Torres F, Lawrence D, Quinn DA, Ghofrani HA (2013) Imatinib mesylate as add-on therapy for pulmonary arterial hypertension: results of the randomized IMPRES study. Circulation 127(10):1128–1138

Humbert M, Sitbon O, Simonneau G (2004) Treatment of pulmonary arterial hypertension. N Engl J Med 351(14):1425–1436

McLaughlin VV, Langer A, Tan M, Clements PJ, Oudiz RJ, Tapson VF, Channick RN, Rubin LJ, Pulmonary Arterial Hypertension (PAH) Quality Enhancement Research Initiative (QuERI) Investigators (2012) Contemporary trends in the diagnosis and management of pulmonary arterial hypertension: an initiative to close the care gap. Chest 143:324–332. doi:10.1378/chest.11-3060

Index

A
AAV-Ang-1. *See* Adeno-associated virus-angiopoietin-1 (AAV-Ang-1)
ACE. *See* Angiotensin converting enzyme (ACE)
Acute respiratory distress syndrome (ARDS), 269
Acute vasodilator testing
  description, 167–168
  drugs, 168
  inhaled iloprost, 168
  PAP and PVR, 168
  pulmonary vasoresponsiveness, 168
Acute vasoreactivity testing, 267
Adeno-associated virus-angiopoietin-1 (AAV-Ang-1), 133
ADMA. *See* Asymmetrical dimethylarginine (ADMA)
Adrenomedullin (AM)
  and PH, 491–492
  structure and expression, 490–491
Adverse cell signaling events, 33
AM. *See* Adrenomedullin (AM)
Ambrisentan, 217–218
Anesthesia, 115
Ang-1. *See* Angiopoietin-1 (Ang-1) overexpression
Ang II. *See* Angiotensin II (Ang II)
Angiopoietin-1 (Ang-1) overexpression
  AAV-Ang-1, 133
  description, 133
  features, 133
  inflammatory changes, 133
Angiopoietins, 90
Angiotensin converting enzyme (ACE), 480
Angiotensin II (Ang II), 480
Angiotensin receptor 1 (AT1R), 480
Animal models
  group 1 PAH, 110–112
  hemodynamic measurements and imaging tools, 114–115
  molecular signaling
    apoptosis resistance, 110
    BMPR-2 expression, 110
    and cellular, 110
    DRP-1 inhibition, 113
    inflammation infiltration, PAH, 113–114
    normoxic activation, HIF-1α, 110, 113
    PASMCs, 110
    potassium channel abnormalities, PPH, 113, 114
  rodent imaging, 115
Anticoagulants, PAH, 156
Anti-endothelial cell (Anti-EC) antibodies, 448
Antifibroblast antibodies, 447–448
Anti-infectious agents
  antihelmintic treatment in schistosomiasis-PAH, 458–459
  antiretroviral therapy in HIV-PAH, 458
Aorto-caval shunt, 140
ARDS. *See* Acute respiratory distress syndrome (ARDS)
Arginase (I and II), 320
Arterial and venous remodeling, PAH.
  See Connective tissue disease (CTD)
Arterial lesions
  complex lesions (*see* Complex lesions)
  intimal fibrosis (*see* Intimal fibrosis)
  levels, anatomical/histological architectural pattern, 60
  medial hypertrophy (*see* Medial hypertrophy)
  occlusive/near-occlusive, 60

Arterial lesions (*cont.*)
    pathological anatomy, 61
    perivascular inflammation and tLTs, 66–67
    typical histological lesions, 61, 62
ASA-STAT trial, 535
Asymmetrical dimethylarginine (ADMA), 283, 319, 321
AT1R. *See* Angiotensin receptor 1 (AT1R)

B
BAY 41-2272
    beneficial effects, 293
    in vitro results, 288
    oral administration, 293–294
    preclinical studies, 288–293
    pulmonary artery infusion, 293
BAY 41-8543
    beneficial effects, preclinical studies, 289–294
    clinical application, 295
    intravenous injection, 294
    right atrial infusion, 294
    YC-1, 294
Beraprost, prostacyclins, 187
Beta-blockers
    animal studies, 542
    carvedilol, 543
    catecholamines, 539
    clinical trials, 543
    molecular action
        adrenoceptors, 540
        beta-adrenoceptors polymorphisms, 540
        cardio-selectivity and vasodilatory properties, 540
        carvedilol activates, 540
        chronic heart failure, 540
        intracellular pathway, β-ARs, 540, 541
        myocardium, 541
        nebivolol and vasodilation, 540
        negative chronotropic effects, 541
        physicians, 540
        supraventricular tachyarrhythmias, 540
Biomarkers, PH
    angiopoietins, 90
    blood biomarkers, 91, 93–94
    BNP and NT-proBNP, 92
    clinical and haemodynamic indices, 86–87
    CPET, 91
    creatinine, 89
    definitions, 78
    diagnosis, 79
    genetic biomarkers, 85–86
    growth and differentiation factor-15, 87–88
    imaging techniques, 92
    inflammatory markers, 89
    miRs, 91
    multiple markers, 92
    natriuretic peptides, 87
    radionuclide imaging, 91–92
    RDW (*see* Red cell distribution width (RDW))
    risk stratification, 79–80
    screening, 79
    specificity and sensitivity, 78
    statistical evaluation (*see* Statistical evaluation, biomarkers)
    susceptibility, 78–79
    temporality and biological gradient, 78
    therapeutic selection and monitoring, 80–81
    uric acid, 90
B lymphocytes, 447
BMP2-PPARg-apolipoprotein E (apoE) pathway, 537–538
BMPR-2-dominant negative transgenic mice
    isoforms, 130–131
    nifedipine, L-type calcium channel blocker, 130
    overexpression, 130
    SMC-specific promotor, 130
BMPR-II. *See* Bone morphogenetic protein receptor type 2 (BMPR-II)
BNP. *See* Brain natriuretic peptide (BNP)
Bone morphogenetic protein receptor type 2 (BMPR-II)
    in human PAECs, 50
    PAH *in vivo*, 38
    serine-threonine kinase, 39
    shear stress, ET-1 and hypoxia, 46
    signaling system
        antiproliferative strategy, 425
        Bmpr2 gene, 425
        kinase domain, 424
        pleiotropic roles, 425
        p38MAPK, 425
        steady-state levels, BMPR-IA, 425
        TGF β family, 424
        transmembrane receptor classes, 424
    signal transduction, 48
Bosentan
    BREATHE-1, 215
    cardiac index, 214–215
    description, 214
    elevation, aminotransferases, 216
    hormonal contraception, 217
    idiopathic PAH, 216
    Kaplan–Meier survival, 215–216
    6MWD, 214

Index

BPD. *See* Bronchopulmonary dysplasia (BPD)
Brain natriuretic peptide (BNP)
  cardiac biomarkers, 81
  Cox analysis, 83
  hypothesis-generating, 82
  natriuretic peptides, 87
  and NT-proBNP levels, 80
Broiler fowl disease, 132
Bronchopulmonary dysplasia (BPD), 266–267

C

Calcitonin gene-related peptide (CGRP)
  and PH, 493
  structure and expression, 492
Calcitonin receptor-like receptor (CRLR), 491
Calcium-channel blockers (CCBs)
  acute vasodilator testing, 167–169
  classification, 164–165
  clinical use
    amlodipine, 171
    diltiazem and nifedipine, 170
    systemic hypotension and edema, 171
    treatment algorithm, 171, 172
    verapamil, 170
  drug interactions, 167
  evidence-based treatment algorithm, 169
  idiopathic and anorexigen, 169
  indications, 166
  mechanism of action, 163–164
  side effects, 166
  SMC and vasodilatation, 162
  vasoconstriction, 167
  vasodilator therapy, 167
cAMP/cGMP pathways
  ABC transporters, 516
  blocking cyclic nucleotides degradation
    PDE1A and PDE1C mRNAs and proteins levels, 518
    PDE1 inhibition, 518
    phosphodiesterases (PDEs) enzymes, 517–518
    vinpocetine, 518
  enhancing cyclic nucleotide production, 516–517
  PASMC, 515
  synthesis, degradation and efflux, 515–516
Cardiomagnetic resonance (CMR), 535
Cardiopulmonary exercise testing (CPET), 91
Cav-1. *See* Caveolin-1 (Cav-1)
Caveolin-1 (Cav-1), 324
CCBs. *See* Calcium-channel blockers (CCBs)
C-C motif ligand (CCL), 441, 449
Cell signaling mechanisms
  cellular responses to hypoxia
    (*see* MicroRNA-mediated regulation)
  eNOS (*see* Endothelial nitric oxide synthase (eNOS))
  ET-1 (*see* Endothelin-1 (ET-1) system)
  mitochondrial dysfunction, 46–48
  PDE (*see* Phosphodiesterase (PDE) inhibition)
  PPAR-γ (*see* Peroxisome proliferator-activated receptor (PPAR-γ))
  prostacyclin signaling, 44–46
  sGC, 42–43
cGMP-dependent protein kinases (PKG)
  Ca^{2+}-activated K^+ channels, 232
  ligand-induced elevation, 232
  myosin light-chain phosphatase, 232
  NO/cGMP/PDE-5 pathway, 232
  N-terminal binding site, PKG1 isozymes, 231
  serine/threonine kinases, 231
  smooth muscle tone, 232
CGRP. *See* Calcitonin gene-related peptide (CGRP)
Chemokines, 441
Cholesterol biosynthesis pathway, 533
Chronic hemolytic anemia
  Doppler echocardiography, 14
  patients, sickle cell disease, 14
  post-capillary PH, 15
  prevalence, pre-capillary PH, 14
Chronic hypoxia rats
  inflammation, 124
  limitations, 125
  mesenchymal precursor and mononuclear cells, 124
  method, 123
  recording, CL and Ppa, 123, 124
  RV, 125
  and Sugen-5416, 128–129
  vascular obstruction/rarefaction, 123
  vascular remodeling, 124
Chronic obstructive pulmonary disease (COPD)
  and eNOS, 283
  NO production, 304
  riociguat, 302
  treatment, riociguat, 297
Chronic thromboembolic pulmonary hypertension (CTEPH)
  BAY 41-8543, 295
  cardiac index, 302
  clinical characteristics and management, patients, 19
  long-term follow-up, 19

Chronic thromboembolic pulmonary
         hypertension (CTEPH) (*cont.*)
    medical and surgical management, 18
    plasma levels, ADMA, 283
    prevalence, 18
    proximal and distal, 18
    randomized control trials, 19
    symptoms and pulmonary haemodynamics,
         302
Ciclosporin A (CsA)
    calcineurin, 458
    calcium synergic action and suppress, 458
    iloprost, 458
    and NFAT, 461
    T-cell production, 457–458
    treatment inhibits PH and RV hypertrophy,
         461
Clinical pharmacology, PDE-5
    cytochrome P-450 (CYP) isoenzyme, 237
    inhibitors, 236
    renal impairment, 238
    sildenafil citrate, 236
    tadalafil, 237, 238
    vardenafil, 236
Complex lesions
    arterial wall alteration within plexiform
         lesions, 65
    cardiac left-to-right shunting, 65
    central plexiform pattern and peripheral
         dilation lesions, 63, 64
    indirect anastomoses, 66
    latter anomaly, dilation, 65
    pathophysiologic significance, 64
    phenotypes, arterial remodeling, 65
    plexiform, 66
    prostacyclin treatment, 64
    shunt hypothesis, 66
    T and B lymphocytes, 65
    vascular compartments, 64
Conditional knockout mice, 130
Congenital heart diseases (CHDs)
    anatomic-pathophysiological classification,
         12, 13
    clinical classification, 12, 14
    Eisenmenger's syndrome, 12
Connective tissue disease (CTD)
    clinical subgroup, PAH, 10
    and inflammatory states, 442
    medial hypertrophy and adventitial
         thickening, 70–71
    muscular vessels, pre-and intra-acinar
         levels, 70
    to PAH, 463–464

PAH, patients, 70
pulmonary edema, 71
Sjögren syndrome, polymyositis/
         rheumatoid arthritis, 11
    and SLE, 442
    systemic scleroderma, 442
    systemic sclerosis, 10, 71
    treatment with vasodilators, 71
COPD. *See* Chronic obstructive pulmonary
         disease (COPD)
COX. *See* Cyclooxygenase (COX) pathway
CPET. *See* Cardiopulmonary exercise testing
         (CPET)
Creatinine, 89
CRLR. *See* Calcitonin receptor-like receptor
         (CRLR)
CsA. *See* Ciclosporin A (CsA)
CTD. *See* Connective tissue disease (CTD)
CTEPH. *See* Chronic thromboembolic
         pulmonary hypertension (CTEPH)
CYC. *See* Cyclophosphamide (CYC)
Cyclic guanosine monophosphate degradation
    cellular and organ functions, 286
    and GTP, 284
    NO signalling, 286
    PDE-5, 284
    vasodilation, 281
    YC-1, 287
Cyclic guanosine nucleotide monophosphate
    phosphodiesterase-5 inhibitors, 333–334
    sGC activators and stimulators, 334–336
Cyclic nucleotides
    efflux (*see* Efflux, cyclic nucleotides)
    and pathogenesis, 514–515
    pathways (*see* cAMP/cGMP pathways)
    PDE5 and MRP4 inhibitors, 523–524
Cyclooxygenase (COX) pathway
    classes, eicosanoids, 44
    isoforms, 44
    pathophysiology, PH, 45
Cyclophosphamide (CYC)
    alkylating agent, 456
    anti-inflammatory function, 456–457
    concomitant corticosteroids, 457
    and glucocorticoids, 463
Cytokines, 441

D

Dehydroepiandrosterone (DHEA), 461
Dendritic cells, 446
DHEA. *See* Dehydroepiandrosterone (DHEA)
Diffuse parenchymal lung disease (DPLD)

collagen vascular diseases, 247
PDE-5 inhibition, 248
Digoxin, PAH, 156
Distal pulmonary arterioles, 33–34
Diuretics, PAH, 154
DPLD. *See* Diffuse parenchymal lung disease (DPLD)
DRP-1. *See* Dynamin-related protein 1 (DRP-1)
Drug-and toxin-induced PAH
 aminorex fumarate, 8
 amphetamines, 10
 benfluorex, 9
 fenfluramine and dexfenfluramine, 8–9
 pre-capillary PH, 10
 SOPHIA study, 9–10
 updated risk factors, 8, 9
Drug interaction profile, ERAs, 213–214
Dynamin-related protein 1 (DRP-1)
 HIF-1α activation, 113
 and mitosis, 113

E
ECs. *See* Endothelial cells (ECs)
ECV304 cells, 537
Efflux, cyclic nucleotides
 ABC transporters
  ATP-binding cassette transporter family, 519
  domain organization, MRP4, 519–520
  extrusion process, 519
  membrane vesicle and mammalian cells, 519
  monophosphorylated nucleotides, 518–519
  NBD, 519
  TMD, 519
  transmembrane proteins, 519
  vascular smooth muscle cells and cardiac myocytes, 518
 MRP4
  cyclic nucleotides homeostasis, 520–522
  vasculature, 522–523
Eisenmenger syndrome associated PAH (ESPAH), 67
Endothelial cells (ECs)
 activation and apoptosis, 442
 anti-EC antibodies, 448
 dysfunction, 440
 and PASMC, 448
Endothelial nitric oxide synthase (eNOS)
 activation, 37
 FAD and FMN, 37
 genetic mediators, 39–40
 G-protein-coupled receptor signal transduction, 37–38
 hypertension sensitivity, 284
 and hypoxia, 38–39
 and oxidant stress, 39
 PAEC/PSMC membranes, intercellular signaling, 37
 riociguat, 297
 uncoupling
  Cav-1, 324
  dihydrobiopterin (BH2), 319–320
  L-arginine, 332–333
  oxygenase domain, 319
  pathologic mechanism, 318, 319
  tetrahydrobiopterin, 331–332
  triggers, 319
  vascular homeostasis, 324
 vascular endothelium, 283
Endothelin B (ET_B) receptors deficiency
 description, 132
 features, 132–133
 limitations, 133
 method, 132
Endothelin receptor antagonists (ERAs)
 ETs (*see* Endothelins (ETs))
 human studies
  ambrisentan, 217–218
  bosentan, 214–217
  macitentan, 218–219
  sitaxentan, 217
 preclinical studies
  classification, 204–205
  ET receptor blockade, 204
  and fibrosis, 208–209
  vascular/cardiac hypertrophy, 206–208
  vasoconstriction/vasodilation, 205
 treatment, PAH
  description, 209
  drug interaction profile, 213–214
  pharmacological profile, 209–211
  safety profile, 211–213
Endothelins (ETs)
 21-amino acid peptides, 200
 pathogenesis, PH
  effects, ET-1, 203–204
  imbalance, endothelial mediators, 202
  upregulation, 204
 production, ET-1, 200–201
 and pulmonary circulation, 202
 receptors, 201–202

Endothelin-1 (ET-1) system
  biological activity, 479–480
  cardiovascular isoform, 479
  Cys1-Cys15 and Cys3-Cys11, 40
  cysteine(s), 40
  and ERAs, 480
  ET_A and ET_B receptors, 40–42
  mammalian cell types, 40
  PAECs and PSMCs, plexiform lesions, 40
  plasma levels, 480
Endothelin-type A (ET_A) receptor, 40–42
Endothelin-type B (ET_B) receptor, 40–42
eNOS. *See* Endothelial nitric oxide synthase (eNOS)
Epidermal growth factor (EGF) signaling system
  ECM, 415
  gefitinib and erlotinib treatments, 415
  pathogenic role, 415
  rat monocrotaline model, PH, 415
  receptors types, 414
  SMC proliferative response, 415
  TK domains autophosphorylate, 414
Epoprostenol
  combination with PAH medication, 184
  continuous intravenous prostacyclin, 183–184
  limitations, 185
  long-term effects, 184
  lung transplantations, PAH, 184
  6MWD, 184
  scleroderma-associated PAH, 184
ERAs. *See* Endothelin receptor antagonists (ERAs)
ESPAH. *See* Eisenmenger syndrome associated PAH (ESPAH)
ET_A. *See* Endothelin-type A (ET_A) receptor
Etanercept, 462
ET_B. *See* Endothelin-type B (ET_B) receptor
ETs. *See* Endothelins (ETs)
Extracellular matrix (ECM) remodeling, 424

F
FAD. *See* Flavin adenine dinucleotide (FAD)
Fawn-hooded rats (FHR)
  BQ-123, 126
  disruption, mitochondrial fusion, 127
  echocardiography, development of PH, 126
  ET-1 and eNOS, 126
  FHR-BN1 consomic rat, 127
  "German brown" albino rats, 125
  hypoxia sensitivity and hyperproliferative phenotype, 126
  limitations, 128
  Long Evan's rats, 125
  PASMC, human, 126
  RV, 128
  SOD2 downregulation, 127
  strains, 125
FHR. *See* Fawn-hooded rats (FHR)
Fibroblast growth factor (FGFs) signaling system
  and FGF2, 413–414
  and FGFRs, 413
  heparin-binding polypeptide mitogens, 413
  protein tyrosine kinase receptors, 413
Fibrosis and ERAs
  bosentan-treated rats, 208–209
  cigarette smoke-induced chronic obstructive disease, 209
  ET_A and ET_B receptors, 208
  evaluation, 208
  pro-fibrotic and pro-inflammatory effects, ET-1, 208
Flavin adenine dinucleotide (FAD), 37, 325
Flavin mononucleotide (FMN), 37
FMN. *See* Flavin mononucleotide (FMN)

G
γ-Herpesvirus 68 (γHV68) reactivation model
  description, 134
  limitations, 135
  PVD lesion, fragmented elastin, 134, 135
  S100A4/Mts mice 1 year, 134
γHV68. *See* γ-Herpesvirus 68 (γHV68) reactivation model
Global knockout mice, 129
Glucocorticoids
  cyclophosphamide, 463
  cytokines, 451
  dexamethasone treatment, 460
  iPAH-PASMC proliferation and migration, 460
  NF-kB, 451
  POEMS syndrome and sarcoidosis, 456
  promote hypokalemia, 464
  prostaglandins synthesis, 456
  small lipophilic compounds, 451
  target B and T cell differentiation and activation, 456
GPCRs. *See* G protein-coupled receptors (GPCRs)
G protein-coupled receptors (GPCRs), 487

Index 563

Growth factor signaling pathways, PAH
  AngII and CTGF, 423
  description, 412
  EGF, 414–415
  FGFs, 413–414
  HGF and NGF, 423
  IGF signaling system, 419–420
  JAK/STAT pathway, 422–423
  mTOR, 420–421
  PDGF, 412–413
  PIGF, 423
  RhoA/Rho-kinase signaling system, 423
  serotoninergic system, 415–418
  VEGF signaling system, 418–419
GTP. *See* Guanosine triphosphate (GTP)
Guanosine triphosphate (GTP), 284–285

H
HAART. *See* Highly active antiretroviral therapy (HAART)
Heart failure due to reduced ejection fraction (HFrEF)
  PH and systolic dysfunction, 245
  primary endpoints, 245–246
  treatment, PDE-5 inhibitors, 246
Heart failure preserved ejection fraction (HFpEF), 246
Hematoxylin–eosin–saffron staining, 61, 63
Hemodynamics
  cardiac output measurement, 117
  fluid-filled catheters, 116–117
  in humans, 115
  measurement, PVR, 115
  mPAP, systolic PAP and LVEDP, 116
  prostacyclins, 182
  pulmonary, 116
  rat, 116, 117
Hemolysis, endothelial dysfunction and vasculopathy
  ADMA, 321
  erythrocytes, 320
  hemoglobin, 321
  NO signaling, 322
  SCD, 321–323
HFpEF. *See* Heart failure preserved ejection fraction (HFpEF)
HFrEF. *See* Heart failure due to reduced ejection fraction (HFrEF)
HHV-8. *See* Human herpes virus 8 (HHV-8)
Highly active antiretroviral therapy (HAART), 443, 458
HIV. *See* Human immunodeficiency virus (HIV)

HIV-associated PAH
  description, 137
  features and limitations, 137
  method, 137
$5HT_{2A}$ receptors
  ketanserin, 374
  SSRIs, 375
  systemic and pulmonary circulation, 374
$5HT_{1B}$ receptors
  human pulmonary arteries, 375
  in PH development, 375–376
  pulmonary circulation, 376
$5HT_{2B}$ receptors, 375
5-HTs. *See* Serotonins (5-HTs)
5-HTTs. *See* Serotonin transporters (5-HTTs)
Human circulatory system, 33
Human herpes virus 8 (HHV-8), 443
Human immunodeficiency virus (HIV), 11, 443
Human PAH
  immunosuppressive agents (*see* Immunosuppressive agents)
  inflammation (*see* Inflammation in human PAH)
Hypoxia
  cellular responses (*see* MicroRNA-mediated regulation)
  downregulation, genes, 123
  eNOS, 38–39
  FHR, 123
  haploinsufficient for HIF-1α, 122
  llamas/yaks, 123
  PH and vascular remodeling, 122

I
Idiopathic and heritable PAH
  BMPR2 mutations, 7–8
  genetic testing, 8
  in women, 8
Idiopathic PAH (IPAH)
  lymphoid neogenesis, lungs, 67
  pulmonary tLTs, 67
  venous lesions, 71
IL. *See* Interleukin (IL)
Iloprost
  inhaled, 188–189
  intravenous, 185
Immunosuppressive agents
  action mechanisms
    anti-infectious agents, 458–459
    CsA, 457–458
    CYC, 456–457

glucocorticoids, 451, 456
MMF, 457
MTX, 457
human PAH
   CTD, 463–464
   iPAH, 463
   POEMS syndrome, 464
   side effects and drug-drug interactions, 464–465
in PAH, 451, 452–455
PH experimental models
   Ciclosporin, 461
   cyclophosphamide, 460
   DHEA, 461
   Etanercept, 462
   glucocorticoids, 460
   hypoxia model, 459
   inflammation in different, 459
   MCT, 459
   Methotrexate, 461
   MMF, 460
   Sirolimus, 462
   Sugen-induced model, 459
Infectious diseases
   HHV-8, 443
   HIV, 443
   schistosomiasis-PAH, 443–445
Inflammation in human PAH
   autoimmune/inflammatory disorders, 440
   CTD, 442
   histological and cytological data, 440–441
   infectious diseases, 443–445
   mediators and biomarkers, 441
   pulmonary vasculature, 450–451
Inflammatory mediators
   and biomarkers in human PAH, 441
   and cells (*see* Pulmonary vascular remodeling)
Inflammatory models
   description, 134
   HIV-associated PAH, 137
   γHV68 (*see* γ-Herpesvirus 68 (γHV68) reactivation model)
   radiation exposure, 137–138
   schistosomiasis, 136
   Stachybotrys chartarum, 136
   viral infections, 134
Inhaled nitric oxide (NO)
   acute pulmonary vasoreactivity, 258
   and acute vasoreactivity testing, 267
   and ARDS, 269
   chronic treatment, PAH
      advantages, 270
      long-term ambulatory breathing, 270
      mechanical ventilation, 269
      pulmonary hemodynamics and cardiac index, 270
      vasodilator therapy in infants, 270
   FDA in adults, 258
   mechanisms, NO-cGMP signaling, 259–260
   neonatal PH
      and developing lung, 261–262
      PPHN (*see* Persistent pulmonary hypertension of the newborn (PPHN))
      and preterm infants, 265–267
   pediatric and adult patients, 258
   physiologic effects
      acute lung injury/chronic lung disease, 260
      colorless and odorless gas, 260
      left ventricular (LV) failure, 261
      met-hemoglobinemia, 261
      "micro-selective" effect, 260
      PAP and PVR, 260
   and postoperative cardiac patients, 268–269
Inhaled nitric oxide (iNO) gas, 328–329
Insulin-like growth factor (IGF) signaling system, 419–420
Interleukin (IL), 441, 448–449, 451
Intimal fibrosis
   anatomical delimitation, 63
   cellularity, 63
   description, 63
   eccentric intimal thickening, 64
   "onion-skin"/"onion-bulb" lesion, 64
   thickening proximal to plexiform lesions, 64
   thrombotic lesions, 64
Intrauterine devices (IUDs), 157
IPAH. *See* Idiopathic PAH (IPAH)
IUDs. *See* Intrauterine devices (IUDs)

J

Janus kinase/signal transducers and activators of transcription (JAK/STAT) pathway
   abnormal activation, 422
   cytokine receptors, 422
   cytoplasmic transcription factors, 422
   dysregulated activation, 422
   family members, 422
   IL-6, 422–423
   inflammation role, 422
   pathogenic role, 422

Index

K
Kaposi's sarcoma-associated herpes virus, 443

L
Left heart disease, 17
Left ventricular dysfunction, PDE-5
  description, 245
  in HFpEF, 246
  in HFrEF, 245–246
Lung diseases and/or hypoxia, 18

M
Macitentan, 218–219
Macrophages, 446–447
Maladaptive changes, pulmonary arterioles, 33
Mammalian target of rapamycin (mTOR) signaling pathway
  activation, 420–421
  Akt, PDGF-induced phosphorylation, 421
  description, 420, 421
  functional domains, 420
  289-kDa serine/threonine protein kinase and PIKK family, 420
  mTORC1, 420
  mTORC2, 420
Management, PH
  anticoagulants, 551–552
  cellular mechanisms, 553
  currently available therapies, 553
  diagnostic approach and treatment strategies, 552
  NIH registry on Primary Pulmonary Hypertension, 552
  right ventricle, 553–554
  systemic vasodilators, 551
  targeted therapies, 553
  vasoconstriction, 551
MAO-A. *See* Monoamine oxidase A (MAO-A)
Mast cells, 446
MCT. *See* Monocrotaline (MCT)
Medial hypertrophy
  clinical and histological findings, 61
  description, 61
  hematoxylin–eosin–saffron staining, 61, 63
  isolated layer, 61
  in vascular compartments, 62
  vasoconstriction, 61
Methotrexate (MTX)
  arthritis and skin manifestations, SLE, 457
  folic acid analogue, 457
  prednisone, 463

MicroRNA-mediated regulation
  description, 50
  HIF-1α-dependent upregulation, 51
  miR-21, 51–52
  miR-17 and miR-328, 51
  pri-mRNA and pre-miRNA, 50
  processing, 50–51
MicroRNAs (miRs), 91
6-Min walk distance (6MWD)
  epoprostenol, 183
  exercise capacity, 189
  and maximal oxygen consumption, 189
miRs. *See* MicroRNAs (miRs)
Mitochondrial dysfunction
  apoptosis-associated proteins, 47
  bioenergetics, 48, 49
  description, 46
  electrons, 47
  electron transport chain complexes I/II, 48
  HIF-1α expression, 47
  hydrogen (hydride), 46–47
  pathological disruptions, 47
  PDGF/IL-6, 48
  PTPC, 47
  pulmonary vascular dysfunction and/or RV hypertrophy, 48
  pulmonary vascular smooth muscle cells, 47
MLC. *See* Myosin light chain (MLC)
MMF. *See* Mycophenolate mofetil (MMF)
MMP inhibitors, 424
Monoamine oxidase A (MAO-A), 368
Monocrotaline (MCT), 356, 357, 359, 451, 459–462
Monocrotaline-induced PAH
  Crotalaria spectabilis, 119
  cytochrome P-450 enzyme CYP3A4, 119
  dehydromonocrotaline, 119
  histological findings, 120
  limitations, 120–122
  method, 119
  platelets, 119–120
  pulmonary vasculitis, 119
  right ventricle (RV), 120
  small preacinar artery 4 days, 120, 121
MRCTs. *See* Multicenter randomized controlled trials (MRCTs)
MRP4
  cyclic nucleotides homeostasis
    adenylate cyclase activation, 521
    caveolins, 521
    CFTR, 521
    drugs transport, 520
    FRET-based cAMP sensor, 521

inhibition, 521
low-affinity cyclic nucleotide exporters, 521
PDEs, 521–522
PMEA and nucleoside monophosphate antiviral agent, 520
vasculature
hypoxia-induced PH, 522
intracellular/extracellular ratios, 522
mRNA and protein, 522
MRP4 KO mice, 522–523
PGVSMCs, 522
SMC, 522
MTX. *See* Methotrexate (MTX)
Multicenter randomized controlled trials (MRCTs)
incidence, BPD, 267
inhaled NO, 267
and PPHN, 270
6MWD. *See* 6-Min walk distance (6MWD)
Mycophenolate mofetil (MMF)
and CYC, 464, 465
immunosuppressive drug, 457
lung disease and skin score, 457
lupus nephritis treatment, 457
macrophages and IL-6 production, 460
and PASMC proliferation, 460
Myosin light chain (MLC), 354

N

NADH. *See* Nicotinamide adenine dinucleotide dehydrogenase (NADH)
Natriuretic peptide receptors (NPRs)
cGMP and CNP, 482
neutral endopeptidase enzymes, 482
NPR-A, NPR-B and NPR-C, 482
Natriuretic peptides (NPS)
amino acid structure and CNP, 481
and ANP and BNP, 481
biological actions, 481–482
endogenous vasoactive peptides, 481
NEP and PH, 485–486
and NPRs, 482
and PH
ANP and BNP, 483
bronchorelaxant activity, 483
lung physiology, 483
NANC innervation, 483
NEP inhibition, 484
NPRs, 483
PDE5i and fibrosis, 484

and pulmonary fibrosis, 483
therapeutic approach, 484
Neonatal PH. *See* Inhaled nitric oxide (NO)
NEP. *See* Neutral endopeptidase (NEP)
Neutral endopeptidase (NEP)
cardio-pulmonary tissue, 486
catabolism, vasoactive peptides, 486
DPP-4, 486
endothelial membrane-bound zinc metallopeptidase, 485
inhibitors, 485–486
organ and cellular distribution, 485
and PH treatment, 486
vasodilators/vasoconstrictors, 485
Nicotinamide adenine dinucleotide dehydrogenase (NADH), 325, 326
Nicotinamide oxidase (NOXs)
FAD, 325
NADH, 325
ROS, 325–326
Nitric oxide (NO)
activation, 284–286
BAY 41-2272, 288
cardiac index, 302
cellular and organ function, 281
CFM-1571, 295
and COPD, 304
disadvantages, 286
donors, 329–330
iNO, 328–329
nitrite, 330–331
oxidative stress, 283
PDE-5 inhibitors, 231
riociguat synergises, 296
signaling
dysregulation, vasodilator systems, 319
L-NMMA, 318–319
NO–sGC–cGMP pathway, 318
sGC, 317–318
YC-1, 287
NO-cGMP pathway, 517
NO-cGMP signaling
direct activation, K+ channels, 259
"endothelium-derived relaxing factor", 259
Hsp90-eNOS interactions, 259
L-amino acid transporters, 260
SMC and PDE, 259
SNO homeostasis, 259
Non-pharmacological measures, PAH
elective surgery, 159
infection control, 158
physical exercise and pulmonary rehabilitation, 157

Index 567

pregnancy and birth control, 157–158
screening, depression and psychosocial support, 158
NPRs. *See* Natriuretic peptide receptors (NPRs)
NPS. *See* Natriuretic peptides (NPS)
NT-pro-BNP, 535

O
Oxidant stress and eNOS, 39
Oxidative metabolism in PAH
    cancer cells, 425
    dichloroacetate, 425
    Krebs cycle, 425
    PDK inhibition, 425
    "Randle cycle" inhibition, 426
    Warburg effect, 425

P
PAAT. *See* Pulmonary artery acceleration time (PAAT)
PAB. *See* Pulmonary artery banding (PAB) model
PACAP. *See* Pituitary adenylate cyclase activating peptides (PACAP)
PAECs. *See* Pulmonary artery endothelial cells (PAECs)
PAH. *See* Pulmonary arterial hypertension (PAH)
PAP. *See* Pulmonary artery pressure (PAP)
PA-SMCs. *See* Pulmonary artery smooth muscles (PA-SMCs)
Pathobiology, PH. *See* Cell signaling mechanisms
Pathological anatomy, PH, 60
Pathophysiology, PH
    biological/molecular factors, disease progression, 35
    chronic changes, RV, 35
    hemodynamic criteria, 34
    impaired pulmonary vascular reactivity, 35, 36
    pathobiological mechanisms, 36
    pulmonary arterioles and homeostatic mechanisms, 34
    pulmonary artery pressure, 37
    pulmonary circulatory performance, 34
    TAPSE, 37
    "two-hit" hypothesis, 34, 35
PDE. *See* Phosphodiesterase (PDE) inhibition
PDE-5. *See* Phosphodiesterase (PDE)-5
PDGF. *See* Platelet-derived growth factors (PDGF)

PEA. *See* Pulmonary endartherectomy (PEA)
Perivascular inflammation
    chemokines, 66
    chemotactic activity, 67
    idiopathic PAH, 66–67
    IPAH and ESPAH, 67
    "scarring" vasculitislike lesions, 67
Permeability transition pore complex (PTPC)
    mitochondrial membrane permeability, 47
    pyridine nucleotides, 47
Peroxisome proliferator-activated receptor (PPAR-γ)
    BMP-RII-interacting proteins, 48, 50
    regulatory effect on genes, 48
Peroxisome proliferator-activated receptor (PPAR) agonists
    animal studies, 538–539
    clinical potential, 539
    description, 537
    molecular action
        isoforms, 537
        nuclear hormone receptor superfamily, 537
        PPARα, 537–538
        PPARg/β-catenin complex, 538
        PPAR-RXR heterodimeric complex, 537, 538
Persistent pulmonary hypertension of the newborn (PPHN)
    clinical
        high-frequency ventilation, 264
        intravenous dilators, 264
        oxidized sGC activators, 265
        selective pulmonary vasodilation, 264
        sildenafil, selective PDE5 inhibitor, 264–265
        treatment with 80 ppm NO, 264
    experimental
        DA in fetal lambs, 263
        in diverse animal models, 263
        functional and structural maturation, lung circulation, 263
        NO-cGMP cascade, 263
        risk for severe asphyxia, 262–263
    pulmonary vasoconstriction, 374
    and SSRIs, 373–374
    vascular resistance, after birth, 373
PH. *See* Pulmonary hypertension (PH)
Pharmacological measures, PAH
    anticoagulants, 156
    digoxin, 156
    diuretics, 154
    oxygen, 154–156

Pharmacological profile, ERAs
  bosentan, ambrisentan and macitentan, 209, 211
  description, 209–211
  tissue-targeting potential, macitentan, 209, 212
  UGTs, 211
PH-ILD. *See* Pulmonary hypertension associated with interstitial lung disease (PH-ILD)
Phosphodiesterase (PDE)-5
  BAY 41-8543, 294
  cGMP, 284, 286
  riociguat, 296
  YC-1, 287
Phosphodiesterase (PDE) inhibition
  Asp-289 and Asp-478, 43
  cGMP-specific kinetic properties, 43, 44
  influence, 43, 45
  PSMCs and RV myocytes, 44
  and sGC, 42–43
  type-5 regulation, cGMP bioactivity, 43
Phosphodiesterase-5 (PDE-5) inhibitors
  clinical efficacy
    adverse effects, 249
    description, 238
    group-1 and group-3, PAH (*see* Pulmonary arterial hypertension (PAH))
    PH and left ventricular dysfunction (*see* Left ventricular dysfunction, PDE-5)
  description, 230
  signaling pathways
    basal activity, sGC, 235
    cGMP-dependent protein kinases (PKG), 231–232
    clinical pharmacology, 236–238
    nitric oxide (NO), 231
    and PDEs, 233–235
    relaxation effect, 235
    sildenafil, tadalafil and vardenafil, 235
Phosphodiesterases (PDEs)
  allosteric binding, 234
  domain structure, 234
  functional domains, 233
  hydrolytic activity, 234
  intracellular cGMP, 233
  NO/cGMP/PKG1 signaling pathway, 233
  physiological importance, 234
  serine (Ser-102), 233–234
  smooth muscle cells, 233
  upregulation, expression and activity level, 234

Pituitary adenylate cyclase activating peptides (PACAP)
  expression, 487
  novel hypothalamic/pituitary-derived hormones, 486
  and PH, 489, 490
  receptors, 487, 488
  transcription, 486–487
PKG. *See* cGMP-dependent protein kinases (PKG)
Platelet-derived growth factors (PDGF)
  cellular proliferation, cultured PSMCs, 51
  IL-6 harvest, PAH patients, 48
  imatinib, 397
  kinase profile, 398
  ligands and receptors, 387
  PAH (*see* Pulmonary arterial hypertension (PAH))
  PEA, 397–398
  pulmonary arterial remodeling, 398
  pulmonary vascular remodeling, 397
  receptor inhibition, PAH
    phase III trial (IMPRES), 395
    phase II trial, 394–395, 396
    preclinical studies, 393–394
    TKI, 392–393
  safety and tolerability, TKI, 398–401
  signaling control, 391, 392
  signal relay downstream
    embryogenesis, 388
    neointima formation, 391
    pulmonary vascular remodeling, 391
    receptor-associated signaling molecules, 389
    receptor's intrinsic tyrosine kinase activity, 387
    roles, PDGF-dependent PI3K and PLCγ, 390
    signaling molecules, 389
    signal transduction, 388
Platelet-derived growth factor (PDGF) signaling system
  autocrine and paracrine mitogen, 412
  autophosphorylation, 412
  homo/heterodimers, 412
  lamb model, 412
  monomeric forms, 412
  mRNA encoding, 413
  PDGFR-β, 413
  receptor, 412
POEMS syndrome, 464
PoPH. *See* Portopulmonary hypertension (PoPH)

Index 569

Portopulmonary hypertension (PoPH)
  defined, 11
  female gender and autoimmune hepatitis, 11–12
  long-term prognosis, 12
  pathologic changes, small arteries, 11
Postoperative cardiac patients
  acute right heart failure, 268
  CHD, 268
  lower PAP and PVR, 268
  pediatric intensive care unit, CPB surgery, 268
  withdrawal PAH, 269
PPAR-γ. *See* Peroxisome proliferator-activated receptor (PPAR-γ)
PPHN. *See* Persistent pulmonary hypertension of the newborn (PPHN)
Preclinical studies, ERAs
  description, 204–205
  and fibrosis, 208–209
  vascular/cardiac hypertrophy, 206–208
  vasoconstriction/vasodilation, 205
Preterm infants and inhaled NO
  BPD, 266–267
  clinical reports, 266
  MRCTs, 267
  placebo-controlled studies, 266
  pulmonary hemodynamics, 265
  RDS, 265
  surfactant therapy, 265
  treatment, PAH, 265
Prostacyclin-cAMP pathway
  adenylyl cyclase (AC), 516
  arachidonic acid, 516
  colforsin, 516
  PGI2, 517
  receptor, 516
Prostacyclins
  beraprost, oral, 187
  description, 178
  differences between analogues, 181–182
  effects, vessel wall and adherent blood cells, 178, 179
  hemodynamic effects, 182–183
  history, 178
  iloprost
    inhaled, 188–189
    intravenous, 185
  intravenous, epoprostenol, 183–185
  pharmacologic properties
    human IP receptor, 179
    intracellular targets, 180
    potassium channels prostacyclin (PGI2), 180
    PPARα and PPARδ, 179

Selexipag, 188
signaling
  arachidonic acid, 44
  hypoxic lung disease, 46
  isoforms, COX, 44–45
  5-lipooxygenase (5-LO) catalysation, 46
  pulmonary vasoconstriction phenotype, 45
  shear stress, ET-1, hypoxia and BMP-RII dysfunction, 46
  TXA_2, 45
therapy management
  and analogues, 190
  practical issues, 191–192
  switch from inhaled to continuous infusion, 192–193
  switch from iv to forms, 192
  treatment hazards, 191
treatment, PAH, 178
treprostinil (*see* Treprostinil)
PSMCs. *See* Pulmonary vascular smooth muscle cells (PSMCs)
PTPC. *See* Permeability transition pore complex (PTPC)
Pulmonary arterial hypertension (PAH)
  anti-proliferative strategies, 426–427
  BMPRII, 440
  BMPR-II signaling system, 424–425
  category of PH, 439
  and CCBs (*see* Calcium-channel blockers (CCBs))
  characterized, 410
  CHDs (*see* Congenital heart diseases (CHDs))
  chronic hemolytic anemia, 14–15
  connective tissue diseases, 10–11
  CTEPH, 302
  definition
  description, 154
  development and progression, 383
  diagnostic approach and treatment strategies, 552
  drug-and toxin-induced, 8–10
  ECM remodeling, 424
  and eNOS, 283
  group 1
    measurement, QOL, 241–242
    multicenter double-blind placebo-controlled trial, 239
    6MWD with sildenafil treatment, 240
    PHIRST-1 study, 241, 242
    sildenafil, 239
    SUPER-1 study, 239, 240
    survival benefit, PDE-5 inhibitor treatment, 243

tadalafil mean changes from baseline, 240–241
vardenail, 242
group 3
  COPD, 247
  double-blind, randomized placebo-controlled trial, 248
  DPLD, 247–248
  parenchymal lung disease, 247
  PDE-5 inhibitors, patients, 247
  treatment, "all comers", 248–249
growth factor signaling pathways (*see* Growth factor signaling pathways, PAH)
HIV infection, 11
idiopathic and heritable, 7–8
incremental pulmonary vasculopathy, 383
irreversible remodeling, pulmonary vasculature, 410
in modern management era, French registry, 15–16
6MWD, 301
myocardial hypertrophy, 427
non-pharmacological measures (*see* Non-pharmacological measures, PAH)
orphan nature, 552
oxidative metabolism, 425–426
oxidative stress, 283
and PATENT, 301
pathobiology, 383–384
pathogenesis, 440
pathogenic role, platelet-derived growth factor, 386–387
PDGF receptor inhibition, 392–396
pediatric
  clinical characteristics and histopathology, 243
  clinical improvement, 245
  EMA endorsement, 244
  lung vasculature, 243
  safety and efficacy, tadalafil, 244–245
  sildenafil, 244
pharmacological measures (*see* Pharmacological measures, PAH)
plasma levels, ADMA, 283
PoPH (*see* Portopulmonary hypertension (PoPH))
and pulmonary endarterectomy, 554
pulmonary vascular remodeling, 383, 445–449
recommendations, general supportive care, 154, 155

Rho-kinase inhibitors, 351–360
schistosomiasis, 12–13
signalling pathways, 553
therapeutic effects, 553
therapies and therapeutic limitations, 384–385
TKI, 383
treatments in animal models, 553
vasculopathy and failing right ventricle, 554
vasoconstriction and vasodilation, disorder YC-1, 287
Pulmonary arterial pressure (PAP)
  acute vasodilator test, 168
  PVR and right ventricular hypertrophy, 170
  vasodilator therapy, 167
Pulmonary artery acceleration time (PAAT)
  PA Doppler signal, 117
  vs. right heart catheterization, 118
Pulmonary artery banding (PAB) model
  description, 139
  features and limitations, 139
  methods, 139
Pulmonary artery endothelial cells (PAECs)
  apoptosis-resistant, 33–34
  and eNOS, 37
  and ET-1, 40
  mitochondria bioenergetics, 46
  perturbations, redox status, 39
Pulmonary artery pressure (PAP)
  adult patients with IPAH, 270
  inhaled NO, 266
  oral vasodilators, 267
  and PVR, 260
Pulmonary artery smooth muscles (PA-SMCs)
  5-HT availability, 369
  hyperplasia, idiopathic PH, 371–372
  S100A4/Mts1-induced human regulation, 375
Pulmonary capillary hemangiomatosis (PCH)
  "angiomatous growth", 70
  anti-α-actin staining, 68
  calcium-encrusting elastic fibers, 69
  fibrous remodeling, intima, 68, 69
  Golde score, 69
  microvessels and capillary hemangiomatosis-like foci, 69, 70
  muscularization, arterioles and medial hypertrophy, 70
  post-capillary obstruction, 70
  pre-septal venules, 68
  septal vein with fibrous, collagen-rich occlusion, 68, 69

thrombotic occlusion, 68–69
tunica media, 69
Venice classification, 68
Pulmonary circulation and endothelins (ETs), 202
Pulmonary endartherectomy (PEA), 397–398
Pulmonary hypertension (PH)
   biomarkers (*see* Biomarkers, PH)
   CTEPH, 18–19
   definitions, 5–6
   disease characterization, 478
   hemodynamic definitions, 5
   5-HTs (*see* Serotonins (5-HTs))
   5-HTTs (*see* Serotonin transporters (5-HTTs))
   left heart disease, 17
   lung diseases and/or hypoxia, 18
   molecular weight signalling peptides, 478
   orphan disease, 4
   PAH (*see* Pulmonary arterial hypertension (PAH))
   patient management and future research, 4
   peptides and pulmonary homeostasis, 479
   polypeptide hormones and paracrine mediators, 478
   "primary"/"secondary", 6
   PVOD and PCH, 6, 16–17
   unclear/multifactorial etiologies, 19–22
   updated clinical classification, 6, 7
   vasoconstrictor peptides (*see* Vasoconstrictor peptides)
   vasodilator peptides (*see* Vasodilator peptides)
Pulmonary hypertension associated with interstitial lung disease (PH-ILD), 302, 303
Pulmonary rehabilitation, 157
Pulmonary vascular remodeling
   antibodies, 447–448
   cell types in inflammatory process, 445–447
   pro-inflammatory mediators
      cytokines and chemokines, 448–449
      growth factors, 449
Pulmonary vascular resistance (PVR)
   acute vasodilator test, 168
   diverse animal models, 263
   human infants with PPHN, 265
   hypoxemia and PPHN-type physiology, 266
   and PAP, 260
   in pulmonary blood flow, 262
   and right ventricular hypertrophy, 170
   vascular injury, 268
   vasodilator therapy, 167
Pulmonary vascular smooth muscle cells (PSMCs), 33–34

Pulmonary veno-occlusive disease (PVOD)
   "angiomatous growth", 70
   anti-α-actin staining, 68
   calcium-encrusting elastic fibers, 69
   characteristics with PAH, 16–17
   clinical examination, radiologic abnormalities, 17
   description, 6
   fibrous remodeling, intima, 68, 69
   Golde score, 69
   microvessels and capillary hemangiomatosis-like foci, 69, 70
   muscularization, arterioles and medial hypertrophy, 70
   and PCH, 16
   post-capillary obstruction, 70
   pre-septal venules, 68
   septal vein with fibrous, collagen-rich occlusion, 68, 69
   thrombotic occlusion, 68–69
   tunica media, 69
   Venice classification, 68
PVOD. *See* Pulmonary veno-occlusive disease (PVOD)
PVR. *See* Pulmonary vascular resistance (PVR)

R
RAMP. *See* Receptor activity modifying protein (RAMP)
Rapamycin, 462
Reactive oxygen and nitrogen species (ROS)
   antimicrobial defense, 324
   and eNOS Uncoupling, 324
   mitochondria, 327–328
   oxidative stress, 324
   radicals, 323
   SOD, 330
   transgenic rat (mRen2)27 models, 337
   trimetazidine, 337
   XOR, 326
Receiver operating characteristic (ROC) analysis, 82, 83, 88
Receptor activity modifying protein (RAMP), 491
Receptors, 5-HTs
   $5HT_{2A}$, 374–375
   $5HT_{1B}$, 375–376
   $5HT_{2B}$, 375
Red cell distribution width (RDW)
   anaemia and hepcidin, 88
   inflammation, 88
   plasma biomarkers, 88

Redox equilibrium and endothelial
    dysfunction, PAH
  arginase (I and II), 320
  eNOS uncoupling, 319–320
  hemolysis, 320–323
  nitric oxide (see Nitric oxide)
  NO signaling, 317–319
  NOXs, 325–326
  ROS (see Reactive oxygen and nitrogen
      species)
  S-nitrosoglutathione reductase inhibitors, 338
  XOR, 326–327
Regisistry to Evaluate Early And Long-term
    PAH disease management
    (REVEAL), 158
RHC. See Right heart catheterization (RHC)
RhoA/Rho-kinase signaling system, 423
Rho-kinase inhibitors, PAH
  cardiovascular fields
    intracellular signaling pathways, 353
    ischemic heart disease, 354
    MLC, 354
    VSMC, 354–355
  cellular functions, cardiovascular diseases,
      352–353
  clinical trial, 357
  fasudil, 360
  hypercontraction, VSMC, 353, 354, 358, 359
  hypertension and ischemic heart disease, 357
  inflammation, 356
  inhalation, 357
  intracellular signaling pathway, 352
  MCT, 356
  myosin phosphatase, 356
  neutrophils and pulmonary arteries, 358
  pleiotropic effects, statins, 352
  pulmonary vascular resistance, 355
  rat model, MCT, 357
  VSMC, 352, 354
Right heart catheterization (RHC)
  accurate diagnosis, PoPH, 11
  diagnosis and confirmation, PAH, 10
  and PH, 5
Riociguat
  cardiopulmonary diseases, 296–297
  clinical trials, 297–300
  COPD, 302
  CTEPH, 302
  discovery, 296
  in vitro results, 296
  left ventricular systolic dysfunction, 303
  PAH, 301
  PH-ILD, 302

ROC analysis. See Receiver operating
    characteristic (ROC) analysis
Rodent models, group 1 PAH
  in animal models (see Animal models)
  BMPR-2-dominant negative transgenic
      mice, 130–131
  catheterization and echocardiography
    anesthesia, 115
    chronic hypoxia and Sugen-5416,
        128–129
    chronic hypoxia rats, 123–125
    cut-down technique, 117
    FHR (see Fawn-hooded rats (FHR))
    hemodynamics (see Hemodynamics)
    hypoxia models, 122–123
    measurement, hemodynamics, 118
    monocrotaline-induced PAH, 119–122
    PAAT, 118
    rat, supine position, 117
    RV function and TAPSE, 118
  clinical characteristics, 110
  description, 107–108
  exercise capacity, Rodent treadmill, 115, 116
  histology, human and rat lung, 108
  hypoxia and/or increased LVEDP, 109
  inflammatory models (see Inflammatory
      models)
  knockout mice
    conditional, 130
    global, 129
  miscellaneous diseases, 109
  mitral stenosis, 109
  monocrotaline model, 109
  PASMCs, 108
  plexiform lesions, 108
  scleroderma patients, 109
  shRNA BMPR-2 knockdown mice
      (see Short hairpin (shRNA)
      BMPR-2 knockdown mice)
  surgical models, 139–140
  transgenic models, BMPR-2 mutation, 129

S
Safety profile, ERAs
  BSEP, 213
  ET_A and ET_B receptors, 212
  hemoglobin concentration and hematocrit,
      213
  liver abnormalities, 212
  mechanism, hepatotoxicity, 212
  peripheral edema observation, 211
SCD. See Sickle cell disease (SCD)

Schistosomiasis
  embolic obstruction, pulmonary arteries, 12
  features and limitations, 136
  hepatosplenic disease, 12
  method, 136
  patients with PAH, 136
  post-capillary PH, 13
  Schistosoma genus, 136
Schistosomiasis-PAH
  description, 443
  hepatosplenic infection, 443–444
  non-exhaustive data, 444–445
  plexiform lesions, 444
  T lymphocytes (Treg) activity, 444
SDH. *See* Subdural hematomas (SDH)
Selective serotonin reuptake inhibitors (SSRIs)
  5-HT bioavailability, 374
  pulmonary circulation, 374
  pulmonary vascular resistance, 374
  risks, PPHN, 375
  treatment, maternal depression, 373
Selexipag, prostacyclins, 188
Serotonin (5-hydroxytryptamine/5-HT)
  amino acid tryptophan, 415–416
  anorectic agent dexfenfluramine, 416–417
  BMPR-II, 417
  distal pulmonary arteries, 416
  endogenous vasoactive indolamine, 415
  5-HTT, 417
  iPAH, 416
  MAO, 416
  raphe nuclei, 416
  receptors, 417–418
  RhoA, 416
  TPH, 416
  transgenic mice types, 417
Serotonin (5-HTT) overexpressing mice, 131
Serotonins (5-HTs)
  appetite suppressants, 367
  effects, pulmonary vascular bed, 366
  FHR, 366
  5-HTTs (*see* Serotonin transporters (5-HTTs))
  intracellular signaling, 368
  PA-SMCs, 367
  in PH
    endogenous vasoactive indoleamine substance, 368
    human and experimental synthesis, 368–370
    MAO-A, 368
    signaling, 376–377
    synthesis and degradation, 368
  platelets, serotonergic components, 366, 367
  pulmonary circulation, 366
  receptors (*see* Receptors, 5-HTs)
Serotonin transporters (5-HTTs)
  description, 370
  idiopathic PH
    gene polymorphism and expression, 372–373
    PA-SMC hyperplasia, 371–372
    Na^+/K^+-ATPase activity, 370
  in PH
    PA-SMC growth, 371
    SM22 promoter, 371
  and PPHN, 373–374
  protein kinase A and C, 370
  treatment, adult patients, 373
sGC. *See* Soluble guanylyl cyclase (sGC)
Short hairpin (shRNA) BMPR-2 knockdown mice
  Ang-1 (*see* Angiopoietin-1 (Ang-1) overexpression)
  broiler fowl disease, 132
  description, 131
  ET_B receptors (*see* Endothelin B (ET_B) receptors deficiency)
  high fat diet, Apo E knockout mice, 133–134
  5-HTT overexpressing mice, 131
shRNA. *See* Short hairpin (shRNA) BMPR-2 knockdown mice
SHR-SP. *See* Spontaneously hypertensive stroke-prone rats (SHR-SP)
Sickle cell disease (SCD)
  hemoglobin, 321
  hemolytic anemia, 322–323
  transgenic mice, 321
Single nucleotide polymorphisms (SNPs), 480
SiPHT, 535
Sirolimus, 462
Sitaxentan, ERAs, 217
SLE. *See* Systemic lupus erythematosus (SLE)
SMCs. *See* Smooth muscle cells (SMCs)
Smooth muscle cells (SMCs)
  biologic effects, NO, 259
  and vascular endothelium, 260
SNPs. *See* Single nucleotide polymorphisms (SNPs)
SOD. *See* Superoxide dismutase (SOD)
SOD2. *See* Superoxide dismutase 2 (SOD2)
Soluble guanylyl cyclase (sGC)
  activators and stimulators, 265
  BAY 58-2667, 43
  BAY 58-2667 and HMR 1766, 334–335
  biologic effects, NO, 259

Soluble guanylyl cyclase (sGC) (cont.)
  cGMP, 318, 326
  cytosolic GTP to cGMP conversion, 42
  drug therapy, 43
  heme-histidine bond, 42
  heme-independent, 42
  metabolism, 329
  NO, 317–318
  PASMC, 336
  pathophysiological concentrations, H_2O_2, 42
  prostacyclin synthase, 330
  stimulators
    acrylamide analogues, 295
    activation, 284–286
    BAY 41-2272, 287–294
    BAY 41-8543, 294–295
    CFM-1571, 295
    cGMP, 284
    genetic factors, 284
    nitric oxide production, 281–283
    oxidative stress, 283–284
    pharmacological stimulation, 286–287
    pulmonary endothelium and vascular smooth muscle, 281
    riociguat, 296–304
    SHR-SP, 295
    YC-1, 287
  transgenic mice deficient, 42
  in vascular SMC, 262
Somatostatin, 493–494
SOPHIA. See Surveillance of Pulmonary Hypertension in America (SOPHIA)
Spontaneously hypertensive stroke-prone rats (SHR-SP), 295
SSRIs. See Selective serotonin reuptake inhibitors (SSRIs)
Stachybotrys chartarum, 136
Statins
  animal models
    atorvastatin, 534
    eNOS and/NO production, 534
    5-HTT, 534–535
    simvastatin, 534
  clinical studies, 535
  fluvastatin and atorvastatin, 536
  HDL-C, 536
  HMG-CoA reductase, 532
  lipophilic statins, 536
  miRNAs, 536
  molecular action
    BMPR2, 534
    cholesterol biosynthesis pathway, 533
    GTPases, 533
    isoprenoids production, 533
    NO production, 534
    ROCK, 533–534
  PAH treatment, 536
  vascular remodelling, 532–533
Statistical evaluation, biomarkers
  BNP, 81, 82
  Cox analysis, 83
  data and/or sample, 81
  data types and statistical analyses, 81, 82
  incident and prevalent disease, 84
  missing data, 84
  multiple testing, 83–84
  over-fitting, 83
  patient population, 81
  potential trial designs, 84–85
  proteomics/metabolomics, 81
  ROC, 82
  severity and progression, disease, 82
  tests, 81
SU-5416. See Sugen-5416 (SU-5416)
Subdural hematomas (SDH), 400–401
Sugen-5416 (SU-5416)
  CH + SU model, 129
  description, 128
  limitations, 129
  method, 128
  PA endothelial cells, 128
  vasodilators, 128
Superoxide dismutase 2 (SOD2)
  CpG islands, 127
  epigenetic silencing, 113
  mitochondrial expression, 127
Superoxide dismutase (SOD), 327, 330, 336, 337
Surgical models
  aorto-caval shunt, 140
  description, 139
  PAB (see Pulmonary artery banding (PAB) model)
Surveillance of Pulmonary Hypertension in America (SOPHIA), 9–10
Systemic lupus erythematosus (SLE), 442

T

TAPSE. See Tricuspid annular plane systolic excursion (TAPSE)
Tertiary lymphoid tissues (tLTs)
  CD^{138+} plasma cells, 67
  CXCL13 and CCL19/CCL21, 67
  lymphoid follicles, 67
  and perivascular inflammation, 67

Index 575

Thromboxane A_2 (TXA$_2$)
  and increased pulmonary vascular sensitivity, 45–46
  pathophysiology, PH, 45
TKI. *See* Tyrosine kinase inhibitors (TKI)
tLTs. *See* Tertiary lymphoid tissues (tLTs)
T Lymphocytes/T cells, 445–446
TNF. *See* Tumor necrosis factor (TNF)
TNF-α. *See* Tumor necrosis factor alpha (TNF-α)
Treprostinil
  inhaled, 190
  intravenous, 186
  subcutaneous, 186–187
Tricuspid annular plane systolic excursion (TAPSE), 37
Tumor necrosis factor (TNF), 447
Tumor necrosis factor alpha (TNF-α), 462
TXA$_2$. *See* Thromboxane A_2 (TXA$_2$)
Tyrosine kinase inhibitors (TKI)
  basal cellular processes, 398
  cardiotoxicity, 399
  imatinib mesylate, 393–394
  induction, PAH, 400
  inhibition, PDGF signaling, 383
  SDH, 400–401

U

Unclear/multifactorial etiologies, PH
  fibrosing mediastinitis, 22
  hematologic disorders, 19
  long-term hemodialysis, 22
  metabolic disorders, 21
  proximal pulmonary arteries, 21
  pulmonary microvascular cytology sampling, 22
  systemic disorders, 20–21

V

Vascular/cardiac hypertrophy, ERAs
  acute PH, 206
  bosentan and macitentan, 207
  chronic models, 206
  chronic thromboembolic PH, 207
  dual and ET$_A$-selective, 208
  monocrotaline-induced PH, rats, 206–207
  overcirculation-induced PH, 207
  pharmacological evaluation, 206
Vascular endothelial growth factor (VEGF) signaling system
  description, 418
  endothelial cell proliferation and promotes, 419
  Flt-1, 418
  KDR/Flk-1, 419
  monocrotaline and chronic hypoxia models, 419
  mRNA levels, 419
  receptors, 419
Vascular smooth muscle cells (VSMC)
  endothelial dysfunction, 355
  eNOS downregulation, 358
  enzyme, 354
  inflammatory cells, 353
  myosin phosphatase, 356
  pulmonary arteries, 359
Vasoactive intestinal peptide (VIP)
  acetylcholine and neuronal NO synthase, 488
  cardiac function and peripheral vasculature, 490
  cytokine-elicited neutrophil accumulation, 489
  endothelium-independent vasodilator, 488
  myocardium, conductance tissue and vascular endothelial cells, 488
  and PACAP peptides, 490
  patient cohort, 489
  peptidase-resistant VIP analogues, 490
  and PH, 489
  pro-inflammatory mediators, 489
  receptors, 487, 488
  smooth muscle, larger airways, 488
  structure and expression, 486
  VPAC1R and VPAC2R, 489
Vasoconstriction/vasodilation, ERAs, 205
Vasoconstrictor peptides
  Ang II, 480–481
  ET-1, 479–480
Vasodilator peptides
  AM (*see* Adrenomedullin (AM))
  CGRP (*see* Calcitonin gene-related peptide (CGRP))
  NPs (*see* Natriuretic peptides (NPs))
  PACAP (*see* Pituitary adenylate cyclase activating peptides (PACAP))

Vasodilator peptides (*cont.*)
 somatostatin, 493–494
 VIP (*see* Vasoactive intestinal peptide (VIP))
Venice classification, 68
Venous and venular lesions
 description, 68
 PVOD/PCH, 68–70

VIP. *See* Vasoactive intestinal peptide (VIP)
VSMC. *See* Vascular smooth muscle cells (VSMC)

X
Xanthine oxidoreductase (XOR), 326–327
XOR. *See* Xanthine oxidoreductase (XOR)

Printed by Printforce, the Netherlands